wave number–micron convers[illegible]

	0	1	2	3	4	5	6	7	8	9									
6.5	1538	1536	1534	1531	1529	1527	1524	1522	1520	1517	[illegible]	[illegible]	[illegible]	[illegible]	[illegible]	[illegible]	[illegible]	[illegible]	[illegible]
6.6	1515	1513	1511	1508	1506	1504	1502	1499	1497	1495	[illegible]	[illegible]	[illegible]	[illegible]	[illegible]	[illegible]	[illegible]	[illegible]	[illegible]
6.7	1493	1490	1488	1486	1484	1481	1479	1477	1475	1473	0	0	1	1	1	1	2	2	2
6.8	1471	1468	1466	1464	1462	1460	1458	1456	1453	1451	0	0	1	1	1	1	2	2	2
6.9	1449	1447	1445	1443	1441	1439	1437	1435	1433	1431	0	0	1	1	1	1	1	2	2
7.0	1429	1427	1425	1422	1420	1418	1416	1414	1412	1410	0	0	1	1	1	1	1	2	2
7.1	1408	1406	1404	1403	1401	1399	1397	1395	1393	1391	0	0	1	1	1	1	1	2	2
7.2	1389	1387	1385	1383	1381	1379	1377	1376	1374	1372	0	0	1	1	1	1	1	2	2
7.3	1370	1368	1366	1364	1362	1361	1359	1357	1355	1353	0	0	1	1	1	1	1	2	2
7.4	1351	1350	1348	1346	1344	1342	1340	1339	1337	1335	0	0	1	1	1	1	1	1	2
7.5	1333	1332	1330	1328	1326	1325	1323	1321	1319	1318	0	0	1	1	1	1	1	1	2
7.6	1316	1314	1312	1311	1309	1307	1305	1304	1302	1300	0	0	1	1	1	1	1	1	2
7.7	1299	1297	1295	1294	1292	1290	1289	1287	1285	1284	0	0	0	1	1	1	1	1	1
7.8	1282	1280	1279	1277	1276	1274	1272	1271	1269	1267	0	0	0	1	1	1	1	1	1
7.9	1266	1264	1263	1261	1259	1258	1256	1255	1253	1252	0	0	0	1	1	1	1	1	1
8.0	1250	1248	1247	1245	1244	1242	1241	1239	1238	1236	0	0	0	1	1	1	1	1	1
8.1	1235	1233	1232	1230	1229	1227	1225	1224	1222	1221	0	0	0	1	1	1	1	1	1
8.2	1220	1218	1217	1215	1214	1212	1211	1209	1208	1206	0	0	0	1	1	1	1	1	1
8.3	1205	1203	1202	1200	1199	1198	1196	1195	1193	1192	0	0	0	1	1	1	1	1	1
8.4	1190	1189	1188	1186	1185	1183	1182	1181	1179	1178	0	0	0	1	1	1	1	1	1
8.5	1176	1175	1174	1172	1171	1170	1168	1167	1166	1164	0	0	0	1	1	1	1	1	1
8.6	1163	1161	1160	1159	1157	1156	1155	1153	1152	1151	0	0	0	1	1	1	1	1	1
8.7	1149	1148	1147	1145	1144	1143	1142	1140	1139	1138	0	0	0	1	1	1	1	1	1
8.8	1136	1135	1134	1133	1131	1130	1129	1127	1126	1125	0	0	0	1	1	1	1	1	1
8.9	1124	1122	1121	1120	1119	1117	1116	1115	1114	1112	0	0	0	1	1	1	1	1	1
9.0	1111	1110	1109	1107	1106	1105	1104	1103	1101	1100	0	0	0	1	1	1	1	1	1
9.1	1099	1098	1096	1095	1094	1093	1092	1091	1089	1088	0	0	0	0	1	1	1	1	1
9.2	1087	1086	1085	1083	1082	1081	1080	1079	1078	1076	0	0	0	0	1	1	1	1	1
9.3	1075	1074	1073	1072	1071	1070	1068	1067	1066	1065	0	0	0	0	1	1	1	1	1
9.4	1064	1063	1062	1060	1059	1058	1057	1056	1055	1054	0	0	0	0	1	1	1	1	1
9.5	1053	1052	1050	1049	1048	1047	1046	1045	1044	1043	0	0	0	0	1	1	1	1	1
9.6	1042	1041	1040	1038	1037	1036	1035	1034	1033	1032	0	0	0	0	1	1	1	1	1
9.7	1031	1030	1029	1028	1027	1026	1025	1024	1022	1021	0	0	0	0	1	1	1	1	1
9.8	1020	1019	1018	1017	1016	1015	1014	1013	1012	1011	0	0	0	0	1	1	1	1	1
9.9	1010	1009	1008	1007	1006	1005	1004	1003	1002	1001	0	0	0	0	0	1	1	1	1
10	1000	990.1	980.4	970.9	961.5	952.4	943.4	934.6	925.9	917.4	.9	1.8	2.7	3.6	4.5	5.5	6.4	7.3	8.2
11	909.1	900.9	892.9	885.0	877.2	869.6	862.1	854.7	847.5	840.3	.8	1.5	2.3	3.0	3.8	4.5	5.3	6.1	6.8
12	833.3	826.4	819.7	813.0	806.5	800.0	793.7	787.4	781.3	775.2	.6	1.3	1.9	2.6	3.2	3.8	4.5	5.1	5.8
13	769.2	763.4	757.6	751.9	746.3	740.7	735.3	729.9	724.6	719.4	.5	1.1	1.6	2.2	2.7	3.3	3.8	4.4	4.9
14	714.3	709.2	704.2	699.3	694.4	689.7	684.9	680.3	675.7	671.1	.5	1.0	1.4	1.9	2.4	2.9	3.3	3.8	4.3
15	666.7	662.3	657.9	653.6	649.4	645.2	641.0	636.9	632.9	628.9	.4	.8	1.3	1.7	2.1	2.5	2.9	3.3	3.8
16	625.0	621.1	617.3	613.5	609.8	606.1	602.4	598.8	595.2	591.7	.4	.7	1.1	1.5	1.8	2.2	2.6	2.9	3.3
17	588.2	584.8	581.4	578.0	574.7	571.4	568.2	565.0	561.8	558.7	.3	.7	1.0	1.3	1.6	2.0	2.3	2.6	3.0
18	555.6	552.5	549.5	546.4	543.5	540.5	537.6	534.8	531.9	529.1	.3	.6	.9	1.2	1.5	1.8	2.0	2.3	2.6
19	526.3	523.6	520.8	518.1	515.5	512.8	510.2	507.6	505.1	502.5	.3	.5	.8	1.1	1.3	1.6	1.8	2.1	2.4
	0	1	2	3	4	5	6	7	8	9	1	2	3	4	5	6	7	8	9

/QC457.D64>C1/
/QC457.D64>C1/

tabulation of infrared spectral data

tabulation of infrared spectral data

david dolphin
the university of british columbia

alexander wick
hoffmann–la roche basel

a wiley–interscience publication
john wiley & sons
new york london sydney toronto

Library of Congress Cataloging in Publication Data

Dolphin, David.

Tabulation of infrared spectral data.

"A Wiley-Interscience publication."

1. Infra-red spectrometry—Tables, etc.

I. Wick, Alexander, 1938– joint author.

II. Title.

QC457.D64 535'.842'0212 76-48994

ISBN 0-471-21780-8

Printed in the United States of America

10 9 8 7 6 5 4 3 2 1

preface

During the past three decades the parallel development of both the theory of infrared spectroscopy and the collection and interpretation of an immense amount of infrared data has allowed chemists to correlate changes in the group frequency of a chromophore with the steric and electronic effects imposed on that chromophore by the rest of the molecule. It was inevitable, however, as such correlations became more generalized, that much of the detailed information contained in a spectrum would be lost when these generalizations were used as interpretive aids.

Indeed, while broad generalizations have considerable pedagogical value, the practical chemist must rely on more specific examples upon which to draw his analogies. In general, this necessitates a foray into the various collections of published spectra or to the original literature. While retrieval of the infrared spectrum of a specific compound is usually straightforward, collecting sufficient examples to provide adequate analogies to the problem in hand can be both a time-consuming and frustrating task. In order that such undertakings may be simplified in the future, we have collected and tabulated the data presented in this book.

So that the reader may make his own generalization concerning his specific problem we briefly describe, at the beginning of each chapter, those steric, electronic, and solute-solvent interactions which bring about changes in the characteristic group frequencies. The tables which follow the brief discussions then show changes in the group frequency as the electronic and steric environment around the group are varied. We have attempted to maintain a consistent order in which these variations are presented from chapter to chapter.

A glance at the chapter covering ketones, page 175, which is the most extensive chapter in the book, shows how these variations have been arranged. Initially the aliphatic side chains of the acyclic system are extended first on one side of the chromophore and then on the other; this is followed by branching of the side chains and then substitution of electronegative groups (halogens, carbon, nitrogen, oxygen, and so on). Aliphatic and then aromatic unsaturation follows, after which these same permutations are applied to the cyclic systems. Finally, specific classes of ketones, such as α- and β-diketones and diazoketones, are covered.

Since solute-solvent interactions and hydrogen bonding can cause significant changes in the group frequency we have included, where appropriate, a discussion of these effects in the introductions to the chapters. In addition, at the beginning of each table, we have shown how various solvents change the group frequency for the simplest members of the series. We have, wherever possible, given the solvent in which the spectrum was measured and the original literature reference. In those few cases when no mention was made of the solvent, *sng* (solvent not given) is indicated.

The intensities of the absorption bands in the tables are strong unless shown as medium (m) or weak (w); shoulders are indicated by an asterisk (*). When group frequencies for two different functional groups are included in the tables, the frequencies relating to the chapter in which the example is given are shown first, followed by the remaining absorptions in decreasing wavenumber.

We have included all of those organic functional groups which have characteristic group frequencies. Starting with X-H single-bond vibrations (including main group hydrides) we continue with C-C vibrations, followed by C-N, C-O, N-N, and N-O vibrations. Since the organic chemist now uses a wide variety of inorganic and organometallic intermediates, we have included chapters on thionyl and phosphoryl compounds, as well as main group and transition metal compounds.

In addition, we have included the complete spectra of some sixty common organic solvents and impurities since the absorptions arising from these contaminants frequently confuse the interpretation of a spectrum.

Finally, we would like to thank all of our colleagues who have encouraged and helped us in this endeavor, as well as Dan Hightower and Penny Pounder for the art work and Marci Hazard and Avril Orloff for the typing and layout.

D. Dolphin   Vancouver

A. E. Wick   Basel

October, 1976

contents

chapter 3 carbon–nitrogen vibrations

chapter 4 carbonyl vibrations

chapter 5 nitrogen–nitrogen vibrations

chapter 6 nitrogen–oxygen vibrations

chapter 7 main group–oxygen vibrations

chapter 8 inorganic vibrations

journal abbreviations

A

Acta Chem Scand	Acta Chemica Scandinavica
Acta Polytec Scand Chem Incl Met	Acta Polytechnica Scandinavica, Chemistry Including Metallurgy Series
Anal Chem	Analytical Chemistry
Anal Chim Acta	Analytica Chimica Acta
Angew	Angewandte Chemie
Angew Chem Int Ed	Angewandte Chemie, International Edition in English
Appl Spectr	Applied Spectroscopy
Arc Pharm	Archiv der Pharmazie (Weinheim)
Arkiv Kemi	Arkiv foer Kemi
Aust JC	Australian Journal of Chemistry

B

B	Chemische Berichte
Ber Bun Physik Chem	Berichte der Bunsen Gesellschaft für Physikalische Chemie
Biochem J	Biochemical Journal
Bol Inst Quim Univ Nacl Auton Mex	Boletim del Instituto de Quimica de la Universidad Nacional Autonoma de Mexico
BCS Japan	Bulletin of the Chemical Society of Japan (Kagaku to Kogyo)
BSC Belge	Bulletin des Societes Chimiques Belges
BSC France	Bulletin de la Societe Chimique de France
Bull Acad Polon	Bulletin de l' Academie Polonaise des Sciences, Serie des Sciences Chimiques

Bull Acad Sci USSR Chemical Science	Bulletin of the Academy of Sciences of the USSR, Division of Chemical Science
Bull Soc Roy Sci Liege	Bulletin de la Societe Royale des Sciences de Liege
C	
Canad J Chem	Canadian Journal of Chemistry
Chem & Ind	Chemistry and Industry (London)
Chem Comm	The Chemical Society, London; Chemical Communications
Chem Pharm Bull	Chemical and Pharmaceutical Bulletin
Col Czech Comm	Collection of Czechoslovak Chemical Communications
CR	Comptes Rendus de L'Académie des Sciences
D	
Disc Farad Soc	Discussions of the Faraday Society
DMS	The Documentation of Molecular Spectroscopy (Butterworth's Scientific Publications, London, England and Verlag Chemie, West Germany)
Doklady Chem	Doklady Akademii Nauk SSSR
E	
Experientia	Experientia
F	
Fortschr Chem Forsch	Fortschritte der Chemischen Forschung
G	
Gazz Chim Ital	Gazzeta Chimica Italiana
H	
Helv	Helvetica Chimica Acta
I	
Inorg & Nucl Chem Letters	Inorganic and Nuclear Chemistry Letters
Inorg Chem	Inorganic Chemistry
Inorg Chim Acta	Inorganica Chimica Acta
Izvest Akad Nauk Khim USSR	Izvestiya Akademii Nauk SSSR, Seriya Khimicheskaya
Izvest Akad Nauk Otdel Khim	Izvestiya Akademii Nauk SSSR, Otdelenie Khimicheskikh Nauk

J

JACS	Journal of the American Chemical Society
J Appl Chem	Journal of Applied Chemistry
JCP	Journal of Chemical Physics
JCS	Journal of the Chemical Society, London
J Gen Chem USSR	Journal of General Chemistry of the USSR (Transl. of Zh. Obshch. Khim)
J Het Chem	Journal of Heterocyclic Chemistry
JINC	Journal of Inorganic and Nuclear Chemistry
J Mol Spectr	Journal of Molecular Spectroscopy
JOC	Journal of Organic Chemistry
J Organometallic Chem	Journal of Organometallic Chemistry
JPC	Journal of Physical Chemistry
J Prakt Chem	Journal für Praktische Chemie

K

Kagaku No Ryoiki	Journal of Japanese Chemistry

M

Monatsh Chem	Monatshefte für Chemie

N

Nature	Nature (London)
Naturwiss	Naturwissenschaften
NKZ	Nippon Kagaku Zasshi (Japanese Journal of Chemistry)

O

Opt & Spec	Optics and Spectroscopy (USSR)

P

Proc Chem Soc	Proceedings of the Chemical Society, London
Proc Roy Soc	Proceedings of the Royal Society of London

Q

Quart Rev	Quarterly Reviews, the Chemical Society, London

R

Rec Trav Chim	Recueil des Travaux Chimiques des Pays-Bas
Rev Roumaine Chim	Revue Roumaine de Chimie

S

SCA	Spectrochimica Acta, Part A
Spec Let	Spectroscopy Letters
Synth	Synthesis

T

T	Tetrahedron
TL	Tetrahedron Letters
Trans Farad Soc	Transactions of the Faraday Society

Y

Yakugak Zasshi	Journal of the Pharmaceutical Society of Japan

Z

Z Anorg Chem	Zeitschrift für Anorganische Chemie
Z Electrochem	Zeitschrift für Electrochemie
Z Naturf	Zeitschrift für Naturforschung
Z Phys Chem (Fr)	Zeitschrift für Physikalische Chemie (Frankfurt am Main)
Z Phys Chem NF	Zeitschrift für Physikalische Chemie, Neue Folge
Zh Fiz Khim	Zhurnal Fizicheskoi Khimii
Zh Obshch Khim	Zhurnal Obshchei Khimii
Zh Organ Khim	Zhurnal Organicheskoi Khimii
Zh Prikl Spektrosk	Zhurnal Prikladnvi Spektroskopii

tabulation of infrared spectral data

1.1

alkanes

Although some of the group frequencies associated with alkanes can be of diagnostic value, they can in general only be used to identify a specific group, since changes in the adjacent molecular structure cause only small changes in the group frequency. Thus only a brief description of the major alkane modes is given.

$-CH_3$	2970-2870 cm^{-1}	Two strong bands arising from the symmetric and asymmetric modes are seen in this region.
	1480-1430 cm^{-1}	A medium-intensity band arising from the asymmetric deformation mode is observed.
	1390-1360 cm^{-1}	A medium-intensity band arises from the symmetric deformation mode.

$\rangle CH_2$	2930-2850 cm^{-1}	Two strong bands arising from the asymmetric and symmetric stretching modes are seen in this region.
	1490-1450 cm^{-1}	A medium-intensity band arising from the methylene scissoring mode is observed.
	750-720 cm^{-1}	A strong band (occasionally a doublet) from the $(-CH_2-)_n$ rocking mode is observed.
$\rangle\!\!-CH$	2900-2880 cm^{-1}	A weak band arises from the C-H stretching mode.
	1360-1320 cm^{-1}	A weak band arising from the C-H deformation mode is found here.

1.2

aromatic systems

Benzene derivatives exhibit a number of characteristic absorptions in their infrared spectra which can be diagnostically useful in determining both the extent and pattern of substitution. When, however, strongly electron-donating or electron-withdrawing groups are present the "characteristic" patterns lose much of their diagnostic value, and caution must be exercised in interpreting the spectra of such compounds.

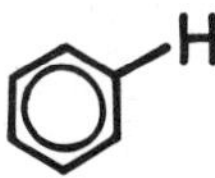

3100-3000 cm^{-1} A sharp band of medium intensity is observed here. When this band is resolved from aliphatic C-H stretching modes, it can be diagnostic of aromatic systems. The band is usually found at 3030 cm^{-1}.

2000-1600 cm^{-1} Exceedingly weak bands arise as overtones of the C-H out-of-plane deformation bands. Figure 1 shows these bands

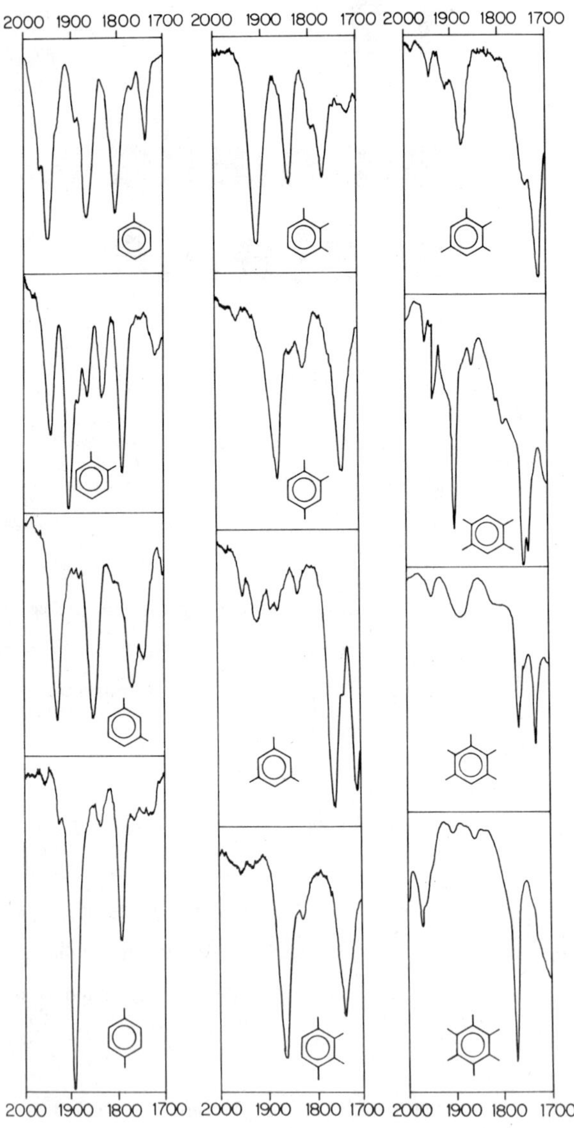

2000 1900 1800 1700 2000 1900 1800 1700 2000 1900 1800 1700
2000 1900 1800 1700 2000 1900 1800 1700 2000 1900 1800 1700

figure 1

for a series of methyl-substituted benzenes. When the substituents are neither strongly electron donating or withdrawing, the patterns in this region may give the correct substitution pattern. As soon, however, as any of the groups strongly perturb the electronic structure of the aromatic ring, the "characteristic pattern" in this region is no longer characteristic and will lead to an incorrect assignment of the substitution pattern.

1625-1440 cm^{-1}

Four bands are usually found in this region and arise from the skeletal vibrations of the benzene nucleus and fall within the following ranges:

1625-1570 cm^{-1}
1610-1560 cm^{-1}
1520-1460 cm^{-1}
1470-1440 cm^{-1}

Some specific examples are given in the tables that follow.

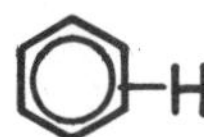

1290-950 cm^{-1}

A series of weak absorptions arise from in-plane C-H deformation modes. In general, these bands are of little diagnostic value.

900-700 cm^{-1}

When neither strongly electron-withdrawing nor donating groups are present in the benzene nucleus, the strong bands between 900 and 700 cm^{-1}, which arise from the C-H out-of-plane bending modes, can be diagnostically useful.

The absorption bands for a number of mono- and disubstituted benzenes are given in the tables that follow. As the benzene nucleus becomes more heavily substituted, the following correlations can be made:

Three adjacent hydrogen atoms
810-750 cm^{-1}

Two adjacent hydrogen atoms
860-800 cm^{-1}

An isolated hydrogen atom
900-860 cm^{-1}

Substituent	Solvent	Monosubstituted benzenes							Ref.
H	liq.	1604	1584	1480		1033		773	1
CH_3	$CHCl_3$	1603		1496	1455	1081	1028	729 (CS_2)	2
CN	$CHCl_3$	1581	1563*	1477	1446	1067	1018	756 (CS_2)	2,3
CHO	$CHCl_3$	1601	1588		1455	1070	1021	748 (liq.)	2,4
$COCH_3$	$CHCl_3$	1601	1585		1448	1075	1022	757 (CS_2)	2,3
$CO.O.CH_2CH_3$	$CHCl_3$	1606	1588	1490*	1451	1070	1028	804 780 (CS_2)	2,3
CF_3	CCl_4	1610	1595	1456	1426			768	5
NH_2	$CHCl_3$	1605		1495	1465		1026	755 (liq.)	2,3
NO_2	$CHCl_3$	1621* 1609	1590*	1478	1458	1068	1021	790 (CS_2)	2,3
OH	$CHCl_3$	1603		1503		1069	1023	750 (CS_2)	2,3
OCH_3	$CHCl_3$	1601	1591*	1494	1453	1077	1018*	750 (CS_2)	2,3
F	CCl_4	1597	1585	1494	1459			752	5
Cl	$CHCl_3$	1587	1562*	1497	1447	1084 1066*	1023	742 (liq.)	2,4
Br	$CHCl_3$	1581	1563*	1477	1446	1067	1018	736 (liq.)	2,4
I	CCl_4	1572	1558	1472	1439			729	5
SH	CCl_4	1585		1480	1443			732	5

1. K. S. Pitzer and D. W. Scott, *JACS* (1943), 803.
2. A. R. Katritzky and J. M. Lagowski, *JCS* (1958), 4155.
3. R. D. Kross, V. A. Fassel and M. Margoshes, *JACS* (1956), 1332.
4. M. Margoshes and V. A. Fassel, *SCA* (1955), *7*, 14.
5. M.-L. Josien and J.-M. Lebas, *BSC France* (1956),53.

Substituents	Solvent	*ortho*-Disubstituted benzenes											Ref.
CH_3 CH_3	$CHCl_3$	1607		1498	1460*			1118	1020	984		748($CHBr_3$)	1,2
CH_3 NH_2	$CHCl_3$	1591		1501	1473	1319*	1155*		1034	985	840	753 (CS_3)	1,3
CH_3 NO_2	$CHCl_3$	1616	1582	1486	1433	1278	1163	1148	1049	950			1
CH_3 OH	$CHCl_3$	1593		1495	1466		1150*		1037	980	840	752 (CS_2)	1,3
CH_3 Cl	$CHCl_3$	1596	1575	1489* 1476	1442			1127	1040			751($CHBr_3$)	1,2
NH_2 NH_2	$CHCl_3$	1596		1505	1463		1150		1033	913	862		1
NH_2 NO_2	$CHCl_3$	1587*	1578	1483	1445		1168 1157*	1105	1017	950	843		1
NO_2 Cl	$CHCl_3$		1590*	1492	1457	1316*	1158	1142	1023		832		1
NO_2 NO_2			1587		1443			1149	1000		862 840	758	4
NO_2 OH	$CHCl_3$	1620	1593 1577*	1479	1457		1161	1115*	1027	950	815		1
NO_2 Cl	$CHCl_3$	1593	1585*	1474	1444	1260		1131	1034	952			1
OH OH	$CHCl_3$	1613		1514	1472		1186* 1173*		1028	916	861 845		1
OH Cl	$CHCl_3$	1599	1590*	1484	1465		1154	1125	1027	927	820		1
Cl Cl	$CHCl_3$	1576		1460	1438	1252		1126	1012	937	816	752($CHBr_3$)	1,2

1. A. R. Katritzky and R. A. Jones, *JCS* (1959), 3670.
2. A. R. H. Cole and H. W. Thompson, *Trans Farad Soc* (1950), 103.
3. D. H. Whiffen and H. W. Thompson, *JCS* (1945), 268.
4. J. G. Grasselli, *Atlas of Spectral Data and Physical Constants for Organic Compounds*, The Chemical Rubber Co., 1973.

Substituents	Solvent	*meta*-Disubstituted benzenes												Ref.
CH_3 CH_3	$CHCl_3$	1611	1596	1483			1165	1094			903	880	773($CHBr_3$)	1,2
CH_3 NH_2	$CHCl_3$		1594*	1491	1467		1166			993	919	868*	774 (CS_2)	1,3
CH_3 NO_2	$CHCl_3$		1594*	1482		1289*		1096	1080		906	880	797	1
CH_3 OH	$CHCl_3$	1616*	1595	1488	1460	1263*	1150		1080	949	925	876	776 (CS_2)	1,3
CH_3 Cl	$CHCl_3$	1600	1578	1476			1160	1098		995		868		1
NH_2 NH_2	$CHCl_3$	1600*		1495			1159					834		1
NH_2 NO_2	$CHCl_3$		1587	1482	1460*		1159	1108	1091	994	927	888		1
NH_2 Cl	$CHCl_3$	1603		1482	1450	1297	1161			993		888		1
NO_2 NO_2	$CHCl_3$	1618	1608	1470*		1266	1158	1091	1065		913	903	808	1
NO_2 OH	$CHCl_3$	1622	1592	1476	1452		1157*	1087	1071	999	928	874	812	1
NO_2 Cl	$CHCl_3$	1602*	1582		1462	1265		1095	1065		913	888		1
OH OH		1613		1493		1299	1176 1149			962		840	775	4
OH Cl	solid	1600*	1589	1467	1435	1241	1153	1085	1067	996		882	767	5
Cl Cl	$CHCl_3$	1580		1460	1413	1285	1155	1109	1071*	997		867	780($CHBr_3$)	1,2

1. A. R. Katritzky and P. Simmons, *JCS* (1959), 2058.
2. A. R. H. Cole and H. W. Thompson, *Trans Farad Soc* (1950), 103.
3. D. H. Whiffen and H. W. Thompson, *JCS* (1945), 268.
4. J. G. Grasselli, *Atlas of Spectral Data and Physical Constants for Organic Compoounds*, The Chemical Rubber Co., 1973.
5. C. F. Pouchert, *The Aldrich Library of Infrared Spectra*, Aldrich 1970, spectrum 488E.

Substituents	Solvent	*para*-Disubstituted benzenes										Ref.
CH_3 CH_3	liq.	1620		1513				1119	1021		793	1
CH_3 NH_2	$CHCl_3$			1512			1177	1122		812		2
CH_3 NO_2	$CHCl_3$	1607	1572	1498*	1419	1297	1177	1109	1016	835		2
CH_3 OH	solid	1612 1600		1509	1430b	1295	1172	1104	1017	840	812b	3
CH_3 Cl	$CHCl_3$			1491				1111	1017	808		2
NH_2 NH_2	nuj.	1635	1514	1465		1265		1129		883 825	800b	4
NH_2 NO_2	$CHCl_3$	1606	1580*					1112			837(kBr)	2
NH_2 Cl	$CHCl_3$	1604*	1578	1491			1170	1119	1004	819		2
NO_2 NO_2	$CHCl_3$			1497	1402*			1102	1011	836		2
NO_2 OH_2	$CHCl_3$	1620 1603					1164			848		2
NO_2 Cl	$CHCl_3$	1609	1582	1479	1422	1278	1170	1109	1012	847		2
OH OH	nuj.	1613	1531	1464		1258		1096	1008	825		5
OH Cl	$CHCl_3$	1592		1490			1170		1008	824		2
Cl Cl	$CHCl_3$			1476	1392			1115*	1012	819		2

1. C. F. Pouchert, *The Aldrich Library of Infrared Spectra*, Aldrich 1970, spectrum 424D.
2. A. R. Katritzky and P. Simmons, *JCS* (1959), 2051.
3. C. F. Pouchert, *The Aldrich Library of Infrared Spectra*, Aldrich 1970, spectrum 489A.
4. C. F. Pouchert, *The Aldrich Library of Infrared Spectra*, Aldrich 1970, spectrum 552D.
5. C. F. Pouchert, *The Aldrich Library of Infrared Spectra*, Aldrich 1970, spectrum 504E.

1.3 amines

The NH-stretching frequencies of amines, like those of the hydroxyl group, depend on the degree of hydrogen bonding, which causes shifts to lower frequencies and broadening of the absorption bands.

The hydrogen bond involving an -NH group is, however, weaker than that involving the hydroxyl group, and a dilute solution of an amine, especially in a nonpolar solvent such as CCl_4, will show only the nonassociated -NH bands.

$-NH_2$	3500-3000 cm^{-1}	Two sharp, medium-intensity bands are seen in this region. The higher frequency band arises from the asymmetric mode and occurs at the high end of the range, whereas the symmetric mode is usually near 3400 cm^{-1}. In dilute solution frequency lowering due to either intra- or intermolecular hydrogen bonding is generally less than 100 cm^{-1}. Nevertheless, when steric requirements

are such that strong hydrogen bonding occurs, -NH frequencies as low as 3100 cm^{-1} may be observed.

Primary amines exhibit the R-NH_2 deformation mode as a band of medium intensity in the region of 1660 to 1575 cm^{-1}, with hydrogen bonding causing only a small increase in frequency.

$\rangle NH$	3500-3300 cm^{-1}	Secondary amines show a single band of medium to weak intensity in this region and a deformation mode between 1660 and 1550 cm^{-1}.
$-\overset{+}{N}H_3$	3350-2850 cm^{-1}	Two strong bands, usually broad and frequently not resolved, are seen in this region although they are usually centered around 3000 cm^{-1}. Medium-intensity bands between 2600 and 2000 cm^{-1} arising from combination bands, and overtones, are frequently observed.

The asymmetric deformation mode occurs as a strong band between 1600 and 1560 cm^{-1}, and the symmetric mode occurs between 1540 and 1500 cm^{-1}.

$\rangle \overset{+}{N}H_2$	3000-2750 cm^{-1}	Two bands (often broad and unresolved) are seen in this region. In addition, a number of medium-intensity

bands between 2700 and 2250 cm^{-1} may be observed. The medium-intensity deformation band occurs between 1630 and 1560 cm^{-1}.

$\rangle\overset{+}{N}H$

The $R_3\overset{+}{N}H$ stretching vibration falls near 2700 cm^{-1}. In addition, a series of medium-intensity bands between 2700 and 2250 cm^{-1} may be observed.

2800-2700 cm^{-1}

The presence or absence of bands in the region from 2800 to 2700 cm^{-1} has proven useful in structural assignments of alkaloids (Bohlmann bands[a]). Thus alkaloids that possess a transdiaxial conformation between a nitrogen lone pair and at least one hydrogen on a neighboring carbon atom show one or more medium- to strong-intensity bands in this region. The absence of such bands indicates an absence of the preceding structural feature.

[a]F. Bohlmann, *B.* (1958) 2157; see also C. N. R. Rao, *Chemical Applications of Infrared Spectroscopy*, Academic, New York, 1963, p. 451.

1.4

alcohols phenols

It is only in the gaseous state that vibrations arising from free hydroxyl groups are seen. Even in exceedingly dilute solution the infrared spectra of hydroxyl groups show that some degree of association has occurred.

Thus the physical condition of any sample containing an hydroxyl group including state, solvent, concentration, and even temperature will affect the position, shape, and intensity of any vibration arising from an hydroxyl group. Because of this it is not possible to assign characteristic frequencies to the hydroxyl group, and in general, one must be satisfied with using the frequencies listed as follows only as a general diagnostic guide to hydroxyl vibrations.

–OH *FREE HYDROXYL GROUPS*

Primary	3650-3600 cm^{-1}
Secondary	3640-3600 cm^{-1}
Tertiary	3640-3590 cm^{-1}

Phenolic	3630-3590 cm^{-1}

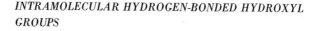

All of the above exhibit a sharp band of medium intensity.

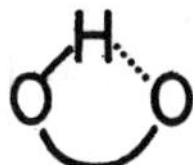

INTRAMOLECULAR HYDROGEN-BONDED HYDROXYL GROUPS

When the hydroxyl group is involved in a hydrogen bond with another group of medium polarity such as another hydroxyl group, a sharp, medium-intensity band is observed between 3600 and 3450 cm^{-1}. When, however, the intramolecular bond occurs in a system such as the enolic form of a β-diketone, or an orthonitrophenol, very broad bands are observed between 3200 and 2500 cm^{-1}. Such bands are weak, because the total intensity is spread over a large area. In general, intramolecular hydrogen bonds do not change upon dilution but may change if the state of the sample, or the solvent, is changed.

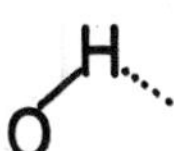

INTERMOLECULAR HYDROGEN-BONDED HYDROXYL GROUPS

A sharp band of medium intensity between 3600 and 3450 cm^{-1} is observed. When more than one hydrogen bond is associated with the particular hydroxyl group a broad band of medium intensity is seen in the region from 3400 to 3200 cm^{-1}. Since the degree of association will depend on the concentration of the sample and the polarity of the solvent, changes in the hydroxyl stretching region will be seen when either of these parameters is changed.

C—O

The C-O stretching frequencies of alcohols give rise to medium to strong absorptions in the region from 1200 to 1050 cm^{-1} and are of little diagnostic value, as are the -OH in-plane bending modes, where the degree of aggregation changes the frequencies.

1.5

main group hydrides

M—H

The structural diversity of the hydrides that are compiled in this chapter does not allow for any useful generalizations. It is apparent, however, that as with the transition metal hydrides (Chapter 8.2, p. 493) the symmetry of the complex and the other coordinated ligands can cause considerable changes in the number and position of the M-H stretching frequencies. In addition, the physical state of the sample or solvent can cause large changes in the stretching frequencies.

Compound	Medium	ν	Ref.
$H_2B(\mu\text{-}H)_2BH_2$	gas	2625 2558 2353 2217w 2134w 1993m	1
$MeHB(\mu\text{-}H)_2BH_2$	gas	2571 2513 2299 2262* 2110 1919	2
$MeHB(\mu\text{-}H)_2BHMe$	gas	2519 2299 2268 2137 1949	3
$H_3B\text{-}CO$	gas	2444 2165 1931m	4
$HB(OMe)_2$	gas	2513	5
$LiBH_4$	nuj.	2320	6
$NaBH_4$	nuj.	2270	6
$Al(BH_4)_3$	nuj.	2559 2493	6

Compound	Medium	ν	Ref.
$Al(BH_4)_3\ Et_2O$	nuj.	2500	6
$LiBH_3CN$	nuj.	2350* 2320 2240* 2220m 2193	7
$NaBH_3CN$	nuj.	2350* 2320 2260* 2240m 2179	7
$i\text{-}Bu_2Al\text{—}H$	C_6H_{12}	1780	8
$LiAlH_4$	nuj.	1810* 1780 1650	9
	THF	1694	9
	Et_2O	1740	9
$NaAlH_4$	nuj.	1650	9
	THF	1680	9

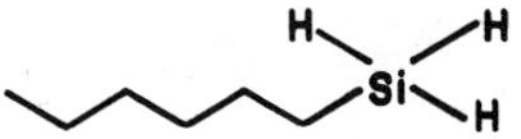

Solvent	Frequency	Ref.
CCl_4	2149	10
CCl_4	2153	10
CCl_4	2158	11
CCl_4	2154	10
CCl_4	2123	10
CCl_4	2142	10
CCl_4	2130	10
CCl_4	2145	10

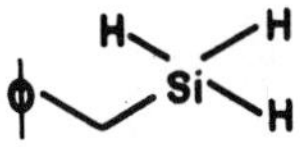

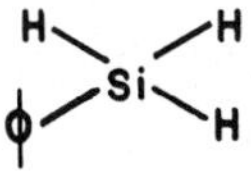

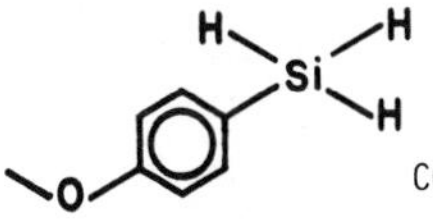

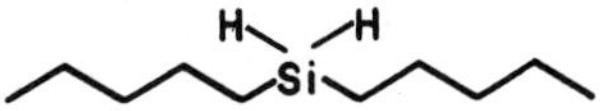

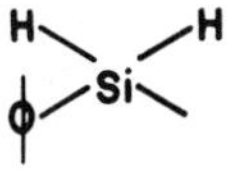

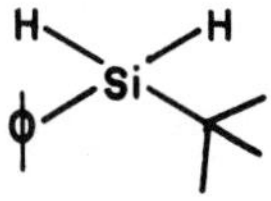

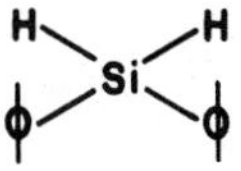

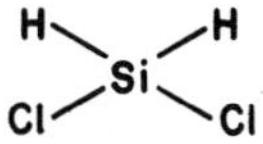

Solvent	Frequency	Ref.
CCl_4	2229	11
CCl_4	2118	11
liq.	2180	13
CCl_4	2106	10
CCl_4	2121	11
CCl_4	2124	11
CCl_4	2122	10
liq.	2135	13
CCl_4	2127	10
CCl_4	2228	11
CCl_4	2168	11

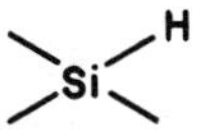

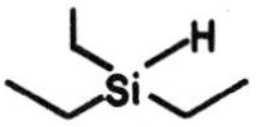

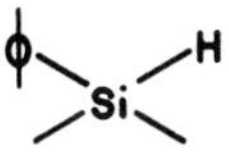

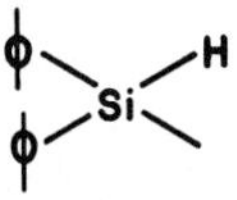

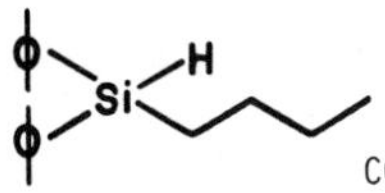

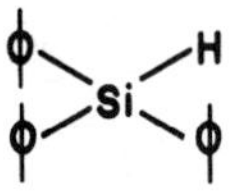

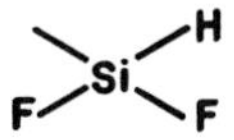

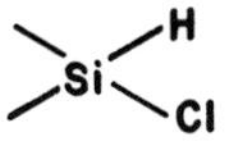

Compound	Medium	ν	Ref.
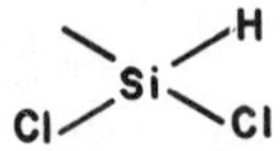 H, Si, Cl, Cl	CCl_4	2214	11
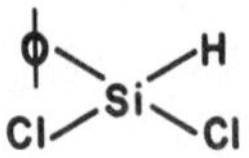 ϕ, H, Si, Cl, Cl	CCl_4	2212	11
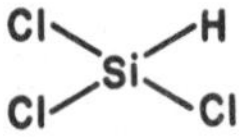 Cl, H, Si, Cl, Cl	CCl_4	2258	11
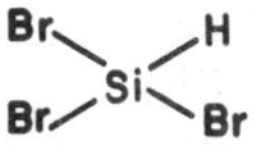 Br, H, Si, Br, Br	CCl_4	2236	11
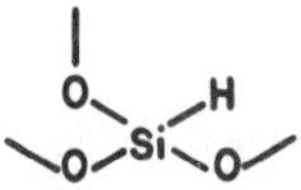 O, H, O, Si, O	CCl_4	2203	11
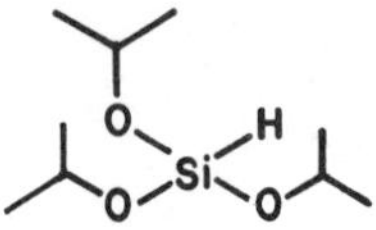 O, H, O, Si, O	CCl_4	2191	11

Compound	Medium	ν	Ref.
PH_3	gas	2327 2421	14
	solid	2370 2268 2193 2075	15
$PH_4^+\ Cl^-$	solid	2433 2398 2079w	15
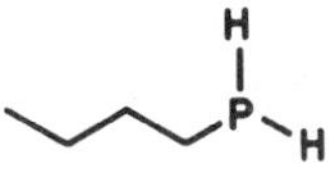 H, P, H	liq.	2286	16
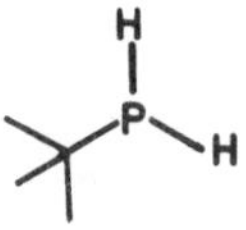 H, P, H	liq.	2282	16
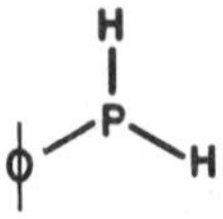 H, ϕ, P, H	liq.	2280	16
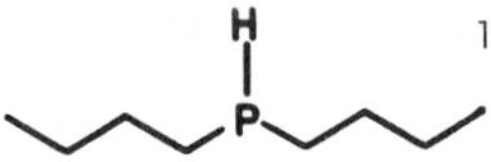 H, P	liq.	2270	16
H, ϕ, P, ϕ	liq.	2270	16
	CCl_4	2282	17
 H, P	gas	2287	18

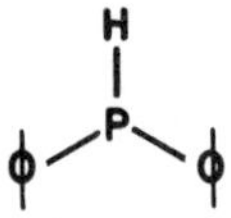

Compound	Medium	ν(cm⁻¹)	Ref.
$Et_3PH^+ Cl^-$	nuj.	2300	19
$Et_3PH^+ Br^-$	KBr	2320	19
$Et_3PH^+ I^-$	KBr	2365m	19
$\phi_3PH^+ Br^-$	KBr	2175mb	19
$\phi_3PH^+ I^-$	KBr	2240mb	19
$(Me_2N)_2P(O)H$	liq.	2320	20
$(MeO)_2P(O)H$	liq.	2434	21
	CCl_4	2439	21
	$CHCl_3$	2443	17
$(EtO)_2P(O)H$	liq.	2434	21
	CCl_4	2433	21
	$CHCl_3$	2439	17
$(i\text{-}PrO)_2P(O)H$	liq.	2434	21
	CCl_4	2429	21
	$CHCl_3$	2437	17
$(\phi CH_2O)(EtO)P(O)H$	liq.	2350	22
$(\phi CH_2O)_2P(O)H$	nuj.	2410	23
$(\phi O)(MeO)P(O)H$	liq.	2440	24
	CCl_4	2444	24
$(\phi O)(i\text{-}PrO)P(O)H$	liq.	2433	24
	CCl_4	2440	24
$(\phi O)_2P(O)H$	CCl_4	2446	17
	$CHCl_3$	2452	17
$(HO)(EtO)P(O)H$	nuj.	2381	22

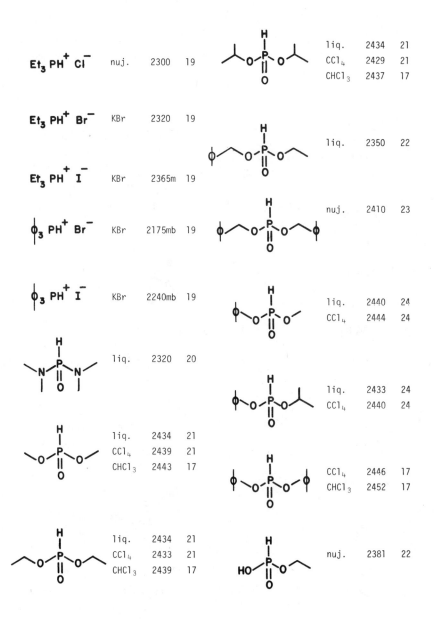

Compound	Medium	ν	Ref.
AsH_3	gas	2185 2122 1005 906	14
ϕAsH_2	liq.	2089	25
ϕ_2AsH	liq.	2071	25
SbH_3	gas	1894 1891 831 782	26
H_2S	gas	2615 1183	27
SH	gas liq.	2597 2550	28 28

Compound	Medium	ν	Ref.
SH	liq.	2564	29
SH	liq.	2574	30
SH	gas liq. CCl_4	2570 2564 2577	31 29 12
ø, ø, ø, SH	CCl_4	2577	12
SH, SH	liq.	2513	32
SH	liq. CCl_4 $CHCl_3$	2575 2590 2579	33 34 35
HO, O, SH	KBr CCl_4	2515 2558	33 33
O, O, SH	liq.	2542	33

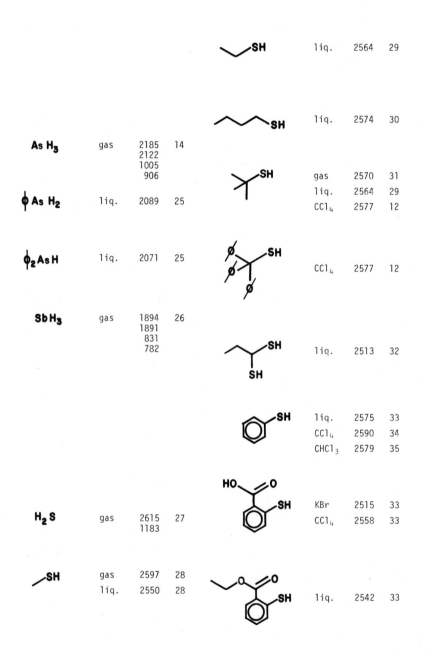

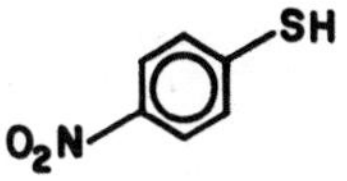

Structure	Medium	Frequency	Ref.
O_2N–⟨ring⟩–SH	CCl_4	2589	34
	$CHCl_3$	2583	35
OH, SH	CCl_4	2595	34
	$CHCl_3$	2588	35
O–⟨ring⟩–SH	CCl_4	2572	34
	$CHCl_3$	2580	35
S, SH	liq.	2538	36
SH, S	liq.	2558	36
O, SH	gas	2550	37
S, SH	liq.	2431	38

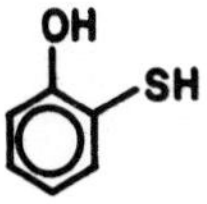

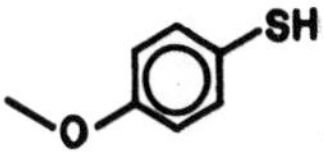

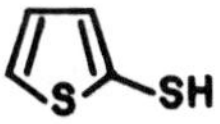

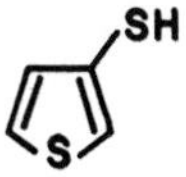

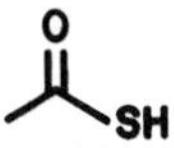

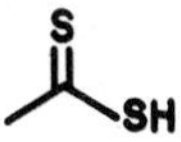

Structure	Medium	Frequency	Ref.
SeH	liq.	2300	39
ϕ SeH	liq.	2300	39
H, H, Sn, H, H	gas	1910 1860	40
	C_6H_{12}	1898	41
H, H, Sn, H	gas	1876	42
	C_6H_{12}	1870	41
H, H, Sn, H	gas	1869	42
	C_6H_{12}	1853	41
H, H, Sn, H	C_6H_{12}	1853	41

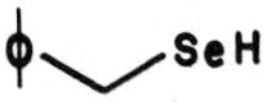

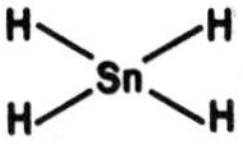

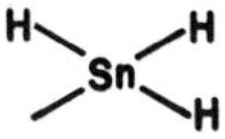

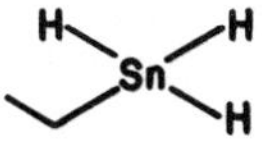

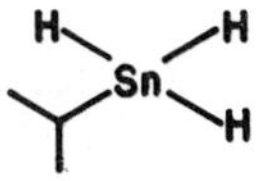

Compound	Medium	ν	Ref
ϕSnH_3	C_6H_{12}	1880	41
$(CH_3)_2SnH_2$	C_6H_{12}	1850	41
ϕ_2SnH_2	C_6H_{12}	1855	41
$(CH_3)_3SnH$	gas	1841	43
	C_6H_{12}	1833	41
$(C_3H_7)_3SnH$	C_6H_{12}	1811	41
$(C_4H_9)_3SnH$	C_6H_{12}	1862	41
ϕ_3SnH	C_6H_{12}	1843	41

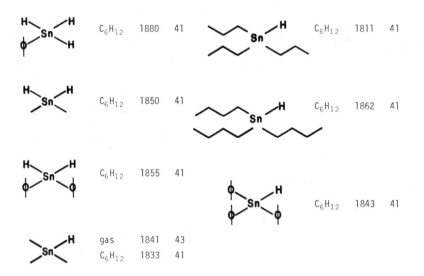

REFERENCES

1. W. C. Price, *JCP* (1948) *16*, 894.
2. W. J. Lehmann, C. O. Wilson, Jr. and I. Shapiro, *JCP* (1960) *32*, 1088.
3. W. J. Lehmann, C. O. Wilson, Jr. and I. Shapiro, *JCP* (1960) *33*, 590.
4. G. W. Bethke and M. K. Wilson, *JCP* (1957) *26*, 1118.
5. W. J. Lehmann, T. P. Onak and I. Shapiro, *JCP* (1959) *30*, 1215.
6. N. Davies, P. H. Bird and M. G. H Wallbridge, *JCS* (1968) *A*, 2269.
7. J. R. Berschied, Jr. and K. F. Purcell, *Inorg Chem* (1970), 624.
8. H. W. Schrötter and E. G. Hoffmann, *Ber Bun Physik Chem* (1964), 627.
9. J. A. Dilts and E. C. Ashby, *Inorg Chem* (1970), 855.
10. R. N. Kniseley, V. A. Fassel and E. E. Conrad, *SCA* (1959) *15*, 651.
11. A. L. Smith and N. C. Angelotti, *SCA* (1959), 412.
12. R. Salinger and R. West, *J Organometallic Chem* (1968) *11*, 631.
13. L. Kaplan, *JACS* (1954), 5880.
14. E. Lee and C. K. Wu, *Trans Farad Soc* (1939), 1366.
15. A. Heinemann, *Ber Bun Physik Chem* (1964), 280.
16. H. Schindlbauer and E. Steininger, *Monatsh Chem* (1961), 868.
17. J. G. David and H. E. Hallam, *JCS* (1966) *A*, 1103.

18. R. I. Wagner, L. D. Freeman, H. Goldwhite and D. G. Rowsell, *JACS* (1967), 1102.
19. M. Van Don Akker and F. Jellinek, *Rec Trav Chim* (1967), 275.
20. R. Burgada and J. Roussel, *BSC France* (1970), 192.
21. C. I. Meyrick and H. W. Thompson, *JCS* (1950), 225.
22. L. W. Daasch and D. C. Smith, *Anal Chem* (1951), 853.
23. L. J. Bellamy and L. Beecher, *JCS* (1952), 1701.
24. D. Houalla and R. Wolf, *BSC France* (1960), 129.
25. H. Stenzenberger and H. Schindlbauer, *SCA* (1970), 1713.
26. W. H. Haynie and H. H. Nielsen, *JCP* (1953) *21*, 1839.
27. H. C. Allen, Jr. and E. K. Plyler, *JCP* (1956) *25*, 1132.
28. I. W. May and E. L. Pace, *SCA* (1968), 1605.
29. W. E. Haines, R. V. Helm, C. W. Bailey and J. S. Ball, *JPC* (1954), 270.
30. R. E. Pennington, D. W. Scott, H. L. Finke, J. P. McCullough, J. F. Messerly, I. A. Hossenlopp and G. Waddington, *JACS* (1956), 3266.
31. I. F. Trotter and H. W. Thompson, *JCS* (1946), 481.
32. T. L. Cairns, G. L. Evans, A. W. Larchar and B. C. McKusick, *JACS* (1952), 3982.
33. A. Wagner, H. J. Becher and K.-G. Kottenhahn, *B* (1956), 1708.
34. J. G. David and H. E. Hallam, *SCA* (1965), 841.
35. C. B. Baddiel and G. J. Janz, *Trans Farad Soc* (1964), 2009.
36. H. D. Martough, *The Chemistry of Heterocyclic Compounds: Thiophene and Its Derivatives* 1952, Interscience.
37. N. Sheppard, *Trans Farad Soc* (1949), 693.
38. R. Mecke and H. Spieseck, *B* (1956), 1110.
39. N. Sharghi and I. Lalezari, *SCA* (1964), 237.
40. L. May and C. R. Dillard, *JCP* (1961) *34*, 694.
41. M. L. Maddoz, N. Flitcroft and H. D. Kaesz, *J Organometallic Chem* (1965) *4*, 50.
42. H. J. Emeléus and S. F. A. Kettle, *JCS* (1958), 2444.
43. C. R. Dillard and L. May, *J Mol Spectr* (1964) *14*, 250.

2.1 olefins

Olefins are characterized in the infrared by two principal absorptions: the C-C double-bond stretching, which occurs between 1600 and 1700 cm^{-1}; and the out-of-plane C-H deformation vibration, which can occur anywhere between 700 and 1000 cm^{-1}.

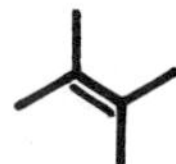

C=C STRETCHING

The intensity of this band varies considerably, with higher intensities being observed when conjugated to a carbonyl group, especially in the S-cis conformer. Nonetheless, the frequency of this absorption varies in a characteristic manner and is diagnostic of the substituents surrounding the double bond. The following tables detail the frequency as a function of olefin substitution. The reader should also check the other chapters for unsaturated systems and note that enamines (Chapter 2.4, p. 92) and enol ethers (Chapter 2.5, p. 96) are covered separately. Other vibrations such as the C-H stretch and C-H in- and out-of-plane deformation vibrations are characteristic of the degree and position of the olefinic substitution pattern and are summarized as follows:

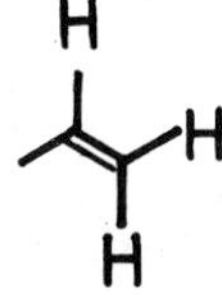

3095-3075 cm^{-1} 3030-3000 cm^{-1}	Both bands are of medium intensity and arise from the C-H stretching mode.
1860-1800 cm^{-1}	A band of medium intensity is frequently found in this region as an overtone of the C-H out-of-plane deformation around 900 cm^{-1}.
1420-1410 cm^{-1} 1300-1285 cm^{-1}	These two bands of medium intensity arise from the C-H in-plane deformation, and their position varies little with olefin substitution.
995-985 cm^{-1}	This strong band arising from the C-H out-of-plane deformation varies in position with olefin substitution. Oxygen substitution can lower the frequency to 950 cm^{-1}, whereas halogen substitution causes a smaller but still significant decrease in frequency.
915-905 cm^{-1}	This second strong absorption, again associated with the C-H out-of-plane deformation, varies little in frequency when conjugated to C-C, but can be

lowered to 810 cm^{-1} when conjugated to a carbonyl group.

	3100-3070 cm^{-1} 2985-2970 cm^{-1}	These two medium-intensity bands arise from asymmetric and symmetric vinylidene C-H stretching modes.
	1800-1750 cm^{-1}	A medium-intensity band which is an overtone of the C-H out-of-plane deformation mode around 900 cm^{-1} is found in this region.
	1420-1410 cm^{-1} 1320-1290 cm^{-1}	These two bands arise from the C-H in-plane deformation mode and vary little in frequency with substitution.
	895-885 cm^{-1}	This strong-intensity band varies considerably with olefin substitution. The changes are recorded in the tables that follow. In general, there is only a small increase in frequency when this chromophore is conjugated to other olefin groups, but conjugation to a carbonyl broup can raise the frequency to 930 to 950 cm^{-1}.

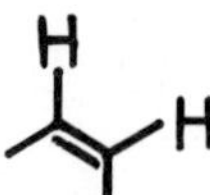

3050-3000 cm^{-1} A medium-intensity band arises from the C-H stretching vibration.

1430-1395 cm^{-1} — This band of medium intensity arising from the in-plane C-H deformation mode, is of a higher frequency than the band associated with the corresponding trans isomer, but is of little diagnostic value.

730-650 cm^{-1} — This medium-intensity band, arising from the C-H out-of-plane deformation, is not characteristic of the olefin substituents. Electronegative groups, especially halogens, cause a shift to 770 to 780 cm^{-1}, as does conjugation to a C-C double bond. Conjugation with a carbonyl group may cause an increase to 820 cm^{-1}.

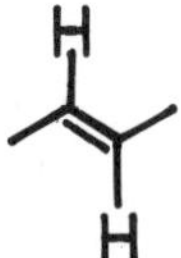

3050-3000 cm^{-1} — A medium-intensity band arising from the C-H stretching vibration is found in this region.

1300-1285 cm^{-1} — This medium-intensity band arises from the C-H in-plane deformation.

990-960 cm^{-1} — A strong band arising from the C-H out-of-plane deformation mode is observed here. The majority of trans olefins absorb at the lower end of the scale (∿ 965 cm^{-1}). Conjugation to C-C causes only a small increase in frequency, whereas conjugation to a carbonyl group can cause the frequency to rise to the higher end of the scale.

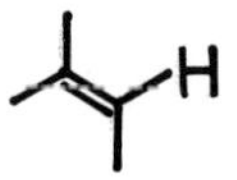

3050-2995 cm^{-1}	A medium-intensity band arises from the C-H stretching vibration.
840-790 cm^{-1}	A strong band arising from the C-H out-of-plane deformation may be seen here. Electronegative groups cause a decrease in frequency toward the lower end of the scale.

Only small changes in any of the preceding absorptions are found in going from gases to condensed phases or in changing from one solvent to another.

	gas	1647	1
	gas	996 919	2
	CS_2	986 908	3
	liq.	1645 994 912	4
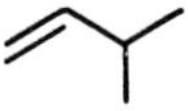	gas	1644 999 912	4
	liq.	999 912	2
	CS_2	996 910	3
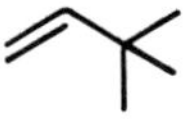	liq.	1645 1000 911	4
	CS_2	999 910	3
 F	CS_2	1652 1630 1426 1413 1289 989 935 925	5
 Cl	CS_2	1649 1643 1429 1412 1289 985 931	5
 Br	CS_2	1647 1638 1409 1294 984 926	5
 I	CS_2	1645 1632 1420 1405 1292 981 920	5

 CN	CCl_4	1645 1420 984 928	6
	CS_2	982 926	3
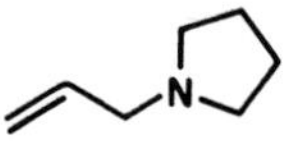 N	liq.	1644	7
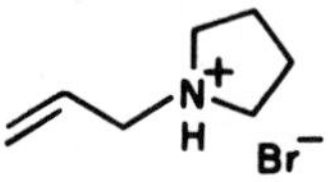 N^+ H Br^-	nuj.	1637	7
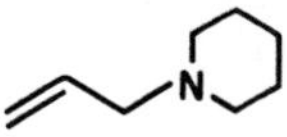 N	liq.	1644	7
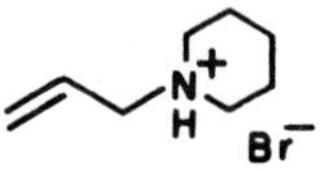 N^+ H Br^-	nuj.	1646	7
 OH	liq.	1642w	8
	CS_2	987 915	3
 O	CS_2	982 919	3
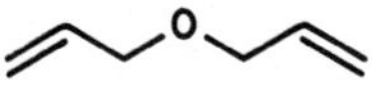 O	CCl_4	1648 988 925	6
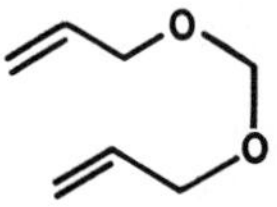 O, O	CCl_4	1645 1420 989 923	6
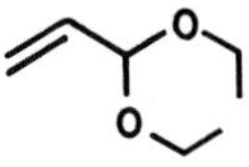 O, O	CCl_4	1647w 986	6
 S	liq.	1634 916	9
	liq.	1661 887 706	4

Structure	Medium	Frequency	Ref.
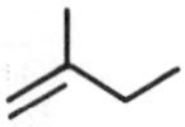 2-methyl-1-butene	gas	1681 801	4
	liq.	1652 892	4
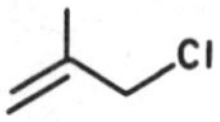 3-chloro-2-methylpropene (Cl)	CS_2	902	3
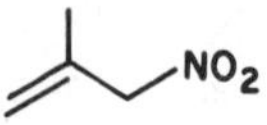 2-methyl-3-nitropropene (NO_2)	liq.	1665 1555 1570 1440 1380	10
	liq.	1661 921	11
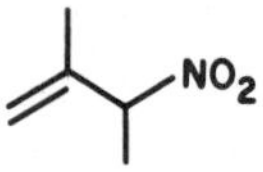 2-methyl-3-nitro-1-butene (NO_2)	liq.	1665 1655 1555 1395 1365	10
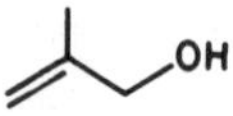 2-methyl-2-propen-1-ol (OH)	CS_2	893	3
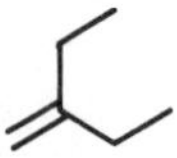 2-ethyl-1-butene	liq.	1647 889 699	4
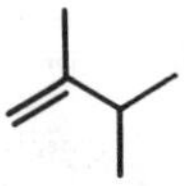 2,3-dimethyl-1-butene	liq.	1639 892	4
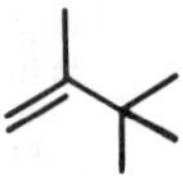 2,3,3-trimethyl-1-butene	liq.	1642 892	4
	CCl_4	890	12
	C_6H_{12}	1640	12
 2-vinylcyclopropanecarboxylic acid (O, OH)	liq.	1641	13
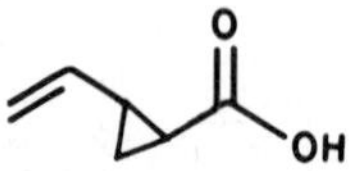 2-vinylcyclopropanecarboxylic acid (O, OH)	liq.	1632	13
 2-vinylcyclopropyl isocyanate (NCO)	liq.	1637	13
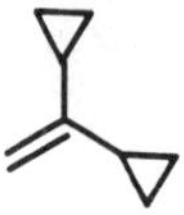 1,1-dicyclopropylethene	liq.	1631 870	14
 cis-2-butene	gas	1660	4
	liq.	1661 675	4
 trans-2-butene	gas	1676	4
	liq.	964	4
 F_3C / CF_3 (cis)	gas	1680	15
 trans (CF_3)	gas	1695	16
 cis-2-pentene	liq.	1658 696	4
trans-2-pentene	liq.	964	4
 trans-1-nitro-2-butene (NO_2)	liq.	1678 962	11
 trans-3-heptene	gas	1668	4
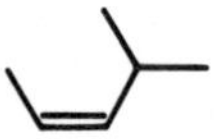 cis-4-methyl-2-pentene	liq.	1655 720	4

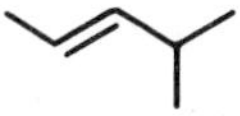	liq.	1670 965	4
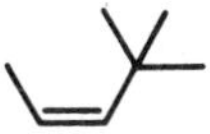	liq.	1653 707	4
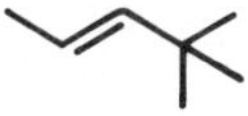	liq.	1675 971	4
	liq.	1653 714	4
	liq.	965	4
	liq.	1650 709	17
	liq.	1670 968	17
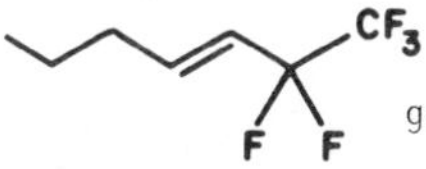 CF_3, F F	gas	1683	18
	liq.	1631 728	4
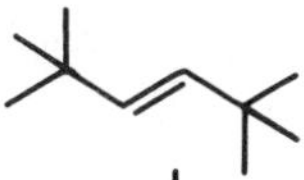	liq.	978	4
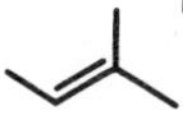	liq.	801	2
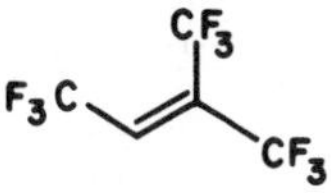 CF_3, F_3C, CF_3	liq.	1751	19
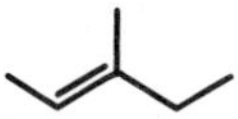	liq.	1675 823 720	4
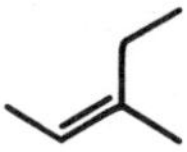	liq.	1675 812	4

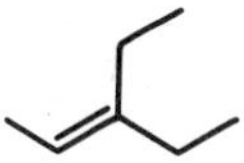	liq.	1667 825	4
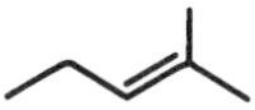	liq.	1675 833	4
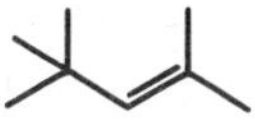	liq.	1664 824	4
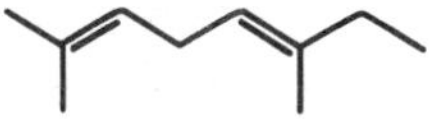	$CHCl_3$	1653 834	20
 F	liq.	925 863	3
 F	gas	1695	16
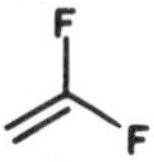 F, F	gas	1728	21
 F, F	gas	1715	22
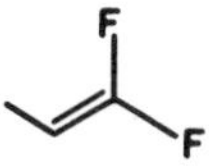 F, F	gas	1745	23
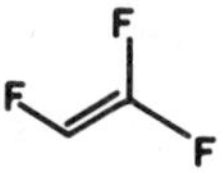 F, F, F	gas	1788	24
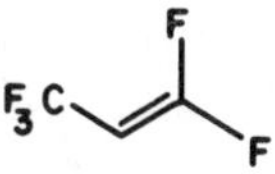 F, F_3C, F	gas	1779	25

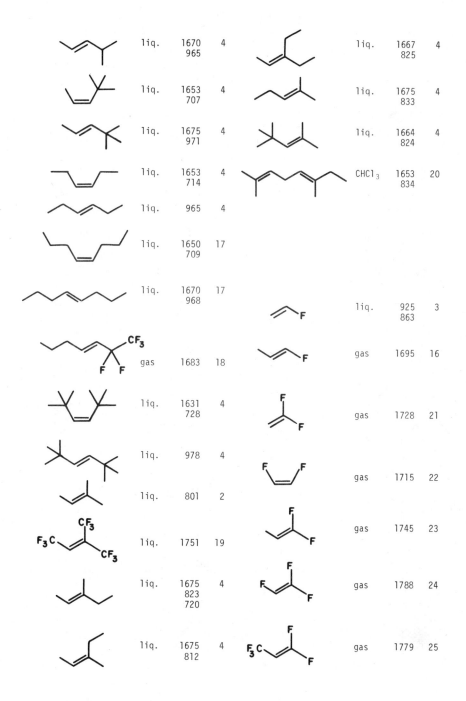

Phase	Frequency	Ref.
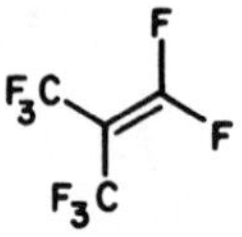 gas	1751	26
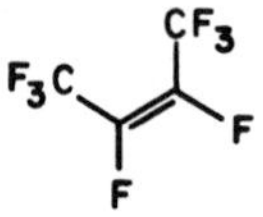 gas	1733	19
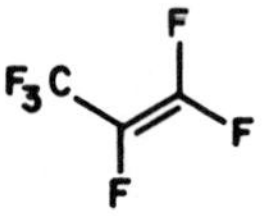 gas	1799	27
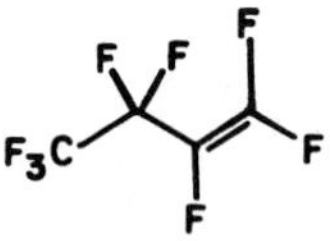 gas	1792	19
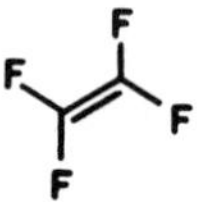 gas	1741 1729	28
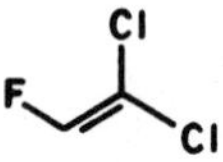 gas	1654	29
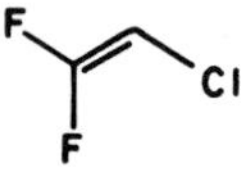 gas	1742	30
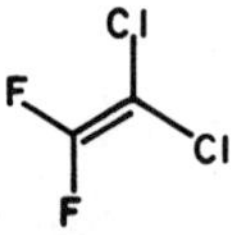 gas	1749	21
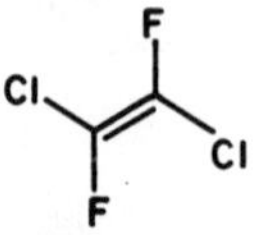 gas	1707	31
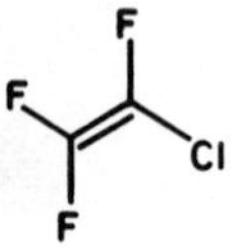 gas	1792	32

Phase	Frequency	Ref.
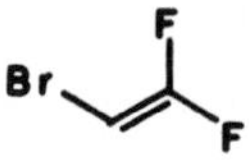 gas	1736	33
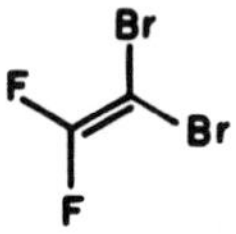 gas	1718	26
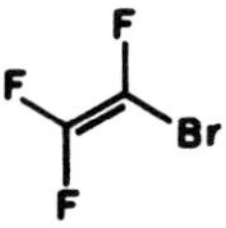 gas	1783	34
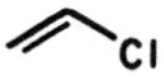 liq.	940 895	2
CS_2	938 894	
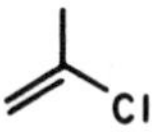 CS_2	875	3
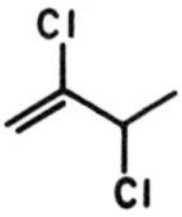 liq.	1640	35
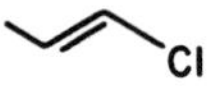 liq.	1195 895 815	36
CS_2	926	3
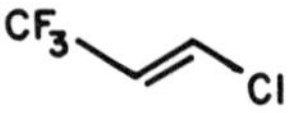 gas	1650	16
 liq.	1639 757	37
 liq.	1642 942	37

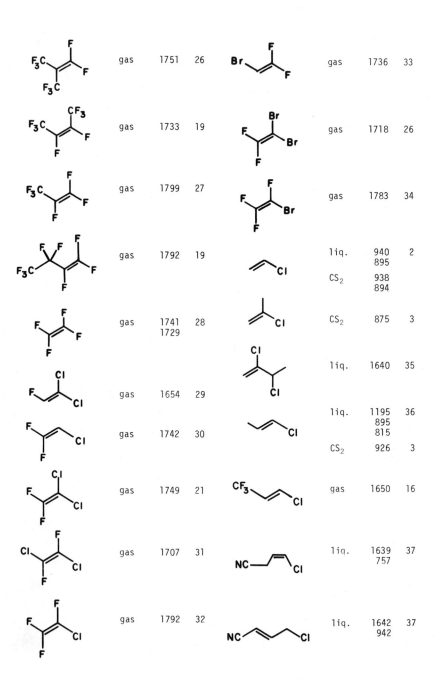

Medium	cm⁻¹	Ref.
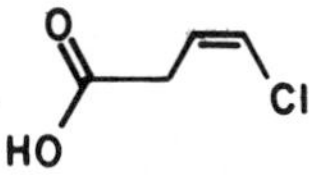 liq.	1639 754	37
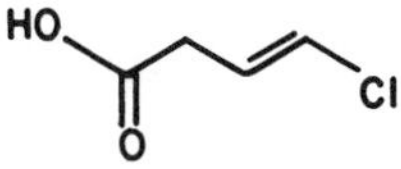 liq.	1639 931	37
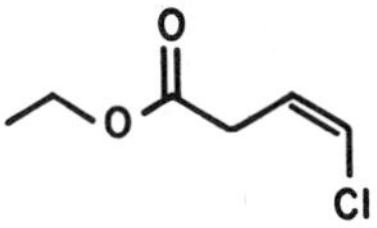 liq.	1639 757	37
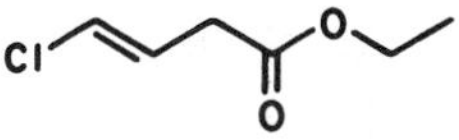 liq.	1639 929	37
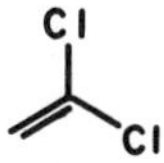 liq.	1614	38
CS_2	867	3
 liq.	1595 1295 845 695	36
 liq.	1658m 1200 895	39
CS_2	892	3
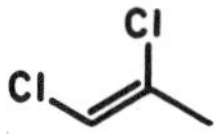 liq.	1614	40
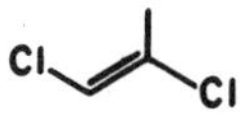 liq.	1615w	40
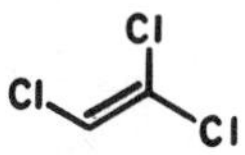 liq.	1587m 1555m 929 840	41

Medium	cm⁻¹	Ref.
 liq.	940 902	2
CS_2	936 898	3
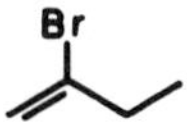 CCl_4	1645	42
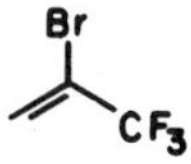 CS_2	929	3
 CCl_4	1668	42
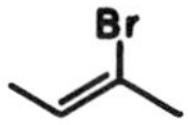 CCl_4	1670	42
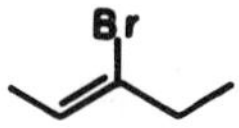 CCl_4	1660	42
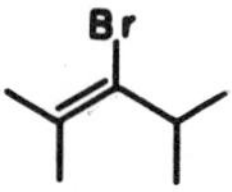 CCl_4	1650	42
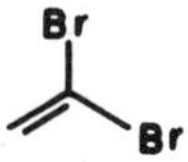 CS_2	877	3
 liq.	1585 1254 748	43
 liq.	1628m 1160 899	43
CS_2	896	3
 liq.	946 909	2
CS_2	943 905	3

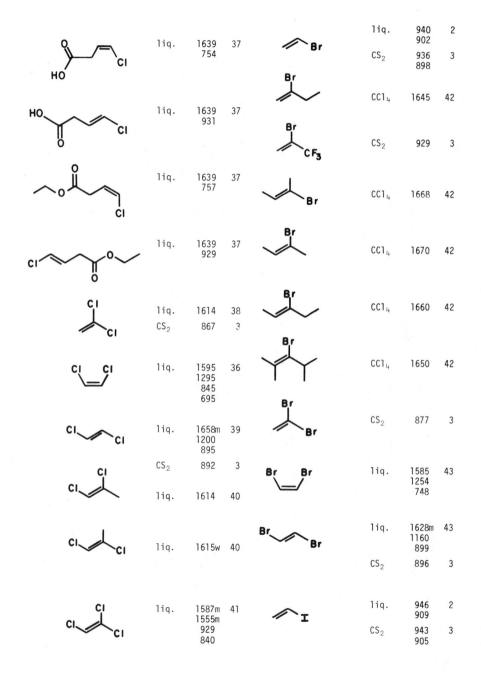

Compound	Medium	Frequency	Ref.
CN	CCl_4	1645 1612 1413 960	6
	CS_2	960	3
	$CHCl_3$	1647 1610	44
Cl, CN	CS_2	916	3
Cl, CN	CS_2	920	3
Br, Br, Br, CN	CCl_4	1541 2224	45
I, I, I, CN	$CHCl_3$	1498 2212	45
CN	CS_2	930	3
	$CHCl_3$	1626	44
CN	CS_2	953	3
	$CHCl_3$	1638	44
CN	$CHCl_3$	1626	44
CN	liq.	1639 2222	46

Compound	Medium	Frequency	Ref.
CN	liq.	1644 2222	46
CN	liq.	1639 2207	46
H_2N, CN	KBr	1646 1600	47
N, CN	liq.	1590	48
CN, CN	CS_2	985	3
NC, CN	$CHCl_3$	1630 1599	44
NC, CN	$CHCl_3$	1612	44
NC, CN	$CHCl_3$	1637 1616	44
CN, CN	$CHCl_3$	1614	44
NC, CN	$CHCl_3$	1645 1615	44
CN, CN	CH_2Cl_2	1613 2237	47
CN, Cl, Cl, CN	KBr	1541	49

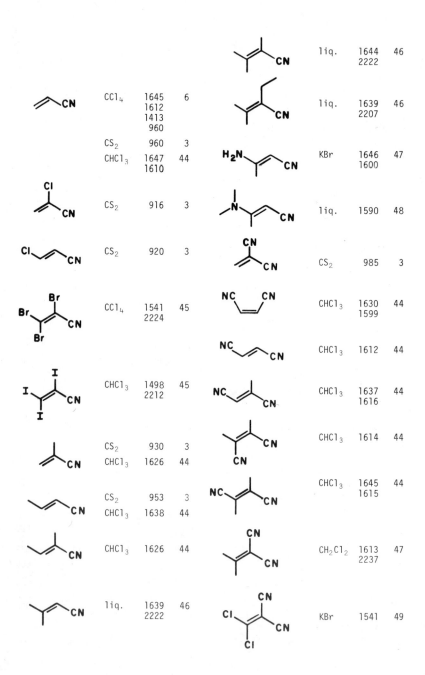

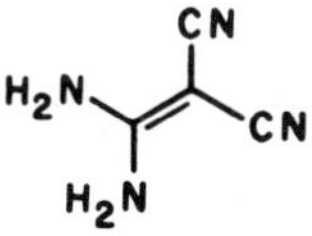 KBr	1636 1545	50
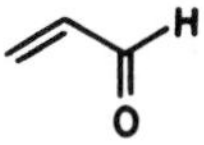 CCl_4	1618 985 971 913	6
CS_2	984 963	3
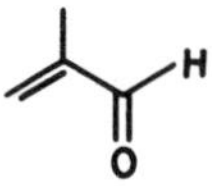 C_2Cl_4	1638	51
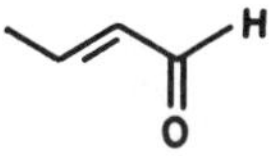 CS_2	964	3
C_2Cl_4	1644	51
$CHCl_3$	1638	52
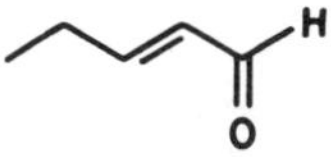 $CHCl_3$	1645	53
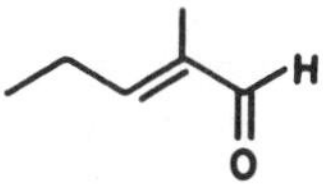 liq.	1620	54
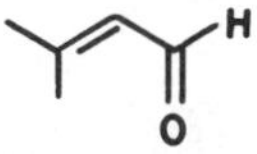 C_2Cl_4	1638 1621	51
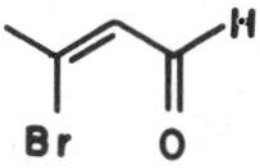 CCl_4	1629	55
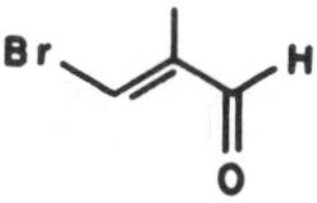 CCl_4	1615	55

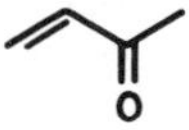 CCl_4	1618 983 953	6
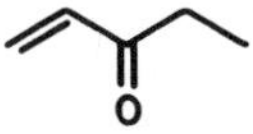 CCl_4	1619 984 955	6
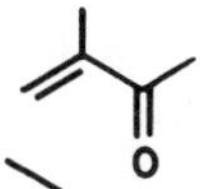 CCl_4	1634 930	6
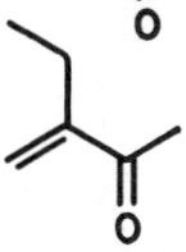 CS_2	931	3
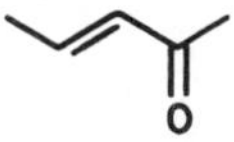 CS_2	938* 972	3
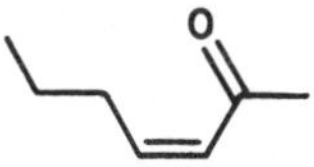 liq.	1623 975 662	56
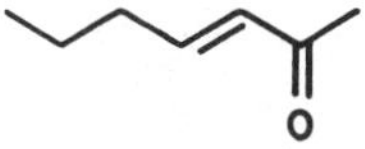 liq.	1631 977	56
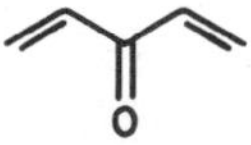 liq.	1660	57
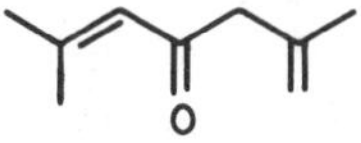 CCl_4	1650* 1626	58
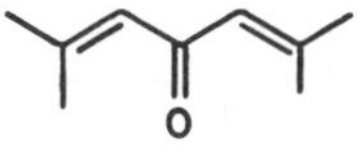 CCl_4	1636 1620	58
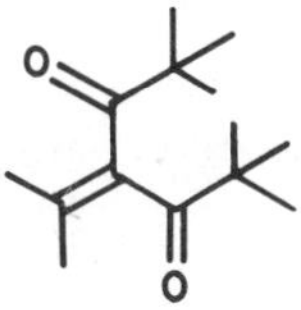 liq.	1645w	59

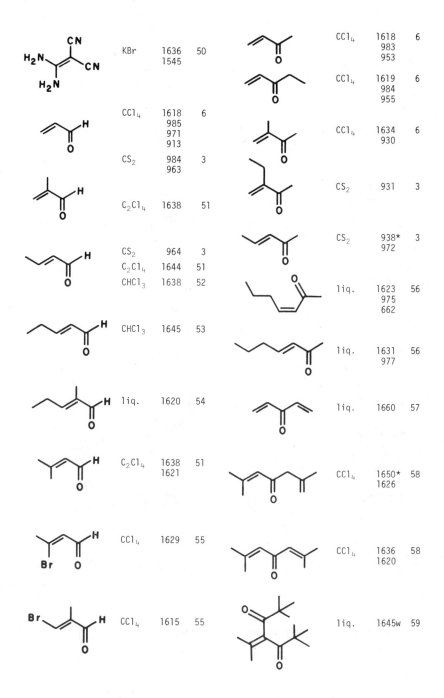

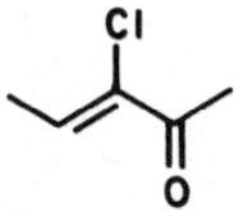

Solvent	ν (cm⁻¹)	Ref.	Solvent	ν (cm⁻¹)	Ref.
$CHCl_3$	1613	60	CCl_4 / CS_2	1639 / 818	66 / 66
liq.	1585	51	nuj. / CCl_4 / CS_2	1653 / 1651 / 970	67 / 66 / 66
CCl_4	1608	61	nuj.	1647	67
liq.	1629	62	nuj.	1621	67
$CHCl_3$	1658	63	$CHCl_3$	1650	47
$CHCl_3$	1635	63	nuj.	1624w	68
CCl_4	1635 / 1615 / 981 / 973 / 930	6	$CHCl_3$	1639w	68
CCl_4	1641	64	CCl_4	1637 / 1404 / 984 / 965	6
nuj.	1618m / 1304 / 909b	65	CCl_4	1640 / 1407 / 984 / 965	6

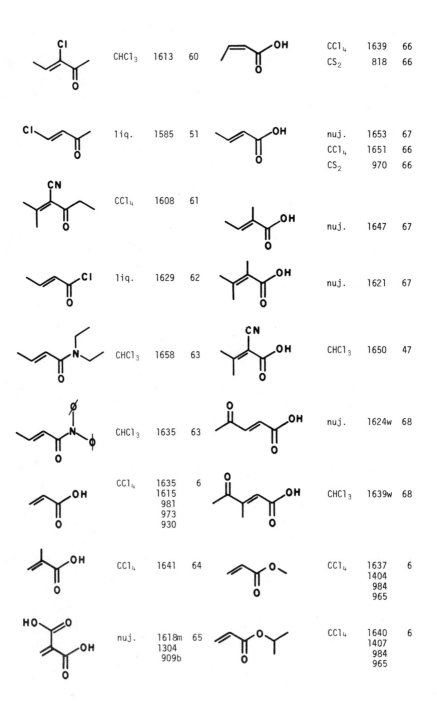

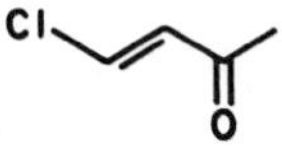

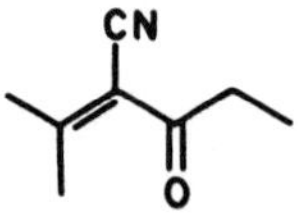

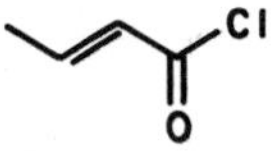

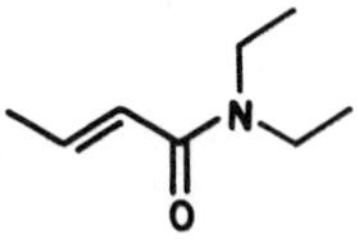

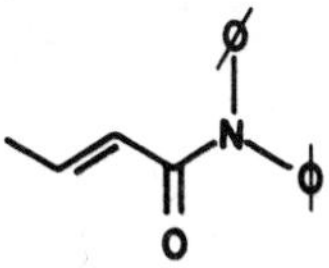

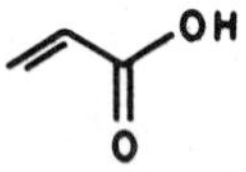

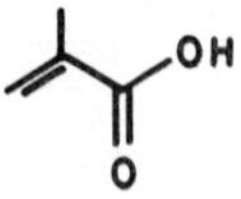

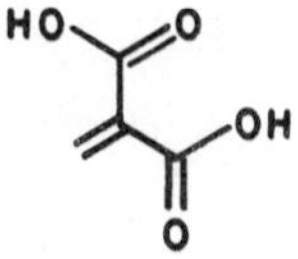

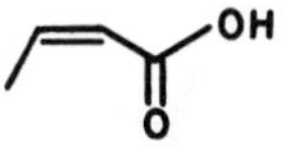

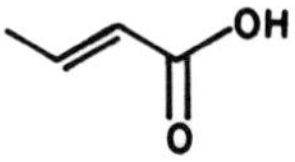

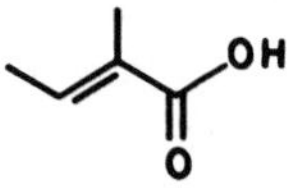

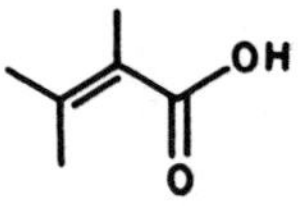

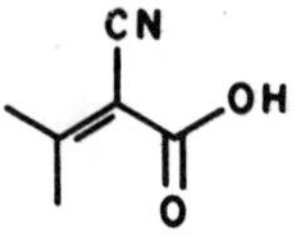

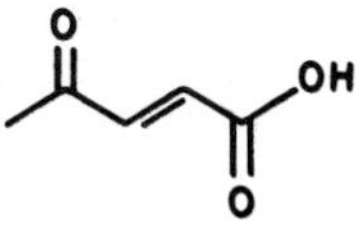

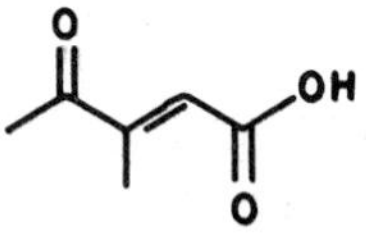

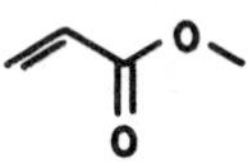

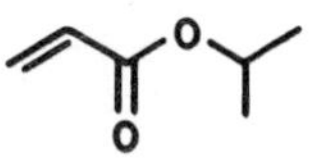

Solvent	Frequency	Ref.
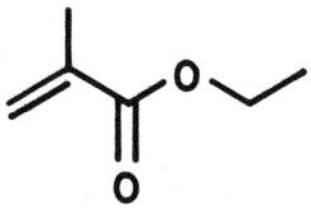 $CHCl_3$	1638	44
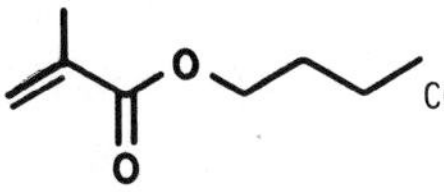 CCl_4	1640 1404 938	6
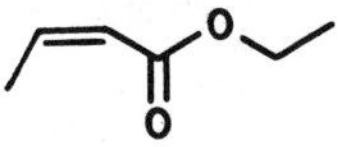 CCl_4 CS_2	1644 812	66 66
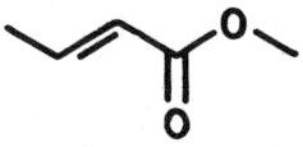 CCl_4 CS_2	1659 969	66 66
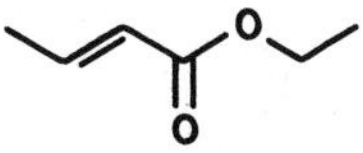 $CHCl_3$	1662	44
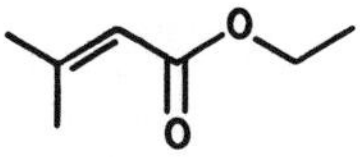 $CHCl_3$	1654	44
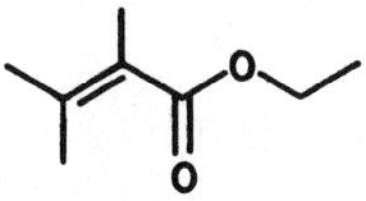 CCl_4	1640	69
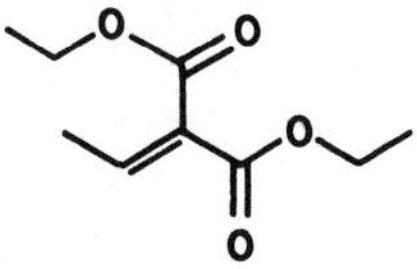 $CHCl_3$	1640	44
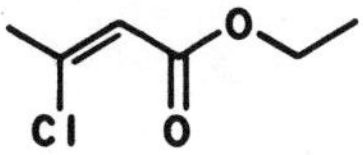 liq.	1631	70

Solvent	Frequency	Ref.
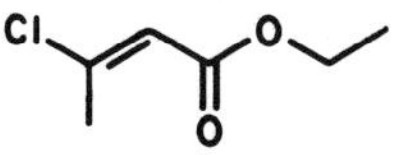 liq.	1637	70
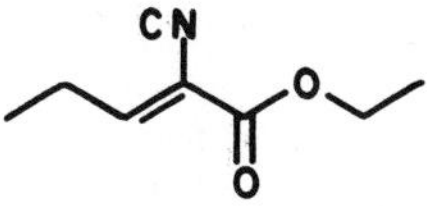 $CHCl_3$	1602	44
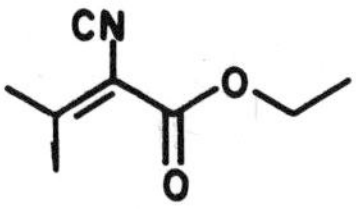 CCl_4	1614	47
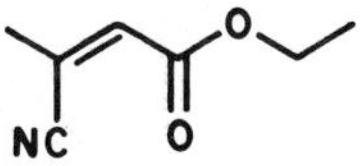 $CHCl_3$	1642	44
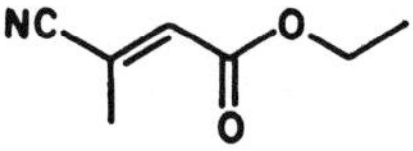 $CHCl_3$	1637	44
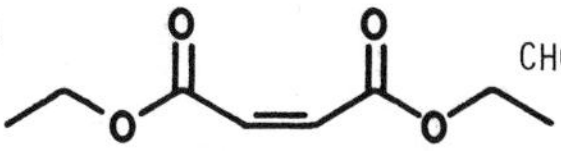 $CHCl_3$	1647	44
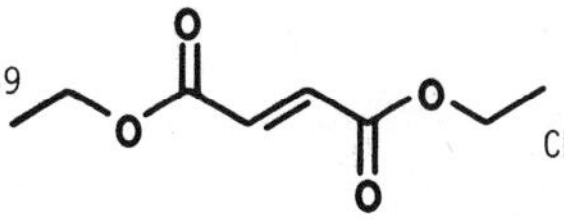 $CHCl_3$	1646	44
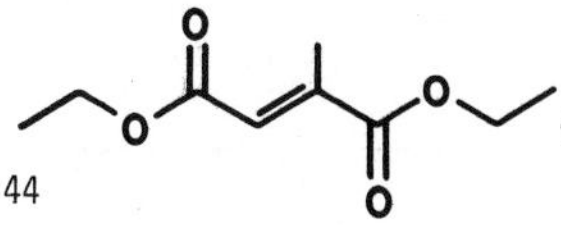 $CHCl_3$	1649	44
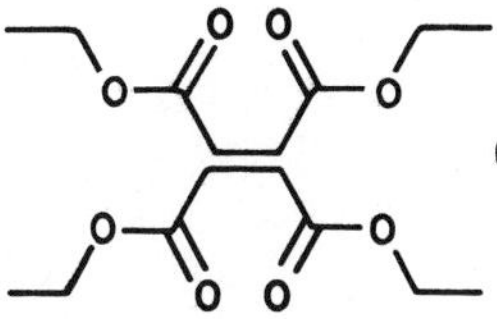 $CHCl_3$	1651	44

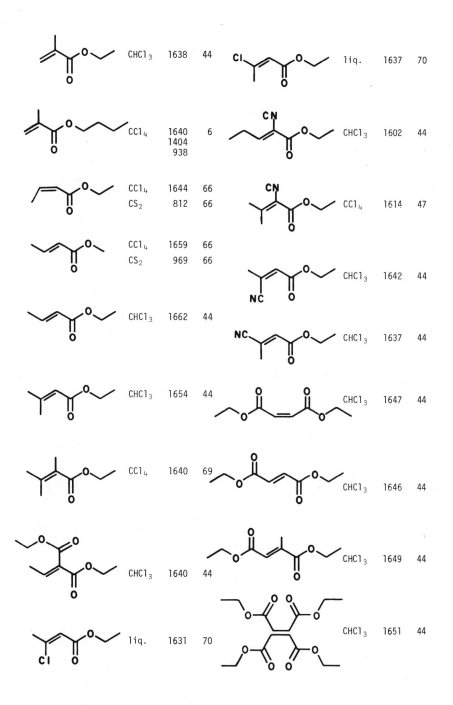

Structure	Medium	Frequencies	Ref.
 NCO	CCl_4	1629	71
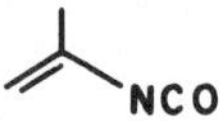 NCO	CCl_4	1651	71
 NCO	CCl_4	1660	71
 NO_2	liq.	1642 965 940	11
 NO_2	liq.	1665 1530 1375 1350	10
 NO_2	liq.	1653 960	11
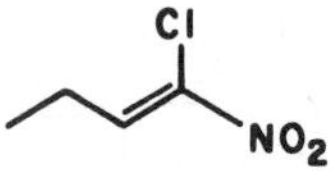 Cl, NO_2	$CHCl_3$	1632 788	11
 NO_2	liq.	1650 1520 1360 1335	10
	$CHCl_3$	1650 823	11
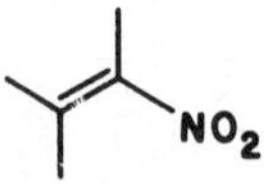 NO_2	liq.	1655 1515 1385 1350	10
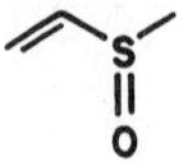 S, O	liq.	1613w 1321w 960 928	9
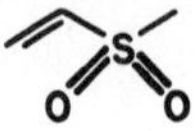 S, O, O	liq.	1626w 961 790	9
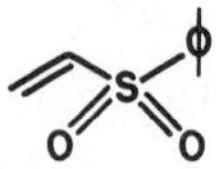 ϕ, S, O, O	liq.	1634w 1299m 976 952m	72
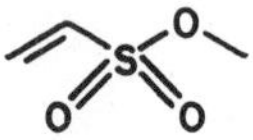 O, S, O, O	liq.	1636 974* 955	73
 $SiCl_3$	liq.	1608 1404 1005 978	74
	liq.	1592 1014 909	75
	CCl_4	1011 907	3
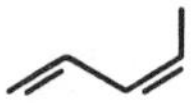	CCl_4	998 908	76
	CS_2	1001 948 896	3
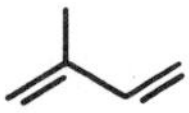	liq.	1656 1613 1016 883	77
	CCl_4	990 910	76
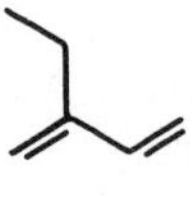	liq.	1633w 1590	78
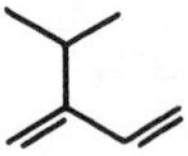	liq.	1632w 1590	78
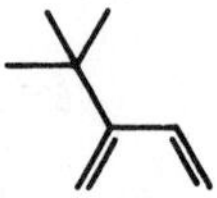	liq.	1645* 1611	78

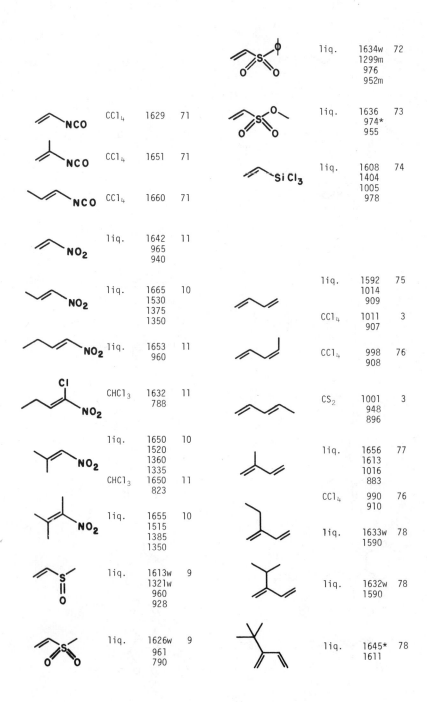

Phase	Frequency	Ref.
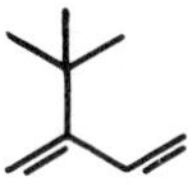 liq.	1625*w 1595*w	78
 CCl_4	883	76
 CCl_4	994 893	76
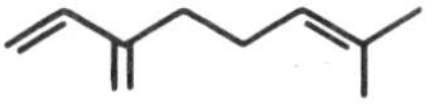 $CHCl_3$	1608m 927 901 834	20
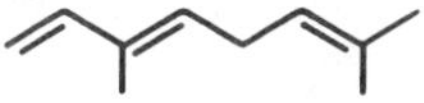 $CHCl_3$	1653m 908 847m	20
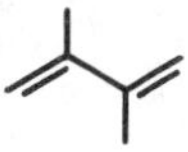 liq.	1637 1603 1016 993 965 936 917 890	79
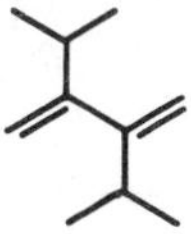 liq.	1625 1600	80
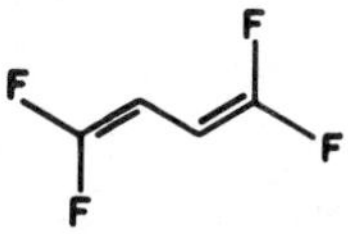 gas	1720	81

Phase	Frequency	Ref.
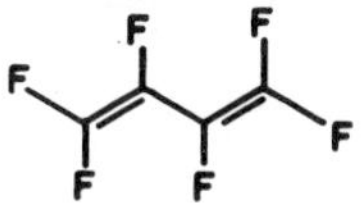 gas	1770	82
 liq.	1634 1580w 1416	83
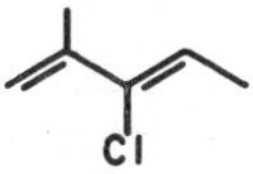 liq.	1633 1613	84
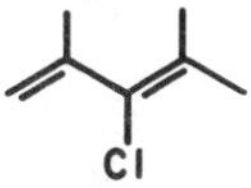 liq.	1640	84
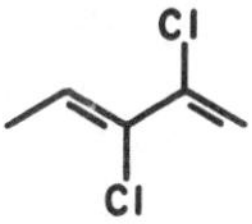 liq.	1608w 1570 891 712	85
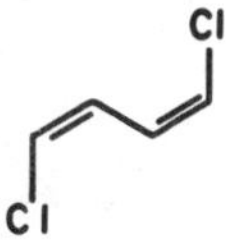 liq.	1567 700	86
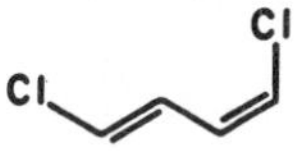 liq.	1567 945 709	86
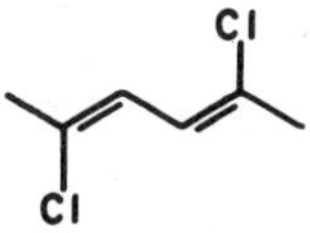 liq.	1610 1080 865	87
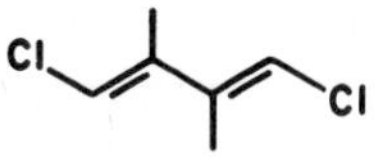 liq.	1580 990 955 790 780 710	87
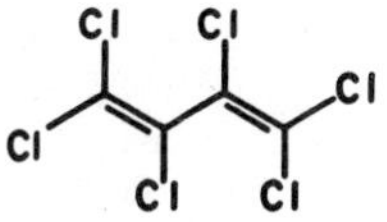 liq.	1608w 1560	88

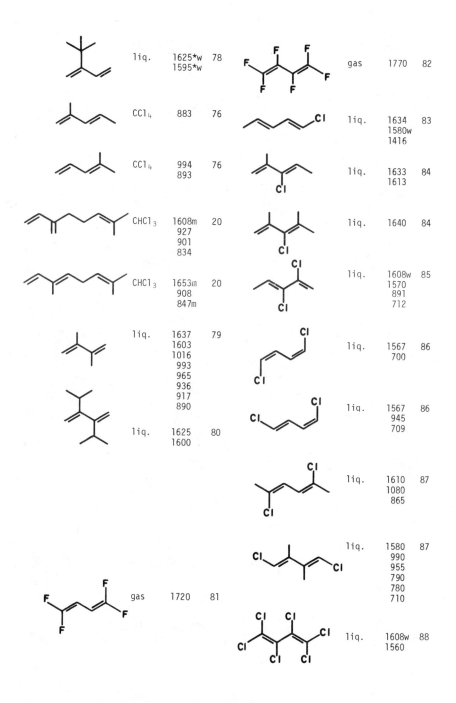

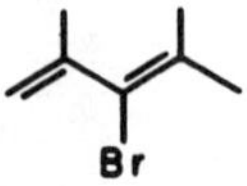

Medium	Frequency (cm^{-1})	Ref.
liq.	1645 1623	84
CCl_4	1583	89
liq.	1634 1587 906 850	84
liq.	1618 1582	84
$CHCl_3$	1642 1608	52
liq.	1633 1591	90
solid	1620 1598m	66
CS_2	829	66
CCl_4	1636 1600m	66
CS_2	996 958 938	66

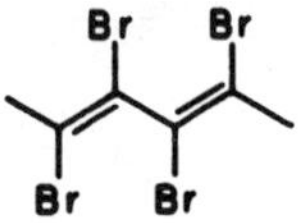

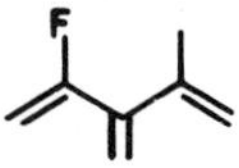

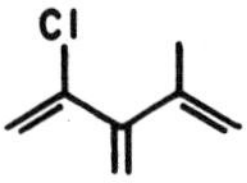

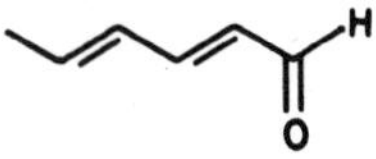

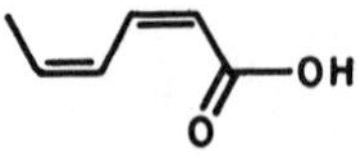

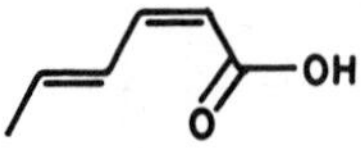

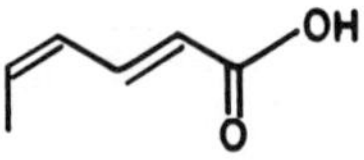

Medium	Frequency (cm^{-1})	Ref.
CCl_4	1627 1601	66
CS_2	990 949	66
nuj.	1639 1613	66
CS_2	998 945	60
KBr	1620	91
CCl_4	1623 1587m	66
CS_2	821	66
CCl_4	1639 1597m	66
CS_2	996 958 938 832	66
CCl_4	1634 1598m	66
CS_2	990 951 936	66
CCl_4	1642 1614	66
CS_2	995 943	66
CCl_4	1620	92
CCl_4	1615	92

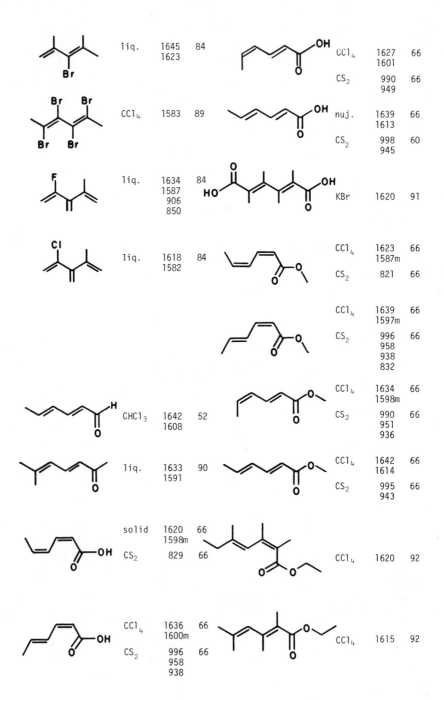

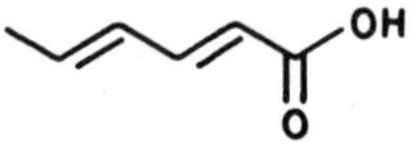

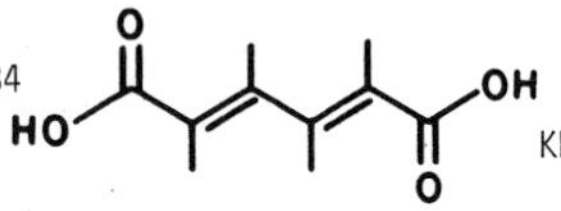

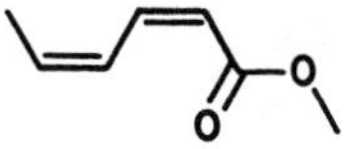

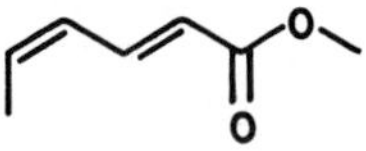

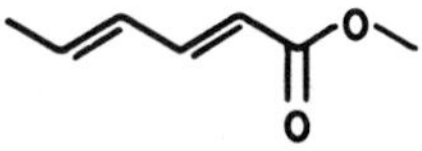

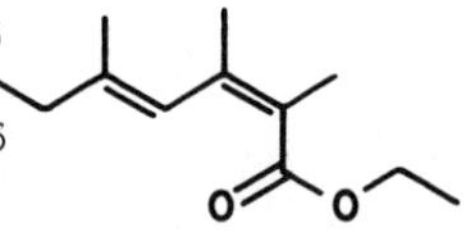

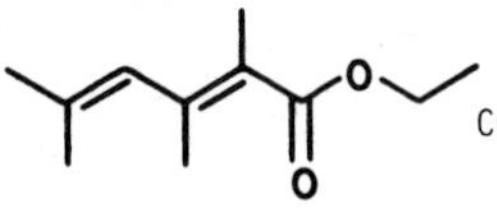

Solvent	Frequency	Ref.
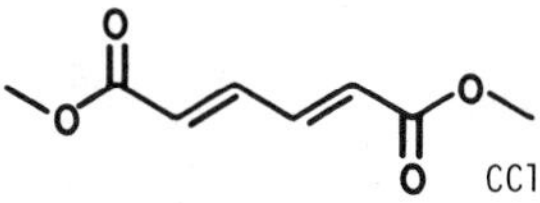 CCl_4	1640	91
 $CHCl_3$	1650 1592	52
 liq.	1693 1639 993 937 838	93
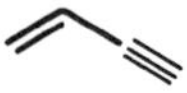 CCl_4	1610 971 928	94
CS_2	972 925	3
 liq.	1608 976 920	94
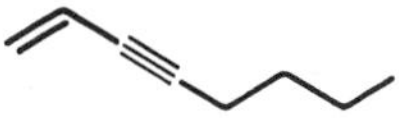 liq.	1640w 1608 1586m 972 911	95
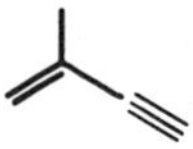 liq.	1621 903	94
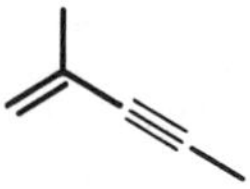 liq.	1616 892	95
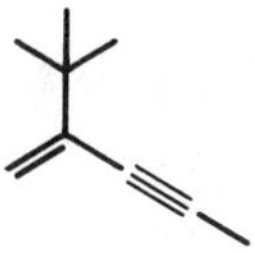 liq.	1643m 1604 892	95

Solvent	Frequency	Ref.
 CS_2	723	66
 liq.	1623 955	95
CS_2	956	66
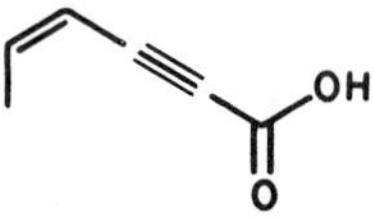 CCl_4	1615	66
CS_2	721	66
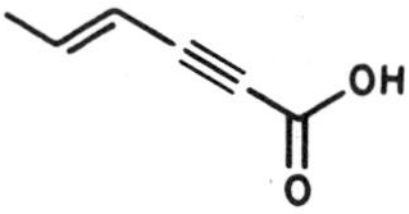 nuj.	1618w 951m	66
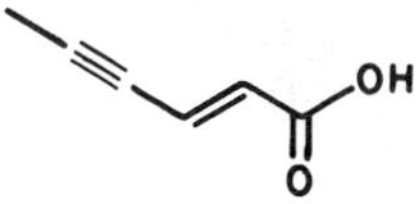 CCl_4	1619	66
CS_2	959	66
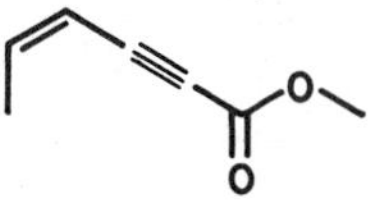 CCl_4	1615	66
CS_2	721	66
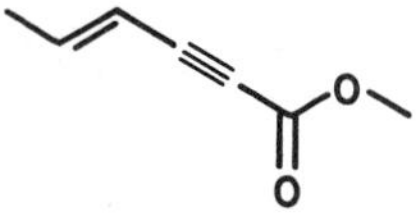 CCl_4	1626	66
CS_2	948	66
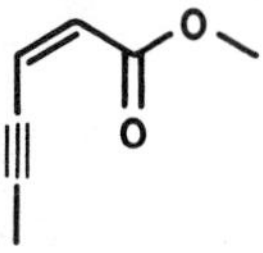 CCl_4	1612	66
CS_2	812	66
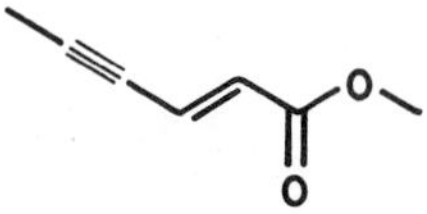 CCl_4	1615	66
CS_2	959	66

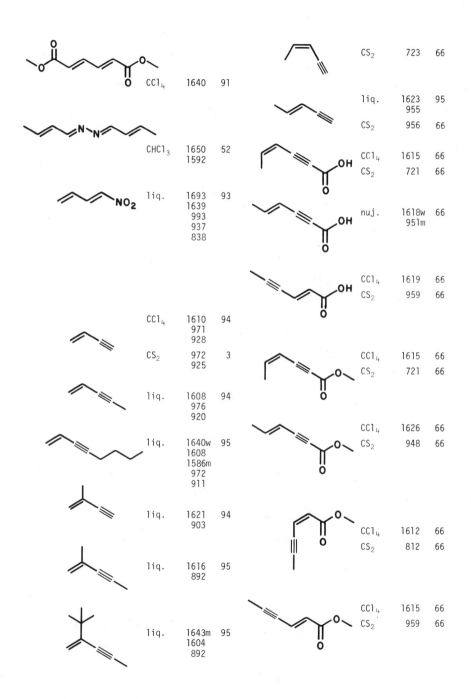

Structure	Medium	Frequencies (cm^{-1})	Ref.
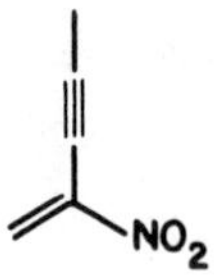 NO$_2$	liq.	1610 940	93
	liq.	1830 1621 1451 987 818	96
	liq.	1800 1621 1011	96
	liq.	941m 899	97
	liq.	1626 1582	98
	liq.	1002 859 730	99
	liq.	1650w 1631 1590	98
	liq.	1002 971	100
	liq.	1653m 1595m	98
	CCl_4	1661	98
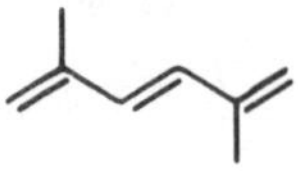	liq.	882	101

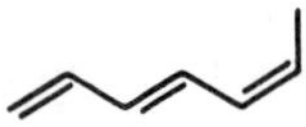

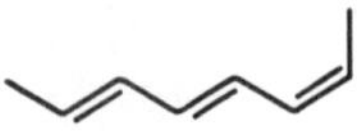

Structure	Medium	Frequencies (cm^{-1})	Ref.
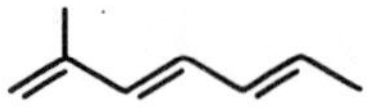	liq.	1012 1006 898	101
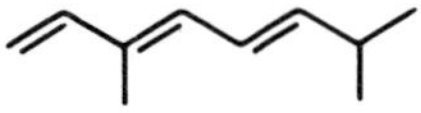	$CHCl_3$	1653	20
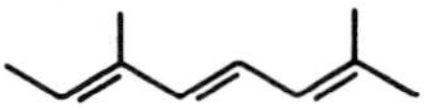	$CHCl_3$	1650 997 956	20
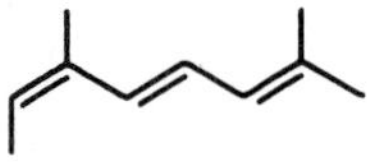	$CHCl_3$	1645m 997 956 844m	20
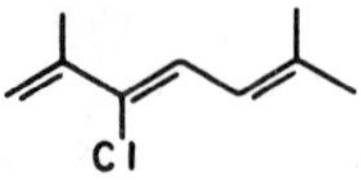 Cl	liq.	1639 1608 951 852	84
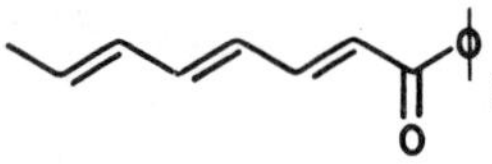 O, φ	KBr	1595 1565	102
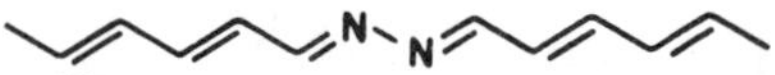 N–N	$CHCl_3$	1624 1573	52
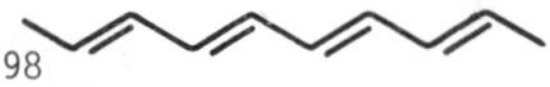	$CHCl_3$	1677 1646 1615	52
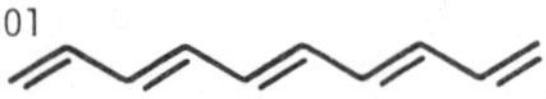	KBr	1802w 1608	103

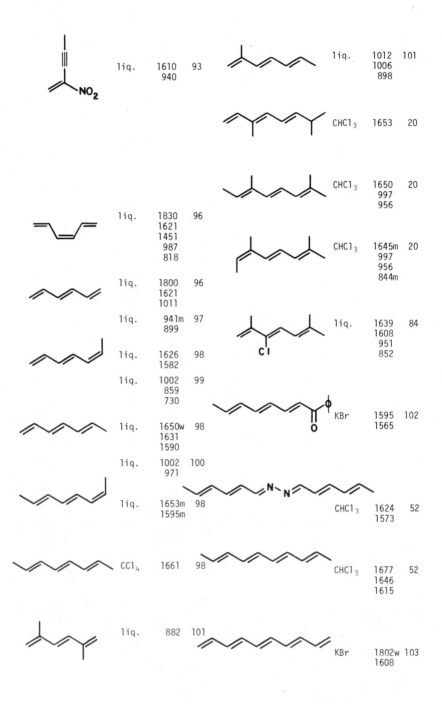

Medium	Frequency	Ref.
 liq.	1664 1635 1611	52
 KBr	1802m 1608 1417 1006 985 899	103
liq.	1636	104
liq.	992 909	105
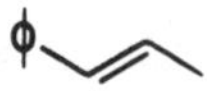 CS_2	959	3
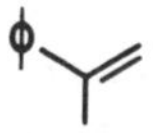 CS_2	890b	3
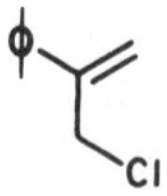 CS_2	907	3
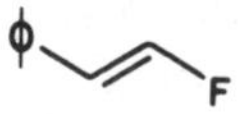 CCl_4	1660	106
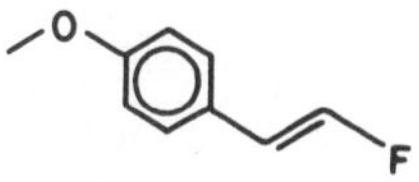 CCl_4	1660	36

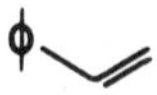

Medium	Frequency	Ref.
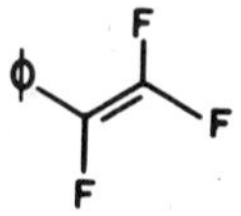 liq.	1765	107
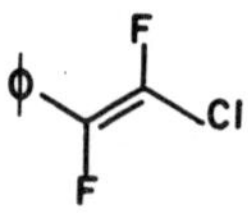 gas	1700	108
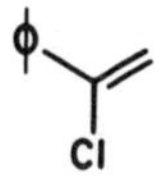 CS_2	877	3
liq.	1626	46
CS_2	962	3
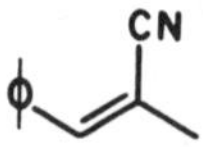 liq.	1626	46
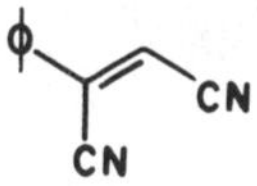 $CHCl_3$	1590	44
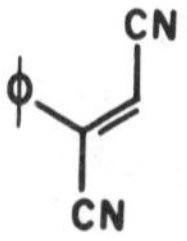 $CHCl_3$	1592	44
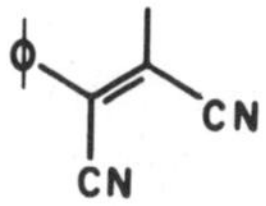 $CHCl_3$	1593	44
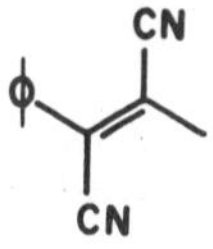 $CHCl_3$	1597	44

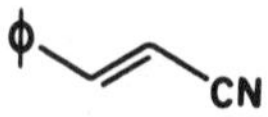

Solvent	Frequency	Ref.
CS_2	972	3
CCl_4	1608	55
CCl_4	1600	55
CCl_4	1628	109
CCl_4	1612	109
CCl_4	1616	109
CCl_4	1610	109
CS_2	984* 975	3
CCl_4	1605	109
CCl_4	1652	110
liq.	1637	111
liq.	1654	112
KBr	1645 1620	112
liq.	1620	112
CCl_4	1690	113
$CHCl_3$	1631	114

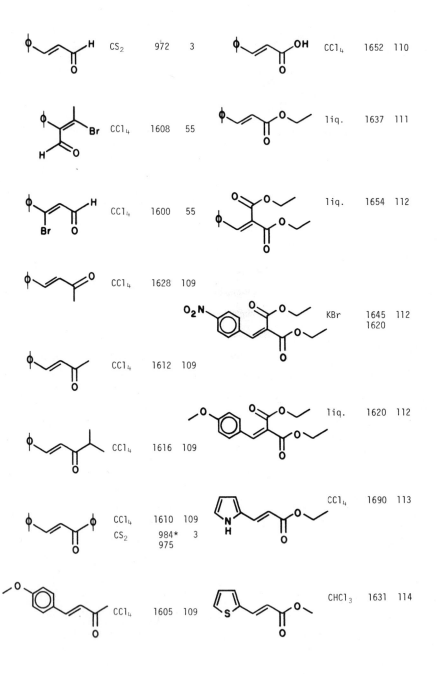

Phase	Frequency	Ref.
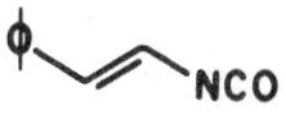 CCl_4	1646	71
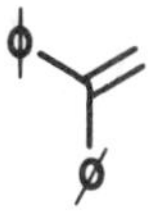 liq.	1608m	115
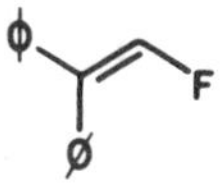 gas	1630	116
 liq.	922 779	117
CCl_4	779 696	118
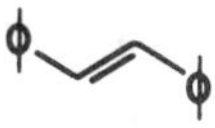 C_6H_{12}	960	118
CCl_4	685	118
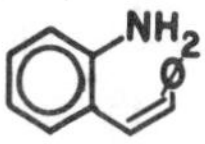 liq.	919 784	117
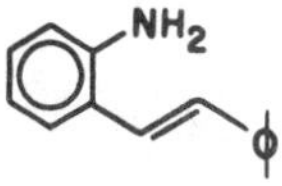 solid	986	117
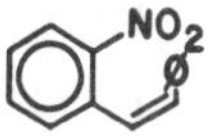 solid	962	117

Phase	Frequency	Ref.
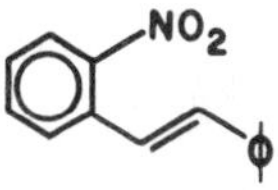 solid	917 791	117
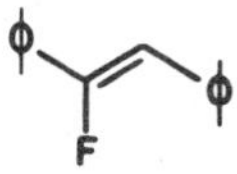 gas	1655	119
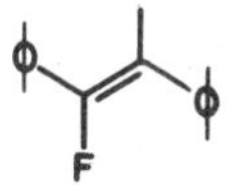 gas	1660	120
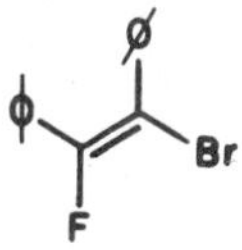 gas	1645	119
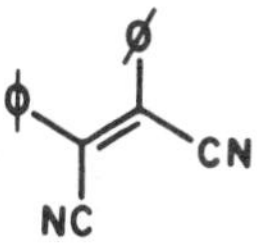 $CHCl_3$	1593	44
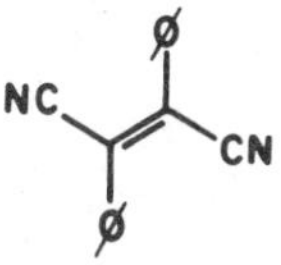 $CHCl_3$	1619	44
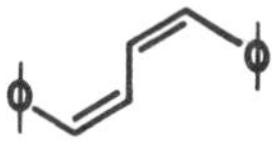 CCl_4	775 704 695	118
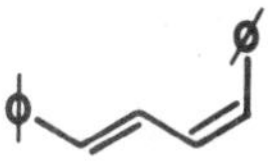 C_6H_{12}	995 945	118
CCl_4	775 698 690	118
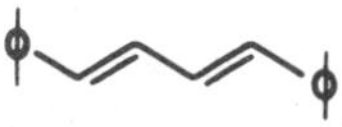 C_6H_{12}	996	118
CCl_4	685	118

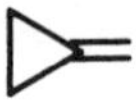

gas	1754 892	122
CCl_4	1730	121
liq.	1761 1637	123
liq.	1727	124
liq.	1776	125
liq.	1798	126
liq.	1786	126
liq.	1786	126
liq.	1779	125
liq.	1776	126

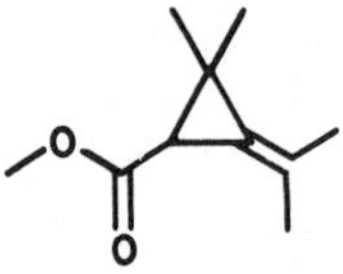

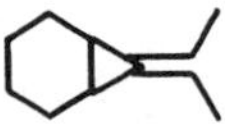

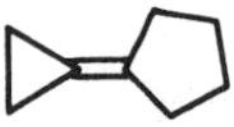

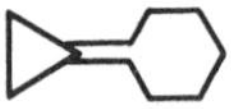

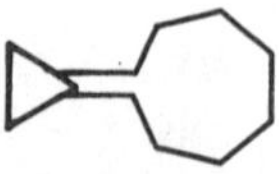

gas	1802 1681 1388 1183	127
CCl_4	1678	121
liq.	1661 880	128
CCl_4	1684 893 1812	129
liq.	1672 864	130
liq.	1669 1595 849	131
liq.	1670 1065 870 820 782	87
liq.	1672 1598 857	128
CCl_4	1672 1598	128

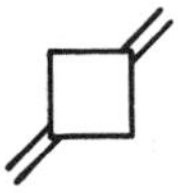

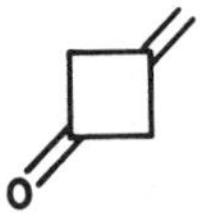

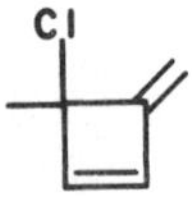

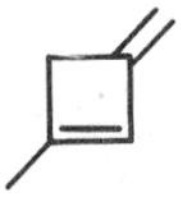

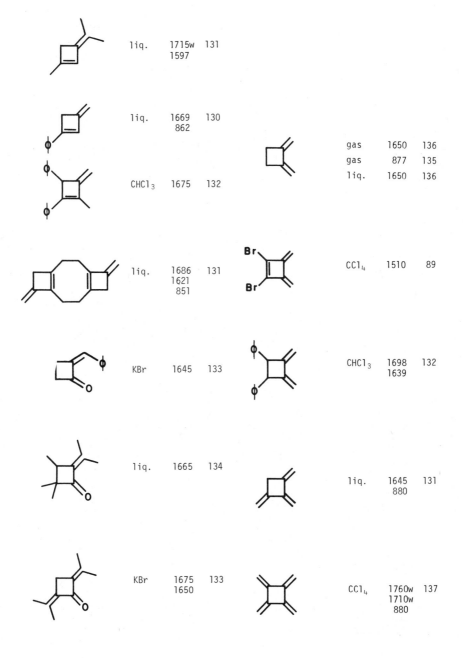

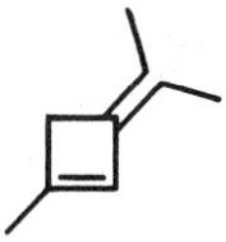

Medium	Frequency	Ref.
liq.	1715w 1597	131
liq.	1669 862	130
$CHCl_3$	1675	132
liq.	1686 1621 851	131
KBr	1645	133
liq.	1665	134
KBr	1675 1650	133

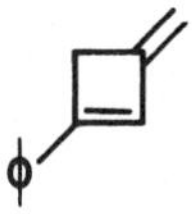

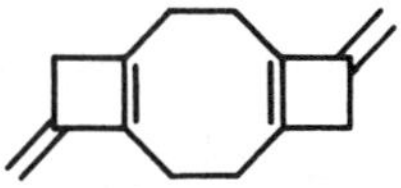

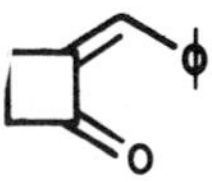

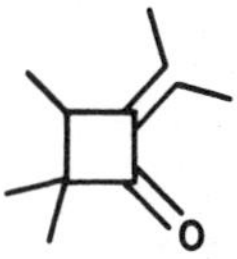

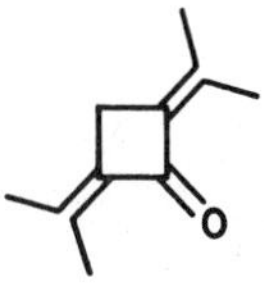

Medium	Frequency	Ref.
gas	1650	136
gas	877	135
liq.	1650	136
CCl_4	1510	89
$CHCl_3$	1698 1639	132

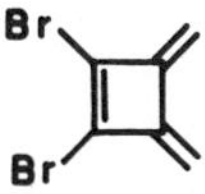

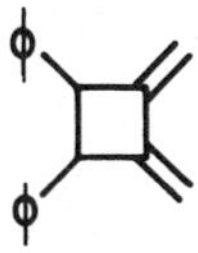

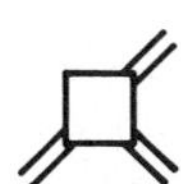

Medium	Frequency	Ref.
liq.	1645 880	131

Medium	Frequency	Ref.
CCl_4	1760w 1710w 880	137

Solvent	Frequency	Ref.
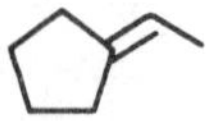 liq.	1672w 1460 1434 804	138
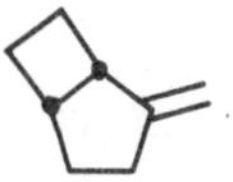 liq.	1650 875	139
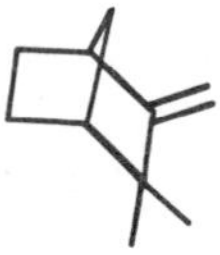 $CHCl_3$	1661m 971m 909 885	20
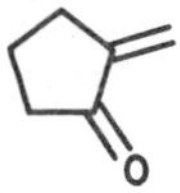 CCl_4	1644	140
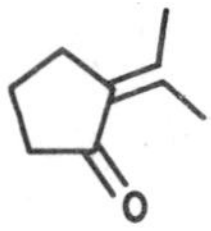 CCl_4	1633	141
$CHCl_3$	1632	140
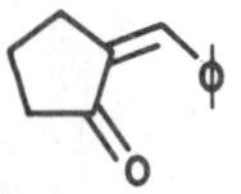 CCl_4	1630 1570w	142
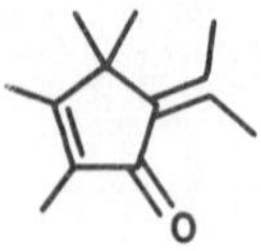 CCl_4	1623	143
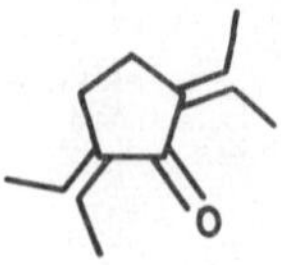 KBr	1635	133
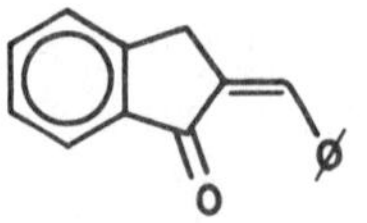 CCl_4	1619	144
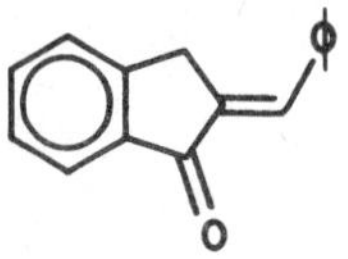 CCl_4	1640	145
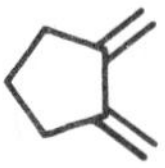 gas	1645w 1625m	146
liq.	1645w 1626 881	146
CCl_4	881	147
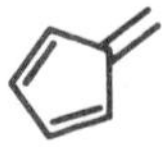 liq.	1644 925 892 765	148
CS_2	1664	149
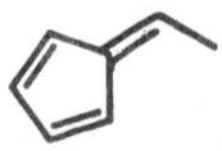 CCl_4	1645	149
CS_2	1645	149
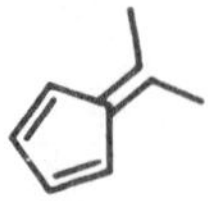 CCl_4	1642	149
CS_2	1642	149

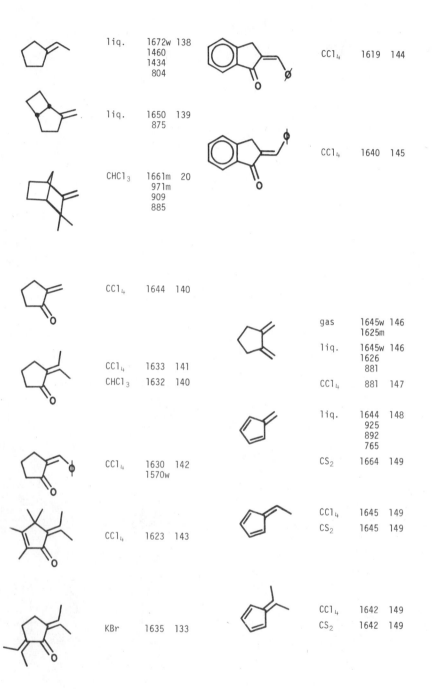

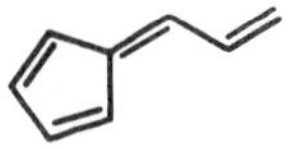

Solvent	cm⁻¹	Ref.
CCl_4	1840m 1800m 1625 1596 888	150
CSe	761	150
CCl_4	1626	149
CS_2	1626	149
CCl_4	1597	149
CS_2	1597	149
CS_2	1629	149
CS_2	1597	149
$CHCl_3$	1655w	151
nuj.	1640	152
$CHCl_3$	1650w 1605w	68
$CHCl_3$	1642w	153
liq.	1650	154
KBr	1642	155
KBr	1629	155
liq.	1642 1445 886	156

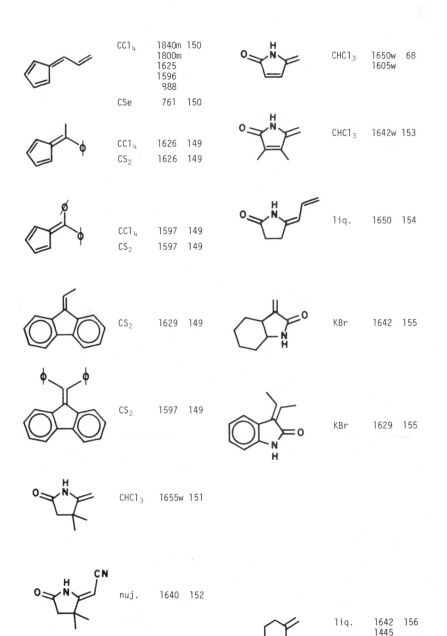

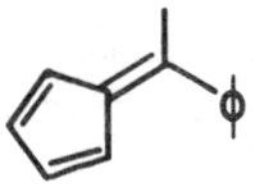

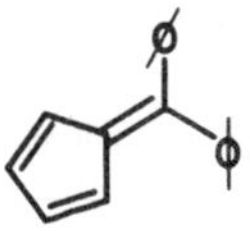

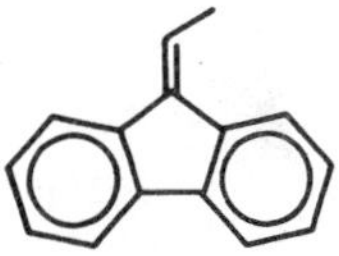

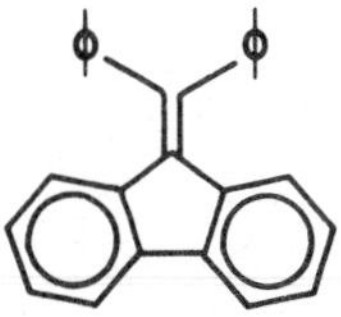

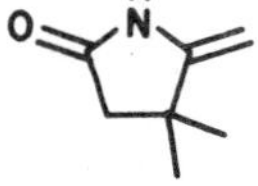

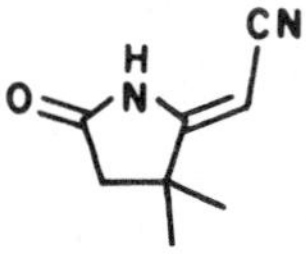

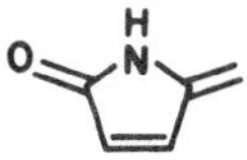

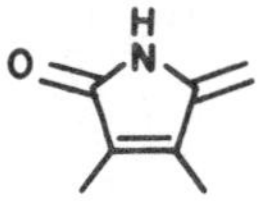

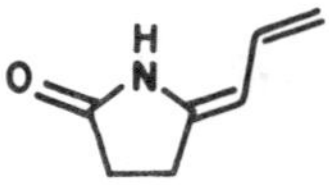

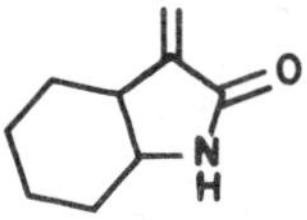

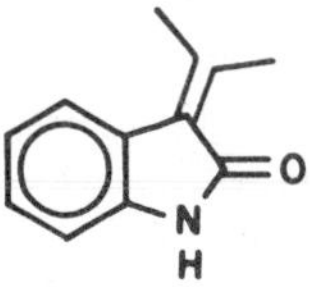

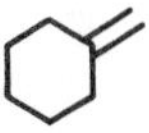

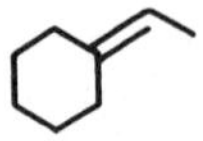

Solvent	Frequencies	Ref.
liq.	1672w, 1445, 814	157
$CHCl_3$	1653, 976, 926, 884	20
$CHCl_3$	1663	158
$CHCl_3$	1684m, 833	20
liq.	1610, 1030m, 847	14
liq.	1653, 1639, 1600, 873	159
liq.	1626, 1550, 1504, 893	159
liq.	1630, 914	160
KBr	1655w, 1620, 830	161
CCl_4	1618	140
liq.	1618	162
CCl_4	1621	163
$CHCl_3$	1616	163
liq.	1575	164
CCl_4	1610	165
CCl_4	1590, 1565	142

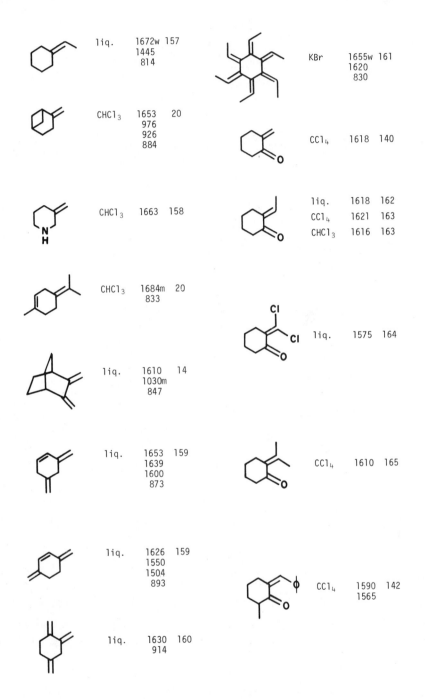

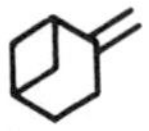

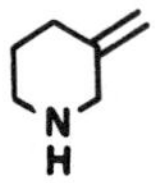

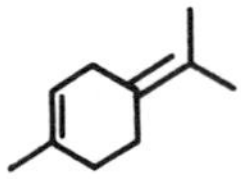

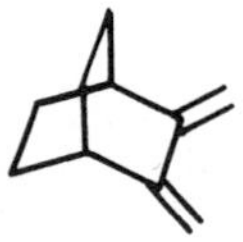

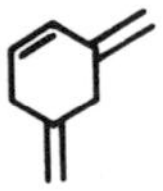

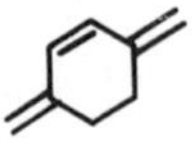

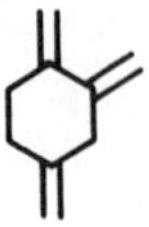

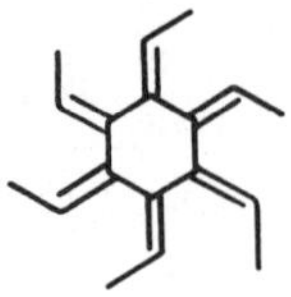

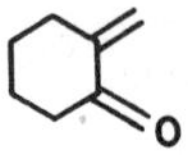

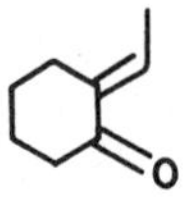

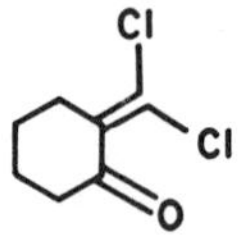

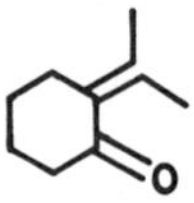

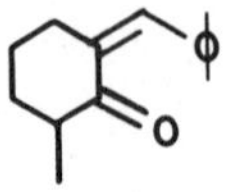

Solvent	Frequency	Ref.
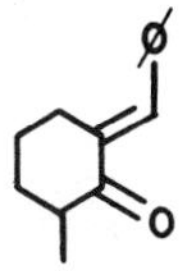 CCl_4	1628 1592	142
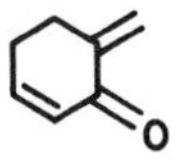 CCl_4	1620 942	166
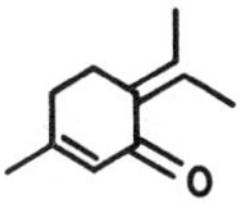 CCl_4	1638 1613	167
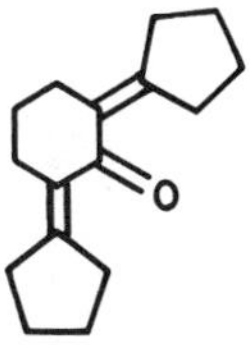 $CHCl_3$	1641	141
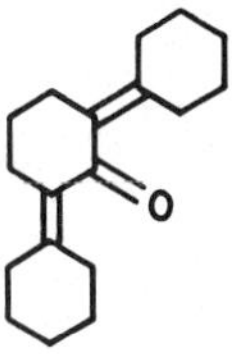 $CHCl_3$	1614	141
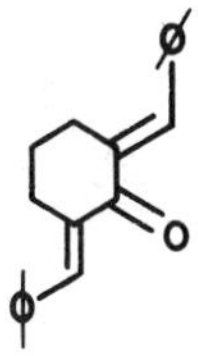 nuj.	1610	168
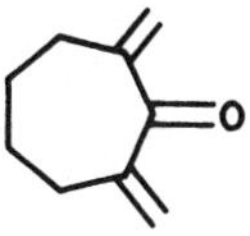 liq.	1610	169
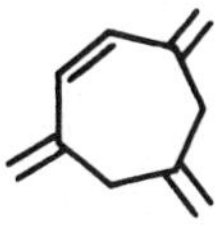 liq.	1639 1626 893	159
liq.	1641	121
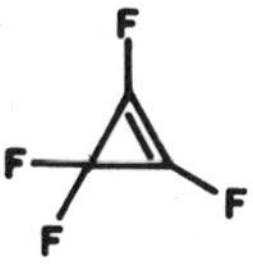 gas	1945	170
 CCl_4	1656	171
 CCl_4	1673	171

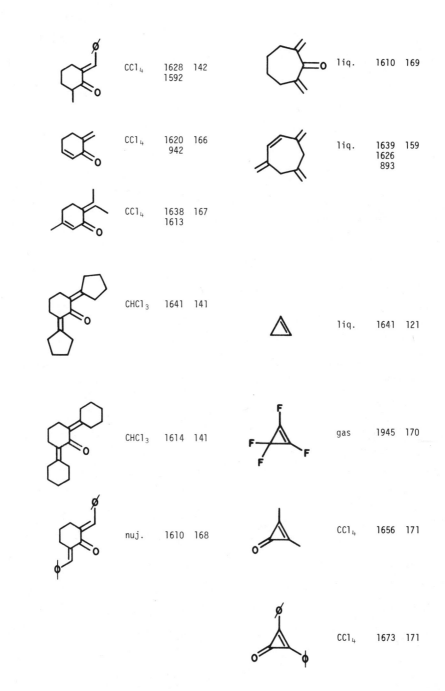

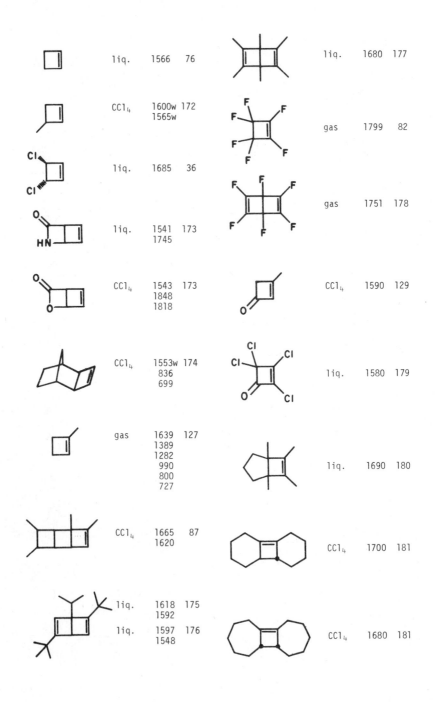

Solvent	Frequency	Ref.
liq.	1566	76
 CCl_4	1600w 1565w	172
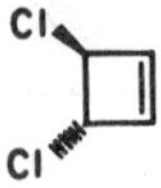 liq.	1685	36
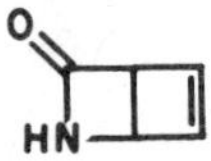 liq.	1541 1745	173
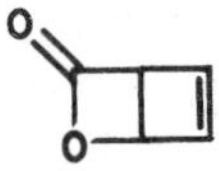 CCl_4	1543 1848 1818	173
 CCl_4	1553w 836 699	174
 gas	1639 1389 1282 990 800 727	127
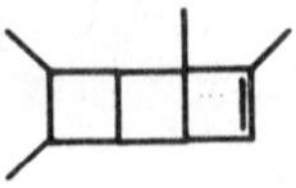 CCl_4	1665 1620	87
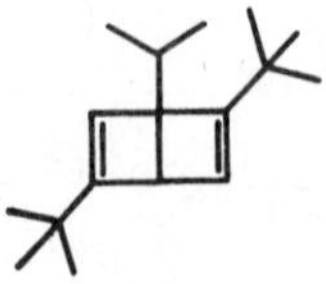 liq.	1618 1592	175
liq.	1597 1548	176
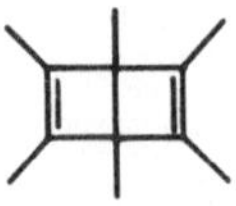 liq.	1680	177
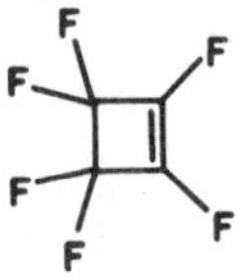 gas	1799	82
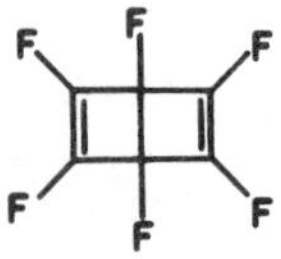 gas	1751	178
 CCl_4	1590	129
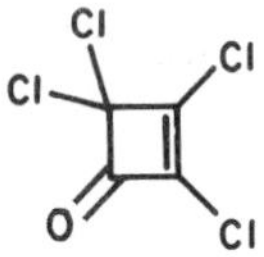 liq.	1580	179
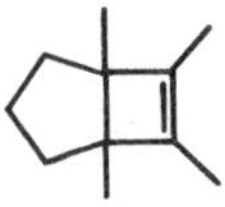 liq.	1690	180
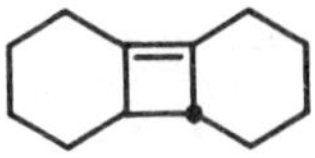 CCl_4	1700	181
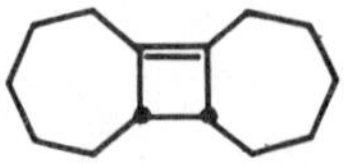 CCl_4	1680	181

Solvent	Frequency	Ref.
CCl_4	1680	181
CCl_4	1670	181
liq.	1612	172
CCl_4	1611	121
CS_2	1612 695	182
liq.	1586 1702	183
liq.	1616	184
liq.	1658	76
liq.	1640m 794	185

Solvent	Frequency	Ref.
liq.	1690m	186
gas	1696	18
gas	1771	18
CCl_4	1620	187
CCl_4	1642w	188
liq.	1603 1562	175
CCl_4	1653w 1768	188

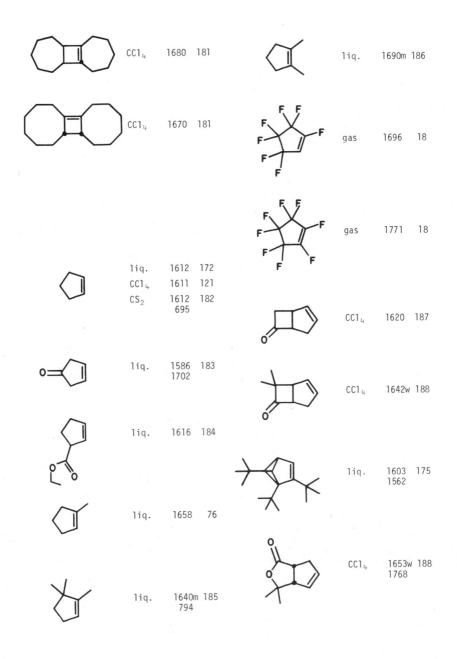

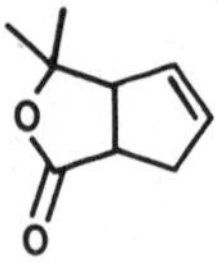

Solvent	cm⁻¹	Ref.
CCl_4	1653w	188
liq.	1600	189
liq.	1605	189
liq.	1575	17
$CHCl_3$	885 859	146
liq.	1634w 1305 650b	190
liq.	1661 1331 876 687	191
$CHCl_3$	1661	20
CCl_4	1661 1631w 1572w 915 896 875	174
CCl_4	1650 1605w 1560 906 893 686	174
liq.	1645	192
KBr	1595	193
$CHCl_3$	1597 1582	194
KBr	1640	195
CCl_4	1628	109
CCl_4	1616	109

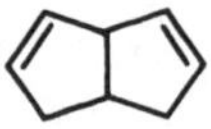

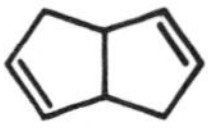

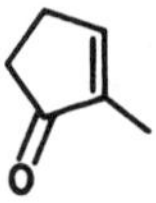

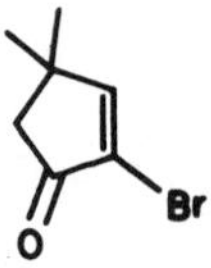

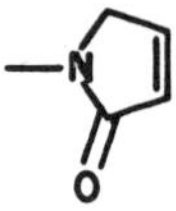

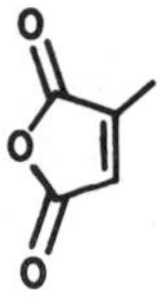

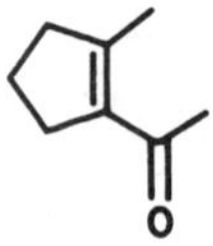

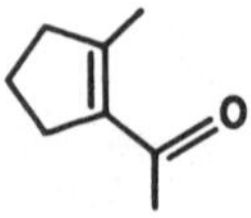

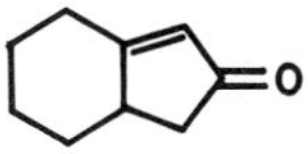	CCl_4	1625	196
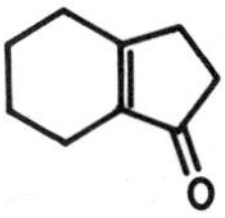	liq.	1626	197
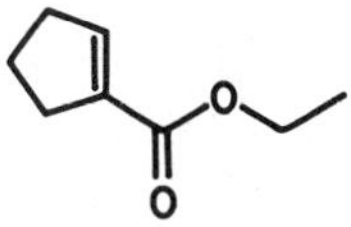	liq.	1620	184
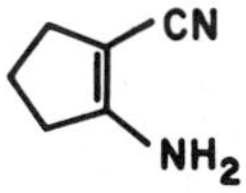 CN, NH_2	$CHCl_3$	1605	198
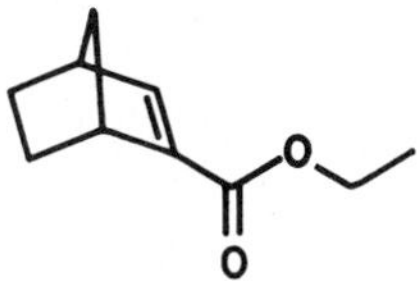	liq.	1595	199
	liq.	1652 1590 959 890 668 662	147

 K^+	KBr	1443 1059	147
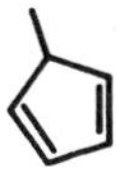	liq.	1630 1507	200
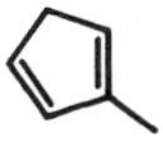	liq.	1619 1526	200
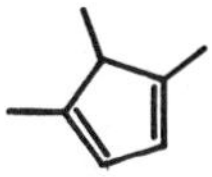	liq.	1625 1575	200
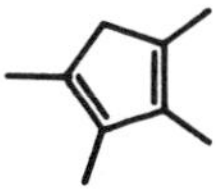	liq.	1658 1613	200
 F F F F F F	gas	1770	201
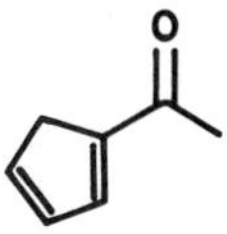	CCl_4	1671	202

Solvent	cm⁻¹	Ref.
liq.	1646	76
CCl_4	1649	121
CS_2	1650 718	182
liq.	1683	203
liq.	1651 1621 700w	204
CCl_4	1655	187
CCl_4	1660	187
liq.	1650 1730	205
CCl_4	1650w	206
gas	1695	82
CS_2	1687 1641 731 723	182
liq.	1661	11
CS_2	1689 1651 724	182
CS_2	1688 1650 725 702	182
liq.	1704 916 858	207
CCl_4	797	208
$CHCl_3$	1647m 982m 916m 898 844w	20
$CHCl_3$	822	20
liq.	1675	203
liq.	1688 1660	135

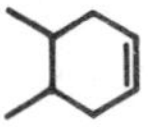

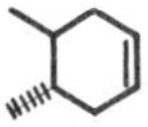

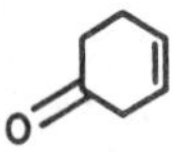

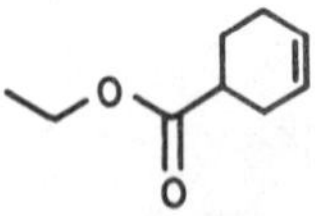

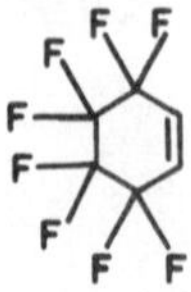

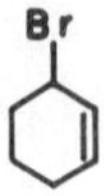

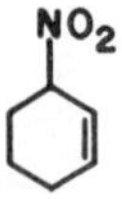

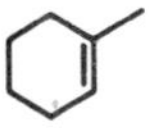

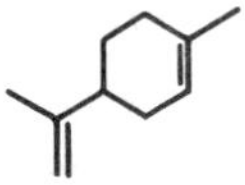

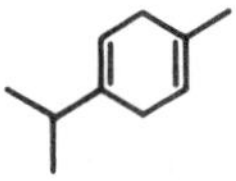

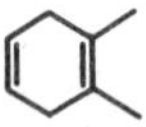

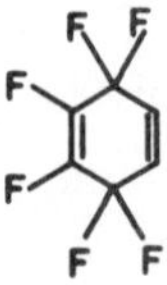

Phase	ν (cm⁻¹)	Ref.
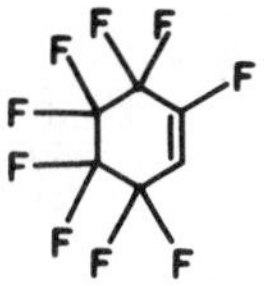 gas	1714	82
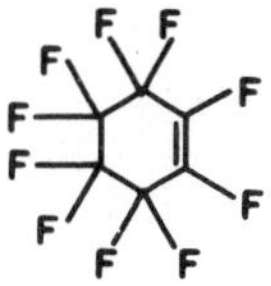 gas	1746	82
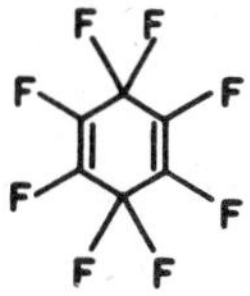 gas	1777w 1741	82
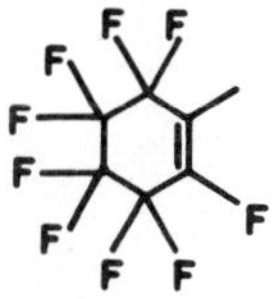 gas	1710	18
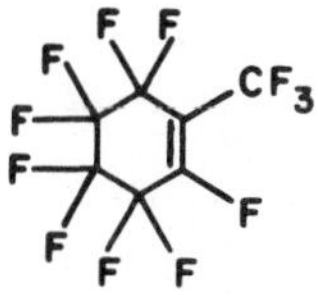 gas	1712	18
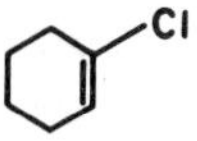 liq.	1656	209
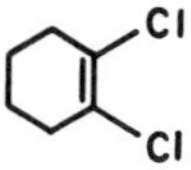 liq.	1649	209
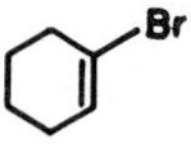 liq.	1650	209
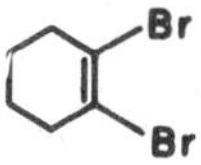 liq.	1636	209
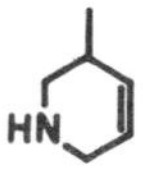 $CHCl_3$	1665	158
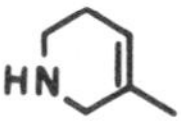 $CHCl_3$	1691	158
 $CHCl_3$	1681w	20
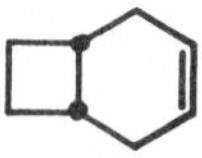 CCl_4	1640	187
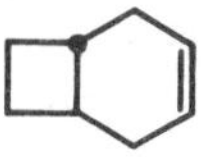 CCl_4	1620	187
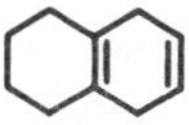 liq.	1660	203
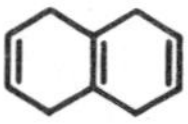 liq.	1662	203
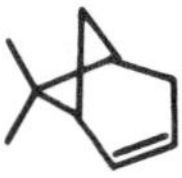 $CHCl_3$	1667	20

Solvent	cm⁻¹	Ref.
 liq.	1614	76
 CS_2	1631 1590	2
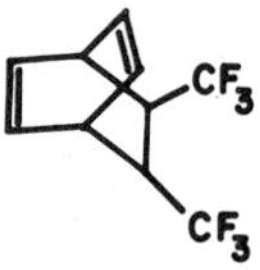 KBr	1682	210
 liq.	1615	189
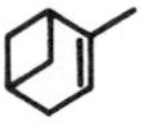 $CHCl_3$	1667w 846	20
 liq.	1626 811	211
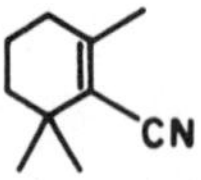 CCl_4	1635	47
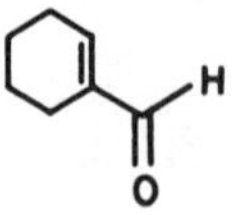 liq.	1637	212

Solvent	cm⁻¹	Ref.
CCl_4	1618	213
CS_2	1651 731 660	182
$CHCl_3$	1618	213
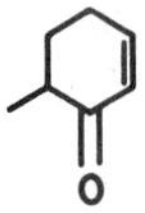 liq.	1626	214
CCl_4	1641	213
$CHCl_3$	1640*	213
CCl_4	1634	213
$CHCl_3$	1632	213
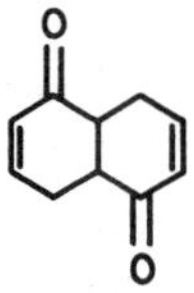 $CHCl_3$	1613	215
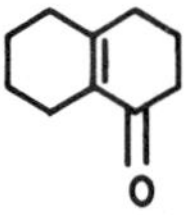 liq.	1632	216
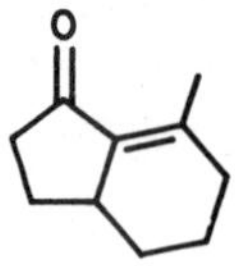 CCl_4	1640	217
 CCl_4	1630	218

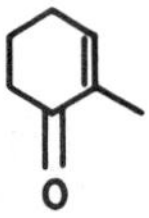

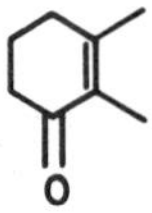

CCl_4	1624	219
liq.	1618	220
KBr	1610	221
liq.	1655	222
CCl_4	1613	109
CCl_4	1650	223
CH_2Cl_2	1639	224

liq.	1650	222
KBr	1645	225
liq.	1639 1587	226
liq.	1672 823	11
KBr	1621	227
$CHCl_3$	1625	229
$CHCl_3$	1615	228
$CHCl_3$	1670 1640	230

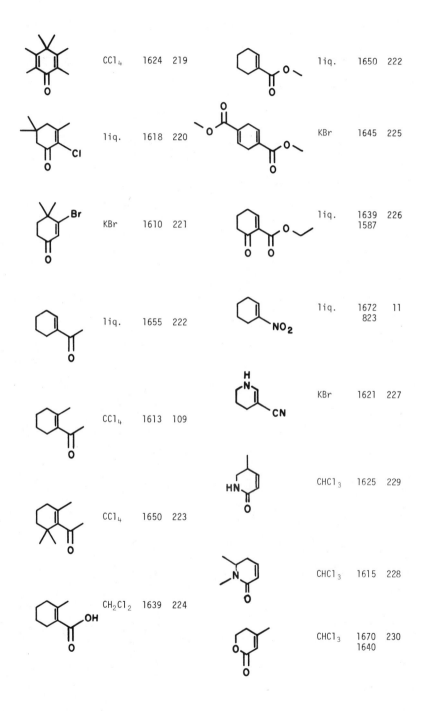

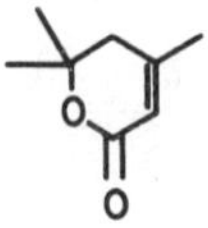 CCl_4	1665	92
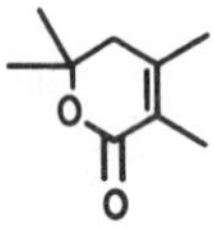 CCl_4	1655	92
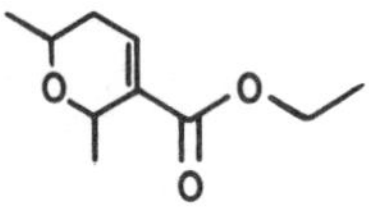 CH_2Cl_2	1653w	208
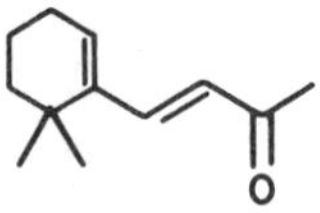 liq.	1685 1605 1460m 1360 1257 980	231
 liq.	1701 1604 675	232
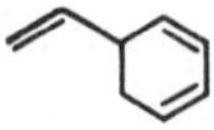 liq.	1640 1575w	189
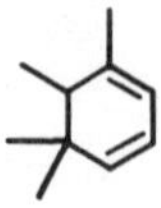 $CHCl_3$	1667	20

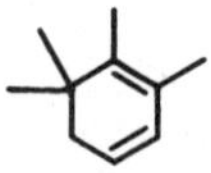 $CHCl_3$	1667m	20
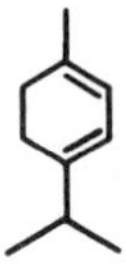 $CHCl_3$	1661 833	20
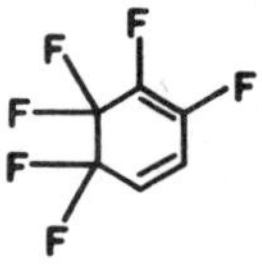 liq.	1734 1635	135
 gas	1746 1712	82
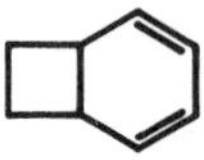 liq.	1645m 1579 850m 813m 777 735m 709	233
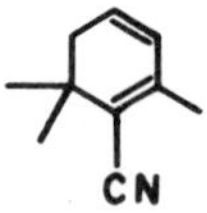 CCl_4	1645	47
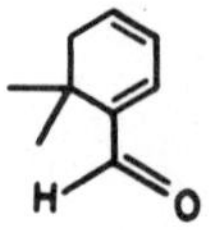 CCl_4	1639 1558	47
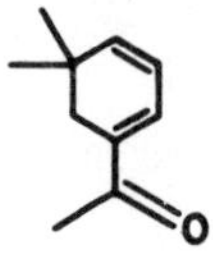 liq.	1600	234

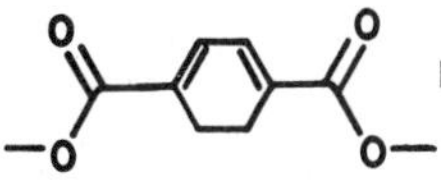

KBr 1645 225

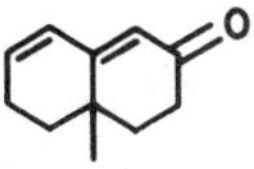

liq. 1618 235
1587

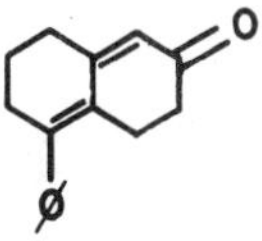

$CHCl_3$ 1615 236

liq. 1650 172
CCl_4 1650 121

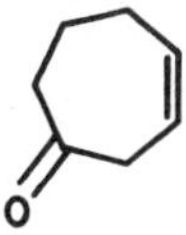

liq. 1660 237
1704

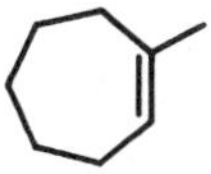

liq. 1673 76

liq. 1665 189

liq. 1650 189

gas 1665 189

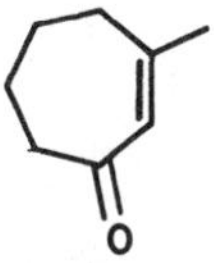

CCl_4 1661 238

CCl_4 1613 218

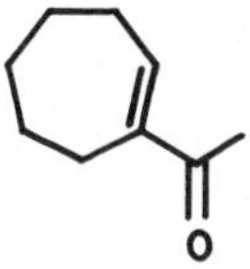

liq. 1640 239

liq. 1610 189

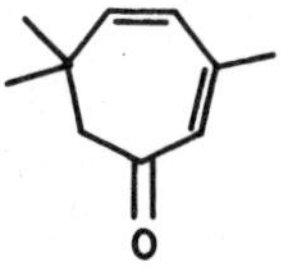

liq. 1575 240

Solvent	Frequency	Ref.
CCl_4	1635	241
$CHCl_3$	1617	242
liq.	1664 700	17
CCl_4	1648	121
liq.	1658	17
liq.	985	2
CCl_4	1653 1710	243
liq.	1631m 680	244
liq.	1648w 1608m 835m 778 742m 701	245
liq.	1680m 1638m 1610m 864 813 775 740 704	233
liq.	1631 796 669b	246
CCl_4	1645	87
CCl_4	1655 1640	87
CCl_4	1560	247
liq.	1642 706	248
liq.	1645 978	248
CCl_4	1640w	180

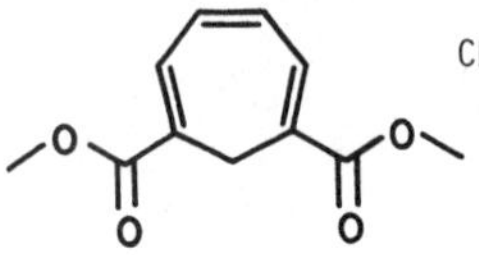

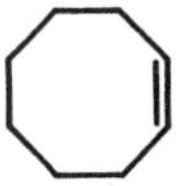

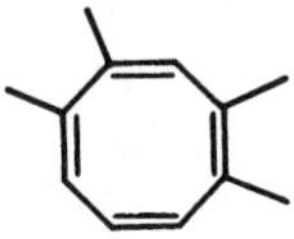

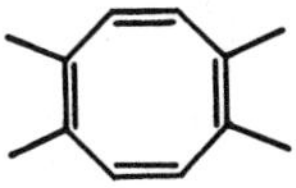

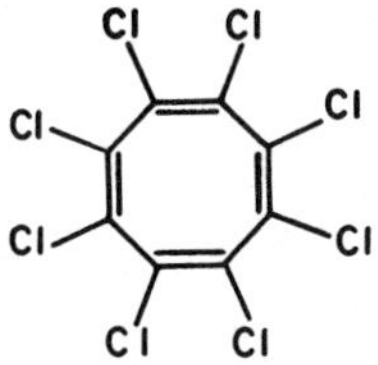

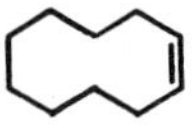

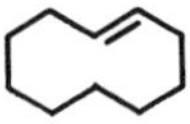

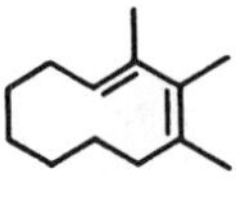

REFERENCES

1. E. B. Wilson, Jr. and A. J. Wells, *JCP* (1941) *9*, 319.
2. J. Hine, J. A. Brown, L. H. Zalkow, W. E. Gardner and M. Hine, *JACS* (1955), 594.
3. W. J. Potts and R. A. Nyquist, *SCA* (1959) *15*, 679.
4. N. Sheppard and D. M. Simpson, *Quart Rev* (1952), 1.
5. R. D. McLachlan and R. A. Nyquist, *SCA* (1958), 103.
6. W. H. T. Davison and G. R. Bates, *JCS* (1953), 2607.
7. N. J. Leonard and V. W. Gash, *JACS* (1954), 2781.
8. C. F. Pouchert, *The Aldrich Library of Infrared Spectra*, Aldrich 1970, spectrum 73E.
9. C. C. Price and R. G. Gillis, *JACS* (1953), 4750.
10. G. Hesse, R. Hatz and H. König, *A* (1967) *709*, 79.
11. J. F. Brown, Jr., *JACS* (1955), 6341.
12. D. Barnard, L. Bateman, A. J. Harding, H. P. Koch, N. Sheppard and G. B. B. M. Sutherland, *JCS* (1950), 915.
13. E. Vogel, R. Erb, G. Lenz and A. A. Bothner-By, *A* (1965) *682*, 1.
14. W. J. Bailey and W. B. Lawson, *JACS* (1957), 1444.
15. R. N. Haszeldine, *JCS* (1952), 2504.
16. R. N. Haszeldine and B. R. Steele, *JCS* (1953), 1199.
17. N. L. Allinger, *JACS* (1958), 1953.
18. J. K. Brown and K. L. Morgan, *Advances in Fluorine Chemistry*, Vol. 4, p. 253, eds. M. Stacey, J. C. Tatlow and A. G. Sharpe; Butterworths, Washington, 1965.
19. T. J. Brice, J. D. LaZerte, L. J. Hals and W. H. Pearlson, *JACS* (1953), 2698.
20. R. T. O'Connor and L. A. Goldblatt, *Anal Chem* (1954), 1726.
21. P. Torkington and H. W. Thompson, *Trans Farad Soc* (1945), 236.
22. H. G. Viehe, *B* (1960), 1697.
23. P. Tarrant and M. R. Lilyquist, *JACS* (1955), 3640.
24. D. E. Mann, N. Acquista and E. K. Plyler, *JCP* (1954) *22*, 1586.
25. R. N. Haszeldine and B. R. Steele, *JCP* (1954), 923.
26. R. Theimer and J. R. Nielsen, *JCP* (1957) *26*, 1374.
27. R. N. Haszeldine, *JCS* (1954), 4026.
28. American Petroleum Institute Infrared Spectrogram No. 1006.
29. C. J. Muelleman, K. Rasaswamy, F. F. Clevland and S. Sundaram, *J Mol Spectr* (1963) *11*, 262.
30. J. R. Nielsen, C. Y. Liang and D. C. Smith, *JCP* (1952) *20*, 1090.

31. D. E. Mann and E. K. Plyler, *JCP* (1957) *26*, 773.
32. J. A. Rolfe and L. A. Woodward, *Trans Farad Soc* (1954), 1030.
33. R. Theimer and J. R. Nielsen, *JCP* (1957) *27*, 264.
34. D. E. Mann, N. Acquista and E. K. Plyler, *JCP* (1954) *22*, 1199.
35. J. A. Landgrebe and L. W. Becker, *JACS* (1968), 395.
36. H. J. Boonstra and L. C. Rinzema, *Rec Trav Chim* (1960), 962.
37. R. Vessière, *BSC France* (1959), 1268.
38. F. Winther and D. O. Hummel, *SCA* (1967), 1839.
39. K. S. Pitzer and J. L. Hollenberg, *JACS* (1954), 1493.
40. H. J. Bernstein and J. Powling, *JACS* (1951), 1843.
41. C. F. Pouchert, *The Aldrich Library of Infrared Spectra*, Aldrich 1970, spectrum 52H.
42. R. Y. Tien and P. I. Abell, *JOC* (1970), 956.
43. J. C. Evans and H. J. Bernstein, *Canad J Chem* (1955), 1171.
44. D. G. I. Felton and S. F. D. Orr, *JCS* (1955), 2170.
45. S. B. Lie, P. Klaboe, E. Kloster-Jensen, G. Hagen and D. H. Christensen, *SCA* (1970), 2077.
46. E. J. Moriconi and C. C. Jalandoni, *JOC* (1970), 3796.
47. C. H. Eugster, L. Leichner and E. Jenni, *Helv* (1963), 543.
48. P. Kurtz, H. Gold and H. Disselnkötter, *A* (1959) *624*, 1.
49. M. Yamaguchi, *NKZ* (1959), 155.
50. E. Allenstein and P. Quis, *B* (1963), 1035.
51. R. Mecke and K. Noack, *B* (1960), 210.
52. E. R. Blout, M. Fields and R. Karplus, *JACS* (1948), 194.
53. K. C. Chan, R. A. Jewell, W. H. Nutting and H. Rapoport, *JOC* (1968), 3382.
54. E. Elkik and P. Vaudescal, *CR* (1965) *261*, 1015.
55. Z. Arnold and A. Holý, *Col Czech Comm* (1961), 3059.
56. V. Theus, W. Surber, L. Colombi and H. Schinz, *Helv* (1955), 239.
57. S. F. Reed, *JOC* (1962), 4116.
58. E. C. Craven and W. R. Ward, *J Appl Chem* (1960) *10*, 18.
59. W. Haefliger and T. Petrzilka, *Helv* (1966), 1937.
60. H. O. House and R. S. Ro, *JACS* (1958), 2428.
61. N. Fuson, M.-L. Josien and E. M. Shelton, *JACS* (1954), 2526.
62. C. F. Pouchert, *The Aldrich Library of Infrared Spectra*, Aldrich 1970, spectrum 324F.
63. M. Yamaguchi, *NKZ* (1960), 1118.
64. M. I. Batuev, A. S. Onishchenko, A. D. Matveeva and N. I. Aronova, *J Gen Chem USSR* (1966), 679.
65. C. F. Pouchert, *The Aldrich Library of Infrared Spectra*, Aldrich 1970, spectrum 232H.

66. J. L. H. Allan, G. D. Meakins and M. C. Whiting, *JCS* (1955), 1874.
67. E. A. Braude and E. A. Evans, *JCS* (1955), 3331.
68. R. Scheffold and P. Dubs, *Helv* (1967), 798.
69. J. C. Sheehan and J. H. Beeson, *JACS* (1967), 362.
70. D. E. Jones, R. O. Morris, C. A. Vernon and R. F. M. White, *JCS* (1960), 2349.
71. M. Sato, *JOC* (1961), 770.
72. C. C. Price and H. Morita, *JACS* (1953), 4747.
73. J. C. Sauer and J. D. C. Wilson, *JACS* (1955), 3793.
74. E. R. Shull, R. A. Thursack and C. M. Birdsall, *JCS* (1956) *24*, 147.
75. American Petroleum Institute Infrared Spectrogram No. 919.
76. R. C. Lord and R. W. Walker, *JACS* (1954), 2518.
77. American Petroleum Institute Infrared Spectrogram No. 452.
78. D. Craig, J. J. Shipman and R. B. Fowler, *JACS* (1961), 2885.
79. American Petroleum Institute Infrared Spectrogram No. 451.
80. Ae. de Groot, B. Evanhius and H. Wynberg, *JOC* (1968), 2214.
81. J. L. Anderson, R. E. Putnam and W. H. Sharkey, *JACS* (1961), 382.
82. J. Burdon and D. H. Whiffen, *SCA* (1958) *12*, 139.
83. C. F. Pouchert, *The Aldrich Library of Infrared Spectra*, Aldrich 1970, spectrum 53E.
84. P. Weyerstahl, D, Klamann, C. Finger, M. Fligge, F. Nerdel and J. Buddrus, *B* (1968), 1303.
85. C. F. Pouchert, *The Aldrich Library of Infrared Spectra*, Aldrich 1970, spectrum 53F.
86. H.-G. Viehe and E. Franchimont, *B* (1964), 602.
87. R. Criegee, W. Eberius and H.-A. Brune, *B* (1968), 94.
88. C. F. Pouchert, *The Aldrich Library of Infrared Spectra*, Aldrich 1970, spectrum 53G.
89. R. Criegee and R. Huber, *B* (1970), 1855.
90. P. A. Stadler, *Helv* (1960), 1601.
91. W. Adam, *B* (1964), 1811.
92. F. Korte and D. Scharf, *B* (1962), 443.
93. A. A. Petrov, K. B. Rall, A. E. Vildavskaya, *Zh Obshch Khim* (1964), 3515.
94. A. A. Petrov, Yu. I. Porfiryeva and G. I. Semenov, *J Gen Chem USSR* (1957), 1167.
95. A. A. Petrov, B. S. Kupin, T. U. Yakovleva and K. S. Mingaleva, *J Gen Chem USSR* (1959), 3732.
96. J. C. H. Hwa, P. L. de Benneville and H. J. Sims, *JACS* (1960), 2537.
97. M. L. Bender and J. Figueras, *JACS* (1953), 6304.
98. K. Alder and H. von Brachel, *A* (1957) *608*, 195.

99. R. B. Woodward, F. Sondheimer, D. Taub, K. Heusler and W. M. McLamore, *JACS* (1952), 4223.
100. W. J. Gensler and J. Casella, Jr., *JACS* (1958), 1376.
101. K. Bowden, B. Lythgoe and D. J. S. Marsden, *JCS* (1959), 1662.
102. M. Kröner, *B* (1967), 3172.
103. F. Sondheimer, D. A. Ben-Efraim and R. Wolovsky, *JACS* (1961), 1675.
104. D. Braun and H.-G. Keppler, *Monatsh Chem* (1963), 1250.
105. J. McKenna, J. K. Norymberski and R. D. Stubbs, *JCS* (1959), 2502.
106. E. Elkik, *BSC France* (1967), 1569.
107. G. V. Kazennikova, T. V. Talalaeva, A. V. Zimin, A. P. Simonov and K. A. Kocheshkov, *Izvest Akad Nauk Khim USSR* (1961), 1063.
108. G. Camaggi, Thesis, Birmingham, England (1964).
109. W. D. Hayes and C. J. Timmons, *SCA* (1968), 323.
110. J. Lecomte and J. Guy, *CR* (1948) *227*, 54.
111. C. F. Pouchert, *The Aldrich Library of Infrared Spectra*, Aldrich 1970, spectrum 758A.
112. J. Zabicky, *JCS* (1961), 683.
113. A. Ya. Yakubovich, E. L. Zaitseva, G. I. Braz and V. P. Bazov, *Zh Obshch Khim* (1962) *32*, 3409.
114. A. R. Katritzky and A. J. Boulton, *JCS* (1959), 3500.
115. C. F. Pouchert, *The Aldrich Library of Infrared Spectra*, Aldrich 1970, spectrum 432C.
116. J. Bornstein, M. S. Blum and J. J. Pratt, Jr., *JOC* (1957), 1210.
117. D. F. DeTar and L. A. Carpino, *JACS* (1956), 475.
118. K. Lunde and L. Zechmeister, *Acta Chem Scand* (1954), 1421.
119. J. Bornstein and M. R. Borden, *Chem & Ind* (1958), 441.
120. J. Bornstein, *Chem & Ind* (1959), 1193.
121. K. B. Wiberg and B. J. Nist, *JACS* (1961), 1226.
122. J. T. Gragson, K. W. Greenlee, J. M. Derfer and C. E. Boord, *JACS* (1953), 3344.
123. American Petroleum Institute Infrared Spectrogram No. 794.
124. J. Meinwald, J. W. Wheeler, A. A. Nimety and J. S. Liu, *JOC* (1965), 1038.
125. M. Tanabe and R. A. Walsh, *JACS* (1963), 3522.
126. J.-P. Vincent, A. Bezaguet and M. Bertrand, *BSC France* (1967), 3550.
127. W. M. Schubert and S. M. Leahy, Jr., *JACS* (1957), 381.
128. F. F. Caserio, Jr., S. H. Parker, R. Piccolini and J. D. Roberts, *JACS* (1958), 5507.
129. P. Dowd and K. Sachdev, *JACS* (1967), 715.
130. D. E. Applequist and J. D. Roberts, *JACS* (1956), 4012.

131. J. K. Williams and W. H. Sharkey, *JACS* (1959), 4269.
132. A. T. Blomquist and Y. C. Meinwald, *JACS* (1959), 667.
133. J.-M. Conia and J.-P. Sandre, *BSC France* (1963), 744.
134. J. Goré, C. Djerassi and J.-M. Conia, *BSC France* (1967), 950.
135. R. Stephens, J. C. Tatlow and E. H. Wiseman, *JCS* (1959), 148.
136. A. T. Blomquist and J. A. Verdol, *JACS* (1955), 1806.
137. G. W. Griffin and L. I. Peterson, *JACS* (1963), 2268.
138. C. F. Pouchert, *The Aldrich Library of Infrared Spectra*, Aldrich 1970, spectrum 30H.
139. R. Schimpf and P. Heimbach, *B* (1970), 2122.
140. R. L. Erskine and E. S. Waight, *JCS* (1960), 3425.
141. M. Horák and P. Munk, *Col Czech Comm* (1959), 3024.
142. A. Hassner and T. C. Mead, *T* (1964) *20*, 2201.
143. J. K. Crandall and D. R. Paulson, *JOC* (1968), 3291.
144. J. Derkosch and W. Kaltenegger, *Monatsh Chem* (1959), 872.
145. B. D. Pearson, R. A. Ayer and N. C. Cromwell, *JOC* (1962), 3038.
146. A. T. Blomquist, J. Wolensky, Y. C. Meinwald and D. T. Longone, *JACS* (1956), 6057.
147. H. P. Fritz, *B* (1959), 780.
148. J. Thiec and J. Wiemann, *BSC France* (1956), 177.
149. J. C. Wood, R. M. Olofson and D. M. Saunders, *Anal Chem* (1958), 1339.
150. M. Neuenschwander, D. Meuche and H. Schaltegger, *Helv* (1963), 1760.
151. R. Scheffold, *Thesis ETH* (1963).
152. W. Häusermann, *Thesis ETH* (1966).
153. R. L. Wineholt, E. Wyss and J. A. Moore, *JOC* (1966), 48.
154. W. Flitsch, *A* (1965) *684*, 141.
155. R. L. Hinman and C. P. Bauman, *JOC* (1964), 2431.
156. C. F. Pouchert, *The Aldrich Library of Infrared Spectra*, Aldrich 1970, spectrum 31B.
157. C. F. Pouchert, *The Aldrich Library of Infrared Spectra*, Aldrich 1970, spectrum 31A.
158. M. Ferles and M. Holík, *Col Czech Comm*
159. R. E. Benson and R. V. Lindsey, Jr., *JACS* (1959), 4250.
160. R. E. Benson and R. V. Lindsey, Jr., *JACS* (1959), 4247.
161. H. Hopff and A. K. Wick, *Helv* (1961), 380.
162. J. K. Crandall, J. P. Arrington and J. Hen, *JACS* (1967), 6208.
163. K. Noack, *SCA* (1962), 1625.
164. J. Libman, M. Sprecher and Y. Mazur, *JACS* (1969), 2062.
165. F. Nerdel, D. Frank and H. Marschall, *B* (1967), 720.
166. I. G. Morris and A. R. Pinder, *JCS* (1963), 1841.

167. Y. R. Naves, *Helv* (1966), 2012.
168. N. J. Leonard and G. C. Robinson, *JACS* (1953), 2714.
169. M. Mühlstaedt, *Naturwiss* (1958), 240.
170. P. B. Sargeant and C. G. Krespan, *JACS* (1969), 415.
171. A. Krebs and B. Schröder. *A* (1968) *709*, 46.
172. S. Pinchas, E. Gil-Av, J. Shabtai and B. Altmann, *SCA* (1965), 783.
173. E. J. Corey and J. Streith, *JACS* (1964), 950.
174. R. R. Sauers, S. B. Schlosberg and P. E. Pfeffer, *JOC* (1968), 2175.
175. K. E. Wilzbach and L. Kaplan, *JACS* (1965), 4004.
176. E. E. van Tamelen and S. P. Pappas, *JACS* (1962), 3789.
177. W. Schäfer, *Angew* (1966), 716.
178. I. Haller, *JACS* (1966), 2070.
179. O. Scherer, G. Hörlein and H. Millauer, *B* (1966), 1966.
180. R. Criegee, G. Bolz and R. Askani, *B* (1969), 275.
181. R. Criegee and H. G. Reinhardt, *B* (1968), 102.
182. H. B. Henbest, G. D. Meakins, B. Nicholls, and R. A. L. Wilson, *JCS* (1957), 997.
183. K. Alder and F. H. Flock, *B* (1956), 1732.
184. H. D. Scharf and F. Korte, *B* (1964), 2425.
185. American Petroleum Institute Infrared Spectrogram No. 214.
186. American Petroleum Institute Infrared Spectrogram No. 325.
187. E. Casadevall, C. Langeau and P. Moreau, *BSC France* (1968), 1514.
188. U. A. Huber and A. S. Dreiding, *Helv* (1970), 495.
189. W. von E. Doering and W. R. Roth, *T* (1963), 715.
190. C. F. Pouchert, *The Aldrich Library of Infrared Spectra*, Aldrich 1970, spectrum 35E.
191. C. F. Pouchert, *The Aldrich Library of Infrared Spectra*, Aldrich 1970, spectrum 34G.
192. K. Spencer, A. L. Hall and C. F. Von Reyn, *JOC* (1968), 3369.
193. H. De Pooter and N. Schamp, *BSC Belge* (1968), 377.
194. E. B. Smith and H. B. Jensen, *JOC* (1967), 3330.
195. K. Nakanishi, *Kagaku-No-Ryoiki* (1959) *13*, 77.
196. R. Fraisse-Jullien and C. Frejaville, *BSC France* (1968), 4449.
197. F. Gautschi, O. Jeger, V. Prelog and R. B. Woodward, *Helv* (1954), 2280.
198. S. Baldwin, *JOC* (1961), 3288.
199. R. A. Finnegan and R. S. McNees, *JOC* (1964), 3234.
200. V. A. Mironov, E. V. Sobolev and A. N. Elizarova, *T* (1963), 1939.
201. R. E. Banks, R. N. Haszeldine and J. B. Walton, *JCS* (1963), 5581.
202. P. Yates, B. L. Shapiro, N. Yoda and J. Fugger, *JACS* (1957), 5756.
203. W. Hückel, E. Vevera and U. Wörffel, *B* (1957), 901.

204. American Petroleum Institute Infrared Spectrogram No. 457.
205. M. Hanack and W. Keberle, *B* (1963), 2937.
206. W. Lwowski and T. W. Mattingly, Jr., *JACS* (1965), 1947.
207. American Petroleum Institute Infrared Spectrogram No. 897.
208. F. E. Bader, *Helv* (1953), 215.
209. G. Chiurdoglu, R. Ottinger, J. Reisse and T. Toussaint, *SCA* (1962), 215.
210. C. G. Krespan, B. C. McKusick and T. L. Cairns, *JACS* (1961), 3428.
211. J. A. Marshall and H. Faubl, *JACS* (1967), 5965.
212. R. Lehr, *Harvard Ph.D. Thesis* (1968).
213. K. Noack, *SCA* (1962), 697.
214. R. H. Wiley and C. H. Jarboe, *JACS* (1956), 624.
215. W. S. Johnson, J. Dolf Bass and K. L. Williamson, *T* (1963) *19*, 861.
216. A. Suzuki and T. Matsumoto, *BCS Japan* (1962) *35*, 2027.
217. H. O. House and M. Schellenbaum, *JOC* (1963), 34.
218. E. W. Garbisch, *JOC* (1965), 2109.
219. H. Hart and D. W. Swatton, *JACS* (1967), 1874.
220. M. Tonoeda, M. Inuzuka, T. Furuta, M. Shinozuka and T. Takahashi, *T* (1968), 959.
221. F. G. Bordwell and K. M. Wellman, *JOC* (1963), 2544.
222. J. Klein, *T* (1964), 465.
223. N. C. Yang and R. A. Finnegan, *JACS* (1958), 5845.
224. L. F. Hatch and H. D. Weiss, *JACS* (1955), 1798.
225. J. C. Kauer, R. E. Benson and G. W. Parshall, *JOC* (1965), 1431.
226. J. E. Brenner, *JOC* (1961), 22.
227. S. E. Ellzey, Jr., C. H. Mack and W. J. Connick, Jr., *JOC* (1967), 846.
228. O. E. Edwards and T. Singh, *Canad J Chem* (1954), 683.
229. K. Gschwend-Steen, *Thesis ETH* (1965).
230. E. Fetz, B. Bohner and Ch. Tamm, *Helv* (1965), 1669.
231. K. R. Farrar, J. C. Hamlet, H. B. Henbest, and E. R. H. Jones, *JCS* (1952), 2657.
232. American Petroleum Institute Infrared Spectrogram No. 459.
233. E. P. Lippincott and R. C. Lord, *JACS* (1957), 567.
234. G. L. Buchanan, R. A. Raphael and I. W. J. Still, *JCS* (1963), 4372.
235. J. A. Marshall and H. Roebke, *JOC* (1966), 3109.
236. H. O. House and R. W. Bashe II, *JOC* (1965), 2942.
237. I. Maclean and R. P. A. Sneeden, *T* (1964) *20*, 31.
238. G. Stork, M. Nussius and B. August, *T* (1968) *Suppl 8*, 105.
239. J. Hanuise and R. R. Smolder, *BSC France* (1967), 2139.
240. W. D. P. Burns, M. S. Carson, W. Cocker and P. V. R. Shannon, *JCS* (1968) *C*, 3073.
241. G. Schröder, *B* (1964), 3140.

242. R. Darms, T. Threlfall, M. Pesaro and A. Eschenmoser, *Helv* (1963), 2893.

243. L. A. Paquette and R. F. Eizember, *JACS* (1967), 6205.

244 C. F. Pouchert, *The Aldrich Library of Infrared Spectra*, Aldrich 1970, spectrum 33A.

245. E. O. Fischer, C. Palm and H. P. Fritz, *B* (1959), 2645.

246. C. F. Pouchert, *The Aldrich Library of Infrared Spectra*, Aldrich 1970, spectrum 33E.

247. R. Criegee and R. Huber, *B* (1970), 1862.

248. A. T. Blomquist R. E. Barge and A. C. Suesy, *JACS* (1952), 3636.

2.2 acetylenes

—≡—H

All terminal acetylenes show a strong, sharp, and characteristic C-H stretching vibration close to 3300 cm^{-1}. The C-H deformation mode occurs as a strong band (two when the molecule is axially symmetric) in the region of 700 to 600 cm^{-1}. A strong overtone of this fundamental vibration may also be seen in the 1400 to 1200 cm^{-1} region but is of little diagnostic value.

—≡—

Monosubstituted acetylenes show a weak C-C triple-bond stretching vibration. The intensity of this mode for disubstituted acetylenes varies considerably: the more symmetric the substitution, the weaker the intensity.

As can be seen from the following tables, the effect of substitution on the acetylenic stretching frequency can be directly related to the electronegativity of the substituent.

Acetylenes coordinated to transition metals are covered in Chapter 8.2, p. 493.

Structure	Phase	Frequency	Ref.
≡	gas	1974	1
D—≡	gas	1853	1
F—≡	gas	2263	1
F—≡—D	gas	1995	1
Cl—≡	gas	2122	1
Cl—≡—D	gas	1988	1
Br—≡	gas	2096	1
I—≡	liq.	2075	2
⁄⁀O—≡	gas	2167	1
	CCl_4	2160	1

Structure	Phase	Frequency	Ref.
⁄⁀S—≡	gas	2057	1
	CCl_4	2051	1
ø⁀S—≡	liq.	2035	3
—≡	CCl_4	2123	1
F⁀—≡	liq.	2135	4
	CCl_4	2148	5
Cl⁀—≡	liq.	2136	4
	CCl_4	2131 2126*	5
Br⁀—≡	liq.	2170w 2125	4
	CCl_4	2126 2121*	5
I⁀—≡	liq.	2128 2110	4

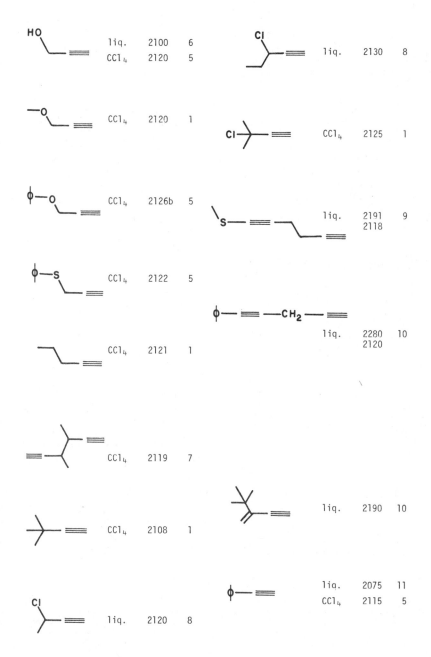

HO	liq.	2100	6
	CCl_4	2120	5
O	CCl_4	2120	1
φ—O	CCl_4	2126b	5
φ—S	CCl_4	2122	5
	CCl_4	2121	1
	CCl_4	2119	7
	CCl_4	2108	1
Cl	liq.	2120	8

Cl	liq.	2130	8
Cl	CCl_4	2125	1
S	liq.	2191 2118	9
φ—≡—CH_2—≡	liq.	2280 2120	10
	liq.	2190	10
φ—≡	liq.	2075	11
	CCl_4	2115	5

Structure	Medium	Frequency	Ref.
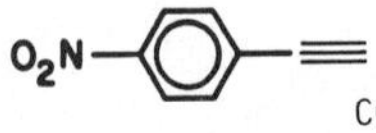 O_2N—C₆H₄—≡	CCl_4	2122	12
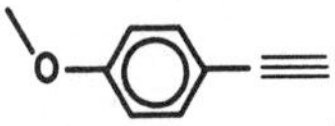 CH₃O—C₆H₄—≡	CCl_4	2116	12
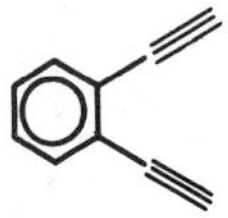 C₆H₄(—≡)₂	CCl_4	2107	13
 Cl—≡—≡	CS_2	2244	14
 Br—≡—≡	CS_2	2238	14
 I—≡—≡	CS_2	2202	14
≡—CN	liq.	2242 2049m	15
	CCl_4	2073	1
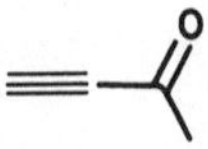 ≡—CHO	liq.	2114	16
	CCl_4	2175	1
≡—C(=O)—	liq.	2114	16
 —≡—CH₂CH₂—≡—	liq.	2232	7
 —≡—Cl	CCl_4	2245 2258	17
 —≡—Br	liq.	2232 2053	18
Cl—≡—Br	gas	2223	19
	CCl_4	2206	20
Cl—≡—I	gas	2191	19
	CCl_4	2177 2160	20
 Br—≡—CH₂CH₂—≡—Br	CCl_4	2223	21
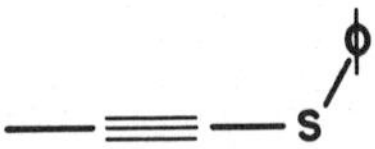 —≡—S—φ	liq.	2190	3
—CH₂CH₂—≡—CH₂—OH	liq.	2240	6

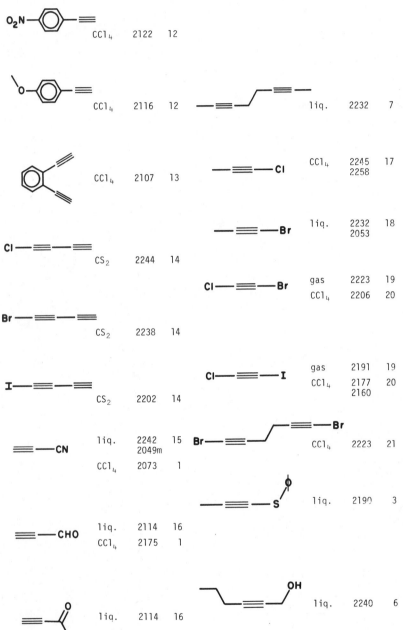

Br
liq. 2260 6

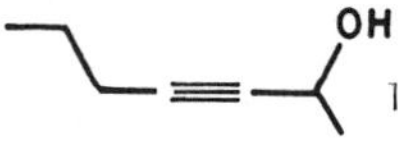

OH
liq. 2260 6

Br
liq. 2210 6

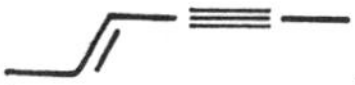

liq. 2270 1607 22

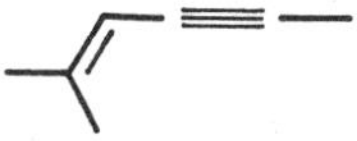

liq. 2270 1613 22

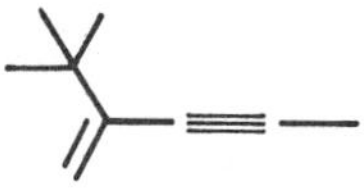

liq. 2240 2206 23

liq. 2204 1611 24

ϕ— ≡ — liq. 2277 2252 25

ϕ— ≡ —Br liq. 2200 26

OH
ϕ— ≡ — liq. 2201 6

OH
ϕ— ≡ — liq. 2220 27

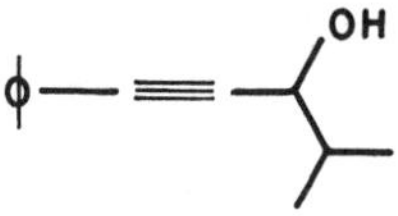

Br
ϕ— ≡ — liq. 2210 6

ϕ— ≡ —CH_2— ≡ —ϕ liq. 2240 2230 10

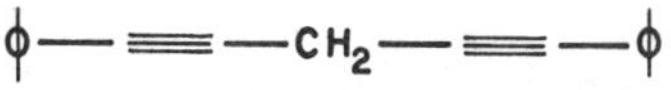

CCl_4 2214 2151 2040 28

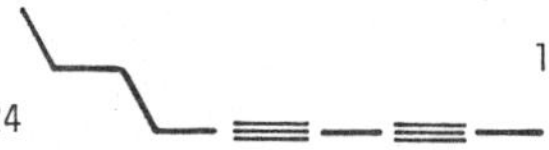

liq. 2260 2175 2140 10

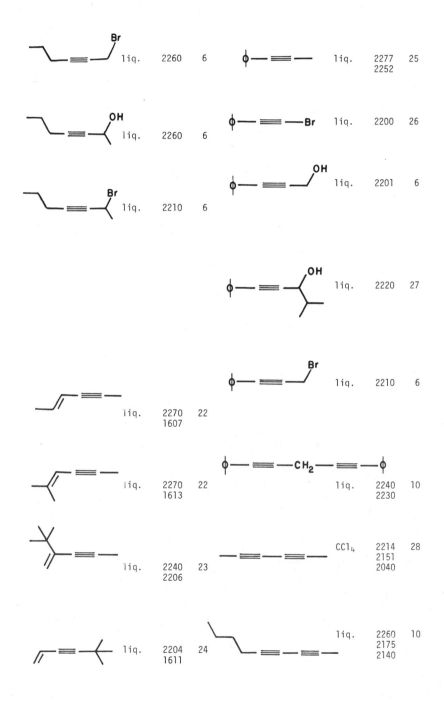

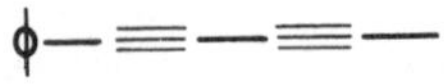

Solvent	cm⁻¹	Ref.
liq.	2247 2202 2164	10
CCl_4	2222	29
CCl_4	2235	30
liq.	2140	31
liq.	2142	31
liq.	2138	31
liq.	2124	31
liq.	2242 2165m	16
KBr	2240	32
liq.	2200w 2140	33
CCl_4	2268	34
CCl_4	2283w 2204	34
CCl_4	2245	34
CCl_4	2312 2220	34
KBr	2200	35
KBr	2220	35
liq.	2220	27

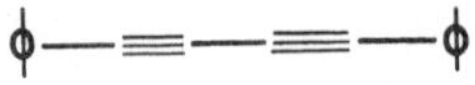

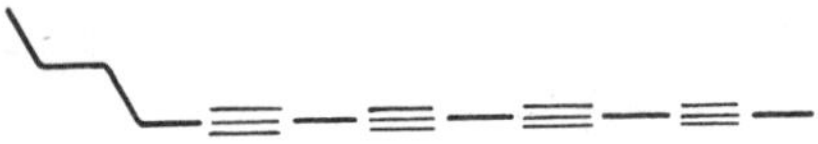

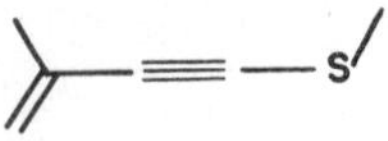

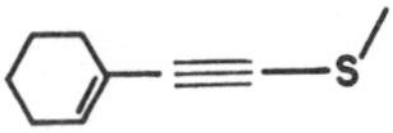

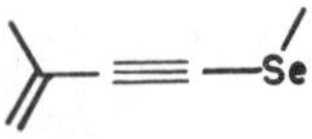

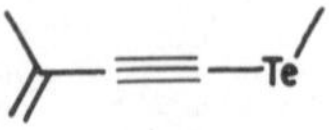

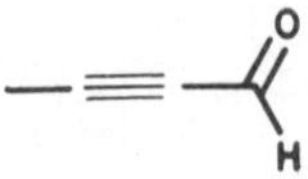

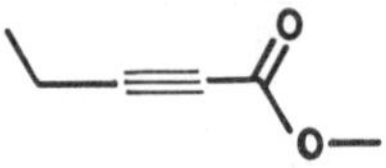

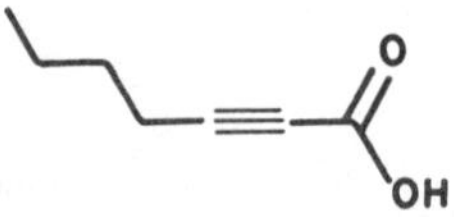

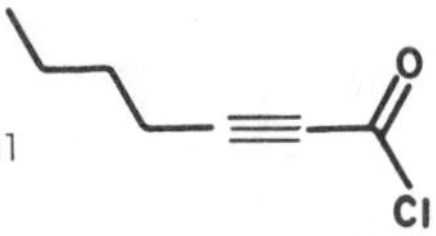

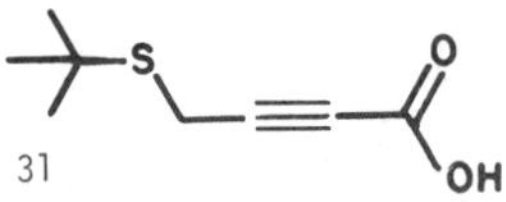

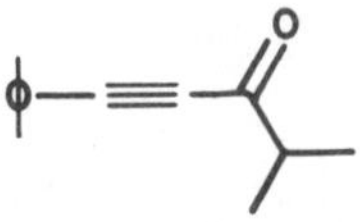

NC—≡—≡—CN
CCl_4 2236m 2066w 15

10
liq. 2208 36

S S
KCl 2225 37

CCl_4 2187 2116 13

Si—≡
CCl_4 2036 1

Si—≡—
liq. 2047 38

Si—≡—Si
CCl_4 2044 38

Si—≡
liq. 2147 1612 1594 24

S═P(—≡—)$_3$
CCl_4 2201 39

S═P(—≡—ϕ)$_3$
CCl_4 2171 39

Ge—≡
CCl_4 2027 1

Ge—≡—
liq. 2046 38

Sn—≡
CCl_4 2013 1

Sn—≡—
gas 2019 38

Sn(—≡—)$_4$
KBr 2170 40

Sn—≡
liq. 2129 1605 24

ϕ—≡—Hg—≡—ϕ
nuj. 2140 26

Pb—≡
CCl_4 2000 1

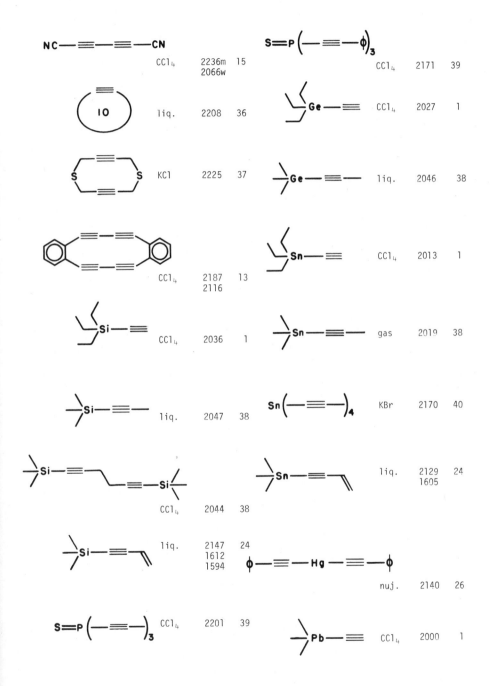

Compound	Phase	cm⁻¹	Ref.
$(C_2H_5)_3Pb-C\equiv C-CH_3$	liq.	2150	40
$\phi_3Pb-C\equiv C-CH_3$	liq.	2150	40
$(C_2H_5)_3Pb-C\equiv C-\phi$	liq.	2200	40
$(CH_3)_3Pb-C\equiv C-\phi$	KBr	2140	40
$\equiv^- Li^+$	nuj.	1876	1
$\equiv^- Na^+$	nuj.	1865	1
$\equiv^- K^+$	nuj.	1856	1
$\equiv^- Rb^+$	nuj.	1850	1
$\equiv^- Cs^+$	nuj.	1840	1
$-\equiv^- Li^+$	nuj.	2052	1
$-\equiv^- Na^+$	nuj.	2032	1
$-\equiv^- K^+$	nuj.	2020	1
$-\equiv^- Rb^+$	nuj.	2019	1
$-\equiv^- Cs^+$	nuj.	2011	1
$\phi-\equiv^- Li^+$	nuj.	2035	22
$\phi-\equiv^- Na^+$	nuj.	2019	1
$\phi-\equiv^- K^+$	nuj.	1999	1
$\phi-\equiv^- Rb^+$	nuj.	1991	1
$\phi-\equiv^- Cs^+$	nuj.	1987	1

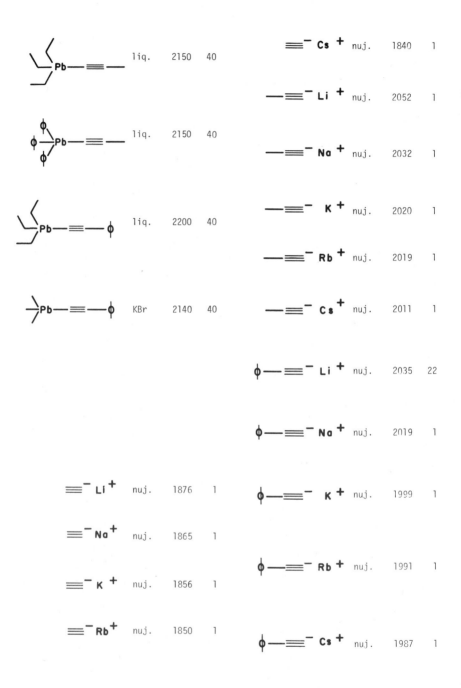

REFERENCES

1. B. Wojtkowiak and R. Queignec, *CR* (B) (1966) *262*, 811.
2. J. K. Brown and J. K. Tyler, *Proc Chem Soc* (1961), 13.
3. W. E. Parham and P. L. Stright, *JACS* (1956), 4783.
4. J. C. Evans and R. A. Nyquist, *SCA* (1963), 1153.
5. R. A. Nyquist and W. J. Potts, *SCA* (1960), 419.
6. J. H. Wotiz, F. A. Miller and R. J. Palchak, *JACS* (1950), 5055.
7. W. D. Huntsman and H. J. Wristers, *JACS* (1967), 342.
8. T. L. Jacobs, W. L. Petty and E. G. Teach, *JACS* (1960), 4094.
9. A. A. Petrov and M. P. Forost, *Zh Organ Khim* (1965), 1550.
10. H. Taniguchi, I. M. Mathai and S. I. Miller, *T* (1966), 867.
11. D. Braun and H.-G. Keppler, *Monatsh Chem* (1963), 1250.
12. A. D. Allen and C. D. Cook, *Canad J Chem* (1963), 1084.
13. O. M. Behr, G. Eglinton, A. R. Galbraith and R. A. Raphael, *JCS* (1960), 3614.
14. P. Klaboe, E. Kloster-Jensen and C. J. Cyvin, *SCA* (1967) *23A*, 2733.
15. A. J. Saggimo, *JOC* (1957), 1171.
16. A. A. Petrov and G. I. Semenov, *J Gen Chem USSR* (1957), 2980.
17. D. W. Davidson and H. J. Bernstein, *Canad J Chem* (1955), 1226.
18. L. F. Hatch and L. E. Kidwell, Jr., *JACS* (1954), 289.
19. E. Kloster-Jensen, *JACS* (1969), 5673.
20. P. Klaboe, E. Kloster-Jensen, D. H. Christensen and I. Johnsen, *SCA* (1970), 1567.
21. R. Criegee and R. Huber, *B* (1970), 1855.
22. I. A. Favorskaia, E. M. Auvinen and Yu. P. Artsybashev, *Zh Obshch Khim* (1958), 1785.
23. I. A. Favorskaia and Yu. P. Artsybashev, *Zh Organ Khim* (1965), 1716.
24. T. V. Yakovleva, A. A. Petrov and V. S. Zavgorodnii, *Opt & Spec* (1962) *12*, 106.
25. S. T. D. Gough and S. Trippett, *JCS* (1962), 2333.
26. G. Eglinton and W. McCrae, *JCS* (1963), 2295.
27. I. A. Favorskaia and A. A. Nikitinia, *Zh Organ Khim* (1965), 2094.
28. Yu. I. Porfir'eva, E. S. Turbanova and A. A. Petrov, *Zh Obshch Khim* (1964), 3966.
29. G. Eglinton and A. R. Galbraith, *JCS* (1959), 889.
30. E. R. H. Jones, G. Lowe and P. V. R. Shannon, *JCS* (1966) *C*, 139.
31. S. I. Radchenko and A. A. Petrov, *Zh Organ Khim* (1965), 2115.
32. J. C. Craig and M. Moyle, *JCS* (1963), 4402.
33. H. J. Boonstra and L. C. Rinzema, *Rec Trav Chim* (1960), 962.

34. L. Lopez, J.-F. Labarre, P. Castan and R. Mathis-Noel, *CR* (1964) *259*, 3483.
35. G. Pourcelot and P. Cadiot, *BSC France* (1966), 3024.
36. A. T. Blomquist, R. E. Barge and A. C. Suesy, *JACS* (1952), 3636.
37. G. Eglinton, I. A. Lardy, R. A. Raphael and G. A. Sim, *JCS* (1964), 1154.
38. R. E. Sacher, W. Davidsohn and F. A. Miller, *SCA* (1970), 1011.
39. G. M. Gogolyubov and A. A. Petrov, *Zh Obshch Khim* (1965), 704.
40. J.-C. Masson, M. Le Quan, W. Chodkiewicz and P. Cadiot, *CR* (1963) *257*, 1111.

2.3 allenes

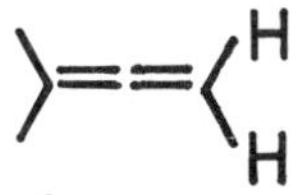

Although the vinylidene groups of olefins have characteristic C-H stretching modes in the region of 3100 to 3070 cm^{-1}, allenes do not show any absorptions in this region. Nevertheless, the out-of-plane deformation mode characteristic of the vinylidene group is seen as a strong absorption between 900 and 800 cm^{-1} in nonsubstituted terminal allenes. The frequency of this band is not, however, of much diagnostic value. The overtone of this band, which occurs in the carbonyl region, is usually of moderate intensity.

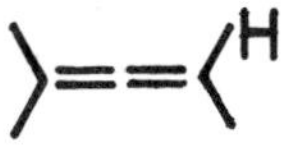

Substituted allenes that still contain a C-H bond show a strong deformation mode in the region of 880 to 860 cm^{-1}.

The cumulative bonds of allenes are vibrationally coupled. This leads to a strong and characteristic absorption between 2000 and 1930 cm^{-1} from the asymmetric mode, and a band between 1100 and 1000 cm^{-1}, which is of little diagnostic value, for the symmetric mode.

When substituted by electron withdrawing groups, terminal allenes often show two bands in the region of the asymmetric vibration. In general, substitution on the allene chromophore has only a small influence on the frequency of the asymmetric vibration.

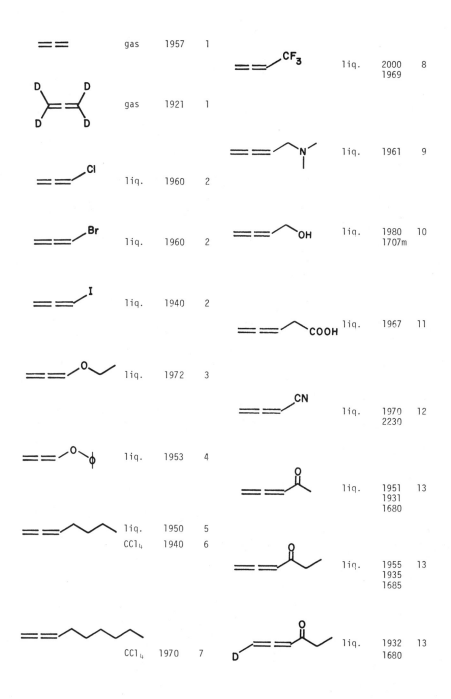

Structure	State	Frequency	Ref.
allene	gas	1957	1
allene-d_4	gas	1921	1
chloroallene	liq.	1960	2
bromoallene	liq.	1960	2
iodoallene	liq.	1940	2
ethoxyallene	liq.	1972	3
phenoxyallene (O–ϕ)	liq.	1953	4
1,2-heptadiene	liq.	1950	5
	CCl_4	1940	6
1,2-nonadiene	CCl_4	1970	7
CF_3-allene	liq.	2000 1969	8
allyl-N(CH3)2	liq.	1961	9
allyl-OH	liq.	1980 1707m	10
allyl-COOH	liq.	1967	11
CN-allene	liq.	1970 2230	12
allenyl methyl ketone	liq.	1951 1931 1680	13
allenyl ethyl ketone	liq.	1955 1935 1685	13
D-allenyl ethyl ketone	liq.	1932 1680	13

Structure	Phase	Frequencies	Ref.
O	liq.	1965 1943 1684	13
O	liq.	1969 1949 1684	13
O	liq.	1960 1927 1666	13
COOH	CS_2	1980	14
	CCl_4	1940 990 900 845	15
	CCl_4	1975 2120	15
ϕ	liq.	1975m 1940	
	liq.	1980	6
	liq.	1960	6
	liq.	1955	6
CN	liq.	1930 1910	17
O	liq.	1959 1710 1677m	13
O, O	liq.	1950m	18
COOH	CCl_4	1950 1930	6

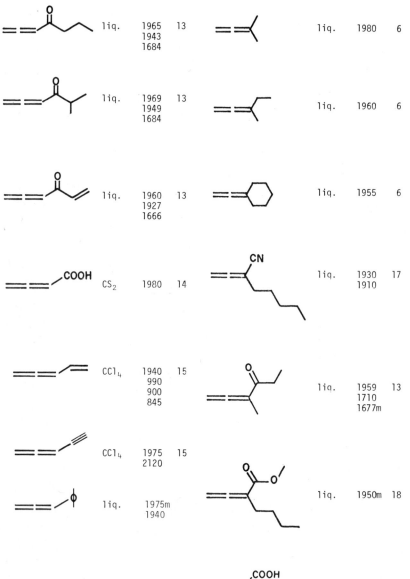

Structure	Phase	Frequencies	Ref.
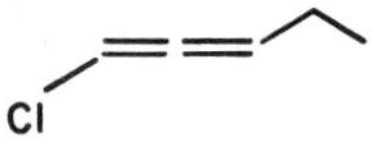 Cl	liq.	1957	19
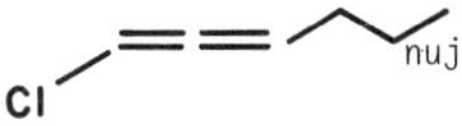 Cl	nuj.	1965 840	20
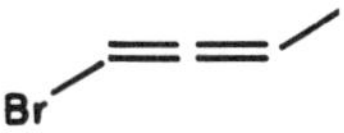 Br	liq.	1950	20
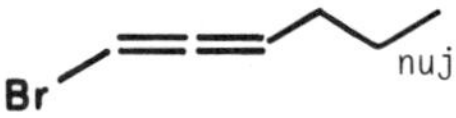 Br	nuj.	1970 830	20
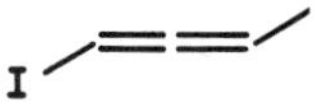 I	liq.	1930 855 800	21
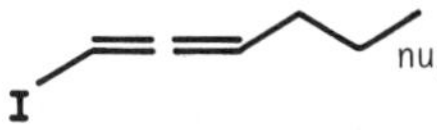 I	nuj.	1950 830	20
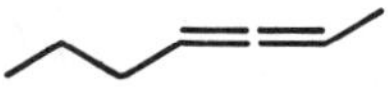	liq.	1964	22
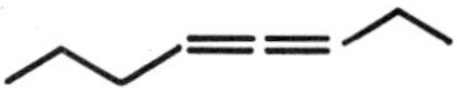	liq.	1964	22
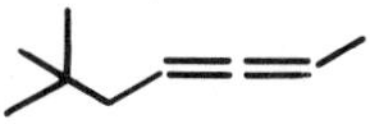	liq.	1964	22
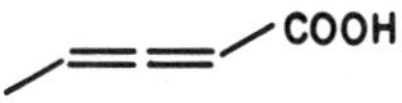 COOH	KBr	1970 1950	23
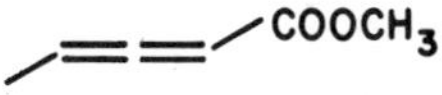 $COOCH_3$	liq.	1970 1960	23
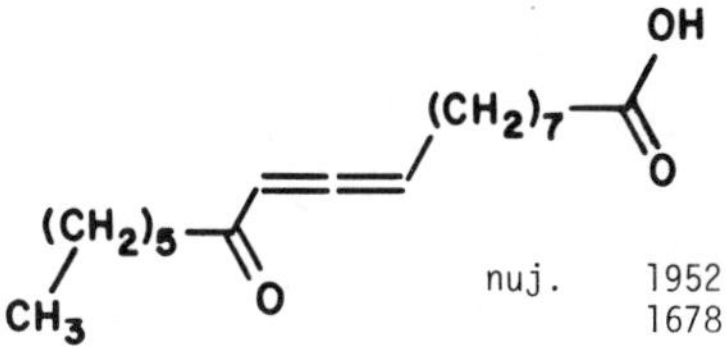 OH, $(CH_2)_7$, O, $(CH_2)_5$, CH_3, O	nuj.	1952 1678	24
 9	CCl_4	1957	7

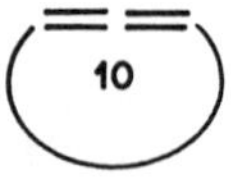

(ring 10)	liq.	1938	25
	CCl_4	1962	7
(ring 11)	CCl_4	1980	7
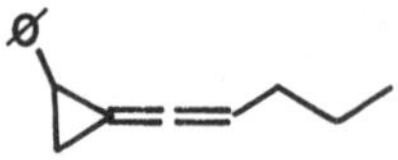	liq.	2020	26
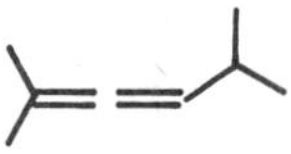	liq.	1980m 1690	10
	CCl_4	1939 1887m 1812m 1760w 1687w	27
Cl	liq.	1950m	28
Cl	liq.	1938m	28
Cl	liq.	1950m	28
Cl	liq.	1955	28
Br	liq.	1950	20
Br	liq.	1950	20
Br	liq.	1950	20
Br	liq.	1940	20

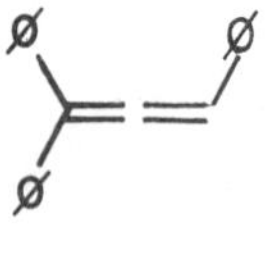

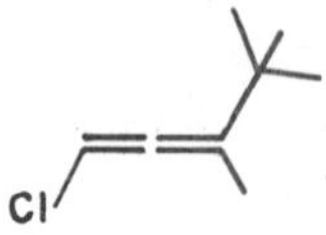

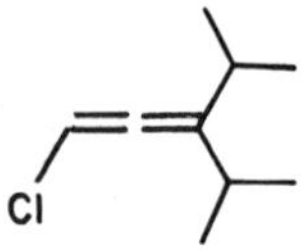

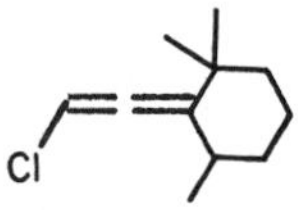

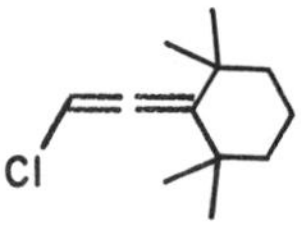

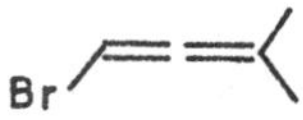

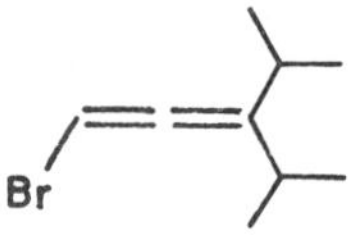

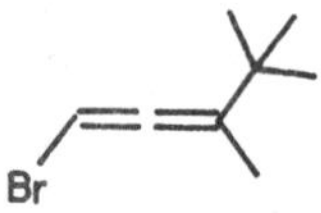

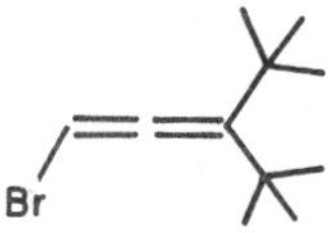

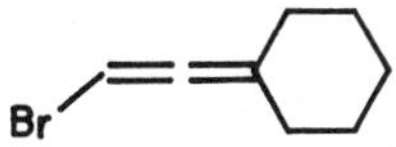	liq.	1950	20
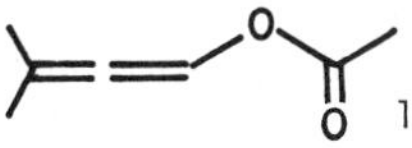	liq.	1987m 1758	29
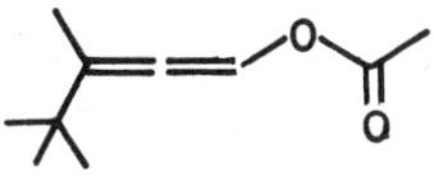	liq.	1979m 1760	29
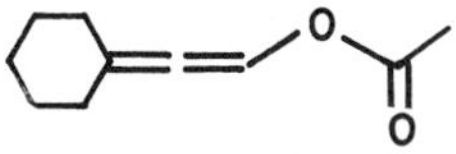	liq.	1977m 1758	29
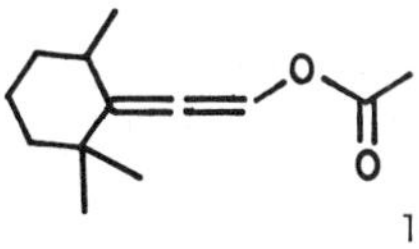	liq.	1970 1950	30
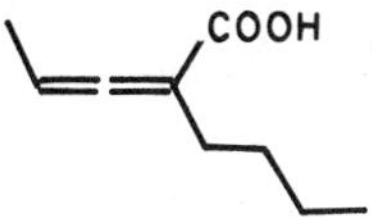	CCl_4	1940	31
	liq.	1960w	32
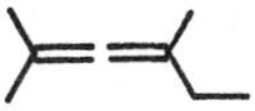	liq.	1960w	32
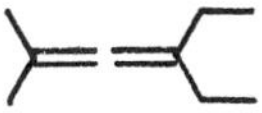	liq.	1950w	32
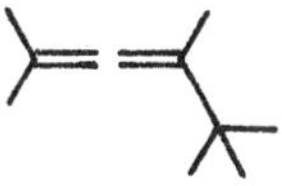	liq.	1960w	32
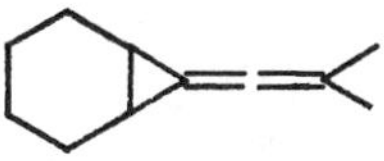	liq.	2020 2000m	26
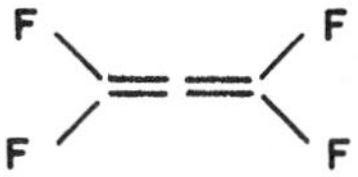	gas	2065	33
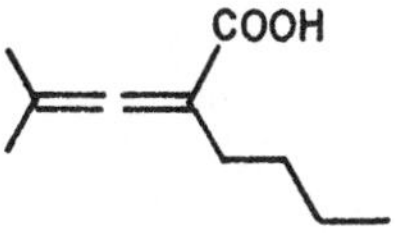	nuj.	1940	6
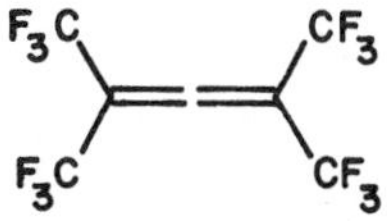	liq.	1992	34

$CH_2=C=CH_2$	gas	1708 1610	35
$F_2C=C=CF_2$	gas	1733	36
$\phi_2C=C=C(\phi)(C_6H_4NO_2)$	KBr	2032	37
$\phi_2C=C=C$(fluorenylidene)	KBr	2032	37

REFERENCES

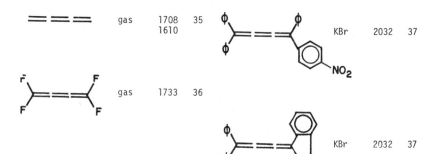

1. R. C. Lord and P. Venkateswarlu, *JCP* (1952) *20*, 1237.
2. T. L. Jacobs and W. F. Brill, *JACS* (1953), 1314.
3. J. F. Arens, *Rec Trav Chim* (1955), 271.
4. L. F. Hatch and H. D. Weiss, *JACS* (1955), 1798.
5. J. H. Wotiz, *JACS* (1951), 693.
6. W. J. Bailey and C. R. Pfeifer, *JOC* (1955), 95.
7. W. R. Moore and H. R. Ward, *JOC* (1962), 4179.
8. R. N. Haszeldine, K. Leedham and B. R. Steele, *JCS* (1954), 2040.
9. V. A. Engelhardt, *JACS* (1956), 107.
10. W. J. Bailey and C. R. Pfeifer, *JOC* (1955), 1337.
11. E. R. H. Jones, G. H. Whitham and M. C. Whiting, *JCS* (1954), 3201.
12. P. Kurtz, H. Gold and H. Disselnkötter, *A* (1959) *624*, 1.
13. J. Chouteau, G. Davidovics, M. Bertrand, J. Le Gras, J. Figarella and M. Santelli, *BSC France* (1964), 2562.
14. G. Eglinton, E. R. H. Jones, G. H. Mansfield and M. C. Whiting, *JCS* (1954), 3197.
15. E. R. H. Jones, H. H. Lee and M. C. Whiting, *JCS* (1960), 341.
16. L. Piaux and M. Gaudemar, *CR* (1955) *240*, 2328.
17. J. H. Wotiz and E. S. Hudak, *JOC* (1954), 1580.
18. J. H. Wotiz, *JACS* (1950), 1639.
19. T. L. Jacobs, W. L. Petty and E. G. Teach, *JACS* (1960), 4094.
20. S. R. Landor, A. N. Patel, P. F. Whiter and P. M. Greaves, *JCS* (1966), 1223.
21. C. S. L. Baker. P. D. Landor, S. R. Landor and A. N. Patel, *JCS* (1965), 4348.

22. A. A. Petrov, T. V. Iakovleva and V. A. Kormer, *Opt & Spec* (1959) *7*, 170.
23. J. C. Craig and M. Moyle, *JCS* (1963), 4402.
24. L. Crombie and A. G. Jacklin, *JCS* (1955), 1740.
25. A. T. Blomquist, R. E. Barge and A. C. Suesy, *JACS* (1952), 3636.
26. S. R. Landor and P. F. Whiter, *JCS* (1965), 5625.
27. T. L. Jacobs, D. Dankner and S. Singer, *T* (1964) *20*, 2177.
28. Y. R. Bhatia, P. D. Landor and S. R. Landor, *JCS* (1959), 24.
29. M. Apparu and R. Glenat, *CR* (1967) *265C*, 400.
30. P. D. Landor and S. R. Landor, *JCS* (1956), 1015.
31. J. H. Wotiz and W. D. Celmer, *JACS* (1952), 1860.
32. T. L. Jacobs and P. Prempree, *JACS* (1967), 6177.
33. T. L. Jacobs and R. S. Bauer, *JACS* (1959), 606.
34. Yu. E. Aronov, Yu. A. Cheburkov and I. L. Knunyants, *Izvest Akad Nauk Khim* (1967), 1758.
35. W. M. Schubert, T. L. Liddicoet and W. A. Lanka, *JACS* (1954), 1929.
36. E. L. Martin and W. H. Sharkey, *JACS* (1959), 5256.
37. R. Kuhn and H. Fischer, *B* (1959), 1849.

2.4

enamines

The medium to strong band of the C-C double bond stretching vibration of enamines falls, for the majority of such compounds, within the narrow range of 1680 to 1630 cm^{-1}. In general this band is more intense than a normal C-C double bond as a result of the polarity of the enamine chromophore.

The positions of the C-H stretching and deformation modes are the same as other olefins and are discussed in detail in Chapter 2.1, p. 26.

Vinylogous amides are discussed separately in Chapter 4.4, p. 263.

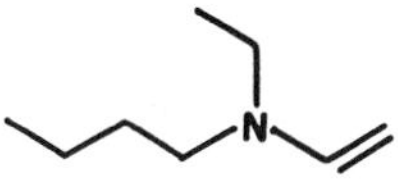

C_6H_{14} 1628 1

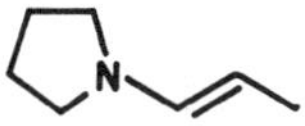

liq. 1644 2
C_6H_{14} 1647 1

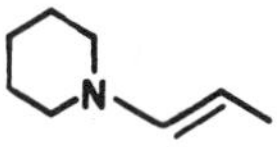

liq. 1644 2
CCl_4 1660 3
C_6H_{14} 1653 1

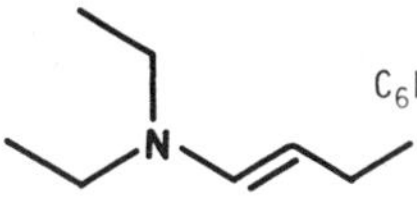

C_6H_{14} 1648 1

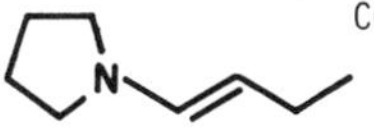

CCl_4 1650 3

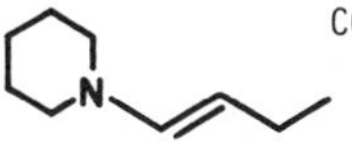

CCl_4 1655 3

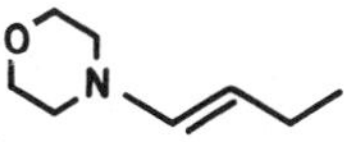

CCl_4 1655 3

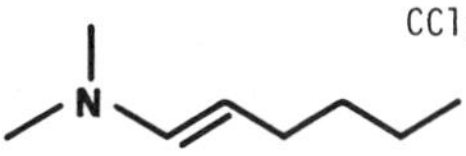

CCl_4 1650 3

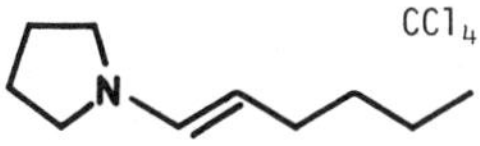

CCl_4 1652 3

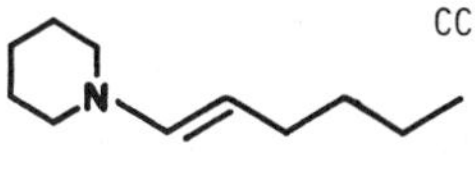

CCl_4 1660 3

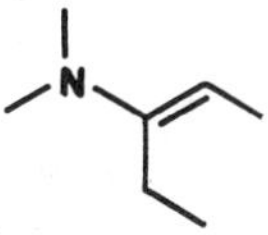

liq. 1640 4

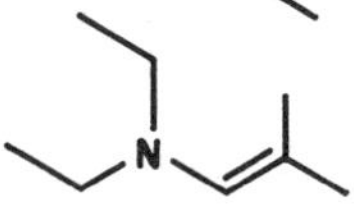

C_6H_{14} 1662m 1

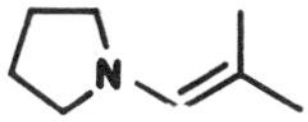

C_6H_{14} 1672 1

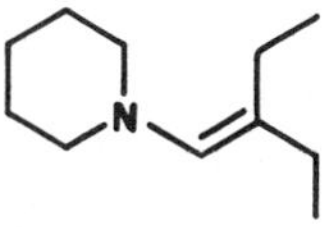

C_6H_{14} 1658m 1

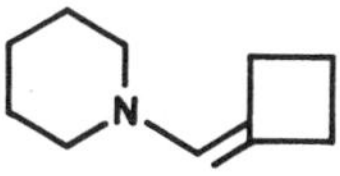

liq. 1668 5

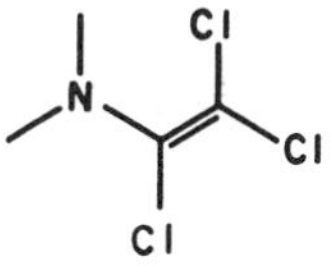

liq. 1616 6

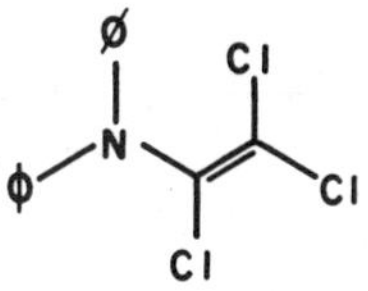

liq. 1590 6

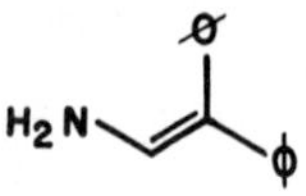 $CHCl_3$	1637	7
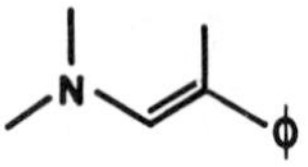 liq.	1639	7
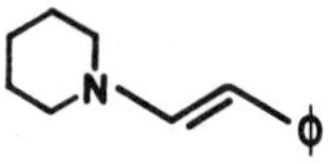 CCl_4	1630	3
liq.	1634	5
C_6H_{14}	1622	1
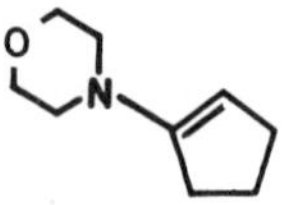 C_6H_{14}	1630	1
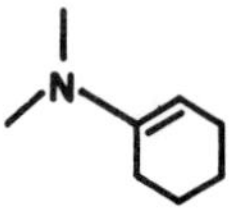 liq.	1640	4
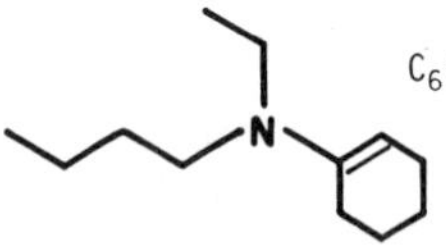 C_6H_{14}	1630	1
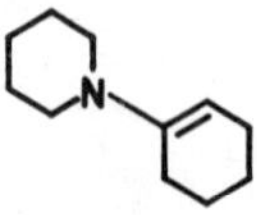 C_6H_{14}	1637	1

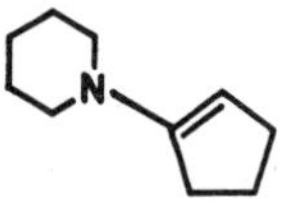

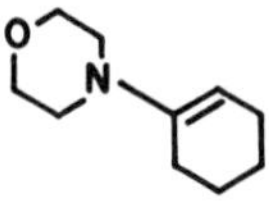 C_6H_{14}	1641	1
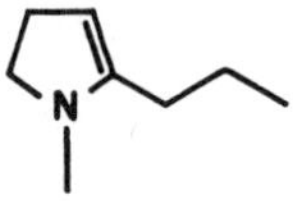 liq.	1665	2
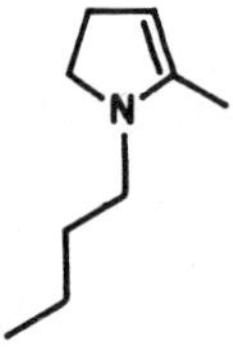 liq.	1639	2
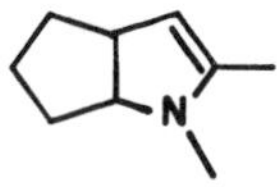 liq.	1640	8
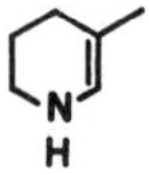 $CHCl_3$	1678	9
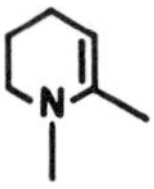 liq.	1649	10
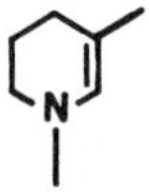 liq.	1673	10
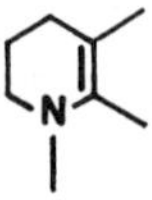 liq.	1657	10

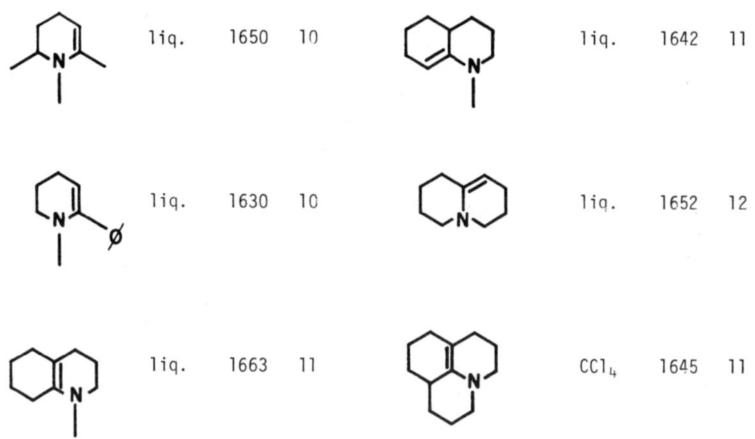

liq. 1650 10
liq. 1642 11
liq. 1630 10
liq. 1652 12
liq. 1663 11
CCl_4 1645 11

REFERENCES

1. G. Opitz, H. Hellmann and H. W. Schubert, *A* (1959) *623*, 112.
2. N. J. Leonard and V. W. Gash, *JACS* (1954), 2781.
3. R. Dulou, E. Elkik and A. Veillard, *BSC France* (1960), 967.
4. R. Burgada and J. Roussel, *BSC France* (1970), 192.
5. K. L. Erickson, J. Markstein and K. Kim, *JOC* (1971), 1024.
6. A. J. Speziale and R. C. Freeman, *JACS* (1960), 903.
7. B. Witkop, *JACS* (1956), 2873.
8. R. Griot and T. Wagner-Jauregg, *Helv* (1959), 121.
9. M. Ferles and M. Holík, *Col Czech Comm* (1966), 2416.
10. N. J. Leonard and F. P. Hauck, Jr., *JACS* (1957), 5279.
11. N. J. Leonard, C. K. Steinhardt and C. Lee, *JOC* (1962), 4027.
12. N. J. Leonard, A. S. Hay, R. W. Fulmer and V. W. Gash, *JACS* (1955), 439.

2.5

enol ethers enol esters

The generalizations that have already been applied to olefins (Chapter 2.1, p. 26) can be applied to enol ethers and enol esters with the exceptions noted as follows. The polar nature of the enolic double bond gives rise to more intense bands than are seen with alkenes. Two bands are often seen in cyclic systems where conformational isomers can exist.

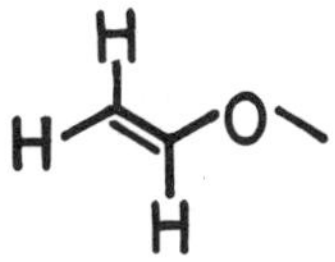

The C-H stretching modes for the variously substituted enol ethers correspond to those seen with other alkenes, except that the bands above 3000 cm^{-1} are usually weaker and their absences do not necessarily indicate the absence of an enol ether or ester.

The other major changes are the lowering of the frequencies of the intense and diagnostically useful bands associated with the out-of-plane C-H bending modes. Thus the band that appears around 1000 cm^{-1} with the vinyl group still occurs between 1000 and 900 cm^{-1} for vinyl ethers, but the frequencies are lower than for the corresponding alkene. These bands often appear as doublets as a result of conformational isomers.

Similarly, the second band that appears between 900 and 800 cm^{-1} in vinyl and vinylidene groups also appears at lower frequencies in these systems. The band is usually closer to 800 cm^{-1} and often occurs as a doublet when conformational isomers are present.

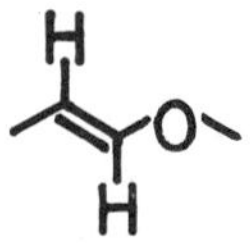

The strong band that occurs between 990 and 960 cm^{-1} with trans-substituted olefins occurs at lower frequencies and is usually found in the region from 940 to 920 cm^{-1}.

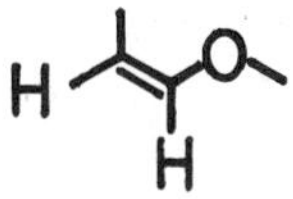

As with alkenes, the out-of-plane C-H deformation for cis substituted derivatives is found between 730 and 650 cm^{-1}. The band is usually found at the higher end of the range but is not diagnostically useful.

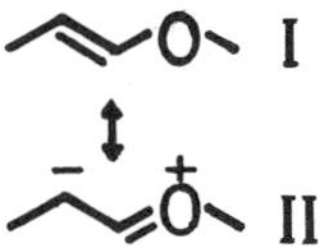

I

II

Delocalization of the oxygen lone pairs into the olefinic double bond (I ⟷ II) would be expected to lower the frequency of the double bond stretching vibration, whereas the field effect of the oxygen is expected to increase its frequency. Neither of these two effects predominates, as can be seen from the following tables, and the double-bond stretching frequency of enols falls in a similar region to those of alkenes.

Solvent	Frequency	Ref.
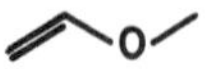 CCl_4	1652m 1639m 1616	1
CS_2	960 813	2
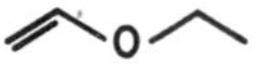 liq.	1610 1320 967 810	3
CCl_4	1634 1608 964 943	4
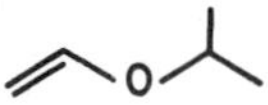 liq.	1614 1322 969 812	3
 liq.	1650* 1633 1611 964 811	5
CCl_4	1653 1638 1613	6
$CHCl_3$	1653 1638 1617	6
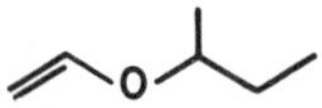 CCl_4	1632 1608 962 941	4
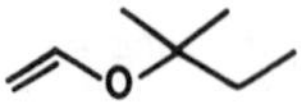 CCl_4	1632 1608 962 941	4

Solvent	Frequency	Ref.
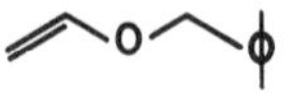 liq.	1611 1320 960 820	3
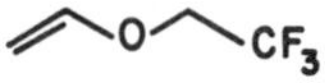 liq.	1645 1626	7
 CCl_4	1637m 1618	1
 CCl_4	989 923	8
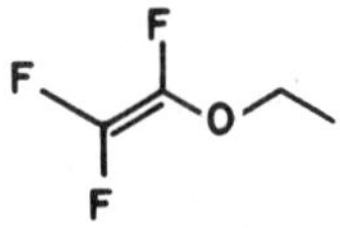 liq.	1832	9
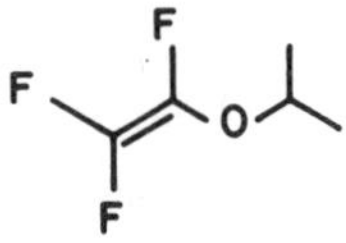 liq.	1825	9
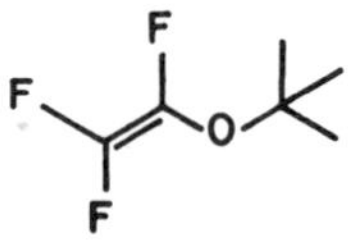 liq.	1818	9
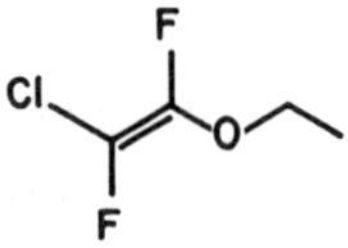 liq.	1754	9

Structure	Solvent	Frequency	Ref.
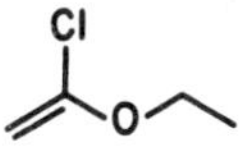 Cl, O	liq.	1640 1590m 1050 980 790	10
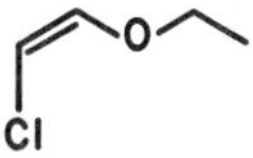 O, Cl	liq.	1645	10
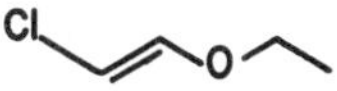 Cl, O	liq.	1605 920 850 825m	10
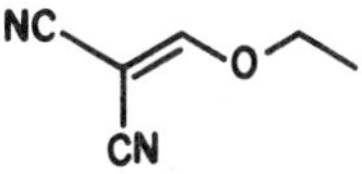 NC, O, CN	$CHCl_3$	1613 2242	11
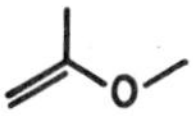 O	CS_2	795	2
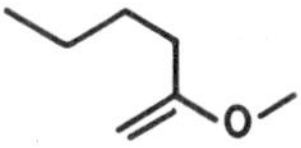 O	liq.	1655 1592 958 793	5
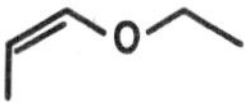 O	CCl_4	1690	12
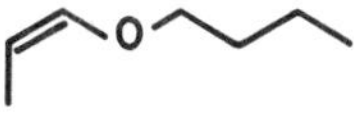 O	CCl_4	1672	1

Structure	Solvent	Frequency	Ref.
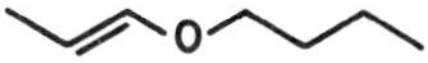 O	CCl_4	1681m 1661	1
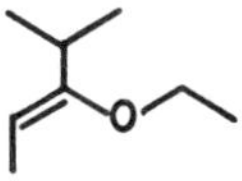 O	CCl_4	1660	12
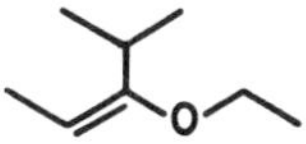 O	CCl_4	1673	12
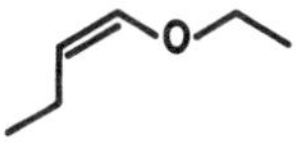 O	CCl_4	1663	12
 O	CCl_4	1670 1652 937 925	12
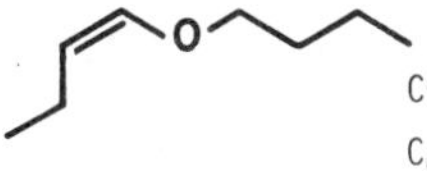 O	CCl_4 C_6H_{12}	1672 723	13 13
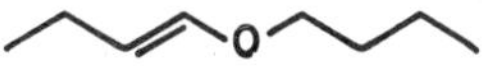 O	CCl_4	1681 1661 933	13
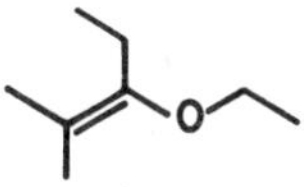 O	CCl_4	1680	12
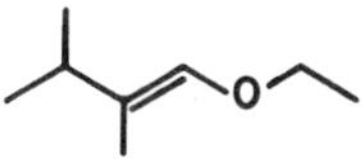 O	CCl_4	1660	12

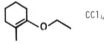

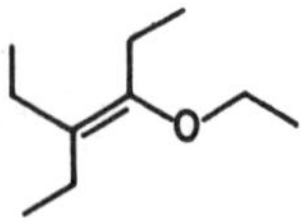	CCl_4	1663	12
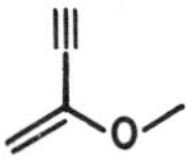	liq.	1603 916 826	14
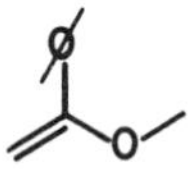	liq.	1608 1314 1303 899	15
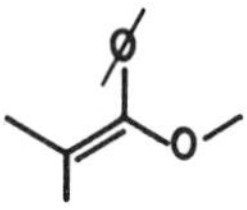	CCl_4	1661	16
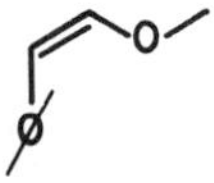	liq.	1650 726	15
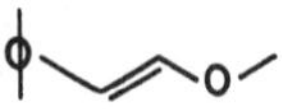	liq.	1642 937	15
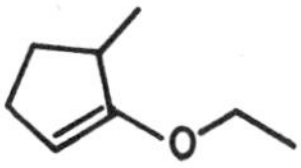	CCl_4	1640	12
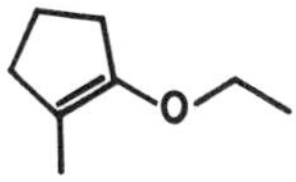	CCl_4	1690	12
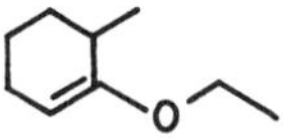	CCl_4	1670	12
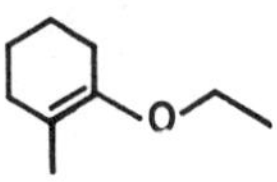	CCl_4	1685	12
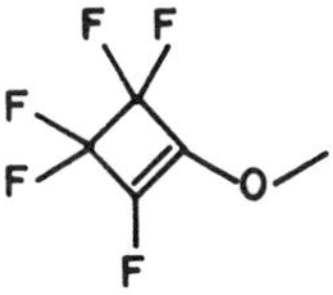	gas	1765	17
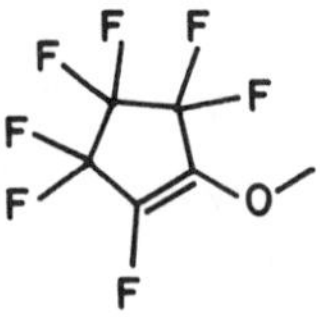	gas	1727	17

 gas	1705	17
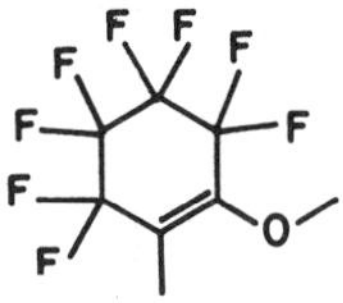 gas	1677	17
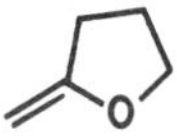 liq.	1669 1590w 882	5
 liq.	1630 715 699	5
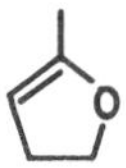 liq.	1677 958 925 891 717	5
 liq.	1615 707	18

 liq.	1644 929 889 835 749 727	5
CCl	1645	1
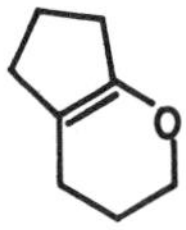 CCl_4	1692	19
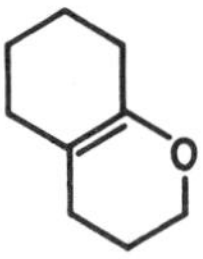 CCl_4	1689	19
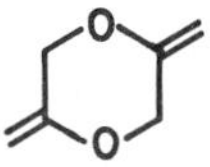 liq.	1671m 1645m 1302 1072 848 823 797	20
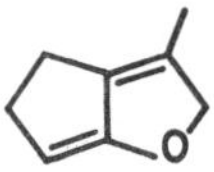 $CHCl_3$	1670 1780 1755	21
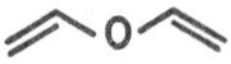 CCl_4	1672w 1641 1623* 1619	1
CCl_4	988 928	8

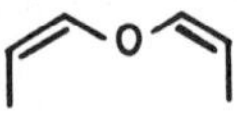	CCl_4	1656	1
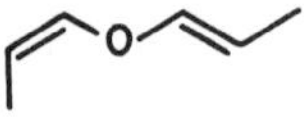	CCl_4	1661	1
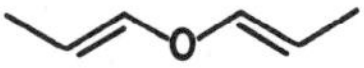	CCl_4	1645	1
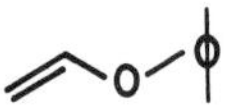	liq.	1644 1313 965 806	3
	CS_2	962* 944 851	2
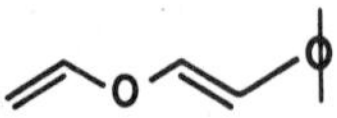	CCl_4	949 868	8
	CCl_4	1686 1634	1
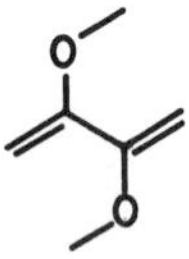	liq.	1592	22
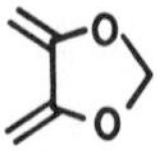	liq.	1658b	22

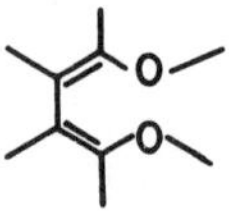	liq.	1668	23
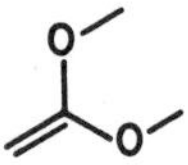	liq.	1639 1036	24
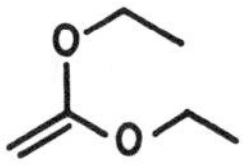	CS_2	711	2
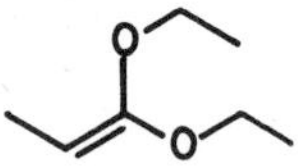	liq.	1675 1018	24
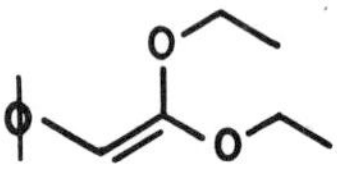	liq.	1634 1022	24
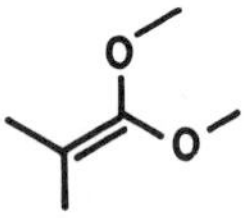	liq.	1633 999	24
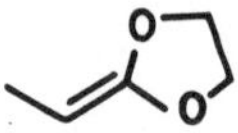	liq.	1656	25

Structure	Phase	Frequency	Ref.
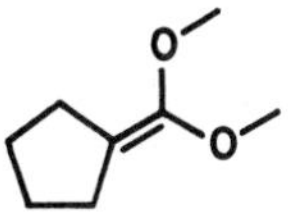 (structure)	liq.	1712 980	24
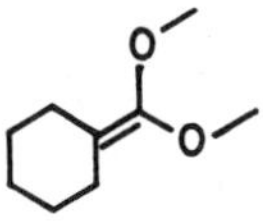 (structure)	liq.	1701 955	24
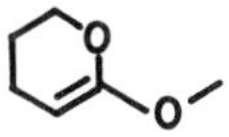 (structure)	liq.	1670	26
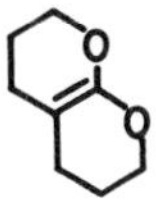 (structure)	liq.	1692	26
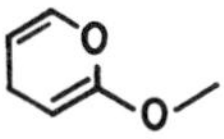 (structure)	liq.	1695 1648	26
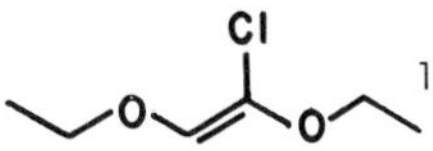 (structure, Cl)	liq.	1750	27
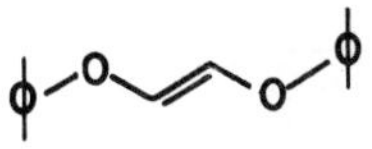 (structure, ϕ)	CCl_4	1690	28
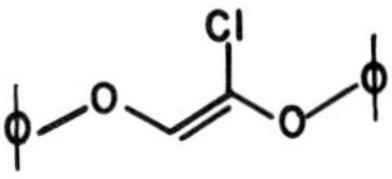 (structure, Cl, ϕ)	liq.	1690	28
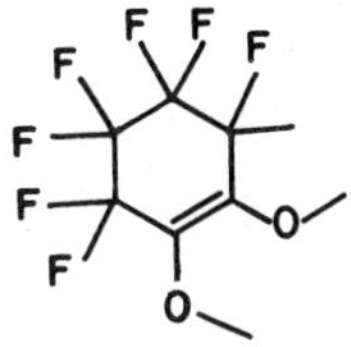 (structure, F)	gas	1678	17
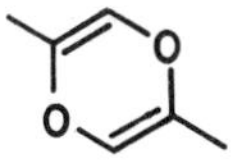 (structure)	liq.	1710* 1680 1387 1352 1236 1142 1001 892 771 725	20
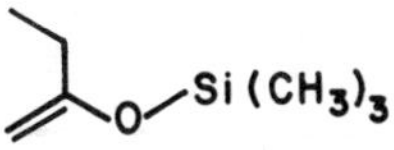 (structure) O–Si(CH$_3$)$_3$	CCl_4	1660w 1635	16

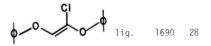

Structure	Solvent	cm⁻¹	Ref.
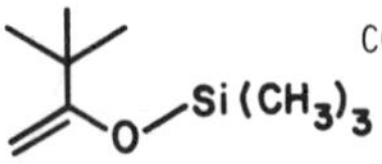 O–Si(CH$_3$)$_3$	CCl_4	1665w 1625	29
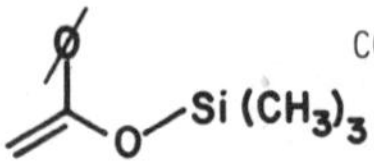 O–Si(CH$_3$)$_3$	CCl_4	1620	29
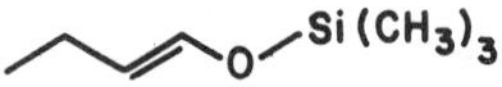 O–Si(CH$_3$)$_3$	CCl_4	1665	29
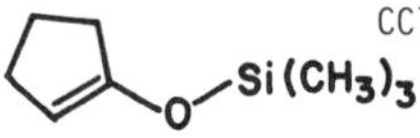 O–Si(CH$_3$)$_3$	CCl_4	1645	29
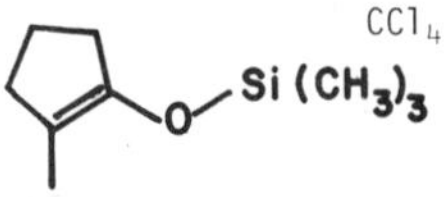 O–Si(CH$_3$)$_3$	CCl_4	1690	29
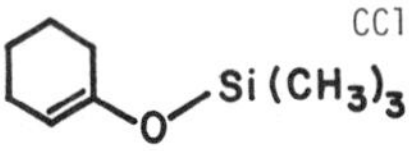 O–Si(CH$_3$)$_3$	CCl_4	1670	29
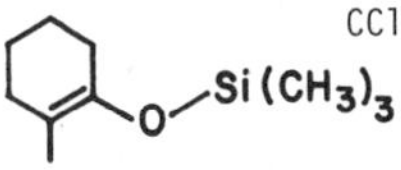 O–Si(CH$_3$)$_3$	CCl_4	1685	29
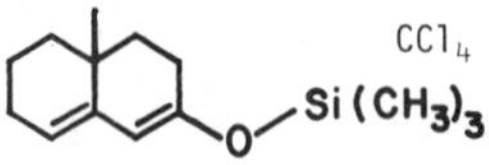 O–Si(CH$_3$)$_3$	CCl_4	1680w 1650 1625	29
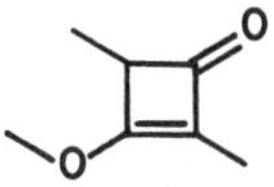 O, O	liq.	1630 1748	30
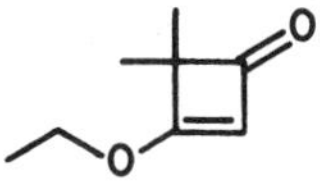 O, O	liq.	1575 1757	31
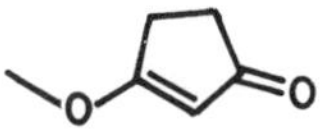 O, O	CCl_4	1680 1600 1705	32
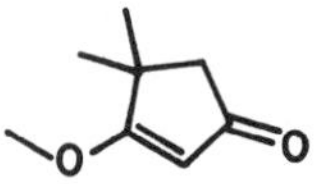 O, O	liq.	1587 1681	33
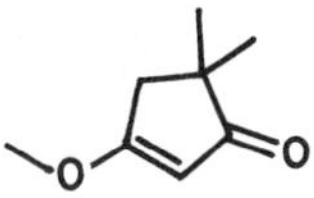 O, O	liq.	1592 1686	33
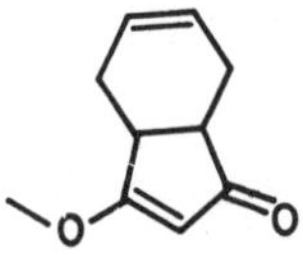 O, O	$CHCl_3$	1595 1685	34

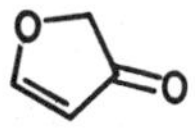

Solvent	cm^{-1}	Ref.
CCl_4	1572 1733 1706	35
CCl_4	1645 1608 1742 1707	36
CCl_4	1610 1712	36
CCl_4	1610 1701	36
$CHCl_3$	1620 1653	37
liq.	1610 1658	38
nuj.	1610 1650	30
CH_2Cl_2	1608 1639	39

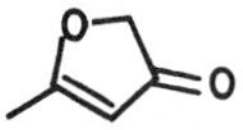

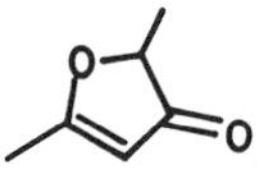

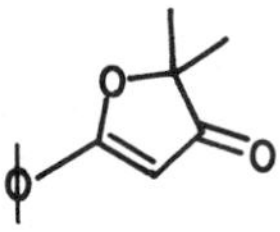

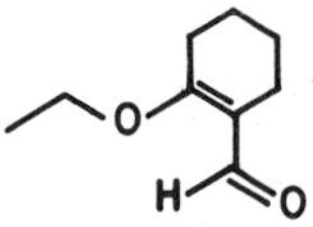

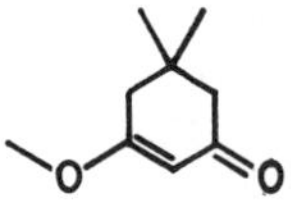

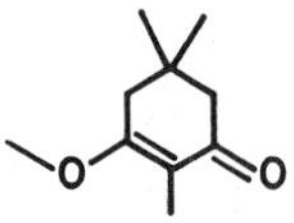

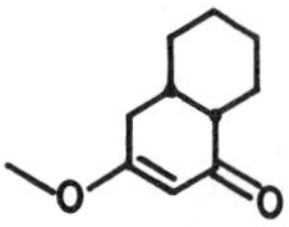

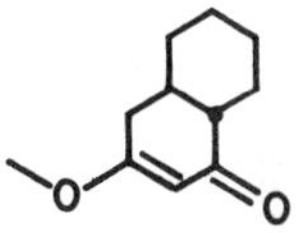

Solvent	cm^{-1}	Ref.
KBr	1605 1639	39
CH_2Cl_2	1603 1639	39
CCl_4	1670	40
KBr	1610 1592 1647	41
liq.	1653	42
C_6H_{12}	1649	43
CS_2	944 876	43
liq.	1645	42
C_6H_{12}	1650	43
CCl_4	1647 948 872	4
CS_2	948 870	43

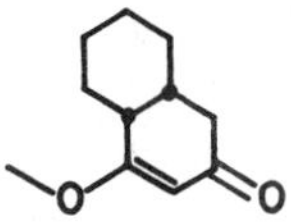

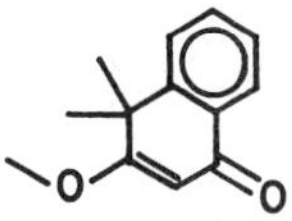

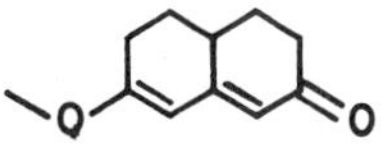

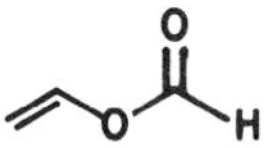

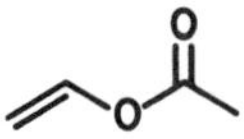

Structure	Solvent	Frequency	Ref.
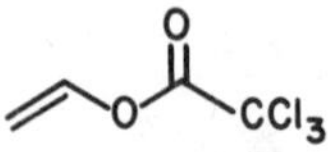 CCl_3	C_6H_{12}	1653	43
	CS_2	936	43
		884	
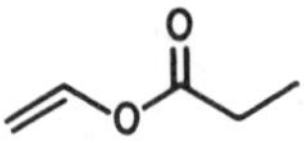	C_6H_{12}	1652	43
	CS_2	947	43
		868	
CN	liq.	1644	25
		1754	
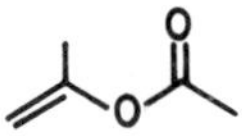	CCl_4	1678	12
		1759	
	CS_2	869	2
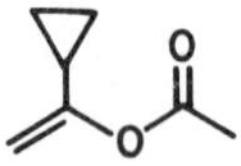	liq.	1660	44
		1760	
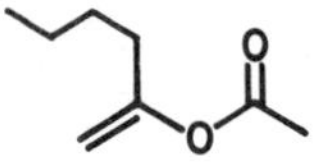	CCl_4	1665	12
		885	
		1755	
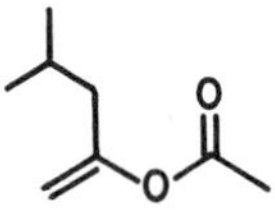	CCl_4	1665	12
		1756	
	liq.	1652	45
		1755	
		1700w	
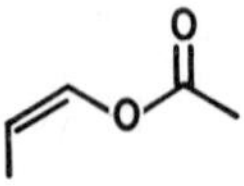	CCl_4	1690	12
		1758	
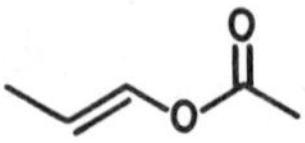	CCl_4	1675	12
		1752	
	liq.	1681	46
		1748	
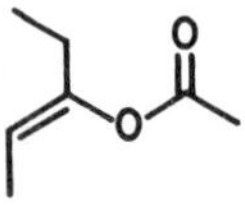	CCl_4	1699	12
		1755	
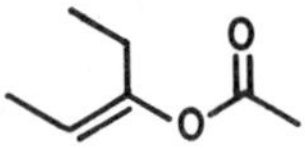	CCl_4	1690	12
		1750	
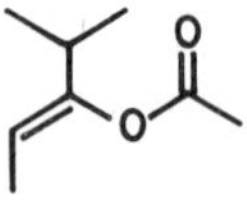	CCl_4	1692	12
		1755	
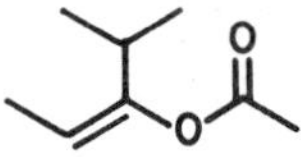	CCl_4	1683	12
		1755	
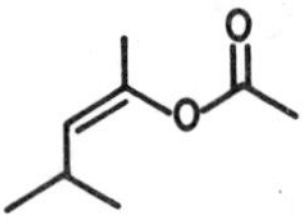	CCl_4	1695	12
		1752	
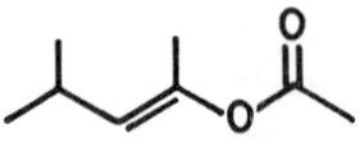	CCl_4	1690	12
		1758	
		1744	
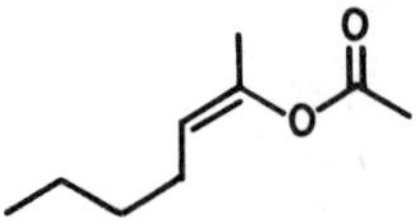	CCl_4	1695	12
		1750	

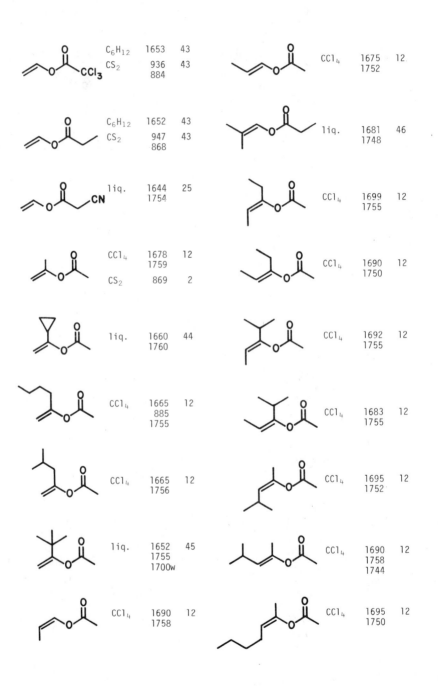

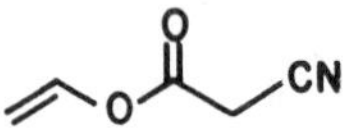

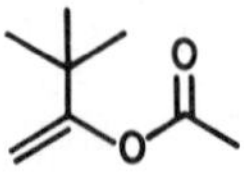

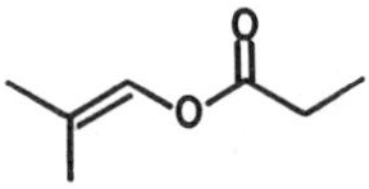

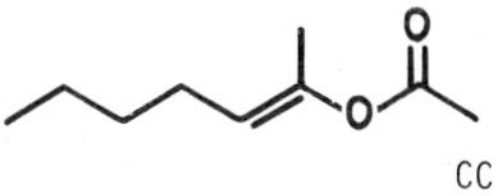	CCl_4	1692 1750	12
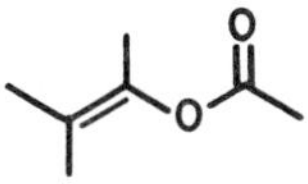	liq.	1710 1765	47
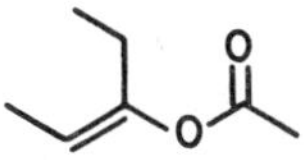	CCl_4	1693 1748	12
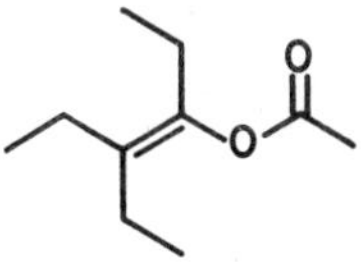	CCl_4	1680 1748	12
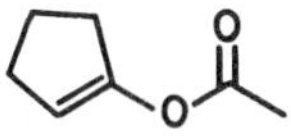	liq.	1689 959 1761	48
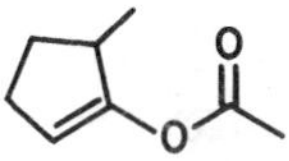	CCl_4	1713 1755	12
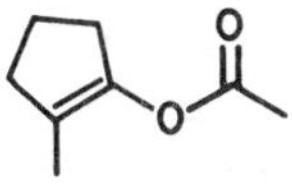	CCl_4	1703 1750	12
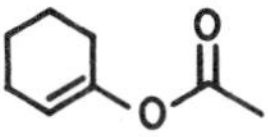	CCl_4	1695	49
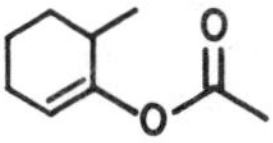	CCl_4	1680 1750	12
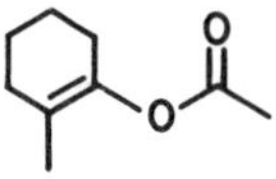	CCl_4	1705 1750	12
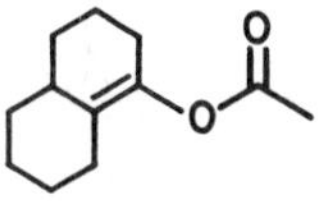	CCl_4	1695 1750	50
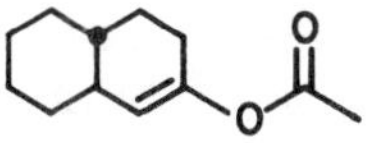	CCl_4	1678 1750	50
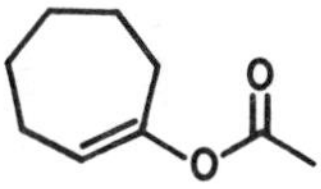	CCl_4	1690	49
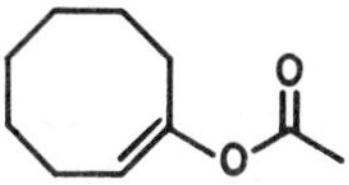	CCl_4	1687	49
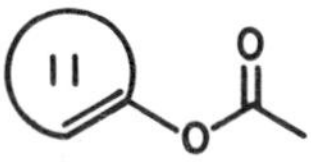	CCl_4	1695 1712	49
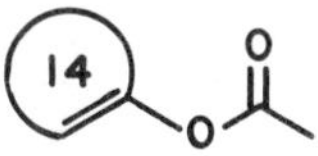	CCl_4	1715	49

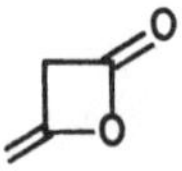

Solvent	Frequencies	Ref.
liq.	1676 960 810 1922 1875 1776 1720	51
CCl_4	1675 1900 1867 1752 1708	51
liq.	1681	52
$CHCl_3$	1640 1780	53
KBr	1661 1600 1754	54
$CHCl_3$	1670	21
$CHCl_3$	1689 1757	55
$CHCl_3$	1669	55
liq.	1696	56

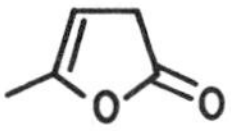

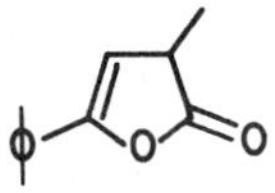

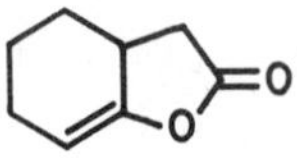

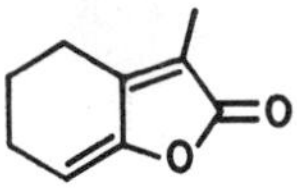

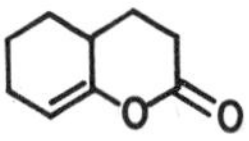

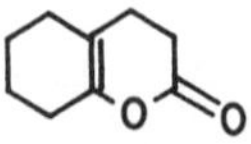

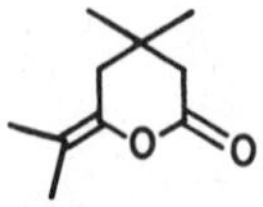

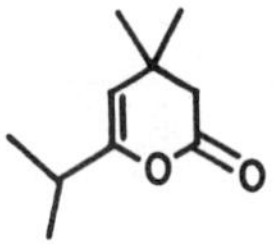

Solvent	Frequencies	Ref.
liq.	1683	56
liq.	1686 1764	57
liq.	1644 1735	58
liq.	1658 1638 1730	58
CCl_4	1647 949 868	4
liq.	1629 1597 1723	58
CCl_4	1647 948 871	4
CS_2	945 869	2

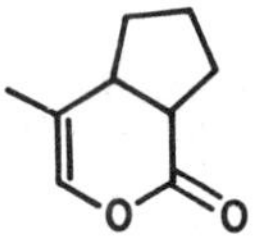

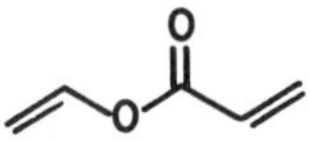

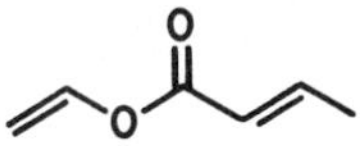

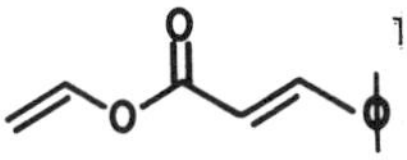

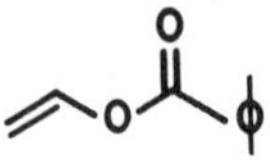

Structure	Solvent	Frequencies	Ref.
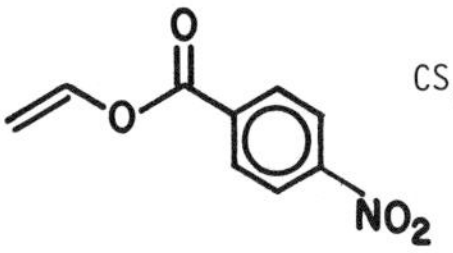	CS_2	944 872	43
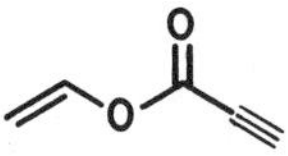	liq.	1644 2120 1721	58
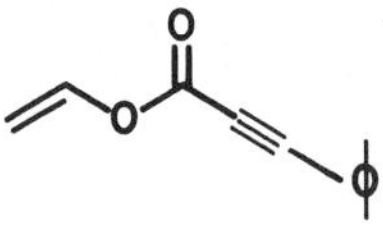	liq.	1639 1592 2213 1714	58
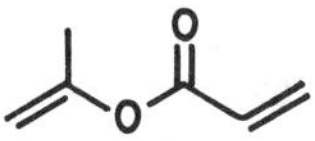	CCl_4	1640 1619 1407 984 965	4
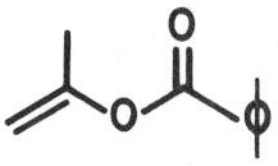	CCl_4	1676 877	4
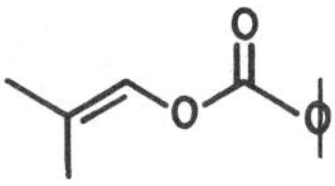	CCl_4	877	8
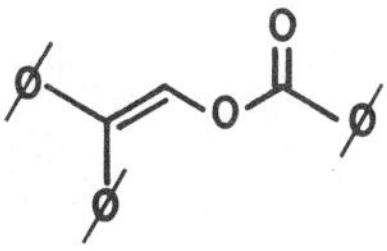	$CHCl_3$	1630 1710	59
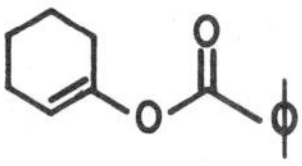	liq.	1695 1608 1739	60
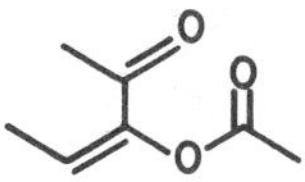	CCl_4	1692 1670 1762	61

Structure	Solvent	Frequencies	Ref.
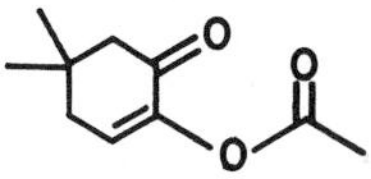	nuj.	1682 1657 1640 1754	62
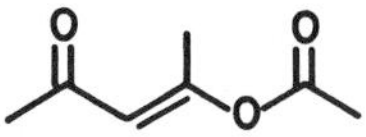	CCl_4	1696 1670w 1762	61
	$CHCl_3$	1695 1664 1631	63
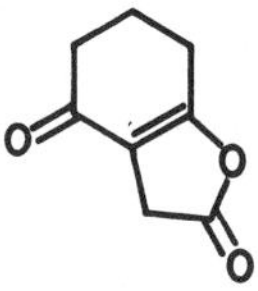	$CHCl_3$	1665 1827	64
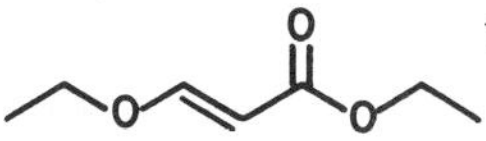	liq.	1652 1632 1724 1708	65
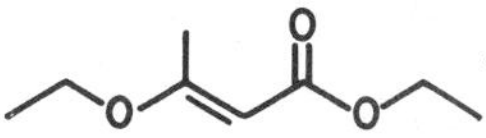	liq.	1650 1742	65
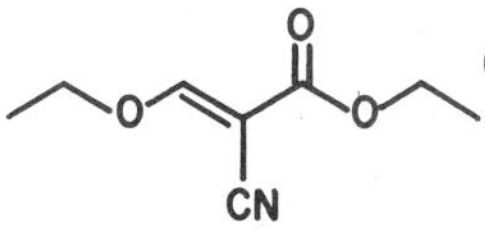	CCl_4	1621 2237 1721	11
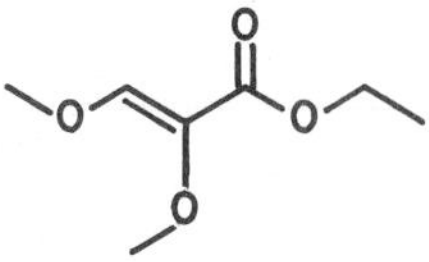	liq.	1660 1722	66

Solvent	Frequencies	Ref.
liq.	1688 1628 1705	67
liq.	1610 1700	68
liq.	1634 1718 1692	67
CH_2Cl_2	1621 1701	69
liq.	1660 1710	70
CCl_4	1660m 1760 1715	61
CCl_4	1678 1780 1720	61
CCl_4	1675m 1795 1725	61
CCl_4	1650w 1785 1715m	61
CCl_4	1737m 1790	61
liq.	1587 1316m 952 855	71
liq.	1585 1260 960 860	3

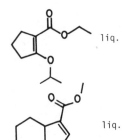

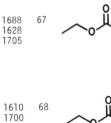

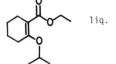

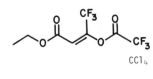

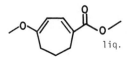

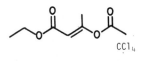

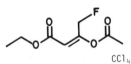

Phase	Frequencies	Ref.
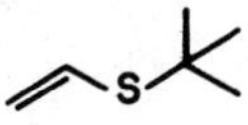 liq.	1580	10
CCl_4	1588	72
CS_2	958 805	72
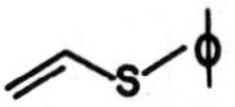 liq.	1600 1272 889 844	73
 liq.	1600 935	10
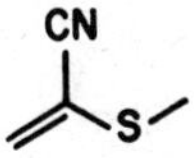 liq.	1577	74
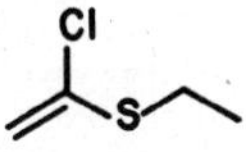 liq.	1590 1130 1080 880 830	10
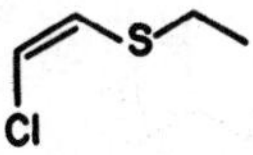 liq.	1560 1290 830	10
 liq.	1200 905 800	10
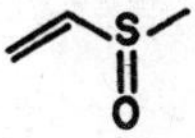 liq.	1613w 960	71
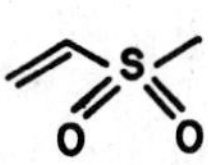 liq.	1626w 962	71
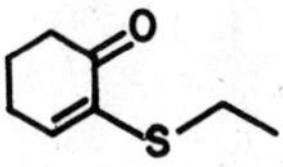 $CHCl_3$	1589	62
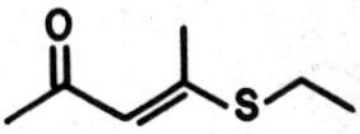 CCl_4	1540 1643	75
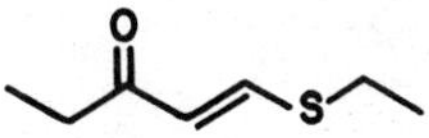 CCl_4	1669 1682	76
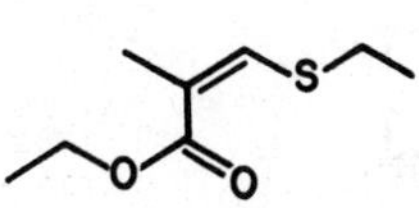 liq.	1592 1701	77
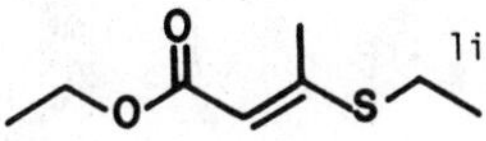 liq.	1577 1689	77
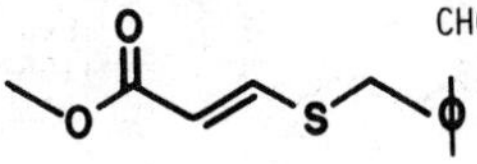 $CHCl_3$	1590 1690	78
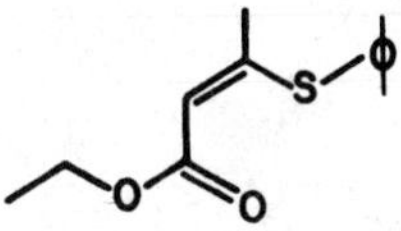 liq.	1595 1580 1698	77

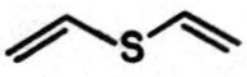

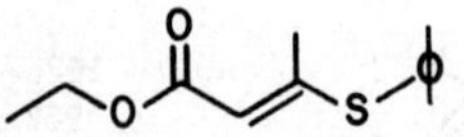

Phase	Bands	Ref.
liq.	1587, 1580, 1695	77
KBr	1565, 1675	79
liq.	1695	80
liq.	1690	80
liq.	1610, 1580, 1110m, 1085m, 1045, 970, 800m	10
liq.	1590, 1105, 855	10
liq.	1590w, 870, 820w	10
liq.	1625, 1340m, 1305m, 1230, 1080	10
liq.	1595, 1320, 1130, 1020, 930, 850m, 825w	10
liq.	1662	81
liq.	1580, 1100, 985, 915, 825	82
liq.	1580w, 1100, 985, 915, 815	82

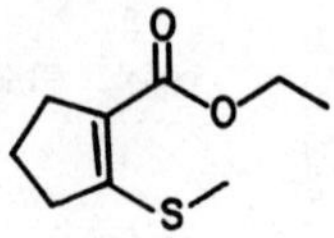

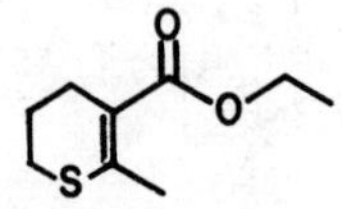

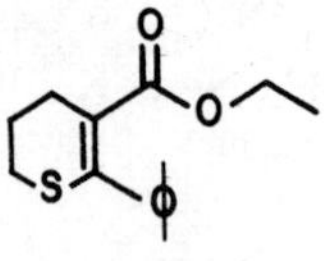

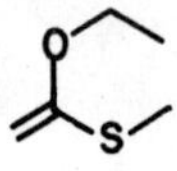

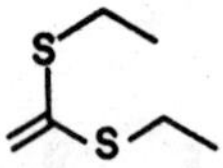

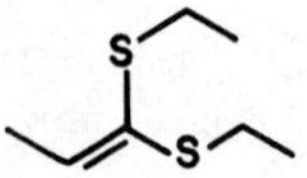

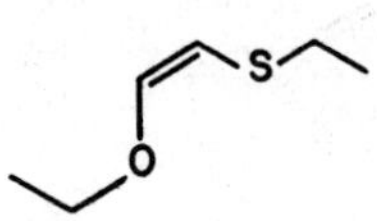

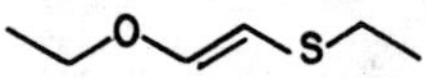

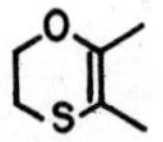

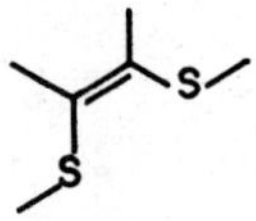

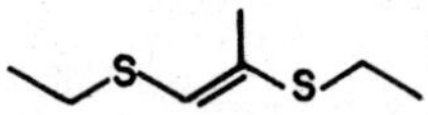

REFERENCES

1. H. Kimmel and W. H. Snyder, *Spec Let* (1971) *4*, 15.
2. W. J. Potts and R. A. Nyquist, *SCA* (1959) *15*, 679.
3. Y. Mikawa, *BCS Japan* (1956), 110.
4. W. H. T. Davison and G. R. Bates, *JCS* (1953), 2607.
5. G. D. Meakins, *JCS* (1953), 4170.
6. E. M. Popov, N. S. Andreev and G. I. Kagan, *Opt & Spec* (1962) *12*, 17.
7. M. L. Brey and P. Tarrant, *JACS* (1957), 6533.
8. R. C. Lord and R. W. Walker, *JACS* (1954), 2518.
9. K. Okuhara, H. Baba and R. Kojima, *BCS Japan* (1962), 532.
10. H. J. Boonstra and L. C. Rinzema, *Rec Trav Chim* (1960), 962.
11. C. H. Eugster, L. Leichner and E. Jenny, *Helv* (1963), 543.
12. H. O. House and V. Kramer, *JOC* (1963), 3362.
13. G. J. Dege, R. L. Harris and J. S. MacKenzie, *JACS* (1959), 3374.
14. A. A. Petrov and G. I. Semonov, *J Gen Chem USSR* (1958), 71.
15. S. I. Miller, *JACS* (1956), 6091.
16. G. A. Russell and E. T. Sabourin, *JOC* (1969), 2336.
17. J. K. Brown and K. J. Morgan, *Advances in Fluorine Chemistry*, Vol. 4, p. 253, eds. M. Stacey, J. C. Tatlow and A. G. Sharpe, Butterworths, Washington, 1965.
18. D. A. Barr and J. B. Rose, *JCS* (1954), 3766.
19. S. J. Etheredge, *JOC* (1966), 1990.
20. R. K. Summerbell and G. J. Lestina, *JACS* (1957), 6219.
21. G. Singh, *Harvard Ph.D. Thesis* (1949).
22. J. B. Miller, *JOC* (1960), 1279.
23. W. Adam, *B* (1964), 1811.
24. S. M. McElvain and R. E. Starn, Jr., *JACS* (1955), 4571.
25. C. O. Parker, *JACS* (1956), 4944.
26. S. M. McElvain and G. R. McKay, Jr., *JACS* (1955), 5601.
27. H. Baganz, *B* (1954), 1725.
28. H. Baganz and P. Klinke, *B* (1955), 1647.
29. H. O. House, L. J. Czuba, M. Gall and H. D. Olmstead, *JOC* (1969), 2324.
30. A. Corbella, G. Jonni, G. Ricca and G. Russo, *Gazz Chim Ital* (1965), 948.
31. R. H. Hasken and J. C. Martin, *JOC* (1962), 3743.
32. H. O. House and G. H. Rasmusson, *JOC* (1963), 27.
33. J. M. Landesberg and D. Kellner, *JOC* (1968), 3374.
34. H. O. House and G. H. Rasmusson, *JOC* (1963), 27.
35. A. Hofmann, W. v. Philipsborn and C. H. Eugster, *Helv* (1965), 1322.

36. R. E. Rosenkranz, K. Allner, R. Good, W. v. Philipsborn and C. H. Eugster, *Helv* (1963), 1259.
37. J. D. Albright and L. Goldman, *JOC* (1966), 273.
38. K. Spencer, A. L. Hall and C. F. Von Reyn, *JOC* (1968), 3369.
39. H. Mühle and Ch. Tamm, *Helv* (1962), 1475.
40. R. D. Campbell and N. H. Cromwell, *JACS* (1957), 3456.
41. J. A. Marshall and N. H. Andersen, *JOC* (1965), 1292.
42. G. E. McManis, *App Spec* (1970) *24*, 495.
43. L. A. Kotorlenko and A. P. Gardenino, *Zh Prakt Spektrosk* (1967) *6*, 620.
44. J. A. Landgrebe and L. W. Becker, *JACS* (1968), 395.
45. F. Merényi and M. Nilsson, *Acta Chem Scand* (1967), 1755.
46. J. Meinwald, J. W. Wheeler, A. A. Nimety and J. S. Liu, *JOC* (1965), 1038.
47. H. Nozaki, Z. Yamaguti, T. Okada, R. Noyori and M. Kawanisi, *T* (1967), 3993.
48. L. Goodman, A. Benitez, C. D. Anderson and B. R. Baker, *JACS* (1958), 6582.
49. N. J. Leonard and F. H. Owens, *JACS* (1958), 6039.
50. H. O. House and H. W. Thompson, *JOC* (1961), 3729.
51. F. A. Miller and G. L. Carlson, *JACS* (1957), 3995.
52. R. S. Rasmussen and R. R. Brattain, *JACS* (1949), 1073.
53. G. Leclerc, C. G. Wermuth and J. Schreiber, *BSC France* (1967), 1302.
54. A. Mondon, H. U. Menz and J. Zander, *B* (1963), 826.
55. J. A. Hartmann, A. J. Tomasewski and A. S. Dreiding, *JACS* (1956), 5662.
56. L. R. Subramanian and G. S. Krishna Rao, *T* (1967), 4167.
57. R. R. Sauers, *JACS* (1959), 925.
58. M. F. Shostakovskii, L. I. Komarova, A. Kh. Filippova and G. V. Ratovskii, *Izvest Akad Nauk Khim* (1967), 2526.
59. H. O. House and D. J. Reif, *JACS* (1955), 6525.
60. M. Gorodetsky and Y. Mazur, *T* (1966) *22*, 3607.
61. R. Filler and S. M. Naqvi, *T* (1963) *19*, 879.
62. M. Tonoeda, M. Inuzuka, T. Furuta, M. Shinozuka and T. Takahashi, *T* (1968), 959.
63. R. S. Rasmussen, D. D. Tunnicliff and R. R. Brattain, *JACS* (1949), 1068.
64. B. M. Goldschmidt, B. L. Van Duuren and C. Mercado, *JCS* (1966) *C*, 2100.
65. I. J. Cantlon, W. Cocker and T. B. H. McMurry, *T* (1961) *15*, 46.
66. A. Schönberg and K. Praefcke, *B* (1966), 196.
67. S. J. Rhoads and R. W. Hasbrouck, *T* (1966) *22*, 3557.
68. F. Korte and K. H. Bückel, *B* (1960), 1025.
69. F. E. Bader, *Helv* (1953), 215.
70. G. M. Iskander and F. Stansfield, *JCS* (1969) *C*, 669.
71. C. C. Price and R. G. Gillis, *JACS* (1953), 4750.

72. K. K. Georgieff and A. Dupré, *Canad J Chem* (1959), 1104.
73. C. C. Price and H. Morita, *JACS* (1953), 4747.
74. K.-D. Gundermann and R. Thomas, *B* (1956), 1263.
75. N. J. Leonard and J. A. Adamcik, *JACS* (1959), 595.
76. P. D. Bartlett and M. Stiles, *JACS* (1955), 2806.
77. D. E. Jones, R. O. Morris, C. A. Vernon and R. F. M. White, *JCS* (1960), 2349.
78. W. Haefliger and T. Petrzilka, *Helv* (1966), 1937.
79. L. O. Ross, L. Goodman and B. R. Baker, *JOC* (1959), 1152.
80. F. Korte and F.-F. Wiese, *B* (1964), 1963.
81. W. E. Parham, J. Heberling and H. Wynberg, *JACS* (1955), 1169.
82. E. I. Heiba and R. M. Dessau, *JOC* (1967), 3837.

3.1

imines iminium salts

The N-H stretching frequency of imines varies considerably with phase, solvent, and concentration, within the range of 3400 to 3000 cm^{-1} and is of little diagnostic value.

The absorption bands associated with the C-N double-bond stretching vibration vary considerably in intensity, but in solution the position of this band, with but a few exceptions, falls within the narrow range of 1670 to 1620 cm^{-1}. There is little correlation between the changes in frequency between condensed phases and various solvents. Individual compounds may show either an increase or decrease in frequency as the phase is changed.

Decrease in ring size and consequent angle strain causes a decrease in the double-bond stretching frequency.

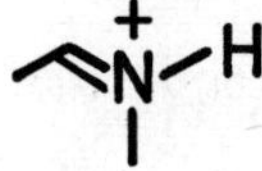

Protonation and formation of the iminium salt causes a small increase (up to 30 cm^{-1}) in the C-N double bond stretching frequency as well as an increase in the intensity of the band.

Solvent	Frequency	Ref.
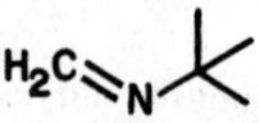 liq.	1652m	1
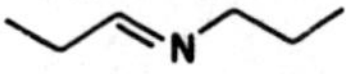 CCl_4	1673	2
$CHCl_3$	1671	2
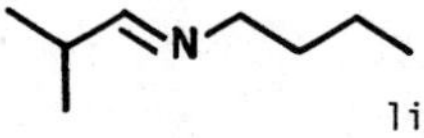 liq.	1668	1
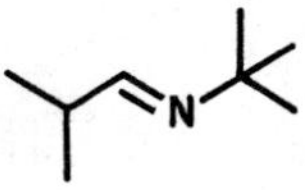 liq.	1670	1
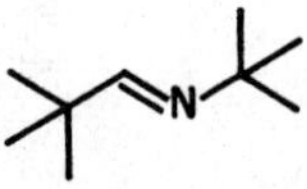 CCl_4	1670	3
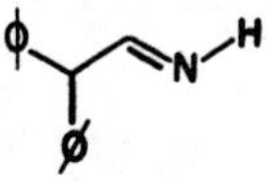 $CHCl_3$	1664	4
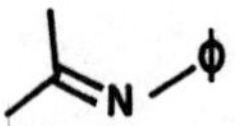 CH_2Cl_2	1649	5
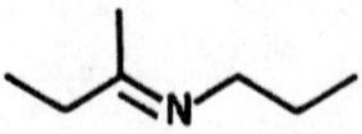 liq.	1660	1
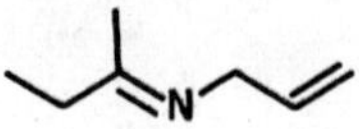 liq.	1658	1
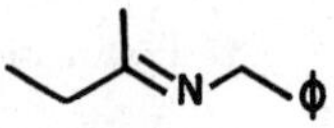 CCl_4	1663	6
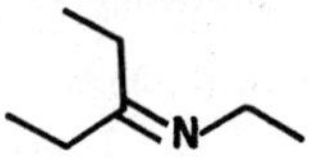 liq.	1660	1
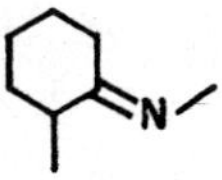 CCl_4	1663	6
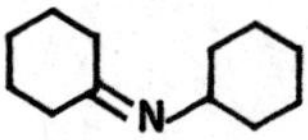 liq.	1647	4
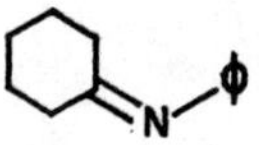 liq.	1667	4
nuj.	1672	4
CCl_4	1658	6
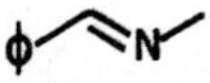 liq.	1605w 1587	7
CH_2Cl_2	1652	5
$CHCl_3$	1654	2
 CH_2Cl_2	1646	5
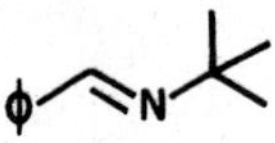 liq.	1638	1

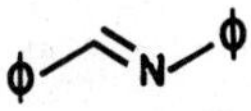

Solvent	cm⁻¹	Ref.
KBr	1629	8
CCl_4	1630	2
CH_2Cl_2	1630	5
$CHCl_3$	1628	2

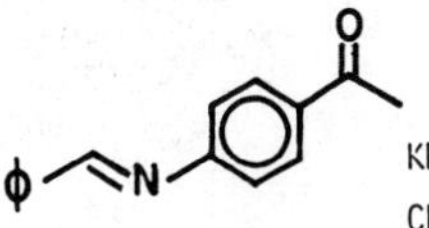

Solvent	cm⁻¹	Ref.
KBr	1600	8
CCl_4	1617	8

Solvent	cm⁻¹	Ref.
KBr	1620	8
$CHCl_3$	1629	8

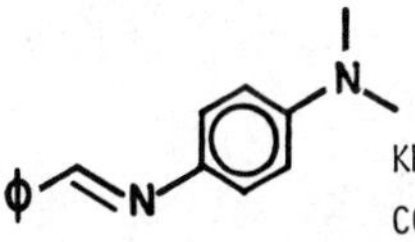

Solvent	cm⁻¹	Ref.
KBr	1630	8
$CHCl_3$	1635	8

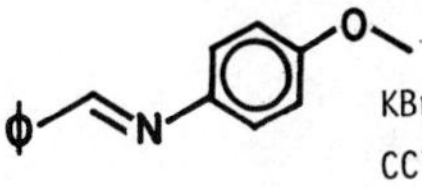

Solvent	cm⁻¹	Ref.
KBr	1616	8
CCl_4	1630	9

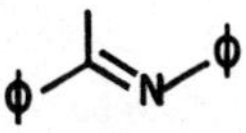

Solvent	cm⁻¹	Ref.
KBr	1619	8
CCl_4	1630	8

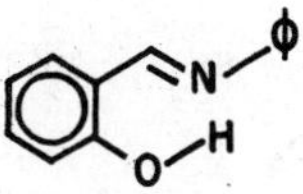

Solvent	cm⁻¹	Ref.
$CHCl_3$	1629	9

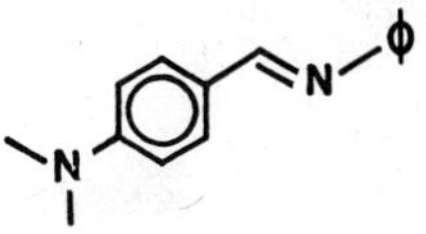

Solvent	cm⁻¹	Ref.
KBr	1627	8
CCl_4	1629	9

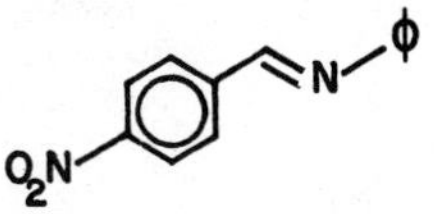

Solvent	cm⁻¹	Ref.
CCl_4	1640	2
$CHCl_3$	1639	2

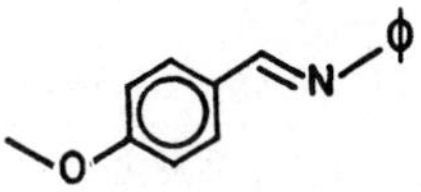

Solvent	cm⁻¹	Ref.
$CHCl_3$	1802	10

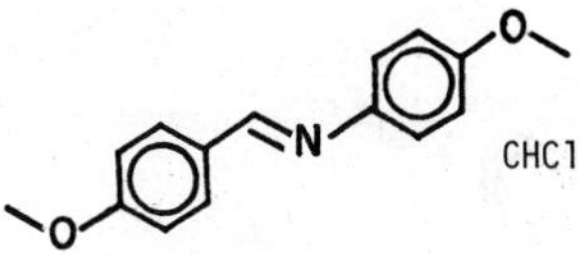

Solvent	cm⁻¹	Ref.
CCl_4	1622	9

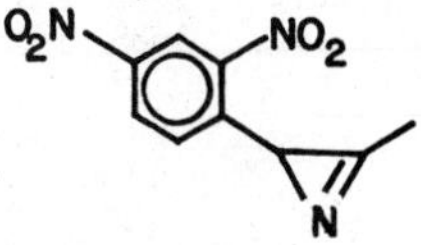

Solvent	cm⁻¹	Ref.
liq.	1617	11

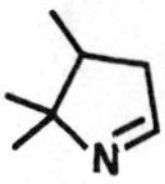

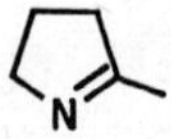

CCl_4	1653	12
liq.	1644	11
liq.	1650	13
nuj.	1616 1572	14
$CHCl_3$	1626 1597 1575m	15
CS_2	1647	4
liq.	1587	13
CCl_4	1642 1585	12

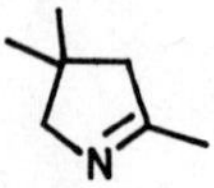

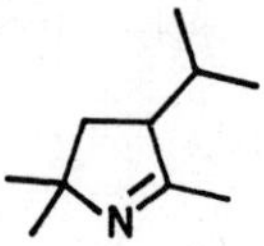

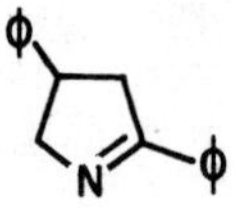

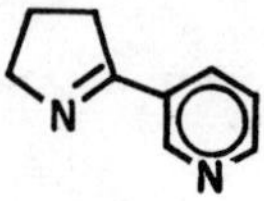

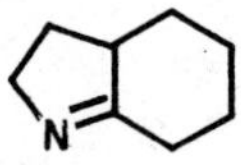

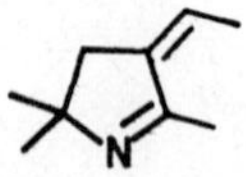

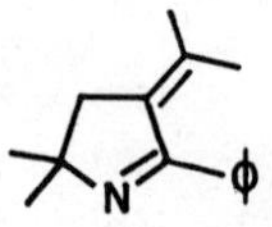

liq.	1645 1562	13
CCl_4	1647 1563	12
liq.	1640m 1575	13
CS_2	1658	4
liq.	1650m	16
nuj.	1675	17
nuj.	1670	17
nuj.	1668	17

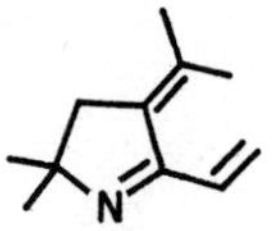

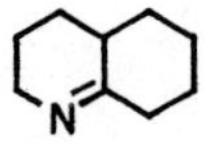

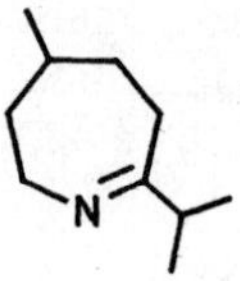

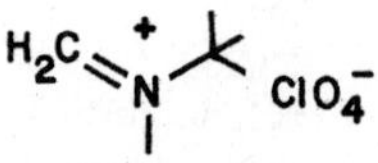

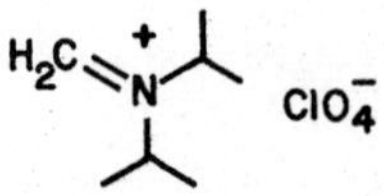

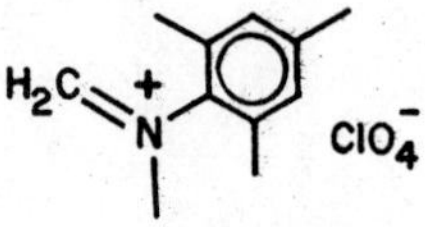

Anion	Medium	ν (cm⁻¹)	Ref.
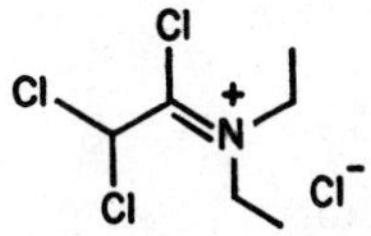 Cl^-	$CHCl_4$	1669, 1658*, 1634*	18
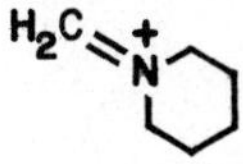 NO_3^-	KBr	1691	19
Cl^-	nuj.	1666	19
ClO_4^-	nuj.	1665	20
$SbCl_6^-$	KBr	1670	19
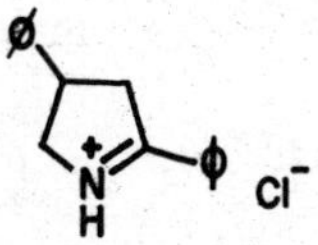 Cl^-	nuj.	1653, 1623	14
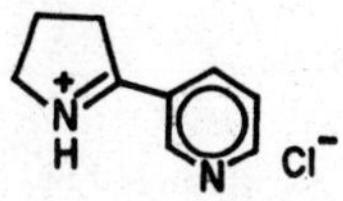 Cl^-	nuj.	1667m, 1585m	15
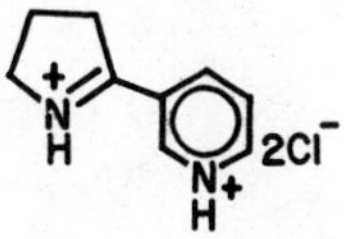 $2Cl^-$	nuj.	1667m, 1634w, 1582, 1558w	15
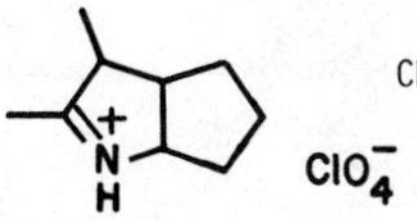 ClO_4^-	$CHCl_3$	1675	21
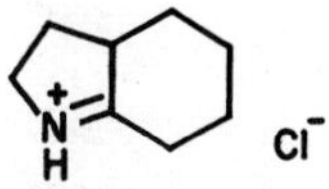 Cl^-	$CHCl_3$	1684	4
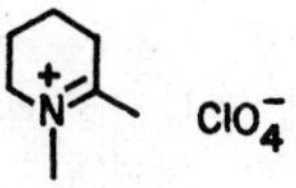 ClO_4^-	nuj.	1690	22
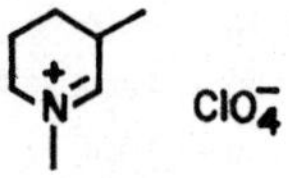 ClO_4^-	nuj.	1698	22
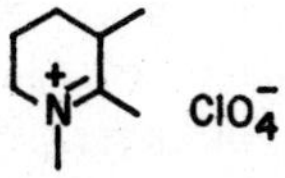 ClO_4^-	nuj.	1682	22
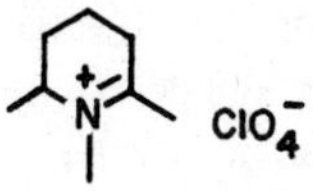 ClO_4^-	nuj.	1679	22
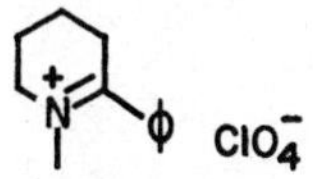 ClO_4^-	nuj.	1691m	22
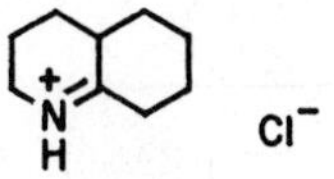 Cl^-	$CHCl_3$	1695	4
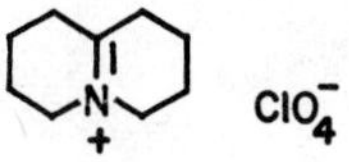 ClO_4^-	nuj.	1696	23

REFERENCES

1. W. D. Emmons, *JACS* (1957), 5739.
2. J. Fabian and M. Legrand, *BSC France* (1956), 1461.
3. J. C. Sheehan and J. H. Beeson, *JACS* (1967), 362.
4. B. Witkop, *JACS* (1956), 2873.
5. W. Krauss and C. Wagner-Bartak, *TL* (1968), 4799.
6. P. W. Hickmott and G. Sheppard, *JCS* (1971) *C*, 1358.
7. G. W. Perold, A. P. Steyn and F. V. K. von Reiche, *JACS* (1957), 462.
8. K. Tabei and E. Saitou, *BSC Japan* (1969), 1440.
9. L. E. Clougherty, J. A. Sousa and G. M. Wyman, *JOC* (1957), 462.
10. D. J. Cram and M. J. Hatch, *JACS* (1953), 33.
11. R. Bonnett, V. M. Clark, A. Giddey and A. Todd, *JCS* (1959), 2087.
12. A. I. Meyers, *JOC* (1959), 1233.
13. A. I. Meyers and J. J. Ritter, *JOC* (1958), 1918.
14. M. C. Kloetzel, J. L. Pinkus and R. M. Washburn, *JACS* (1957), 4222.
15. B. Witkop, *JACS* (1954), 5597.
16. J. H. Boyer and F. C. Canter, *JACS* (1955), 3287.
17. H. Volz and H. H. Kiltz, *TL* (1970), 1917.
18. A. J. Speziale and R. C. Freeman, *JACS* (1960), 909.
19. G. Opitz, H. Hellmann and H. W. Schubert, *A* (1959) *623*, 117.
20. N. J. Leonard and K. Jann, *JACS* (1960), 6418.
21. R. Griot and T. Wagner-Jauregg, *Helv* (1959), 121.
22. N. J. Leonard and F. P. Hauck, Jr., *JACS* (1957), 5279.
23. N. J. Leonard, A. S. Hay, R. W. Fulmer and V. W. Gash, *JACS* (1955), 439.

3.2

nitriles cyanides

$—C \equiv N$

The C-N triple-bond vibrational frequencies of organic nitriles fall within very narrow ranges. Saturated aliphatic nitriles absorb between 2285 and 2240 cm^{-1}, and α-halogen substitution causes a small decrease in frequency. Conjugated unsaturated systems absorb in the region of 2250 to 2200 cm^{-1}, unless the double bond is substituted by a heteroatom when delocalization of the heteroatom lone pairs into the nitrile causes a further decrease in the vibrational frequency.

Aromatic nitriles also absorb within the narrow range of 2240 to 2220 cm^{-1}, irrespective of the nature and position of additional substituents.

Changes in frequency that result from change in state or solvent are both small and unpredictable. Aliphatic nitriles absorb at higher frequencies in $CHCl_3$ than in CCl_4, the biggest difference (28 cm^{-1}) being found with acetonitrile, whereas no difference is found between these two solvents for butyronitrile. Unsaturated aliphatic nitriles show the opposite behavior, higher frequencies being measured in CCl_4 compared to $CHCl_3$. No significant difference is found between aromatic nitriles in these two solvents.

The intensity of the nitrile absorption is extremely variable and can change as a function of solvent. Although unsaturated and aryl nitriles generally show a more intense absorption than their saturated aliphatic counterparts, many nitriles have such weak absorptions as to be virtually undetectable.

Compound	Medium	cm^{-1}	Ref.
HCN	gas	2097	1
	CCl_4	2096	1
	$CHCl_3$	2095	1
DCN	gas	1925	1
	$CHCl_3$	1917	1
FCN	gas	2290 1077	2
ClCN	gas	2219 714w	3
	liq.	2206 730	4
BrCN	liq.	2191 568	4
	solid	2194 573m	3
NC—CN	gas	2159	5
	solid	2165 2167	5
Na CN	nuj.	2085	6
KCN	nuj.	2076	6
Cu CN	nuj.	216C	6
Ag CN	nuj.	2138	6
CN	CCl_4	2255	7
	$CHCl_3$	2283	8

Compound	Medium	cm^{-1}	Ref.
CN	CCl_4	2249	7
	$CHCl_3$	2257	8
CN	CCl_4	2253	7
	$CHCl_3$	2253	7
CN	CCl_4	2248	7
	$CHCl_3$	2249	7
CN	CCl_4	2253	7
ϕ CN	CCl_4	2253	7
NC CN	$CHCl_3$	2273	7
NC CN	$CHCl_3$	2256	7
F CN	$CHCl_3$	2256	9
Cl CN	$CHCl_3$	2259	9
Br CN	$CHCl_3$	2255	9

Structure labels	Solvent	ν	Ref.
I, CN	$CHCl_3$	2248	9
Cl_3C, CN	$CHCl_3$	2250	7
H_2N, CN	$CHCl_3$	2248	9
N, CN	CCl_4	2232	7
O_2N, CN, CN	liq.	2262	10
HO, CN	$CHCl_3$	2257	9
O, O, CN	$CHCl_3$	2267	11
O, O, O, O, NC, CN	$CHCl_3$	2259	11
CN, NO_2, NO_2	liq.	2257 1600	12
CN, OH	$CHCl_3$	2249	9
CN, OH	$CHCl_3$	2243	7
CN	liq.	2242	13
CN	CCl_4	2245	14
ϕ, CN	CCl_4	2250	15
CN	CCl_4	2235	14
O, CN	CCl_4	2248	16
O, CN	liq.	2250	17

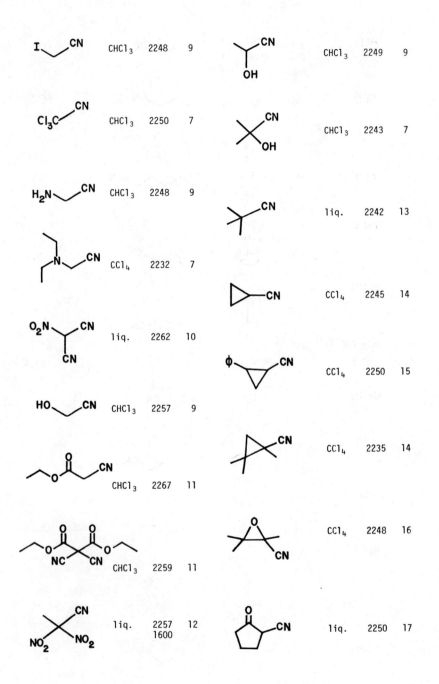

Solvent	Frequencies	Ref.
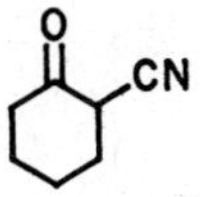 $CHCl_3$	2240 2190	18
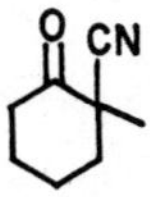 $CHCl_3$	2240	18
liq.	2248	19
CCl_4	2230	7
$CHCl_3$	2224	8
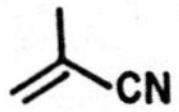 $CHCl_3$	2224	8
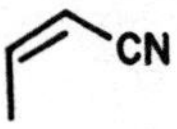 $CHCl_3$	2217	8.
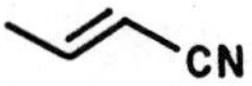 $CHCl_3$	2220	8
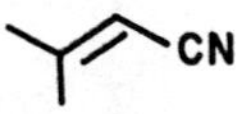 $CHCl_3$	2216	8
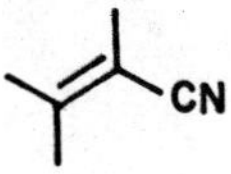 CCl_4	2218	14

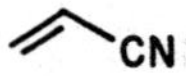

Solvent	Frequencies	Ref.
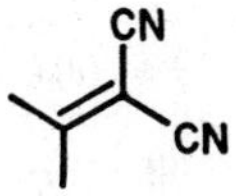 CH_2Cl_2	2237 1613	20
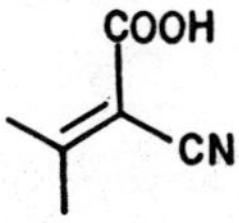 $CHCl_3$	2212 1605	20
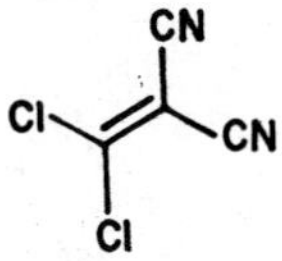 KBr	2220 1541	21
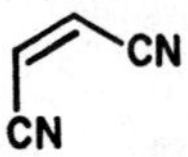 $CHCl_3$	2251 2231	11
 $CHCl_3$	2240	11
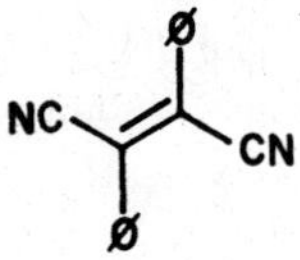 $CHCl_3$	2224	11
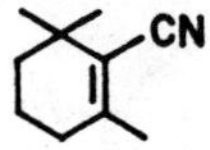 CCl_4	2222 1635	20
 sng	2217	22
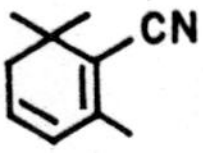 CCl_4	2203 1645	20
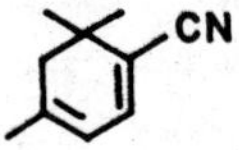 CCl_4	2210 1647	20

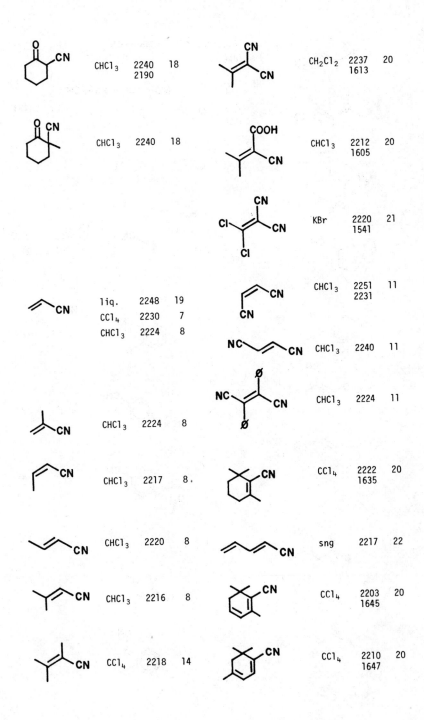

Structure	Medium	Frequency	Ref.
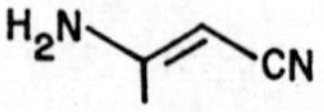 H_2N, CN	KBr	2174 1646 1600	20
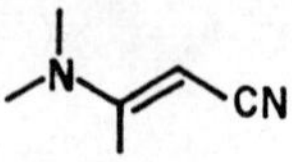 N, CN	liq.	2190 1590	23
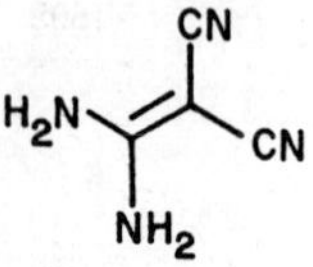 CN, H_2N, CN, NH_2	KBr	2188 1636 1545	24
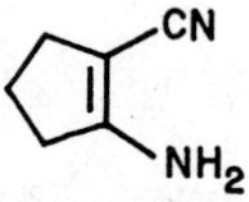 CN, NH_2	$CHCl_3$	2189 1605	25
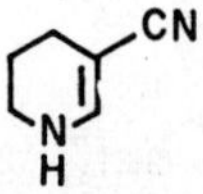 CN, NH	KBr	2179 1621	26
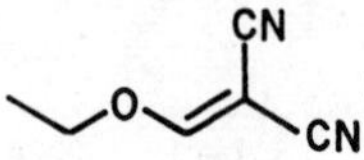 CN, O, CN	$CHCl_3$	2242 1613	20
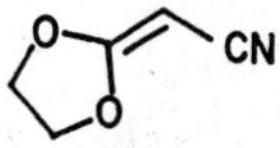 O, O, CN	liq.	2195 1656	27
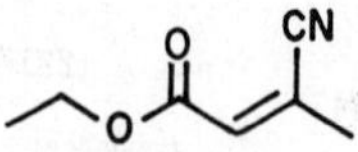 O, CN, O	$CHCl_3$	2229	11
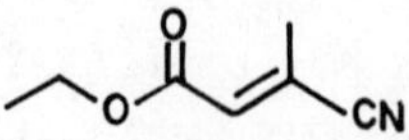 O, O, CN	$CHCl_3$	2228	11

Structure	Medium	Frequency	Ref.
 ≡—CN	liq.	2242	28
 NC—≡—≡—CN	CCl_4	2237	28
ϕ—CN	CCl_4	2232	29
	$CHCl_3$	2231	30
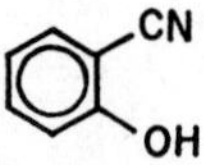 CN, OH	$CHCl_3$	2227	30
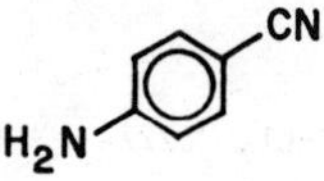 H_2N, CN	$CHCl_3$	2221	30
O_2N, CN	CCl_4	2238	29
	$CHCl_3$	2237	30
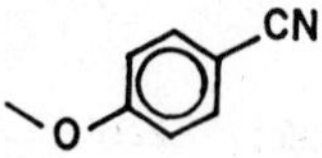 O, CN	CCl_4	2228	29
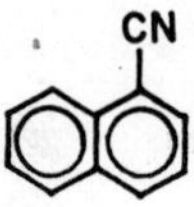 CN	CCl_4	2222	31

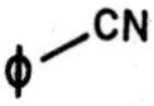

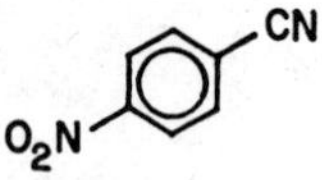

Compound	Phase	ν (cm^{-1})	Ref.
CN	CCl_4	2227	31
S CN	CCl_4	2226	32
CN N	$CHCl_3$	2239	30
O CN	liq.	2235	33
O CN	liq.	2235	33

Compound	Phase	ν (cm^{-1})	Ref.
O NC CN	liq.	2230 1711	34
N CN H	liq.	2237	35
N CN	liq.	2222	35
	$CHCl_3$	2218	25
N CN	$CHCl_3$	2209	25
N —CN	nuj.	2215	36

REFERENCES

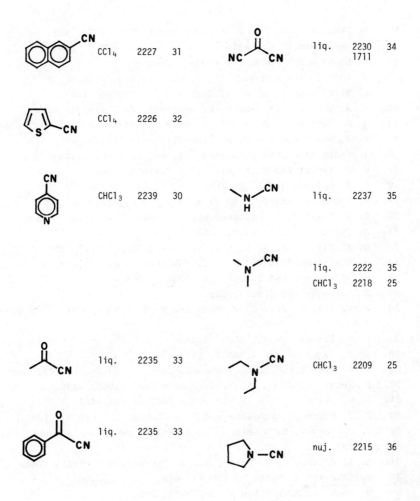

1. G. L. Caldow and H. W. Thompson, *Proc Roy Soc* (1960) *A254*, 1.
2. R. E. Dodd and R. Little, *SCA* (1960), 1083.
3. W. O. Freitag and E. R. Nixon, *JCP* (1956) *24*, 109.
4. J. Wagner, *Z Physik Chem* (1943) *193A*, 55.
5. F. D. Verderame, J. W. Nebgen and E. R. Nixon, *JCP* (1963) *39*, 2274.
6. M. F. Amr El-Sayed and R. K. Sheline, *JINC* (1958) *6*, 187.
7. J. P. Jesson and H. W. Thompson, *SCA* (1959) *13*, 217.

8. N. Sheppard and G. B. B. M. Sutherland, *JCS* (1947), 453.
9. P. Sensi and G. G. Gallo, *Gazz Chim Ital* (1955), 224.
10. W. Ruske and E. Ruske, *B* (1958), 2505.
11. D. G. I. Felton and S. F. D. Orr, *JCS* (1955), 2170.
12. L. W. Kissinger and H. E. Ungnade, *JOC* (1960), 1471.
13. J. J. McBride, Jr. and H. C. Beachell, *JACS* (1952), 5247.
14. G. W. Cannon, A. A. Santilli and P. Shenian, *JACS* (1959), 1660.
15. R. J. Mohrbaker and N. H. Cromwell, *JACS* (1957), 401.
16. J. Cantacuzène and D. Ricard, *BSC France* (1967), 1587.
17. C. F. Hammer and R. A. Hines, *JACS* (1955), 3649.
18. L. J. Bellamy and L. Beecher, *JCS* (1954), 4487.
19. A. Rosenberg and J. P. Devlin, *SCA* (1965), 1613.
20. C. H. Eugster, L. Leichner and E. Jenny, *Helv* (1963), 543.
21. M. Yamaguchi, *NKZ* (1959), 155.
22. R. E. Kitson and N. E. Griffith, *Anal Chem* (1952), 334.
23. P. Kurtz, H. Gold and H. Disselnkötter, *A* (1959) *624*, 1.
24. E. Allenstein and P. Quis, *B* (1963), 1035.
25. S. Baldwin, *JOC* (1961), 3288.
26. S. E. Ellzey, Jr., C. H. Mack and W. J. Connick, Jr., *JOC* (1967), 846.
27. C. O. Parker, *JACS* (1956), 4944.
28. A. J. Saggiomo, *JOC* (1957), 1171.
29. M. R. Mander and H. W. Thompson, *Trans Farad Soc* (1957), 1402.
30. P. Sensi and G. G. Gallo, *Gazz Chim Ital* (1955), 235.
31. M. W. Skinner and H. W. Thompson, *JCS* (1955), 487.
32. J. J. Peron, P. Saumagne and J. M. Lebas, *SCA* (1970), 1651.
33. C. S. Marvel, N. O. Brace, F. A. Miller and A. R. Johnson, *JACS* (1949), 34.
34. F. A. Miller, B. M. Harney and J. Tyrrell, *SCA* (1971), 1003.
35. M. Kuhn and R. Mecke, *B* (1960), 618.
36. H. Bock and H. tom Dieck, *B* (1966), 213.

3.3

isonitriles isocyanides

$-N \equiv C$

The N-C triple-bond stretching frequency of isonitriles falls within the narrow range of 2180 to 2100 cm^{-1}. Conjugation to both alkenyl and aryl residues causes a small decrease in the frequency of the intense band.

Small changes in frequency are observed between condensed phases and various solvents, but there appears to be no systematic change.

The frequency of isonitrile groups bonded to transition metals depends on the degree of back bonding from the metal *d*-orbitals into the empty π^*-orbitals of the isonitrile. The extent of such back bonding depends not only on the metal and its oxidation state but on the other ligands coordinated to the metal; the stretching frequencies of metal-coordinated isonitriles are found at values both higher and lower than those observed for the non-coordinated system.

Structure	Medium	ν	Ref.
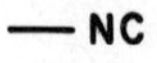 —NC	liq.	2183	1
	CCl_4	2169	2
	$CHCl_3$	2142	2
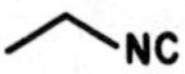 NC	liq.	2160	1
	$CHCl_3$	2160	3
 NC	liq.	2146	4
	$CHCl_3$	2151	3
 φ NC	liq.	2146	4
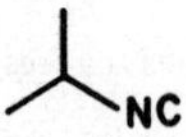 NC	liq.	2140	4
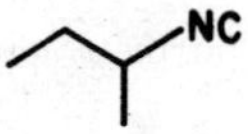 NC	CCl_4	2125	5
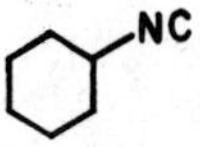 NC	liq.	2138	4
 NC	liq.	2110	6
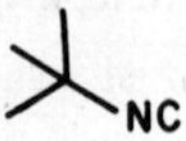 NC	liq.	2134	4
	CCl_4	2127	5
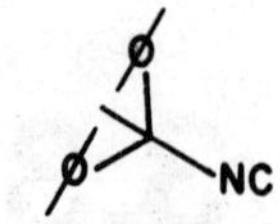 φ φ NC	CCl_4	2120	6
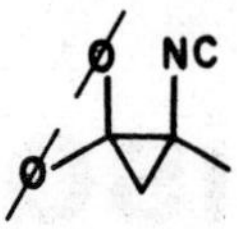 φ φ NC	CCl_4	2120	6
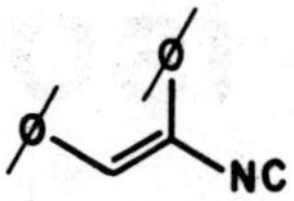 φ φ NC	liq.	2105	6
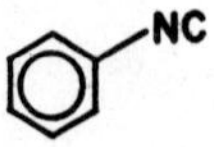 NC	liq.	2117	4
	CCl_4	2133	2
	$CHCl_3$	2136	2
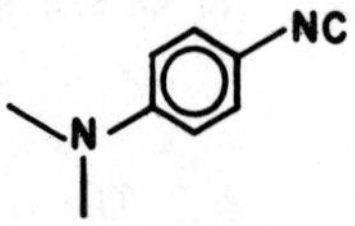 N NC	KBr	2124	4
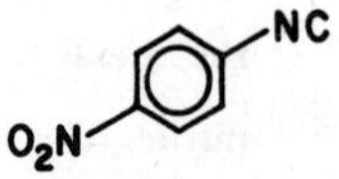 O_2N NC	KBr	2116	4
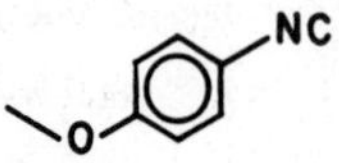 O NC	CCl_4	2128	2
	$CHCl_3$	2137	5

Compound	Phase	Frequency	Ref.
2-naphthyl-NC	KBr	2122	4
$(CH_3)_3SiNC$	liq.	2169	7
$Cr(\phi NC)_6$	$CHCl_3$	2070 2012 1965	2
$Fe(CO)_4CH_3NC$	$CHCl_3$	2213	2
$Fe(CO)_4\phi NC$	$CHCl_3$	2165	2
$Ni(\phi NC)_4$	$CHCl_3$	2050 1990	2

REFERENCES

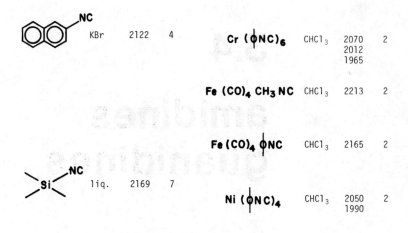

1. W. Gordy and D. Williams, *JCP* (1936) *4*, 85.
2. F. A. Cotton and F. Zingales, *JACS* (1961), 351.
3. N. Sheppard and G. B. B. M. Sutherland, *JCS* (1947), 453.
4. I. Ugi and R. Meyr, *B* (1960), 239.
5. J. Casanova, N. D. Werner and R. E. Schuster, *JOC* (1966), 3473.
6. H. M. Walborsky and G. E. Niznik, *JOC* (1972), 187.
7. J. J. McBride, Jr. and H. C. Beachell, *JACS* (1952), 5247.

3.4

amidines guanidines

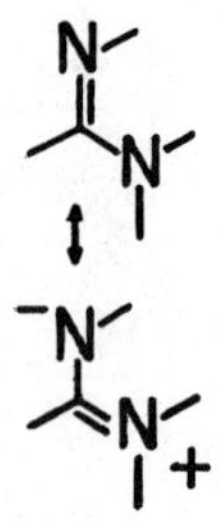

Delocalization of the nitrogen lone pair into the imine double bond will lower the frequency, whereas the field effect of the nitrogen will increase the frequency of the C-N double-bond stretching vibration. However, unlike imino ethers where these two effects balance each other, the lower electronegativity of nitrogen both increases the extent of lone-pair delocalization and decreases the field effect, with the result that guanidines absorb at frequencies lower than the corresponding imines and imino ethers.

Substitution on the amino-nitrogen by an electron withdrawing group causes an increase in the C-N double-bond stretching frequency, whereas electron donating groups, including alkyl, lower it. Conversely, electron withdrawing groups on the imine-nitrogen lower the C-N double-bond stretching frequency.

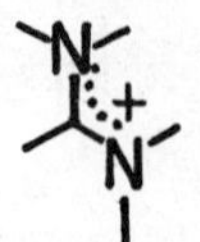

Protonation of the amidine chromophore, as with imines, causes an increase in the stretching frequency of the C-N double bond.

Solvent	cm⁻¹	Ref.
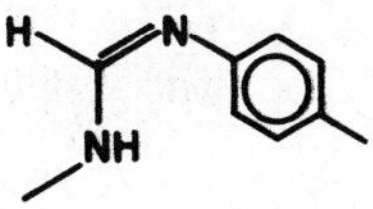 $CHCl_3$	1626 1600	1
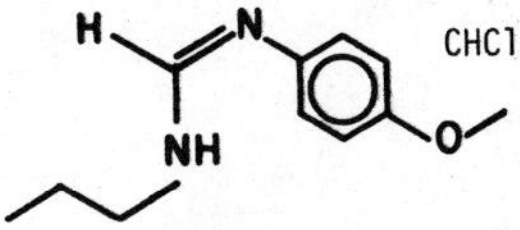 $CHCl_3$	1620 1600m	1
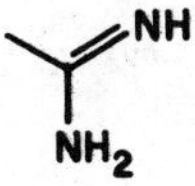 liq.	1650 1608	2
CCl_4	1624	3
$CHCl_3$	1618	3
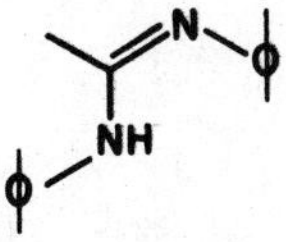 $CHCl_3$	1655 1633	4
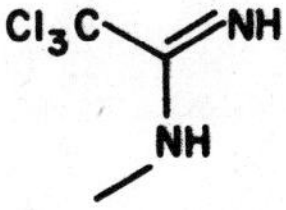 KBr	1635 1522	5
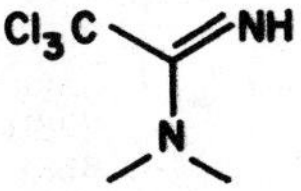 liq.	1603	5
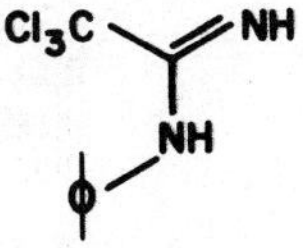 KBr	1657 1584	5
CCl_4	1651	3
$CHCl_3$	1634	3
CCl_4	1621	3
$CHCl_3$	1614	3
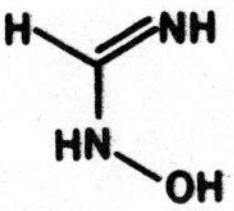 KBr	1678	6
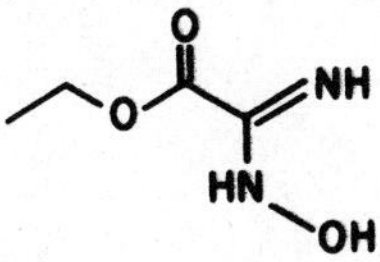 $CHCl_3$	1664 1724	6
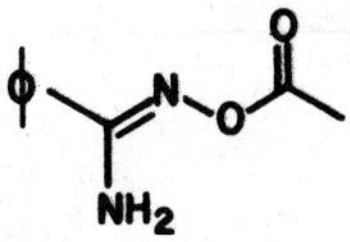 $CHCl_3$	1638 1761	7

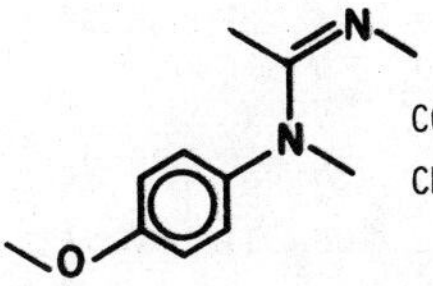

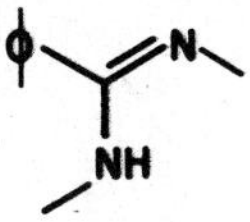

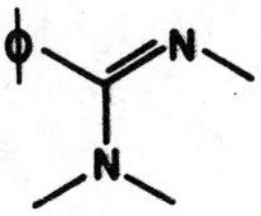

Structure	Medium	cm⁻¹	Ref.
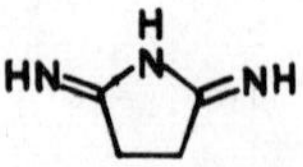 HN=(pyrrolidine, N–H)=NH	nuj.	1621 1767w 1681	8
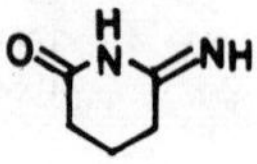 O=(piperidine, N–H)=NH	nuj.	1661 1816w 1697	8
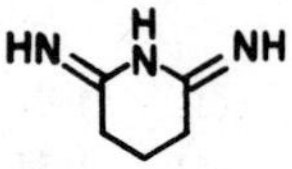 HN=(piperidine, N–H)=NH	nuj.	1688 1603	8
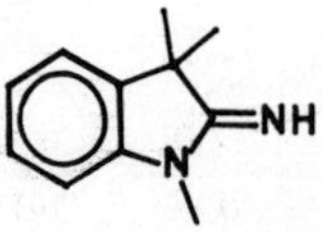 3,3-dimethyl-1-methyl-indoline =NH	CH_2Cl_2	1653	9
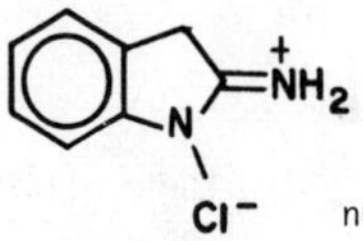 1-methyl-indoline $=\overset{+}{N}H_2$ Cl^-	nuj.	1709	9
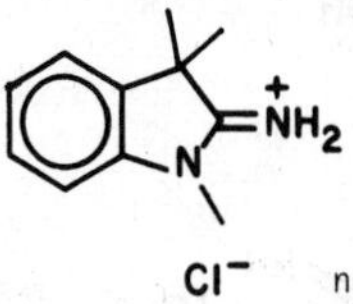 3,3-dimethyl-1-methyl-indoline $=\overset{+}{N}H_2$ Cl^-	nuj.	1706	9
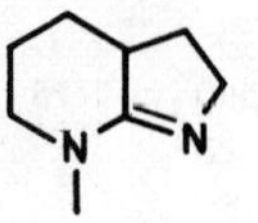 N-methyl bicyclic amidine (N, =N)	liq. $CHCl_3$	1615 1610	10 10
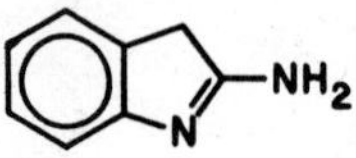 indolenine –NH_2 (N)	CH_2Cl_2	1642	9
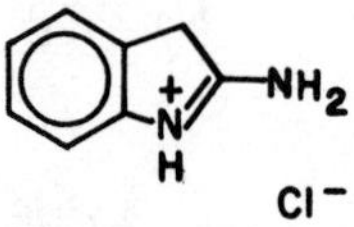 indolenine –NH_2 ($\overset{+}{N}$–H) Cl^-	nuj.	1704	9
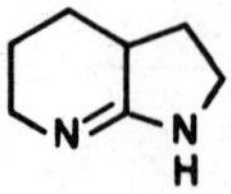 bicyclic amidine (N=, N–H)	KBr $CHCl_3$	1652 1635	10 10
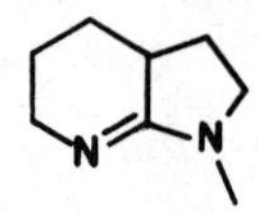 bicyclic amidine (N=, N–CH₃)	liq. $CHCl_3$	1657 1652	10 10
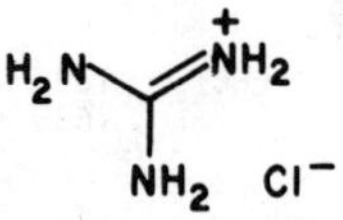 H_2N–C(NH_2)=$\overset{+}{N}H_2$ Cl^-	KBr	1653	11
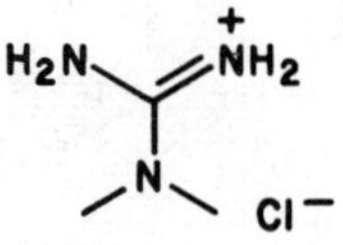 H_2N–C(N(CH₃)₂)=$\overset{+}{N}H_2$ Cl^-	KBr	1654 1634	11

Compound	Medium	Frequency	Ref.
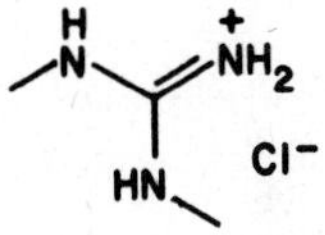 H, N, +, NH_2, Cl^-, HN	KBr	1680 1593	11
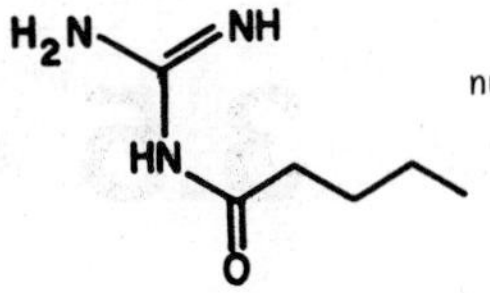 H_2N, NH, HN, O	nuj.	1630 1580 1730 1710	12
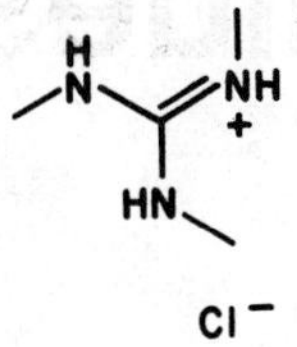 H, N, NH, +, HN, Cl^-	KBr	1641	11
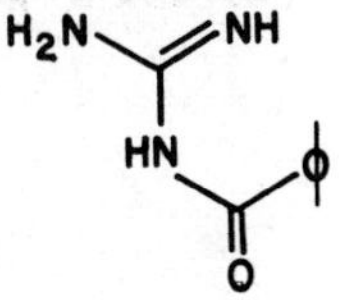 H_2N, NH, HN, O, φ	nuj.	1670 1580 1780	12

REFERENCES

1. J. Fabian, V. Delaroff and M. Legrand, *BSC France* (1956), 287.
2. M. Davies and A. E. Parsons, *Z Phys Chem (Fr)* (1959) *20*, 34.
3. J. Fabian and M. Legrand, *BSC France* (1956), 1461.
4. D. Prevoršek, *BSC France* (1958), 788.
5. J. C. Grivas and A. Taurins, *Canad J Chem* (1959), 795.
6. H. E. Ungnade and L. W. Kissinger, *JOC* (1958), 1794.
7. C. L. Bell, C. N. V. Nambury and L. Bauer, *JOC* (1964), 2873.
8. J. A. Elvidge, R. P. Linstead and A. M. Salaman, *JCS* (1959), 208.
9. J. Kebrele and K. Hoffmann, *Helv* (1956), 116.
10. M. M. Robison, F. P. Butler and B. L. Robison, *JACS* (1957), 2573.
11. T. Goto, K. Nakanishi and M. Ohashi, *BCS Japan* (1957), 723.
12. P. A. Boivin, W. Bridges and J. L. Boivin, *Canad J Chem* (1954), 242.

3.5 carbodiimides

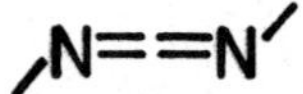

The exceedingly intense asymmetric stretching mode falls within a very narrow and thus characteristic range between 2260 and 2100 cm^{-1}. In general, only one band is seen, but some aryl-substituted carbodiimides may show two or more bands in this region.

The symmetric vibration for the cumulative bonds of carbodiimides falls within the range of 1300 to 1200 cm^{-1} and is of no diagnostic value.

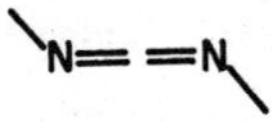 liq.	2140	1
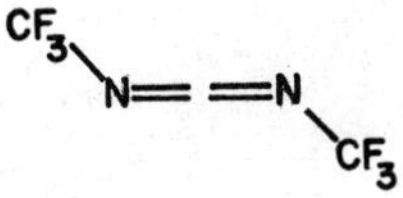	2212	2
liq.	2128	4
CCl_4	2138	3
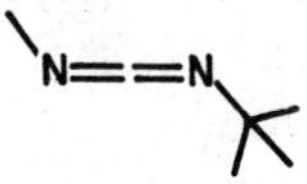 liq.	2130	4
liq.	2116	4
CCl_4	2128	3
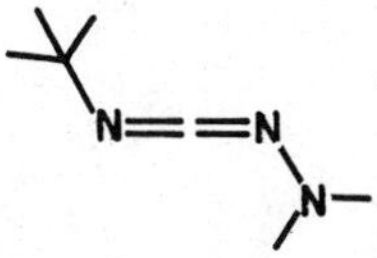 liq.	2090	5
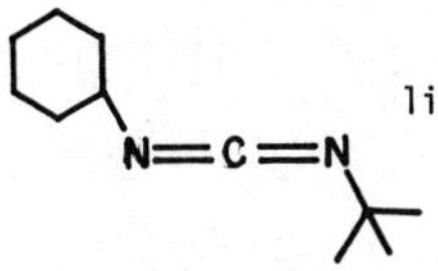 liq.	2160	6
liq.	2121	4
CCl_4	2130	3
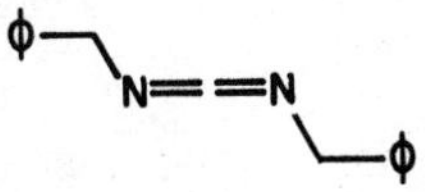 CCl_4	2140	3
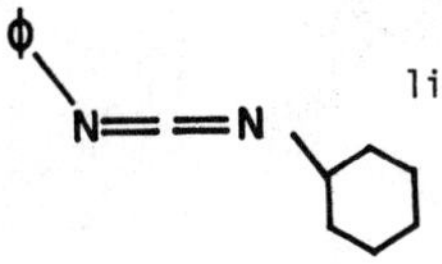 liq.	2150	5
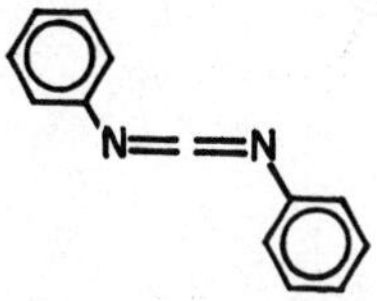 liq.	2150	7
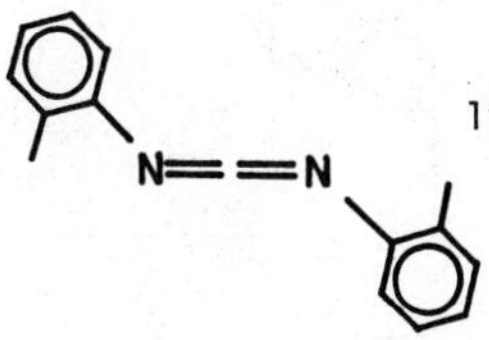 liq.	2148	4
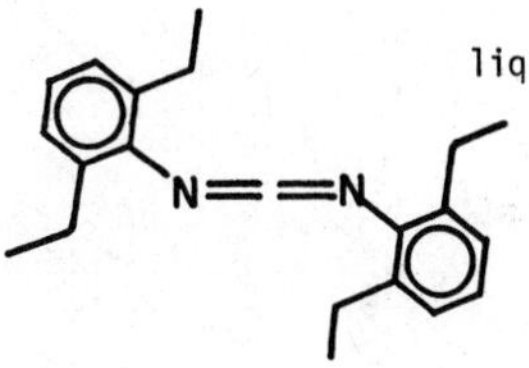 liq.	2168	4

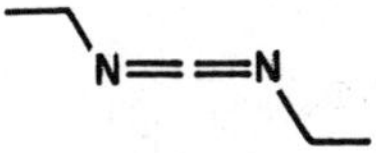

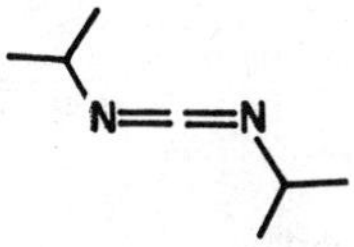

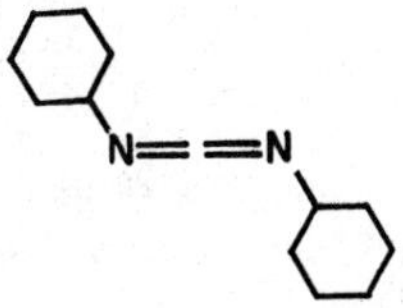

Structure	Medium	cm⁻¹	Ref.
Di-p-tolylcarbodiimide	CCl_4	2145 2120	3
Bis(p-nitrophenyl)carbodiimide (NO_2)	KBr	2160 2100w	8
Bis(p-methoxyphenyl)carbodiimide	CCl_4	2142 2120	3
Di-1-naphthylcarbodiimide	CCl_4	2152 2100	3
H_3Si–N=C=N–SiH_3	liq.	2260	9
$(CH_3)_3Si$–N=C=N–$Si(CH_3)_3$	liq.	2190	9
$(CH_3)_3Si$–N=C=N–$Si\phi_3$	liq.	2190	9
ϕ_3Si–N=C=N–$Si\phi_3$	KBr	2180	9
$(EtO)_3Si$–N=C=N–$Si(OEt)_3$	liq.	2230	9
n-Butyl–N=C=N–$SO_2\phi$	liq.	2198	10
Isopropyl–N=C=N–$SO_2\phi$	liq.	2174	10
tert-Butyl–N=C=N–$SO_2\phi$	liq.	2165	10

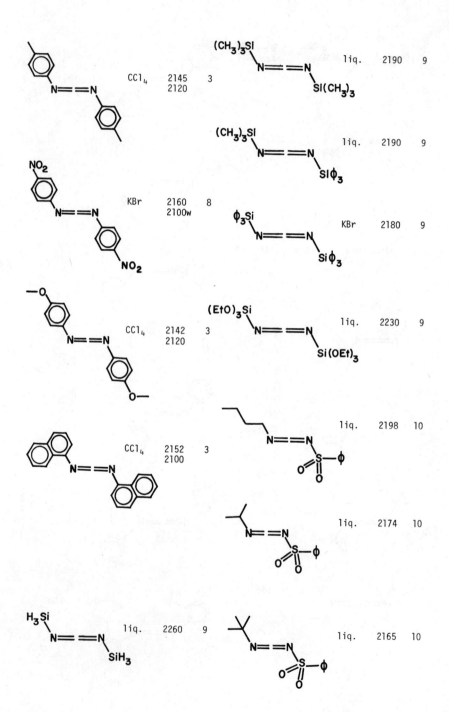

Compound	Phase	Frequency	Ref.
$H_3Ge{-}N{=}C{=}N{-}GeH_3$	solid	2140 2063	11
$(C_2H_5)_3Sn{-}N{=}C{=}N{-}Sn(C_2H_5)_3$	liq.	2165 2102 2090	13
$(CH_3)_3Sn{-}N{=}C{=}N{-}Sn(CH_3)_3$	C_6H_6	2160 2060	12

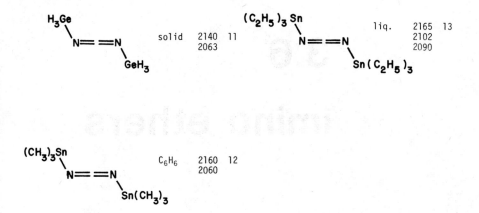

REFERENCES

1. G. Rapi and G. Sbrana, *Chem Comm* (1968), 128.
2. R. A. Mitsch and P. H. Ogden, *Chem Comm* (1967), 59.
3. G. D. Meakins and R. J. Moss, *JCS* (1957), 993.
4. P. H. Mogul, *USAEC* (1967), 1S-T-160.
5. W. S. Wadsworth, Jr. and W. D. Emmons, *JOC* (1964), 2816.
6. T. Saegusa, Y. Ito and T. Shimizu, *JOC* (1970), 3995.
7. E. Haruki, T. Inaike and E. Imoto, *BSC Japan* (1965), 1806.
8. A. E. Wick, unpublished results.
9. J. Pump, E. G. Rochow and U. Wannagat, *Monatsh Chem* (1963), 588.
10. R. Neidlein, W. Haussmann and E. Heukelbach, *B* (1966), 1252.
11. S. Cradock and E. A. V. Ebsworth, *JCS* (1968) *A*, 1423.
12. O. J. Scherer and R. Schmitt, *B* (1968), 3302.
13. V. F. Gerega, Yu. I. Dergunov, E. A. Kuz'mina, Yu. A. Aleksandrov and Yu. I. Mushkin, *J Gen Chem USSR* (1969), 1307.

3.6
imino ethers

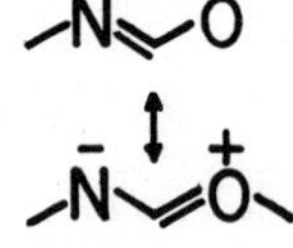

Delocalization of the nonbonding electrons on oxygen into the imino double bond is expected to cause a decrease in the frequency of the C-N double-bond vibration, whereas the field effect of the electronegative oxygen should operate in the opposite direction. In general, these two effects balance each other, and imino ethers, like imines, absorb between 1700 and 1600 cm^{-1}.

The absorption band is usually strong, and a doublet, because of rotational isomerization, is occasionally seen. Protonation causes only small changes in the C-N stretching frequency.

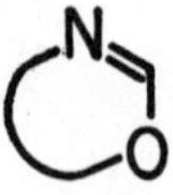

As with imines (Chapter 3.1, p. 116) decrease in ring size, and consequent angle strain, causes a decrease in the stretching frequency.

Solvent	Frequency	Ref.
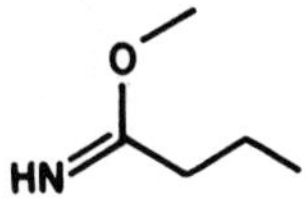 CCl_4	1680	1
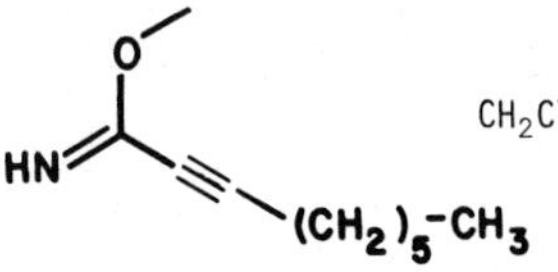 CH_2Cl_2	1621	2
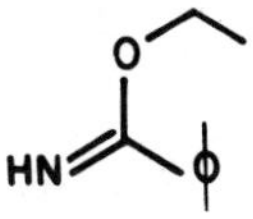 liq.	1637 1605w 1580m	3
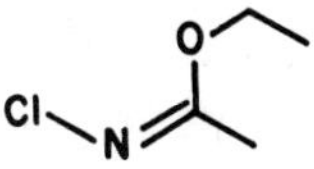 CCl_4	1616	4
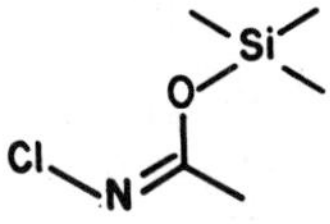 CCl_4	1609	4
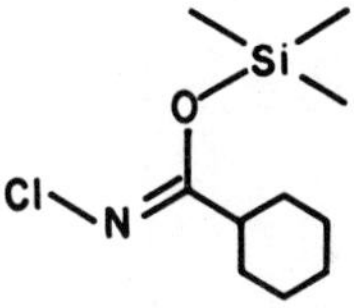 CCl_4	1600	4
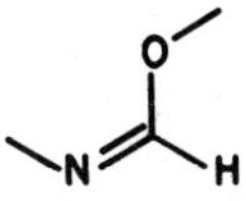 liq.	1675	5

Solvent	Frequency	Ref.
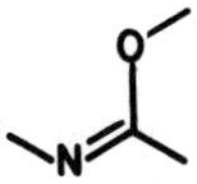 liq.	1690	5
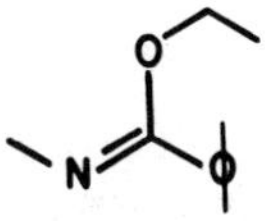 CCl_4	1675 1603	3
CCl_4	1665	6
CH_2Cl_2	1669	7
$CHCl_3$	1663	6
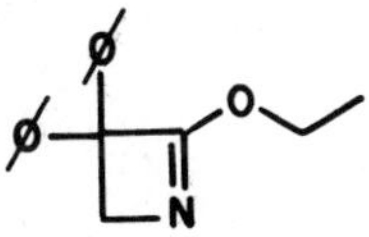 nuj.	1620	8
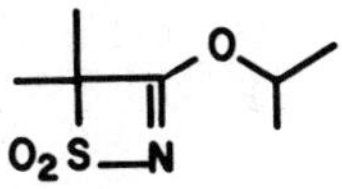 nuj.	1570	9
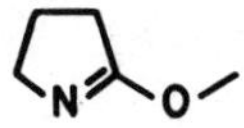 liq.	1660	10
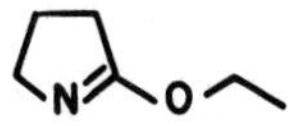 $CHCl_3$	1640	11
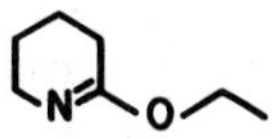 $CHCl_3$	1670	11
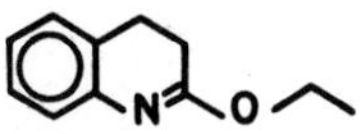 $CHCl_3$	1632	12

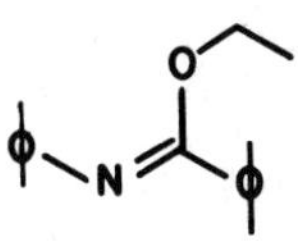

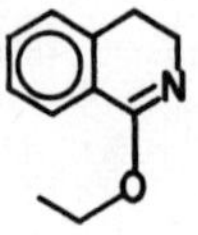

Solvent	Frequency	Ref.	Solvent	Frequency	Ref.
$CHCl_3$	1648	12	$CHCl_3$	1647	15
$CHCl_3$	1680m	12	liq.	1670	16
liq.	1670	13	liq.	1670	16
liq.	1625 1612 1535	14	liq.	1656m	17
$CHCl_3$	1705	11	liq.	1650	16
$CHCl_3$	1645	15	liq.	1665	18
$CHCl_3$	1660	15	liq.	1650	18
$CHCl_3$	1666	15	liq.	1668	18

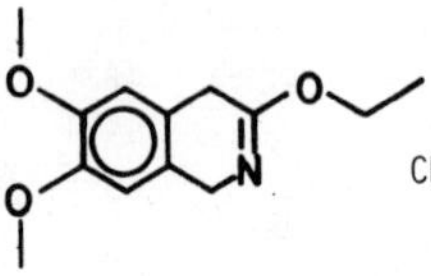

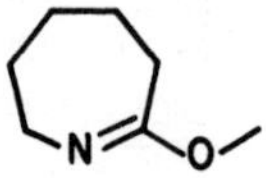

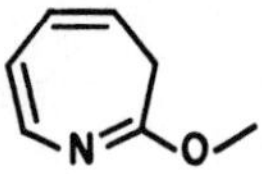

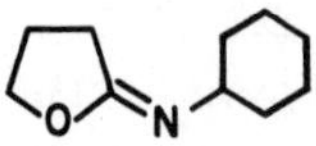

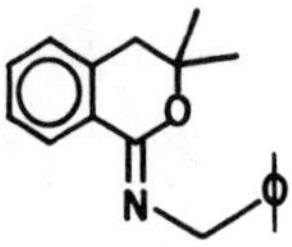

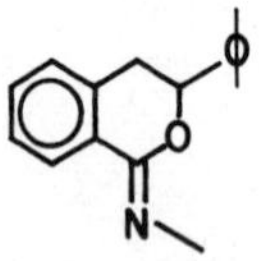

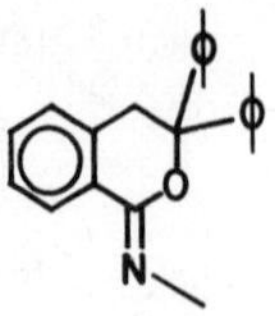

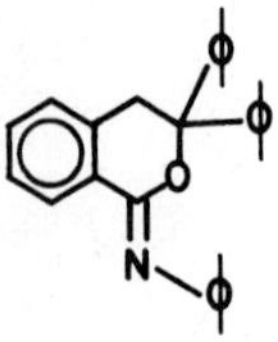

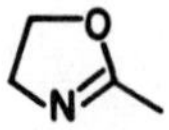

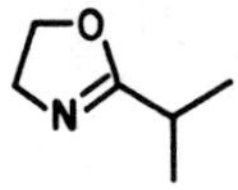

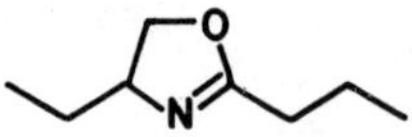

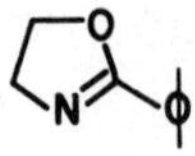

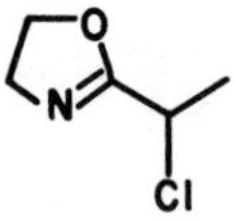

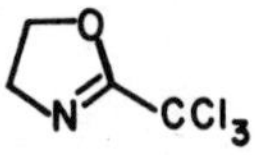

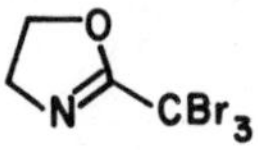

Medium	Frequency	Ref.
liq.	1660	19
CCl_4	1643 1598	21
liq.	1673	16
liq.	1542 1502	22
liq.	1670	19
liq.	1642 1722	13
$CHCl_3$	1667	20
KBr	1825 1815 1650 1605w	23
liq.	1660	19
KBr	1675 1600m 1825m 1785 1740m	23
liq.	1650	19
KBr	1660 1600m 1805 1770m	23
liq.	1650	19
KBr	1660 1600m 1785 1745m	23
liq.	1647	19
$CHCl_3$	1645 1605 1756b	24

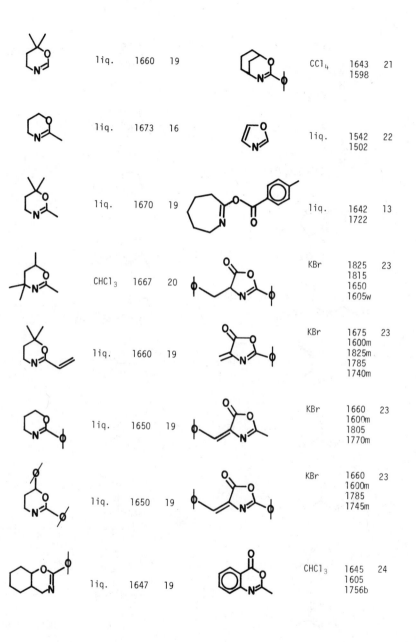

Solvent	Frequency	Ref.
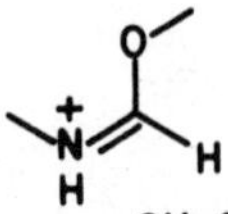 liq.	1695	5
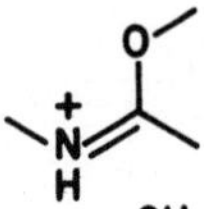 liq.	1680	5
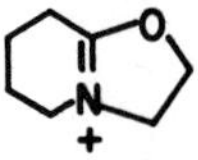 nuj.	1691	25
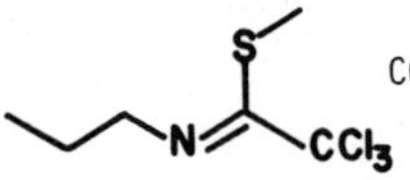 CCl_4	1585	26
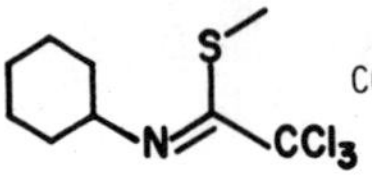 CCl_4	1620	26
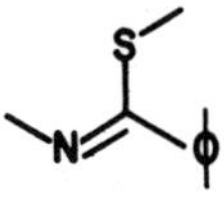 $CHCl_3$	1622	27
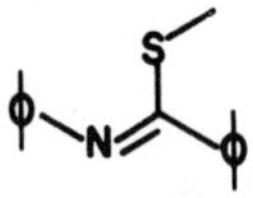 $CHCl_3$	1611	27
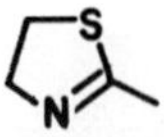 liq.	1639	28
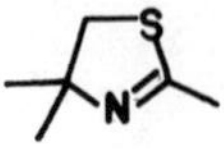 liq.	1641	29
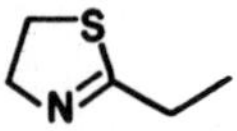 liq.	1631	28
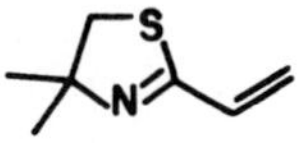 liq.	1587	29
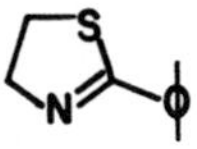 liq.	1613 1605m	28
CCl_4	1616	30
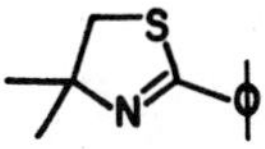 liq.	1615	29
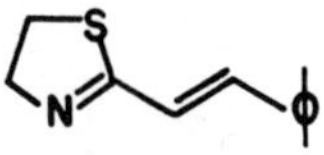 KBr	1587	28
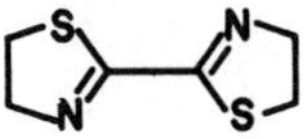 liq.	1593	28
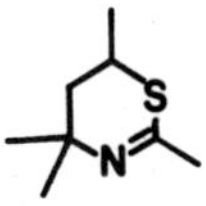 $CHCl_3$	1637	20
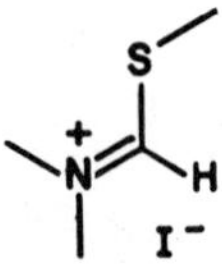 $CHCl_3$	1607	27
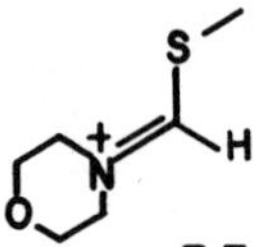 $CHCl_3$	1581	27

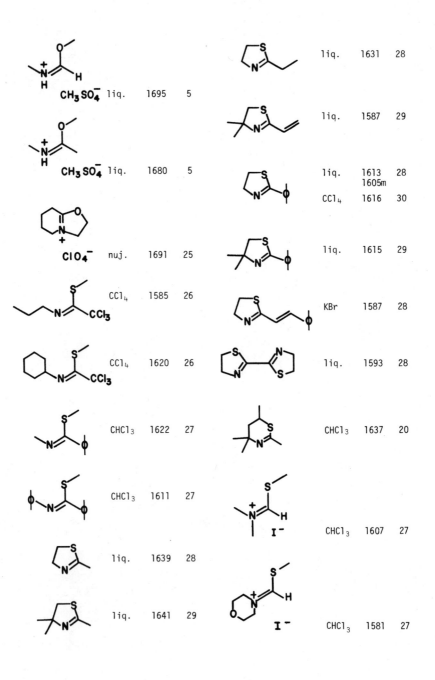

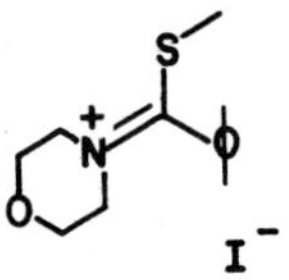

I⁻ $CHCl_3$ 1580 27

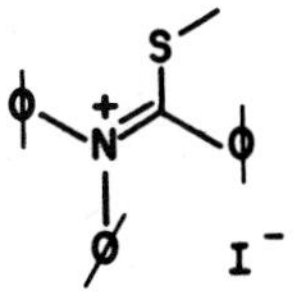

I⁻ $CHCl_3$ 1510 27

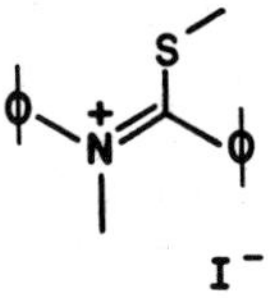

I⁻ $CHCl_3$ 1562 27

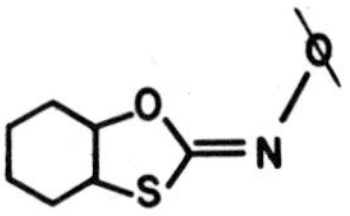

KBr 1634 31

REFERENCES

1. A. Ya. Yakubovich, E. L. Zaitseva, G. I. Braz and V. P. Bazov, *Zh Obshch Khim* (1962) *32*, 3409.
2. C. A. Grob and B. Fischer, *Helv* (1955), 1794.
3. E. S. Hand and W. P. Jenks, *JACS* (1962), 3505.
4. L. Birkofer and H. Dickopp, *B* (1968), 2585.
5. H. Bredereck, F. Effenberger and E. Henseleit, *B* (1965), 2754.
6. J. Fabian and M. Legrand, *BSC France* (1956), 1461.
7. H. U. Daeniker, *Helv* (1964), 33.
8. G. Pifferi, P. Consonni, G. Pelizza and E. Testa, *J Het Chem* (1967), 619.
9. B. J. R. Nicolaus, E. Bellasio and E. Testa, *Helv* (1963), 450.
10. A. E. Wick, P. A. Bartlett and D. Dolphin, *Helv* (1971), 513.
11. H. Peter, M. Brugger, J. Schreiber and A. Eschenmoser, *Helv* (1963), 577.
12. F. Schenker, unpublished results.
13. W. Z. Heldt, *JACS* (1958), 5880.
14. E. Vogel, R. Erb, G. Lenz and A. A. Bothner-By, *A* (1965) *682*, 1.
15. D. M. Bailey and C. G. De Grazia, *JOC* (1970), 4088.
16. W. Seeliger and W. Thier, *A* (1966) *698*, 158.
17. H. R. Nace and E. P. Goldberg, *JACS* (1953), 3646.
18. E. Aufderhaar and W. Seeliger, *A* (1967) *701*, 166.

19. W. Seeliger and W. Diepers, *A* (1966) *697*, 171.
20. A. I. Meyers, *JOC* (1961), 218.
21. R. R. Burford, F. R. Hewgill and P. R. Jefferies, *JCS* (1957), 2937.
22. H. Bredereck and R. Bangert, *B* (1964), 1414.
23. F. Micheel and B. Schleppinghoff, *B* (1955), 763.
24. G. N. Walker, *JACS* (1955), 6698.
25. N. J. Leonard, K. Conrow and R. R. Sauers, *JACS* (1958), 5185.
26. C. Metzger and R. Wegler, *B* (1968), 1136.
27. J. D. S. Goulden, *JCS* (1953), 997.
28. W. Otting and F. Drawert, *B* (1955), 1469.
29. A. I. Meyers and J. J. Ritter, *JOC* (1958), 1918.
30. A. I. Meyers, *JOC* (1959), 1233.
31. L. Goodman, A. Benitez, C. D. Anderson and B. R. Baker, *JACS* (1958), 6582.

3.7
oximes

N—OH

The hydroxyl stretching frequency in dilute solution occurs between 3660 and 3500 cm^{-1}. As with all hydroxyl stretching frequencies they are neither characteristic nor diagnostic, and hydrogen bonding can cause considerable changes in both the position and shape of the band (see Chapter 1.4, p. 14).

C=N

The C-N double-bond stretching frequency in oximes is generally weak. Conjugation to both alkenyl and carbonyl groups causes a lowering of the frequency as do electronegative groups attached directly to the double bond. The hybridization of the carbon atom of the double bond affects the frequency, as with ketonic carbonyl groups an increase in the *p*-character of the C-N bond causes an increase in the stretching frequency. Thus cyclobutanone oxime absorbs at 1709, cyclopentanone oxime at 1686, and cyclohexanone oxime at 1667 cm^{-1}.

The C-N double-bond stretching frequency varies with both concentration and solvent. Thus in concentrated solutions, where intermolecular hydrogen bonding can occur, lower frequencies are observed than in dilute solution. Similarly, samples whose spectra are measured in nujol or KBr usually exhibit lower stretching frequencies. In general, there is little difference between $CHCl_3$ and CCl_4 as solvents.

N—OH

N
|
OH

Too few examples of syn and anti isomers have been reported to discern any major trends, and unless one isomer can undergo a dipolar interaction (*e.g.*, hydrogen bonding), the difference in frequency between the two isomers is small.

N—O

The N-O single-bond stretching vibration occurs as a strong absorption between 1000 and 900 cm^{-1} but has no diagnostic value.

Phase	ν (cm⁻¹)	Ref.
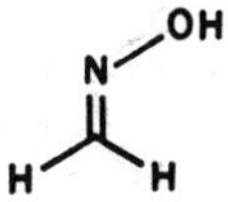 gas	1642	1
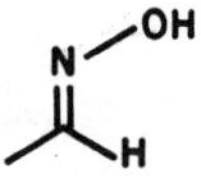 nuj.	1669	2
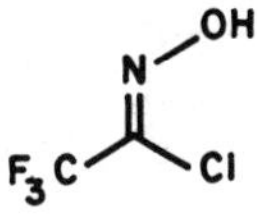 liq.	1637	3
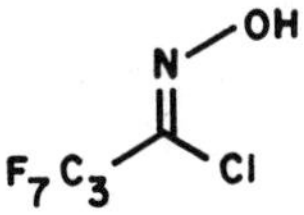 liq.	1633	3
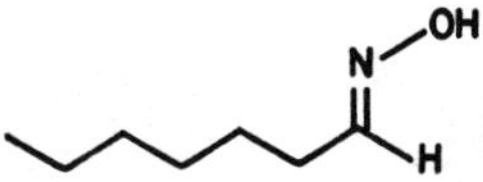 solid	1670	4
liq.	1603w 1582m	5
α nuj.	1630	6
β nuj.	1645	6
$CHCl_3$	1651	7
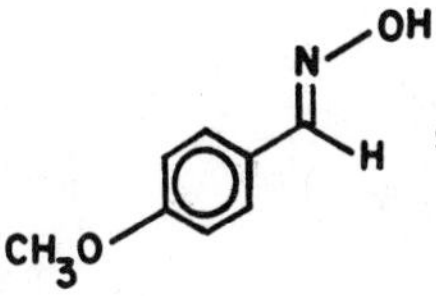 solid	1637	4
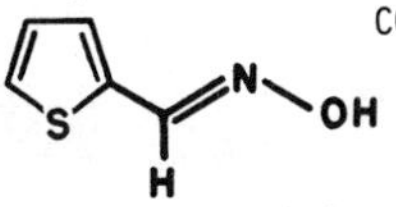 CCl_4	1634b	8
nuj.	1658	9
CCl_4	1674 1703	9
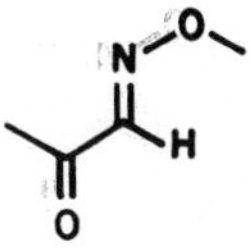 liq.	1588 1696	9
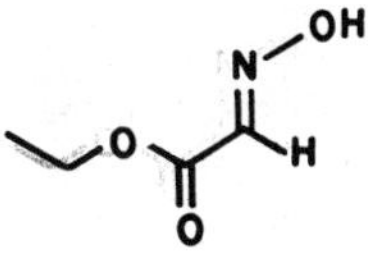 $CHCl_3$	1629 1727	10
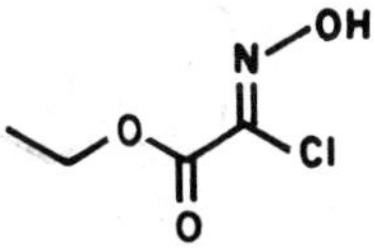 $CHCl_3$	1603 1739	10
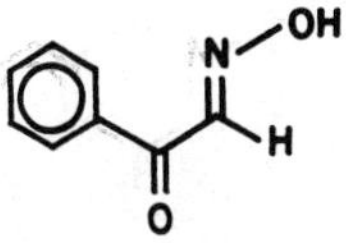 KBr	1650* 1670	11
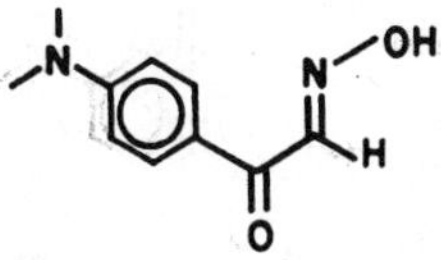 KBr	1610	11
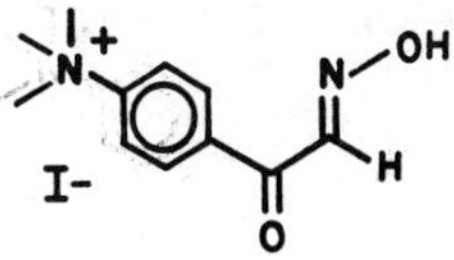 KBr	1645 1690	11
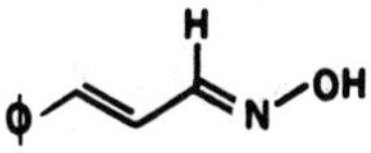 $CHCl_3$	1630 1620	12

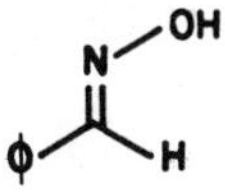

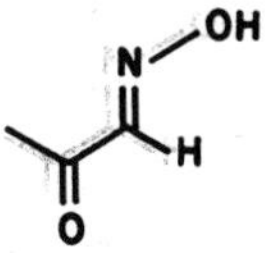

Solvent	ν	Ref.
CCl_4	1675	13
CCl_4	1661	14
$CHCl_3$	1660	14
nuj.	1630	6
CCl_4	1630	14
$CHCl_3$	1630	14
liq.	1603m 1577m	5
liq.	1603* 1580m	5
KBr	1608m 1585m	5
liq.	1605 1572m	5

Solvent	ν	Ref.
nuj.	1665	9
CCl_4	1686* 1698	9
liq.	1610 1695	9
$CHCl_3$	1623 1718	10
$CHCl_3$	1623 1721	10
$CHCl_3$	1603 1739	10
$CHCl_3$	1629 1745	10
CH_2Cl_2	1631	15

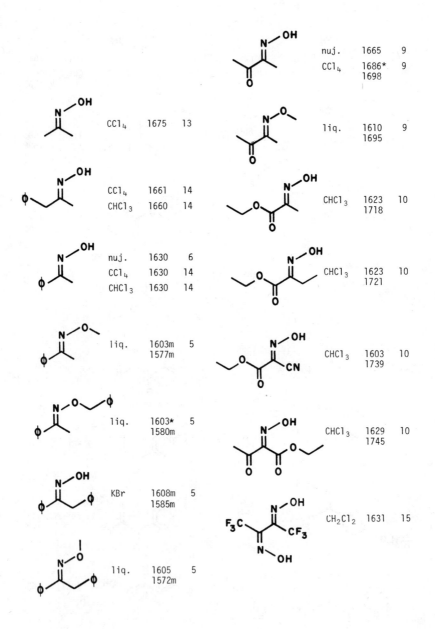

Solvent	Frequency	Ref.
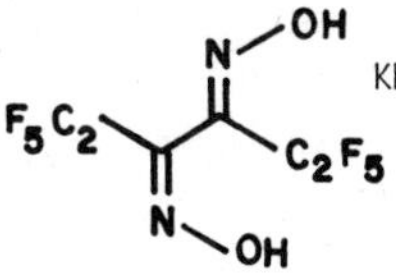 KBr	1625	15
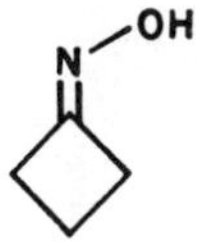 CCl_4	1730* 1709	16
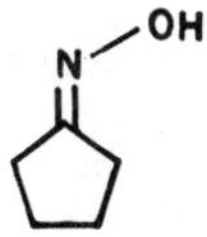 CCl_4	1686	16
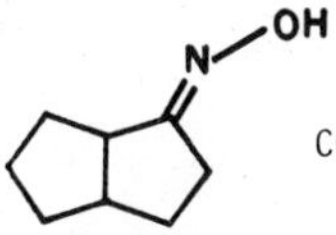 CCl_4	1677	16
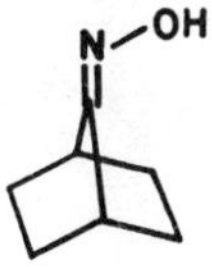 CCl_4	1690	16
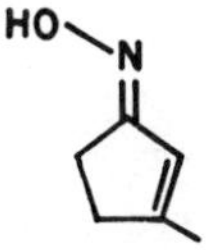 $CHCl_3$	1655 1643 1625	17
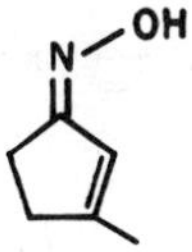 $CHCl_3$	1645 1620	17
nuj.	1662	6
CCl_4	1667	16
$CHCl_3$	1656	14
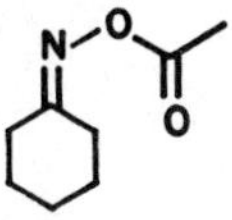 liq.	1648 1755	18
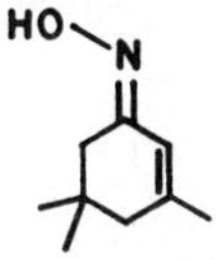 $CHCl_3$	1645	17
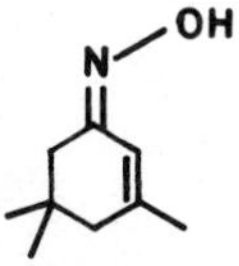 $CHCl_3$	1637	17
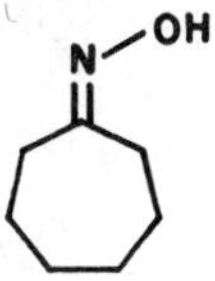 CCl_4	1652	16
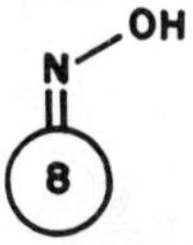 CCl_4	1647	16
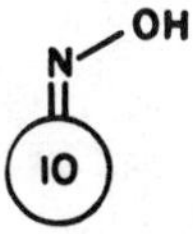 CCl_4	1650	16
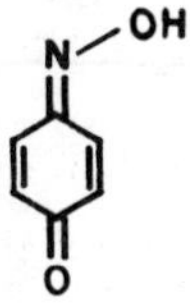 nuj.	1555 1628	19

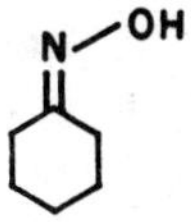

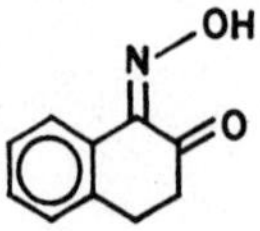 nuj.	1526 1618	19
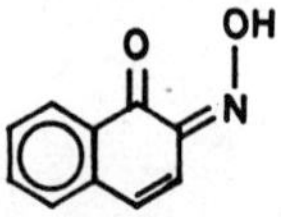 nuj.	1550 1668	19
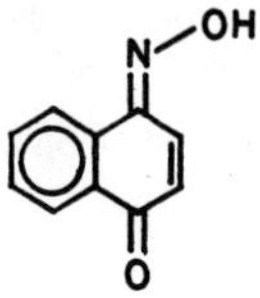 nuj.	1577 1630	19
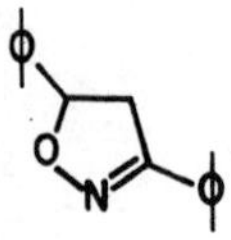 KBr	1597m 1570m	5
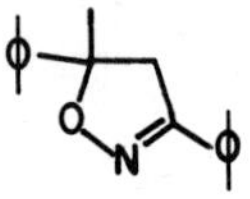 KBr	1592m 1567m	5
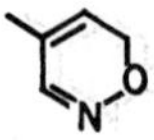 liq.	1665 1560	20
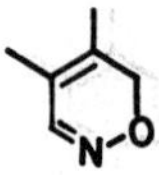 liq.	1675 1570	20
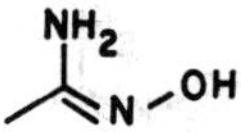 KBr	1661 1590	21
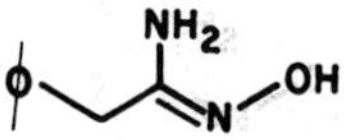 $CHCl_3$	1670	7
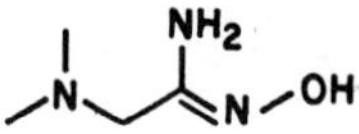 KBr	1664 1618	21

REFERENCES

1. S. Califano and W. Lüttke, *Z Phys Chem NF* (1956) *6*, 83.
2. C. F. Pouchert, *The Aldrich Library of Infrared Spectra*, Aldrich 1970, spectrum 1040A.
3. L. W. Kissinger, W. E. McQuistion and M. Schwartz, *T* (1963) *Suppl I*, 137.
4. F. Mathis, *CR* (1951) *232*, 505.
5. G. W. Perold, A. P. Steyn and F. V. K. von Reiche, *JACS* (1957), 462.
6. A. Palm and H. Werbin, *Canad J Chem* (1953), 1004.
7. C. L. Bell, C. N. V. Nambury and L. Bauer, *JOC* (1964), 2873.

8. J. J. Peron, C. Saumagne and J. M. Lebas, *SCA* (1970), 1651.
9. Y. Kuroda and M. Kimura, *BSC Japan* (1963), 464.
10. H. E. Ungnade and L. W. Kissinger, *JOC* (1958), 1794.
11. J. J. Norman, *Canad J Chem* (1962), 2023.
12. G. Duyckaerts, *Bull Soc Roy Sci Liége* (1952), 196.
13. L. H. Cross and A. C. Rolfe, *Trans Farad Soc* (1951), 354.
14. J. Fabian and M. Legrand, *BSC France* (1956), 1461.
15. L. W. Kissinger, W. E. McQuistion, M. Schwartz and L. Goodman, *T* (1963) *Suppl I*, 131.
16. S. Bank and W. D. Closson, *T* (1968), 381.
17. K. Gschwend, Thesis #3745 ETH, Zurich, 1965.
18. W. Z. Heldt, *JACS* (1958), 5880.
19. D. Hadži, *JCS* (1956), 2725.
20. A Schönberg and K. Junghans, *B* (1966), 531.
21. C. S. Hollander, R. A. Yoncoskie and P. L. deBenneville, *JOC* (1958), 1112.

3.8

nitrones imine oxides

The dipolar nature of the nitrone group gives rise to strong absorptions in the range of 1620 to 1550 cm^{-1}. As might be expected, substitution on the carbon atom of the C-N double bond decreases the contribution from resonance form II and increases the nitrone stretching frequency. Similarly, electron donating groups on a C-phenyl substituent raise the frequency, whereas electron withdrawing groups on the phenyl lower it, compared to the unsubstituted phenyl derivative.

As with imines (Chapter 3.1, p. 116) a decrease in ring size, and consequent angle strain, results in a decrease in the stretching frequency of cyclic nitrones.

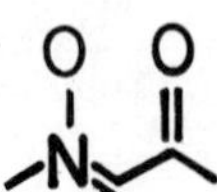

In systems where the oxygen lone pairs of the nitrone group can be delocalized, as in β-keto nitrones, the nitrone frequency is lowered considerably.

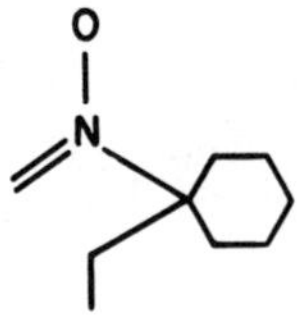

Solvent	Frequency	Ref.
liq.	1550	1
liq.	1585	2
$CHCl_3$	1658 1527	3
KBr	1592	5
CCl_4	1587 1172	4
KBr	1580	5
KBr	1580 1563	5
KBr	1592	5
nuj.	1595 1577 1548	6
nuj.	1592 1550	6
nuj.	1613 1534	6
nuj.	1600 1538	6
KBr	1575	7
KBr	1600 1555	7
liq.	1582	5

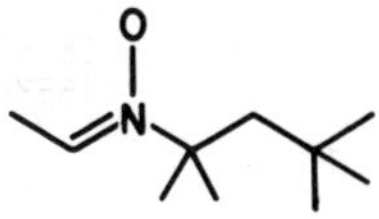

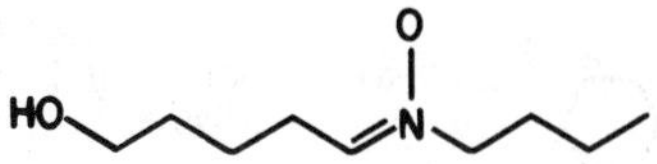

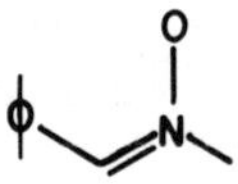

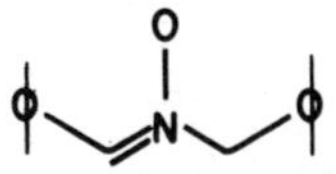

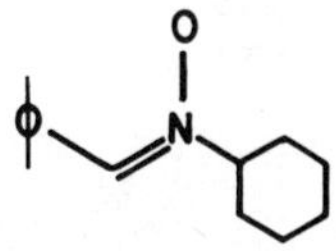

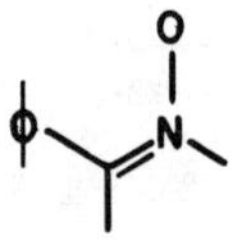

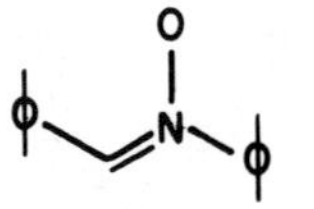

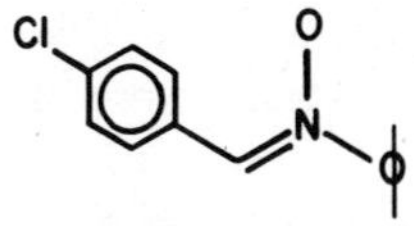

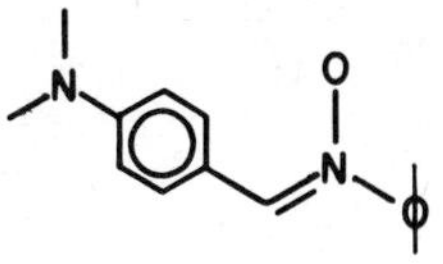

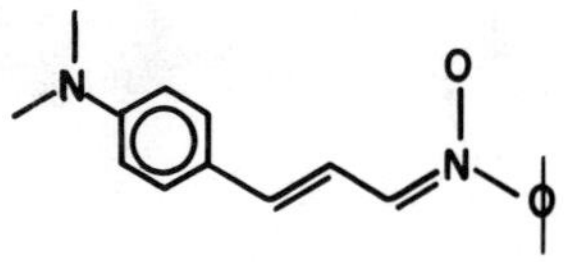

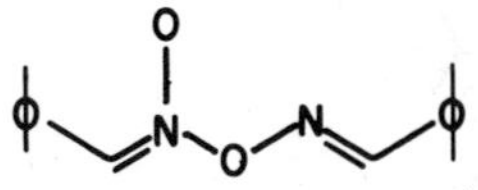

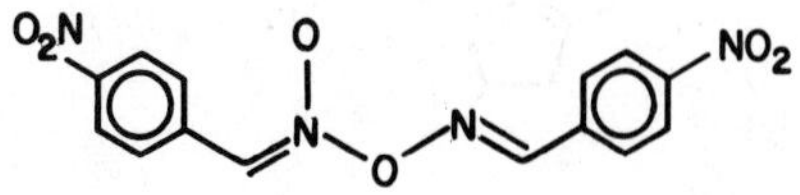

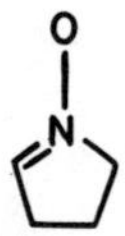

Structure	Phase	Frequency	Ref.
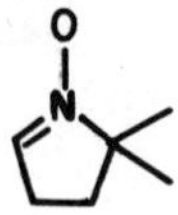 1-oxide, 2,2-dimethyl	liq.	1573	8
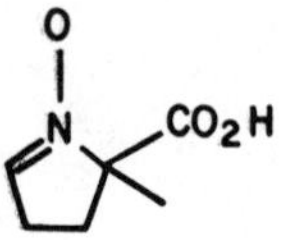 1-oxide, 2-methyl-2-CO_2H	liq.	1585	8
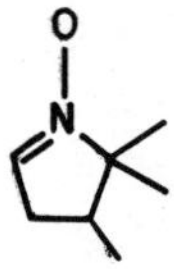 1-oxide, 2,2,3-trimethyl	liq.	1572	8
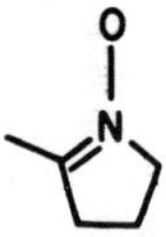 1-oxide, 5-methyl	liq.	1612	9
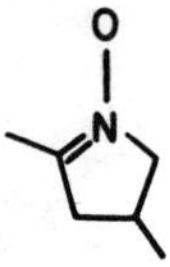 1-oxide, 5-methyl-3-methyl	liq.	1618	9
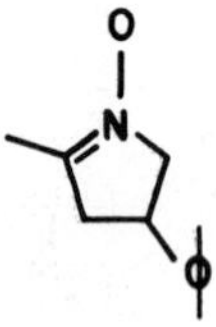 1-oxide, 5-methyl-3-ϕ	liq.	1621	9
 1-oxide, 5-methyl-3,3-dimethyl	liq.	1613	8
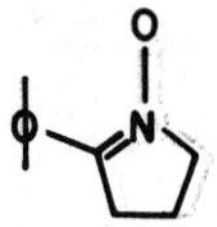 1-oxide, 5-ϕ	KBr	1575 1563	5
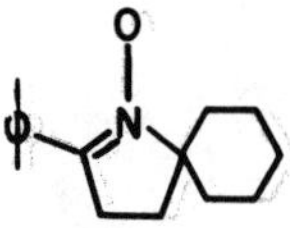 1-oxide, 5-ϕ, spirocyclohexane	nuj.	1567 1546	10
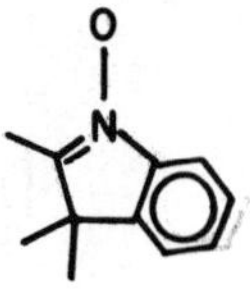 indolenine N-oxide	CCl_4	1625 1545 1470	11
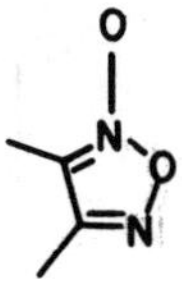 furoxan, dimethyl	KBr	1605 1534	7
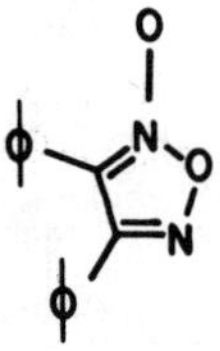 furoxan, diphenyl (ϕ)	KBr	1582 1567	7

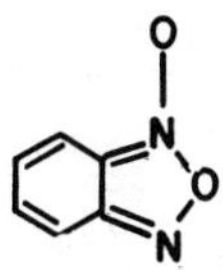	KBr	1605 1576	7
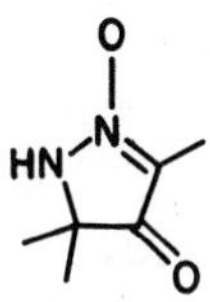	CCl_4	1588 1730	12
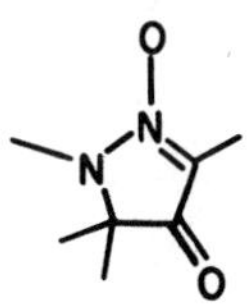	CCl_4	1550 1705	12
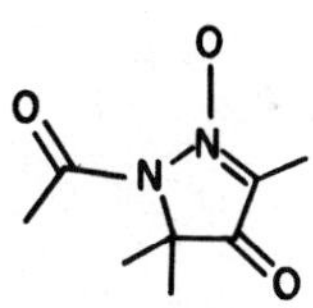	CCl_4	1570 1710	12
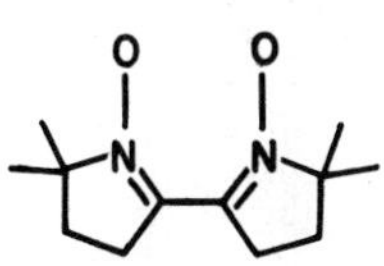	$CHCl_3$	1509 1503	13
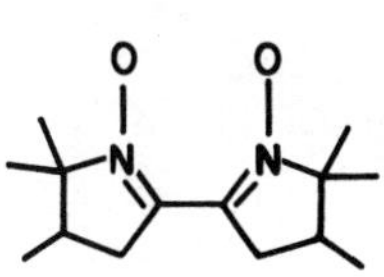	nuj.	1504	13

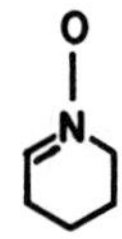	$CHCl_3$	1625	14
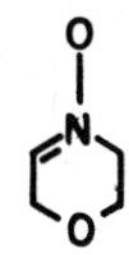	$CHCl_3$	1628	14
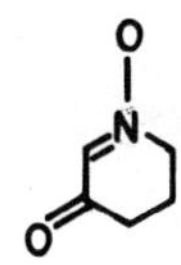	nuj.	1555 1610	9
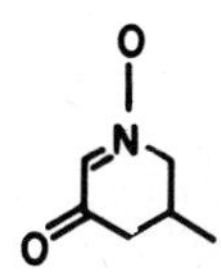	nuj.	1556 1660	9
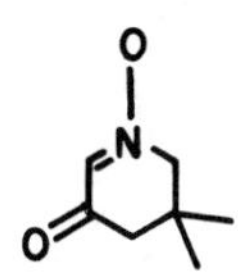	nuj.	1546 1653	15
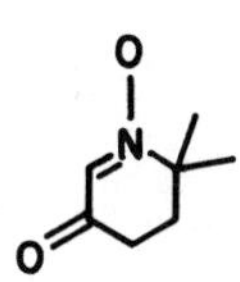	nuj.	1524 1663	9
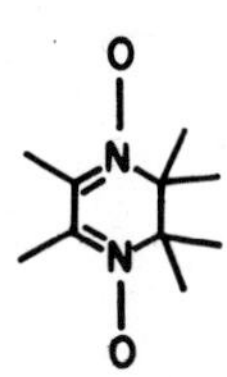	KCl	1540 1300	16

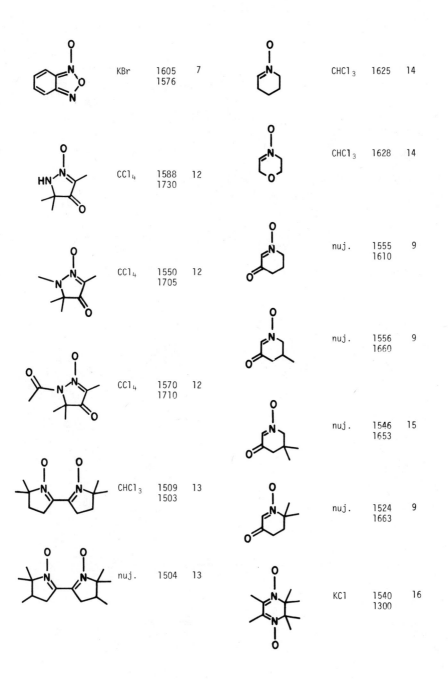

REFERENCES

1. J. E. Baldwin, A. K. Quershi and B. Sklarz, *Chem Comm* (1968), 373.
2. J. E. Baldwin, R. G. Pudussery, B. Sklarz and M. K. Sultan, *Chem Comm* (1968), 1361.
3. A. A. R. Sayigh and H. Ulrich, *JOC* (1962), 4662.
4. H. Shindo and B. Umezawa, *Chem Pharm Bull* (1962), 492.
5. J. Thesing and W. Sirrenberg, *B* (1959), 1748.
6. E. Boyland and R. Nery, *JCS* (1963) 3141
7. H. Kropf and R. Lambeck, *A* (1966) *700*, 18.
8. R. Bonnett, R. F. C. Brown, V. M. Clark, I. O. Sutherland, and A. Todd, *JCS* (1959), 2094.
9. R. F. C. Brown, L. Subrahmanyan and C. P. Whittle, *Aust JC* (1967), 339.
10. M. C. Kloetzel, F. L. Chubb, R. Gobran and J. L. Pinkus, *JACS* (1961), 1128.
11. M. Mousseron-Canet and J.-P. Boca, *BSC France* (1967), 1296.
12. J. P. Freeman, *JOC* (1962), 2881.
13. R. F. C. Brown, V. M. Clark, M. Lamchen and A. Todd, *JCS* (1959), 2116.
14. J. F. Elsworth and M. Lamchen, *JCS* (1968) *C*, 2423.
15. R. F. C. Brown, V. M. Clark and A. Todd, *JCS* (1959), 2105.
16. M. Lamchen and T. M. Mittag, *JCS* (1966) *A*, 2300.

3.9

nitrile oxides

$$-\overset{}{C}\equiv\overset{+}{N}-\overset{-}{O} \quad \text{I}$$

$$\updownarrow$$

$$-\overset{-}{C}=\overset{+}{N}=O \quad \text{II}$$

Oxidation of a nitrile to a nitrile oxide causes only a small increase in the frequency of the C-N triple-bond stretching vibration. As with nitriles, the range of frequencies for nitrile oxides is narrow.

Contributions from the resonance form II are stabilized by electron withdrawing groups on the carbon, and such groups cause a small decrease in the stretching frequency.

N—O

The N-O single-bond stretching frequency falls between 1400 and 1200 cm^{-1} and is of no diagnostic value.

Compound	Medium	cm^{-1}	Ref.
HCNO	gas	2190 1251	1
ONC−CNO	CCl_4	2190 1235	2
−CNO	CCl_4	2315 1319	3
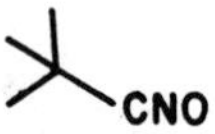 CNO	sng	2273	4
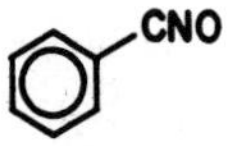 CNO	CCl_4	2288 1712w	5
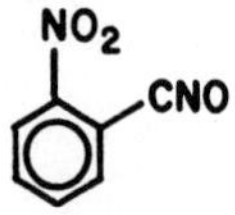 NO_2 CNO	CCl_4	2304 1536	5
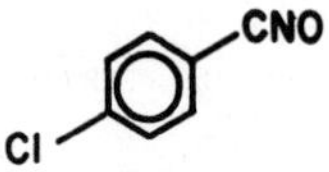 CNO Cl	CCl_4	2292 1377	6
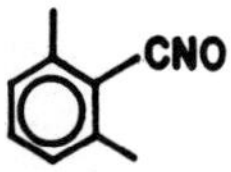 CNO	CCl_4	2288 1349	7
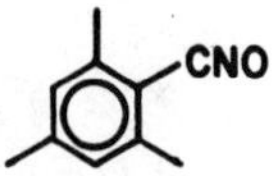 CNO	CCl_4	2287 1348	6
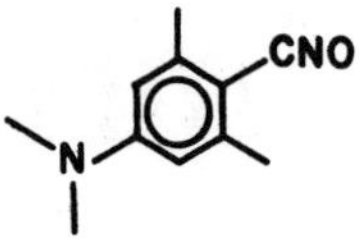 CNO N	CCl_4	2287 1341	8
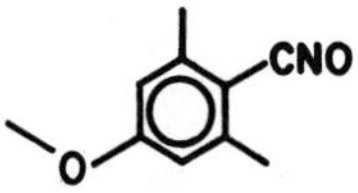 CNO O	CCl_4	2290 1351	7
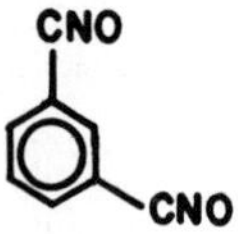 CNO CNO	nuj.	2300 1345	7
 CNO ONC	nuj.	2283 1345	7

REFERENCES

1. W. Beck and K. Feldl, *Angew* (1966), 746.
2. C. Grundmann, V. Mini, J. M. Dean and H.-D. Frommeld, *A* (1965) *687*, 191.
3. W. G. Isner and G. L. Humphrey, *JACS* (1967), 6442.
4. G. Zinner and A. H. Günther, *Angew* (1964), 440.
5. R. H. Wiley and B. J. Wakefield, *JOC* (1960), 546.
6. S. Califano, R. Moccia, R. Scarpati and G. Speroni, *JCP* (1957) *26*, 1777.
7. M. Yamakawa, T. Kubota and H. Akazawa, *BCS Japan* (1967), 1600.
8. M. Yamakawa, T. Kubota, H. Akazawa and I. Tanaka, *BCS Japan* (1968), 1046.

3.10

cyanates thiocyanates

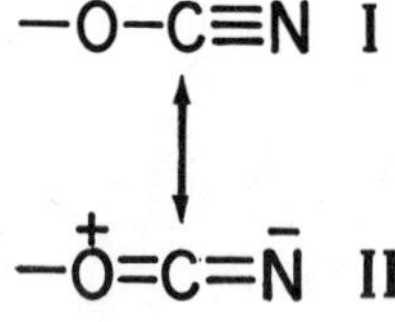

Both cyanates and thiocyanates can be represented by the resonance forms I and II, and as would be expected, electron withdrawing groups bonded to oxygen or sulfur lower the asymmetric vibrational stretching frequencies. Thus a phenyl substituent causes a lowering of the frequency, whereas further substitution causes only additional small changes.

The position of the medium-intensity band of cyanates and thiocyanates varies little with state or solvent.

The cyanate ion absorbs at 2170 cm^{-1} and the thiocyanate ion at 2020 cm^{-1}.

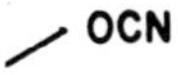 OCN	CCl_4	2307m 2236 2216*w	1
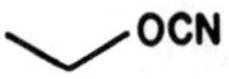 OCN	CCl_4	2266m 2247 2197*w	1
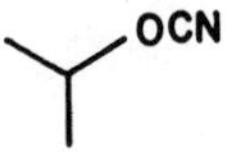 OCN	CCl_4	2282 2255 2226m 2202*w	1
 OCN	CCl_4	2266m 2247 2197*w	1
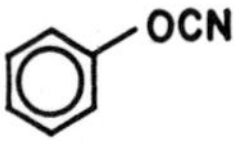 OCN	liq.	2222	2
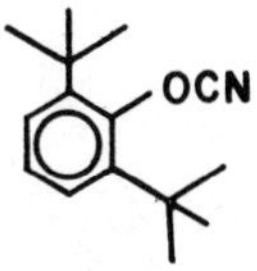 OCN	CCl_4	2247	3

KSCN	nuj.	2041	4
	KBr	2053	4
Ag SCN	nuj.	2169	4
	KBr	2083	4
Hg $(SCN)_2$	nuj.	2090	5
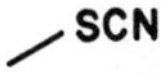 SCN	liq.	2141	6
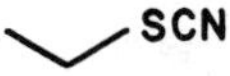 SCN	liq.	2141	6
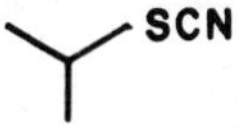 SCN	liq.	2153	7
 SCN	liq.	2137	6
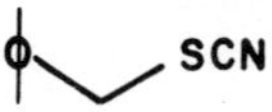 ϕ SCN	liq.	2141	6

Compound	Medium	cm⁻¹	Ref.
SCN	liq.	2175	8
SCN, N	CCl_4	2164	9
	$CHCl_3$	2161	9
SCN, O_2N	CCl_4	2174	9
NC, S, CN	KBr	2190m	10
		2179	
		2132m	
SCN, N, H	liq.	2139	11
SCN, N	liq.	2137	11
NC, S, S, CN	CS_2	2171	12
Se CN	liq.	2154	13

REFERENCES

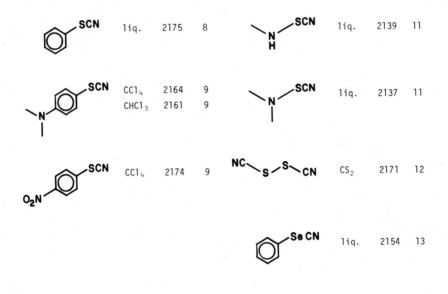

1. N. Groving and A. Holm, *Acta Chem Scand* (1965), 443.
2. E. Grigat and R. Pütter, *B* (1964), 3012.
3. H. Hoyer, *B* (1961), 1042.
4. P. Kinell and B. Strandberg, *Acta Chem Scand* (1959), 1607.
5. P. C. H. Mitchell and R. J. P. Williams, *JCS* (1960), 1912.
6. E. Lieber, C. N. R. Rao and J. Ramachandran, *SCA* (1958) *13*, 296.
7. R. P. Hirschmann, R. N. Kniseley and V. A. Fassel, *SCA* (1964), 809.
8. N. S. Ham and J. B. Willis, *SCA* (1960), 393.
9. G. L. Caldow and H. W. Thompson, *SCA* (1958) *13*, 212.
10. D. A. Long and D. Steele, *SCA* (1963), 1731.
11. M. Kuhn and R. Mecke, *B* (1960), 618.
12. M. J. Nelson and A. D. E. Pullin, *JCS* (1960), 604.
13. E. Bulka, D.-D. Ahlers and E. Tuček, *B* (1967), 1367.

4.1
aldehydes

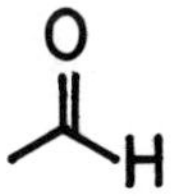

O
H

Aldehydes show two weak, but characteristic, bands in the region of 2900 to 2700 cm^{-1} that arise from C-H stretching vibrations. Frequently one of these two bands falls in the narrow range of 2730 to 2710 cm^{-1}. The C-H deformation mode is usually weak and of little diagnostic value, with reported frequencies varying from 1000 to 750 cm^{-1}.

O

In all respects the influence of neighboring substituents on the carbonyl stretching frequency of aldehydes qualitatively parallels those observed with ketonic carbonyl groups, and reference should be made to Chapter 4.2, p. 175 for a discussion of these effects.

As with ketones changes in state and polarity of the solvent cause small, but consistent and characteristic, changes in the carbonyl stretching frequency. In general, solutions in CCl_4 absorb at 10 cm^{-1} lower than the gaseous state and $CHCl_3$ solutions absorb at 10 cm^{-1} lower than those in CCl_4.

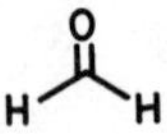

Structure	Solvent	ν	Ref.
	gas	1744	1
	CCl_4	1733	2
	$CHCl_3$	1725	3
	liq.	1729	4
	CCl_4	1738	2
	liq.	1735	4
	CCl_4	1729	2
	CCl_4	1729	5
	CCl_4	1729	2
	CCl_4	1715	6
ϕ	CCl_4	1728	5
Cl	CCl_4	1742	7
	gas	1752	1

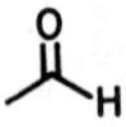

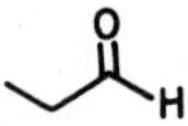

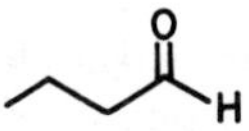

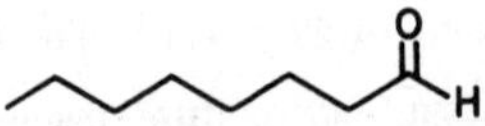

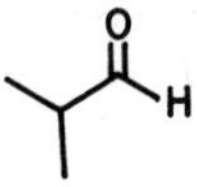

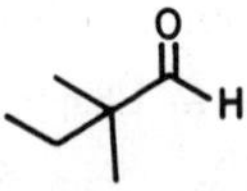

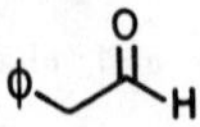

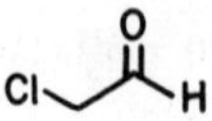

Structure	Solvent	ν	Ref.
Cl, Cl	CCl_4	1748	7
	gas	1761	1
F_3C	gas	1788	1
Cl_3C	CCl_4	1768	7
	gas	1778	1
Br_3C	CCl_4	1745 1707w	8
	liq.	1700	9
	liq.	1682	9
	liq.	1715	10
	CCl_4	1715	6

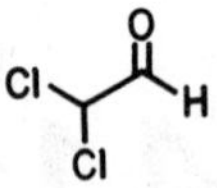

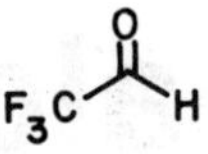

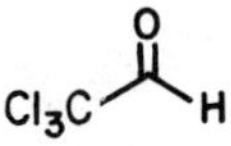

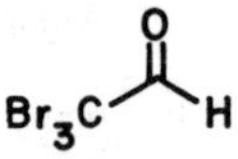

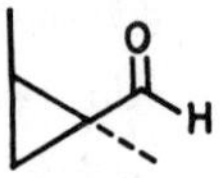

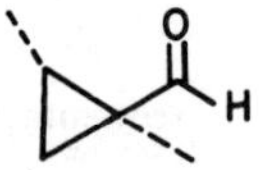

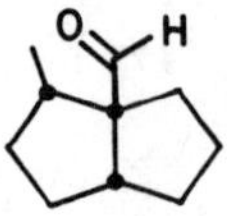

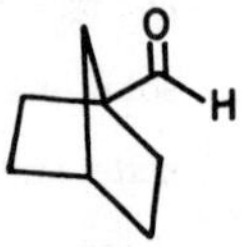

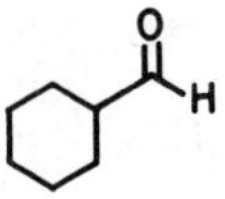	CCl_4	1729	5
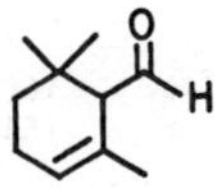	$CHCl_3$	1714	11
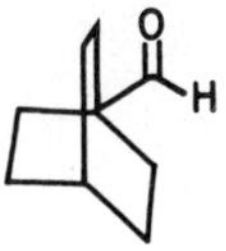	CCl_4	1715	6
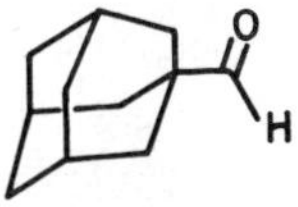	KBr	1722	12
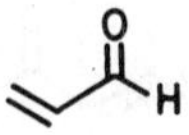	CCl_4	1704	2
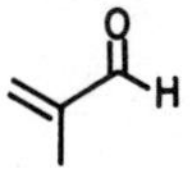	C_2Cl_4	1702	13
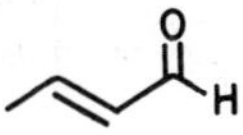	CCl_4	1696	2
	$CHCl_3$	1685	14
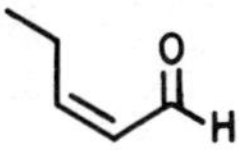	CS_2	1685	15

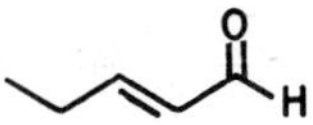	CS_2	1692	15
	$CHCl_3$	1685 1645	16
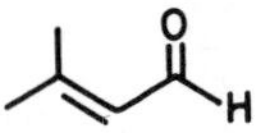	C_2Cl_4	1686	17
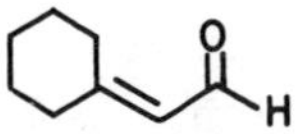	$CHCl_3$	1686	18
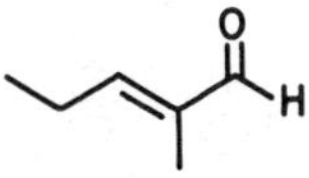	liq.	1690 1620	19
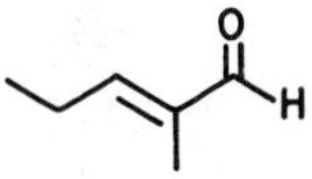	$CHCl_3$	1685 1645	16
	C_2Cl_4	1693	13
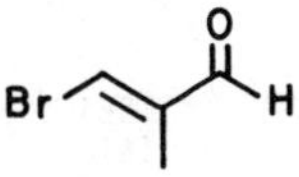	CCl_4	1706 1615	20
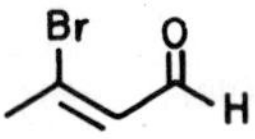	CCl_4	1692 1629	20
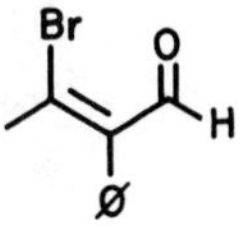	CCl_4	1693 1608	20

Solvent	Frequency	Ref.
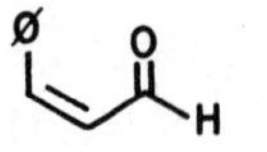 $CHCl_3$	1684	21
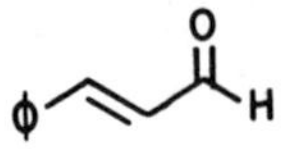 $CHCl_3$	1674	21
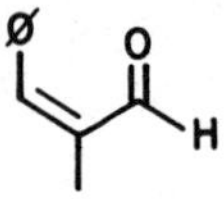 $CHCl_3$	1687	21
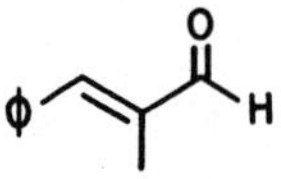 $CHCl_3$	1679	21
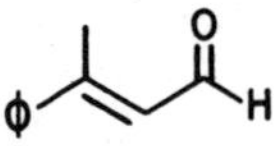 $CHCl_3$	1658	21
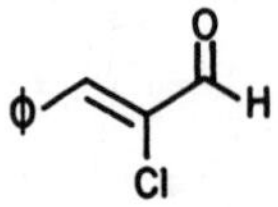 liq.	1685	19
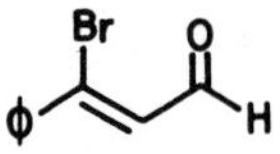 CCl_4	1681 1600	20
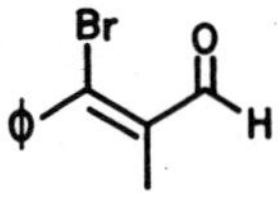 CCl_4	1684 1618	20
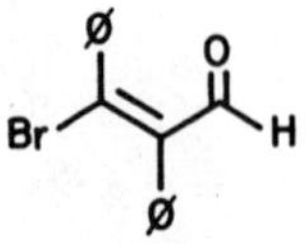 $CHCl_3$	1681 1611	20
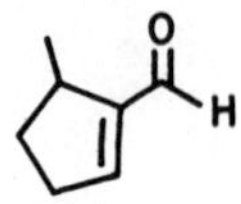 liq.	1680	22

Solvent	Frequency	Ref.
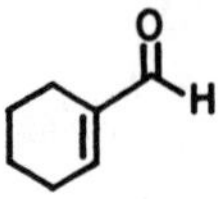 liq.	1672 1637	23
CCl_4	1690	5
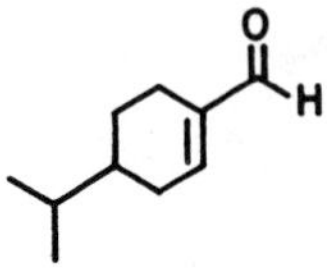 liq.	1685 1649	24
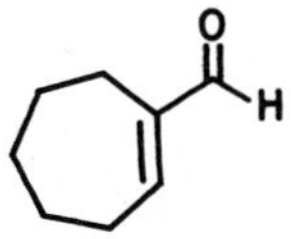 CCl_4	1690	5
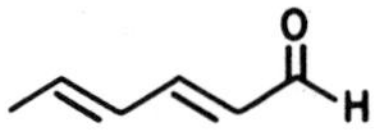 $CHCl_3$	1677	14
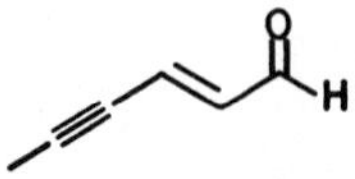 CS_2	1689	25
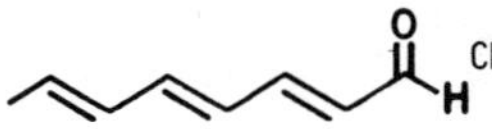 $CHCl_3$	1674	14
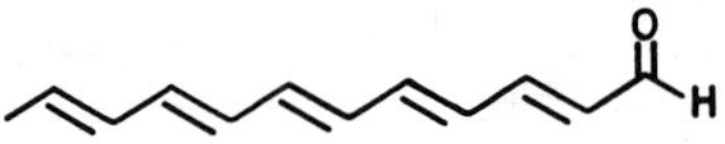 $CHCl_3$	1664	14
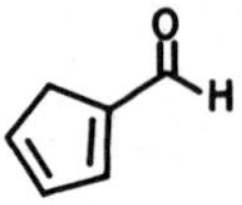 CCl_4	1717 1684* 1676 1668	26
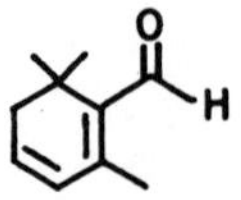 CCl_4	1672 1639 1558	27
CS_2	1672 1639 1558	27

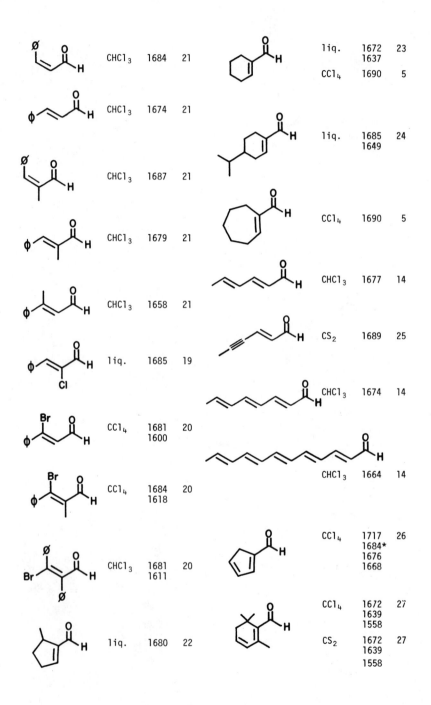

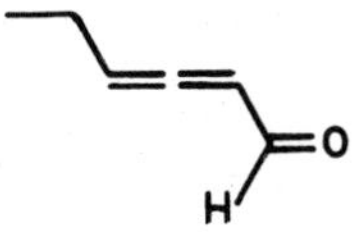

liq. 1688, 1932 — 28

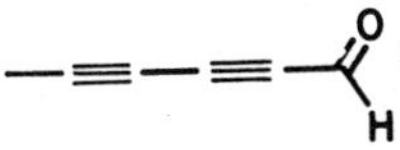

CH_2Cl_2 1657, 2228 — 29

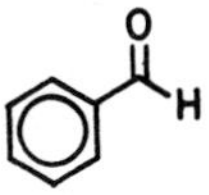

liq. 1705 — 4
CCl_4 1710 — 30

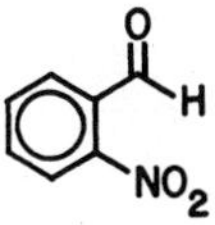

CCl_4 1706 — 31

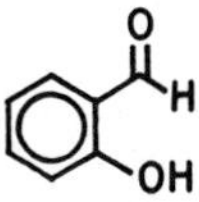

CCl_4 1670 — 32

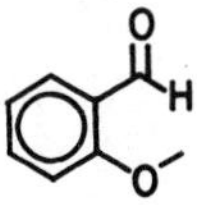

liq. 1679, 1666 — 33

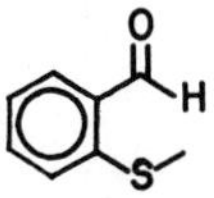

liq. 1700 — 34

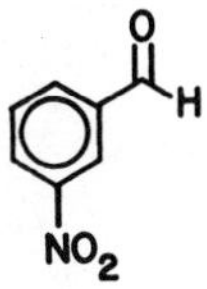

CCl_4 1714 — 30

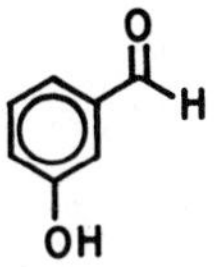

$CHCl_3$ 1679 — 30

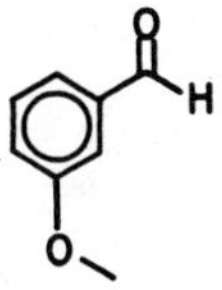

CCl_4 1712 — 30

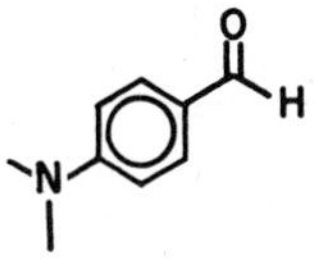

CCl_4 1696 — 30

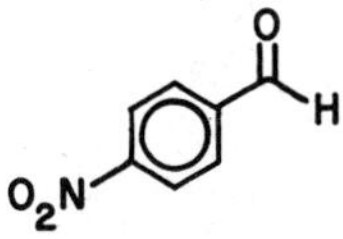

CCl_4 1717 — 30
$CHCl_3$ 1710 — 30

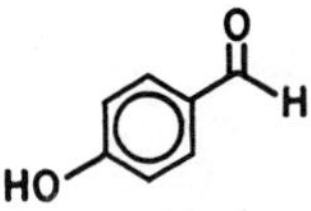

$CHCl_3$ 1690 — 30

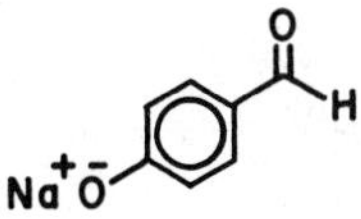

nuj. 1639 — 35

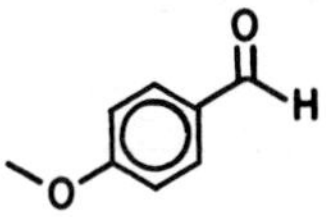

CCl_4 1704 — 30

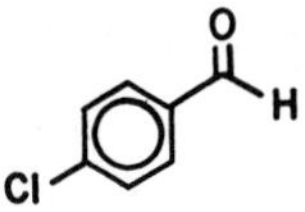

CCl_4 1712 — 30

Solvent	ν (cm⁻¹)	Ref.
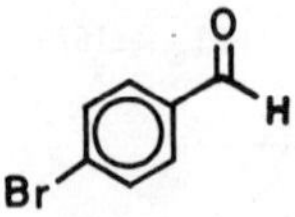 CCl_4	1715	30
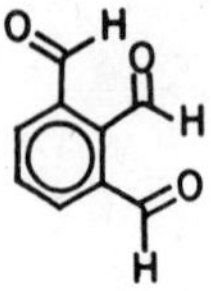 nuj.	1704	36
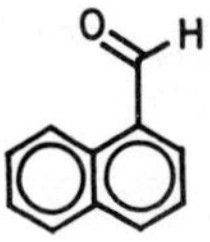 CCl_4	1700	37
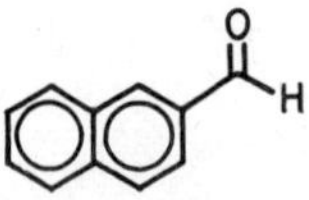 CCl_4	1702	37
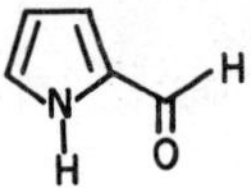 CCl_4	1665	38
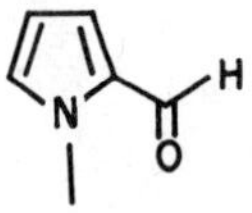 CCl_4	1672	38
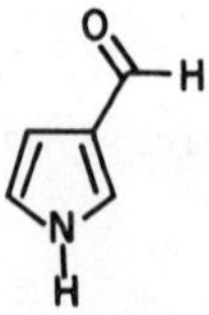 CCl_4	1683	39
CCl_4	1700 1683	40
$CHCl_3$	1692 1676	40
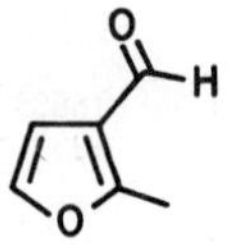 CCl_4	1721 1695	41
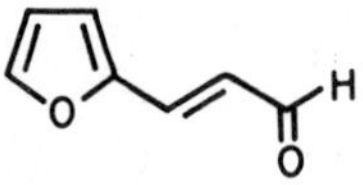 $CHCl_3$	1677 1629	14
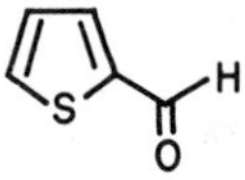 $CHCl_3$	1673	13
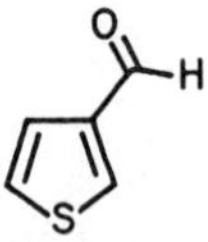 $CHCl_3$	1691	13
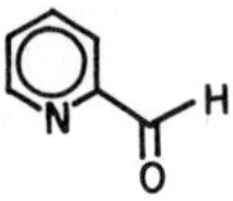 $CHCl_3$	1717	42
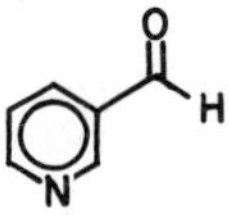 $CHCl_3$	1712	42
 $CHCl_3$	1721	42
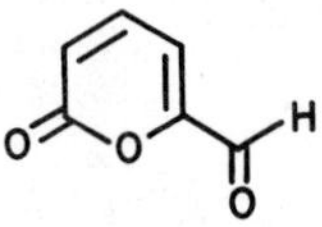 $CHCl_3$	1745 1708 1625	43

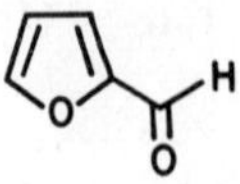

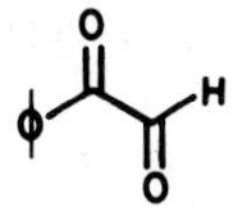

Solvent	Frequency	Ref.
$CHCl_3$	1732 1704	44
KBr	1700	45
$CHCl_3$	1755 1686	44
nuj.	1674 1540	46

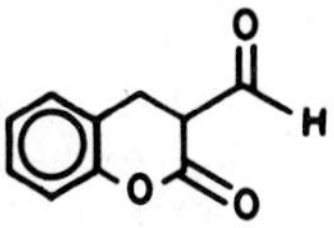

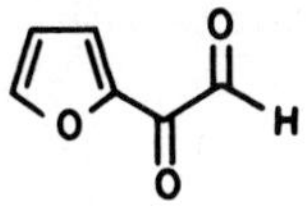

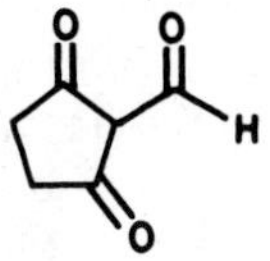

REFERENCES

1. D. F. DeTar and L. A. Carpino, *JACS* (1956), 475.
2. D. Cook, *JACS* (1958), 49.
3. D. Dolphin and A. E. Wick, unpublished results.
4. E. J. Hartwell, R. E. Richards and H. W. Thompson, *JCS* (1948), 1436.
5. J. F. King and B. Vig, *Canad J Chem* (1962), 1023.
6. D. E. Applequist and L. Kaplan, *JACS* (1965), 2194.
7. L. J. Bellamy and R. L. Williams, *JCS* (1958), 3465.
8. G. Lucazeau and A. Novak, *SCA* (1969), 1615.
9. J.-L. Ripoll and J.-M. Conia, *TL* (1965), 979.
10. P. Beslin, R. Bloch, G. Moinet and J.-M. Conia, *BSC France* (1969), 508.
11. R. Joos, *Thesis ETH* (1970).
12. K. Bott, *Angew* (1968), 970.
13. S. Gronowitz and A. Rosenberg, *Arkiv Kemi* (1955), *8*, 23.
14. E. R. Blout, M. Fields and R. Karplus, *JACS* (1948), 194.
15. D. A. Thomas and W. K. Warburton, *JCS* (1965), 2988.
16. K. C. Chan, R. A. Jewell, W. H. Nutting and H. Rapoport, *JOC* (1968), 3382.
17. R. Mecke and K. Noack, *B* (1960), 210.
18. N. A. Milas and C. P. Priesing, *JACS* (1957), 6295.
19. E. Elkik and P. Vaudescal, *CR* (1965), *261*, 1015.
20. Z. Arnold and A. Holý, *Col Czech Comm* (1961), 3059.
21. J. Elguero, R. Jacquier and C. Marzin, *BCS France* (1967), 3005.
22. F. Korte, J. Falbe and A. Zschocke, *T* (1959), *6*, 201.
23. R. Lehr, *Harvard Ph.D. Thesis* (1968).
24. P. A. Stadler, *Helv* (1960), 1601.

25. I. Bell, E. R. H. Jones and M. C. Whiting, *JCS* (1958), 1313.
26. K. Hafner, G. Schulz and K. Wagner, *A* (1964), *678*, 39.
27. C. H. Eugster, L. Leichner and E. Jenny, *Helv* (1963), 543.
28. M. Winter, *Helv* (1963), 1749.
29. R. K. Bentley, U. Graf, E.R. H. Jones, R. A. M. Ross, V. Thaller and R. A. Vere Hodge, *JCS* (1969), *C*, 683.
30. J. L. Mateos, M. J. Cerecer and R. Cetina, *Bol Inst Quim Univ Nacl Auton Mex* (1960) *12(2)*, 59.
31. W. J. Forbes, *Canad J Chem* (1962), 1891.
32. I. M. Hunsberger, D. Lednicer, H. S. Gutowsky, D. L. Bunker and P. Taussig, *JACS* (1955), 2466.
33. J. Depireux, *BSC Belge* (1957), 218.
24. B. Eistert, W. Schade and H. Selzer, *B* (1964), 1470.
35. L. A. Cohen, *JOC* (1957), 1333.
36. F. González-Sánchez, *SCA* (1958), *12*, 17.
37. I. M. Hunsberger, *JAC* (1950), 5626.
38. W. Herz and J. Brasch, *JOC* (1958), 1513.
39. M. K. A. Khan, K. J. Morgan and D. P. Morrey, *T* (1966), 2095.
40. W. Suetaka, *Gazz Chim Ital* (1956), 783.
41. J. S. Scarpa, M. Ribi and C. H. Eugster, *Helv* (1966), 858.
42. A. R. Katritzky, A. M. Monro, J. A. T. Beard, D. P. Dearnaley and N. J. Earl, *JCS* (1958), 2182.
43. J. T. Kurek and G. Vogel, *J Het Chem* (1968), 275.
44. D. J. Cosgrove, D. G. H. Daniels, J. K. Whitehead and J. D. S. Goulden, *JCS* (1952), 4821.
45. F. Korte and K. H. Büchel, *B* (1960), 1025.
46. B. D. Akehurst and J. R. Bartels-Kieth, *JCS* (1957), 4798.

4.2 ketones

The carbonyl groups of ketones are characterized by a strong absorption from the C-O double-bond stretching vibration. The carbonyl stretching frequency is the only characteristic absorption in the infrared spectra of ketones and is influenced, in an additive manner, by both electronic and steric effects exerted by neighboring groups.

The force constant of the double bond is a function of the hybridization of the carbon atom of the carbonyl group, which is in turn related to the C-C(O)-C bond angle. Thus although acetone absorbs at 1719 cm^{-1} in CCl_4, di-tert-butyl ketone absorbs at 1689 cm^{-1} in the same solvent. Here the lowering in the frequency arises from the larger C-C(O)-C bond angle, which results from the steric interaction of the two bulky groups.

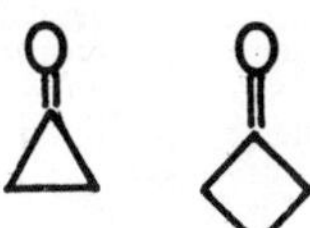

Similar changes in frequency are observed with cyclic ketones. Cyclopropanone absorbs at 1813 cm^{-1}, cyclobutanone at 1788 cm^{-1}, cyclopentanone at 1751 cm^{-1}, and cyclohexanone at 1718 cm^{-1}. Coulson[a] argues that as the

[a]C. A. Coulson and W. Moffitt, *Phil. Mag.* (1949) *40,* 1.

$\overset{O}{\text{C-C-C}}$ bond angle decreases, the carbonyl carbon has greater *p*-orbital character in the orbitals of the ring with a consequent increase in the s-orbital character of the C-O σ-bond. This in turn increases the force constant of the carbonyl group and causes an increase in the frequency of the carbonyl stretching vibration.

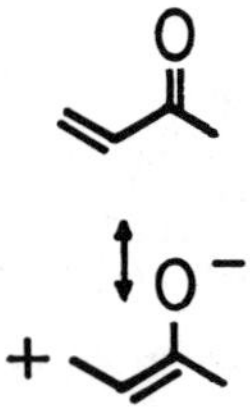

I

II

A decrease in the carbonyl stretching frequency results from the conjugation of a carbonyl group with unsaturated chromophores. Thus conjugation to an olefinic double bond causes a decrease in the carbonyl stretching frequency of about 35 cm^{-1}. Contributions from the resonance forms of the type II lead to a decrease in the force constant of the carbonyl group (more single-bond character) and a decrease in the frequency. Further γ-δ conjugation has only a small effect on any further lowering of the frequency, whereas conjugation on both sides of the carbonyl group tends to be additive. Thus methyl vinyl ketone absorbs at 1684 cm^{-1}, hexa-3,5-diene-2-one at 1681 cm^{-1}, and divinyl ketone at 1672 cm^{-1}. When, however, the contribution from the dipolar resonance form is increased, as is the case, for example, with vinylogous amides III $\longleftrightarrow$ IV (see Chapter 4.4, p. 263) an additional decrease in frequency is observed. Thus 4-amino-but-3-ene-2-one absorbs at 1664 cm^{-1}. Similar increases and decreases are seen with aromatic ketones when the dipolar form is either stabilized or destabilized.

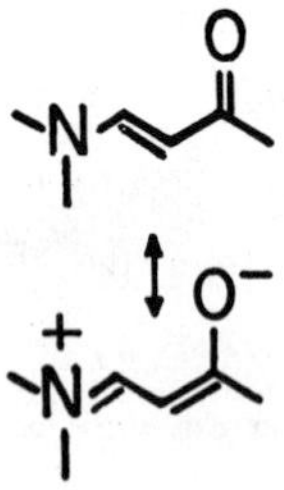

III

IV

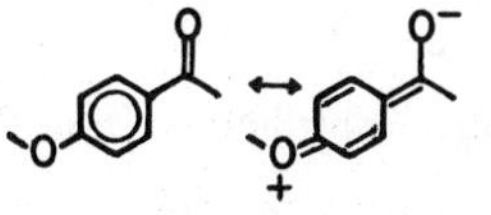

1682 cm^{-1} 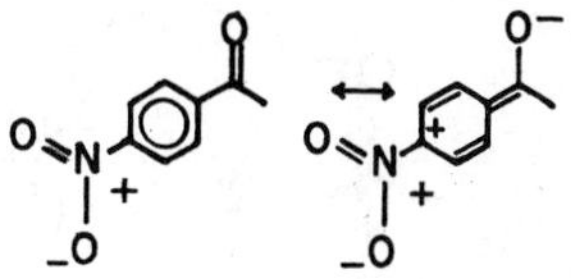 1702 cm^{-1}

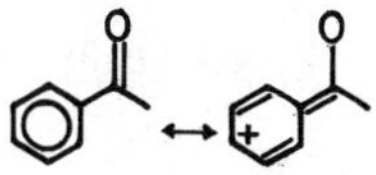

1692 cm^{-1}

Not only does the extent of the conjugated system and its substitution pattern affect the carbonyl stretching frequency, but the orientation of the conjugated systems also has an effect. This is readily seen in the case of *exo* and *endo* α-β unsaturated cyclohexanones.

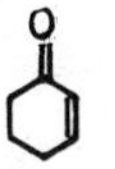

1691 cm^{-1} 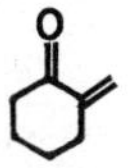 1697 cm^{-1}

But even in acyclic systems S-cis and S-trans conformations have different vibrational frequencies.

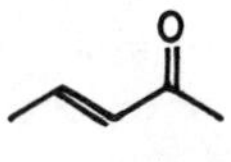

s-cis 1699 cm^{-1} s-cis 1690 cm^{-1}
s-trans 1701 cm^{-1} s-trans 1675 cm^{-1} 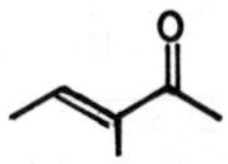
1682 cm^{-1}

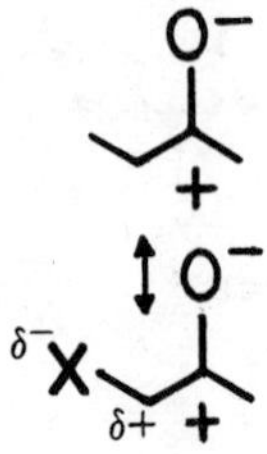

Electronegative groups α-to the carbonyl group destabilize the contribution from the polar resonance forms, that is, V is more stable than VI such that an α-halo grouping can cause an increase in the carbonyl stretching frequency. Thus α-chloroacetone absorbs at 1752 cm^{-1}. However, α-fluoroacetone, which absorbs at 1721 cm^{-1} is essentially unaltered compared to acetone; moreover, although 2,6-dibromocyclohexanone absorbs at 1750 cm^{-1} when both halogen atoms are equatorial, the corresponding diaxial isomer absorbs at 1718 cm^{-1}, the same as cyclohexanone. Clearly, then, the simple inductive effect of the electronegative group cannot, by itself, account for the increases in frequencies that are observed. Observations such as these prompted Bellamy and Williams[b] and Jones and Sandorfy[c] to suggest that the increase in frequency was a spatially dependent effect which operated through space, and Dewar[d] has extended these concepts of the field effect. The importance of the spatial orientation of the field can be seen in α-chloroacetone, which shows two carbonyl absorptions at 1752 cm^{-1} and 1726 cm^{-1}. The higher frequency band arises from the "eclipsed" conformation (VII) in which the dihedral angle between the C-O and the C-Cl bonds is small. This conformation corresponds to an α-halocyclohexanone with an equatorial halogen. The lower carbonyl frequency of α-chloroacetone arises from the

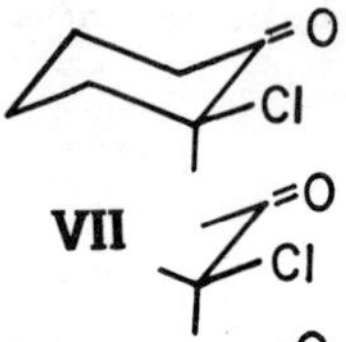

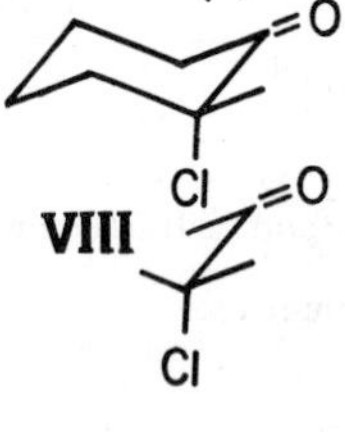

[b] L. J. Bellamy and R. L. Williams, *JCS* (1957) 4294.

[c] R. N. Jones and C. Sandorfy, in *Chemical Applications of Spectroscopy,* ed. W. West, Interscience, New York, 1956.

[d] M. J. S. Dewar, *The Molecular Orbital Theory of Organic Chemistry,* McGraw-Hill, New York, 1969.

staggered conformation (VIII), which corresponds to the axial α-halocyclohexanone, where the dihedral angle between the C-O and carbon-halogen bonds is much larger.

SOLVENTS AND HYDROGEN BONDING

Solvation, including self-association, can cause a lowering of the carbonyl frequency from that observed in the gas phase. In the gas phase acetone absorbs at 1744 cm^{-1}, whereas the neat liquid absorbs at 1715 cm^{-1}. The frequency is raised to 1719 cm^{-1} in CCl_4 and lowered to 1712 cm^{-1} in $CHCl_3$. The lowering of the carbonyl frequency as the polarity of the solvent increases is observed with the majority of ketones, and in general, the frequency in $CHCl_3$ is about 10 cm^{-1} lower than that measured in CCl_4, with the value for the neat ketone falling between the two.

Although relatively small changes in frequency are observed between one solvent and another, much larger changes can be observed when the carbonyl group becomes hydrogen bonded. This effect can be large, especially when an intramolecular bond is formed. For example, acetophenone absorbs at 1692 cm^{-1}, whereas *o*-hydroxyacetophenone absorbs at 1646 and *o*-methoxyacetophenone at 1667 cm^{-1}. An extreme example of this phenomenon is shown by enolizable β-diketones. Thus, although dimethyl acetylacetone shows the expected carbonyl frequency at 1725 cm^{-1}, acetylacetone shows only one broad absorption at 1610 cm^{-1}. This arises from the intramolecularly

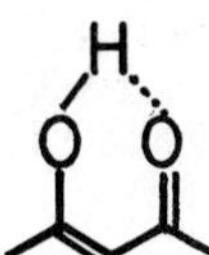

hydrogen-bonded enolic form. It is not, however, possible to assign this absorption to either the hydrogen-bonded carbonyl, or the C-C double bond. Rather the absorption must be considered as that arising from the whole conjugated chromophore.

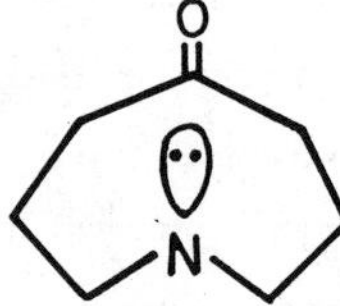

Transannular interactions with heteroatoms can also cause changes in the carbonyl stretching frequency. In six-membered rings such interactions are small, especially when the ring adopts a chair conformation. Thus 4-azacyclohexanone absorbs at 1710 cm^{-1}, 4-oxacyclohexanone at 1715 cm^{-1}, and cyclohexanone at 1718 cm^{-1}. In larger ring systems where the molecule can adopt a conformation in which transannular interactions can occur, the carbonyl group and its vibrational frequency can be perturbed. Thus 5-methyl-5-azacyclooctanone absorbs at 1683 cm^{-1} compared to cyclooctanone, which absorbs at 1704 cm^{-1}. This lowering can be attributed to an interaction between the lone pair on nitrogen and the carbonyl groups. Note, however, that such interactions do not always lower the frequency of the carbonyl groups, since in 5-oxacyclooctanone the frequency is raised compared to that of the parent ketone.

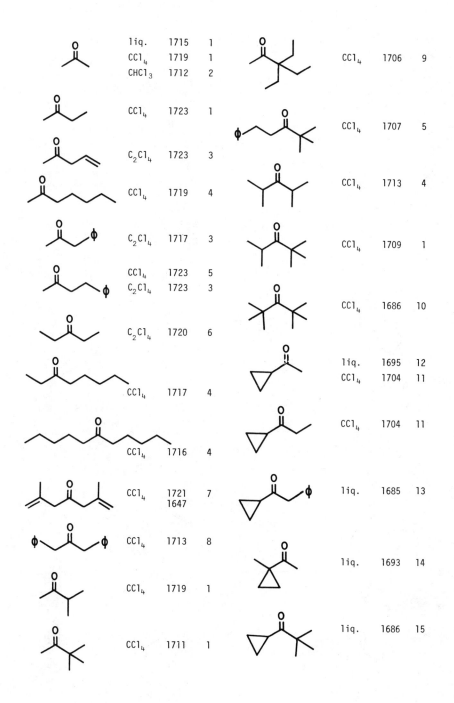

liq.	1715	1
CCl_4	1719	1
$CHCl_3$	1712	2
CCl_4	1723	1
C_2Cl_4	1723	3
CCl_4	1719	4
C_2Cl_4	1717	3
CCl_4	1723	5
C_2Cl_4	1723	3
C_2Cl_4	1720	6
CCl_4	1717	4
CCl_4	1716	4
CCl_4	1721 1647	7
CCl_4	1713	8
CCl_4	1719	1
CCl_4	1711	1

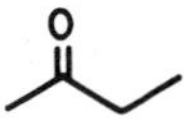

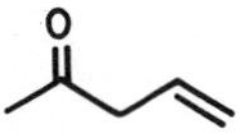

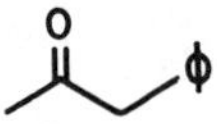

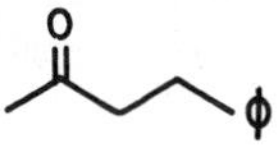

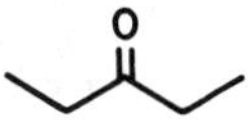

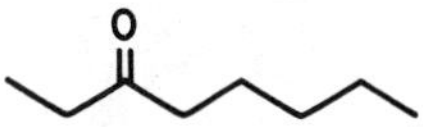

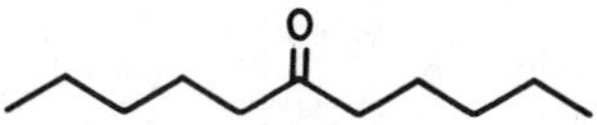

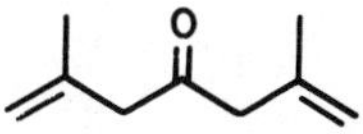

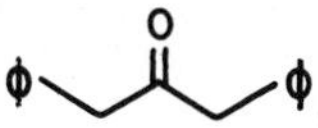

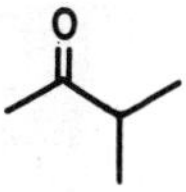

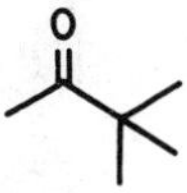

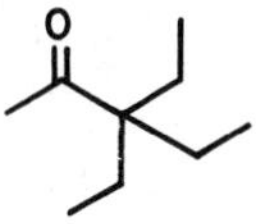

CCl_4	1706	9
CCl_4	1707	5
CCl_4	1713	4
CCl_4	1709	1
CCl_4	1686	10
liq.	1695	12
CCl_4	1704	11
CCl_4	1704	11
liq.	1685	13
liq.	1693	14
liq.	1686	15

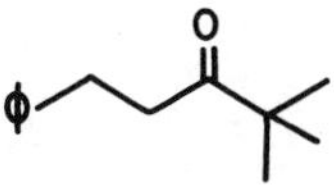

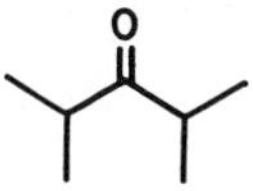

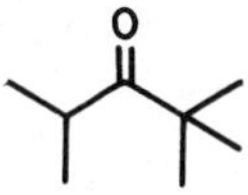

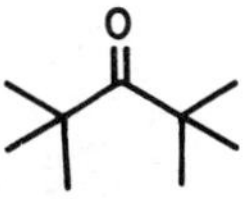

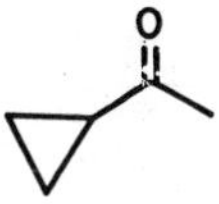

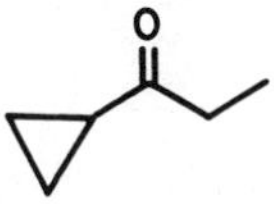

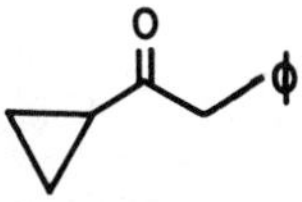

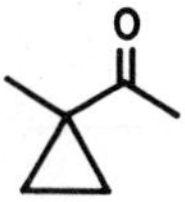

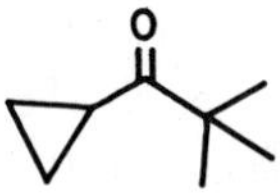

Structure	Phase	Frequency	Ref.
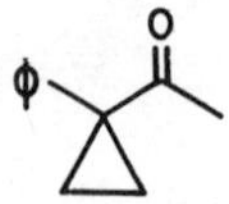 ϕ, O	liq.	1706	14
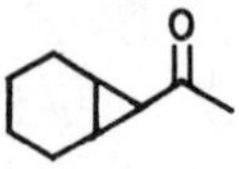 O	liq.	1695	14
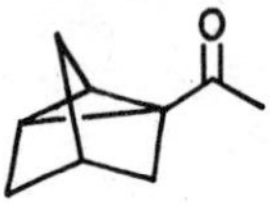 O	liq.	1680	14
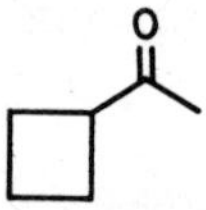 O	liq.	1706	12
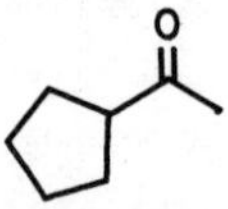 O	liq.	1704	12
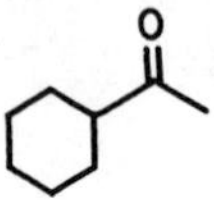 O	liq.	1704	12
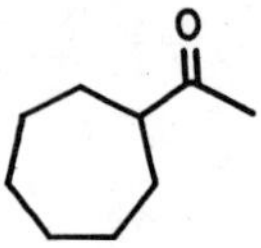 O	liq.	1710	16
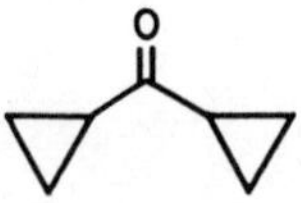 O	CCl_4	1694	17
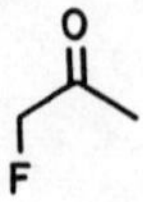 O, F	liq.	1721	18
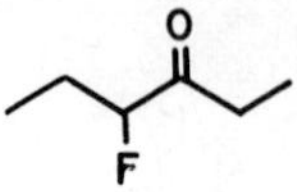 O, F	liq.	1725	19

Structure	Phase	Frequency	Ref.
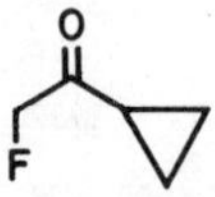 O, F	CCl_4	1740	20
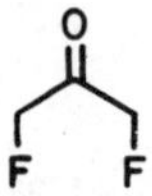 O, F, F	liq.	1750	18
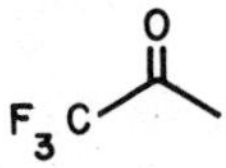 O, F_3C	gas	1780	21
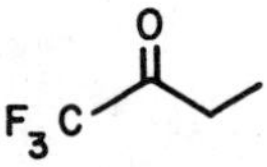 O, F_3C	CCl_4	1764	22
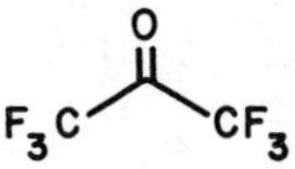 O, F_3C, CF_3	gas	1809	23
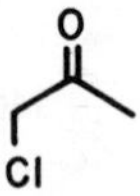 O, Cl	CCl_4	1752 1726	1
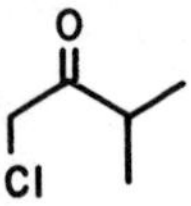 O, Cl	CCl_4	1740 1720	1
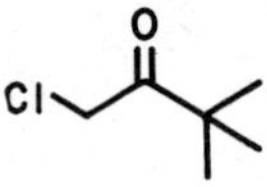 Cl, O	CCl_4	1732 1713	1
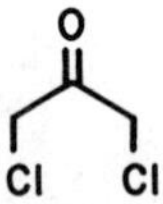 O, Cl, Cl	CCl_4	1746 1730	1
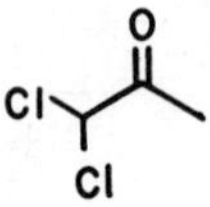 Cl, O, Cl	CCl_4	1743 1724	1
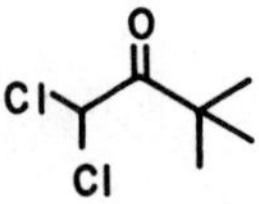 Cl, O, Cl	CCl_4	1745 1734	1

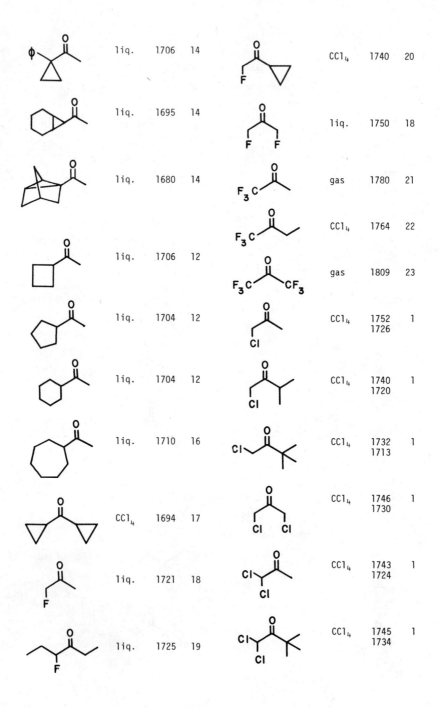

Solvent	Frequency	Ref.
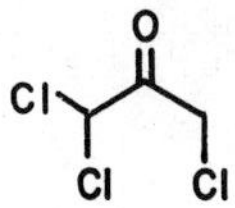 CCl_4	1746 1730	1
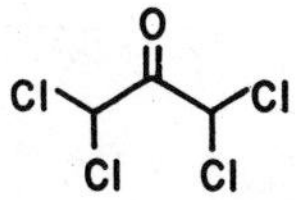 CCl_4	1774 1764	1
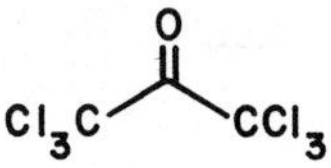 CCl_4	1780 1751	1
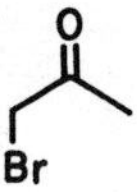 liq.	1739 1733 1728 1721 1710	24
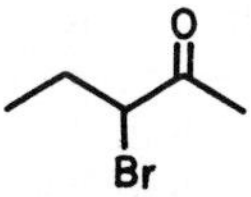 liq.	1725	24
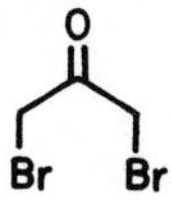 liq.	1756 1740 1725 1710	24
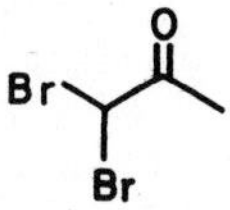 liq.	1741 1733 1708	24
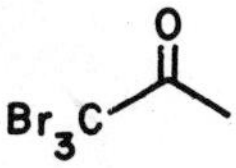 liq.	1748	24
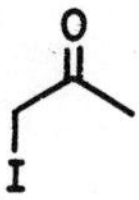 liq.	1724 1704	25
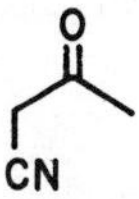 liq.	1739 2262	26
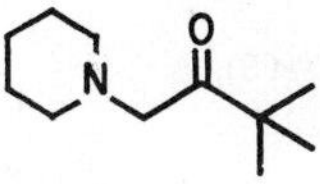 liq.	1713	27
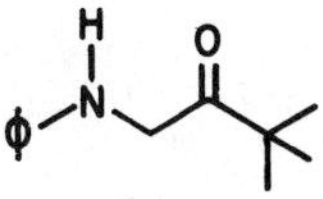 nuj	1705	27
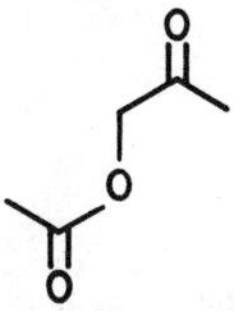 $CHCl_3$	1733 1747	2
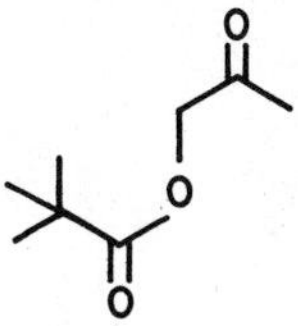 $CHCl_3$	1723 1736	2
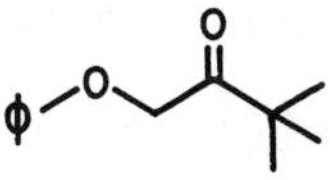 liq.	1721	27
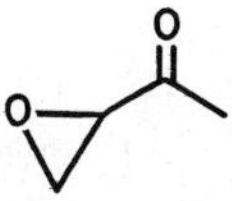 liq.	1718	28
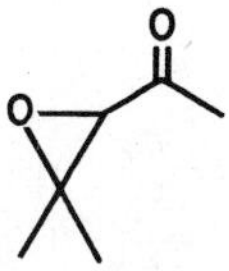 liq.	1718	28
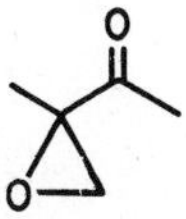 liq.	1721	28
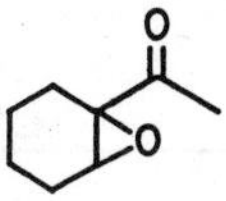 CCl_4	1714	29
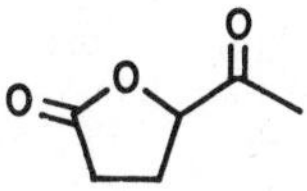 CCl_4	1725 1795	30

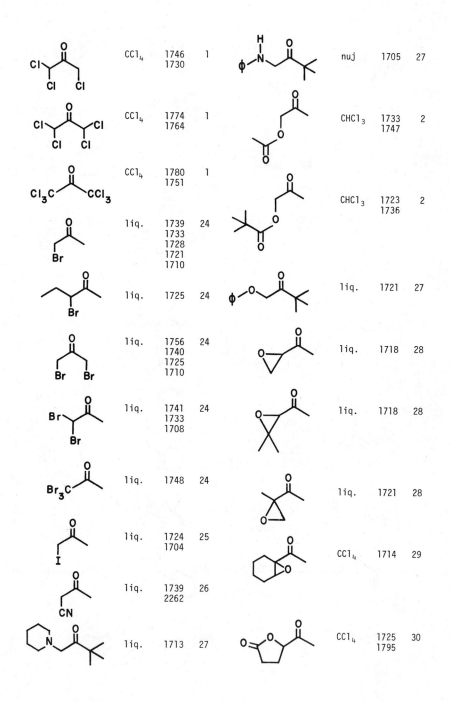

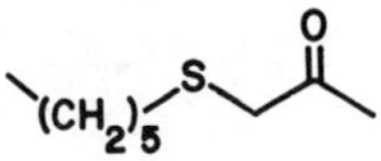	liq.	1706	27
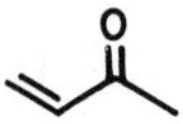	CCl_4	1684	11
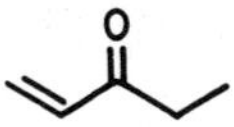	C_2Cl_4	1707 1690	3
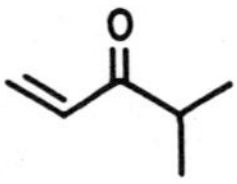	$CHCl_3$	1704 1675	11
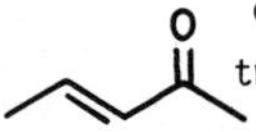 cis	C_2Cl_4	1699	3
trans	C_2Cl_4	1701 1682	3
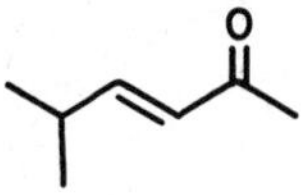	C_2Cl_4	1702 1683	3
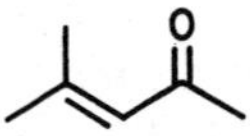	liq.	1694	14
	CCl_4	1693	31
	CS_2	1692	31
	$CHCl_3$	1685	31
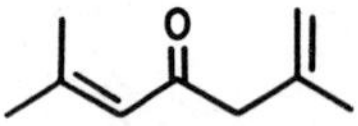	CCl_4	1693 1650* 1626	32
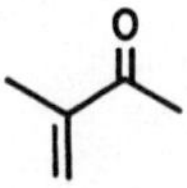	C_2Cl_4	1624	3
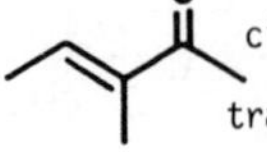 cis	CCl_4	1690	33
trans	CCl_4	1675	33

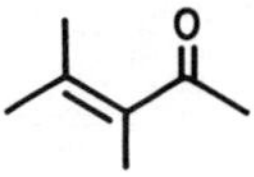 CCl_4	1687	31
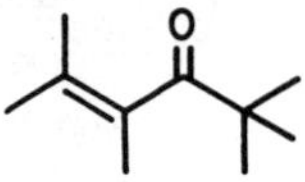 CCl_4	1681	34
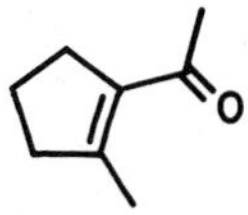 CCl_4	1658 1628	5
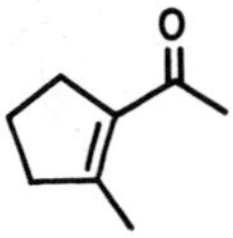 CCl_4	1681 1616	5
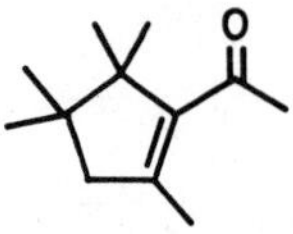 liq.	1667 1619	35
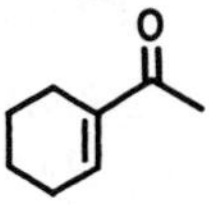 liq.	1685 1655	36
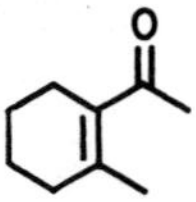 CCl_4	1685 1613	5
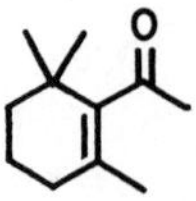 CCl_4	1693 1650	28
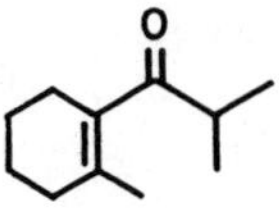 CCl_4	1688 1638	37
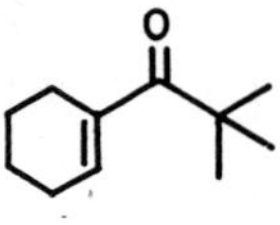 CCl_4	1694	38

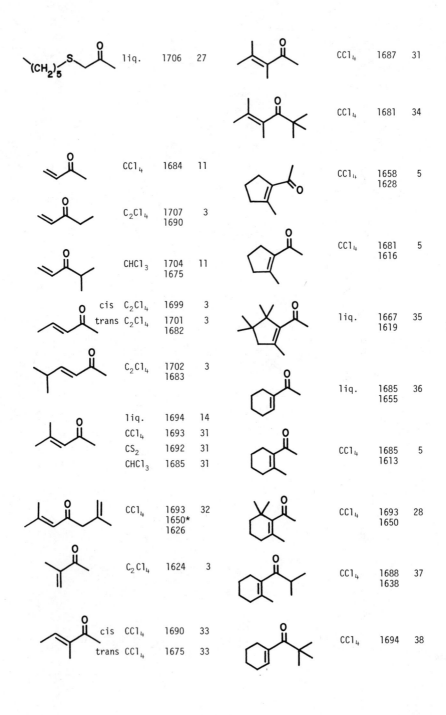

Solvent	Frequency	Ref.
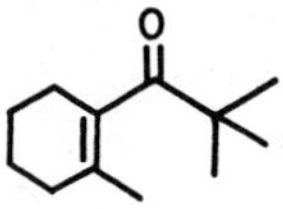 CCl_4	1684	38
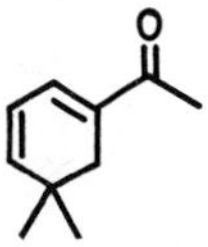 liq.	1660 1600	39
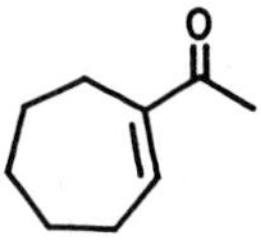 liq.	1670 1640	16
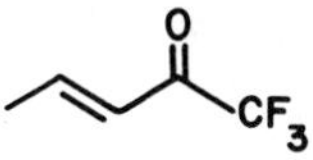 liq.	1773	22
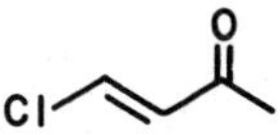 liq.	1691* 1677 1585	3
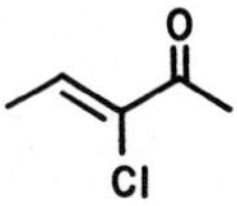 CCl_4	1681 1613	33
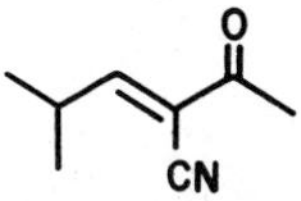 CCl_4	1712 2212 1608	4
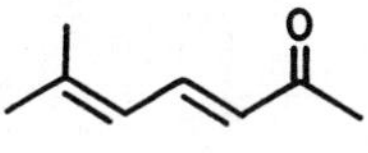 liq.	1675 1633 1591	40
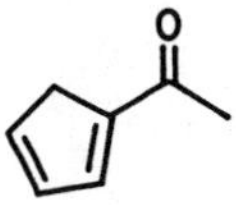 CCl_4	1701 1671	41
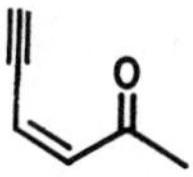 CS_2	1671	42
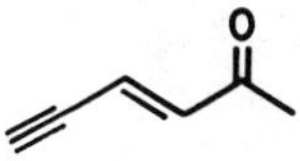 CS_2	1697 1680	42

Solvent	Frequency	Ref.
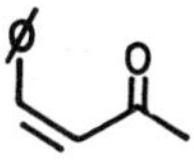 $CHCl_3$	1688	43
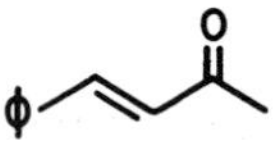 CCl_4	1697 1612	5
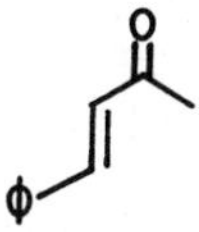 CCl_4	1674 1628	5
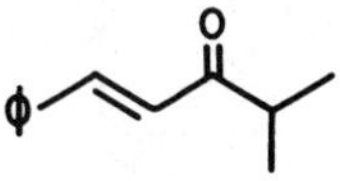 CCl_4	1694 1616	5
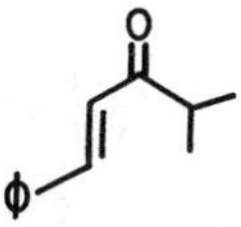 CCl_4	1671 1628	5
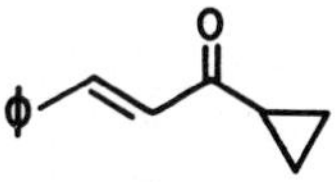 CCl_4	1685 1662	4
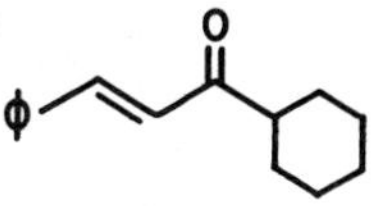 CCl_4	1690 1663	4
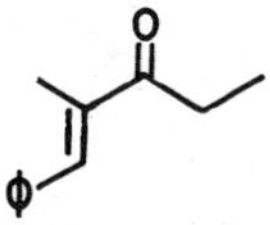 CCl_4	1676 1628	5
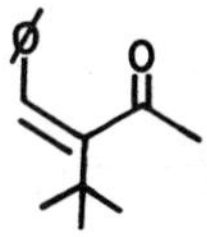 CCl_4	1687 1612	5
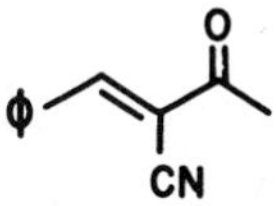 CCl_4	1712 2217 1592 1570	1

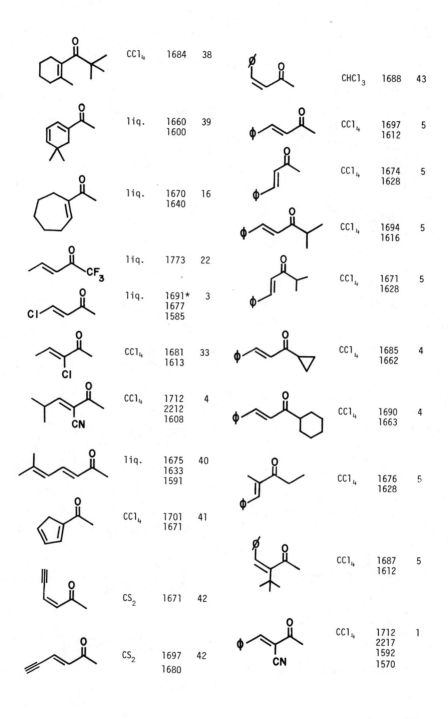

Solvent	Frequencies	Ref.
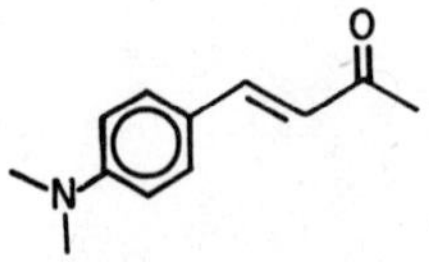 $CHCl_3$	1657	44
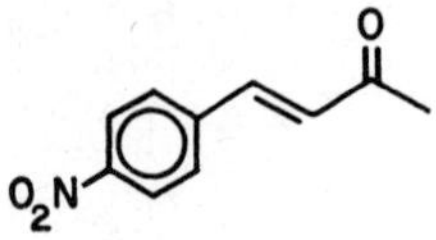 CCl_4	1692	45
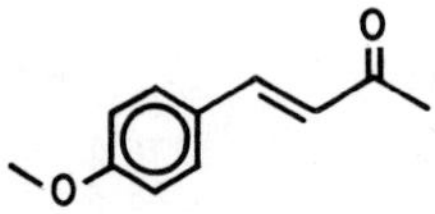 CCl_4	1693 1605	5
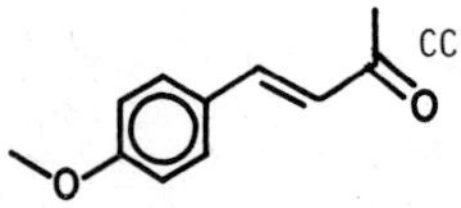 CCl_4	1670 1626	5
 CCl_4	1678 2222	46
 liq.	1667 2212	47
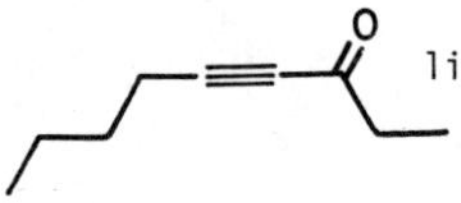 liq.	1667 2222	47
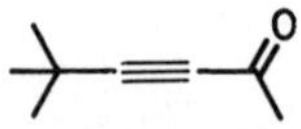 liq.	1670 2220 2180	48
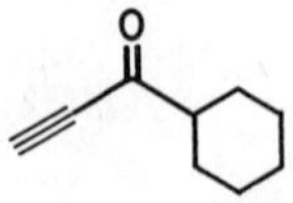 liq.	1665 2085	49
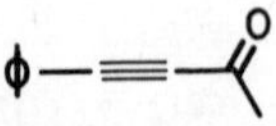 CCl_4	1675 2203	46
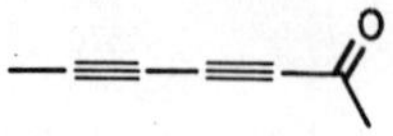 KBr	1690 2250 2160m	50
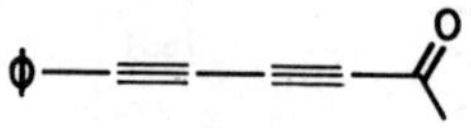 KBr	1670 2220 2150m	50
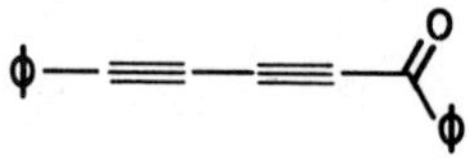 KBr	1635 2210 2140w	50
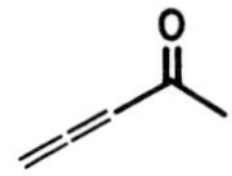 liq.	1680 1951 1931	51
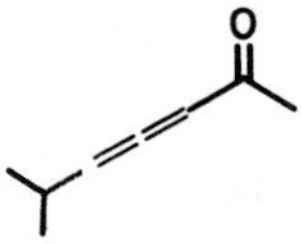 $CHCl_3$	1676 1957	52
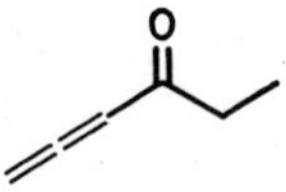 liq.	1685 1955 1931	51
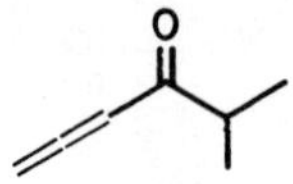 liq.	1666 1960 1927	51
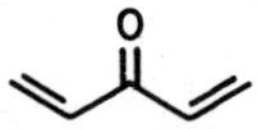 liq.	1672 1660	53
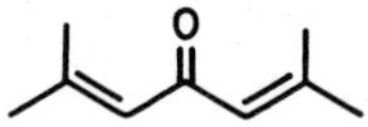 CCl_4	1678 1636 1620	32
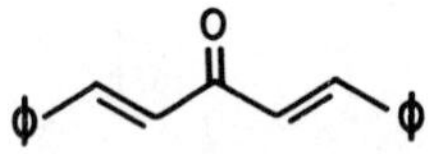 KBr	1660	54
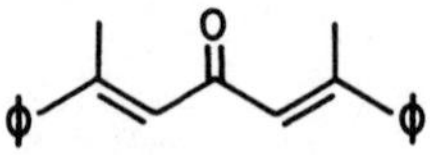 $CHCl_3$	1626	55
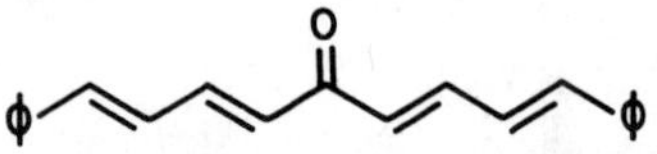 KBr	1610	54

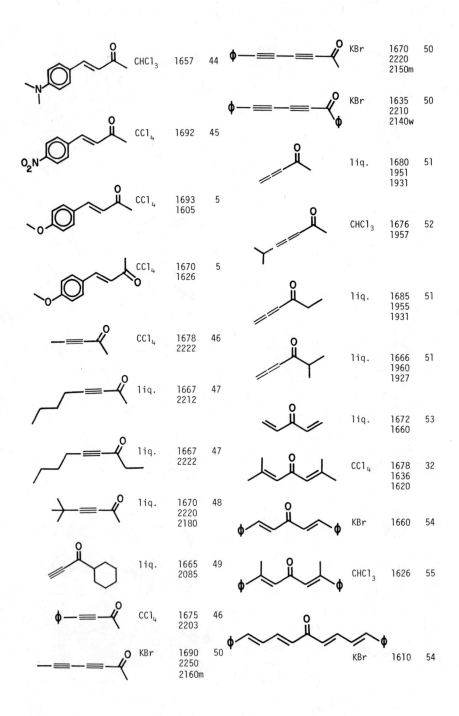

Solvent	Frequency	Ref.
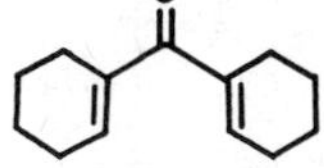 CCl_4	1639	56
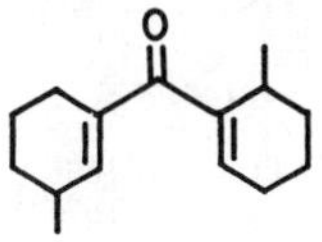 CCl_4	1642 1631	57
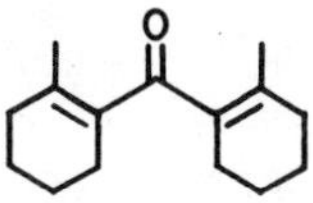 CCl_4	1661m 1626	57
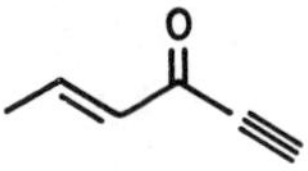 liq.	1650 2102	49
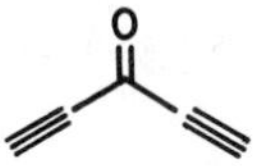 liq.	1635 2100	58
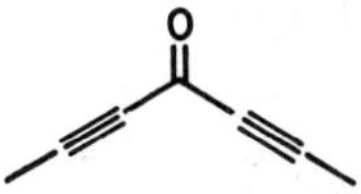 CCl_4	1639 2222	46
KBr	1625 2245 2220	50
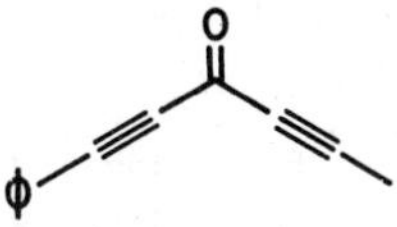 KBr	1625 2240 2210	50
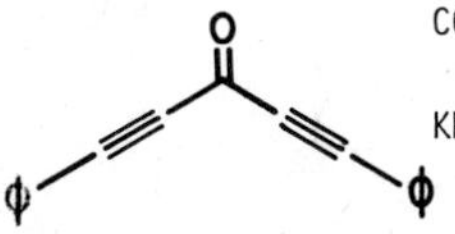 CCl_4	1623 2217	46
KBr	1610 2205 2106	50
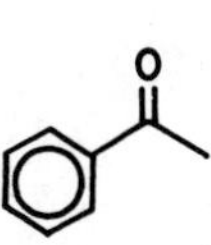 CCl_4	1692	1
liq.	1687	59

Solvent	Frequency	Ref.
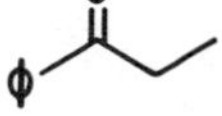 CCl_4	1692	4
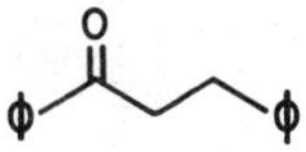 CCl_4	1692	5
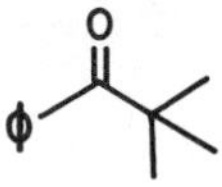 liq.	1680	60
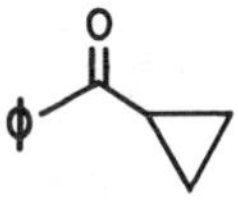 CCl_4	1677	4
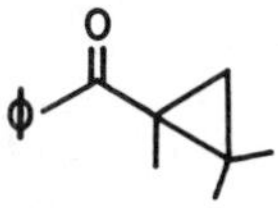 CCl_4	1696	11
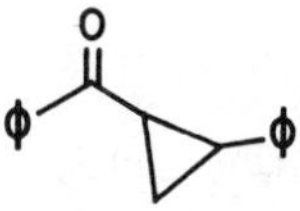 CCl_4	1669	61
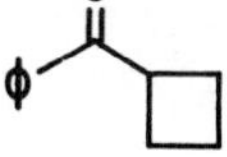 CCl_4	1686	4
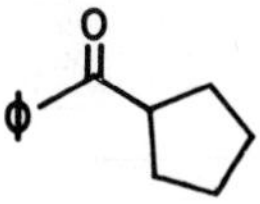 CCl_4	1687	4
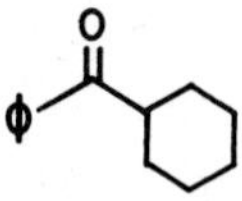 CCl_4	1686	4

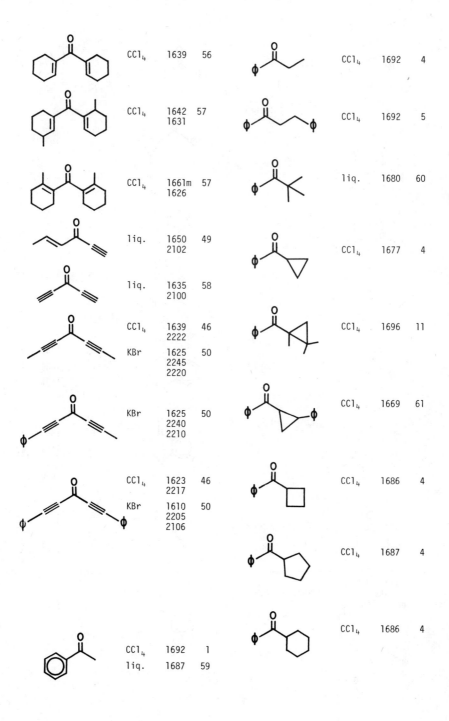

Solvent	ν	Ref.
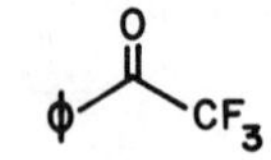 CCl_4	1721	22
liq.	1724	22
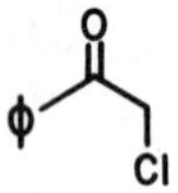 CCl_4	1715 1696	1
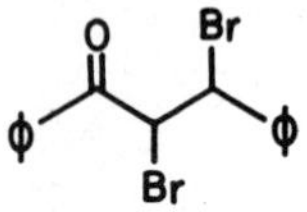 KBr	1680	62
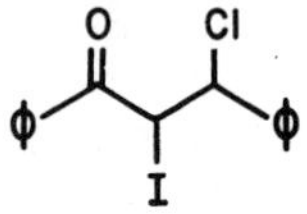 KBr	1671	62
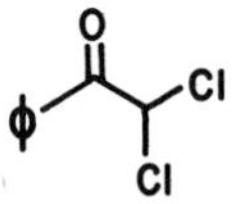 CCl_4	1716 1692	1
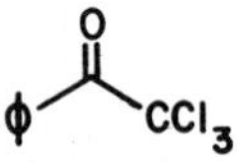 liq.	1715	63
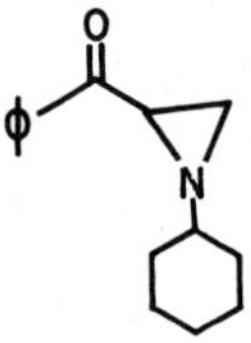 CCl_4	1695 1673	64
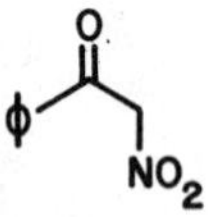 KBr	1695 1605 1560 1330	65
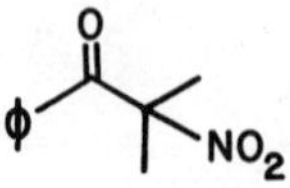 $CHCl_3$	1689	66
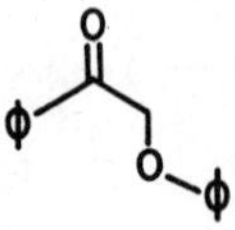 nuj.	1694	67
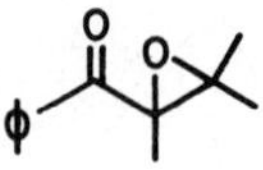 CCl_4	1679	29
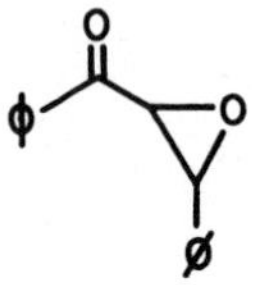 CCl_4	1696 1677*	68
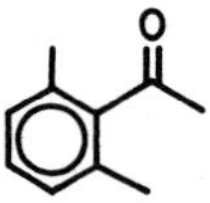 C_2Cl_4	1705	3
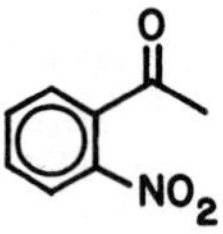 $CHCl_3$	1704	69
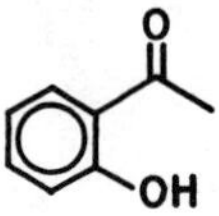 CCl_4	1646	70
$CHCl_3$	1634	71
liq.	1635	72
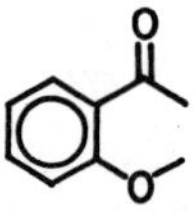 $CHCl_3$	1667	71
liq.	1649	72
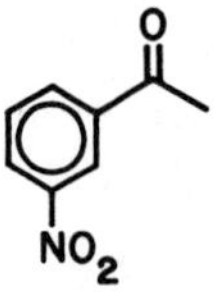 CCl_4	1701	4
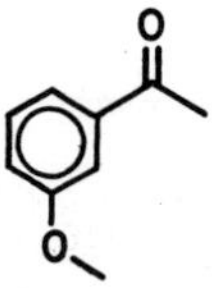 liq.	1681	72

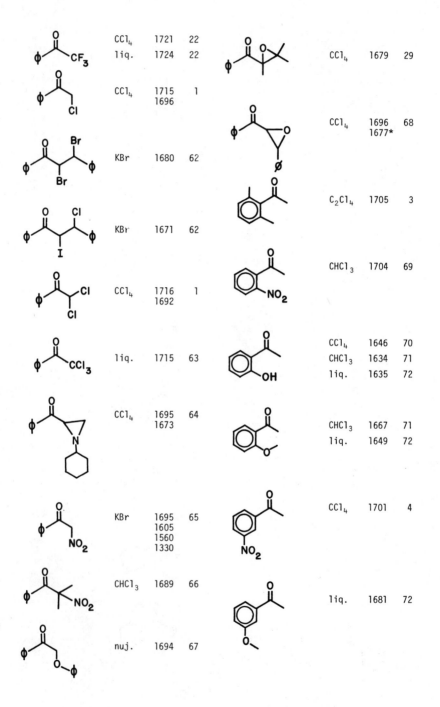

Solvent	Frequency	Ref.
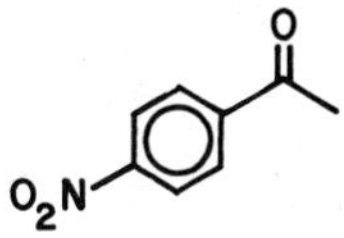 CCl_4	1702	4
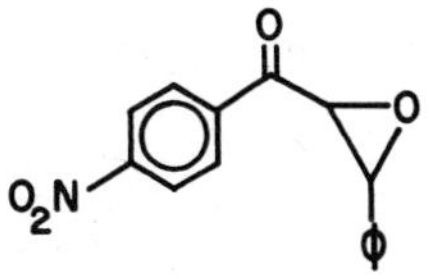 $CHCl_3$	1698	68
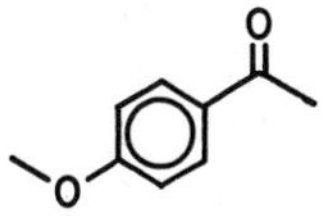 CCl_4	1684	4
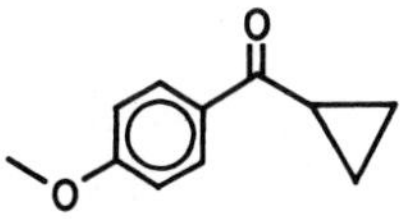 CCl_4	1663	14
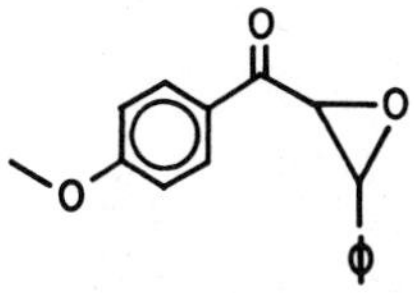 CCl_4	1691 1673*	68
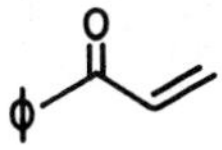 CCl_4	1678	61
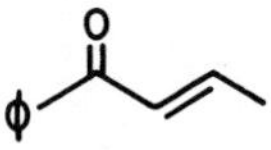 CCl_4	1673	45
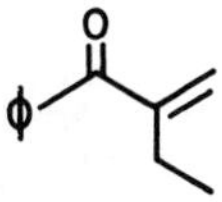 $CHCl_3$	1645 1625*	73
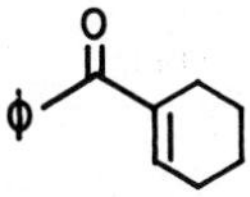 CCl_4	1655	56

Solvent	Frequency	Ref.
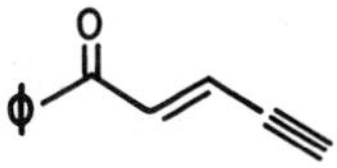 CCl_4	1690 2240 1620	74
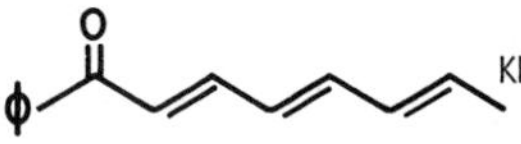 KBr	1650 1595 1565	75
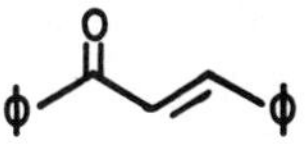 CCl_4	1670 1610	5
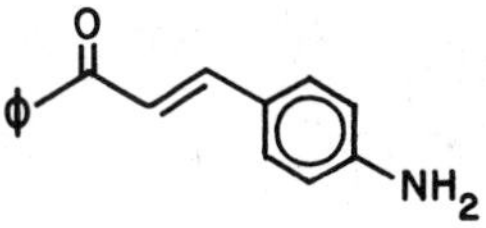 CCl_4	1642	76
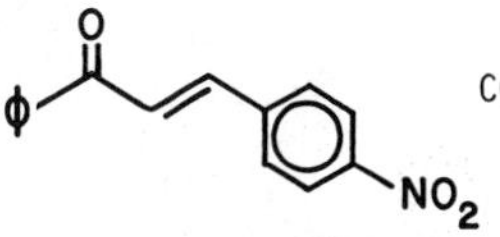 CCl_4	1650	77
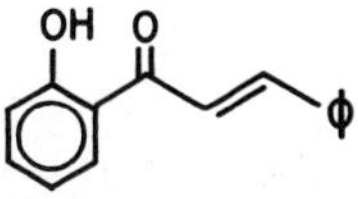 $CHCl_3$	1637	71
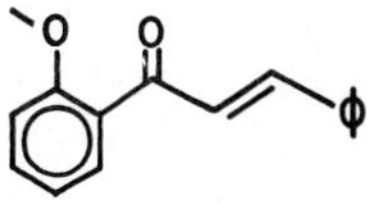 $CHCl_3$	1650	71
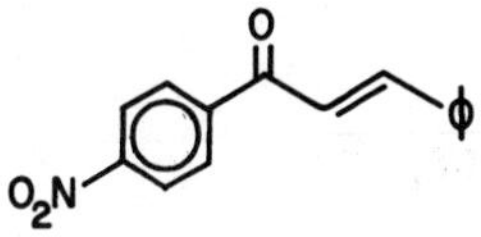 KBr	1660	68
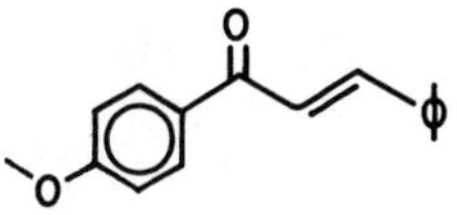 $CHCl_3$	1656	71

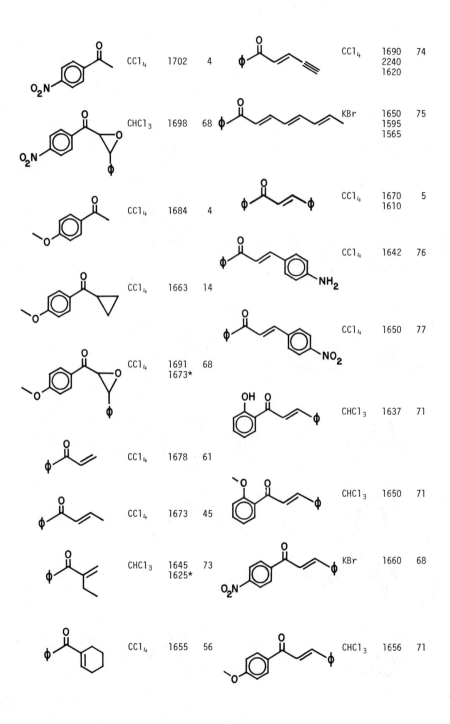

Solvent	ν	Ref.
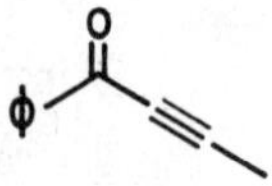 CCl_4	1645 2212	46
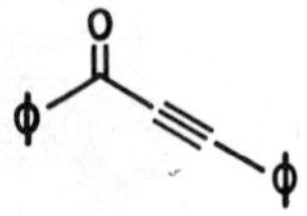 CCl_4	1634 2203	46
CCl_4	1666	6
$CHCl_3$	1651	6
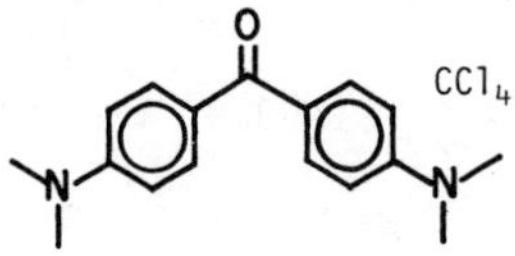 CCl_4	1639	4
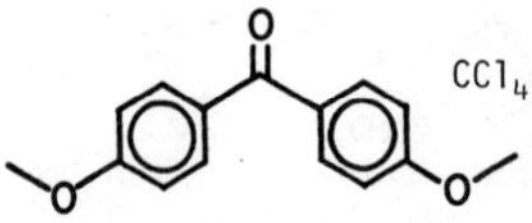 CCl_4	1648	78
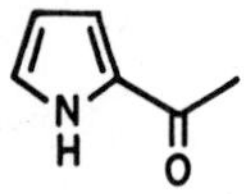 $CHCl_3$	1642	79
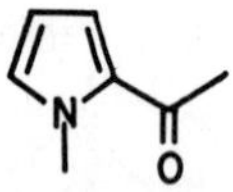 liq.	1660	80
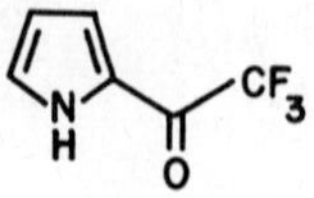 CCl_4	1667	81
CCl_4	1670	82
KBr	1645	82
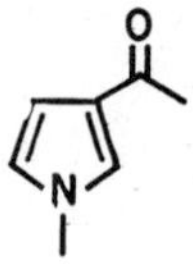 liq.	1660	80
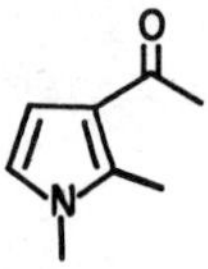 KBr	1660	82
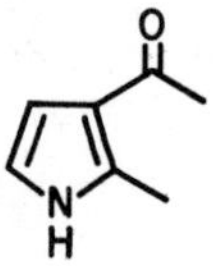 $CHCl_3$	1637	79
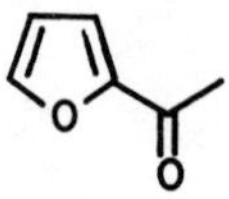 $CHCl_3$	1681 1570	83
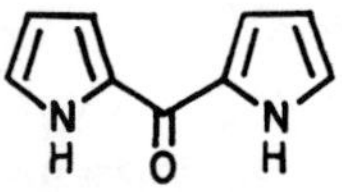 $CHCl_3$	1597 1575 1500	84
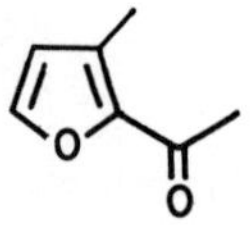 liq.	1670	85
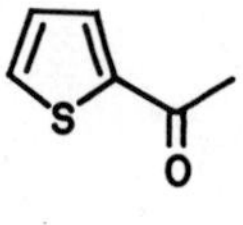 liq.	1667	86

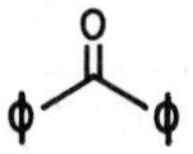

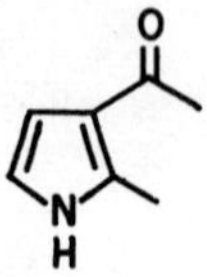

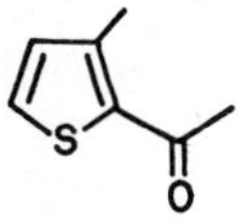	liq.	1664	87
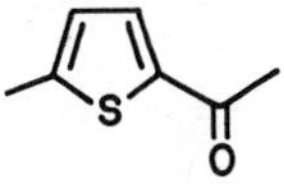	liq.	1655	88
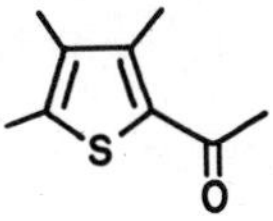	liq.	1656	89
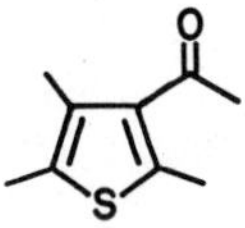	liq.	1664	90
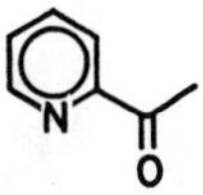	$CHCl_3$	1695	91
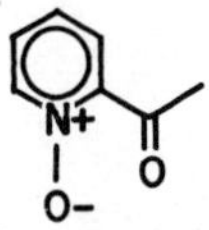	$CHCl_3$	1691	91
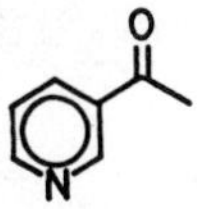	$CHCl_3$	1690	91
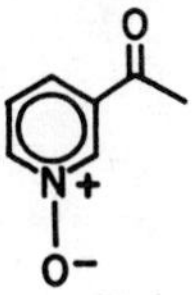	$CHCl_3$	1707	91
	CHCl	1700	91
	$CHCl_3$	1692	91
	CH_2Cl_2	1813	92
	CH_2Cl_2	1850 1822	92
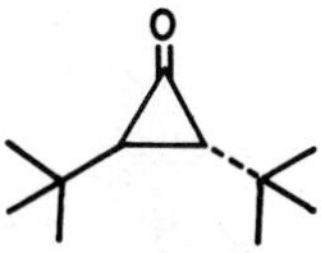	CCl_4	1822	93
	CCl_4	1879 1851 1654	94
	$CHCl_3$	1815	95
	CH_2Cl_2	1843 1823	92

Solvent	ν (cm⁻¹)	Ref.
 CCl_4	1869 1851 1656	94
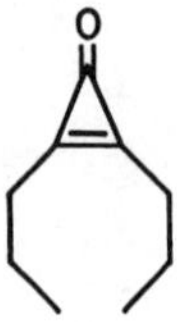 CCl_4	1840 1630	96
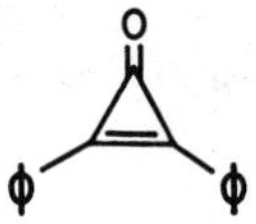 CCl_4	1861 1845* 1673	94
 CCl_4	1788	97
CCl_4	1788	97
liq.	1780	97
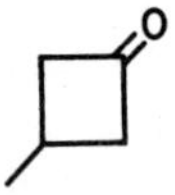 CCl_4	1789	97
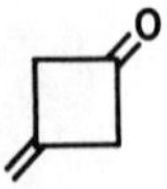 CCl_4	1812 1684 893	98
 CCl_4	1782	97
 CCl_4	1793	97
 CCl_4	1798	97
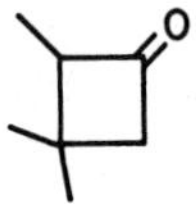 CCl_4	1780	97
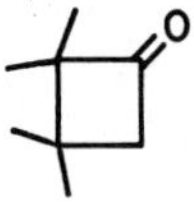 CCl_4	1790	92
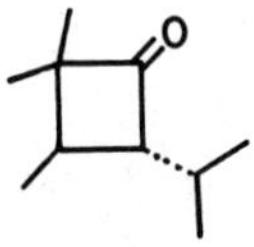 CCl_4	1770	99
 liq.	1776	100
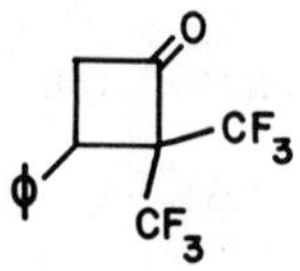 liq.	1824	101
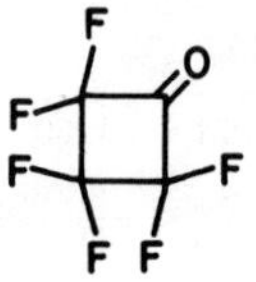 liq.	1852	102

ϕ ϕ

CF_3 CF_3 ϕ

F F F F F F

Structure	Solvent	ν (cm^{-1})	Ref.
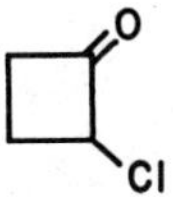	CCl_4	1805	103
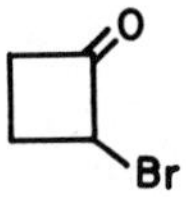	CCl_4	1804	97
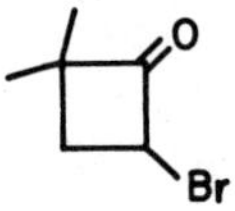	CCl_4	1797	97
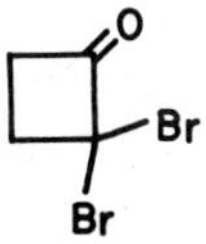	CCl_4	1810	97
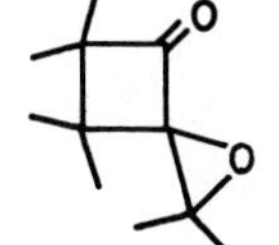	CCl_4	1776	104
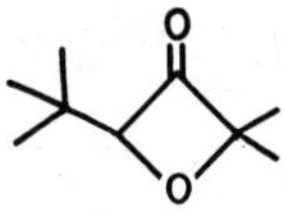	liq.	1818	105
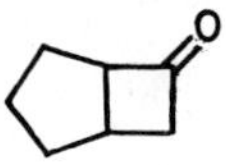	CCl_4	1730 1620	106
	CCl_4	1779 1770	107
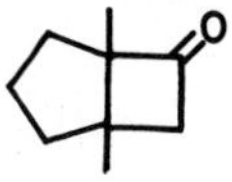	CCl_4	1773 1765	107

Structure	Solvent	ν (cm^{-1})	Ref.
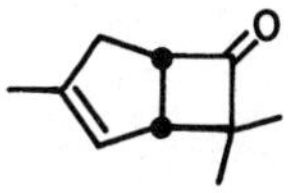	CCl_4	1778 1642w	101
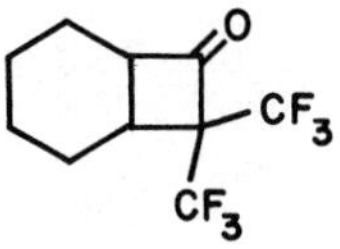	liq.	1804	108
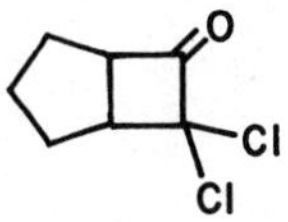	liq.	1800	109
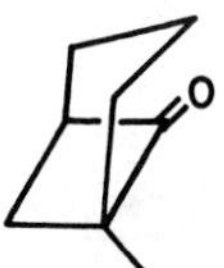	CCl_4	1772 1769	107
	CCl_4	1763	107
	CCl_4	1769	110
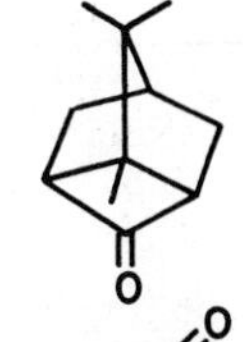	CCl_4	1770	111
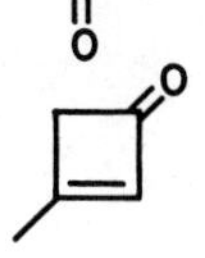	CCl_4	1770 1590	98

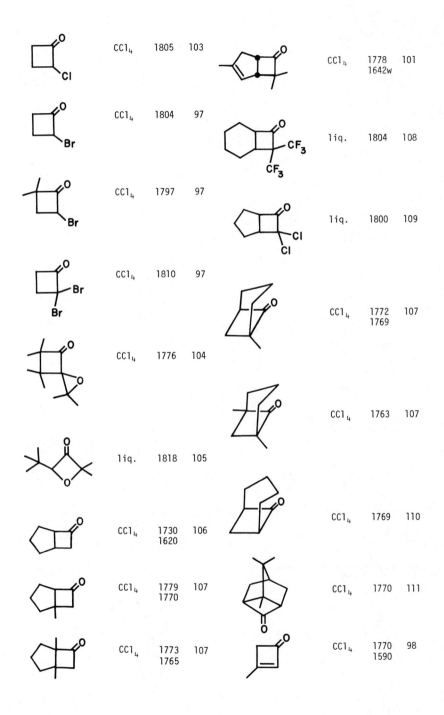

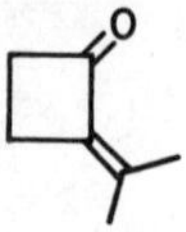

Phase	Frequency	Ref.
liq.	1745	112
CCl_4	1741 1667	99
liq.	1739 1665	99
KBr	1735 1645	112
liq.	1800 1580	113
liq.	1800 1540	113
$CHCl_3$	1780 1610 1600 1590	114
CCl_4	1785 1620	114
KBr	1725 1675 1650	112
CCl_4	1751	115
liq.	1744	59
$CHCl_3$	1730	2
liq.	1740	116
CCl_4	1710	117
liq.	1735	116
$CHCl_3$	1725	118
liq.	1702 1586	119

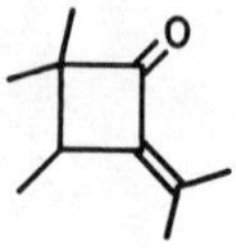

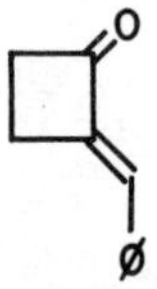

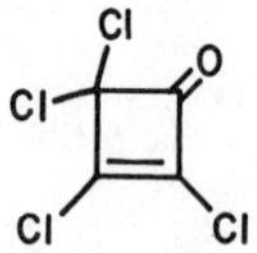

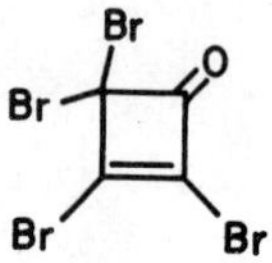

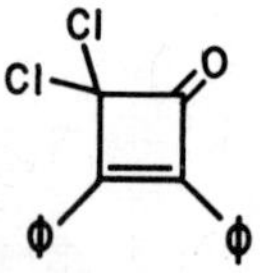

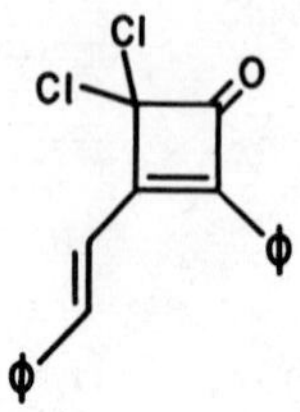

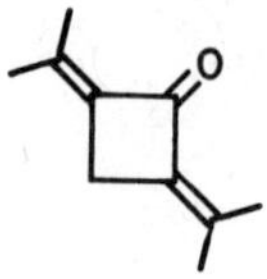

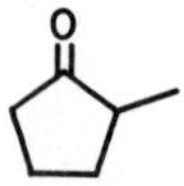

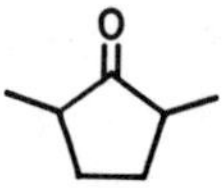

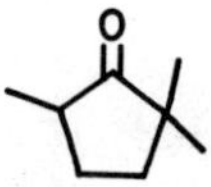

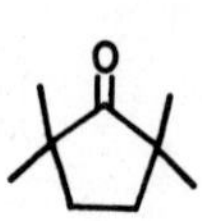

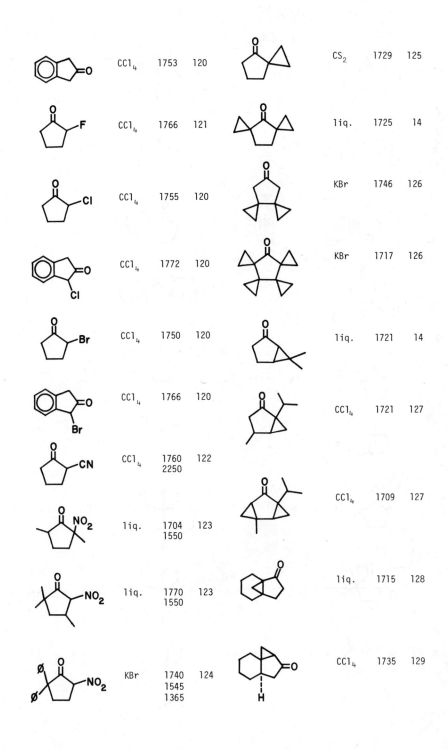

Solvent	ν (cm⁻¹)	Ref.
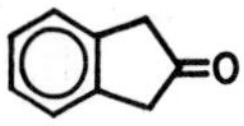 CCl_4	1753	120
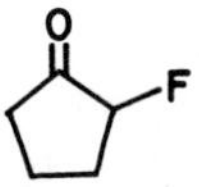 CCl_4	1766	121
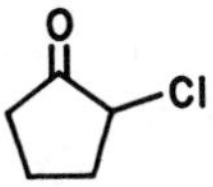 CCl_4	1755	120
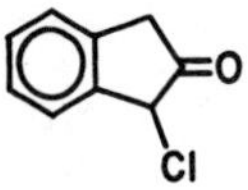 CCl_4	1772	120
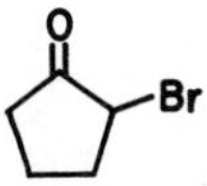 CCl_4	1750	120
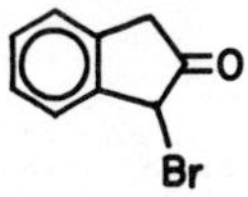 CCl_4	1766	120
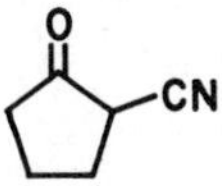 CCl_4	1760 2250	122
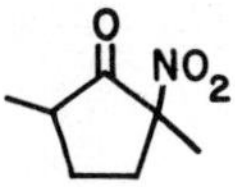 liq.	1704 1550	123
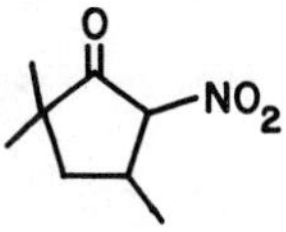 liq.	1770 1550	123
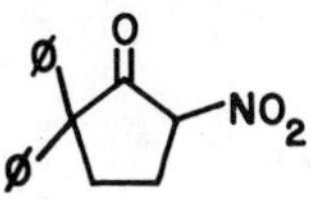 KBr	1740 1545 1365	124

Solvent	ν (cm⁻¹)	Ref.
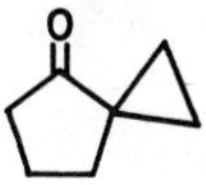 CS_2	1729	125
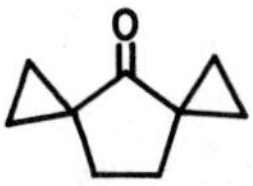 liq.	1725	14
 KBr	1746	126
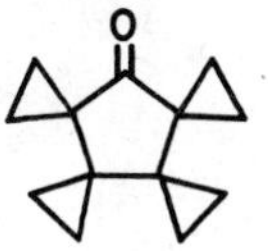 KBr	1717	126
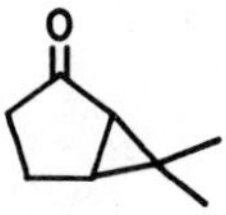 liq.	1721	14
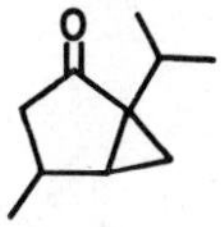 CCl_4	1721	127
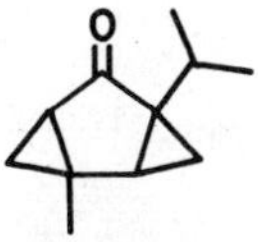 CCl_4	1709	127
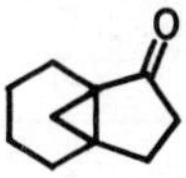 liq.	1715	128
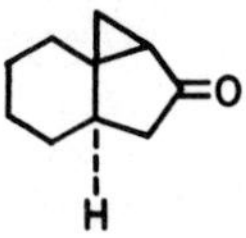 CCl_4	1735	129

Solvent	ν	Ref.
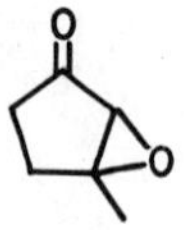 CCl_4	1760	130
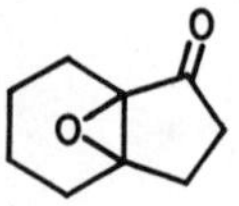 $CHCl_3$	1742	131
 liq.	1763* 1739 1700*	132
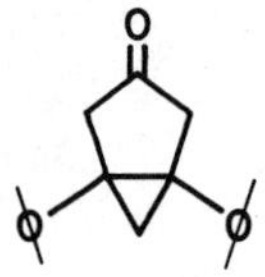 $CHCl_3$	1732	133
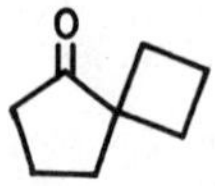 CCl_4	1739	134
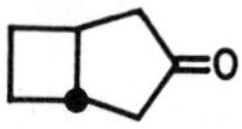 CCl_4	1751	135
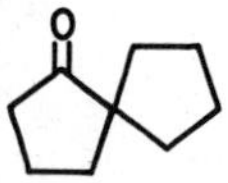 CCl_4	1738	136
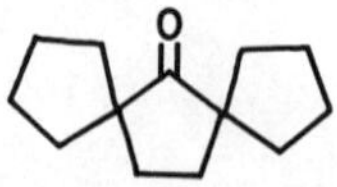 CCl_4	1726	136
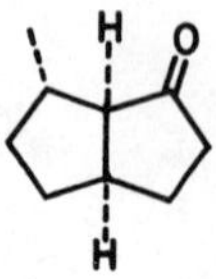 CCl_4	1736	137
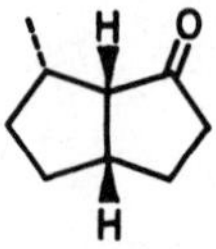 CCl_4	1733	138
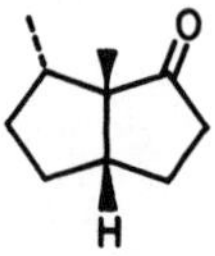 CCl_4	1735	138
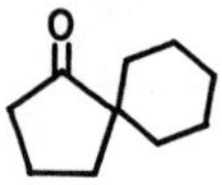 CCl_4	1738	117
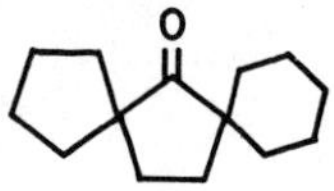 CCl_4	1729	136
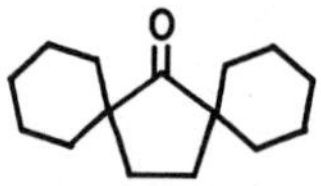 CCl_4	1724	136
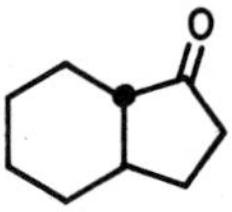 CCl_4	1740	139
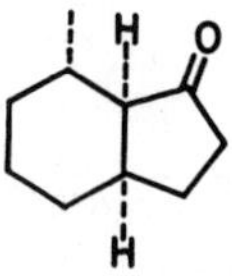 CCl_4	1736	137
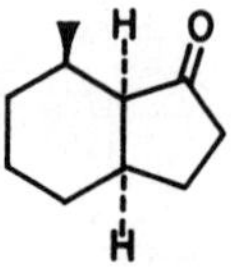 CCl_4	1735	137
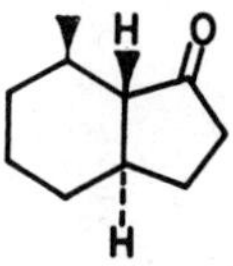 CCl_4	1737	137

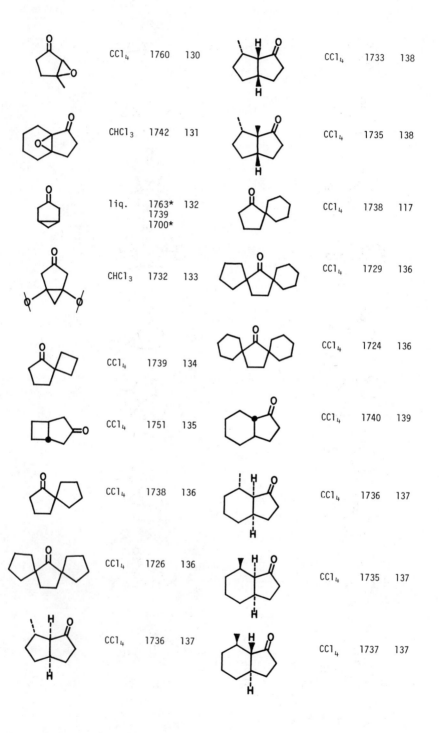

Solvent	Frequency	Ref.
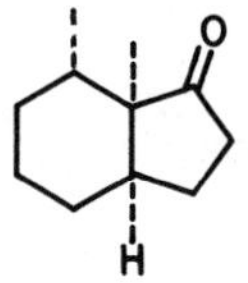 CCl_4	1730	137
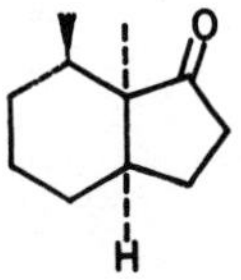 CCl_4	1734	137
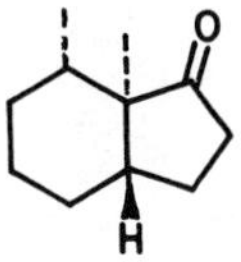 CCl_4	1730	137
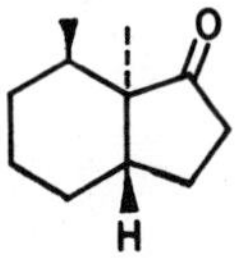 CCl_4	1731	137
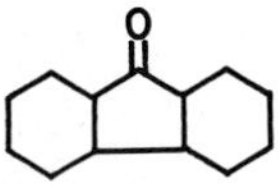 CCl_4	1745	56
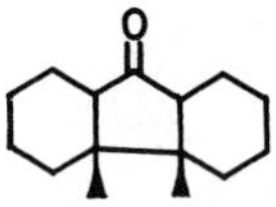 CCl_4	1739	57
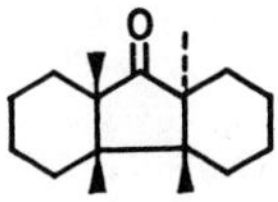 CCl_4	1742	57
 CS_2	1781 1755	140
 liq.	1862m 1792 1779m	141

Solvent	Frequency	Ref.
 liq.	1790	142
 CCl_4	1782	143
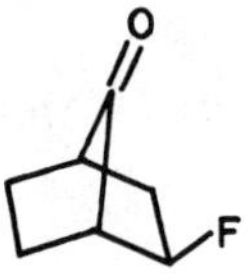 CS_2	1785	144
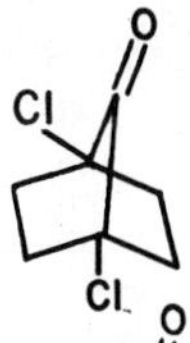 CS_2	1815	145
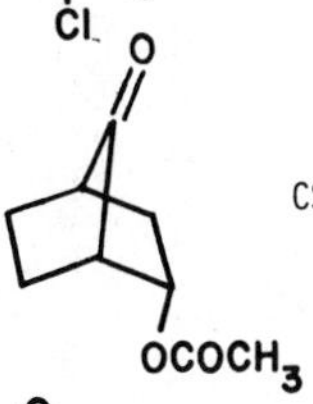 CS_2	1781	144
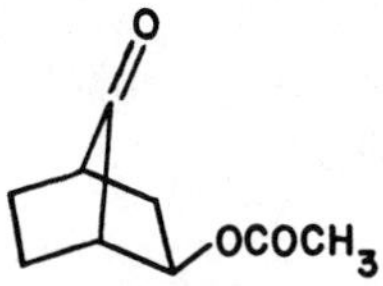 CS_2	1783	144
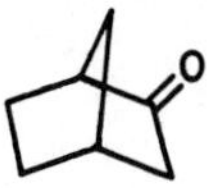 CS_2	1750	144
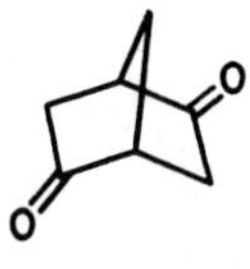 CCl_4	1761 1739	146

F

Cl

Cl

$OCOCH_3$

$OCOCH_3$

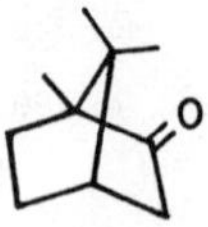 CCl_4	1744	120
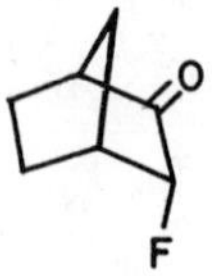 CS_2	1766	144
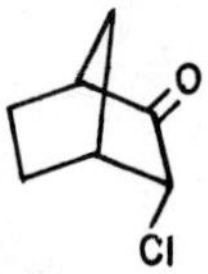 CS_2	1763	144
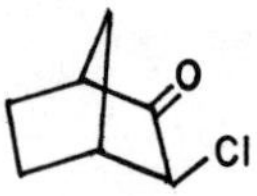 CS_2	1764	144
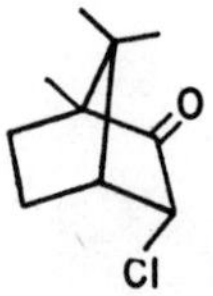 CCl_4	1763	120
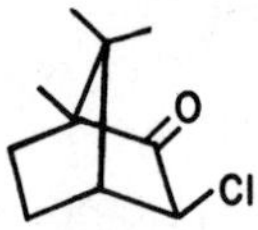 CCl_4	1762	120
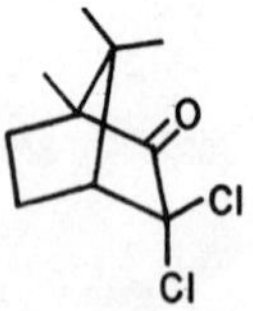 CCl_4	1774	120
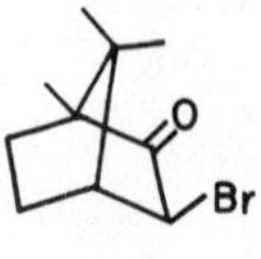 CCl_4	1760	120

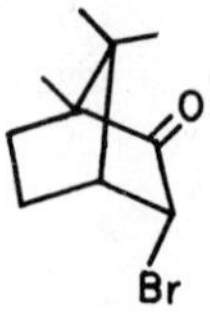 CCl_4	1758	120
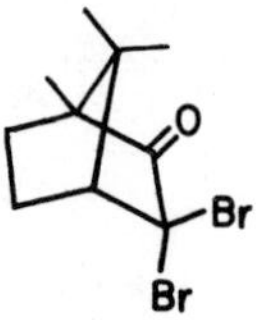 CCl_4	1766	120
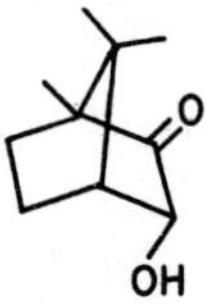 CS_2	1750	144
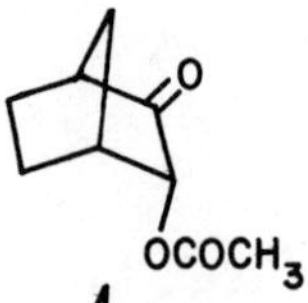 CS_2	1764	144
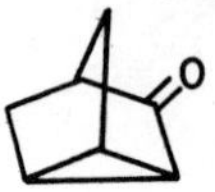 CCl_4	1753	127
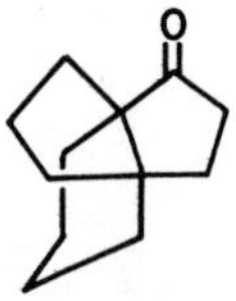 CCl_4	1733	147
 CCl_4	1746	127
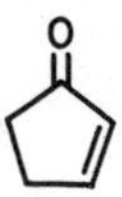 CCl_4	1715	127

Solvent	cm⁻¹	Ref.
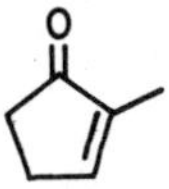 liq.	1704 1645	148
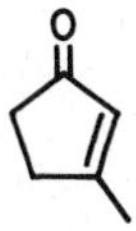 $CHCl_3$	1710	149
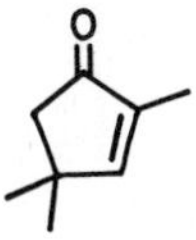 liq.	1706 1645	148
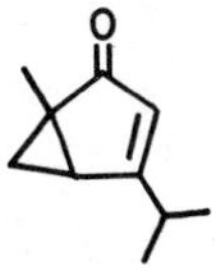 liq.	1695	150
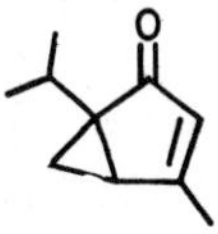 CCl_4	1701	127
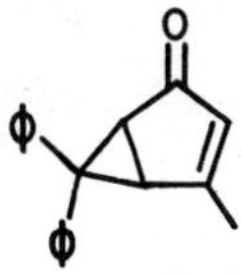 CCl_4	1694	127
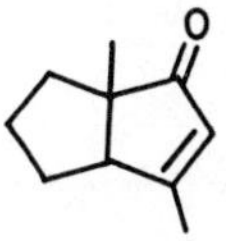 liq.	1709 1626	151
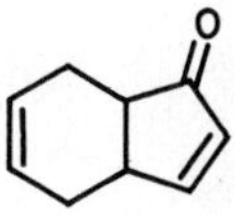 CCl_4	1710 1595	139

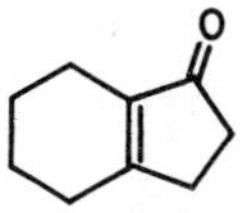

Solvent	cm⁻¹	Ref.
liq.	1661 1626	152
CCl_4	1700 1645	153
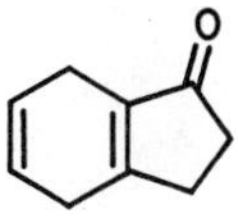 CCl_4	1700 1675 1635	139
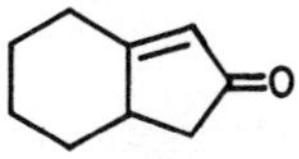 CCl_4	1715 1625	129
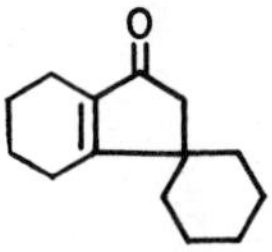 CH_2Cl_2	1695 1642	57
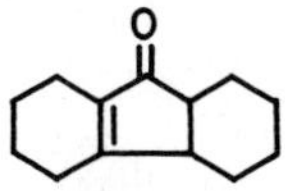 liq.	1704 1645	56
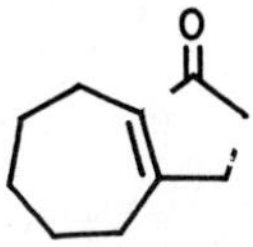 liq.	1705 1645	154
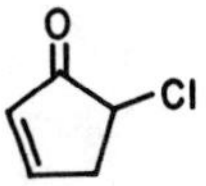 liq.	1740* 1725	155
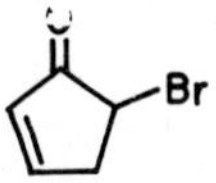 CCl_4	1730	156
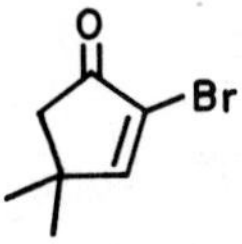 KBr	1720 1595	157

Solvent	Frequency (cm⁻¹)	Ref.
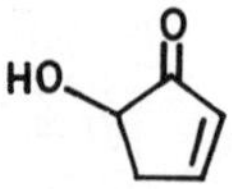 liq.	1725	156
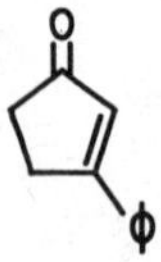 $CHCl_3$	1690	149
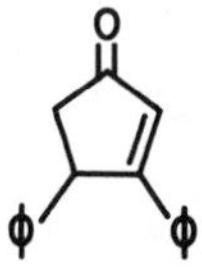 CCl_4	1718 1695	158
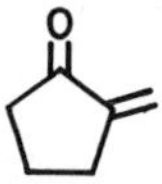 CCl_4	1727 1644	159
$CHCl_3$	1703 1632	160
CCl_4	1706 1633	159
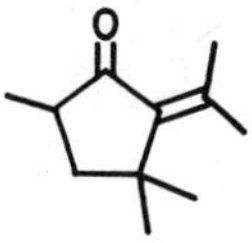 CCl_4	1709 1623	104
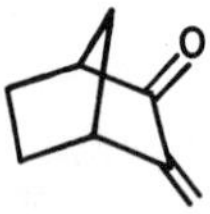 CCl_4	1739	161
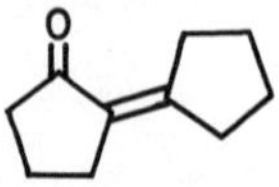 liq.	1704	162
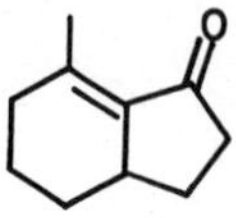 CCl_4	1705 1640	163
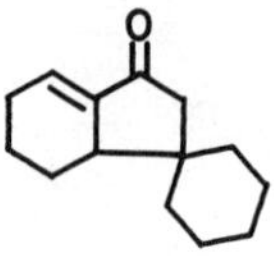 CH_2Cl_2	1715 1653	57
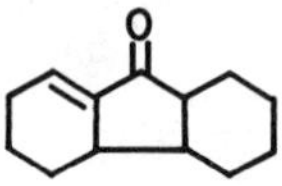 CCl_4	1724 1658	56
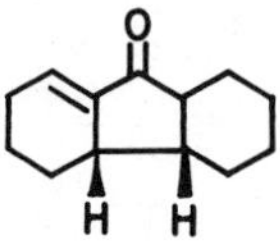 liq.	1730 1661	56
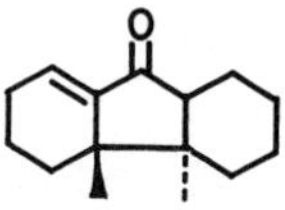 CCl_4	1724 1653	57
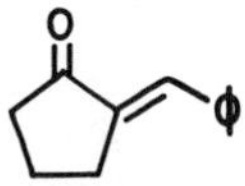 CCl_4	1722 1630 1570w	164
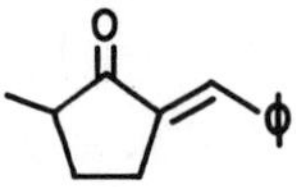 CCl_4	1708 1620 1565	164
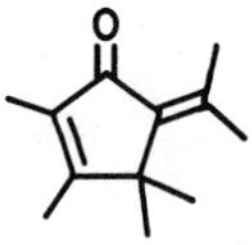 CCl_4	1689 1623	104
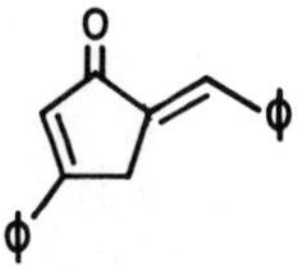 CCl_4	1699	4
CCl_4	1712	165
KBr	1718	166

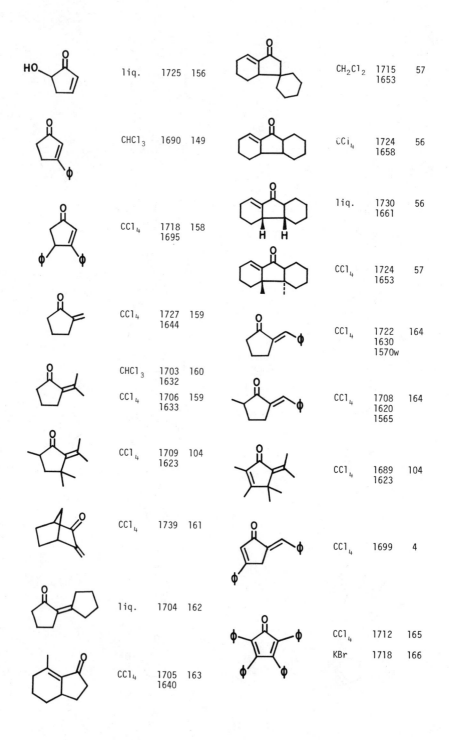

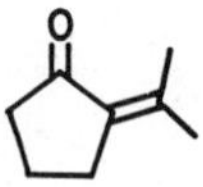

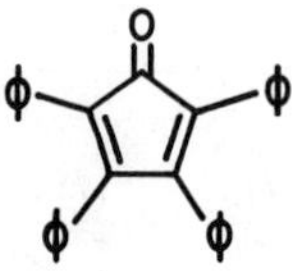

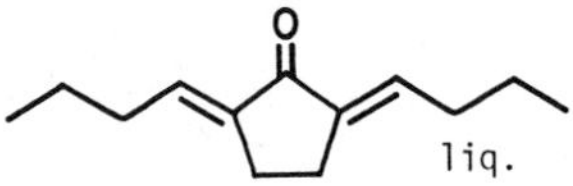	liq.	1684	167
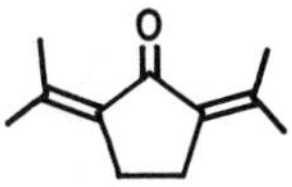	KBr	1692 1635	112
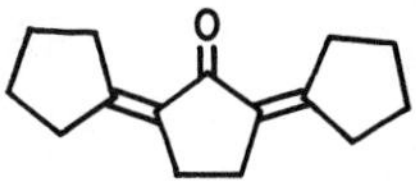	$CHCl_3$	1692 1641	160
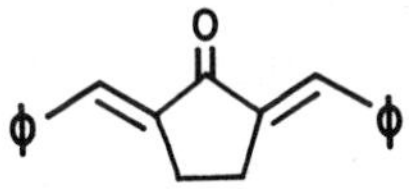	$CHCl_3$	1692	168
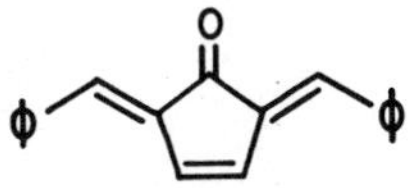	$CHCl_3$	1706	168
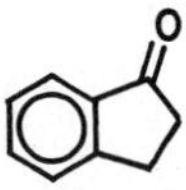	CCl_4 $CHCl_3$	1721 1708	168 169
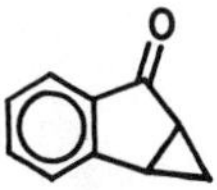	liq.	1723	14
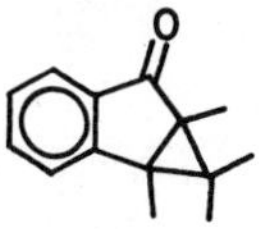	CCl_4	1699 1606	170
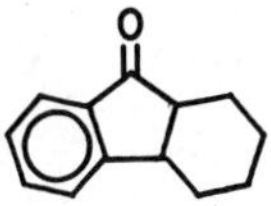	CCl_4	1723	171
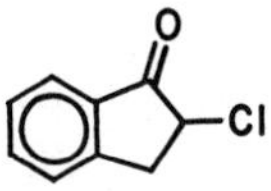	CCl_4	1737	120
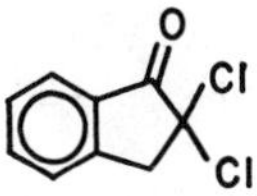	CCl_4	1755	120
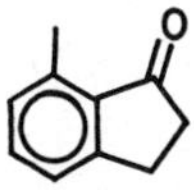	CCl_4	1705	172
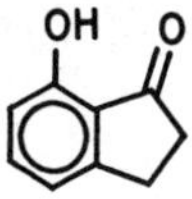	$CHCl_3$	1681	169
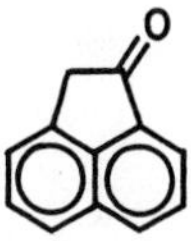	CCl_4	1724	169
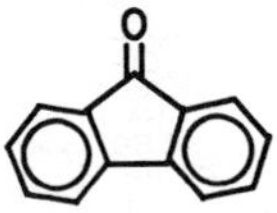	CCl_4	1720	165
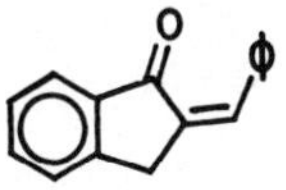	CCl_4	1700 1619	173
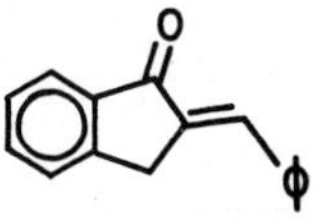	CCl_4	1706 1640	174
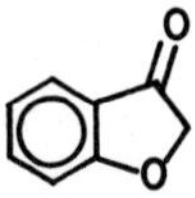	nuj.	1712	67

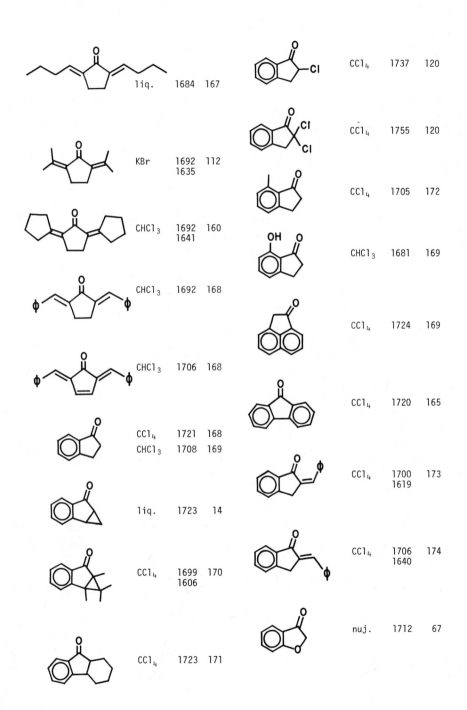

Solvent	ν (cm⁻¹)	Ref.
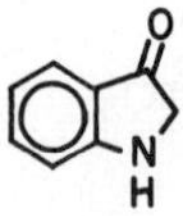 KBr	1743	175
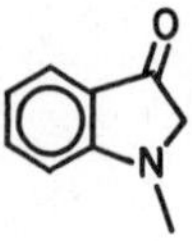 KBr	1733	175
 CCl_4	1718	115
$CHCl_3$	1711	176
liq.	1714	59
 KBr	1710	177
 liq.	1730 1650	178
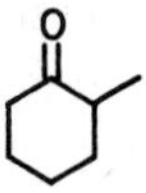 CCl_4	1715	179
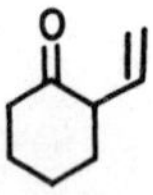 liq.	1712 1639	180
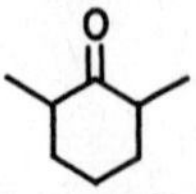 CCl_4	1714	179
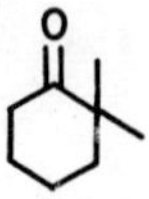 CCl_4	1702	179
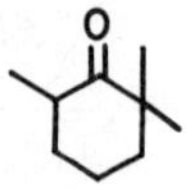 CCl_4	1705	181
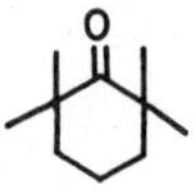 CCl_4	1690	181
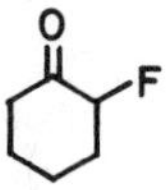 CCl_4	1738	121
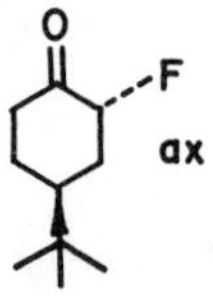 CCl_4 (ax)	1735	182
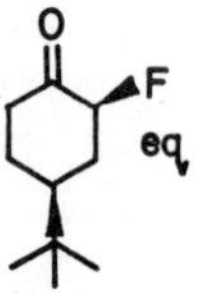 CCl_4 (eq)	1745	182
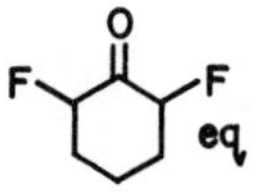 CCl_4 (eq)	1772	183
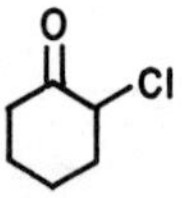 CCl_4	1722	184
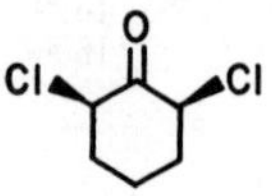 CCl_4	1787w 1753 1730w	185
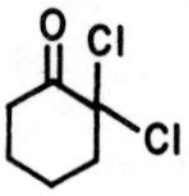 CCl_4	1745	186

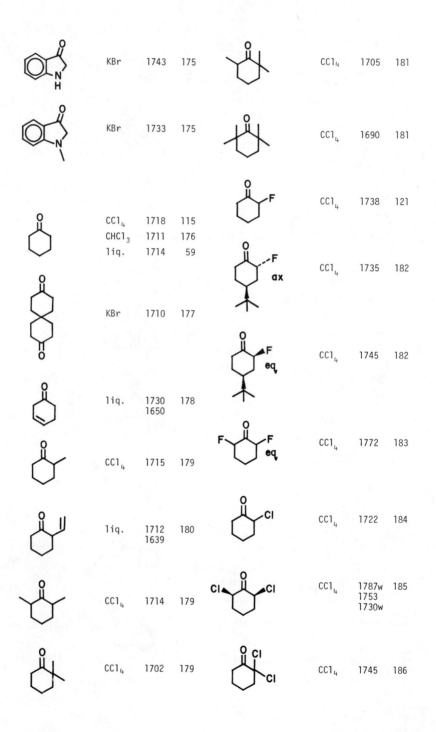

Solvent	cm⁻¹	Ref.	Solvent	cm⁻¹	Ref.
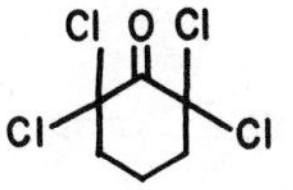 CCl_4	1766	186	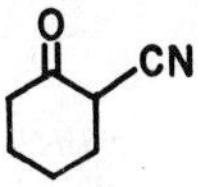 $CHCl_3$	1735	188
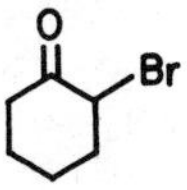 CCl_4	1716	184	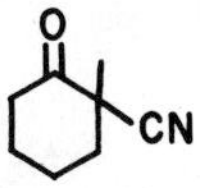 $CHCl_3$	1721	189
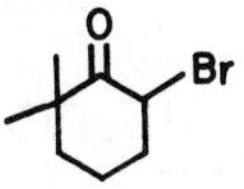 CCl_4	1723 1700	179	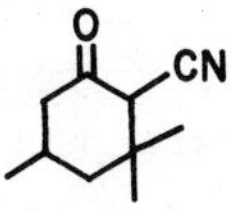 CCl_4	1733 2250 2208	190
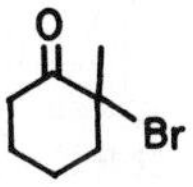 CCl_4	1723 1700	179	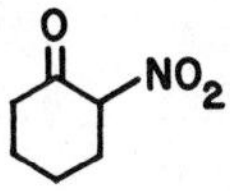 liq.	1739 1550	123
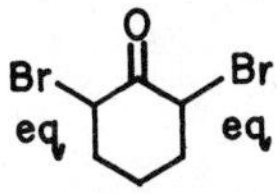 CCl_4	1750	187	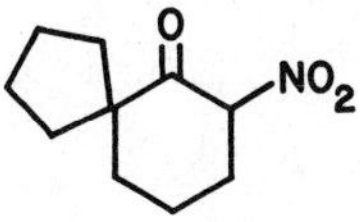 KBr	1710 1545 1395	124
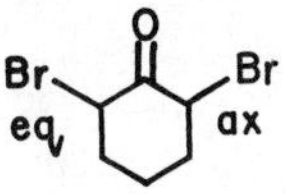 CCl_4	1732	187	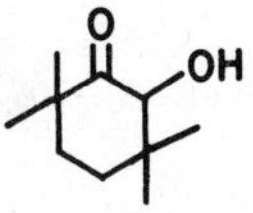 nuj.	1705	191
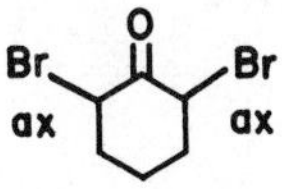 CCl_4	1718	183	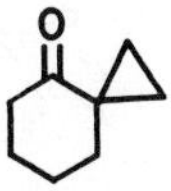 liq.	1685	14
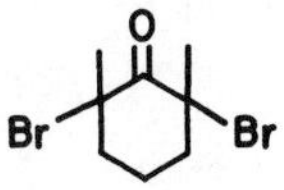 CCl_4	1722	179	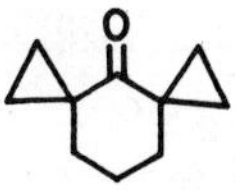 liq.	1680	14
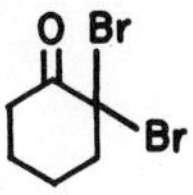 CCl_4	1727	187	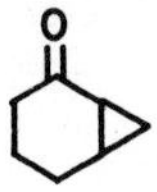 liq.	1695	14

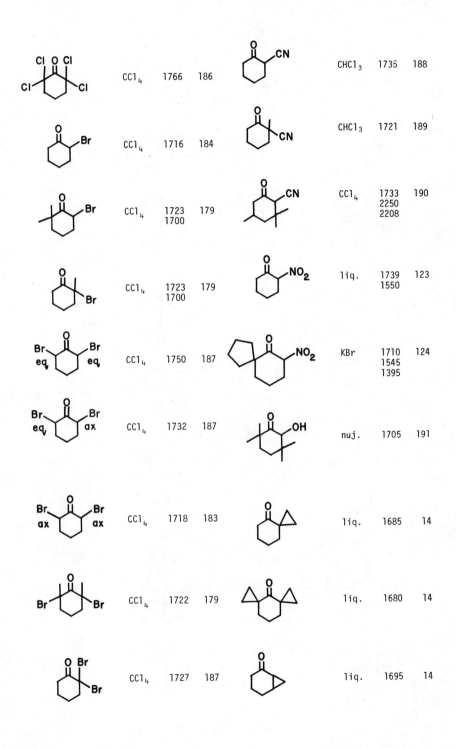

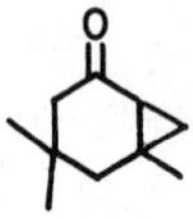 liq.	1698	14
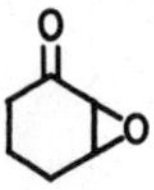 liq.	1718	28
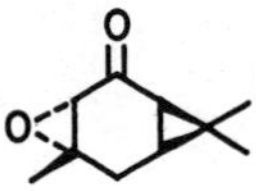 liq.	1680	192
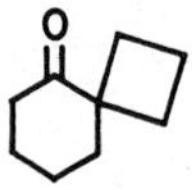 CCl_4	1709	134
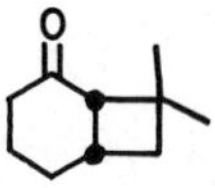 liq.	1698	135
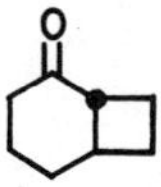 liq.	1704	135
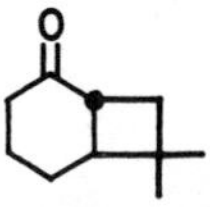 liq.	1721	193
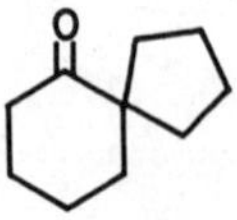 CCl_4	1701	186
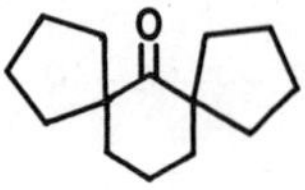 CCl_4	1693	136

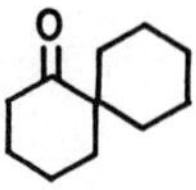 CCl_4	1708	117
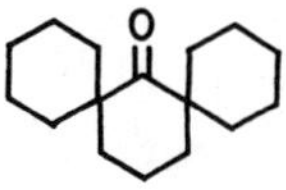 CCl_4	1691	136
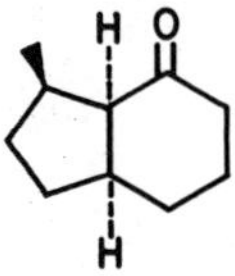 CCl_4	1701	194
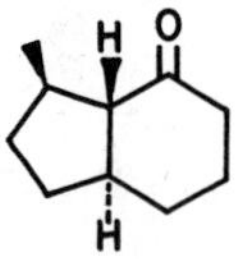 CCl_4	1711	194
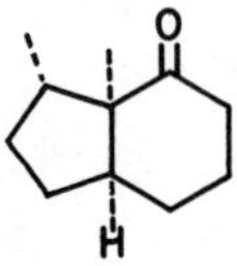 CCl_4	1697	194
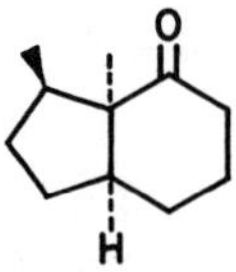 CCl_4	1701	194
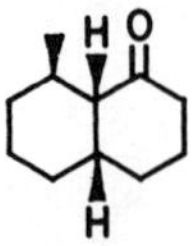 CCl_4	1705	138
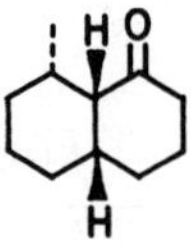 liq.	1700	138

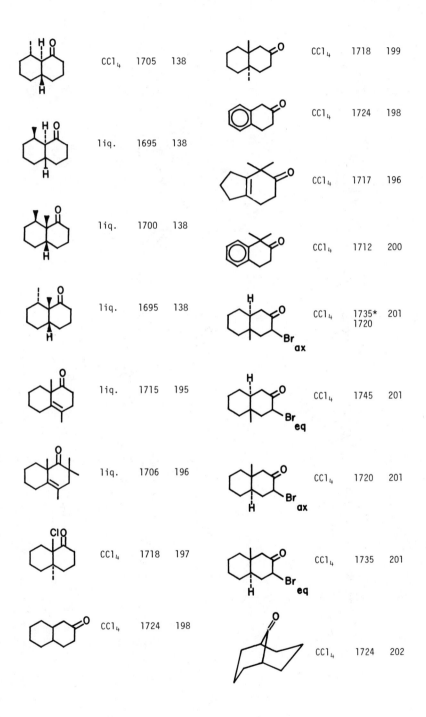

Structure	Solvent	Frequency	Ref.
	CCl_4	1705	138
	liq.	1695	138
	liq.	1700	138
	liq.	1695	138
	liq.	1715	195
	liq.	1706	196
	CCl_4	1718	197
	CCl_4	1724	198
	CCl_4	1718	199
	CCl_4	1724	198
	CCl_4	1717	196
	CCl_4	1712	200
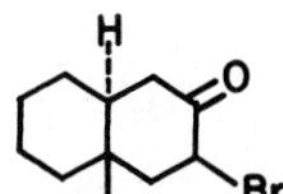 Br ax	CCl_4	1735* 1720	201
Br eq	CCl_4	1745	201
Br ax	CCl_4	1720	201
Br eq	CCl_4	1735	201
	CCl_4	1724	202

Solvent	ν (cm⁻¹)	Ref.
CCl_4	1670	127
CCl_4	1732 1723m	203
CCl_4	1731	204
nuj.	1739	205
CS_2	1761 1748	205
KBr	1718	205
KBr	1724	205

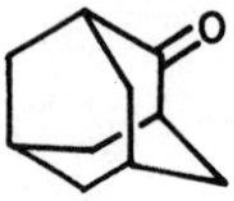

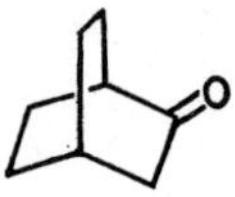

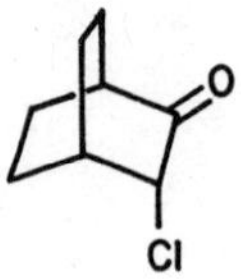

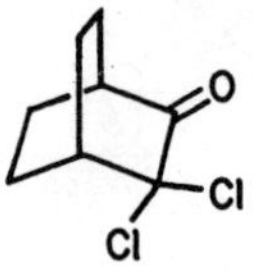

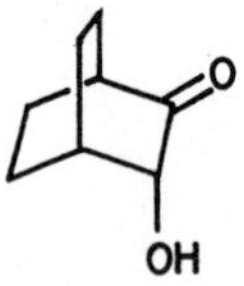

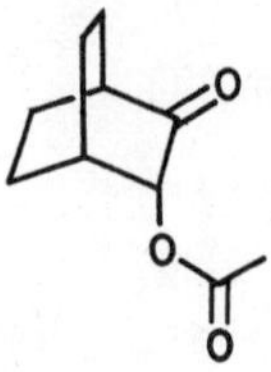

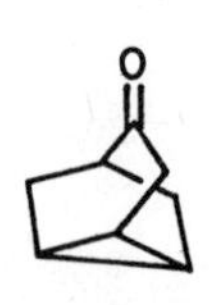

Solvent	ν (cm⁻¹)	Ref.
CS_2	1730	206
CS_2	1739	207
CCl_4	1717	204
CCl_4	1699	147
liq.	1716	208
nuj.	1725	208
$CHCl_3$	1710	209
liq.	1716	208

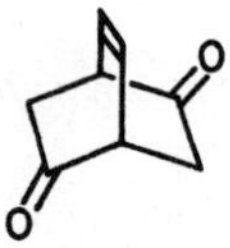

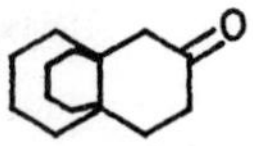

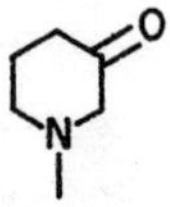

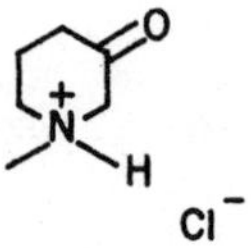

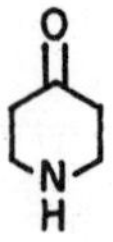

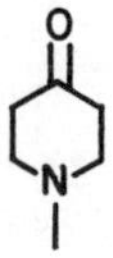

Structure	Solvent	ν (cm^{-1})	Ref.
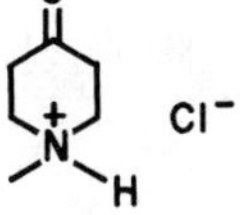	nuj.	1733b	208
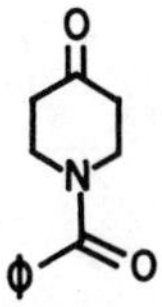	liq.	1715 1625	210
	liq.	1715	211
	liq.	1710	212
	CCl_4	1716	213
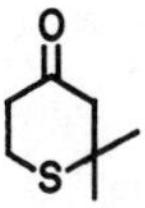	liq.	1718	214
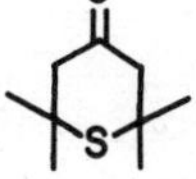	liq.	1710	214

Structure	Solvent	ν (cm^{-1})	Ref.
	CCl_4	1725	215
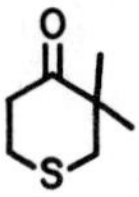	liq.	1740	214
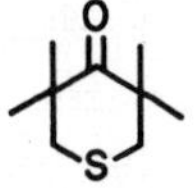	liq.	1690	216
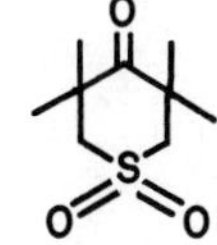	KBr	1707 1692	216
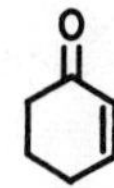	CCl_4 C_2Cl_4	1691 1691	217 3
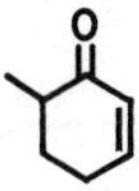	liq. CCl_4	1680 1626 1684	218 219
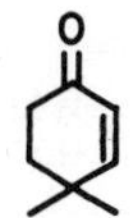	CCl_4	1680	220

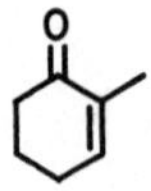

Solvent	ν	Ref.	Solvent	ν	Ref.
CCl_4	1681	219	KBr	1689	222
liq.	1660	192	KBr	1695 1610	222
liq.	1695 1645	192	KBr	1712 1613	222
C_2Cl_4	1680	3	CCl_4	1695 2247 1642	190
CCl_4 C_2Cl_4	1667 1674	209 3	nuj.	1665 1640	221
liq.	1695 1658	192	nuj.	1682 1754 1657 1640	221
C_2Cl_4	1673	3	liq.	1669 1608	221
liq.	1690 1618	221	$CHCl_3$	1671 1589	221

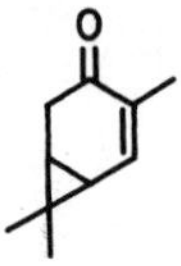

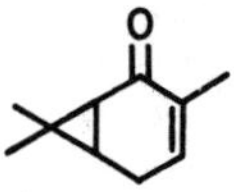

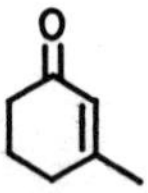

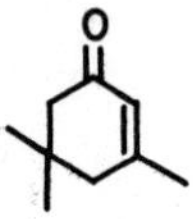

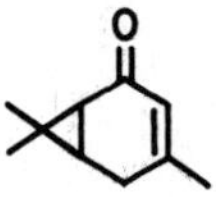

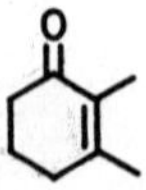

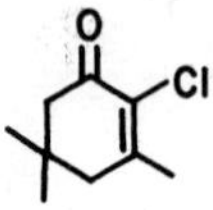

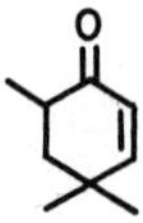

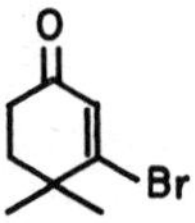

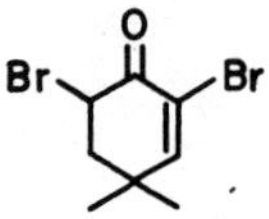

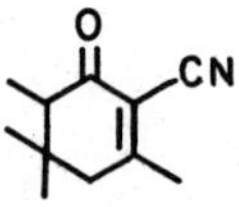

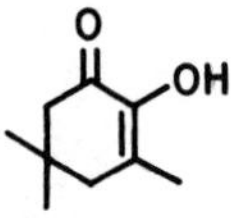

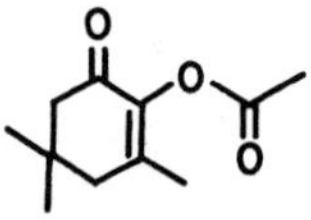

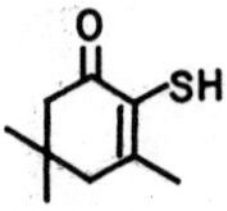

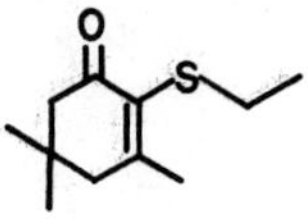

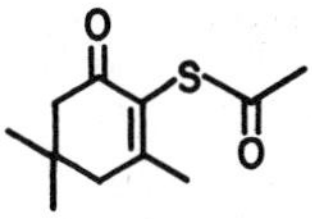	nuj.	1673 1704 1693 1596	221
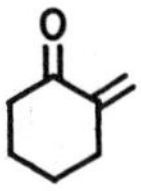	CCl_4	1697 1618	159
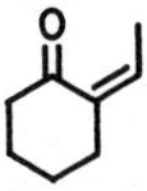	liq. CCl_4	1684 1626 1692	180 31
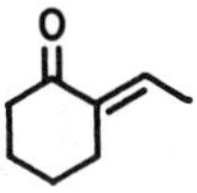	liq. C_2Cl_4	1684 1618 1693	180 3
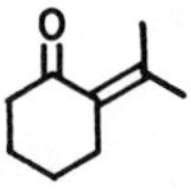	CCl_4	1680 1610	107
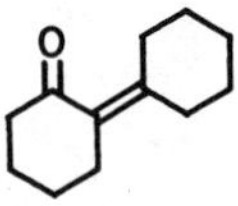	CCl_4	1670	223
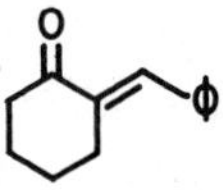	CCl_4	1670 1588 1562m	164
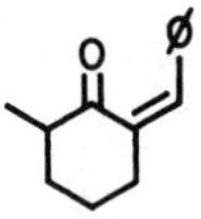	CCl_4	1692 1628 1592	164
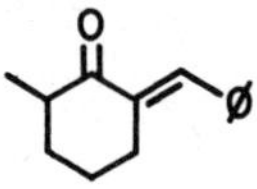	CCl_4	1660 1590 1565	164
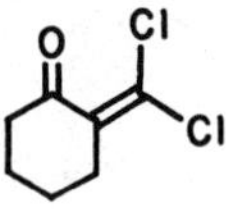 Cl Cl	liq.	1700 1575	224
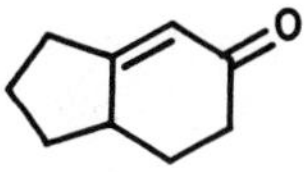	CCl_4	1667 1630	196
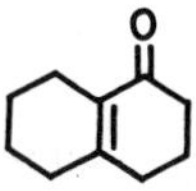	liq.	1660 1632	225
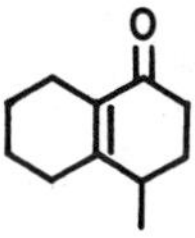	liq.	1661	195
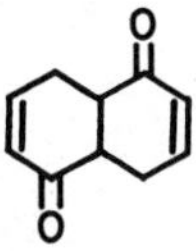	$CHCl_3$	1672 1613	226
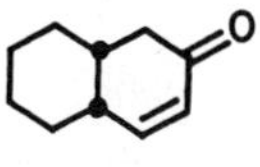	CH_2Cl_2	1678 1618	227
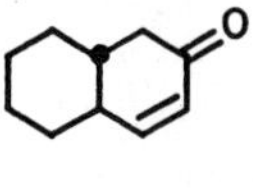	CH_2Cl_2	1669	227
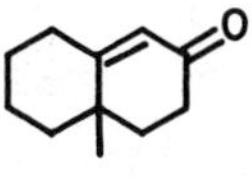	CCl_4	1685	199
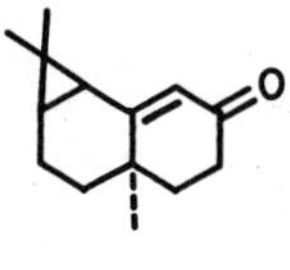	CCl_4	1667	127

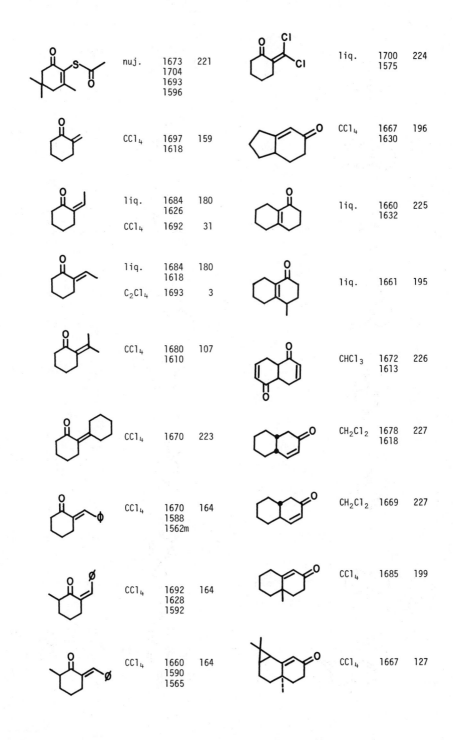

Solvent	Frequency	Ref.
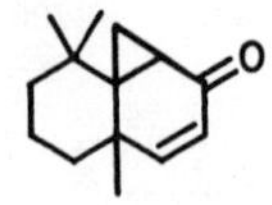 CCl_4	1672	127
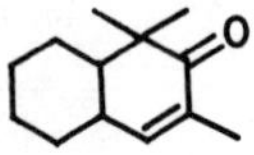 CCl_4	1665	228
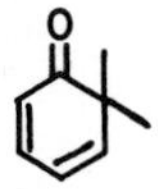 CCl_4	1670	173
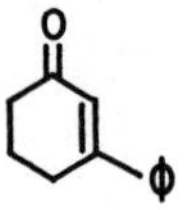 $CHCl_3$	1660	149
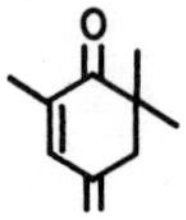 CCl_4	1669	229
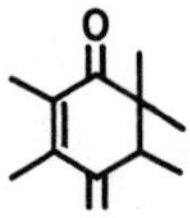 CCl_4	1667	230
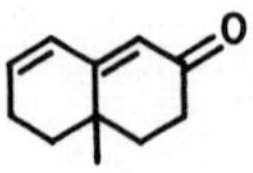 liq.	1664 1618 1587	231
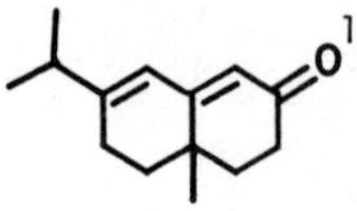 liq.	1670 1645 1605	232
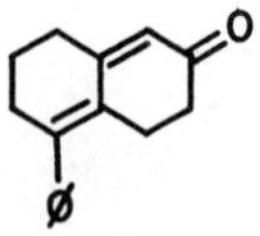 $CHCl_3$	1660 1615	233
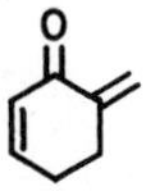 CCl_4	1675 1620 942	234
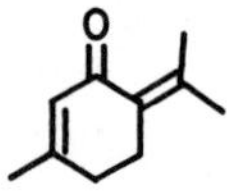 CCl_4	1662 1638 1613	235
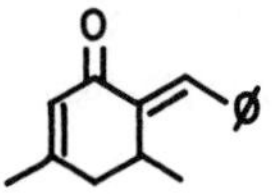 CCl_4	1664	4
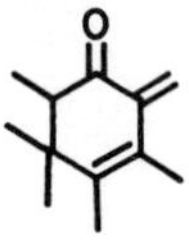 CCl_4	1700	230
 CCl_4	1664 1630	220
 $CHCl_3$	1640	236
 KBr	1685* 1660 1615	177

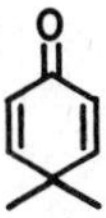

Solvent	ν (cm⁻¹)	Ref.
CCl_4	1669	219
KBr	1610 1578	237
CCl_4	1653 1624	230
KBr	1672 1610	222
KBr	1689 1613	222
CCl_4	1667	219
$CHCl_3$	1659 1614	160
nuj.	1663 1610	238

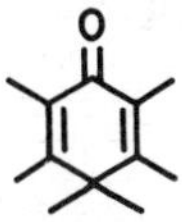

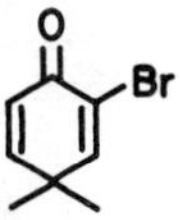

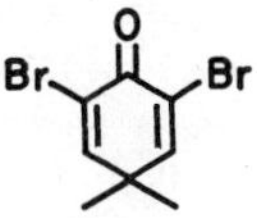

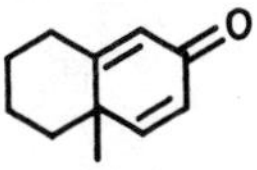

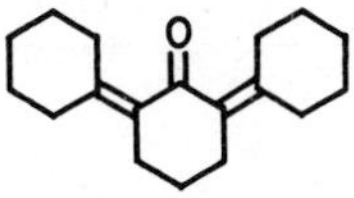

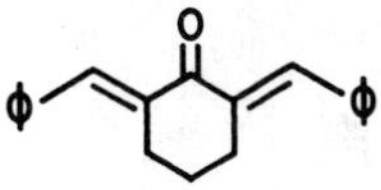

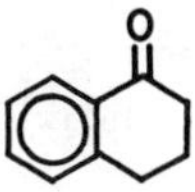

Solvent	ν (cm⁻¹)	Ref.
CCl_4 $CHCl_3$	1687 1680	169 169
CCl_4	1689	239
CCl_4	1685	200
liq.	1667	14
liq.	1681	171
CCl_4 nuj.	1690 1677	200 200
CCl_4	1697	239
CCl_4	1715	239

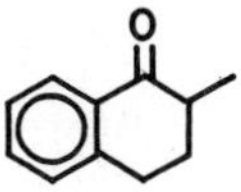

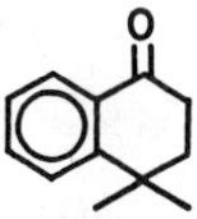

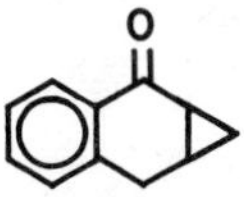

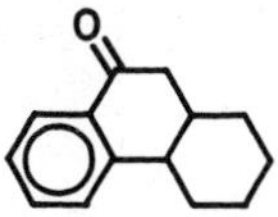

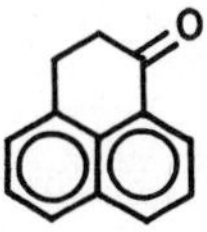

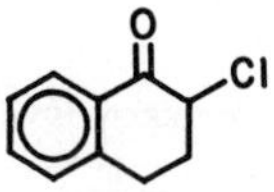

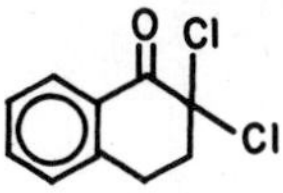

Solvent	Frequency	Ref.
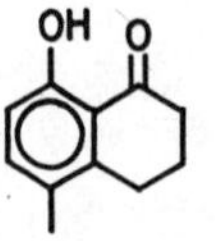 $CHCl_3$	1644	169
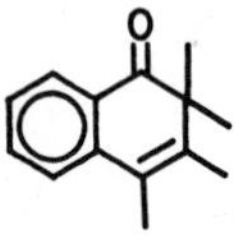 liq.	1674 1633 1600	170
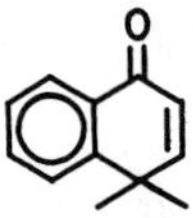 CCl_4	1665	200
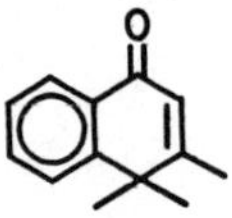 liq.	1678	240
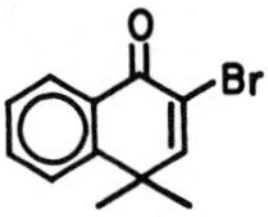 CCl_4	1665	200
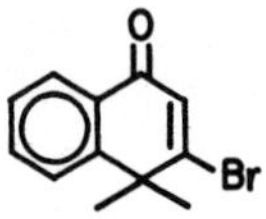 CCl_4	1673	200
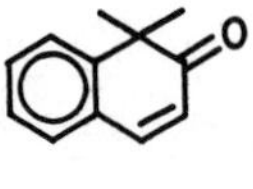 CCl_4	1662	200
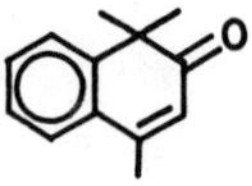 liq.	1652 1621 1600	170
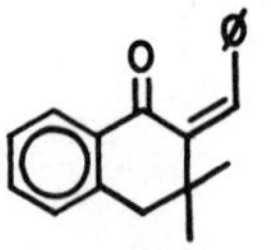 CCl_4	1695 1615	173
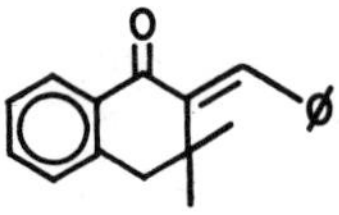 CCl_4	1705 1627	173
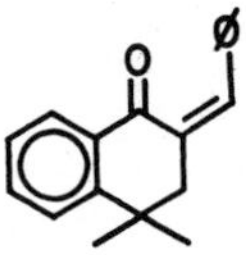 CCl_4	1676 1602	241
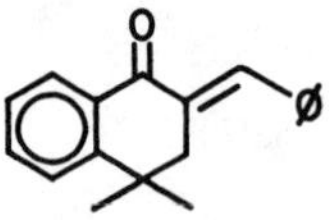 CCl_4	1673 1610	241
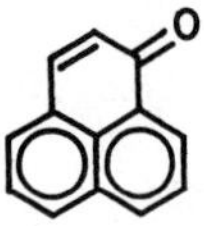 CCl_4 nuj.	1646 1636	200 200
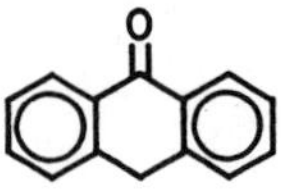 CS_2	1665	242
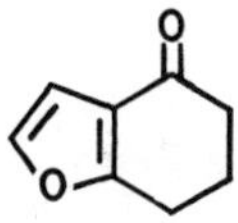 liq.	1675	82
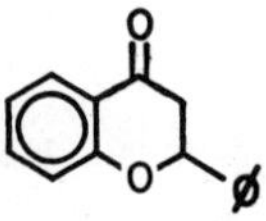 nuj.	1680	72

Structure	Solvent	Frequency	Ref.
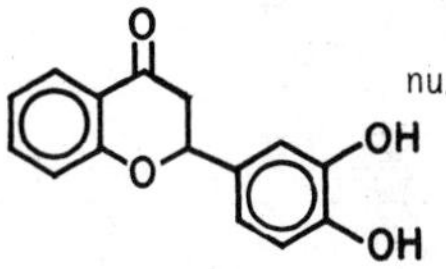 OH, OH	nuj.	1665	72
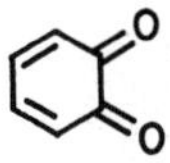	CCl_4	1669	243
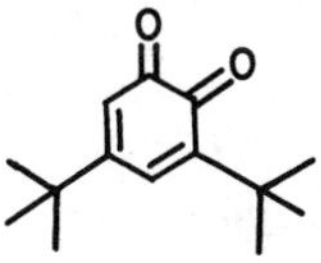	CCl_4	1689 1664	244
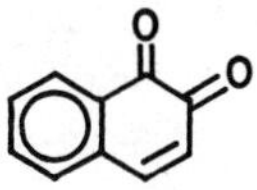	CCl_4	1678	243
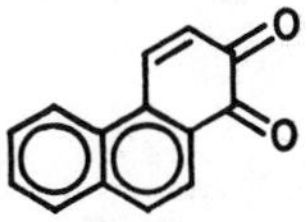	CCl_4	1677	243
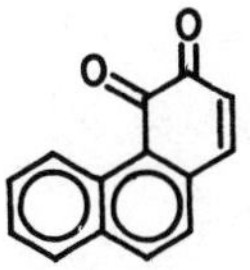	CCl_4	1668	243
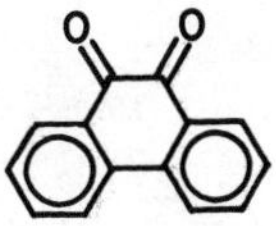	CCl_4	1684	243

Structure	Solvent	Frequency	Ref.
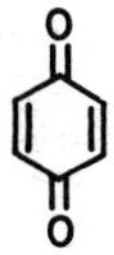	CS_2	1669	245
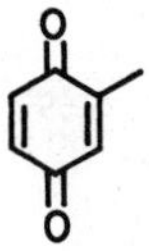	CS_2	1664	245
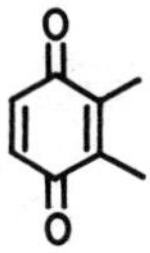	CS_2	1661	245
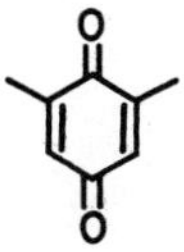	CS_2	1658	245
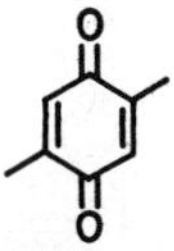	CS_2	1661	245
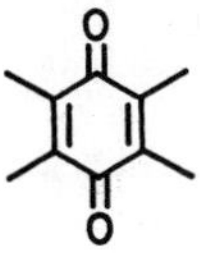	CS_2	1642	245
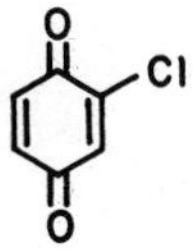 Cl	CS_2	1675 1658	245

Solvent	ν	Ref.
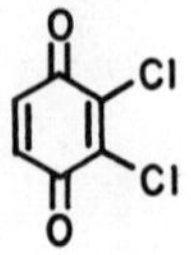 CS_2	1681	245
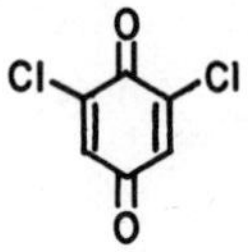 CS_2	1695 1653	245
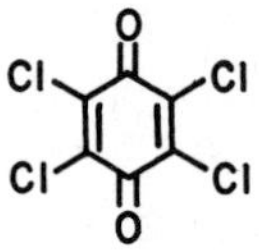 nuj.	1681	245
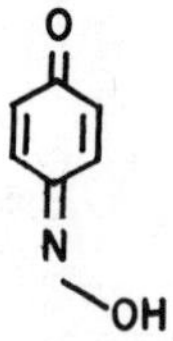 nuj.	1628 1555	246
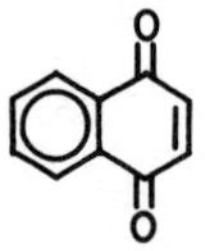 CCl_4	1675	243
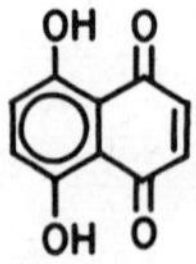 CCl_4	1623	243
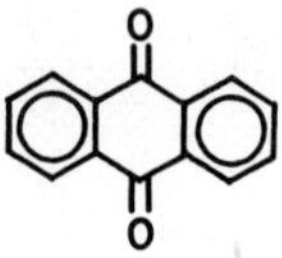 CCl_4	1678	247
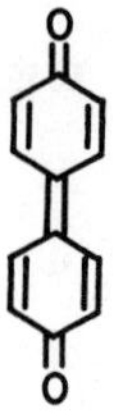 $CHCl_3$	1635	4
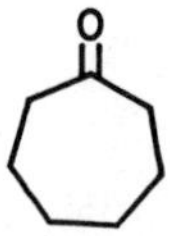 CCl_4	1706	115

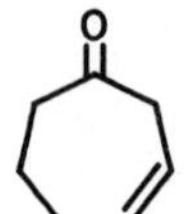

Solvent	ν	Ref.
liq.	1704 1660	248
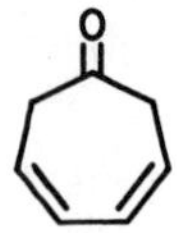 CCl_4	1709	220
 CCl_4	1706 1658	249
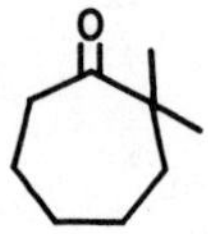 CCl_4	1706	117

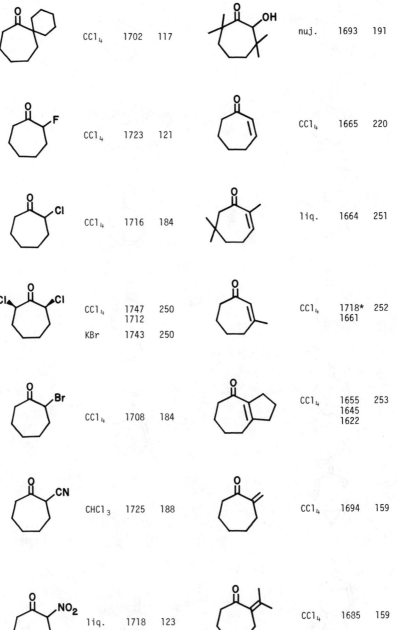

Solvent	Frequency	Ref.
CCl_4	1702	117
CCl_4	1723	121
CCl_4	1716	184
CCl_4	1747 1712	250
KBr	1743	250
CCl_4	1708	184
$CHCl_3$	1725	188
liq.	1718 1553	123
nuj.	1693	191
CCl_4	1665	220
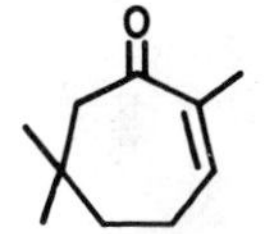 liq.	1664	251
CCl_4	1718* 1661	252
CCl_4	1655 1645 1622	253
CCl_4	1694	159
CCl_4	1685	159

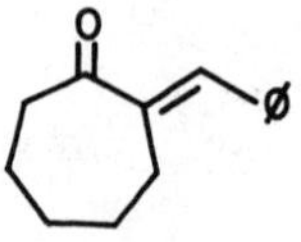

Solvent	ν (cm⁻¹)	Ref.
CCl_4	1680	223
liq.	1668	254
CCl_4	1665	255
liq.	1653 1590	192
liq.	1631 1575	192
CCl_4	1647 1613	220
liq.	1650 1603	256
liq.	1650 1608	249
liq.	1680 1610	257
nuj.	1672 1623 1612	238
CS_2	1672	258
$CHCl_3$	1638	259
$CHCl_3$	1623 1563	249
CCl_4	1635	260
CCl_4	1610* 1597	260
CCl_4	1618	260

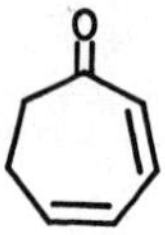

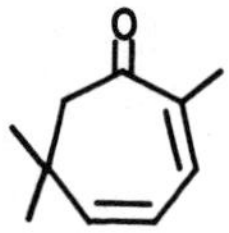

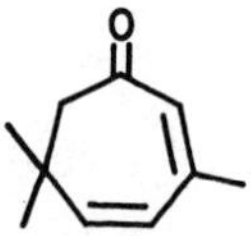

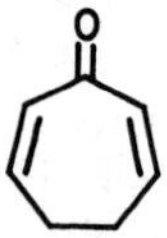

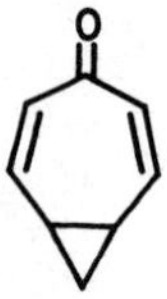

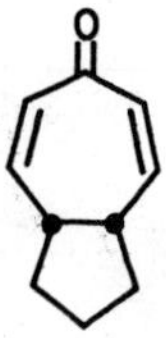

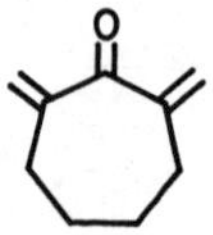

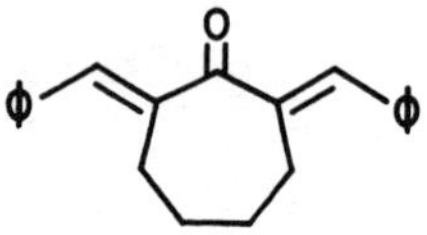

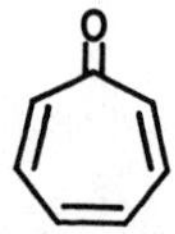

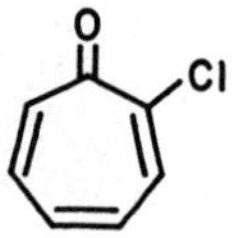

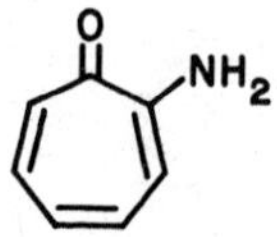

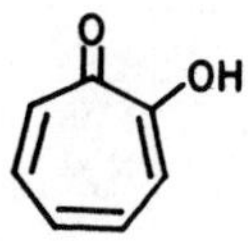

Solvent	Frequency	Ref.
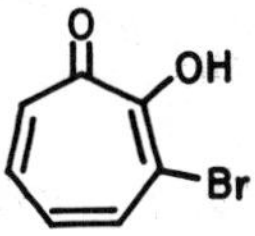 CCl_4	1607	260
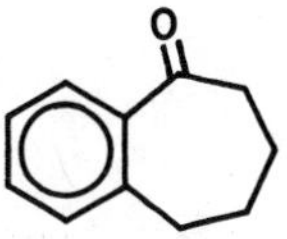 $CHCl_3$	1672	169
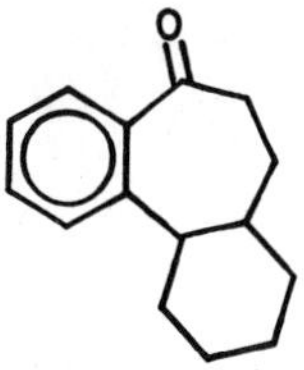 liq.	1676	171
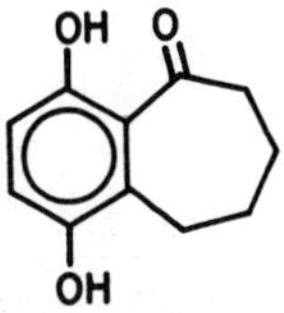 $CHCl_3$	1637	169
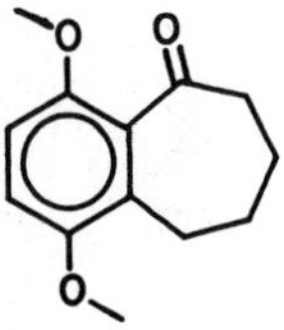 $CHCl_3$	1691	169
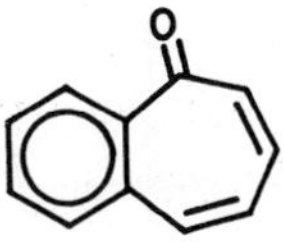 $CHCl_3$	1642	153
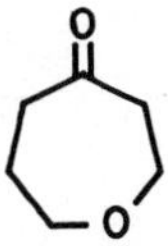 liq.	1703	261
 CCl_4	1711	262
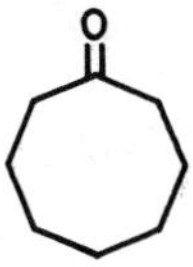 CCl_4	1704	115
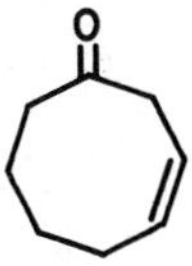 CCl_4	1710 1653	263
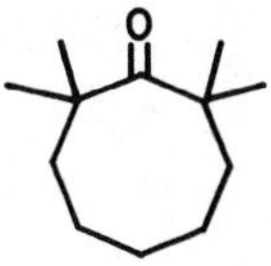 nuj.	1692	191
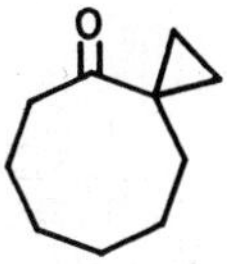 liq.	1681	14
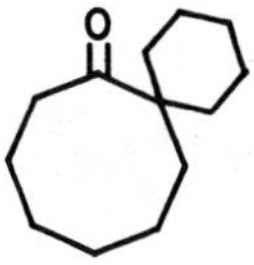 CCl_4	1698	117

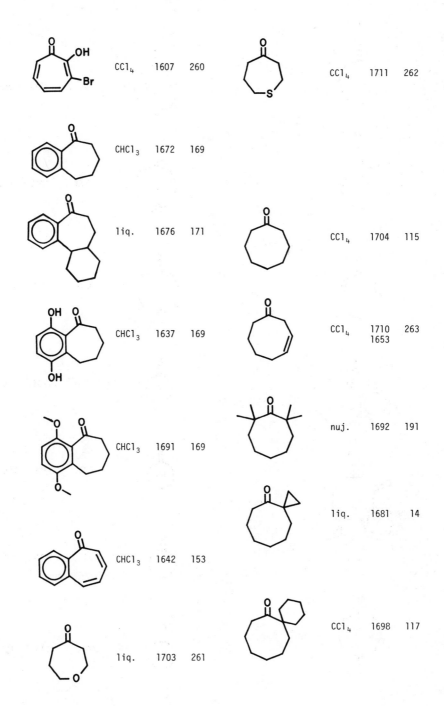

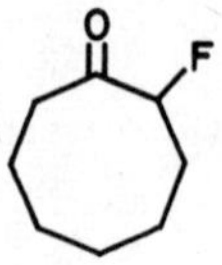

Solvent	cm⁻¹	Ref.
CCl_4	1715	121
CCl_4	1664	220
CCl_4	1640	220
CCl_4	1675	264
liq.	1667	171
CCl_4	1683	265
CCl_4	1714	262
liq.	1701	266
CCl_4	1710	215
CCl_4	1703	115
CCl_4	1705	115
CCl_4	1713	115
CCl_4	1713	115

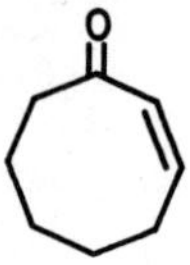

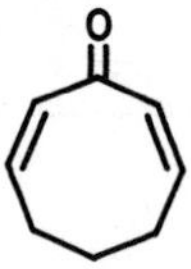

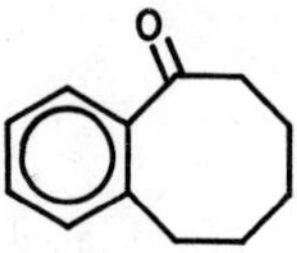

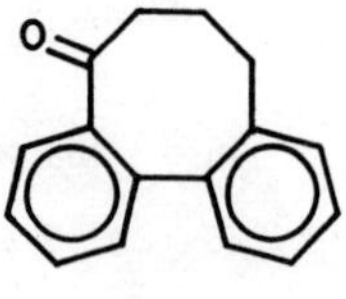

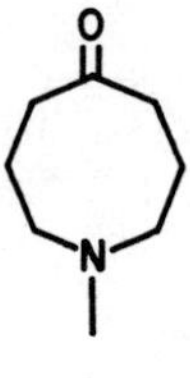

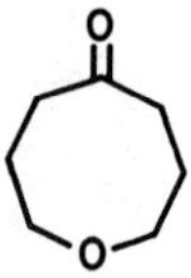

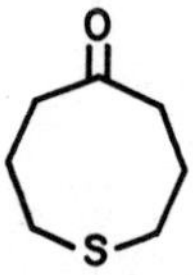

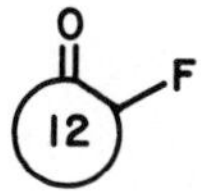

CCl_4	1718	121
CCl_4	1715	115
CCl_4	1757w 1718 1681w	267
liq.	1721	268
liq.	1710	269
liq.	1700	269
$CHCl_3$	1678	270
CCl_4	1679	217
$CHCl_3$	1681	271

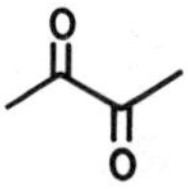

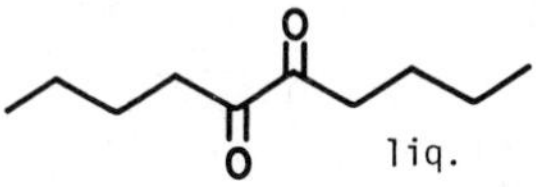

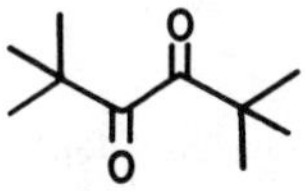

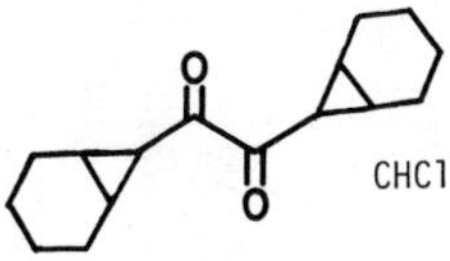

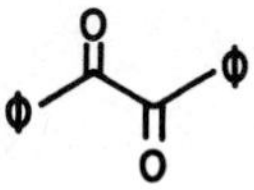

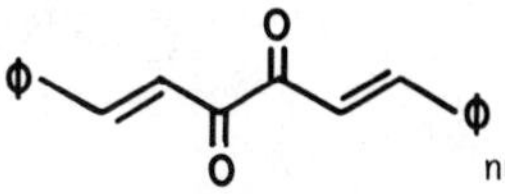

nuj.	1667	238
$CHCl_3$	1658	272
KBr	1735 1640	273
liq.	1785 1763	274
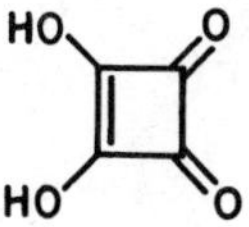	1818 1639	275
KBr	1790	274
CCl_4	1800 1610 1595	114

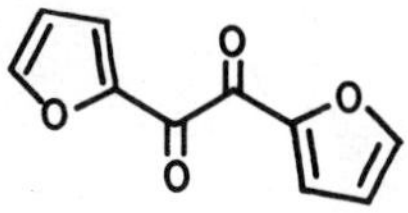

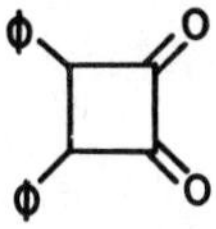

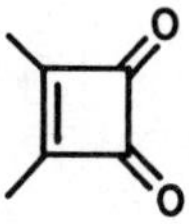

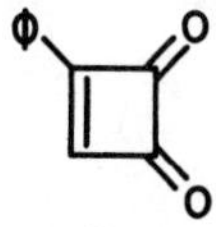

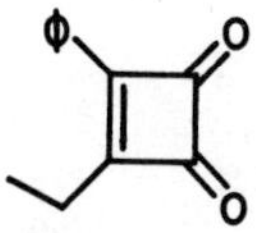

Structure	Solvent	Frequency	Ref.
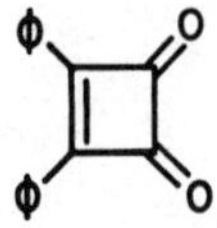	KBr	1785 1776 1760*	274
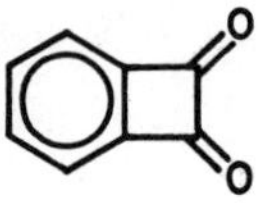	CS_2	1810 1780	274
	$CHCl_3$	1700 1655	276
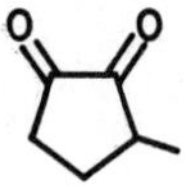	KBr	1709 1659	277
	CCl_4	1755	278
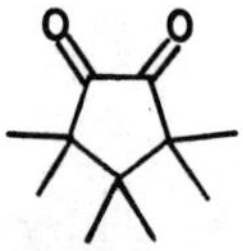	CCl_4	1739	104
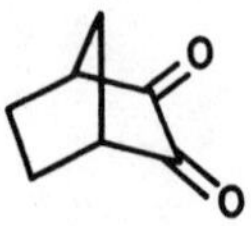	CCl_4	1771 1760	279
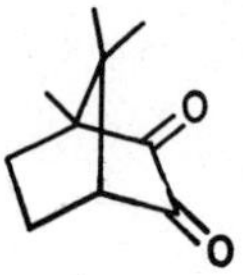	CCl_4	1776 1760	279
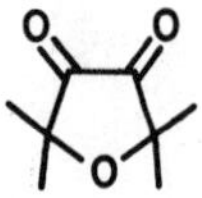	CCl_4	1765	280
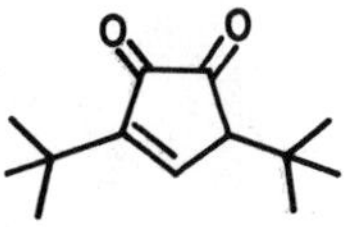	CCl_4	1779 1715	281
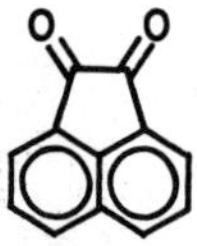	CS_2	1780 1736	242
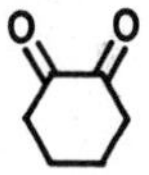	$CHCl_3$	1735	282
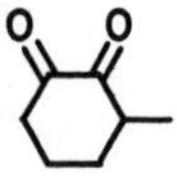	CCl_4	1723 1677	283
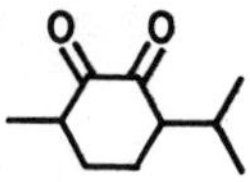	nuj.	1668 1645	284
	CCl_4	1667	285
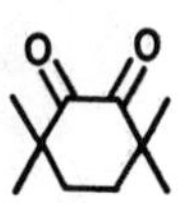	nuj.	1709	286

Solvent	Frequencies	Ref.
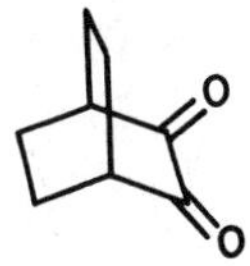 CCl_4	1760, 1730	286
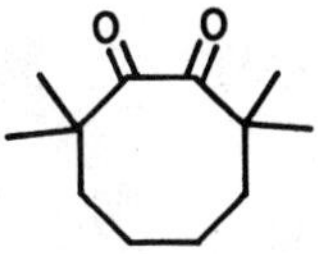 liq.	1697	191
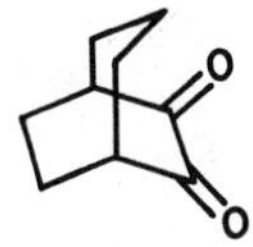 KBr	1710	287
 nuj.	1663, 1602	238
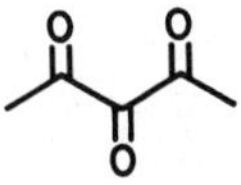 liq.	1722	289
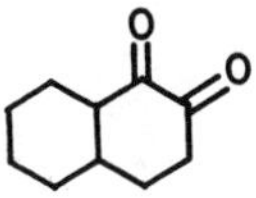 CCl_4	1670, 1642	288
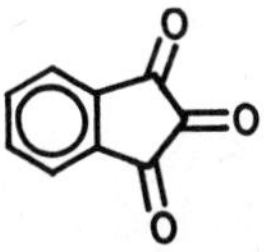 KBr	1770, 1750, 1725, 1595m	290
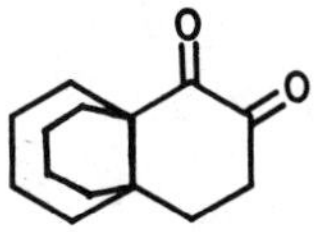 CCl_4	1679, 1648	147
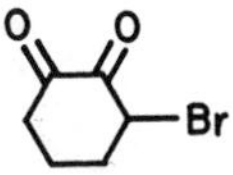 KBr	1680, 1643	277
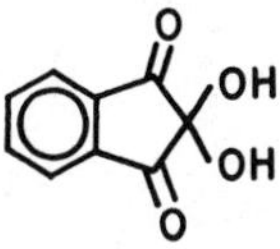 KBr	1750, 1685, 1585	290
CCl_4	1649	200
CCl_4	1715, 1600*, 1580	291
liq.	1710	191
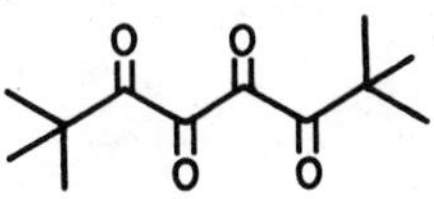 KBr	1730, 1690	292
nuj.	1685, 1602	238
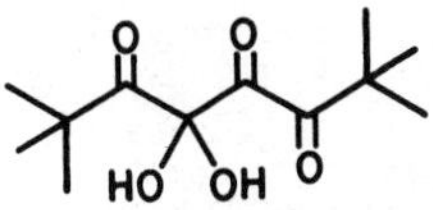 KBr	1730m, 1700	292

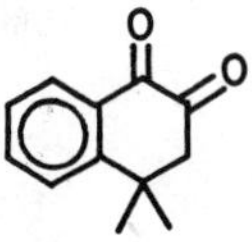

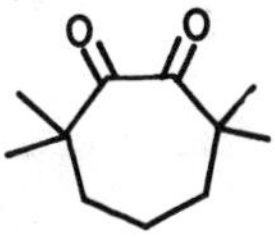

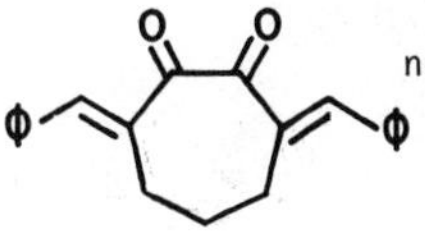

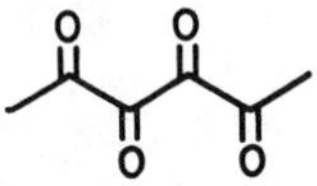

Solvent	Frequencies	Ref.
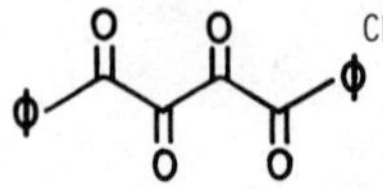 CH_2Cl_2	1769 1761 1605 1553	291
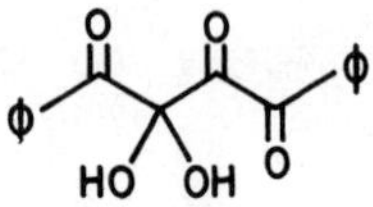 KBr	1735m 1690 1660	292
CCl_4	1610	293
$CHCl_3$	1724 1608	189
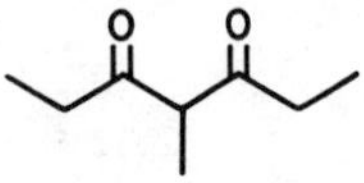 KBr	1709 1598	294
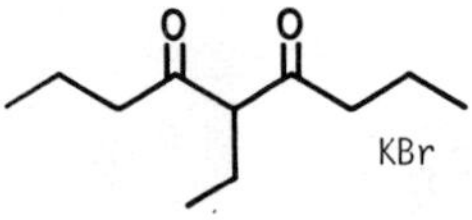 KBr	1700 1590	294
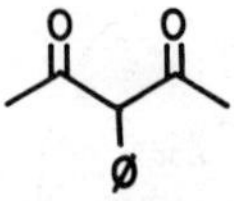 nuj.	1610	295
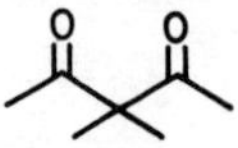 liq.	1725 1703	296
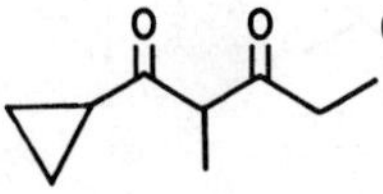 CCl_4	1725 1688 1642 1600	297
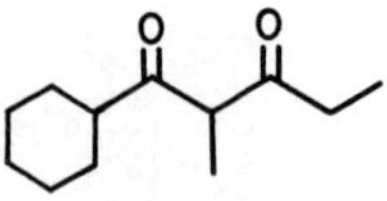 CCl_4	1728 1700 1658 1600	297
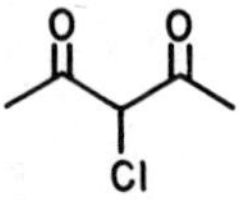 CCl_4	1610	293
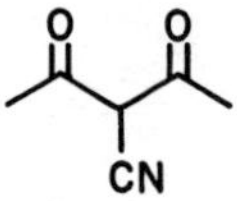 CCl_4	1607b 2220	298
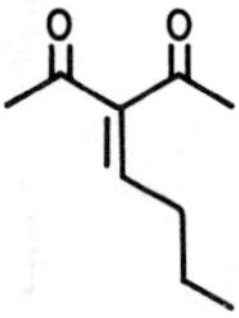 $C_2H_4Cl_2$	1711 1670 1635* 1609 1560	295
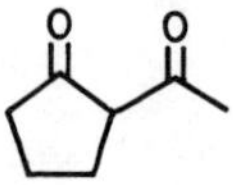 liq.	1740 1704 1660 1610	299
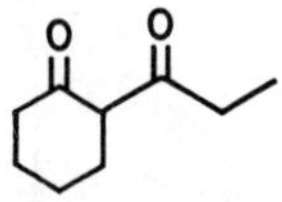 liq.	1718m 1706m 1600b	300
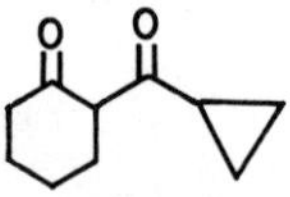 CCl_4	1725 1692 1640 1608	297
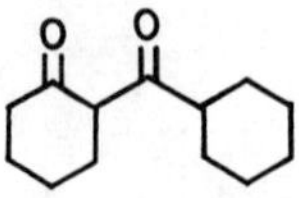 CCl_4	1725 1720 1640 1610	297
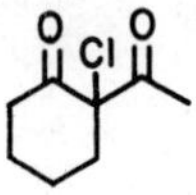 liq.	1745 1725	301

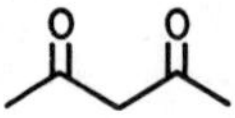

Solvent	Frequency	Ref.
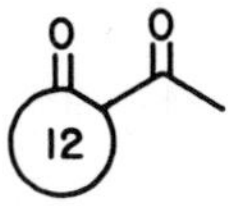 C_2Cl_4	1730* 1704 1595	203
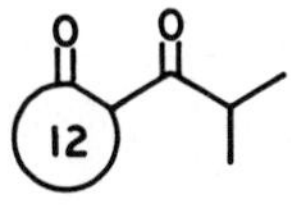 C_2Cl_4	1723* 1696	302
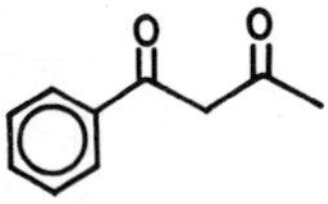 CCl_4	1714w 1604b	45
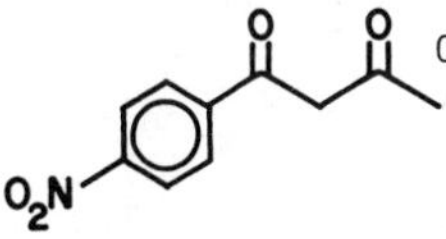 CCl_4	1602	45
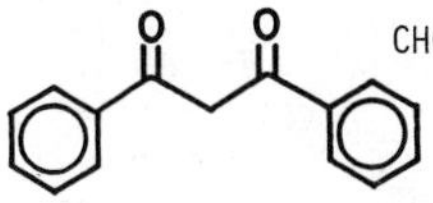 $CHCl_3$	1600	189
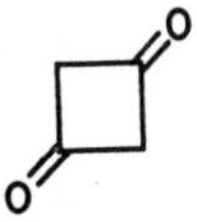 CCl_4	1766	303
$CHCl_3$	1755 1570w	304
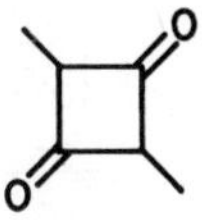 $CHCl_3$	1740 1565	305
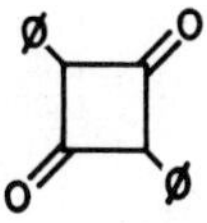 nuj.	1742 1635 1602 1565 1549	295

Solvent	Frequency	Ref.
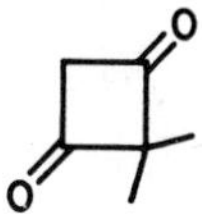 KBr	1883b 1701	306
$CHCl_3$	1754	306
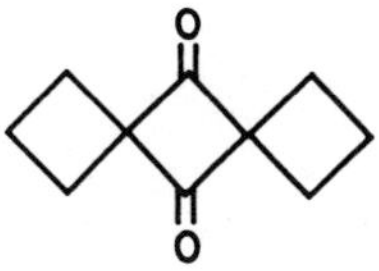 CCl_4	1733	307
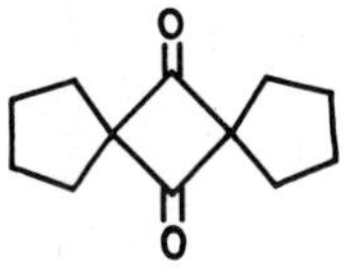 CCl_4	1748	307
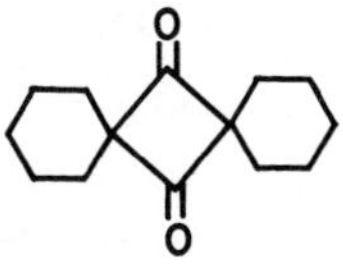 CCl_4	1748	307
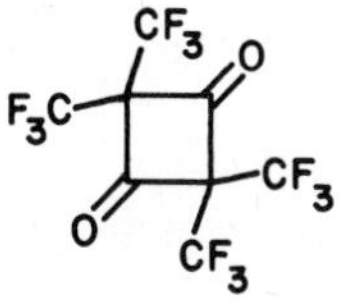 liq.	1730	308
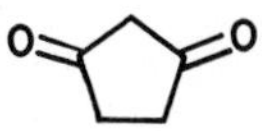 nuj.	1755 1725 1635 1590	305
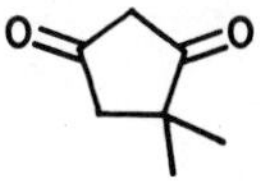 $C_2H_4Cl_2$	1740 1714 1630	295
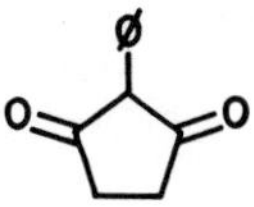 nuj.	1657 1610 1560	295

Solvent	Frequency	Ref.
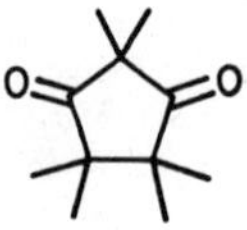 CCl_4	1724	104
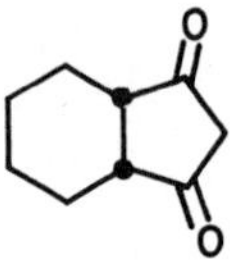 $CHCl_3$	1725 1635 1590b	139
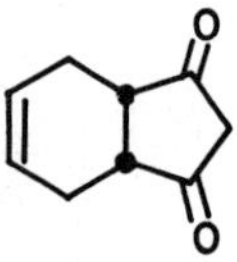 KBr	1635 1585	139
 $CHCl_3$	1745 1715	309
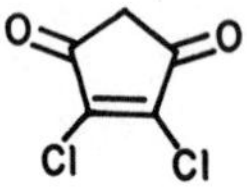 CCl_4	1767 1733	310
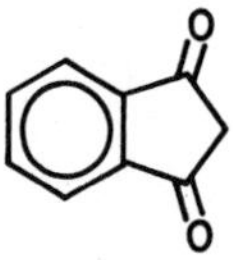 KBr	1750 1710	311
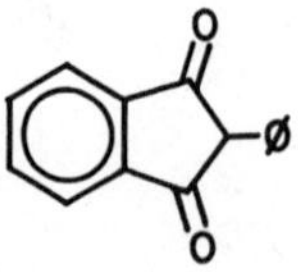 $CHCl_3$	1750 1717	312
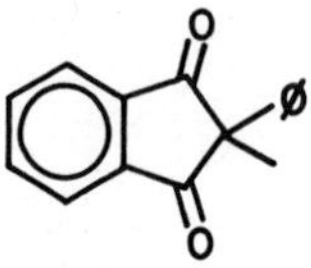 $CHCl_3$	1748 1713	312
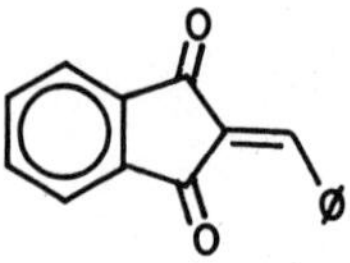 $CHCl_3$	1737 1696	311
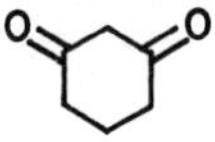 $CHCl_3$	1630 1607	313
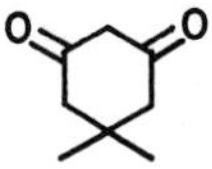 $CHCl_3$	1733 1706 1603	189
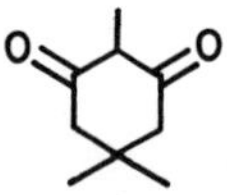 $CHCl_3$	1748 1706 1639 1575	305
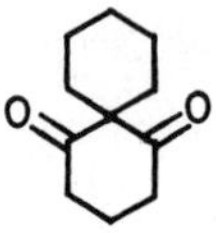 CCl_4	1700 1680	223
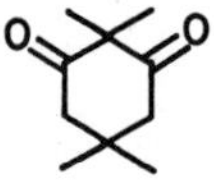 $CHCl_3$	1730 1702	314
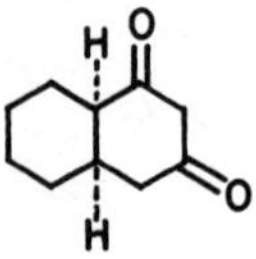 CH_2Cl_2	1721m 1698 1600b	315
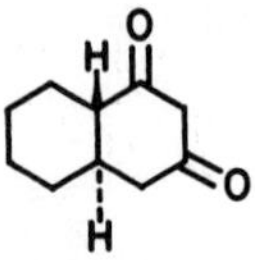 CH_2Cl_2	1724m 1701 1600	315

Solvent	Frequencies	Ref.
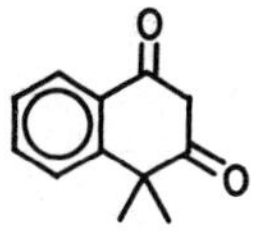 CCl_4	1649	200
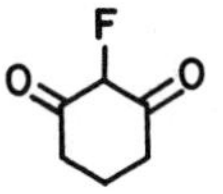 CCl_4	1755 1725 1645 1625	121
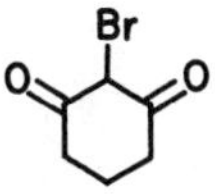 $CHCl_3$	1672 1654 1596	313
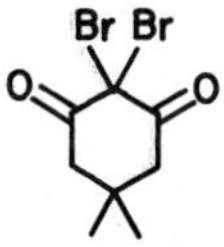 KBr	1745 1725	157
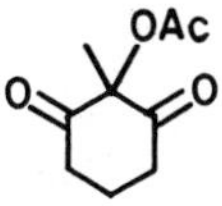 KBr	1718 1736	148
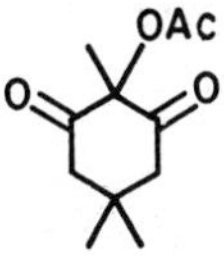 KBr	1712 1754*	148
 $CHCl_3$	1749 1712	316
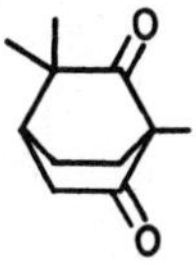 CS_2	1712	316
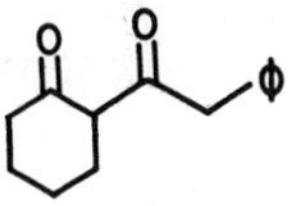 CCl_4	1724 1683 1601bm	317
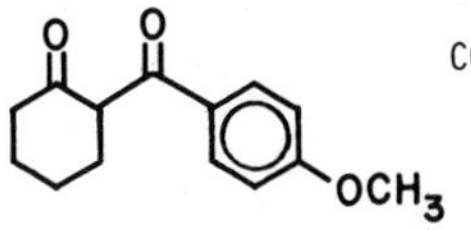 CCl_4	1717 1679 1602 1510m	317
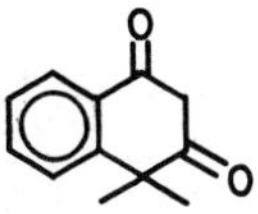 CCl_4	1663	200
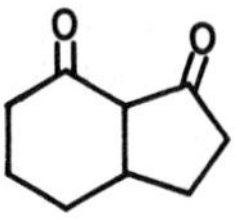 liq.	1761m 1692 1626	318
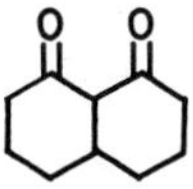 CCl_4	1613b	318
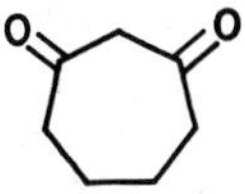 liq.	1724 1709	319
CCl_4	1728 1704	248
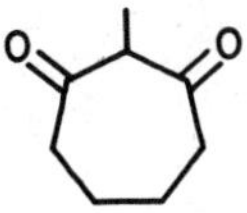 $CHCl_3$	1724 1695	320
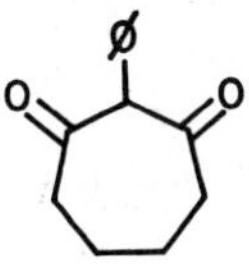 CCl_4	1720* 1695	223
 C_2Cl_4	1706* 1695 1608	302

Solvent	Frequency	Ref.
 C_2Cl_4	1730* 1704 1595	302
 C_2Cl_4	1724* 1699	302
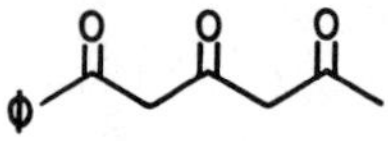 KBr	1603 1570	321
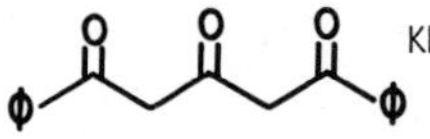 KBr	1603 1570	321
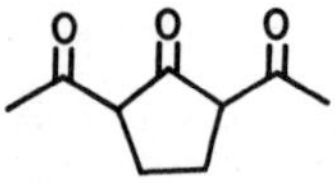 liq.	1725 1670 1605	322
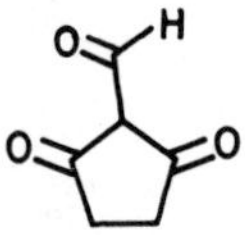 nuj.	1674 1540	323
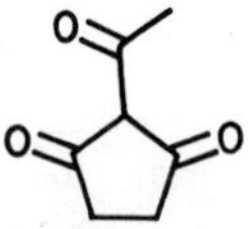 CCl_4	1710 1635 1595	324
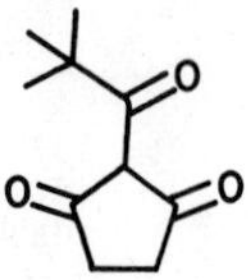 $CHCl_3$	1692 1590 1562	325
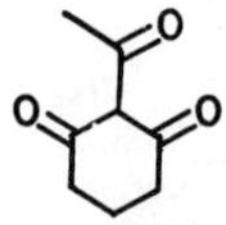 $CHCl_3$	1680 1565	326
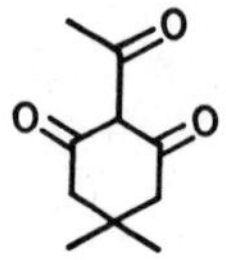 $CHCl_3$	1656 1555	326
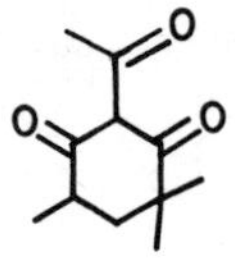 $CHCl_3$	1650 1543	326
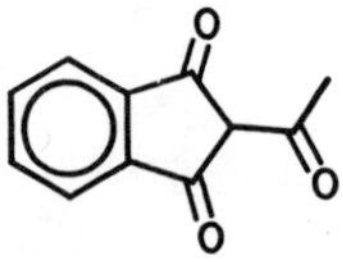 nuj.	1706 1658 1629	327
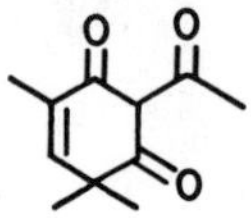 $CHCl_3$	1660 1630 1540	326
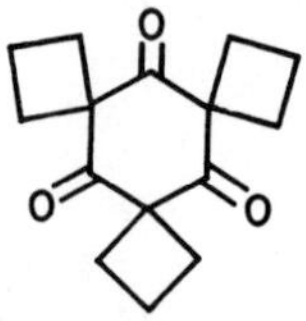 CCl_4	1701	307
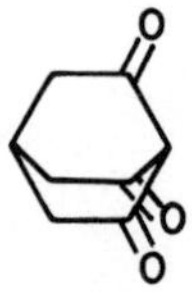 KBr	1770 1725	328
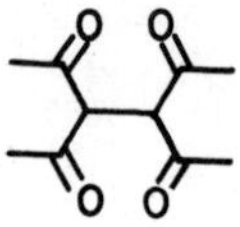 KBr	1595	329

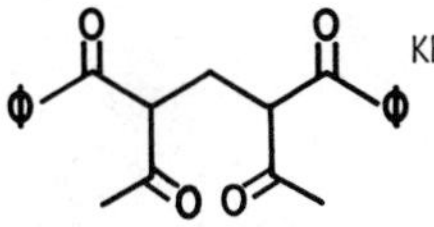

Solvent	Frequency	Ref.
KBr	1692	330
KBr	1704 1686 1672 1595	329
$CHCl_3$	1590 1397	331
$CHCl_3$	1575 1370	331
$CHCl_3$	1586 1373	331
$CHCl_3$	1605 1389	331
$CHCl_3$	1580 1390	331
$CHCl_3$	1524 1391	331
$CHCl_3$	1577 1375	331
$CHCl_3$	1712	2
KBr	1680	332
$CHCl_3$	1731 1721	2
$CHCl_3$	1708	2
$CHCl_3$	1749 1712	316
KBr	1640	332
$CHCl_3$	1690	333

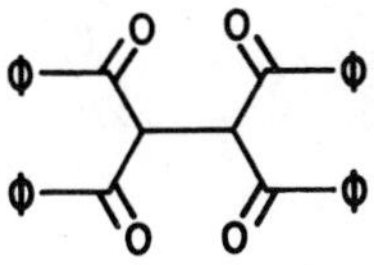

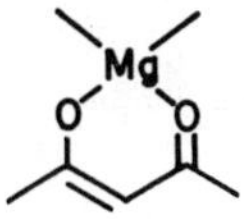

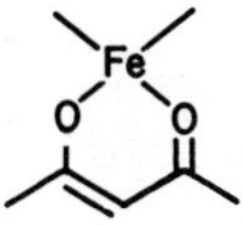

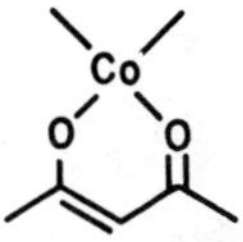

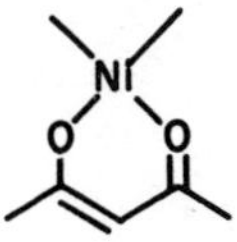

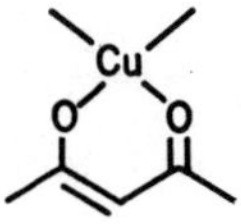

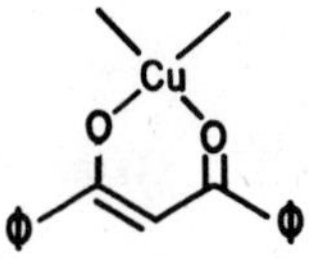

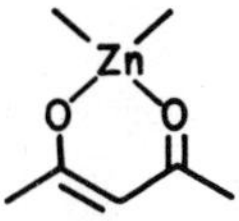

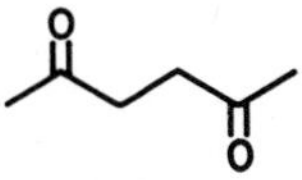

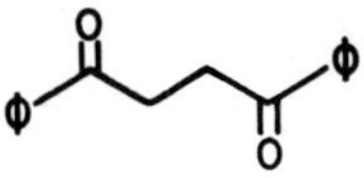

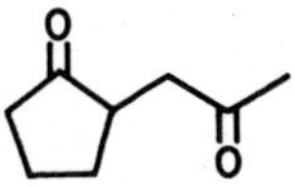

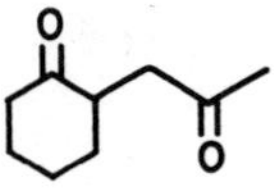

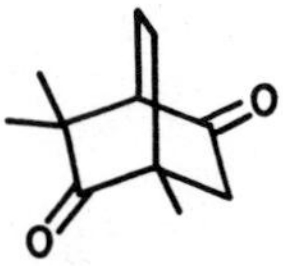

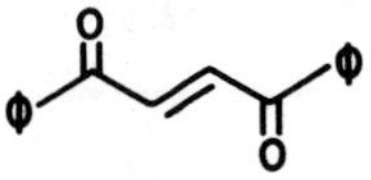

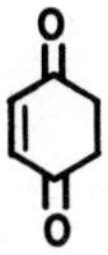

Solvent	cm⁻¹	Ref.
$CHCl_3$	1700	333
$CHCl_3$	1685	333
nuj.	1645 1613	192
liq.	1680	334
nuj.	1705	335
nuj.	1660	335
CCl_4	1675	4
CCl_4	1707	4

Solvent	cm⁻¹	Ref.
$CHCl_3$	1689	336
CCl_4	1783 1693	196
CCl_4	1736 1689	196
CCl_4	1645	337
CCl_4	1637	337
CCl_4	1645	338
liq. CCl_4	1616 1618	337 338
CCl_4	1618	339

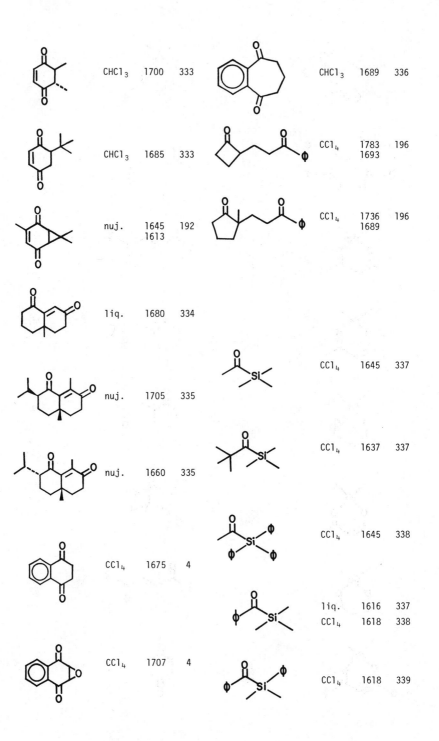

Solvent	Frequency	Ref.
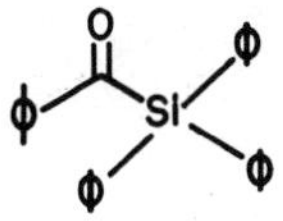 CCl_4	1618	338
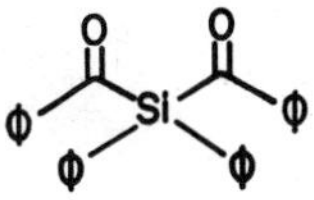 CCl_4	1675 1618	339
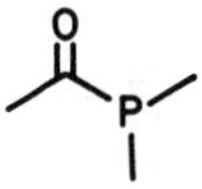 liq.	1758 1660	340
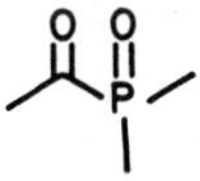 liq.	1760 1723	340
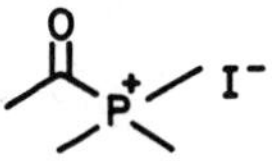 KBr	1722	340
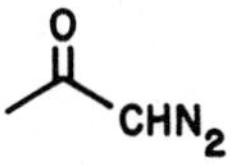 CH_2Cl_2	1647 2092	41
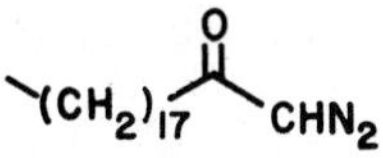 CH_2Cl_2	1642 2088	41
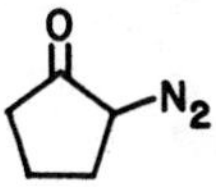 liq.	1672 2086	341
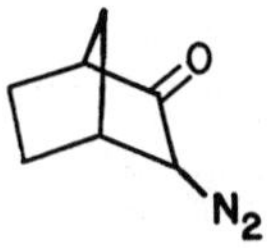 CS_2	1695	144
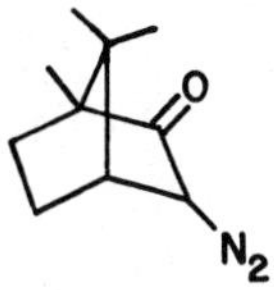 CH_2Cl_2	1684	41
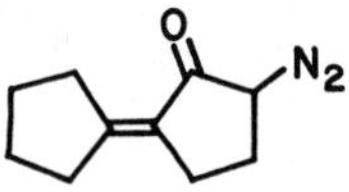 KBr	1675 2075 1634	341
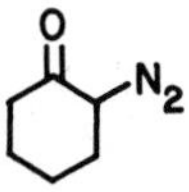 liq.	1634 2083	341
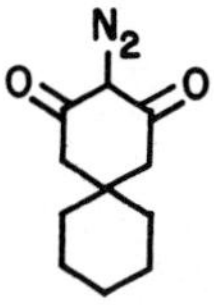 KBr	1672 1639 2193 2151	342
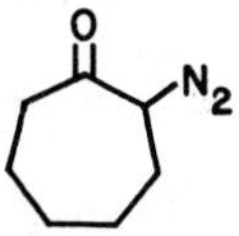 liq.	1631 2079	341
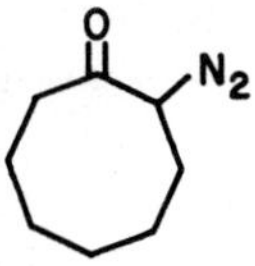 liq.	1631 2081	341
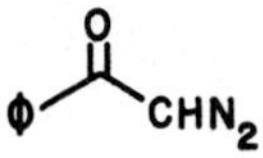 CH_2Cl_2	1621 2092	41

Compound	Solvent	ν	Ref.
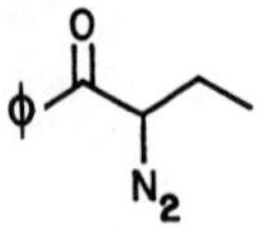	CH_2Cl_2	1613	41
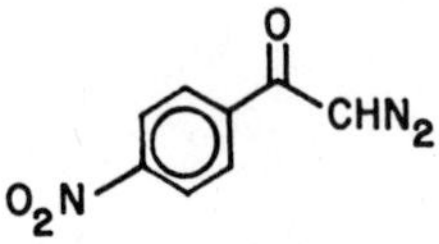	CH_2Cl_2	1629 2101	41
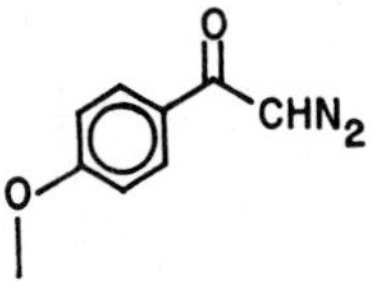	CH_2Cl_2	1621 2101	41
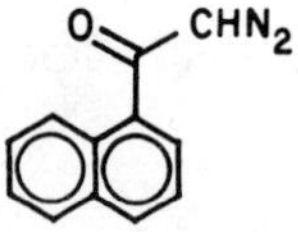	CH_2Cl_2	1629 1613 2088	41
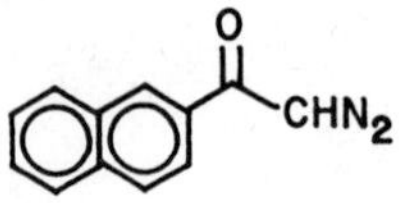	CH_2Cl_2	1613 2092	41
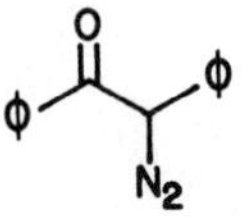	CH_2Cl_2	1623 2062	41
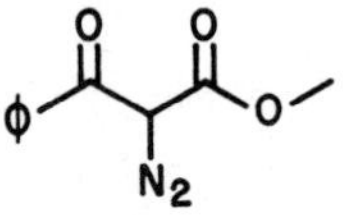	CCl_4	1633 1728	343
	nuj.	1616 1716	344
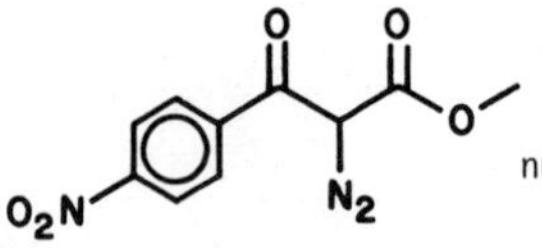	nuj.	1726 1632	344

REFERENCES

1. L. J. Bellamy and R. L. Williams, *JCS* (1957), 4294.
2. Y. Mazur and F. Sondheimer, *Experientia* (1960), 181.
3. R. Mecke and K. Noack, *B* (1960), 210.
4. N. Fuson, M.-L. Josien and E. M. Shelton, *JACS* (1954), 2526.
5. W. D. Hayes and C. J. Timmons, *SCA* (1968), 323.
6. H. W. Thompson and D. A. Jameson, *SCA* (1958) *13*, 236.
7. K. J. Crowley, R. A. Schneider and J. Meinwald, *JCS* (1966) 571.
8. J. DePireux, *BSC Belge* (1957), 218.
9. J. E. Dubois, M. Chastrette and E. Schunk, *BCS France* (1967), 2011.
10. P. D. Bartlett and M. Stiles, *JACS* (1955), 2806.
11. G. W. Cannon, A. A. Santilli and P. Shenian, *JACS* (1959), 1660.
12. S. L. Friess and R. Pinson, Jr., *JACS* (1952), 1302.
13. M. Hanack and H. M. Ensslin, *A* (1966) *697*, 100.
14. J.-L. Pierre, R. Barlet and P. Arnaud, *SCA* (1967) *23A*, 2297.

15. S. E. Wiberley and S. C. Bunce, *Anal Chem* (1952), 623.
16. J. Hanuise and R. R. Smolders, *BSC France* (1967), 2139.
17. H. Hart and O. E. Curtis, Jr., *JACS* (1956), 112.
18. E. D. Bergmann and S. Cohen, *JCS* (1958), 2259.
19. E. D. Bergmann, S. Cohen, E. Hoffman and Z. Rand-Meir, *JCS* (1961), 3452.
20. M. Hanack and H. Eggensperger, *B* (1963), 1341.
21. L. J. Bellamy and R. L. Williams, *JCS* (1957), 861.
22. C. E. Griffin, *SCA* (1960), 1464.
23. C. V. Berney, *SCA* (1965), 1809.
24. C. Rappe, *Arkiv Kemi* (1964) *21*, 503.
25. B. R. Cook and G. A. Crowder, *JCP* (1967) *47*, 1700.
26. H. Dahn and H. Hauth, *Helv* (1964), 1424.
27. N. J. Leonard and S. Gelfand, *JACS* (1955), 3272.
28. N. C. Yang and R. A. Finnegan, *JACS* (1958), 5845.
29. J. Cantacuzène and A. Keramat, *BSC France* (1968), 4540.
30. H. Machleidt, *A* (1963) *667*, 24.
31. N. Noack, *SCA* (1962), 1625
32. E. C. Craven and W. R. Ward, *J Appl Chem* (1960) *10*, 18.
33. H. O. House and R. S. Ro, *JACS* (1958), 2428.
34. M. S. Newman and A. Arkell, *JOC* (1959), 385.
35. N. Bacon, S. Brewis, G. E. Usher, and E. S. Waight, *JCS* (1961), 2255.
36. J. Klein, *T* (1964), 465.
37. J. K. Groves and N. Jones, *T* (1969), 223.
38. E. A. Braude and C. J. Timmons, *JCS* (1955), 3766.
39. G. L. Buchanan, R. A. Raphael and I. W. J. Still, *JCS* (1963), 4372.
40. P. A. Stadler, *Helv* (1960), 1601.
41. P. Yates, B. L. Shapiro, N. Yoda and J. Fugger, *JACS* (1957), 5756.
42. I. Bell, E. R. H. Jones and M. C. Whiting, *JCS* (1958), 1313.
43. J. Elguero, R. Jacquier and C. Marzin, *BSC France* (1967), 3005.
44. N. J. Leonard, J. C. Little and A. J. Kresge, *JACS* (1957), 6436.
45. J. U. Lowe, Jr., and L. N. Ferguson, *JOC* (1965), 3000.
46. M. Fontaine, J. Chauvelier and P. Barchewitz, *BSC France* (1962), 2145.
47. R. B. Davis and D. H. Scheiber, *JACS* (1956), 1675.
48. J. W. Wilson and V. S. Stubblefield, *JACS* (1968), 3423.
49. M. Barrelle and R. Glenat, *BSC France* (1967), 453.
50. E. Müller and A. Segnitz, *Synth* (1970), 147.

51. J. Chouteau, G. Davidovics, M. Bertrand, J. Le Gras, J. Figarella and M. Santelli, *BSC France* (1964), 2562.
52. W. E. Willy and W. E. Thiessen, *JOC* (1970), 1235.
53. S. F. Reed, *JOC* (1962), 4116.
54. H. Brockmann and B. Franck, *Naturwiss* (1955), 70.
55. P. Yates, N. Yoda, W. Brown and B. Mann, *JACS* (1958), 202.
56. R. Lehr, *Harvard Ph.D. Thesis* (1968).
57. D. B. Kurland, *Harvard Ph.D. Thesis* (1967).
58. F. Wille and R. Strasser, *B* (1961), 1606.
59. E. J. Hartwell, R. E. Richards and H. W. Thompson, *JCS* (1948), 1436.
60. J. L. Adelfang, P. H. Hess and N. H. Cromwell, *JOC* (1961), 1402.
61. R. J. Mohrbacher and N. H. Cromwell, *JACS* (1957), 401.
62. F. G. Weber, *T* (1969), 4283.
63. A. Winston, J. P. M. Bederka, W. G. Isner, P. C. Juliano and J. C. Sharp, *JOC* (1965), 2784.
64. N. H. Cromwell, R. W. Bambury and J. L. Adelfang, *JACS* (1960), 4241.
65. R. D. Campbell and F. J. Schultz, *JOC* (1960), 1877.
66. E. H. White and W. J. Considine, *JACS* (1958), 626.
67. M. Yamaguchi, *NKZ* (1957), 1543.
68. H. O. House and G. D. Ryerson, *JACS* (1961), 979.
69. M. Yamaguchi, *NKZ* (1959) 155.
70. C. J. W. Brooks and J. F. Morman, *JCS* (1961), 3372.
71. M. Yamaguchi, *NKZ* (1957), 1236.
72. H. L. Hergert and E. F. Kurth, *JACS* (1953), 1622.
73. H. O. House, D. J. Reif and R. L. Wasson, *JACS* (1957), 2490.
74. F. Bohlmann and K.-M. Kleine, *B* (1962), 39.
75. M. Kröner, *B* (1967), 3172.
76. N. H. Cromwell and G. D. Mercer, *JACS* (1957), 6201.
77. N. H. Cromwell and G. D. Mercer, *JACS* (1957), 3815.
78. E. J. Moriconi, W. F. O'Connor and W. F. Forbes, *JACS* (1960), 5454.
79. U. Eisner and R. L. Erskine, *JCS* (1958), 971.
80. A. Ermili, A. J. Castro and P. A. Westfall, *JOC* (1965), 339.
81. W. D. Cooper, *JOC* (1958), 1382.
82. E. Bisagni, J.-P. Marquet, J. André-Louisfert, A. Cheutin and F. Feinte, *BSC France* (1967), 2796.
83. R. E. Rosenkranz, K. Allner, R. Good, W. V. Philipsborn and C. H. Eugster, *Helv* (1963), 1259.
84. H. Rapoport and C. D. Willson, *JACS* (1962), 630.
85. P. A. Finan and G. A. Fothergill, *JCS* (1963), 2723.

86. DMS 2257.
87. DMS 2272.
88. DMS 2259.
89. DMS 2240.
90. DMS 2588.
91. A. R. Katritzky, A. M. Monro, J. A. T. Beard, D. P. Dearnaley and N. J. Earl, *JCS* (1958), 2182.
92. N. J. Turro and W. B. Hammond, *T* (1968), 6017.
93. J. F. Pazos and F. D. Greene, *JACS* (1967), 1030.
94. A. Krebs and B. Schrader, *A* (1968) *709*, 46.
95. W. B. Hammond and N. J. Turro, *JACS* (1966), 2880.
96. R. Breslow and R. Peterson, *JACS* (1960), 4426.
97. J. Goré, *Ph.D. Thesis Caen* (1964).
98. P. Dowd and K. Sachdev, *JACS* (1967), 715.
99. J. Goré, C. Djerassi and J.-M. Conia, *BSC France* (1967), 950.
100. J. Meinwald, J. W. Wheeler, A. A. Nimetz and J. S. Liu, *JOC* (1965), 1038.
101. D. C. England and C. G. Krespan, *JOC* (1970), 3300.
102. D. C. England, *JACS* (1961), 2205.
103. J.-M. Conia and J.-L. Ripoll, *BSC France* (1963), 768.
104. J. K. Crandall and D. R. Paulson, *JOC* (1968), 3291.
105. J. K. Crandall, W. H. Machleder and M. J. Thomas, *JACS* (1968), 7346.
106. E. Casadevall, C. Largeau and P. Moreau, *BSC France* (1968), 1514.
107. F. Nerdel, D. Frank and H. Marschall, *B* (1967), 720.
108. U. A. Huber and A. S. Dreiding, *Helv* (1970), 495.
109. W. T. Brady and O. H. Waters, *JOC* (1967), 3703.
110. C. D. Gutsche and T. D. Smith, *JACS* (1960), 4067.
111. P. Yates and A. G. Fallis, *TL* (1968), 2493.
112. J.-M. Conia and J.-P. Sandré, *BSC France* (1963), 744.
113. O. Scherer, G. Hörlein and H. Millauer, *B* (1966), 1966.
114. E. V. Dehmlow, *B* (1967), 3829.
115. T. Bürer and Hs. H. Günthard, *Helv* (1956), 356.
116. H. O. House and B. M. Trost, *JOC* (1965), 2502.
117. A. P. Krapcho and J. E. McCullough, *JOC* (1967), 2453.
118. B. A. Gubler, *Thesis ETH* (1965).
119. K. Alder and F. H. Flock, *B* (1956), 1732.
120. C. W. Shoppee, R. H. Jenkins and G. H. R. Summers, *JCS* (1958), 3048.
121. H. Machleidt and V. Hartmann, *A* (1964) *679*, 9.

122. H. O. House, P. P. Wickham and H. C. Müller, *JACS* (1962), 3139.
123. H. Feuer and P. M. Pivawer, *JOC* (1966), 3152.
124. A. Hassner, J. M. Larkin and J. E. Dowd, *JOC* (1968), 1733.
125. R. Mayer and H.-J. Schubert, *B* (1958), 768.
126. J. L. Ripoll and J.-M. Conia, *TL* (1969), 979.
127. J.-P. Pete, *BSC France* (1967), 357.
128. J. R. Williams and H. Ziffer, *Chem Comm* (1967), 194.
129. R. Fraisse-Jullien and C. Frejaville, *BSC France* (1968), 4449.
130. H. O. House and R. L. Wasson, *JACS* (1957), 1488.
131. R. Joos, *Thesis ETH* (1970).
132. S. Winstein and J. Sonnenberg, *JACS* (1961), 3235.
133. W. Broser and D. Rahn, *B* (1967), 3472.
134. S. J. Etheredge, *JOC* (1966), 1990.
135. J. Meinwald, J. J. Tufariello and J. J. Hurst, *JOC* (1964), 2914.
136. J. Brugidou and H. Christol, *BSC France* (1968), 1141.
137. J.-M. Conia and G. Moinet, *BSC France* (1969), 500.
138. P. Beslin, R. Bloch, G. Moinet and J.-M. Conia, *BSC France* (1969), 508.
139. H. O. House and G. H. Rasmusson, *JOC* (1963), 27.
140. H. K. Hall, Jr. and R. Zbinden, *JACS* (1958), 6428.
141. P. G. Gassman and P. G. Pape, *JOC* (1964), 160.
142. J. Haywood-Farmer, R. E. Pincock and J. I. Wells, *T* (1966), 2007.
143. J. J. Hurst and G. H. Whitham, *JCS* (1963), 710.
144. J. Dolde, *Thesis Tübingen* (1966), p. 49.
145. W. G. Dauben, J. L. Chitwood and K. V. Scherer, Jr., *JACS* (1968), 1014.
146. J. Meinwald, J. K. Crandall and P. G. Gassman, *T* (1962) *18*, 815.
147. G. Snatzke and G. Zanati, *A* (1965) *684*, 62.
148. T. A. Spencer, A. L. Hall and C. F. von Reyn, *JOC* (1968), 3369.
149. H. O. House and R. L. Wasson, *JACS* (1957), 1488.
150. H. E. Smith and R. H. Eastman, *JACS* (1957), 5500.
151. S. B. Kulkarni and Sukh Dev, *T* (1968), 545.
152. F. Gautschi, O. Jeger, V. Prelog and R. B. Woodward, *Helv* (1954), 2280.
153. H. H. Rennhard, G. DiModica, W. Simon, E. Heilbronner and A. Eschenmoser, *Helv* (1957), 957.
154. C. D. Gutsche, I. Y. C. Tao and J. Kozma, *JOC* (1967), 1782.
155. B. W. Ponder and D. R. Walker, *JOC* (1967), 4136.
156. K. Hafner and K. Goliasch, *B* (1961), 2909.
157. H. De Pooter and N. Schamp, *BSC Belges* (1968), 377.

158. P. Yates and L. L. Williams, *JACS* (1958), 5896.
159. R. L. Erskine and E. S. Waight, *JCS* (1960), 3425.
160. M. Horák and P. Munk, *Col Czech Comm* (1959), 3024.
161. W. F. Erman, *JOC* (1967), 765.
162. DMS 1546.
163. H. O. House and M. Schellenbaum, *JOC* (1963), 34.
164. A. Hassner and T. C. Mead, *T* (1964), 2201.
165. M.-L. Josien and N. Fuson, *CR* (1953) *236*, 1879.
166. H. Güsten, G. Kirsch and D. Schulte-Frohlinde, *T* (1968), 4393.
167. O. B. Edgar and D. H. Johnson, *JCS* (1958), 3925.
168. M. Yamaguchi, *NKZ* (1960), 1118.
169. V. C. Farmer, N. F. Hayes and R. H. Thomson, *JCS* (1956), 3600.
170. H. Hart and R. K. Murray, Jr., *JOC* (1967), 2448.
171. W. M. Schubert and W. A. Sweeney, *JACS* (1955), 4172.
172. H. O. House and G. H. Rasmusson, *JOC* (1963), 31.
173. J. Derkosch and W. Kaltenegger, *Monatsh Chem* (1959), 872.
174. B. D. Pearson, R. A. Ayer and N. H. Cromwell, *JOC* (1962), 3038.
175. S. J. Holt, A. E. Kellie, D. G. O'Sullivan and P. W. Sadler, *JCS* (1958), 1217.
176. J. Allinger and N. L. Allinger, *T* (1958) *2*, 64.
177. G. Farger and A. S. Dreiding, *Helv* (1966), 552.
178. M. Hanack and W. Keberle, *B* (1963), 2937.
179. E. J. Corey, J. H. Topie and W. A. Wozniak, *JACS* (1955), 5415.
180. J. K. Crandall, J. P. Arrington and J. Hen, *JACS* (1967), 6208.
181. E. G. Cummins and J. E. Page, *JCS* (1957), 3847.
182. J. Cantacuzène and R. Jantzen, *T* (1970), 2429.
183. J. Cantacuzène and M. Atlani, *T* (1970), 2447.
184. E. J. Corey, *JACS* (1953), 2301.
185. J. P. Bervelt, R. Ottinger, P. A. Peters, J. Reisse and G. Chiurdoglu, *SCA* (1968) *24A*, 1411.
186. E. J. Corey and H. J. Burke, *JACS* (1955), 5418.
187. E. J. Corey, *JACS* (1953), 3297.
188. M. E. Kuehne, *JACS* (1959), 5400.
189. L. J. Bellamy and L. Beecher, *JCS* (1954), 4487.
190. C. H. Eugster, L. Leichner and E. Jenny, *Helv* (1963), 543.
191. N. J. Leonard and P. M. Mader, *JACS* (1950), 5388.
192. W. D. P. Burns, M. S. Carson, W. Cocker and P. V. R. Shannon, *JCS* (1968) *C*, 3073.
193. E. J. Corey, J. D. Bass, R. LeMahieu and R. B. Mitra, *JACS* (1964), 5570.
194. J.-M. Conia and P. Beslin, *BSC France* (1969), 483.

195. P. S. Wharton and C. E. Sundin, *JOC* (1968), 4255.
196. G. L. Buchanan, A. C. W. Curran, J. M. McCrae and G. W. McLay, *T* (1967), 4729.
197. H. O. House and G. A. Frank, *JOC* (1965), 2948.
198. A. W. Baker, *SCA* (1965), 1603.
199. M. Yanagita and K. Yamakawa, *JOC* (1956), 500.
200. R. D. Campbell and N. H. Cromwell, *JACS* (1957), 3456.
201. A. Casadevall, E. Casadevall and M. Lasperas, *BSC France* (1968), 4506.
202. C. S. Foote and R. B. Woodward, *T* (1964), 687.
203. P. Von R. Schleyer and R. D. Nicholas, *JACS* (1961), 182.
204. R. Zbinden and H. K. Hall, Jr., *JACS* (1960), 1215.
205. R. N. McDonald and R. N. Steppel, *JOC* (1970), 1250.
206. J. T. Lumb and G. H. Whitham, *T* (1965), 499.
207. C. A. Grob and A. Weiss, *Helv* (1960), 1390.
208. R. E. Lyle, R. E. Adel and G. G. Lyle, *JOC* (1959), 342.
209. D. Dolphin and A. E. Wick, unpublished data.
210. R. A. Johnson, M. E. Herr, H. C. Murray and G. S. Fonken, *JOC* (1968), 3187.
211. G. R. Owen and C. B. Reese, *JCS* (1970) *C*, 2401.
212. H. E. Shook, Jr. and L. D. Quin, *JACS* (1967), 1841.
213. N. J. Leonard, T. L. Brown and T. W. Milligan, *JACS* (1959), 504.
214. P. Y. Johnson and G. A. Berchtold, *JOC* (1970), 584.
215. N. J. Leonard and C. R. Johnson, *JOC* (1962), 282.
216. AE. De Groot, J. A. Boerma, J. De Valk and H. Wynberg, *JOC* (1968), 4025.
217. L. J. Bellamy and R. J. Pace, *SCA* (1963), 1831.
218. R. H. Wiley and C. H. Jarboe, *JACS* (1956), 624.
219. S. Inayama, *Chem Pharm Bull* (1956) *4*, 198.
220. E. W. Garbisch, Jr., *JOC* (1965), 2109.
221. M. Tomoeda, M. Inuzuka, T. Furuta, M. Shinozuka, and T. Takahashi, *T* (1968), 959.
222. F. G. Bordwell and K. M. Wellman, *JOC* (1963), 2544.
223. H. O. House and R. L. Wasson, *JACS* (1956), 4394.
224. J. Libman, M. Sprecher and Y. Mazur, *JACS* (1969), 2062.
225. A. Suzuki and T. Matsumoto, *BCS Japan* (1962), 2027.
226. W. S. Johnson, J. Dolf Bass and K. L. Williamson, *T* (1963), 861.
227. H. Mühle and Ch. Tamm, *Helv* (1962), 1475.
228. P. Bey and G. Ourisson, *BSC France* (1968), 1402.
229. J. J. Beereboom, *JOC* (1966), 2026.
230. H. Hart and D. W. Swatton, *JACS* (1967), 1874.

231. J. A. Marshall and H. Roebke, *JOC* (1966), 3109.
232. D. W. Theobald, *T* (1966), 2869.
233. H. O. House and R. W. Bashe II, *JOC* (1965), 2942.
234. I. G. Morris and A. R. Pinder, *JCS* (1963), 1841.
235. Y.-R. Naves, *Helv* (1966), 2012.
236. R. Baird and S. Winstein, *JACS* (1957), 4238.
237. R. Gompper, R. R. Schmidt and E. Kutter, *A* (1965) *684*, 37.
238. N. J. Leonard and G. C. Robinson, *JACS* (1953), 2714.
239. C. L. Stevens, J. J. Beereboom, Jr. and K. G. Rutherford, *JACS* (1955), 4590.
240. J. W. Huffman and T. W. Bethea, *JOC* (1965), 2956.
241. D. N. Kevill, E. D. Weiler and N. H. Cromwell, *JOC* (1964), 1276.
242. M.-L. Josien and N. Fuson, *BSC France* (1952), 389.
243. M.-L. Josien, N. Fuson, J.-M. Lebas and T. M. Gregory, *JCP* (1953) *21*, 331.
244. J.-C. Salfeld, *B* (1960), 737.
245. P. Yates, M. I. Ardao and L. F. Fieser, *JACS* (1956), 650.
246. D. Hadži, *JCS* (1956), 2725.
247. M. St. C. Flett, *JCS* (1948), 1441.
248. I. Maclean and R. P. A. Sneeden, *T* (1965), 31.
249. O. L. Chapman and T. H. Koch, *JOC* (1966), 1042.
250. R. Borsdorf, W. Flamme, H. Kumpfert and M. Mühlstädt, *T* (1968), 65.
251. J. R. B. Campbell, A. M. Islam and R. A. Raphael, *JCS* (1956), 4096.
252. G. Stork, M. Nussim and B. August, *T Suppl* (1966) *8*, 105.
253. H. Christol, F. Plénat and C. Reliaud, *BSC France* (1968), 1566.
254. A. P. ter Borg and H. Kloosterziel, *Rec Trav Chim* (1963), 1189.
255. S. Moon and C. R. Ganz, *JOC* (1970), 1241.
256. O. L. Chapman and R. A. Fugiel, *JACS* (1969), 215.
257. M. Mühlstädt, *Naturwiss* (1958), 240.
258. N. J. Leonard, L. A. Miller and J. W. Berry, *JACS* (1957), 1482.
259. E. Kloster-Jensen, N. Tarköy, A. Eschenmoser and E. Heilbronner, *Helv* (1956), 786.
260. W. Treibs and P. Grossman, *B* (1959), 267.
261. S. Olsen and R. Bredoch, *B* (1958), 1589.
262. N. J. Leonard, T. W. Milligan and T. L. Brown, *JACS* (1960), 4075.
263. L. A. Paquette and R. F. Eizember, *JACS* (1967), 6205.
264. W. M. Schubert, W. A. Sweeney and H. K. Latourette, *JACS* (1954), 5462.
265. N. J. Leonard and M. Öki, *JACS* (1955), 6245.

266. R. A. Benkeser and R. F. Cunico, *JOC* (1967), 395.
267. R. S. Rasmussen, D. D. Tunicliff and R. R. Brattain, *JACS* (1949), 1068.
268. H. M. Randall, R. G. Fowler, N. Fuson and J. R. Dangl, *Infrared Determination of Organic Structures* (1949), D. Van Nostrand, New York.
269. N. J. Leonard, H. A. Laitinen and E. H. Mottus, *JACS* (1953), 3300.
270. S. Bien and D. Ovadia, *JOC* (1970), 1028.
271. H. E. Baumgarten and J. E. Dirks, *JOC* (1958), 900.
272. A. W. Johnson, J. R. Quayle, T. S. Robinson, N. Sheppard and A. R. Todd, *JCS* (1951), 2633.
273. W. Ried and D. P. Schäfer, *B* (1969), 4193.
274. A. Treibs and K. Jacob, *A* (1966) *699*, 153.
275. S. Cohen, J. R. Lacher and J. D. Park, *JACS* (1959), 3480.
276. H. O. House, B. M. Trost, R. W. Magin, R. G. Carlson, R. W. Franck and G. H. Rasmusson, *JOC* (1965), 2513.
277. K. Sato, S. Suzuki and Y. Kojima, *JOC* (1967), 339.
278. N. J. Leonard and J. C. Little, *JACS* (1958), 4111.
279. K. Alder, H. K. Schäfer, H. Esser, H. Krieger and R. Reubke, *A* (1955) *593*, 23.
280. E. Tamada, *NKZ* (1958), 494.
281. J.-C. Salfeld and E. Baume, *B* (1960), 745.
282. K. Kotera, *Yakugaku Zasshi* (1960), 1281.
283. C. J. W. Brooks, G. Eglinton and D. S. Magrill, *JCS* (1961), 308.
284. R. J. W. LeFèvre, F. Maramba and R. L. Werner, *JCS* (1953), 2496.
285. W. Reusch and R. LeMahieu, *JACS* (1964), 3068.
286. N. B. Haynes, D. Redmore and C. J. Timmons, *JCS* (1963), 2420.
287. J. Braband, M. Mühlstädt and G. Mann, *T* (1970), 3667.
288. H. O. House and H. W. Thompson, *JOC* (1961), 3729.
289. W. H. Urry, Mei-Shu H. Pai and C. Y. Chen, *JACS* (1964), 5342.
290. A. R. Lepley and J. P. Thelman, *T* (1966), 101.
291. D. F. Martin, M. Shamma and W. C. Fernelius, *JACS* (1958), 4891.
292. L. Horner and F. Maurer, *B* (1968), 1783.
293. S. Bratož, D. Hadži and G. Rossmy, *Trans Farad Soc* (1956), 464.
294. H. Musso and K. Figge, *A* (1963) *668*, 1.
295. É. Ya. Gren, A. K. Grinvalde and G. Ya. Vanag, *J Gen Chem USSR* (1966), 17.
296. H. Nozaki, Z. Yamaguti, T. Okada, R. Noyori and M. Kawanisi, *T* (1967), 3993.
297. V. C. Petrus, *Thesis U of Montpelier* (1965).

298. K. L. Wierzchowski and D. Shugar, *SCA* (1965), 943.
299. A. W. Allan and R. P. A. Sneeden, *T* (1962) *18*, 821.
300. I. A. Kaye and R. S. Matthews, *JOC* (1964), 1341.
301. C. W. Hussey and A. R. Pinder, *JCS* (1961), 3525.
302. S. Hünig, H.-J. Buysch, H. Hoch and W. Lendle, *B* (1967), 3996.
303. F. A. Miller, F. E. Kiviat and I. Matsubara, *SCA* (1968) *24A*, 1523.
304. H. H. Wasserman and E. V. Dehmlow, *JACS* (1962), 3786.
305. A. Corbella, G. Jommi, G. Ricca and G. Russo, *Gazz Chim Ital* (1965), 948.
306. R. H. Hasek, P. G. Gott and J. C. Martin, *JOC* (1964), 2510.
307. J. L. E. Erickson, F. E. Collins, Jr. and B. L. Owens, *JOC* (1966), 480.
308. D. C. England and C. G. Krespan, *JOC* (1970), 3322.
309. C. H. DePuy and E. F. Zaweski, *JACS* (1959), 4920.
310. A. Roedig and H. Ziegler, *B* (1961), 1800.
311. A. Mustafa and A. H. E. Harhash, *JACS* (1956), 1649.
312. Yu. N. Sheinker, B. E. Zaitsev and M. E. Perel'son, *Izvest Akad Nauk Otdel Khim* (1964), 2114.
313. S. N. Ananchenko, I. V. Berezin and I. V. Torgov, *Izvest Akad Nauk Otdel Khim* (1960), 1644.
314. B. Eistert and W. Reiss, *B* (1954), 92.
315. Ch. Tamm and R. Albrecht, *Helv* (1960), 768.
316. D. Y. Curtin and R. R. Fraser, *JACS* (1959), 662.
317. R. D. Campbell and H. M. Gilow, *JACS* (1960), 5426.
318. I. A. Kaye and R. S. Matthews, *JOC* (1963), 325.
319. B. Eistert, F. Haupter and K. Schank, *A* (1963) *665*, 55.
320. R. Selvarajan, J. P. John, K. V. Narayanan and S. Swaminathan, *T* (1966), 949.
321. R. J. Light and C. R. Hauser, *JOC* (1960), 538.
322. M. Vandewalle and S. Dewaele, *BSC Belge* (1967), 468.
323. B. D. Akehurst and J. R. Bartels-Keith, *JCS* (1957), 4798.
324. F. Merényi and M. Nilsson, *Acta Chem Scand* (1963), 1801.
325. F. Merényi and M. Nilsson, *Acta Chem Scand* (1967), 1755.
326. W. R. Chan and C. H. Hassall, *JCS* (1956), 3495.
327. F. M. Dean, R. A. Eade, R. Moubasher and A. Robertson, *JCS* (1957), 3497.
328. W. Theilacker and E. Wegner, *A* (1963) *664*, 125.
329. D. F. Martin, W. C. Fernelius and M. Shamma, *JACS* (1959), 130.
330. D. F. Martin and W. C. Fernelius, *JACS* (1959), 1509.
331. L. J. Bellamy and R. F. Branch, *JCS* (1954), 4491.

332. A. Kreutzberger and P. A. Kalter, *JOC* (1960), 554.
333. D. D. Chapman, W. J. Musliner and J. W. Gates, *JCS* (1969) *C*, 124.
334. S. K. Malhotra, J. J. Hostynek and A. F. Lundin, *JACS* (1968), 6565.
335. D. W. Theobald, *T* (1964), 1455.
336. J. F. Garden and R. H. Thomson, *JCS* (1957), 2851.
337. E. J. Corey, D. Seebach and R. Freedman, *JACS* (1967), 434.
338. A. G. Brook, M. A. Quigley, G. J. P. Peddle, N. V. Schwartz and C. M. Warner, *JACS* (1960), 5102.
339. A. G. Brook and G. J. D. Peddle, *Canad J Chem* (1963), 2351.
340. R. G. Kostyanovsky, V. V. Yakshin, and S. L. Zimont, *T* (1968), 2995.
341. M. Regitz and J. Rüter, *B* (1968), 1263.
342. B. Eistert and G. Heck, *A* (1965) *681*, 123.
343. J. H. Looker and D. N. Thatcher, *JOC* (1957), 1233.
344. J. H. Looker and C. H. Hayes, *JOC* (1963), 1342.

4.3 amides

(structure: $CH_3C(=O)NH_2$)	3550-3450 cm^{-1} 3450-3350 cm^{-1}	Medium-intensity bands arising from the symmetric and asymmetric N-H stretching vibrations are found here. These bands result from nonassociated primary amides and are seen in dilute solution.
	3350-3000 cm^{-1}	In solution several medium-intensity bands may be observed resulting from various types of associated $-NH_2$ groups. With solids (KBr and nujol) this complex pattern frequently collapses to two broad absorptions in this region.
	1650-1580 cm^{-1}	The amide II band is of medium intensity and arises from the $-NH_2$ deformation mode. In the solid state the band occurs at the high-frequency end of the scale, whereas in dilute solution the

frequency falls toward the low end of the range. In concentrated solution bands arising from both free and associated -NH_2 deformation modes may be observed.

3475-3400 cm^{-1} A medium-intensity band arises from the stretching mode of the free -NH group of the S-trans conformer.

3450-3400 cm^{-1} A medium-intensity band arising from the stretching mode of the free -NH group in the S-cis conformation may be found here. This band is frequently weaker than the corresponding band of the S-trans isomer.

3375-3000 cm^{-1} A medium-intensity band at the higher end of the range and several weaker bands toward the lower end of the range may be observed in condensed phases and concentrated solutions.

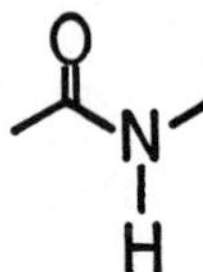

1580-1510 cm^{-1} A medium-intensity band resulting from the coupling of the -NH deformation and C-N stretching modes in a trans-coplanar conformation is seen here. This amide II band appears at higher frequencies in the solid state or concentrated solutions, and moves toward lower frequencies in dilute solution. Bands due to both free and associated groups may appear in concentrated solution.

1315-1200 cm^{-1} — The amide III band is generally weak and results from a coupling of the C-N stretching and -NH deformation modes.

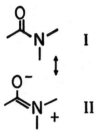

Most of the changes in frequency that arise from various substitution patterns with amides can be rationalized in terms of the resonance structures I and II which contribute to the amide chromophore. Delocalization of the lone pair of electrons on nitrogen into the carbonyl group is considerable, and this accounts for an activation energy (18 Kcal $mole^{-1}$ with DMF) for rotation about the C-N bond. This delocalization decreases the force constant of the carbonyl group with consequent lowering of the carbonyl stretching frequencies of amide carbonyl groups compared to ketonic carbonyl groups.

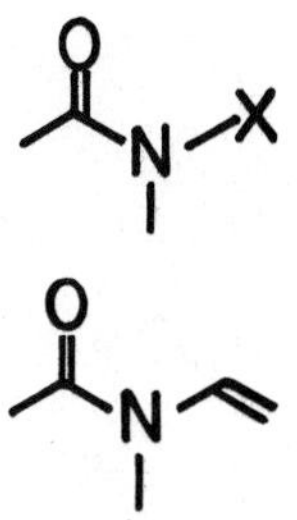

Substitution on the amide nitrogen of electronegative groups or unsaturated groups that are capable of lowering the degree of lone-pair delocalization cause an increase in the carbonyl stretching frequency. Such increases are observed with nitroso, nitro, halogen, alkenyl, and aryl substituents. An extreme example of this is N-acetylpyrrole, which absorbs at 1732 cm^{-1} in KBr. In this case, the lone pair of electrons on nitrogen are part of the aromatic sextet and are not available for delocalization into the carbonyl group, with the result that this amide shows a typical ketonic stretching frequency.

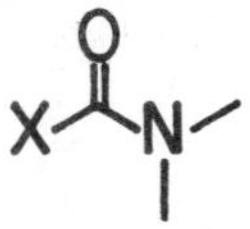

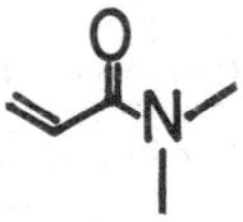

Substitutions on the carbonyl side of the amide chromophore have similar, though smaller, effects to substitution on ketones (Chapter 4.2, p. 175). Thus α-halogens cause an increase in frequency, whereas conjugation with alkenyl and aryl groups cause a decrease in frequency.

The association of amides causes marked decreases in the carbonyl stretching frequencies. Thus formamide, which absorbs at 1742 cm^{-1} in the vapor phase, suffers successive decreases in frequency (1722 cm^{-1} in CCl_4; 1709 cm^{-1} in $CHCl_3$; 1672 cm^{-1} as a liquid) as solute-solvent interaction increases. Similar decreases in frequency are observed with all amides as the polarity of the solvent increases; neat liquids or samples in KBr or nujol invariably absorb at lower frequencies and often exhibit broad bands. In concentrated solutions two carbonyl bands, which result from free and associated molecules, may be observed.

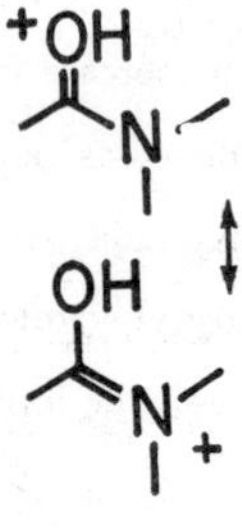

III

IV

Protonation[a] of amides gives the resonance stabilized species (III ⟷ IV) and results in a lowering of the carbonyl stretching frequency.

[a] R. Stewart, L. J. Muenster and J. T. Edward, *Chem. and Ind.*, (1961) 1906.

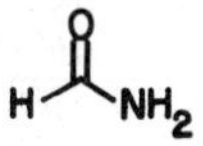

Structure	Medium	ν (cm⁻¹)	Ref.
O / H / NH_2	liq.	1672	3
	CCl_4	1722	1
	$CHCl_3$	1709	2
O / NH_2	KBr	1684 1641m 1595 1461m	4
	CCl_4	1714	1
	$CHCl_3$	1675 1592 1385	4
ϕ / O / NH_2	nuj.	1634	6
	$CHCl_3$	1678	5
F / O / NH_2	KBr	1665b	7
O / F_3C / NH_2	$CHCl_3$	1750	2
Cl / O / NH_2	KBr	1655 1640 1608	7
	$CHCl_3$	1695	2
Cl / O / NH_2 / Cl	KBr	1671	7
	$CHCl_3$	1716	2

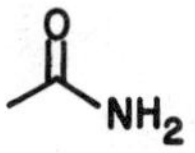

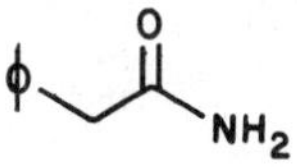

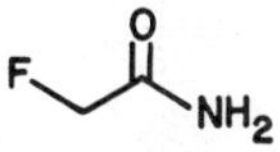

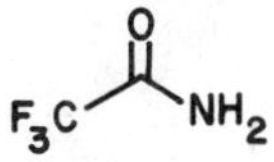

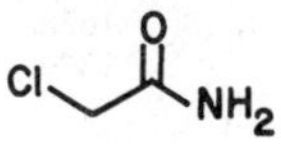

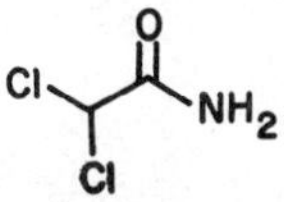

Structure	Medium	ν (cm⁻¹)	Ref.
O / Cl_3C / NH_2	$CHCl_3$	1732	2
I / O / NH_2	KBr	1633	7
NC / O / NH_2	nuj.	1694	8
O / O / NH_2	$CHCl_3$	1691	2
O / NH_2 · HCl	nuj.	1770m 1718 1665w 1608m 1474	4
O / NH_2	KBr	1654b	7
	$CHCl_3$	1687	2
O / NH_2 / Br	KBr	1661b	7
O / NH_2 / OH	KBr	1653b 1637b	7
O / F_3C / NH_2 / F F	nuj.	1600	9

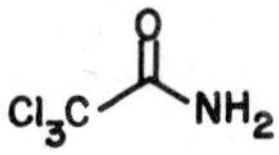

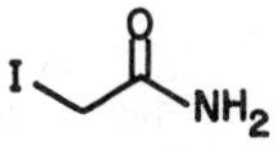

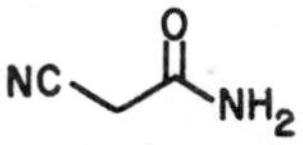

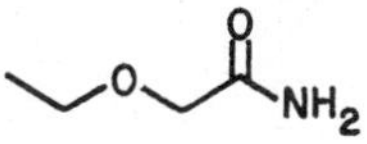

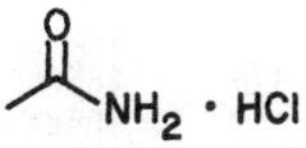

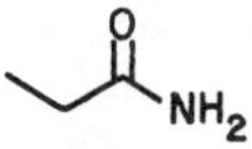

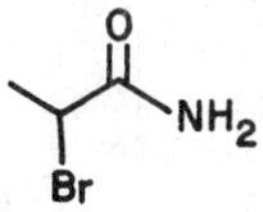

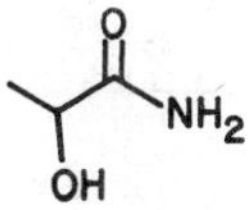

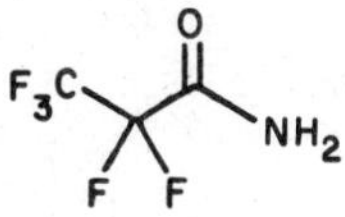

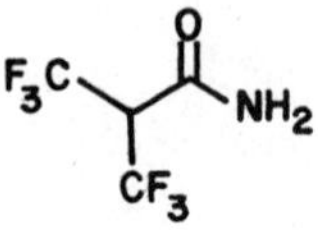

Structure	Solvent	Frequency	Ref.
F_3C, CF_3, O, NH_2	CCl_4	1742	10
O, NH_2	nuj.	1631	12
	$CHCl_3$	1679	11
O, NH_2	KBr	1655	13
	$CHCl_3$	1664	13
O, NH_2	KBr	1660* 1645* 1631	7
O, NH_2	nuj.	1695 1645*	14
	$CHCl_3$	1686 1647m	15
O, NH_2	KBr	1677 1617b	7
O, NH_2	nuj.	1656 1642*	16
	KBr	1658 1598	7
ϕ, O, NH_2	KBr	1658 1627m 1602	7
	CHCl	1675	17
O, NH_2	KBr	1655 1631 1603	7
ϕ, O, NH_2	nuj.	1672	18
O, NH_2	KBr	1650 1632 1600	7
O, NH_2	KBr	1658 1600	7
O, NH_2	KBr	1678	19
$CH_3-(CH_2)_5-C\equiv C-C(=O)NH_2$	CH_2Cl_2	1678 1590	20
O, NH_2	KBr	1656 1623	21
	$CHCl_3$	1678	2

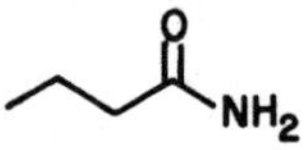

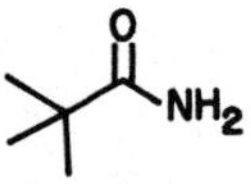

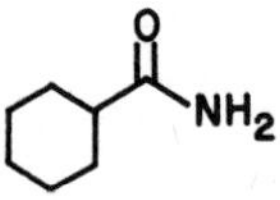

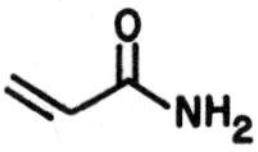

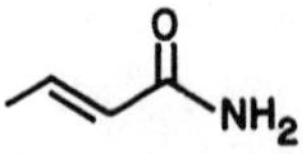

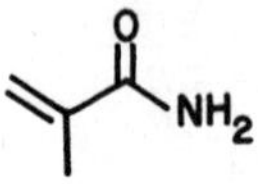

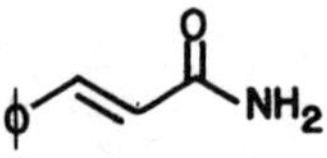

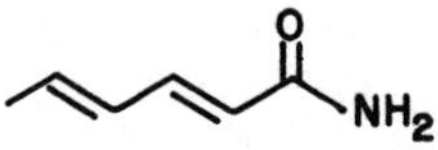

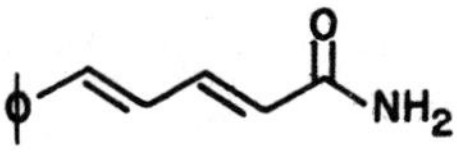

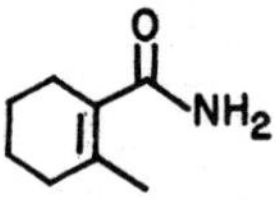

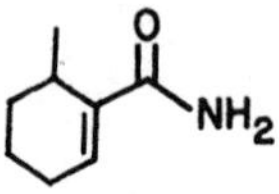

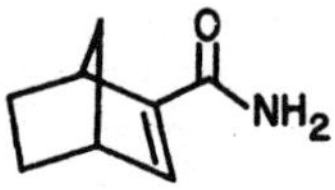

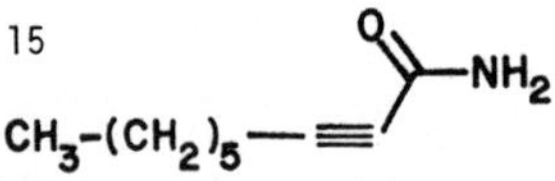

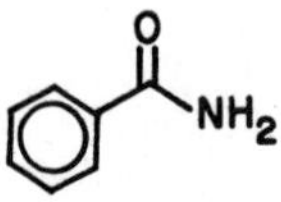

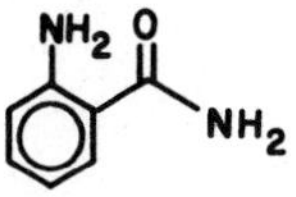

KBr	1650b 1620b 1603*	7
KBr	1672 1627	7
KBr	1632b 1610b 1600b	7
KBr	1670 1620m 1602m 1735	7
nuj.	1669	22
$CHCl_3$	1690	22
nuj.	1645	22
$CHCl_3$	1675	22
KBr	1653 1615	7
KBr	1660* 1651 1608b	7

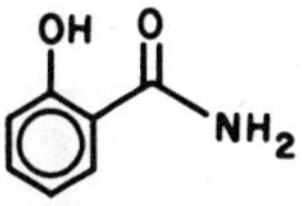

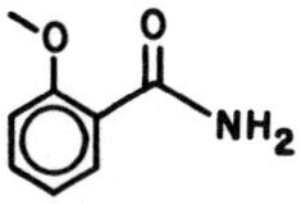

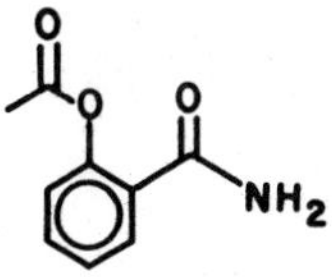

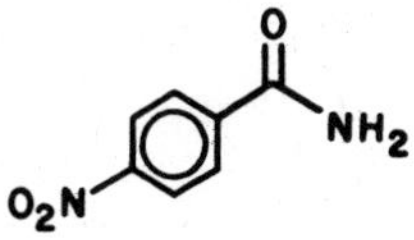

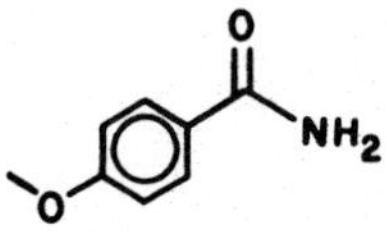

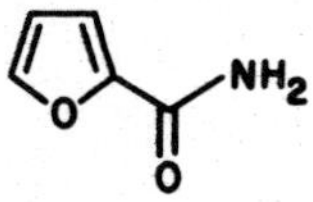

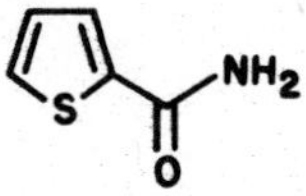

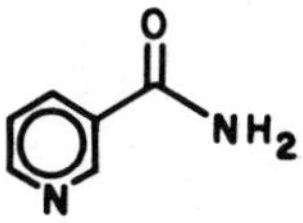

nuj.	1695* 1684	22
$CHCl_3$	1686	22
nuj.	1684	22
$CHCl_3$	1692	22
nuj.	1653	23
KBr	1656 1608	24
KBr	1658 1532	24
KBr	1691m 1654 1535	25
nuj.	1730m 1672	26
nuj.	1684 1647	27
nuj.	1684b 1656* 1733m	28

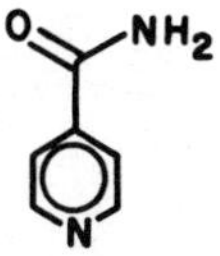

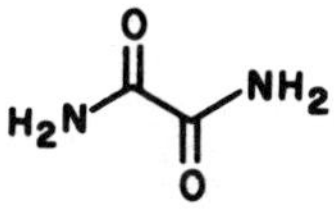

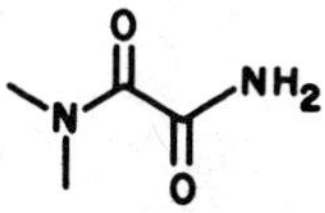

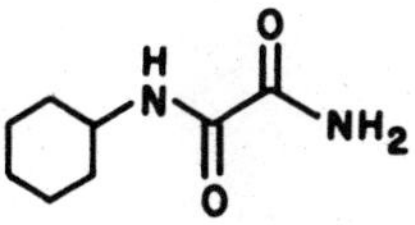

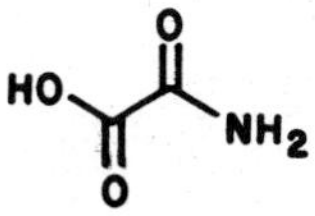

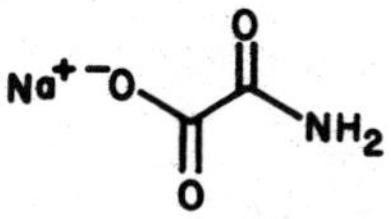

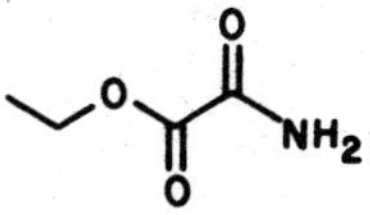

Compound	Medium	ν (cm⁻¹)	Ref.
$H_2NCOCH_2CONH_2$	nuj.	1656	8
$H_2NCOCHFCONH_2$	KBr	1700	29
$H_2NCOCF_2CONH_2$	KBr	1695 1620	29
$\phi COCH_2CONH_2$	nuj.	1661b	30
$H_2NCO(CH_2)_2CONH_2$	nuj.	1678 1637	31
cyclopropane-1,2-dicarboxamide	KBr	1660b 1605	7
$H_2NCOCH{=}CHCONH_2$	KBr	1692 1629	7
$H_2NCO(CH_2)_4CONH_2$	nuj.	1654	32

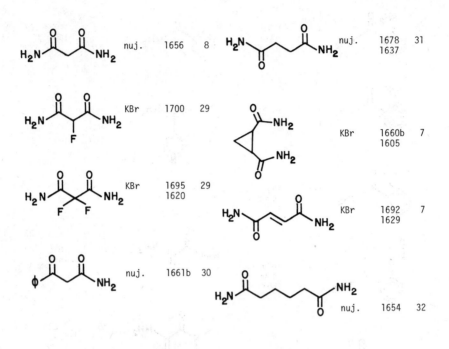

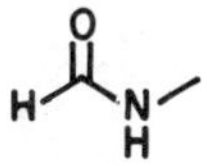

Solvent	ν	Ref.
liq.	1672	33
liq.	1637	33
liq.	1667	33
nuj.	1664	34
$CHCl_3$	1692	34
KBr	1653	36
CCl_4	1688	35
CCl_4	1687	37
CH_2Cl_2	1671	37
$CHCl_3$	1667	37
CCl_4	1690	35
CCl_4	1688	35
liq.	1656 1645	38

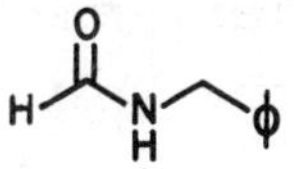

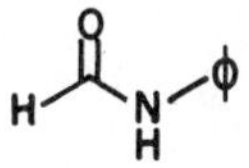

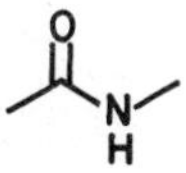

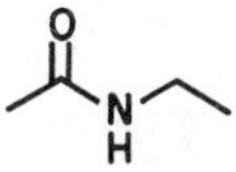

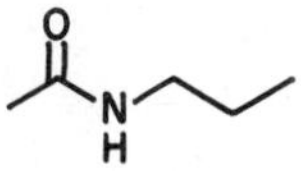

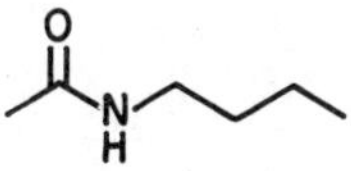

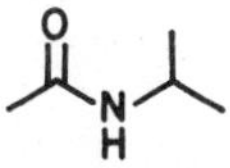

Solvent	ν	Ref.
CCl_4	1687	35
liq.	1664	39
CCl_4	1688	35
CCl_4	1689	40
KBr	1715 1590	41
KBr	1715 1585	41
KBr	1675	42
nuj.	1658	34
CCl_4	1705	43
$CHCl_3$	1691	44

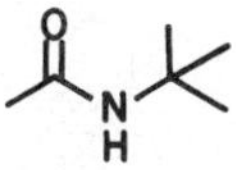

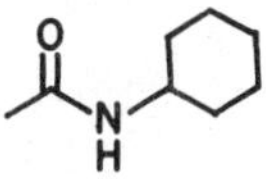

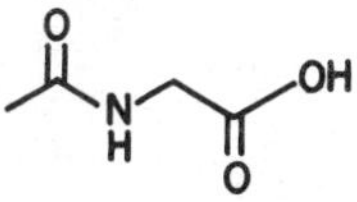

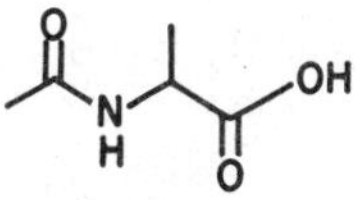

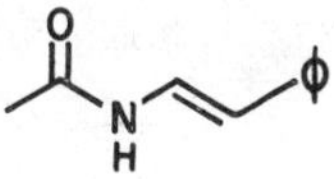

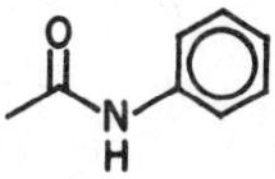

Solvent	ν (cm⁻¹)	Ref.
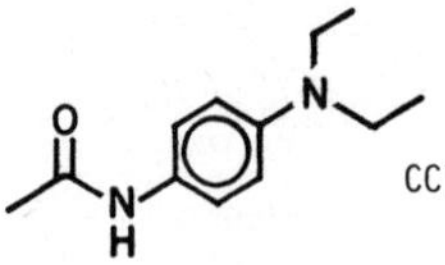 CCl_4	1693	43
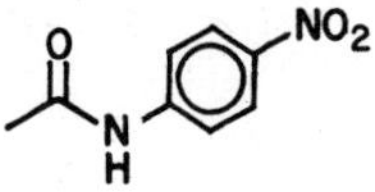 CCl_4	1718	43
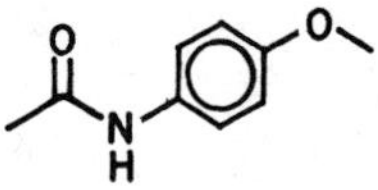 $CHCl_3$	1683	44
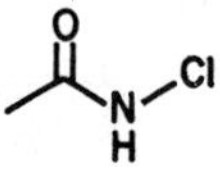 $CHCl_3$	1705	45
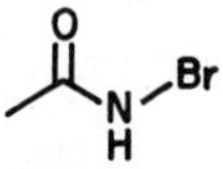 $CHCl_3$	1705	45
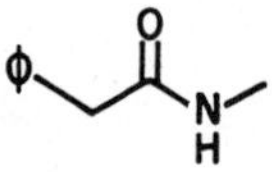 CCl_4	1656	8
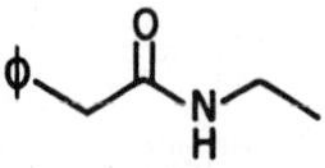 nuj.	1630	46
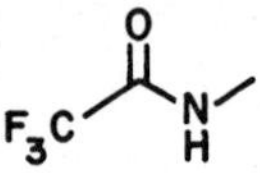 CCl_4	1741	35
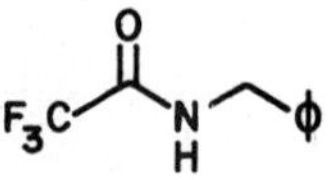 nuj.	1706	29
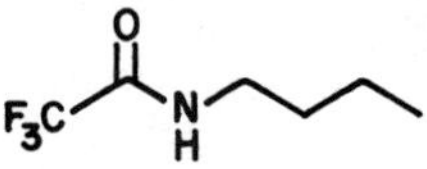 liq.	1718	33
CCl_4	1736	38
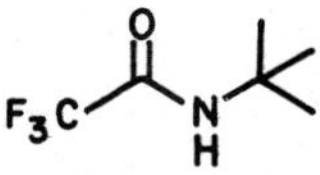 CCl_4	1736	35
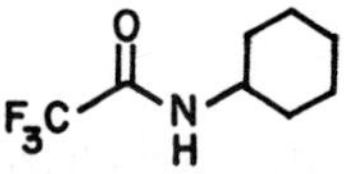 CCl_4	1718	47
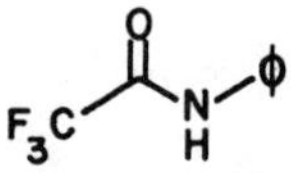 nuj.	1698	33
CCl_4	1734	47
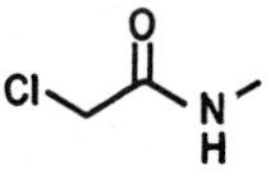 liq.	1660	48
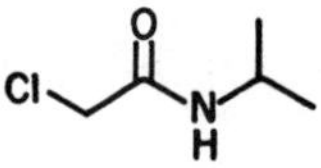 CCl_4	1682	35
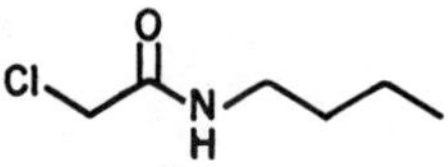 liq.	1661 1653	38
CCl_4	1686	38

Solvent	Frequency	Ref.
CCl_4	1684	35
CCl_4	1692	35
CCl_4	1707	35
liq.	1692 1675	38
CCl_4	1706	38
CCl_4	1702	35
CCl_4	1713	35
CCl_4	1728	35
CCl_4	1726	35
CCl_4	1725	35
CCl_4	1731	35
CCl_4	1723	43
CCl_4	1738	43
CCl_4	1680	35
CCl_4	1700	35
CCl_4	1712	35
CCl_4	1689	35

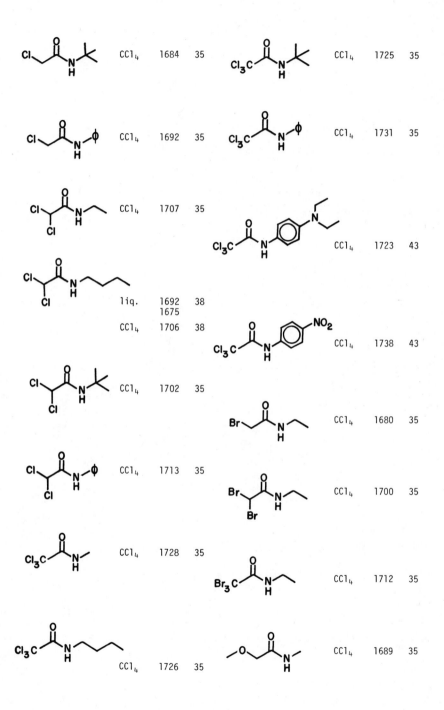

Solvent	Frequency	Ref.
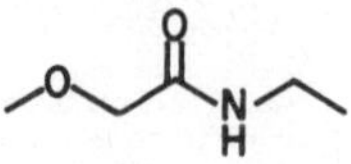 CCl_4	1689	35
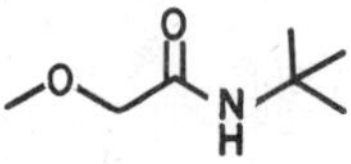 CCl_4	1690	35
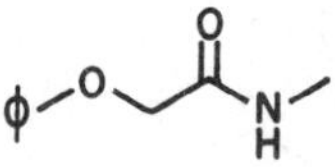 CCl_4	1691	35
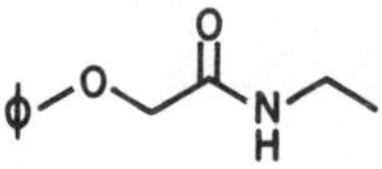 CCl_4	1691	35
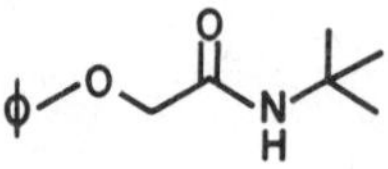 CCl_4	1689	35
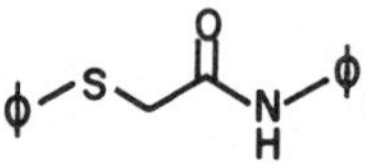 nuj.	1660 1515	49
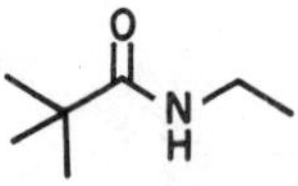 CCl_4	1644	50
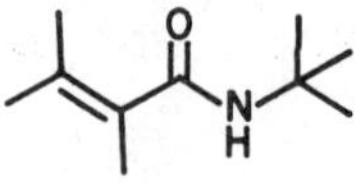 CCl_4	1670	51
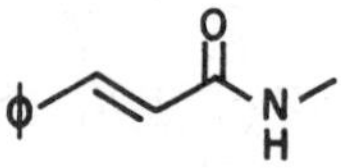 nuj.	1660	52
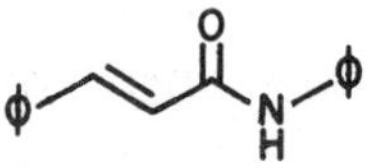 $CHCl_3$	1675	17
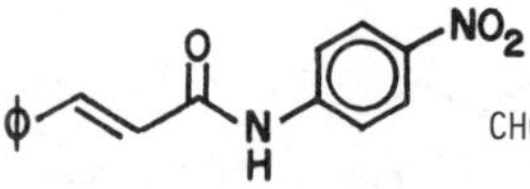 $CHCl_3$	1686	17
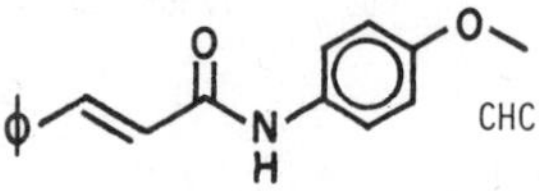 $CHCl_3$	1669	17
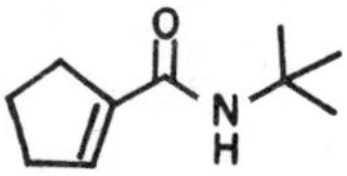 CCl_4	1670 1630	53
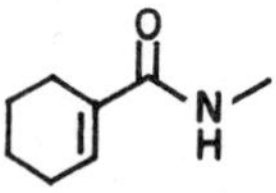 KBr	1640	54
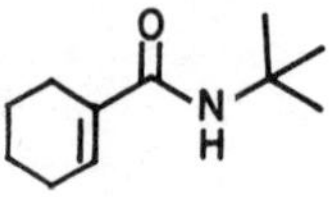 CCl_4	1670 1640	55
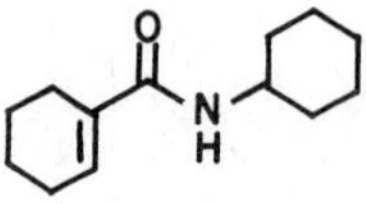 CCl_4	1657 1612	56

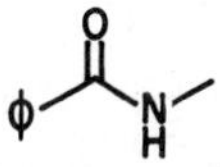

$CHCl_3$ 1660 44

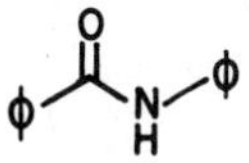

nuj. 1656 34
CCl_4 1688 57
$CHCl_3$ 1674 34

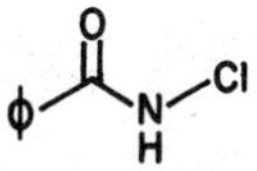

KBr 1691 58

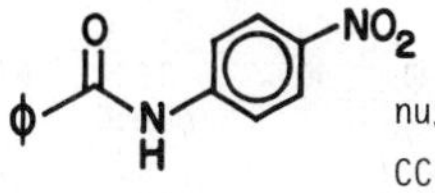

nuj. 1660 59
CCl_4 1695 57

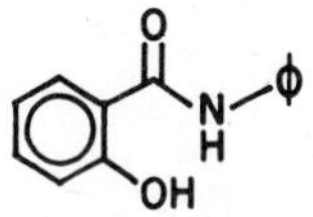

CCl_4 1659 60

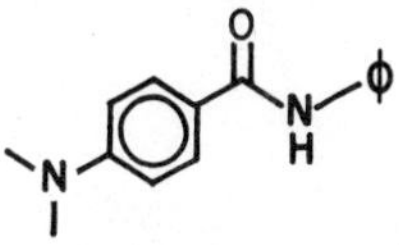

nuj. 1646 59
C_2Cl_4 1665 59

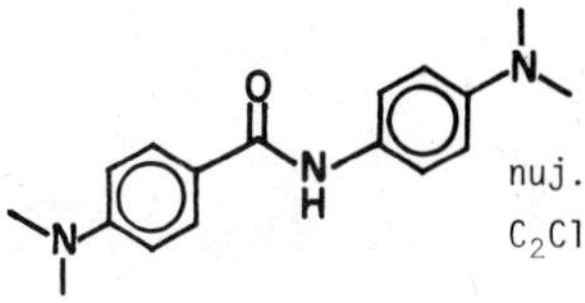

nuj. 1635 59
C_2Cl_4 1660 59

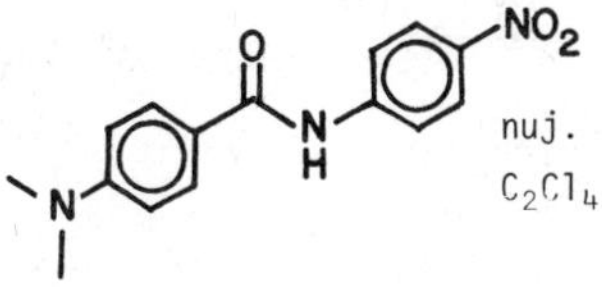

nuj. 1655 59
C_2Cl_4 1680 59

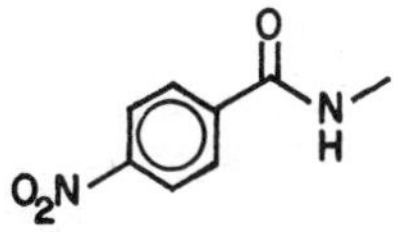

$CHCl_3$ 1671 44

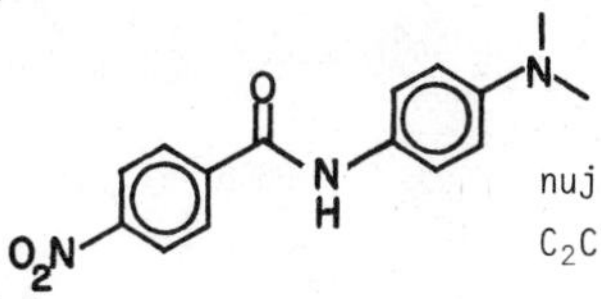

nuj. 1645 59
C_2Cl_4 1675 59

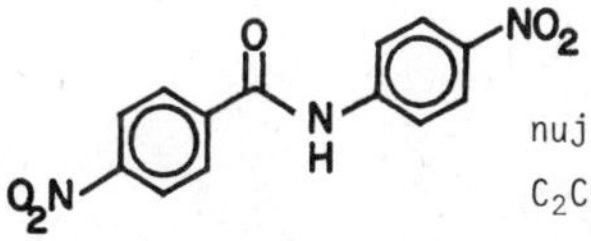

nuj. 1685 59
C_2Cl_4 1698 59

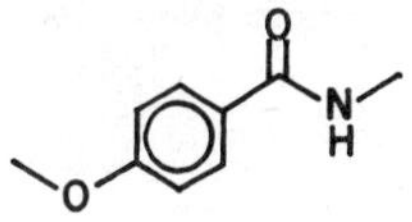

CCl_4 1676 35

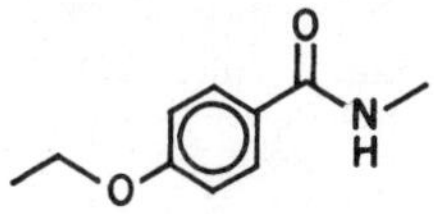

$CHCl_3$ 1655 44

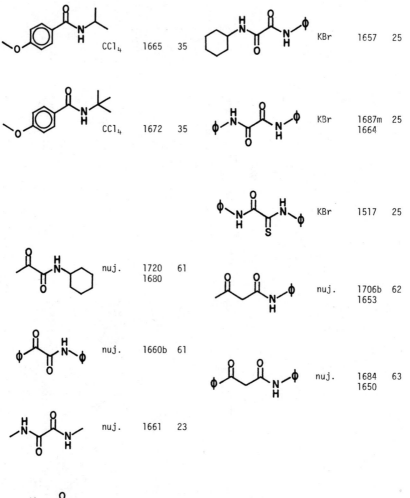

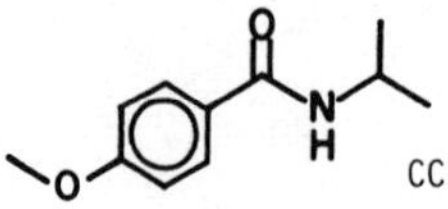	CCl_4	1665	35
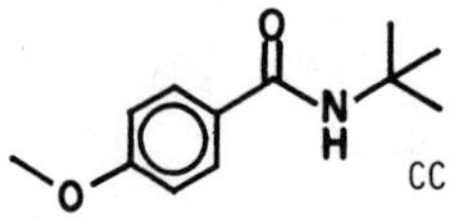	CCl_4	1672	35
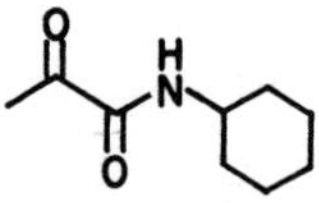	nuj.	1720 1680	61
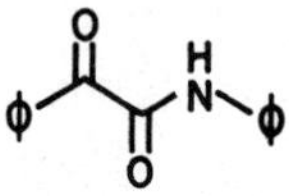	nuj.	1660b	61
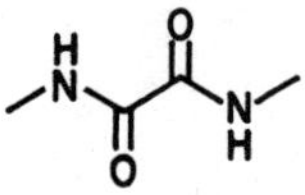	nuj.	1661	23
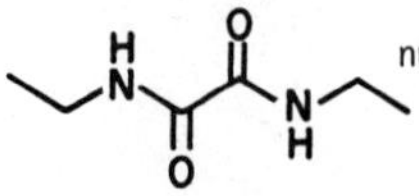	nuj.	1656	23
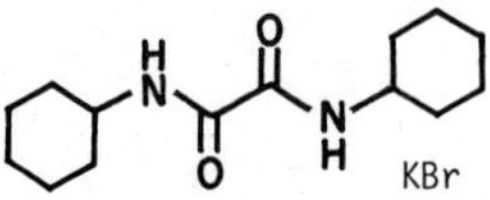	KBr	1645	25
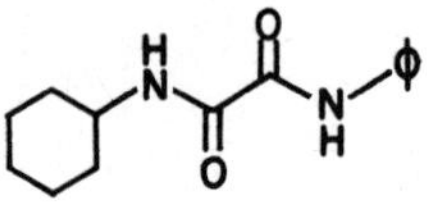	KBr	1657	25
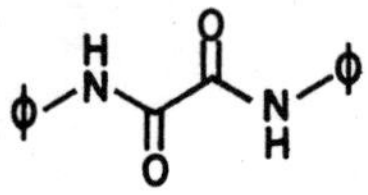	KBr	1687m 1664	25
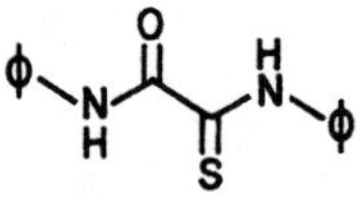	KBr	1517	25
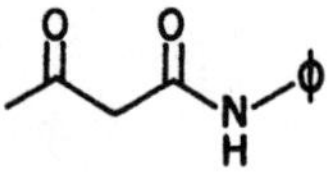	nuj.	1706b 1653	62
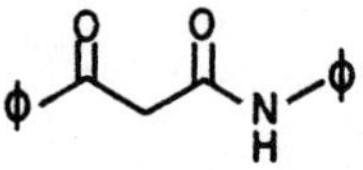	nuj.	1684 1650	63

Solvent	ν (cm⁻¹)	Ref.
liq.	1670	65
CCl_4	1688	44
$CHCl_3$	1673	64
liq.	1672	29
CCl_4	1684	44
$CHCl_3$	1663	66
liq.	1678	29
$CHCl_3$	1660	66
liq.	1715m 1658	67
liq.	1678	68
CCl_4	1739	69
CCl_4	1751* 1742	69

Solvent	ν (cm⁻¹)	Ref.
liq.	1645	70
CCl_4	1653	44
$CHCl_3$	1633	66
KBr	1765w 1692 1653w 1605w	71
liq.	1645	33
CCl_4	1650	72
$CHCl_3$	1628	66
CCl_4	1647	44
liq.	1690	73
liq.	1642	75
CCl_4	1652	74
$CHCl_3$	1625	66
CS_2	1681 1622	76

Solvent	ν (cm⁻¹)	Ref.
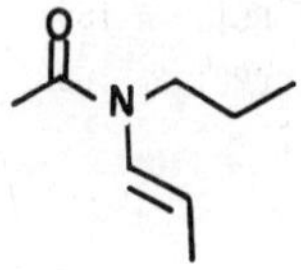 CS_2	1675 1650	76
nuj.	1692	77
KBr	1692	78
CCl_4	1671	74
$CHCl_3$	1653	11
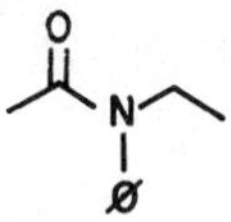 CCl_4	1667	74
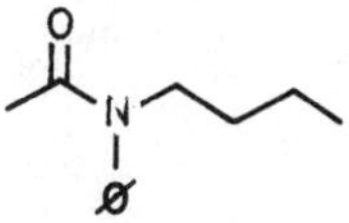 CCl_4	1667	74
CCl_4	1679	44
$CHCl_3$	1661	17
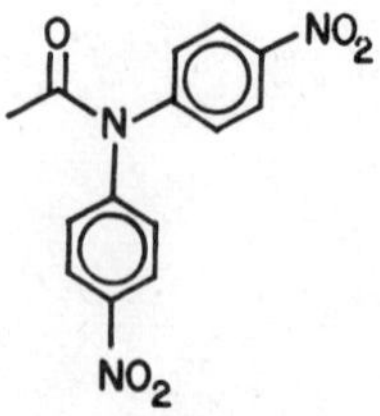 $CHCl_3$	1684	17
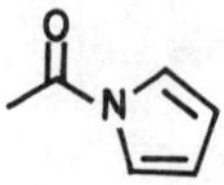 KBr	1732	79
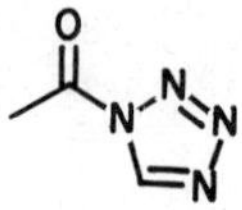 KBr	1779	79
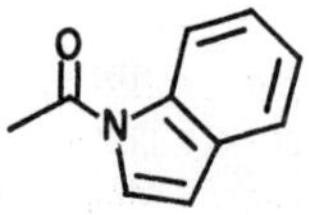 CCl_4	1711	80
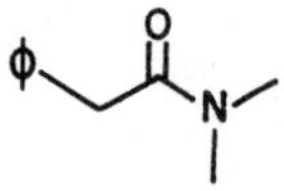 CCl_4	1653	8
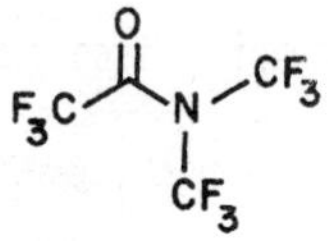 liq.	1799	66
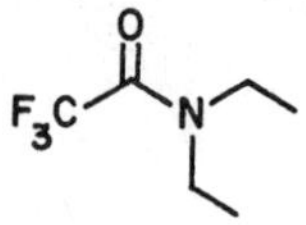 liq.	1698	33
liq.	1689	38
CCl_4	1692	38
liq.	1658	38
CCl_4	1656	38

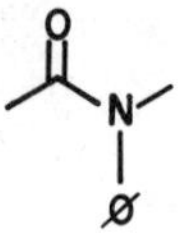

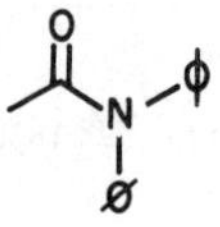

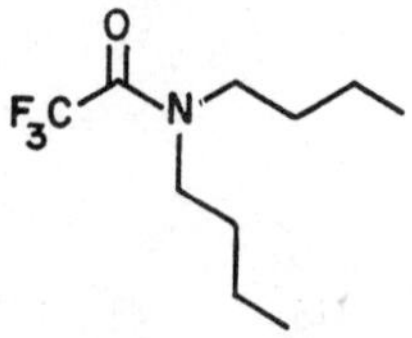

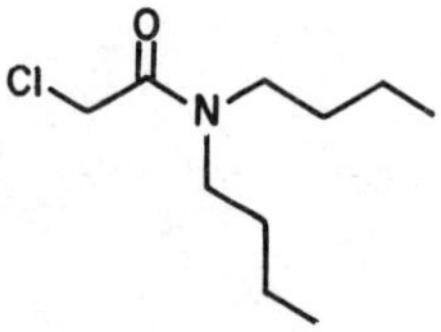

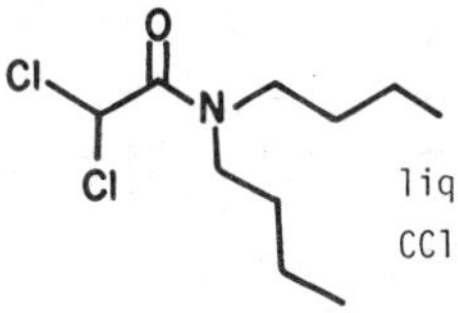

Solvent	Frequency	Ref.
liq.	1667	38
CCl_4	1684 1658	38
CS_2	1689	81
$CHCl_3$	1686	81
nuj.	1635	82
CCl_4	1660	72
$CHCl_3$	1633	66
$CHCl_3$	1624	66
liq.	1701 1664	83
$CHCl_3$	1605	77
$CHCl_3$	1605	77
$CHCl_3$	1668 1635	17
$CHCl_3$	1639	17
$CHCl_3$	1639	17
$CHCl_3$	1645	17

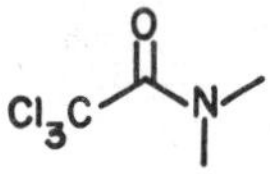

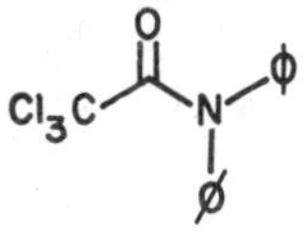

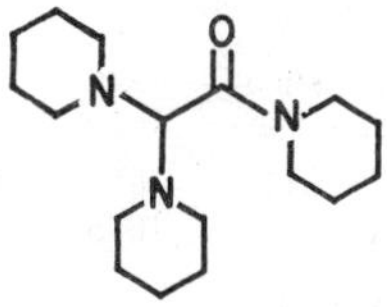

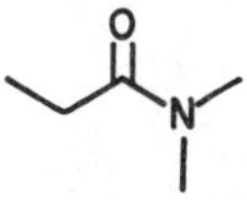

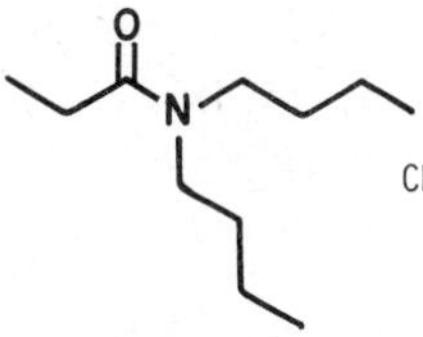

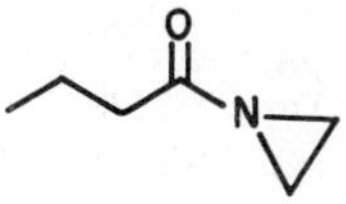

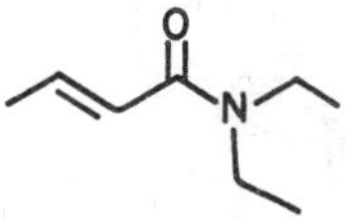

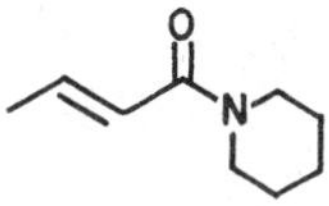

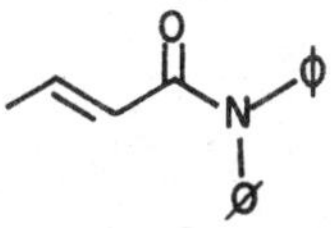

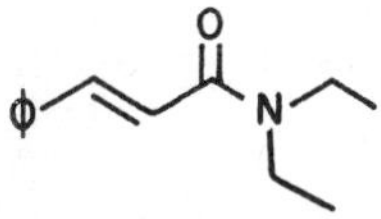

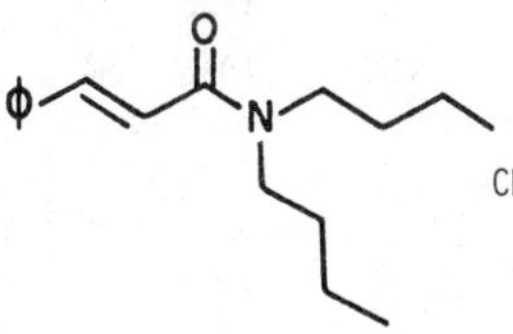

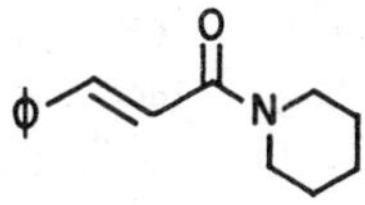

Solvent	Frequency	Ref.
$CHCl_3$	1656	17
liq.	1625 1715	86
KBr	1656	42
$CHCl_3$	1663	17
$CHCl_3$	1641	87
KBr	1645	54
KBr	1670 1625	88
$CHCl_3$	1690 1625	88
CH_2Cl_2	1629	20
KBr	1686 1650	88
$CHCl_3$	1650	88
CCl_4	1644	72
$CHCl_3$	1645	84
liq.	1630	85
CCl_4	1641	74
liq.	1578	89
CCl_4	1721	89

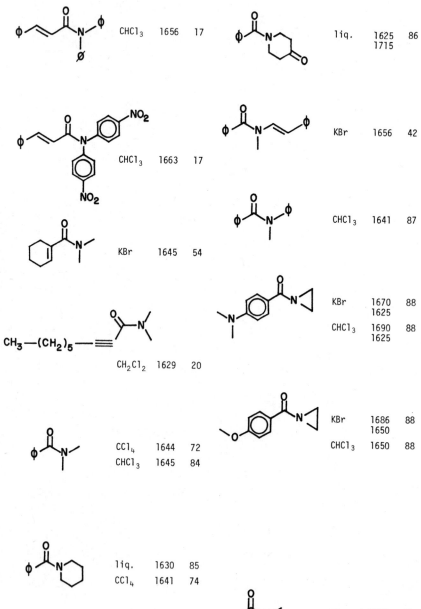

Solvent	Frequency	Ref.
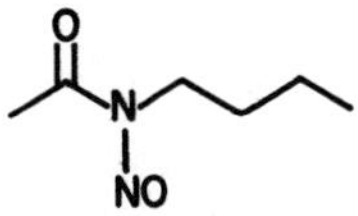 CCl_4	1739 1515	40
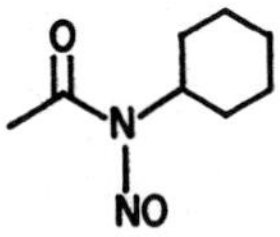 CCl_4	1745 1520	40
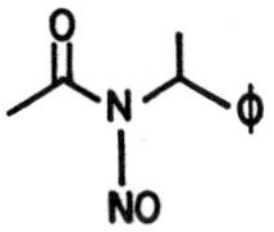 CCl_4	1733 1515	40
liq.	1562	89
CCl_4	1730	89
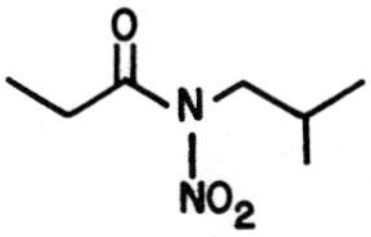 CCl_4	1724	89
liq.	1565	89
CCl_4	1745	89
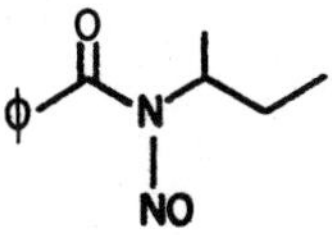 CCl_4	1712 1518	40

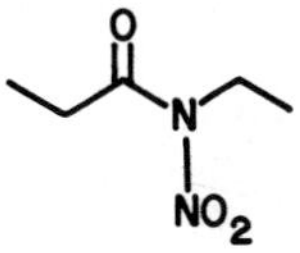

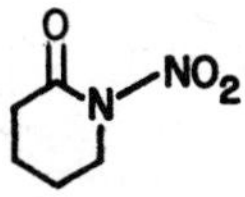

REFERENCES

1. D. Cook, *JACS* (1958), 49.
2. T. L. Brown, J. F. Regan, R. D. Schuetz and J. C. Sternberg, *JPC* (1959), 1324.
3. C. F. Pouchert, *The Aldrich Library of Infrared Spectra*, Aldrich (1970), spectrum 332A.
4. E. Spinner, *SCA* (1959) *15*, 95.
5. A. E. Kellie, D. G. O'Sullivan and P. W. Sadler, *JCS* (1956), 3809.
6. C. F. Pouchert, *The Aldrich Library of Infrared Spectra*, Aldrich (1970), spectrum 789C.
7. A. Jart, *Acta Polytec Scand Chem Ind Met* (1965) *42*, 55.
8. H. M. Randall, R. G. Fowler, N. Fuson and J. R. Dangl, *Infrared Determination of Organic Structures* (1949), D. Van Nostrand Company, New York.
9. D. R. Husted and A. H. Ahlbrecht, *JACS* (1953), 1605.

10. E. Halpern and J. Goldenson, *JPC* (1956), 1372.
11. R. N. Jones and C. Sandorfy in *Chemical Applications of Spectroscopy*, ed. W. West (1956), Wiley (Interscience), New York.
12. C. F. Pouchert, *The Aldrich Library of Infrared Spectra*, Aldrich (1970), spectrum 332D.
13. G. T. Tisue, S. Linke and W. Lwowski, *JACS* (1967), 6303.
14. C. F. Pouchert, *The Aldrich Library of Infrared Spectra*, Aldrich (1970), spectrum 333C.
15. N. Jonathan, *J Mol Spectr* (1961) *6*, 205.
16. C. F. Pouchert, *The Aldrich Library of Infrared Spectra*, Aldrich (1970), spectrum 333D.
17. M. Yamaguchi, *NKZ* (1957), 1236.
18. K. E. Schulte, J. Reisch and R. Hobl, *Arc Pharm* (1960), 687.
19. R. A. Finnegan and R. S. McNees, *JOC* (1964), 3234.
20. C. A. Grob and B. Fischer, *Helv* (1955), 1794.
21. M. Mashima, *BCS Japan* (1962) *35*, 332.
22. S. Yoshida, *Chem Pharm Bull* (1963), 628.
23. J. Chouteau, *BSC France* (1953), 1148.
24. M. Mashima, *BCS Japan* (1962) *35*, 2020.
25. B. Milligan, E. Spinner and J. M. Swan, *JCS* (1961), 1919.
26. C. F. Pouchert, *The Aldrich Library of Infrared Spectra*, Aldrich (1970), spectrum 339A.
27. C. F. Pouchert, *The Aldrich Library of Infrared Spectra*, Aldrich (1970), spectrum 339C.
28. C. F. Pouchert, *The Aldrich Library of Infrared Spectra*, Aldrich (1970), spectrum 339D.
29. E. D. Bergmann, S. Cohen and I. Shahak, *JCS* (1959), 3286.
30. C. F. Pouchert, *The Aldrich Library of Infrared Spectra*, Aldrich (1970), spectrum 792G.
31. C. F. Pouchert, *The Aldrich Library of Infrared Spectra*, Aldrich (1970), spectrum 339H.
32. DMS 1011.
33. J. H. Robson and J. Reinhart, *JACS* (1955), 498.
34. E. J. Forbes, K. J. Morgan and J. Newton, *JCS* (1963), 835.
35. R. A. Nyquist, *SCA* (1963), 509.
36. M. Mashima, *BCS Japan* (1962) *35*, 423.
37. W. Klemperer, M. W. Cronyn, A. H. Maki and G. C. Pimentel, *JACS* (1954), 5846.
38. H. Letaw, Jr. and A. H. Gropp, *JCP* (1953), 1621.
39. C. F. Pouchert, *The Aldrich Library of Infrared Spectra*, Aldrich (1970), spectrum 335A.

40. E. H. White, *JACS* (1955), 6008.
41. F. Micheel and B. Schleppinghoff, *B* (1955), 763.
42. H. Böhme and G. Berg, *B* (1966), 2127.
43. R. A. Nyquist, *SCA* (1963), 1595.
44. H. W. Thompson and D. A. Jameson, *SCA* (1958) *13*, 236.
45. A. Gierer, *Z Naturf* (1953) *8b*, 644 and 654.
46. J. Denkosch, K. Schlögl and H. Woidich, *Monatsh Chem* (1957), 35.
47. E. J. Bourne, S. H. Henry, C. E. M. Tatlow and J. C. Tatlow, *JCS* (1952), 4014.
48. S. Mizushima, T. Shimanouchi, I. Ichishima, T. Miyazawa, I. Nakagawa and T. Araki, *JACS* (1956), 2038.
49. P. Baudet, M. Calin and E. Cherbuliez, *Helv* (1965), 2023.
50. DMS 1005.
51. J. C. Sheehan and J. H. Beeson, *JACS* (1967), 362.
52. A. Chatterjee, S. Bose and S. K. Srimany, *JOC* (1959), 687.
53. J. C. Sheehan and J. H. Beeson, *JACS* (1967), 366.
54. J. Klein, *T* (1964), 465.
55. J. C. Sheehan and I. Lengyel, *JACS* (1964), 746.
56. A. Lukasiewicz and J. Lesińska, *T* (1965), 3247.
57. W. Geiger, *SCA* (1966), 495.
58. R. Gompper, *B* (1960), 198.
59. J. Reichel, R. Bacaloglu and W. Schmidt, *Rev Roumaine Chim* (1964) *9*, 294.
60. D. Welti, *SCA* (1966), 281.
61. Y. Ito, M. Okano and R. Oda, *T* (1966), 447.
62. C. F. Pouchert, *The Aldrich Library of Infrared Spectra*, Aldrich (1970), spectrum 791A.
63. C. F. Pouchert, *The Aldrich Library of Infrared Spectra*, Aldrich (1970), spectrum 791D.
64. R. L. Adelman, *JOC* (1964), 1837.
65. DMS 1001.
66. J. A. Young, T. C. Simmons and F. W. Hoffmann, *JACS* (1956), 5637.
67. C. F. Pouchert, *The Aldrich Library of Infrared Spectra*, Aldrich (1970), spectrum 337G.
68. C. F. Pouchert, *The Aldrich Library of Infrared Spectra*, Aldrich (1970), spectrum 338D.
69. R. A. Nyquist and W. J. Potts, *SCA* (1961), 679.
70. C. F. Pouchert, *The Aldrich Library of Infrared Spectra*, Aldrich (1970), spectrum 336F.
71. M. J. Janssen, *SCA* (1961), 475.
72. C. D. Schmulbach and R. S. Drago, *JPC* (1960), 1956.

73. T. A. Foglia, L. M. Gregory and G. Maerker, *JOC* (1970), 3779.
74. L. J. Bellamy and R. J. Pace, *SCA* (1963), 1831.
75. C. F. Pouchert, *The Aldrich Library of Infrared Spectra*, Aldrich (1970), spectrum 338E.
76. H. Breederveld, *Rec Trav Chim* (1960), 401.
77. M. Yamaguchi, *NKZ* (1960), 1118.
78. W. Ziegenbein and W. Franke, *B* (1957), 2291.
79. W. Otting, *B* (1956), 1940.
80. H. A. Staab, *B* (1957), 1320.
81. A. J. Speziale and R. C. Freeman, *JACS* (1960), 903.
82. L. Duhamel, P. Duhamel and G. Plé, *BSC France* (1968), 4423.
83. C. F. Pouchert, *The Aldrich Library of Infrared Spectra*, Aldrich (1970), spectrum 337E.
84. L. J. Bellamy, *JCS* (1955), 4221.
85. E. H. Hoffmeister and D. S. Tarbell, *T* (1965), 35 and 2857.
86. R. A. Johnson, M. E. Herr, M. C. Murray and G. S. Fonken, *JOC* (1968), 3187.
87. A. R. Katritzky and R. A. Jones, *JCS* (1959), 2067.
88. G. R. Boggs and J. T. Gerig, *JOC* (1969), 1484.
89. E. H. White, M. C. Chen, and L. A. Dolak, *JOC* (1966), 3038.

4.4

vinylogous amides

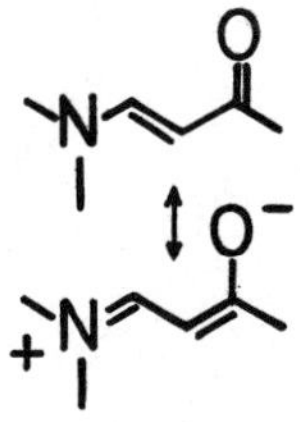

I

II

As with amides, contributions from the dipolar resonance structure (II) play an important role in the properties of vinylogous amides. Indeed the extent of delocalization of the nitrogen lone pair is of the same magnitude as in amides, such that the range of carbonyl frequencies is comparable to those of amides.

In addition to this, the spectra are complicated by a dynamic cis-trans equilibrium (III ⇌ IV). In solution both forms may be observed, but the cis isomer predominates.

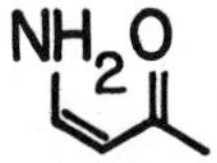

III

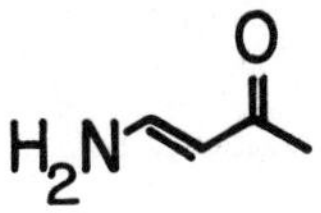

IV

The trans isomers, with -NH_2 or -NHR groups, show two or three broad bands in the region of 3350 to 3100 cm^{-1} in the solid phase as a result of intermolecular associations. Solutions of the trans isomer rapidly equilibrate with the cis isomer, and the cis isomer shows two intense, sharp bands in the region of 3550 to 3430 cm^{-1}, that result from the symmetric and asymmetric nonbonded -NH_2 vibrations. In concentrated solutions dimerization occurs, and the bands broaden and fall in frequency.

A band of medium intensity in the region of 1625 to 1560 cm^{-1} is found for both the cis and trans isomers that corresponds to the amide II band (see amide Chapter 4.3, p. 241) and arises from the NH_2 deformation mode.

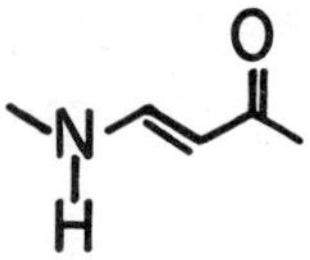

Few examples of secondary vinylogous amides have been reported, but bands arising from the free and associated -NH group are seen in the region of 3500 to 3000 cm^{-1}. Only in very dilute solutions are the nonassociated bands seen, otherwise broad bands at the lower end of the range are observed.

A band of medium intensity between 1580 and 1510 cm^{-1} is seen for both the cis and trans isomers which corresponds to the amide III band of secondary amides (Chapter 4.3, p. 241).

The major contributions of resonance forms I and II do not allow discrete absorptions to be assigned to the carbonyl and olefinic groups. Rather, the two vibrations can be considered as coupled, and the resulting absorptions are characteristic of the enamino chromophore.

Two bands, one occurring between 1700 and 1600 cm^{-1} and the other between 1585 and 1475 cm^{-1}, are found for both the cis and trans isomers. It is important to note that the position of these bands varies with the physical state and concentration of the sample.

Solvent	Frequency	Ref.
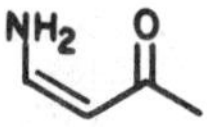 liq.	1640 1603 1501	1
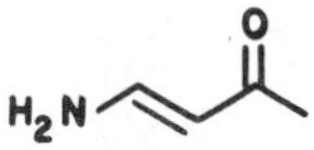 liq.	1664* 1622* 1559* 1543	1
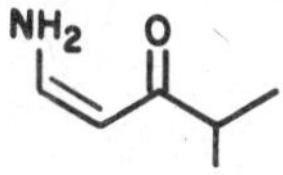 liq.	1634 1606* 1498	1
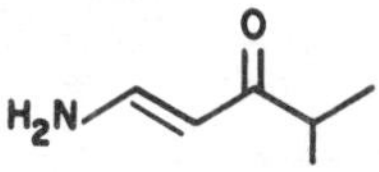 liq.	1662* 1619* 1558*	1
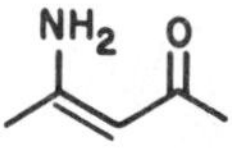 nuj.	1625	2
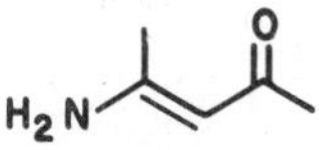 liq.	1700 1625 1540	2
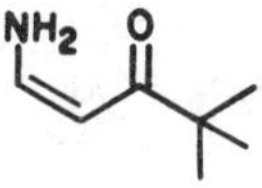 liq.	1636 1599 1493	1
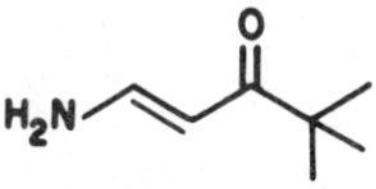 liq.	1658 1563* 1542* 1528*	1
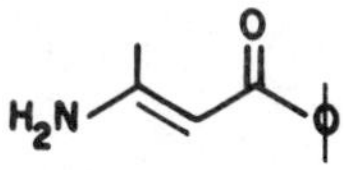 nuj.	1625b 1593 1568b	2
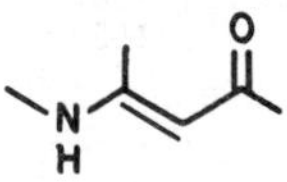 nuj.	1628 1595	2

Solvent	Frequency	Ref.
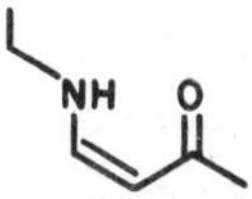 C_2Cl_4	1643 1577 1485	3
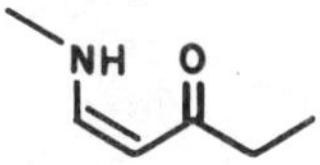 CCl_4	1644 1579 1487	3
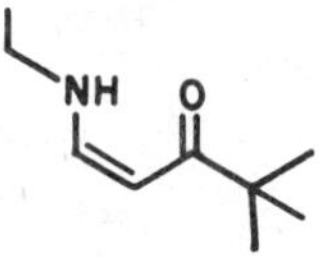 C_2Cl_4	1634 1526 1483	3
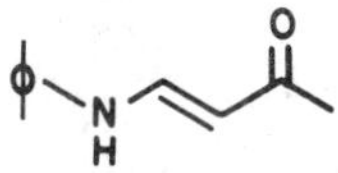 nuj.	1612 1588 1569	2
KBr	1640	4
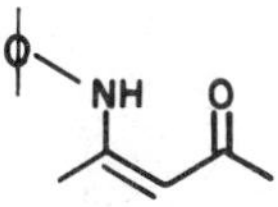 nuj.	1612	2
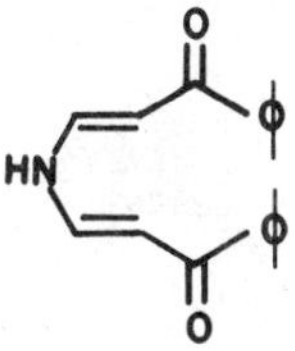 KBr	1648	4
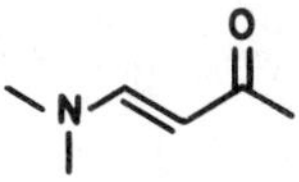 KBr	1631 1584	5
C_2Cl_4	1673m 1623m 1586	6

Solvent	Frequency	Ref.
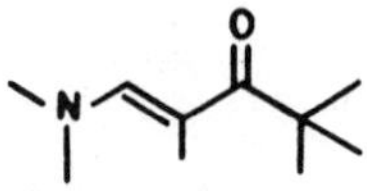 C_2Cl_4	1665m 1584	6
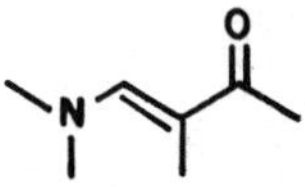 C_2Cl_4	1651m 1553	6
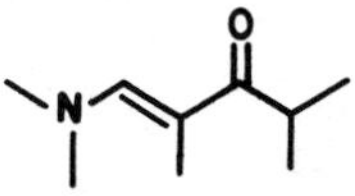 liq.	1625 1590b	5
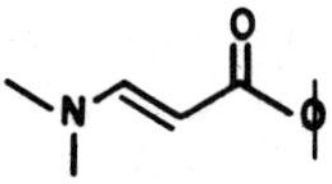 CCl_4	1653 1604 1591 1568	7
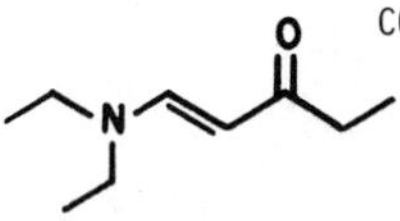 CCl_4	1664 1616 1574	7
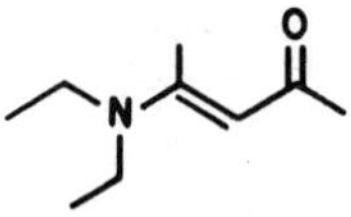 CCl_4	1643 1540	7
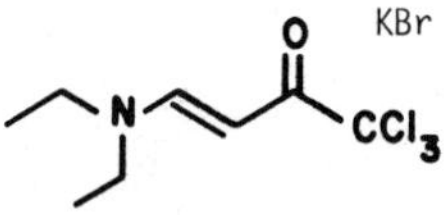 KBr	1675 1663	8
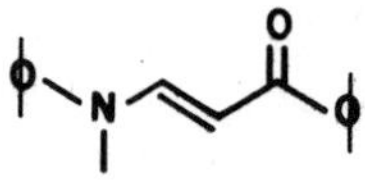 KBr	1657	4
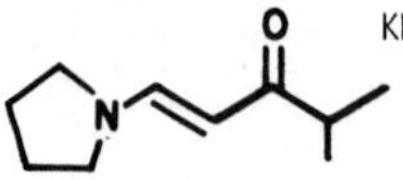 KBr	1650 1610 1570	5
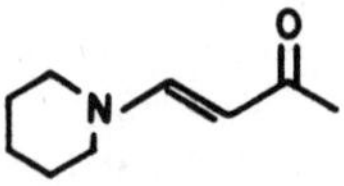 C_2Cl_4	1668m 1615m 1575	6
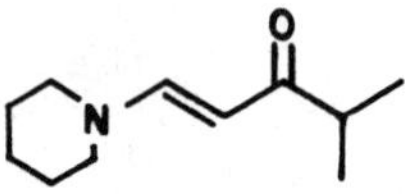 liq.	1650 1612 1570b	9
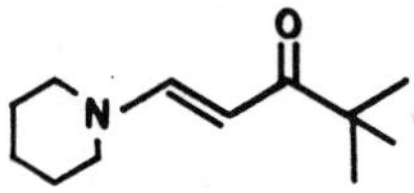 C_2Cl_4	1662m 1575	6
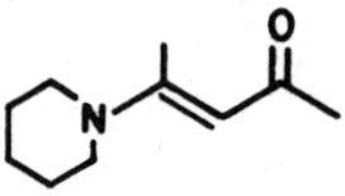 CCl_4	1623 1535	7
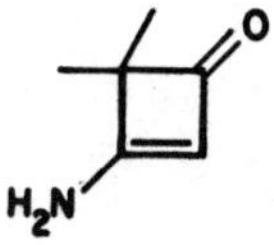 KBr	1672 1556b	10
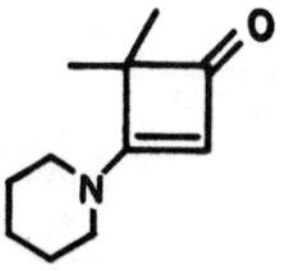 KBr	1739 1592	10
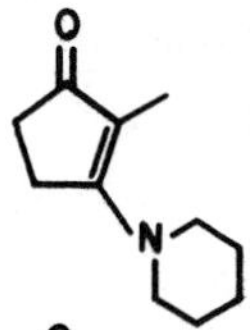 KBr	1661 1572	11
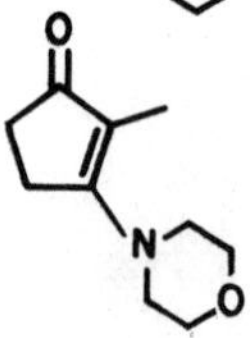 KBr	1661 1558	11

Solvent	Frequencies	Ref.
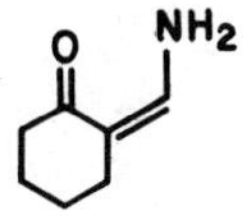 $CHCl_3$	1643 1585 1477	1
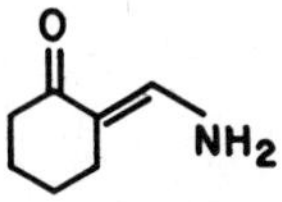 nuj.	1656 1483b	1
$CHCl_3$	1659* 1616m 1543mb	1
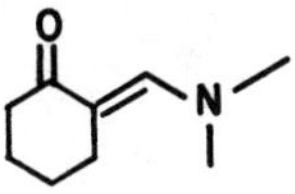 C_2Cl_4	1659m 1557	6
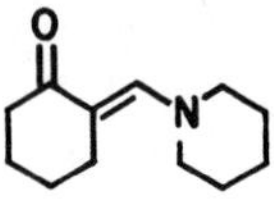 C_2Cl_4	1654 1545	5
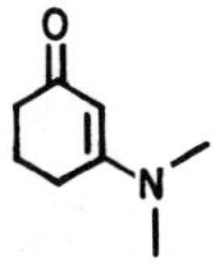 C_2Cl_4	1635m 1571	6
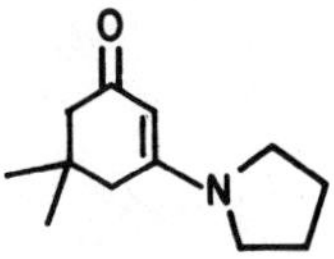 nuj.	1607 1560	5
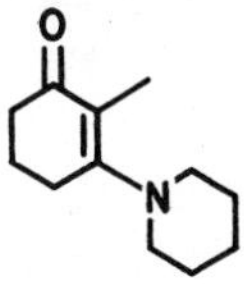 CCl_4	1619 1573	7
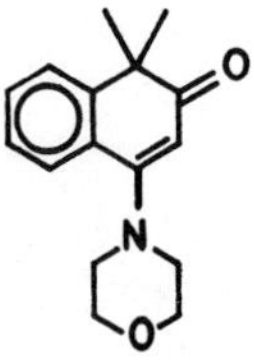 CCl_4	1670	12
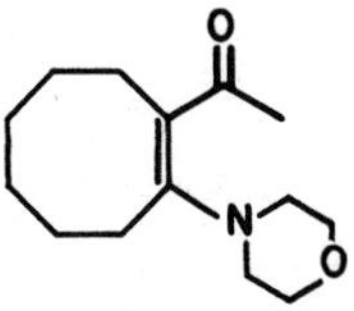 KBr	1590 1525	5
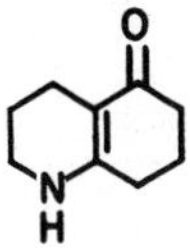 CH_2Cl_4	1615 1580	13
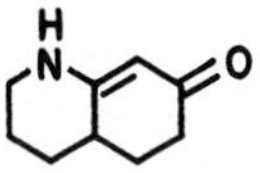 CH_2Cl_4	1610 1580	13
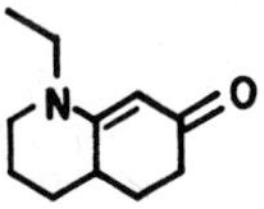 CH_2Cl_4	1600 1550	5
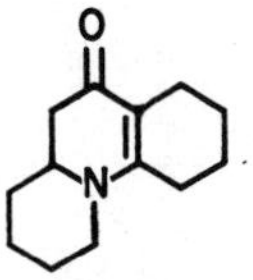 CCl_4	1640 1544	5
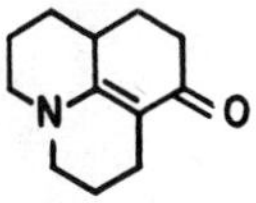 CCl_4	1625 1550	14
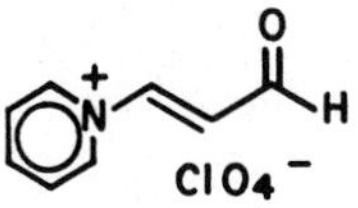 KBr	1690 1628	4
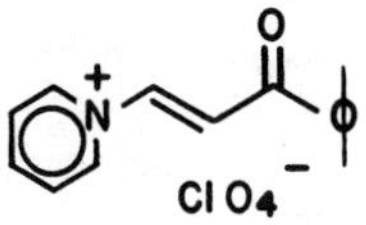 KBr	1680	4

ClO_4^-

ClO_4^-

REFERENCES

1. J. Dabrowski, *SCA* (1963), 475.
2. H. F. Holtzclaw, Jr., J. P. Collman and R. M. Alire, *JACS* (1958), 1100.
3. J. Dabrowski and U. Dabrowski, *B* (1968), 3392.
4. G. W. Fischer, *B* (1970), 3470.
5. S. Hünig and H. Hoch. *Fortschr Chem Forsch* (1970) *14*, 235.
6. J. Dabrowski and K. Kamieńska-Trela, *SCA* (1966), 211.
7. N. J. Leonard and J. A. Adamcik, *JACS* (1959), 595.
8. R. R. Fraser and R. B. Swingle, *T* (1969), 3469.
9. G. Opitz and F. Zimmermann, *A* (1963) *662*, 178.
10. R. H. Hasek, P. G. Gott and J. C. Martin, *JOC* (1964), 2510.
11. J. J. Panouse and C. Sannié, *BSC France* (1956), 1374.
12. R. D. Campbell and N. H. Cromwell, *JACS* (1957), 3457.
13. C. A. Grob and H. J. Wilkens, *Helv* (1967), 725.
14. Z. Valenta, P. Deslongschamps, R. A. Ellison and K. Wiesner, *JACS* (1964), 2532.

4.5

imides diacylamines

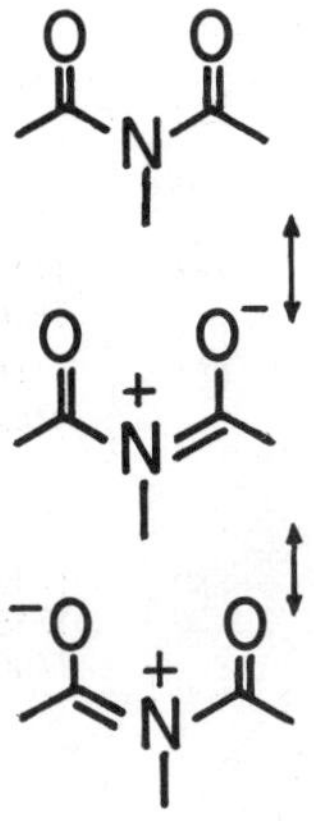

I

II

III

Both cyclic and acyclic imides exhibit bands in the 3500 to 3000 cm^{-1} region. These absorptions correspond to those discussed for secondary amides (Chapter 4.3, p. 241) and lactams (Chapter 4.6, p. 278). Cyclic imides, which are constrained to a cis-NH to carbonyl conformation, do not show an amide II band. However, acyclic imides, which may take up an -NH to carbonyl trans conformation, show an amide II band in the region of 1600 to 1500 cm^{-1}.

Contributions from the resonance forms I, II, and III result in a planar structure for imides, and coupling between the two carbonyl stretching vibrations leads to symmetric and asymmetric modes. This results in the appearance of two strong bands in the carbonyl region.

As with amides and lactams, self-association, or hydrogen bonding to solvent, causes a decrease in the frequency of both the symmetric and asymmetric modes.

Medium	Frequency	Ref.
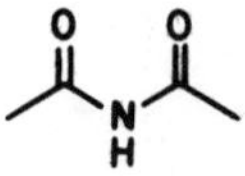 CCl_4	1747 1720 1698	1
nuj.	1734 1505	2
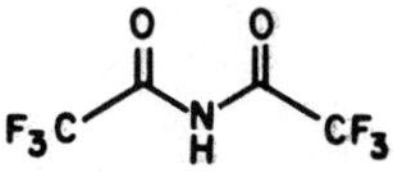 nuj.	1802 1748	4
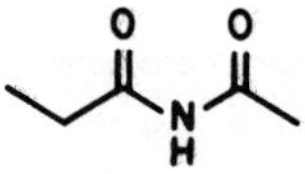 CCl_4	1747 1718 1698	1
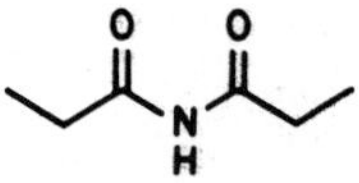 CCl_4	1747 1718 1694	1
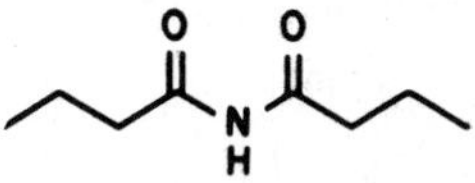 CCl_4	1743 1715 1690	1
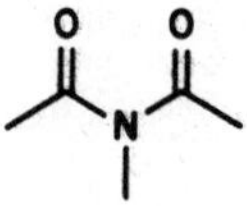 KBr	1736 1706 1689	5
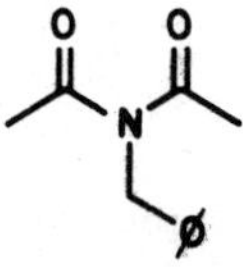 $CHCl_3$	1727* 1712* 1698	6
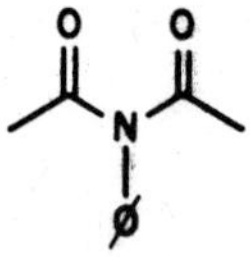 $CHCl_3$	1724* 1704 1698*	6
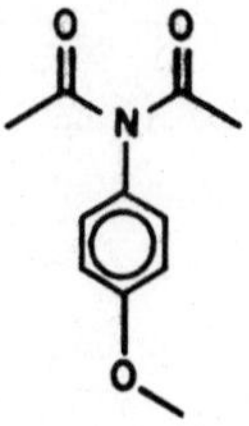 $CHCl_3$	1727* 1706 1701*	6

Medium	Frequency	Ref.
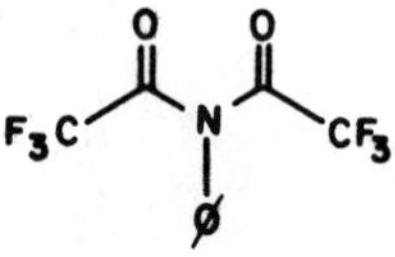 CCl_4	1734	7
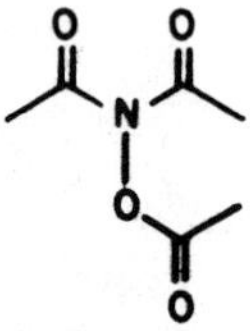 liq.	1810	8
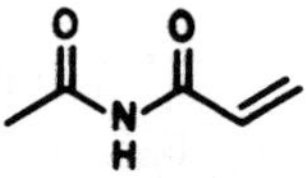 KBr	1738 1695	3
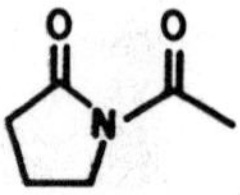 KBr	1730 1686	5
CCl_4	1745 1702	9
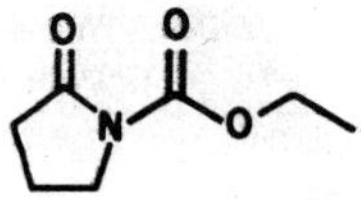 KBr	1786 1745 1706	5
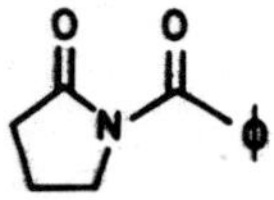 KBr	1692	5
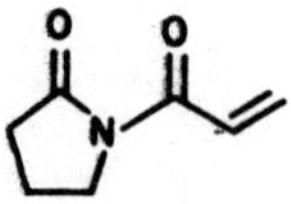 KBr	1739 1678	3
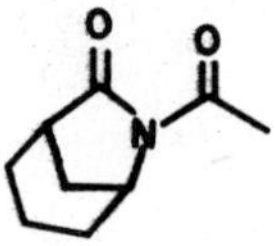 CCl_4	1746 1697	9

Solvent	cm⁻¹	Ref.
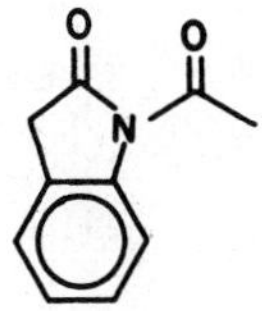 $CHCl_3$	1770 1689	6
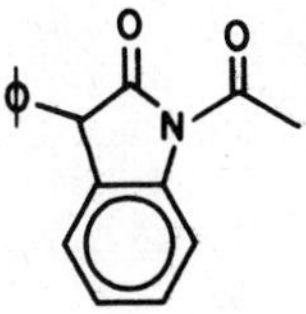 nuj.	1770 1712	10
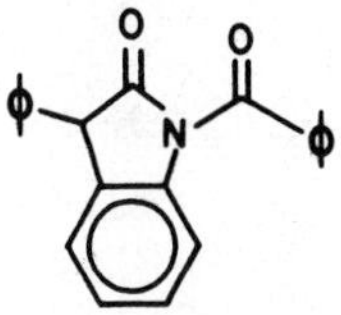	1761 1695	10
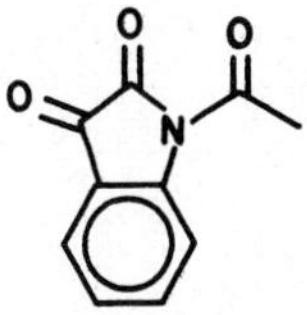 $CHCl_3$	1775 1742 1721	11
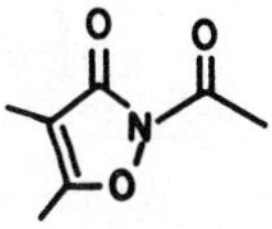 $CHCl_3$	1731* 1706b 1677 1604m	12
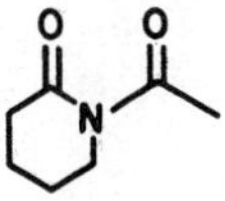 KBr	1686	5
CCl_4	1701	9
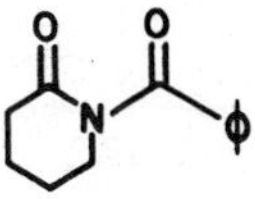 KBr	1739 1684	5
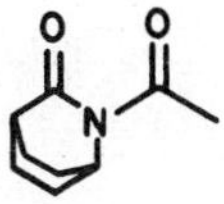 CCl_4	1719 1700	9

Solvent	cm⁻¹	Ref.
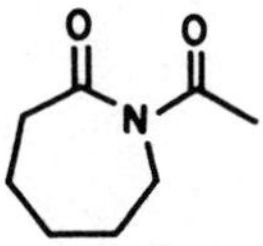 liq.	1700 1670	13
KBr	1686	5
CCl_4	1701	9
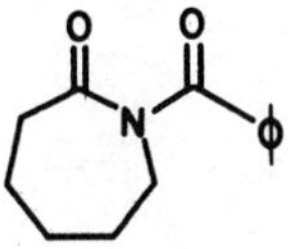 KBr	1695* 1672	5
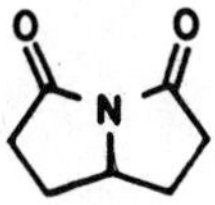 KBr	1765 1690	14
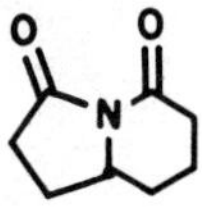 KBr	1757 1667	14
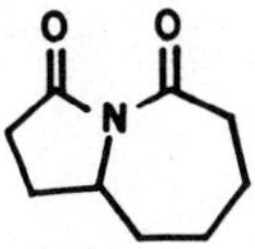 KBr	1756 1668	14
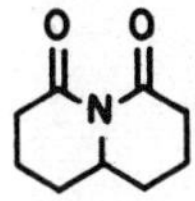 KBr	1737 1657	14
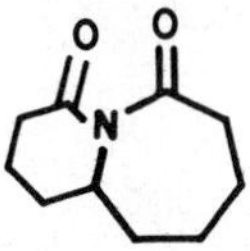 KBr	1729 1656	14
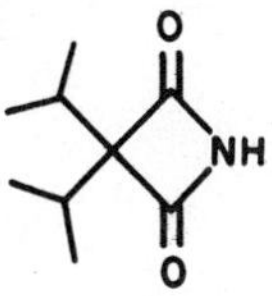 nuj.	1740	15
CCl_4	1761	15

Solvent	Frequency	Ref.
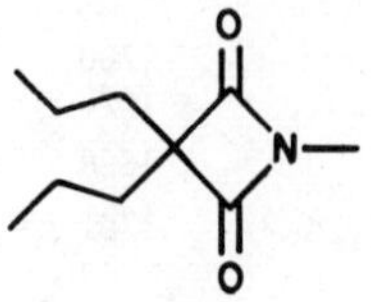 CCl_4	1761	15
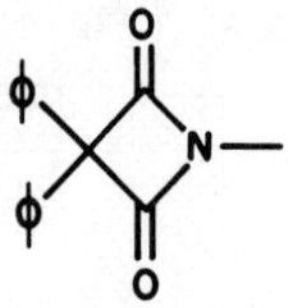 KBr	1735b	16
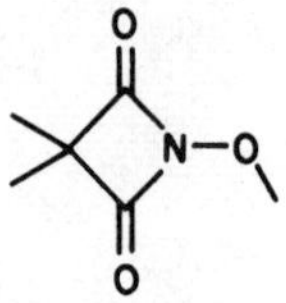 liq.	1870w 1754 1681w	17
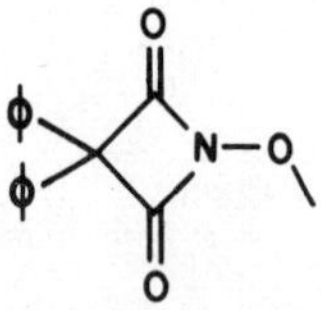 KBr	1893w 1874w 1754	17
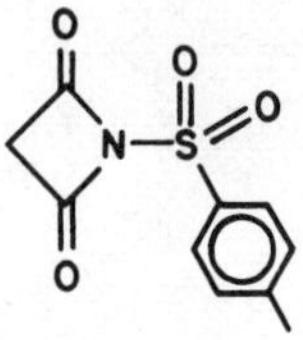 KBr	1929 1780 1730	18
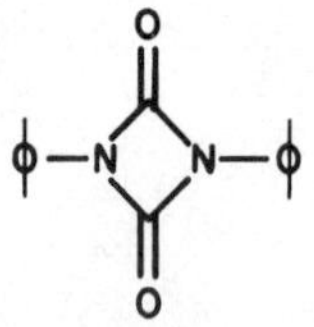 CCl_4	1783	15

Solvent	Frequency	Ref.
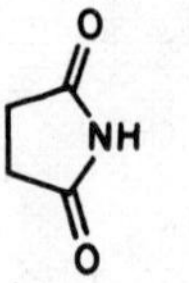 KBr	1771w 1698	9
CCl_4	1753 1727	9
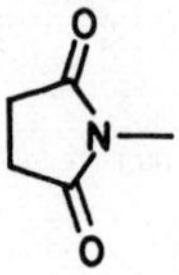 KBr	1760w 1690	9
CCl_4	1721 1705	9
$CHCl_3$	1701	19
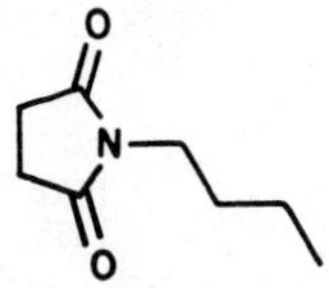 liq.	1767w 1698	20
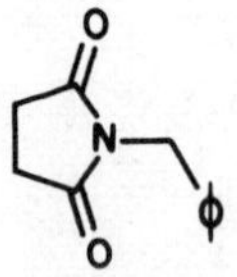 KBr	1769w 1700	20
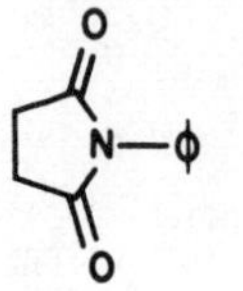 KBr	1770w 1700	20
CS_2	1786w 1725	21
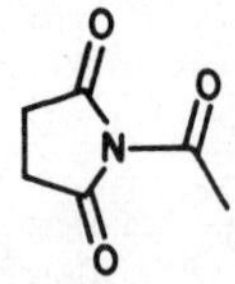 KBr	1783 1724 1681	5
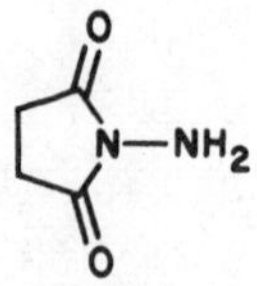 $CHCl_3$	1786w 1722w 1711	22
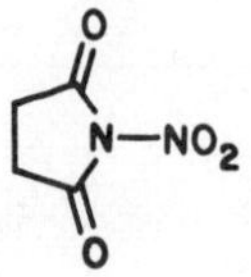 $CHCl_3$	1795 1739	23

Solvent	Frequencies	Ref.	Solvent	Frequencies	Ref.
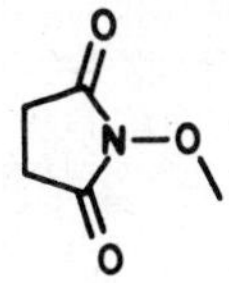 $CHCl_3$	1779 1715	23	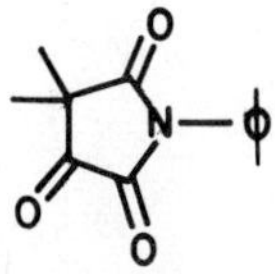 nuj.	1776 1712 1695 1678	24
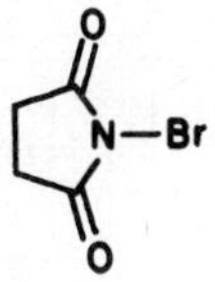 $CHCl_3$	1783 1733	6	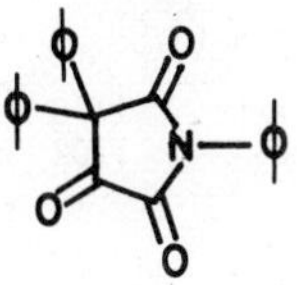 nuj.	1761w 1701	24
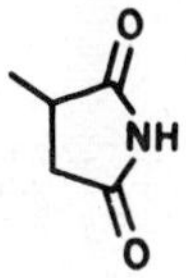 KBr	1760w 1700	20	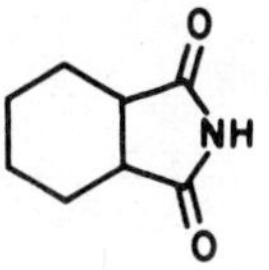 KBr	1748 1692	25
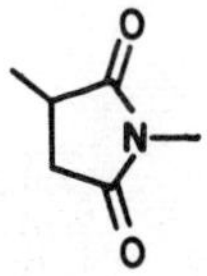 liq.	1765w 1695	20	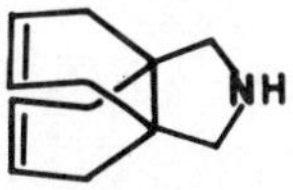 $CHCl_3$	1778 1720	26
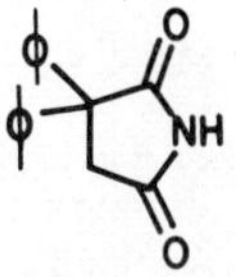 CCl_4	1735	15	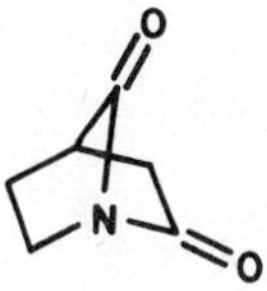 CCl_4	1737 1715w	9
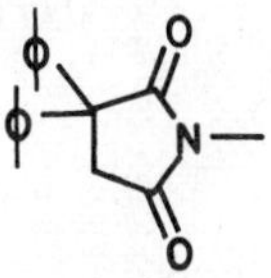 CCl_4	1712	15	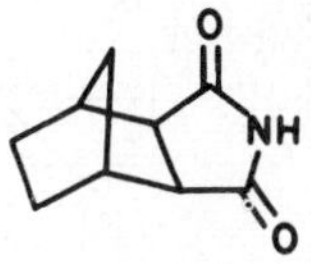 KBr	1773 1684	27
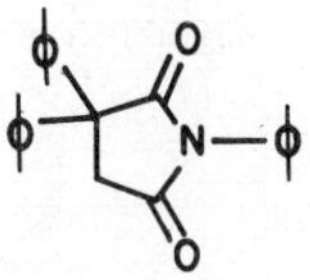 CCl_4	1722	15	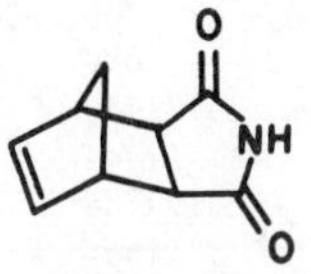 nuj.	1786 1724	28
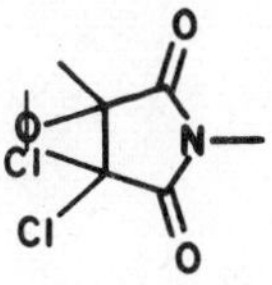 CCl_4	1732	15	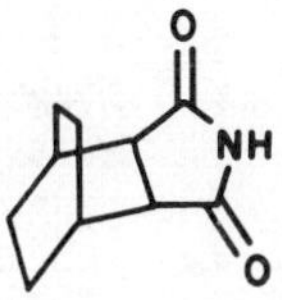 KBr	1754 1695	25

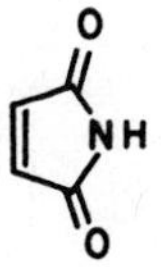 KBr	1770 1702	29
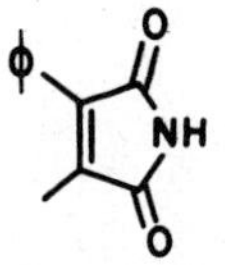 KBr	1786 1724	30
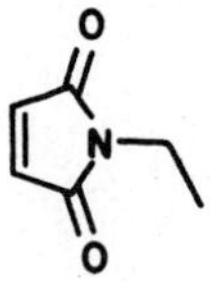 $CHCl_3$	1712	31
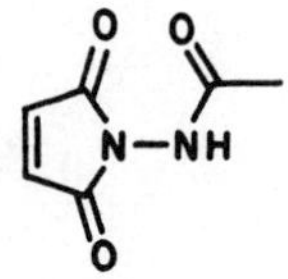 nuj.	1779 1704 1681	32
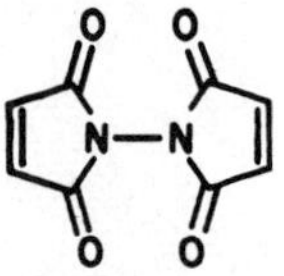 nuj.	1789	33
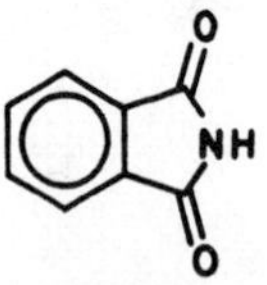 nuj.	1700 1680	34
KBr	1774 1749 1724	35
CHCl	1778w 1735	36
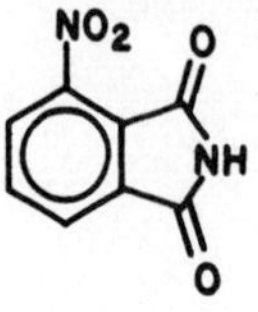 $CHCl_3$	1778w 1737	36

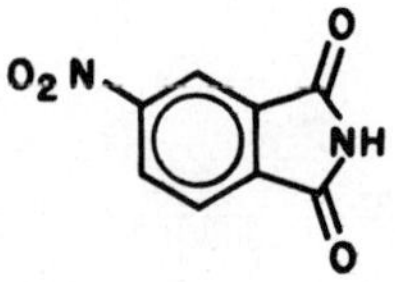 $CHCl_3$	1780w 1736	36
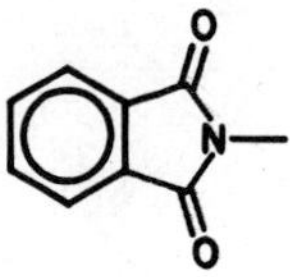 $CHCl_3$	1772 1712	36
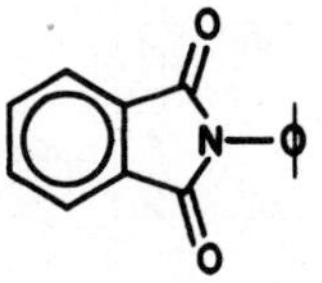 CCl_4	1782w 1738 1719	15
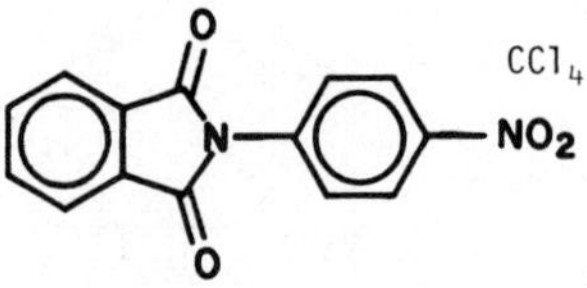 CCl_4	1791w 1754 1739 1718	15
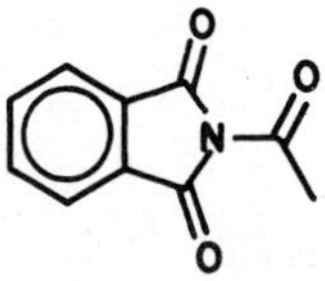 $CHCl_3$	1818 1779	23
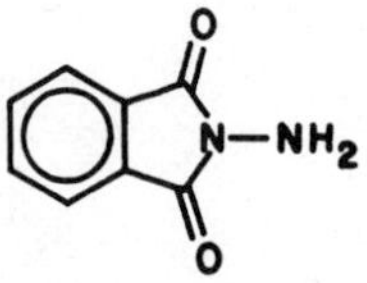 nuj.	1776 1715	37
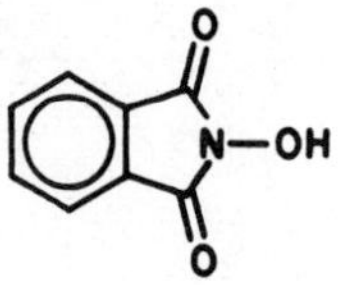 nuj.	1795 1745 1715	34

Solvent	Frequency	Ref.
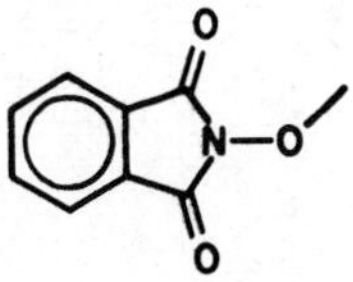 nuj.	1790 1735	34
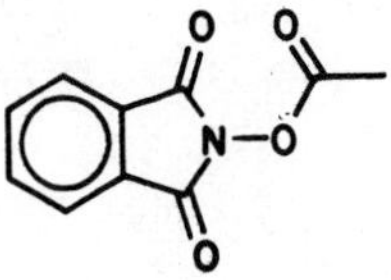 nuj.	1792 1751	38
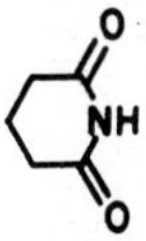 CCl_4	1742w 1730 1718	9
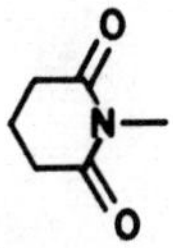 KBr	1718 1670	14
CCl_4	1729w 1686	9
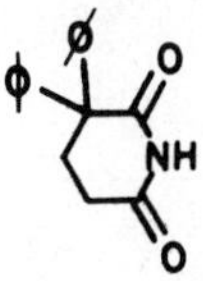 CCl_4	1735w 1717	15
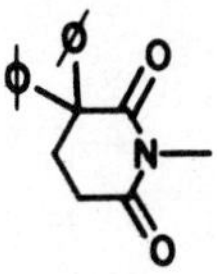 CCl_4	1730w 1685	15
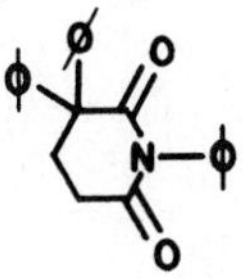 CCl_4	1740w 1696	15

Solvent	Frequency	Ref.
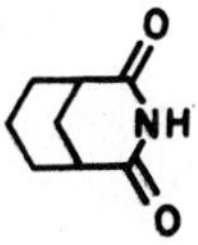 CCl_4	1730w 1714	9
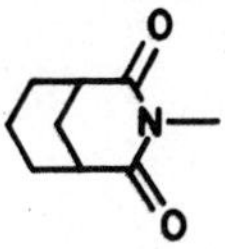 CCl_4	1728w 1679	9
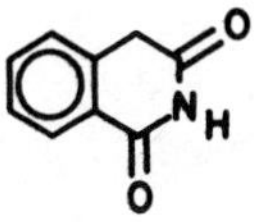 KBr	1706 1684	39
$CHCl_3$	1721 1704	39
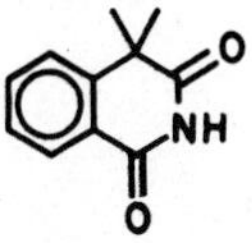 CCl_4	1718 1709	39
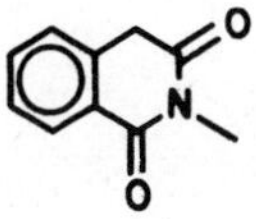 KBr	1718 1667	39
CCl_4	1727 1681	39
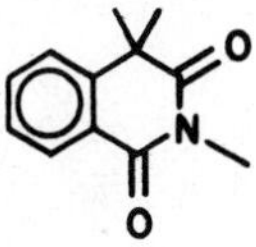 KBr	1712 1667	39
CCl_4	1721 1678	39
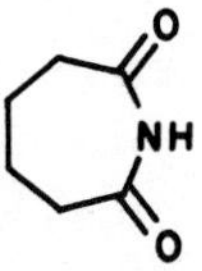 KBr	1700	40

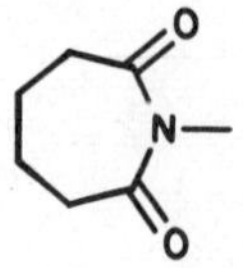

Phase	Frequency	Ref.
KBr	1715 1664	14
KBr	1742	3
KBr	1722w 1672	9
CCl_4	1728w 1697	9
KBr	1749	3
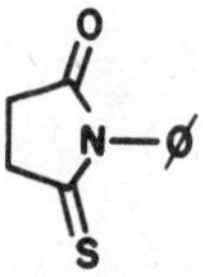 KBr	1730	29
KBr	1698	3
KBr	1471	41

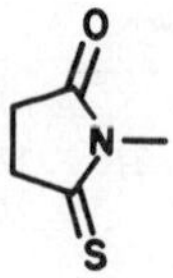

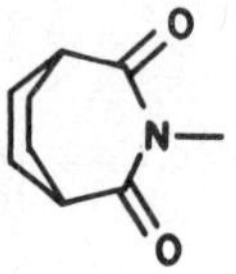

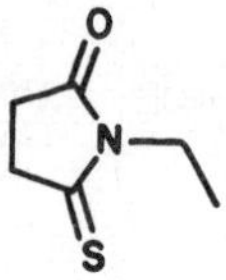

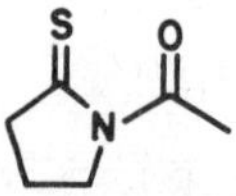

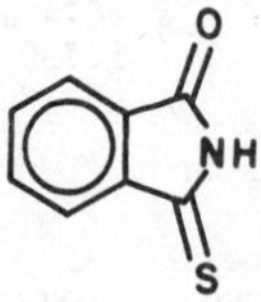

REFERENCES

1. T. Uno and K. Machida, *BCS Japan* (1967), 1226.
2. T. Uno and K. Machida, *BCS Japan* (1961), 551.
3. P. Bassignana, G. Cogrossi, S. Franco and G. Polla Mattiot, *SCA* (1965), 677.
4. W. C. Firth, Jr., *JOC* (1968), 441.
5. C. M. Lee and W. D. Kumler, *JACS* (1962), 565.
6. R. A. Abramovitch, *JCS* (1957), 1413.
7. E. J. Bourne, S. H. Henry, C. E. M. Tatlow and J. C. Tatlow, *JCS* (1952), 4014.
8. J. P. Freeman, *JACS* (1958), 5954.
9. H. K. Hall, Jr. and R. Zbinden, *JACS* (1958), 6428.
10. J. M. Bruce and F. K. Sutcliffe, *JCS* (1957), 4789.
11. S. J. Holt, A. E. Kellie, D. G. O'Sullivan and P. W. Sadler, *JCS* (1958), 1217.
12. L. Bauer, C. N. V. Nambury and C. L. Bell, *T* (1964) *20*, 165.

13. W. Z. Heldt, *JACS* (1958), 5880.
14. W. Flitsch, *B* (1964), 1548.
15. C. Fayat and A. Foucaud, *BSC France* (1970), 4491.
16. C. Metzger and R. Wegler, *B* (1968), 1120.
17. G. Zimmer and R. Moll, *B* (1966), 1292.
18. E. Mundlos and R. Graf, *A* (1964) *677*, 108.
19. E. B. Smith and H. B. Jensen, *JOC* (1967), 3330.
20. J. Falbe and F. Korte, *B* (1962), 2680.
21. R. C. Cookson, J. E. Page and M. E. Trevett, *JCS* (1954), 4028.
22. J. Gut, A. Nováček and P. Fiedler, *Col Czech Comm* (1968), 2087.
23. L. W. Kissinger and H. E. Ungnade, *JOC* (1958), 815.
24. G. S. Skinner and R. E. Ludwig, *JACS* (1956), 4656.
25. C. M. Lee and W. D. Kumler, *JOC* (1961), 4586.
26. G. Snatzke and G. Zanati, *A* (1965) *684*, 62.
27. E. J. Moriconi and W. C. Crawford, *JOC* (1968), 370.
28. W. S. Worrall, *JACS* (1960), 5707.
29. P. Bassignana, C. Cogrossi, G. Polla Mattiot and M. Gillio-Tos, *SCA* (1962), 809.
30. R. L. Wineholt, E. Wyss and J. A. Moore, *JOC* (1966), 48.
31. K. C. Tsou, R. J. Barrnett and A. M. Seligman, *JACS* (1955), 4613.
32. H. Feuer and J. P. Asunskis, *JOC* (1962), 4684.
33. H. Feuer and H. Rubinstein, *JACS* (1958), 5873.
34. A. Calder, A. R. Forrester and R. H. Thomson, *JCS* (1969) *C*, 512.
35. R. Gompper, *B* (1960), 198.
36. A. E. Kellie, D. G. O'Sullivan and P. W. Sadler, *JCS* (1956), 3809.
37. L. A. Carpius, *JACS* (1957), 98.
38. D. E. Ames and T. F. Grey, *JCS* (1955), 3518.
39. A. L. Bluhm, *SCA* (1958) *13*, 93.
40. N. Tokura, R. Tada and K. Yokoyama, *BCS Japan* (1961), 1812.
41. M. E. Baguley and J. A. Elvidge, *JCS* (1957), 709.

4.6 lactams

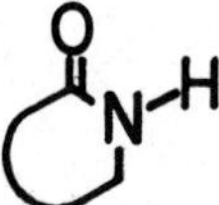

Amide functions contained in eight-membered, or smaller, lactams are constrained to maintain a cis conformation between the NH and carbonyl groups. This cis conformation leads to the absence of the amide II band (see Chapter 4.3, p. 241) and results in strong intermolecular hydrogen bonding, such that dimers are found even in dilute solution.

As with associated secondary amides a number of bands are seen in the region of 3375 to 3000 cm^{-1}, with the strongest band (of medium intensity) being found at the higher end of the range. In very dilute solution non-associated NH vibrations may be found in the region of 3450 to 3400 cm^{-1}.

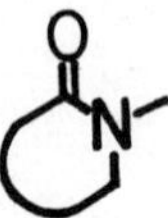

The factors influencing the carbonyl stretching frequencies of lactams are the same as those which operate for acyclic amides (Chapter 4.3, p. 241) with the added perturbation of ring size. Six, and larger, membered rings exhibit frequencies similar to those of their acyclic counterparts, whereas the smaller, higher strained, rings systems absorb at higher frequencies.

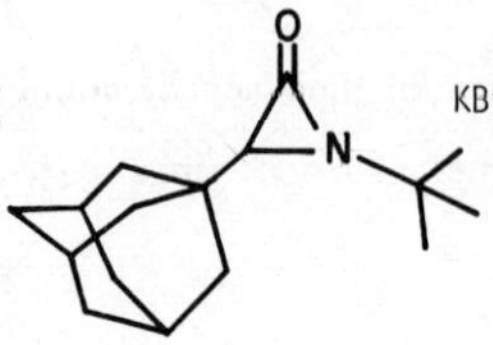

Solvent	ν	Ref.
KBr	1820	1
CCl_4	1849	2
$CHCl_3$	1847	2
$CHCl_3$	1830	3
CCl_4	1837	4
$CHCl_3$	1840 1675w	4
CCl_4	1835 1675	5

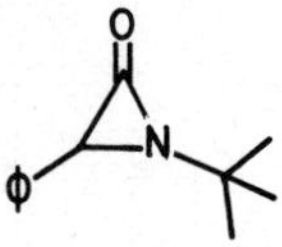

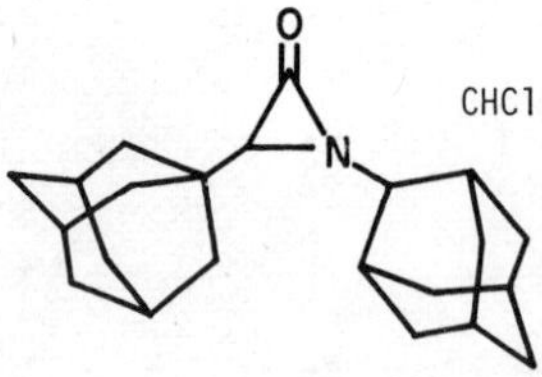

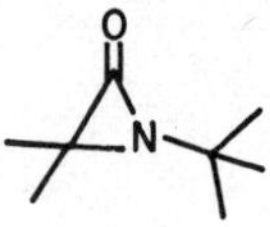

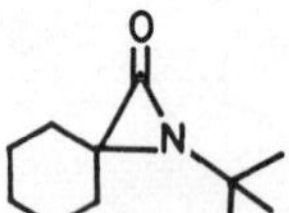

Solvent	ν	Ref.
$CHCl_3$	1764	6
$CHCl_3$	1764	6
nuj.	1786	7
CCl_4	1748	8
nuj.	1724	9
CCl_4	1761	8
$CHCl_3$	1745	6

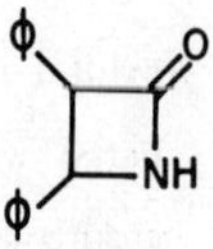

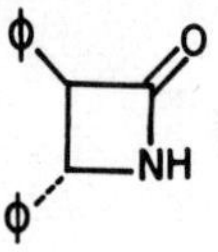

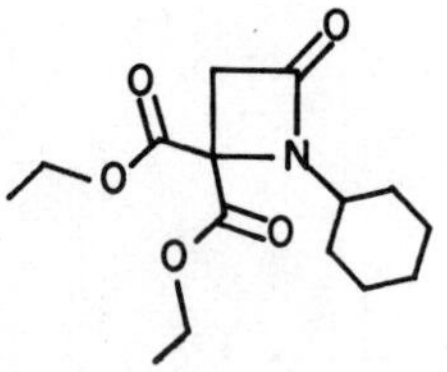

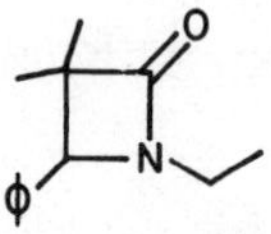

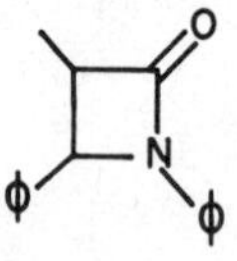

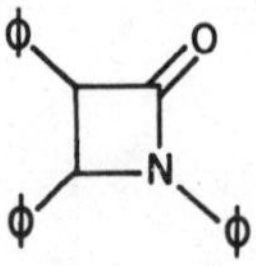

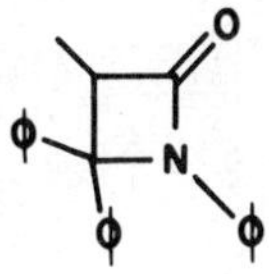

Medium	Frequency	Ref.
nuj.	1727	9
$CHCl_3$	1745	8
liq.	1742 1541	10
CCl_4	1828 1541	10
liq.	1748	11
CH_2Cl_2	1748	11
KBr	1739	11
$CHCl_3$	1754	11
CS_2	1776 1757	12
nuj.	1776	13
CCl_4	1783	3
nuj.	1779	14
liq.	1691	15
CCl_4	1702	16
$CHCl_3$	1690	17

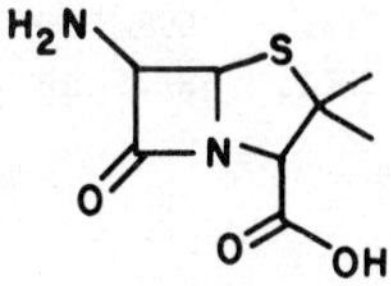

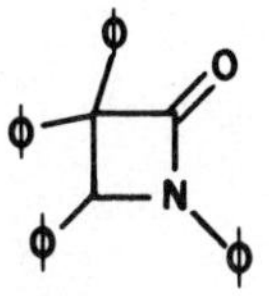

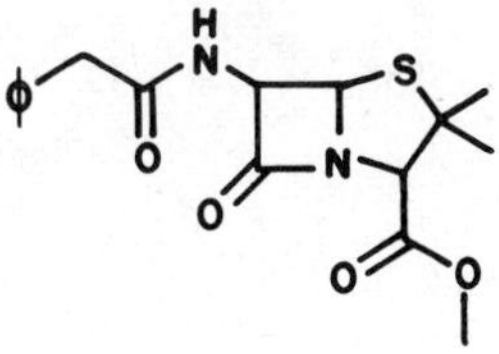

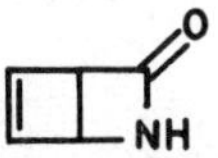

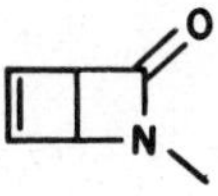

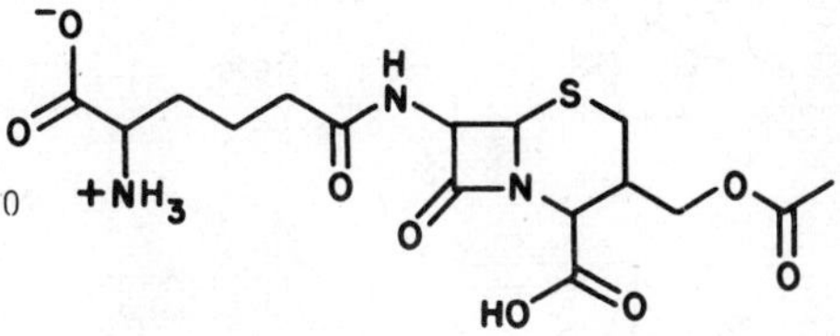

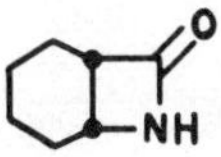

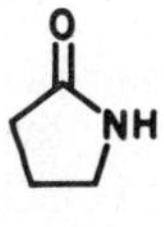

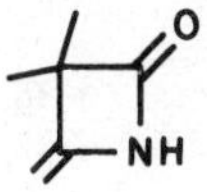

Solvent	ν (cm⁻¹)	Ref.
liq.	1688	15
CCl_4	1698	15
$CHCl_3$	1674	18
KBr	1652	19
liq.	1678	20
liq.	1690	21
liq.	1690	22
KBr	1664	23
liq.	1680	24
liq.	1686	25
KBr	1690	22
KBr	1700	22
KBr	1690	22
KBr	1675	22
nuj.	1718 1661* 1645	26
$CHCl_3$	1695	27
$CHCl_3$	1770 1712	28

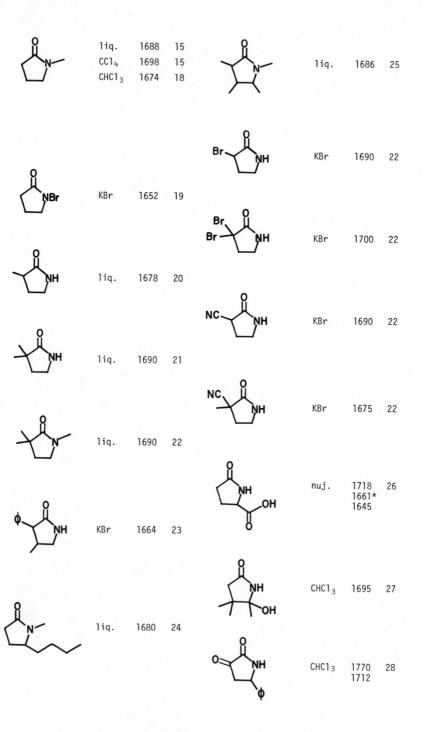

Solvent	ν (cm⁻¹)	Ref.
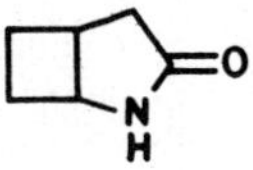 CCl_4	1718 1690	29
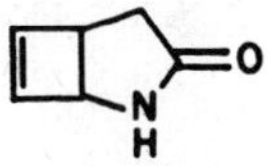 CCl_4	1727 1695	29
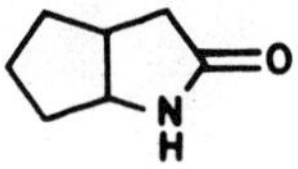 liq.	1700	30
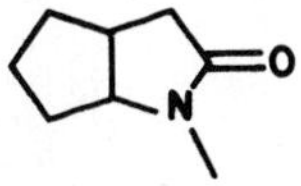 liq.	1685	30
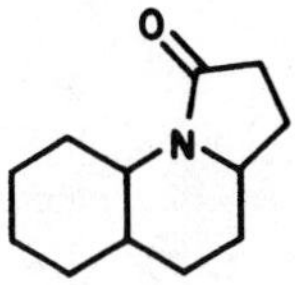 $CHCl_3$	1688* 1680 1586w	31
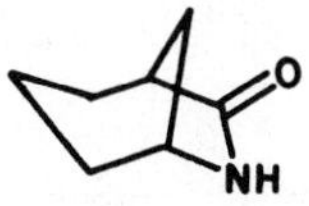 CCl_4	1723	15
 nuJ.	1689 1653 1603	32
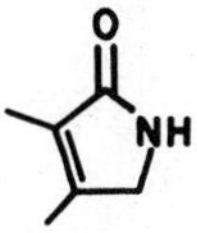 KBr	1670	33
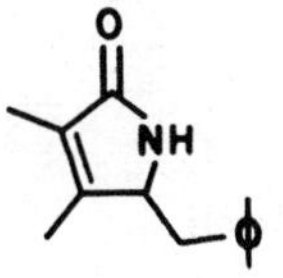 KBr	1680	33
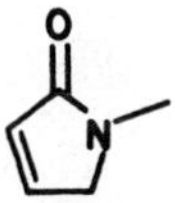 $CHCl_3$	1675 1597 1582	34
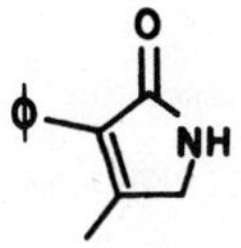 KBr	1686	23
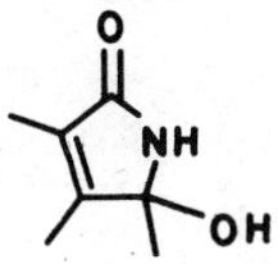 CCl_4	1680	25
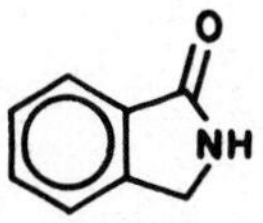 KBr	1686 1661	35
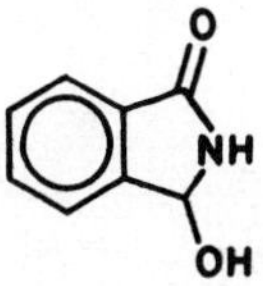 KBr	1695	36

Solvent	ν (cm^{-1})	Ref.
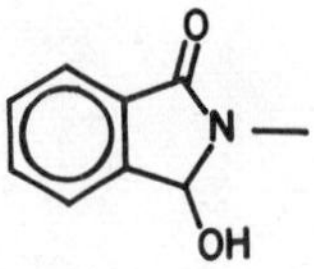 KBr	1675	36
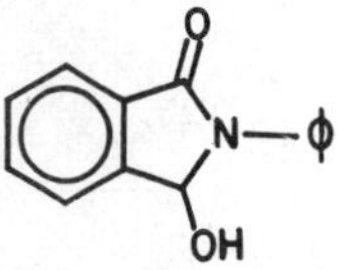 KBr	1720	36
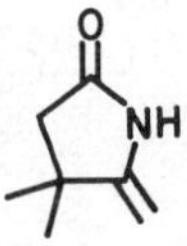 $CHCl_3$	1710 1690 1655w	27
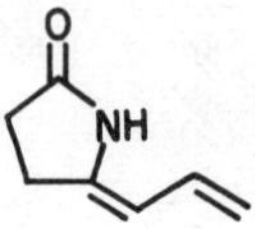 liq.	1730 1650	24
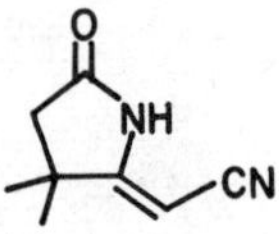 nuj.	1750 1640	37
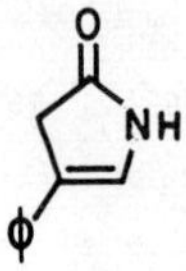 CH_2Cl_2	1705	38
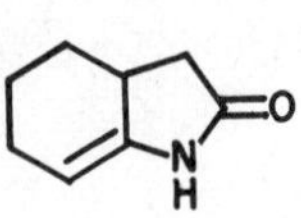 $CHCl_3$	1690	39
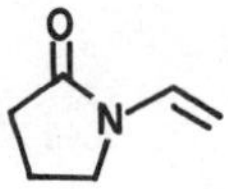 liq.	1695	40
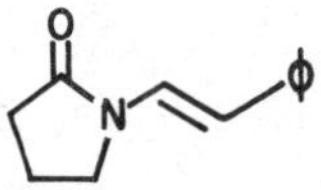 KBr	1692	41
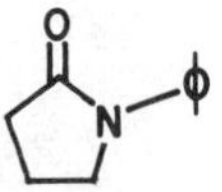 CCl_4 $CHCl_3$	1708 1684 1620	42 31
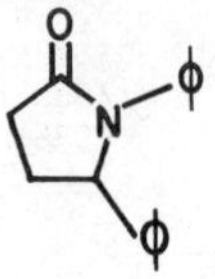 KBr	1705	30
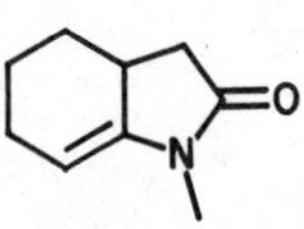 liq.	1672	39
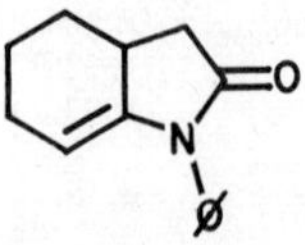 $CHCl_3$	1680	39
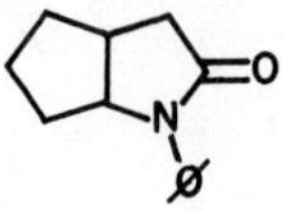 KBr	1685	30

Solvent	Frequency	Ref.
KBr	1683	30
nuj.	1704* 1692	44
$CHCl_3$	1725 1708	43
$CHCl_3$	1700	43
KBr	1709 1675	45
$CHCl_3$	1754 1722	43
$CHCl_3$	1718 1700	43
$CHCl_3$	1758	43
$CHCl_3$	1739	43
$CHCl_3$	1755 1740	46
$CHCl_3$	1741	46
$CHCl_3$	1700 1650w 1605w	47
$CHCl_3$	1698 1642w	47
KBr	1670	33
CH_2Cl_2	1675	38

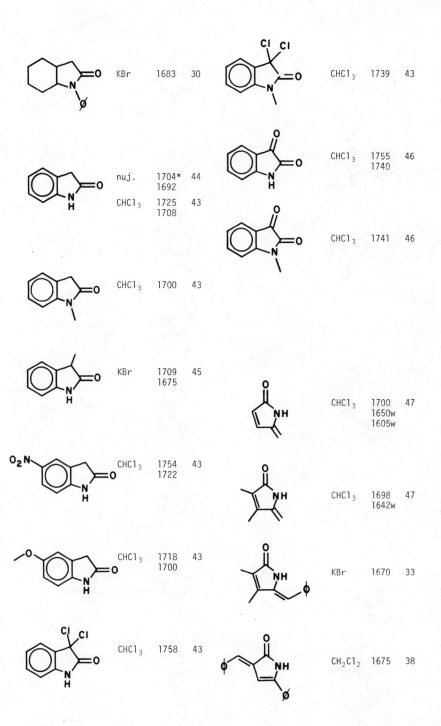

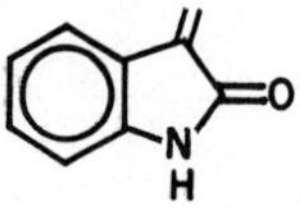 KBr	1724 1709 1642	45
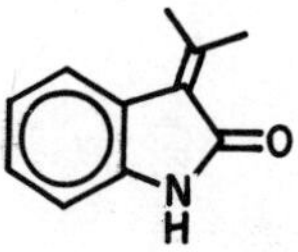 KBr	1689 1629	45
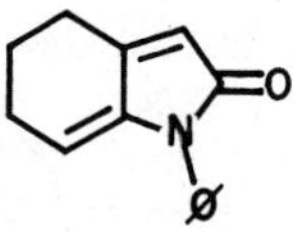 CCl_4	1700 1650 1600	48
KBr	1651	15
CCl_4	1673	15
$CHCl_3$	1667	17
CCl_4	1658	49
$CHCl_3$	1669 1631w	50
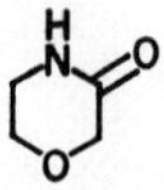 nuj.	1661	51
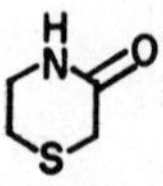 nuj.	1639b	52

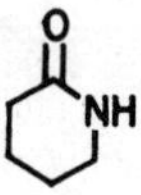

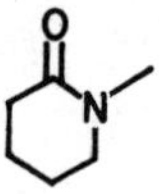

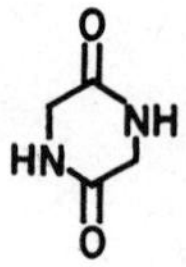 nuj.	1696	53
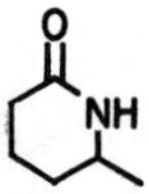 nuj.	1647	54
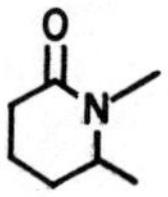 liq.	1637	55
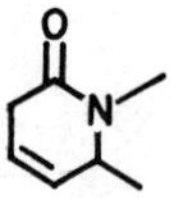 liq.	1647	54
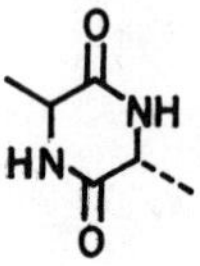 KBr	1686 1658	56
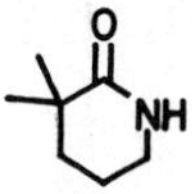 KBr	1670	22
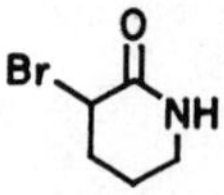 KBr	1635	22
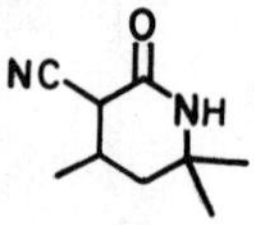 $CHCl_3$	1672 2257w	57

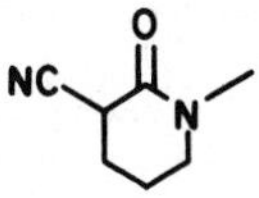 liq.	1650	22
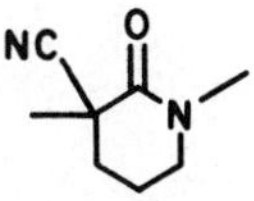 liq.	1650	22
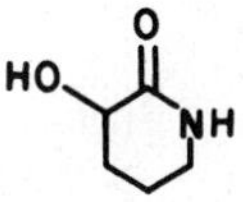 nuj.	1639	58
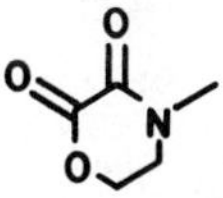 $CHCl_3$	1700 1778	59
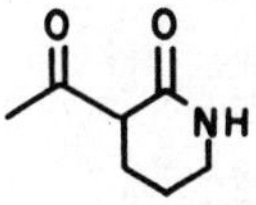 $CHCl_3$	1653 1713	60
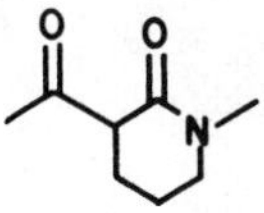 $CHCl_3$	1630 1710	60
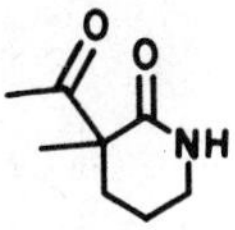 $CHCl_3$	1655 1705	61
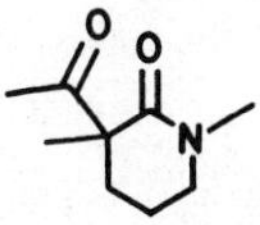 $CHCl_3$	1625 1705	61
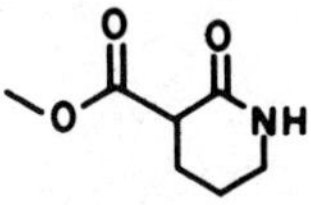 nuj.	1647 1724	62
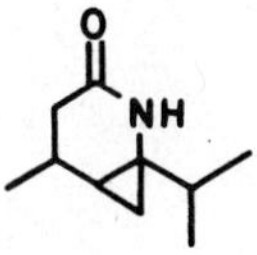 nuj.	1656	63
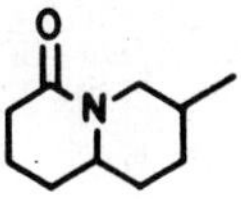 CCl_4	1630	64
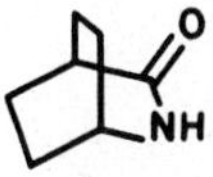 CCl_4	1695	49
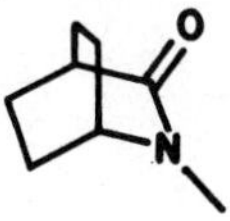 CCl_4	1678	49
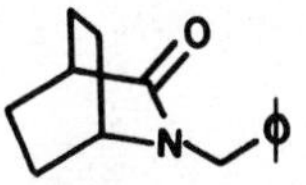 nuj.	1650	65
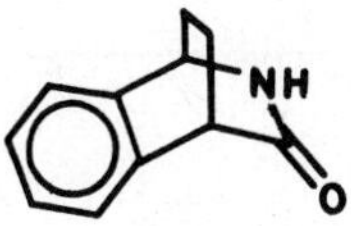 nuj.	1656b	66

Solvent	cm⁻¹	Ref.
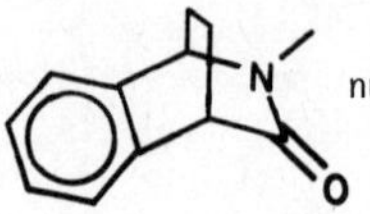 nuj.	1667	66
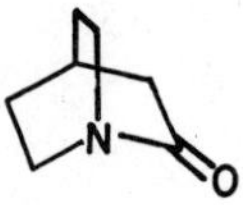 CCl_4	1750	49
KBr	1799	49
CCl_4	1762	49
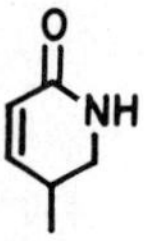 $CHCl_3$	1675 1625	67
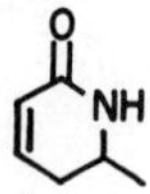 $CHCl_3$	1675 1631 1608	55
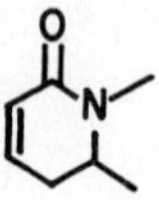 $CHCl_3$	1666 1615	35
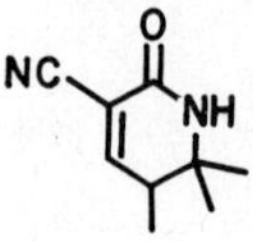 KBr	1667 2222	57
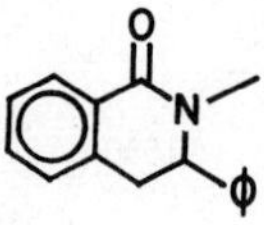 $CHCl_3$	1640	68
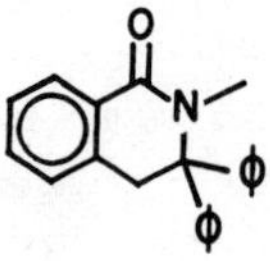 $CHCl_3$	1635	68
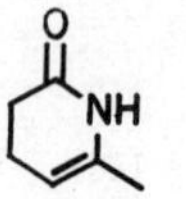 $CHCl_3$	1673	29
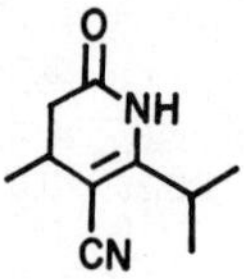 nuj.	1689 1634	63
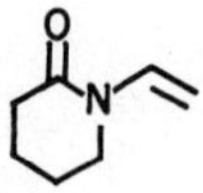 CCl_4	1666 1634	69
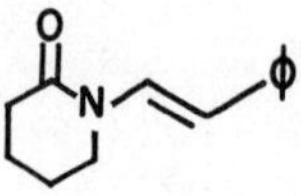 KBr	1656	70

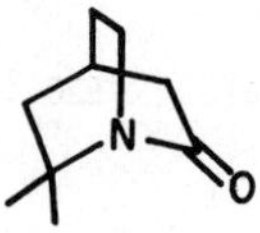

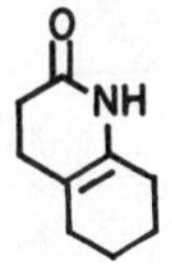	sng	1680 1640	71
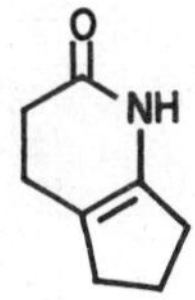	sng	1690 1640	71
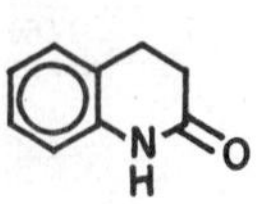	$CHCl_3$	1688* 1676	72
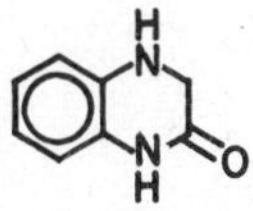	KBr $CHCl_3$	1665 1685	73 73
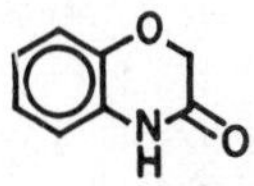	KBr $CHCl_3$	1705 1095	73 73
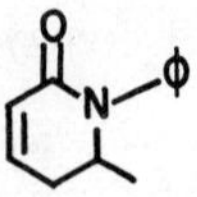	liq.	1661 1623 1605	54
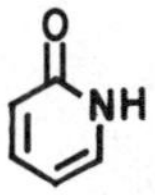	CCl_4 $CHCl_3$	1682m 1658 1620m 1544w 1654	74 75
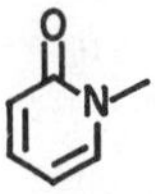	$CHCl_3$	1665	76
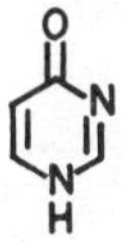	KBr	1716	77
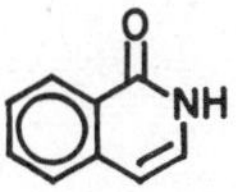	$CHCl_3$	1658	75
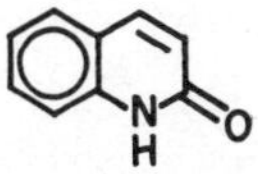	$CHCl_3$	1656	75
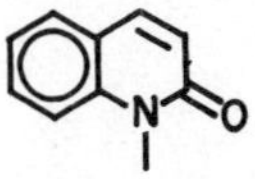	CS_2	1656	78
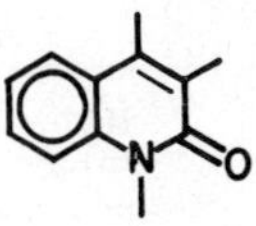	CS_2	1634	78
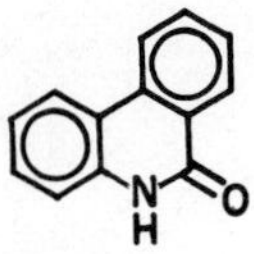	nuj.	1653	79

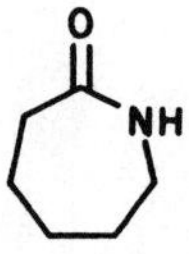

CCl_4	1671	15
$CHCl_3$	1660	17
CCl_4	1652	15
$CHCl_3$	1634	50
KBr	1630	22
CCl_4	1657	15
liq.	1661 1613	80

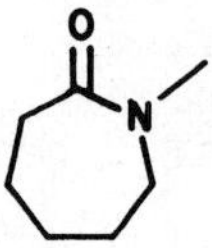

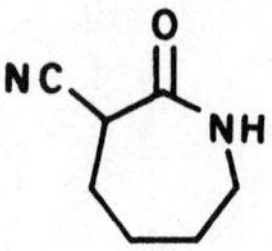

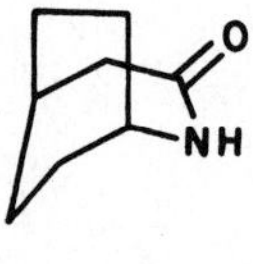

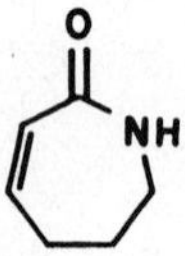

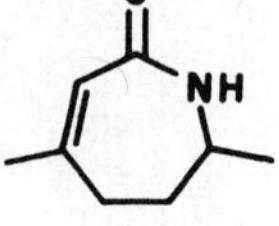

$CHCl_3$	1657 1615	29
$CHCl_3$	1658 1622	29
CCl_4	1661 1627	69
CCl_4	1668	41
CCl_4	1676	41
$CHCl_3$	1657	29
$CHCl_3$	1657	29

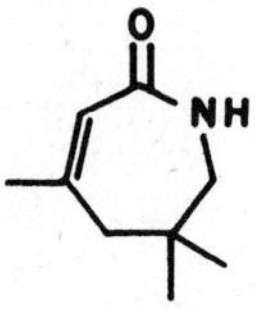

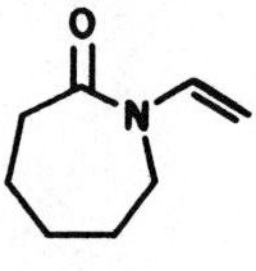

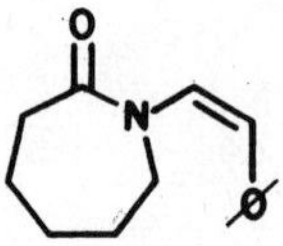

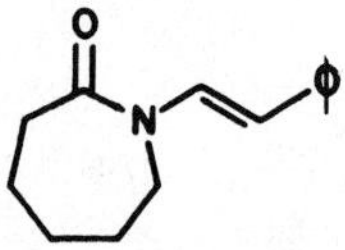

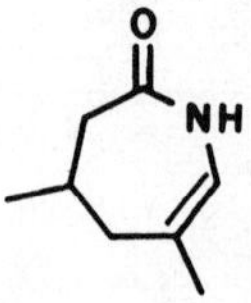

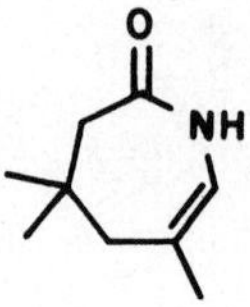

Compound	Medium	Frequency	Ref.
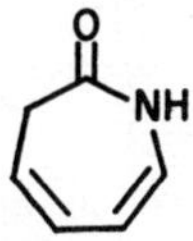	KBr	1680 1639 1593	29
	CCl_4	1670 1637 1593	29
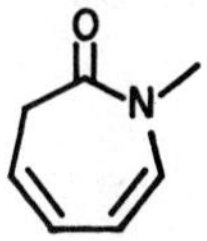	liq.	1667 1639 1593	29
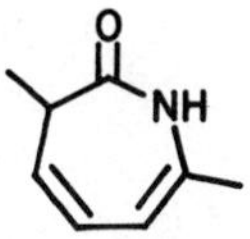	KBr	1665 1638 1595	29
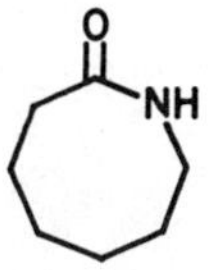	$CHCl_3$	1658	17
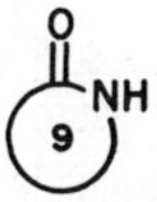	$CHCl_3$	1648 1515	17
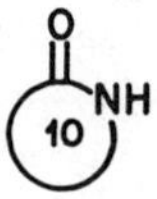	$CHCl_3$	1652 1515	17
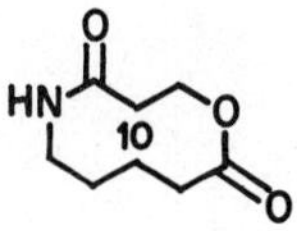	$CHCl_3$	1692 1672w 1729	81

Compound	Medium	Frequency	Ref.
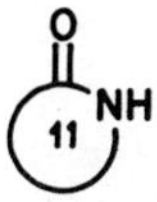	$CHCl_3$	1656 1520	17
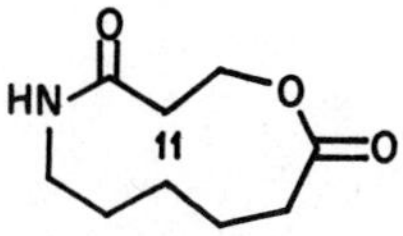	CCl_4	1695 1748	81
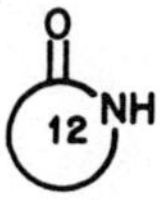	$CHCl_3$	1656 1520	17
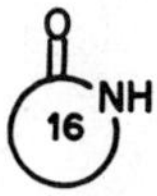	$CHCl_3$	1657 1510	17
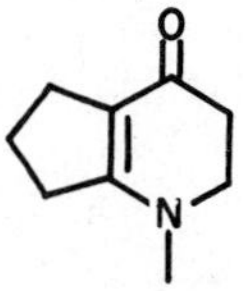	liq.	1621 1553	82
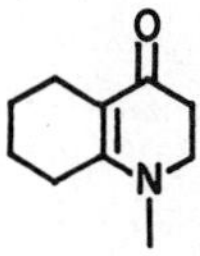	liq.	1639 1577	82
	$CHCl_3$	1638	75
	$CHCl_3$	1664 1575	76

Structure	Solvent	cm⁻¹	Ref.
	$CHCl_3$	1645	75
ϕ ϕ N H	KBr	1618 1575 1529	83
N H ϕ	KBr	1623 1577 1527	83

REFERENCES

1. K. Bott, *Angew Chem Int Ed* (1967), 946.
2. H. E. Baumgarten, *JACS* (1962), 4975.
3. E. R. Talaty, A. E. Dupuy, Jr. and A. E. Cancienne, Jr., *J Het Chem* (1967), 657.
4. J. C. Sheehan and I. Lengyel, *JACS* (1964), 1356.
5. J. C. Sheehan and I. Lengyel, *JACS* (1964), 746.
6. O. L. Chapman and W. R. Adams, *JACS* (1967), 4243.
7. J. C. Sheehan and A. K. Bose, *JACS* (1951), 1761.
8. H. T. Clarke, J. R. Johnson and Sir R. Robinson, *Chemistry of Penicillin* (1949), Princeton Univ Press, (a) p. 405, (b) p. 397.
9. A. M. Van Leusen and J. F. Arens, *Rec Trav Chim* (1959), 551.
10. E. J. Corey and J. Streith, *JACS* (1964), 950.
11. H. Bestian, H. Biener, K. Clauss and H. Heyn, *A* (1968) *718*, 94.
12. E. J. Moriconi and J. F. Kelly, *JACS* (1966), 3657.
13. C. J. Pouchert, *The Aldrich Library of Infrared Spectra*, Aldrich (1970), spectrum 350F.
14. E. P. Abraham and G. G. F. Newton, *Biochem J* (1961) *79*, 377.
15. H. K. Hall, Jr. and R. Zbinden, *JACS* (1958), 6428.
16. L. J. Bellamy and R. J. Pace, *SCA* (1963), 1831.
17. R. Huisgen, H. Brade, H. Walz and I. Glogger, *B* (1957), 1437.
18. J. Gut, A. Nováček and P. Fiedler, *Col Czech Comm* (1968), 2087.
19. A. E. Parsons, *J Mol Spectr* (1961) *6*, 201.
20. C. M. Lee and W. D. Kumler, *JACS* (1961), 4593.
21. R. Bonnett, V. M. Clark and A. Todd, *JCS* (1959), 2102.
22. F. Korte and H. Wamhoff, *B* (1964), 1970.
23. R. L. Wineholt, E. Wyss and J. A. Moore, *JOC* (1966), 48.
24. W. Flitsch, *A* (1965) *684*, 141.

25. D. Seebach, *B* (1963), 2723.
26. C. J. Pouchert, *The Aldrich Library of Infrared Spectra*, Aldrich (1970), spectrum 350C.
27. R. Scheffold, *Thesis ETH* (1963).
28. H. H. Wasserman and R. C. Koch, *JOC* (1962), 35.
29. E. Vogel, R. Erb, G. Lenz and A. A. Bothner-By, *A* (1965) *682*, 1.
30. A. Bertho and G. Rödl, *B* (1959), 2218.
31. H. E. Baumgarten, R. B. Beckerbauer and M. R. De Brunner, *JOC* (1961), 1539.
32. W. S. Worall, *JACS* (1960), 5707.
33. H. Plieninger, H. Bauer and A. R. Katritzky, *A* (1962) *654*, 165.
34. E. B. Smith and H. B. Jensen, *JOC* (1967), 3330.
35. A.-H. Khuthier and J. C. Robertson, *JOC* (1970), 3760.
36. W. Flitsch, *B* (1970), 3205.
37. W. Häusermann, *Thesis ETH* (1966).
38. Y. S. Rao and R. Filler, *JCS* (1963), 4996.
39. A. Bertho and J. F. Schmidt, *B* (1964), 3284.
40. DMS 2531.
41. W. Ziegenbein and W. Franke, *B* (1957), 2291.
42. G. R. Proctor and R. H. Thomson, *JCS* (1957), 2302.
43. A. E. Kellie, D. G. O'Sullivan and P. W. Sadler, *JCS* (1956), 3809.
44. C. J. Pouchert, *The Aldrich Library of Infrared Spectra*, Aldrich (1970), spectrum 802H.
45. R. L. Hinman and C. P. Bauman, *JOC* (1964), 2431.
46. D. G. O'Sullivan and P. W. Sadler, *JCS* (1956), 2202.
47. R. Scheffold and P. Dubs, *Helv* (1967), 798.
48. M. Prasad and C.-G. Wermuth, *BSC France* (1967), 1386.
49. H. Pracejus, M. Kehlen, H. Kehlen and H. Matschiner, *T* (1965), 2257.
50. N. Ogata, *BCS Japan* (1961), 245.
51. C. J. Pouchert, *The Aldrich Library of Infrared Spectra*, Aldrich (1970), spectrum 351A.
52. C. J. Pouchert, *The Aldrich Library of Infrared Spectra*, Aldrich (1970), spectrum 352A.
53. R. Schwyzer, B. Iselin, W. Rittel and P. Sieber, *Helv* (1956), 872.
54. A. J. Verbiscar and K. N. Campbell, *JOC* (1964), 2472.
55. O. E. Edwards and T. Singh, *Canad J Chem* (1954), 683.
56. H. Brockmann and H. Musso, *B* (1956), 241.
57. C. H. Eugster, L. Leichner and E. Jenny, *Helv* (1963), 543.

58. C. J. Pouchert, *The Aldrich Library of Infrared Spectra*, Aldrich (1970), spectrum 351H.
59. G. Drefahl, M. Hartmann and A. Skurk, *B* (1966), 2716.
60. K. Hoegerle and H. Erlenmeyer, *Helv* (1956), 1203.
61. K.-J. Ploner, H. Wamhoff and F. Korte, *B* (1967), 1675.
62. C. J. Pouchert, *The Aldrich Library of Infrared Spectra*, Aldrich (1970), spectrum 352B.
63. W. D. Burrows and R. H. Eastman, *JACS* (1957), 3756.
64. F. Bohlmann, E. Winterfeldt, P. Studt, H. Laurent, G. Boroschewski and K.-M. Kleine, *B* (1961), 3151.
65. L. H. Werner and S. Ricca, Jr., *JACS* (1958), 2733.
66. G. N. Walker and D. Alkalay, *JOC* (1971), 491.
67. K. Gschwend-Steen, *Thesis ETH* (1965).
68. D. M. Bailey and C. G. DeGrazia, *JOC* (1970), 4088.
69. P. P. Shorygin, T. N. Shkurina, M. F. Shostakovskii, F. P. Sidel' Kovskaya and M. G. Zelenskaya, *Bull Acad Sci USSR Chemical Science* (1959), 2103.
70. E. H. White, M. C. Chen and L. A. Dolak, *JOC* (1966), 3038.
71. J. J. Vill, T. R. Steadman and J. J. Godfrey, *JOC* (1964), 2780.
72. H. E. Baumgarten, P. L. Creger and R. L. Zey, *JACS* (1960), 3977.
73. D. G. O'Sullivan and P. W. Sadler, *JCS* (1957), 2916.
74. R. A. Coburn and G. O. Dudek, *JPC* (1968), 1177.
75. S. F. Mason, *JCS* (1957), 4874.
76. A. R. Katritzky and R. A. Jones, *JCS* (1960), 2947.
77. R. Gompper, *B* (1960), 198.
78. D. J. Cook, R. S. Yunghans, T. R. Moore and B. E. Hoogenboom, *JOC* (1957), 211.
79. H. Gilman and J. Eisch, *JACS* (1957), 5479.
80. F. J. Donat and A. L. Nelson, *JOC* (1957), 1107.
81. V. K. Antonov, A. M. Shkrob, V. I. Shchelokov and M. M. Shemyakin, *TL* (1963), 1353.
82. A. I. Meyers, A. H. Reine, J. C. Sircar, K. B. Rao, S. Singh, H. Weidmann and M. Fitzpatrick, *J Het Chem* (1968), 151.
83. R. J. Light and C. R. Hauser, *JOC* (1960), 539.

4.7

carboxylic acids

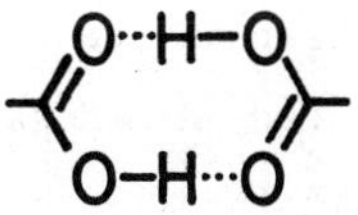

In organic solvents and even to some extent in the vapor phase, carboxylic acids exist as dimers because of strong intermolecular hydrogen bonding. In the vapor phase or in dilute nonpolar solutions the monomeric carboxylic acid shows a medium-intensity O-H stretching frequency between 3600 and 3500 cm^{-1}. Frequently, however, the monomeric band is not seen, and even when it is observed it is dominated by a number of medium-intensity absorptions in the range of 3000 to 2500 cm^{-1}. Although the origin of these absorptions is still uncertain, they are both characteristic and diagnostic for carboxylic acids.

—OH

The O-H out-of-plane deformation occurs in the region of 1000 to 900 cm^{-1}. It is of variable intensity, usually broad, and of little diagnostic value. Similarly, the C-O stretching vibration and the coupled C-O and O-H modes that fall in the region from 1450 to 1200 cm^{-1} have little diagnostic value.

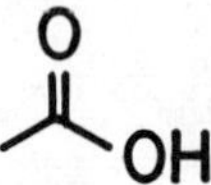

The carbonyl stretching frequency of the dimeric carboxylic acids is usually from 30 to 60 cm^{-1} lower in frequency than that of the monomeric species. When spectra are measured on solids or neat liquids, essentially no monomeric species are detectable. In dilute solutions the monomer may be observed, usually as a weak absorption or shoulder at wavelengths higher than the major carbonyl absorption of the dimeric species.

In both the monomeric and dimeric carboxylic acids, as with esters, two opposing phenomena influence the carbonyl stretching frequency. Delocalization of the hydroxyl lone pairs into the carbonyl group causes a lowering of the frequency, compared to that of a ketonic carbonyl group, whereas the electronegative oxygen destabilizes the dipolar contribution to the carbonyl group, which will raise the frequency of the carboxyl carbonyl. As with esters the latter effect dominates, and monomeric carboxylic acids absorb at higher frequencies than ketones and at frequencies comparable to the corresponding ester. However, the influence of hydrogen bonding lowers the carbonyl stretching frequency to that of the corresponding aldehyde.

The influence of conjugation and neighboring electronegative groups parallels those noted for ketones (Chapter 4.2, p. 175). When intramolecular hydrogen bonding can occur, as with salicyclic or anthranilic acids, considerable lowering of the carbonyl stretching frequency is observed.

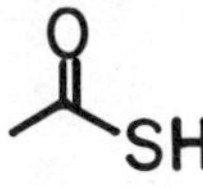

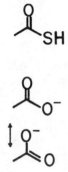

Thio carboxylic acids, in a manner analogous to thiol esters (Chapter 4.10, p. 332), generally absorb at frequencies lower than the corresponding carboxylic acid.

Carboxylate anions have two equivalent C-O bonds and can be represented by two equivalent resonance structures. The carboxylate anion, like the isoelectronic nitro group, shows two intense bands. The higher frequency band results from the asymmetric mode, and the lower frequency from the symmetric mode. As with the nitro group, the frequency shifts of the asymmetric vibration are of more diagnostic value than those of the symmetric vibration. Qualitatively, the changes observed with carboxylate anions parallel those seen with the acids themselves. Association with the counter ion can, however, cause considerable frequency changes.

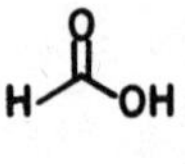

Phase	ν (cm⁻¹)	Ref.
liq.	1727	1
CCl_4	1756w 1724	1
liq.	1759w 1718	1
CCl_4	1768w 1717	1
liq.	1787w 1736	3
CCl_4	1712	2
liq.	1701	4
CCl_4	1709	2
liq.	1722	5
liq.	1716	5
CCl_4	1715	6
KBr	1686	7
CCl_4	1704	6

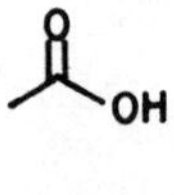

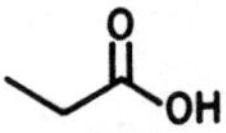

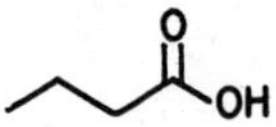

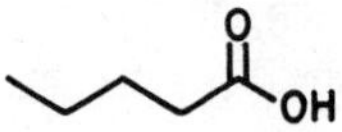

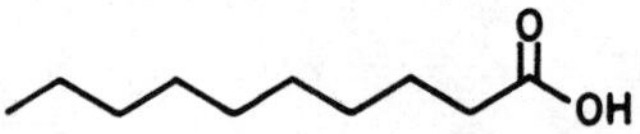

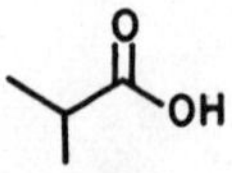

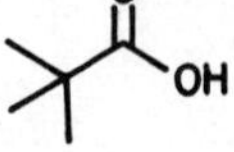

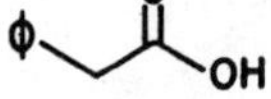

Phase	ν (cm⁻¹)	Ref.
CCl_4	1712	8
KBr	1670	7
CCl_4	1693	7
KBr	1695	9
KBr	1675	7
CCl_4	1696	7
nuj.	1686	10
KBr	1670	7
CCl_4	1706	7
KBr	1646	7
CCl_4	1654	7
KBr	1633	7
CCl_4	1654	7

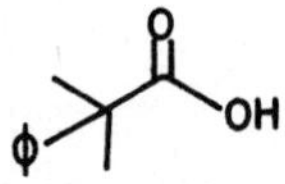

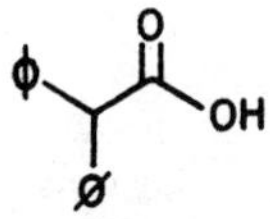

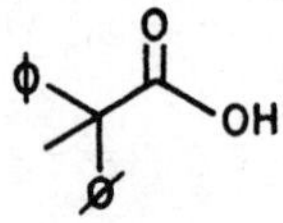

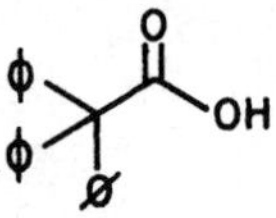

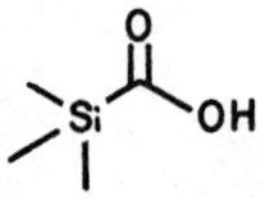

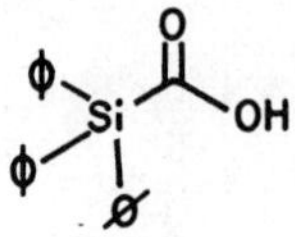

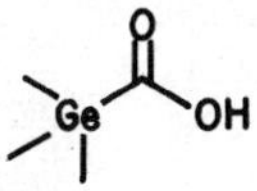	KBr CCl_4	1650 1648	7 7
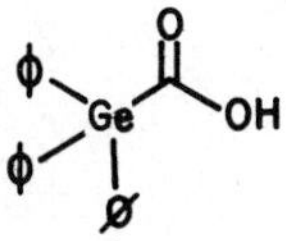	KBr CCl_4	1642 1663	7 7
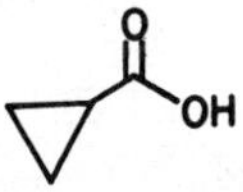	liq.	1696	11
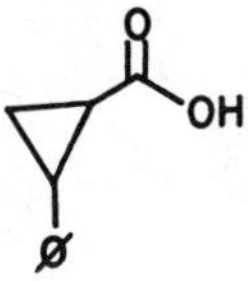	CCl_4	1694	12
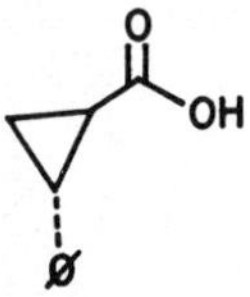	CCl_4	1688	12
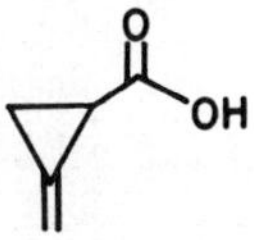	$CHCl_3$	1712	13
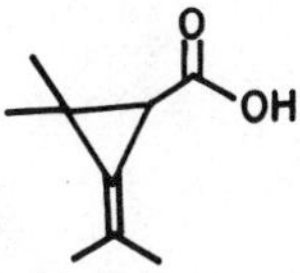	$CHCl_3$	1695	14

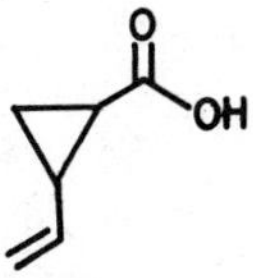	liq.	1695 1632	15
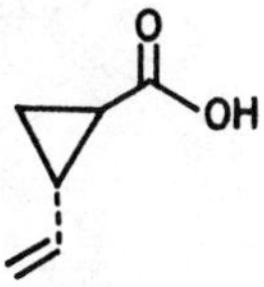	liq.	1695 1641	15
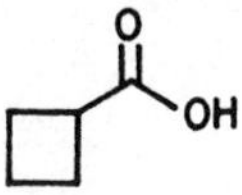	liq.	1701	16
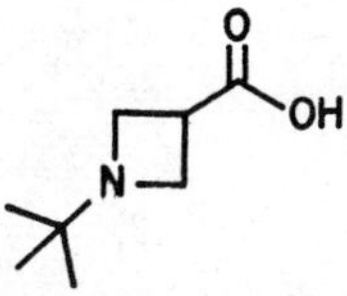	KBr	1600	9
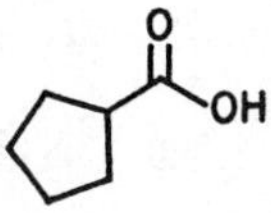	liq.	1701	17
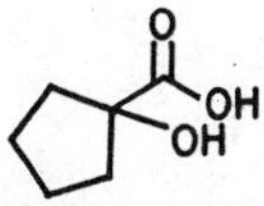	nuj.	1712	18
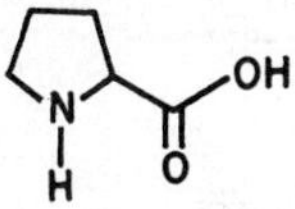	KBr	1618 1555 1527*	19
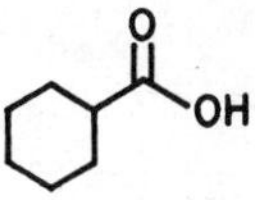	solid	1718	20

Compound	Medium	ν (cm⁻¹)	Ref.
cyclohex-3-enecarboxylic acid	CCl_4	1705	21
cycloheptanecarboxylic acid	liq.	1698	22
cycloundecanecarboxylic acid (11)	liq.	1698	23
cubanecarboxylic acid	CCl_4	1692	24
adamantane-1-carboxylic acid	nuj.	1689	25
FCH_2COOH	CCl_4	1790w 1735 1715	26
$F_2CHCOOH$	liq.	1739 1700	26
	CCl_4	1794w 1764	27
F_3CCOOH	liq.	1776	26
	CCl_4	1813w 1780	27
	$CHCl_3$	1783	28
F_3CCF_2COOH	liq.	1780	9
$ClCH_2COOH$	CCl_4	1791w 1737	27
$Cl_2CHCOOH$	liq.	1799w 1739	1
	CCl_4	1784w 1744	27
Cl_3CCOOH	liq.	1742	1
	CCl_4	1789w 1752	27
	$CHCl_3$	1742	28
$BrCH_2COOH$	CCl_4	1772w 1726	27
$CH_3CHBrCOOH$	CCl_4	1721	29
$CH_3CH_2CHBrCOOH$	CCl_4	1720	2

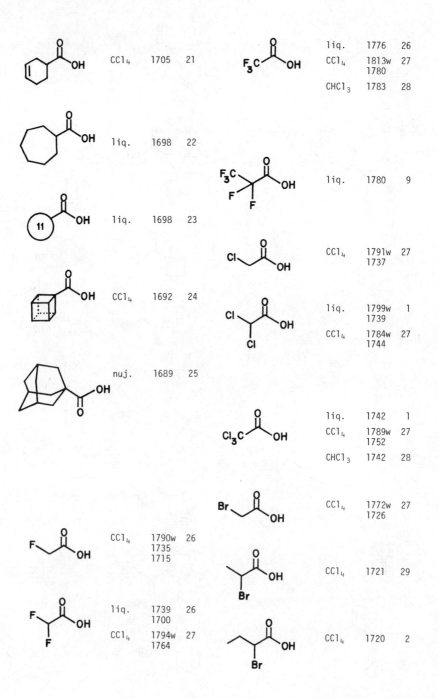

Solvent	Frequency	Ref.
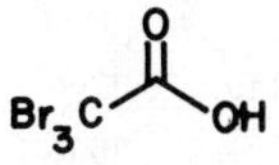 CCl_4	1772w 1735	27
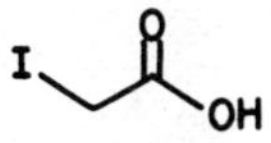 CCl_4	1772w 1713	27
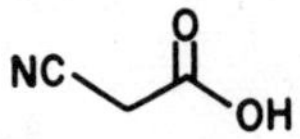 KBr	1730	9
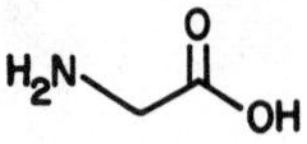 KBr	1605b 1515	19
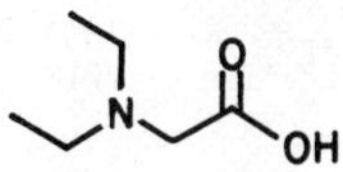 CCl_4	1786	30
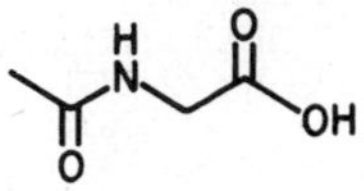 KBr	1715 1590	31
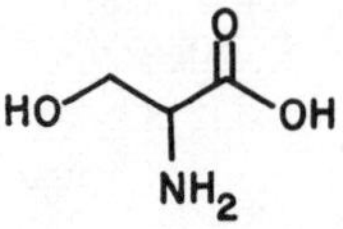 KBr	1600b 1468	19
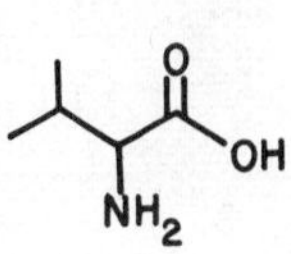 KBr	1613* 1575 1499	19
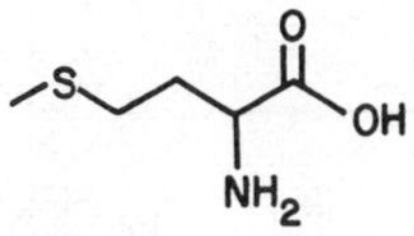 KBr	1626* 1580b 1508	19
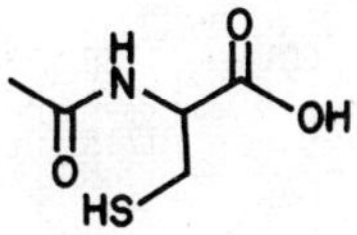 nuj.	1718 1587	32
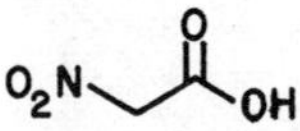 CCl_4	1740	2
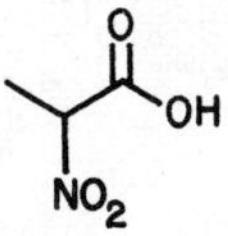 CCl_4	1736	2
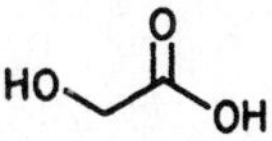 CCl_4	1765 1720	33
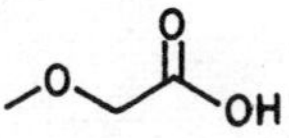 CCl_4	1791 1761 1731	34
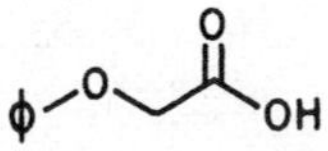 CCl_4	1791 1763 1737	34
liq.	1718	4
CCl_4	1755 1715	33
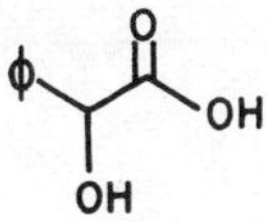 nuj.	1721	35

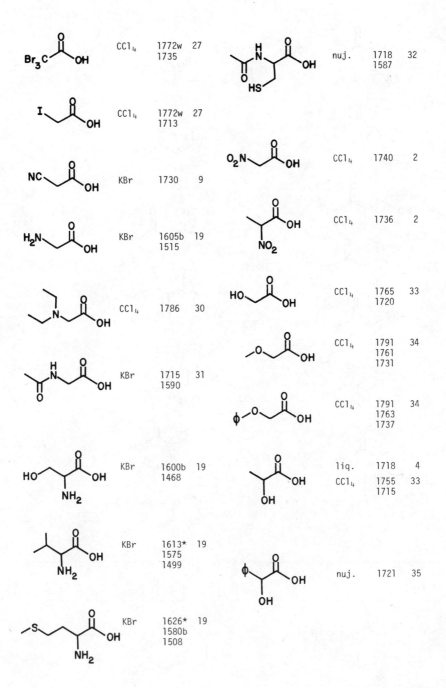

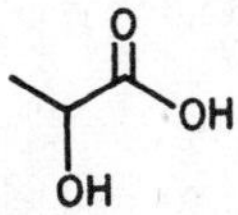

Solvent	Frequency	Ref.
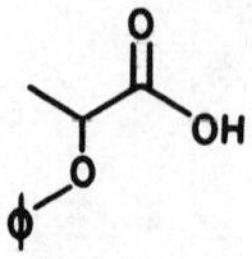 CCl_4	1786, 1754, 1716	34
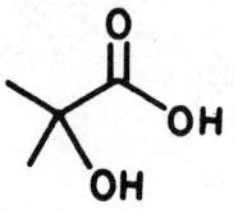 liq.	1721	4
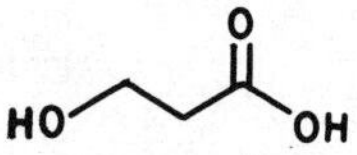 liq.	1709	4
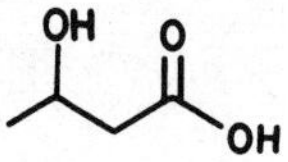 CCl_4	1748, 1720	33
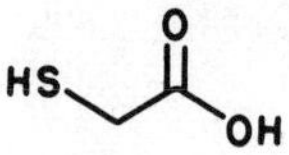 liq.	1710	9
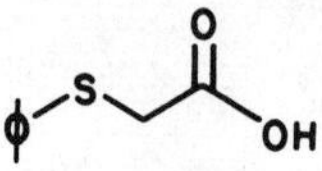 nuj.	1701	36
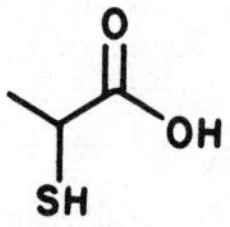 CCl_4	1700	37
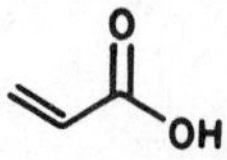 liq.	1730*m, 1706	9
liq.	1700	9
CCl_4	1665, 1641	38
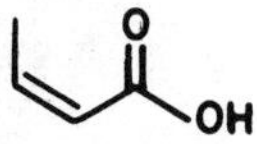 CCl_4	1692	39
liq.	1700	9
CCl_4	1694	39
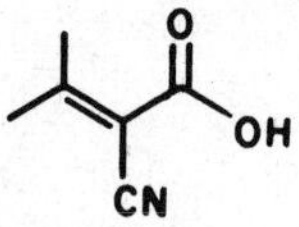 $CHCl_3$	1701, 1650, 2212	40
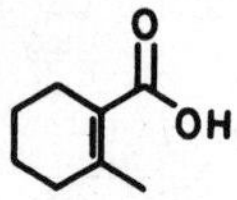 CH_2Cl_2	1742, 1701, 1639	41
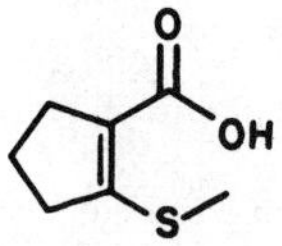 KBr	1650, 1560	42
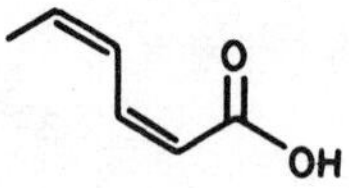 CCl_4	1691	39
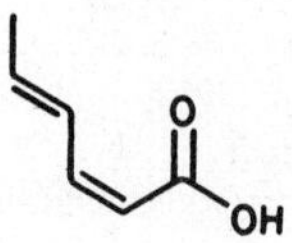 CCl_4	1687	39
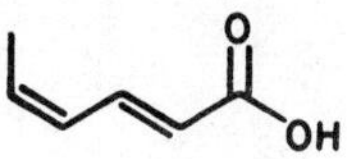 CCl_4	1683	39

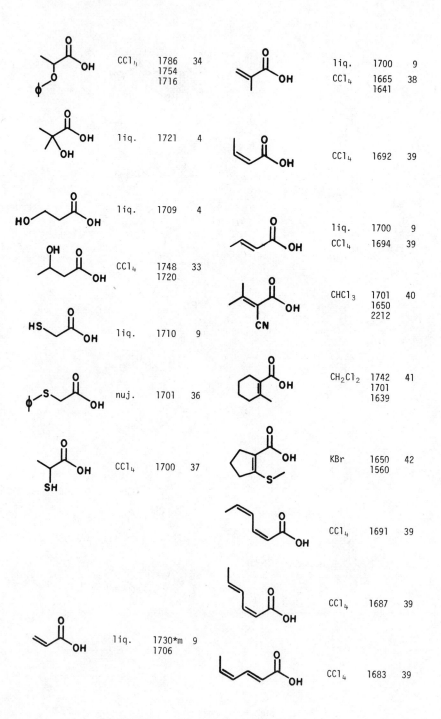

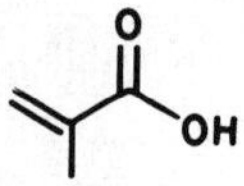

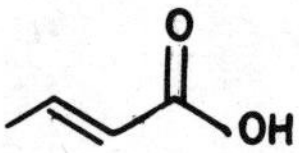

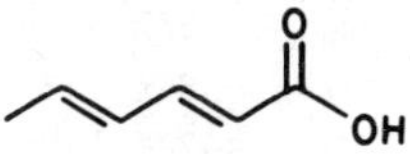

CCl_4 1702 39

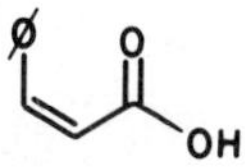

CCl_4 1709 43
1650

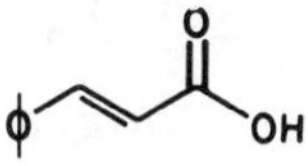

KBr 1693 44
CCl_4 1699 43
1652

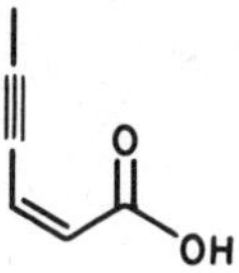

CCl_4 1702 39

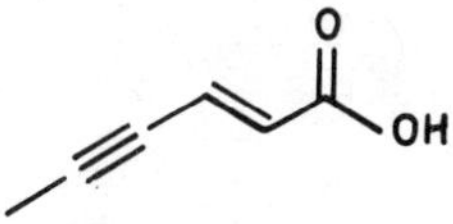

CCl_4 1685 39

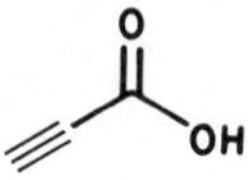

CCl_4 1698 45
2130

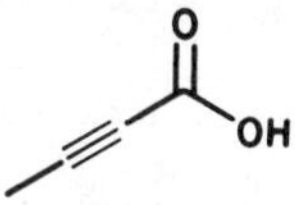

CCl_4 1684 39

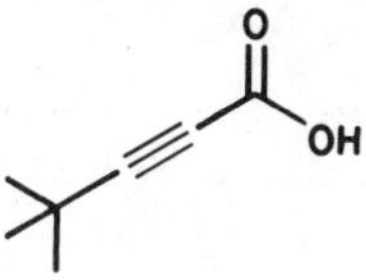

liq. 1670 46
2210

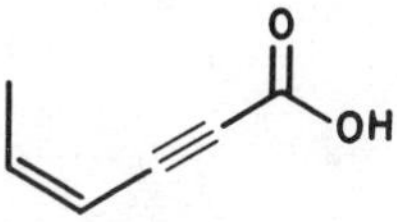

CCl_4 1692 27

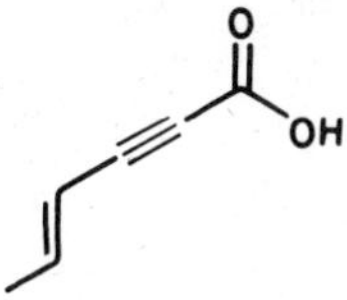

CCl_4 1667 39

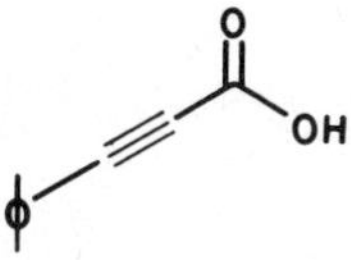

CCl_4 1692 43
2220

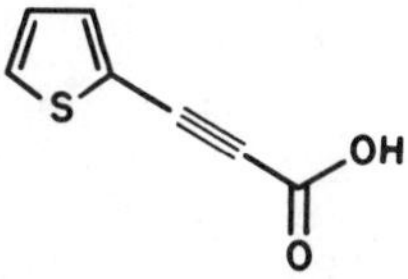

CCl_4 1686 47

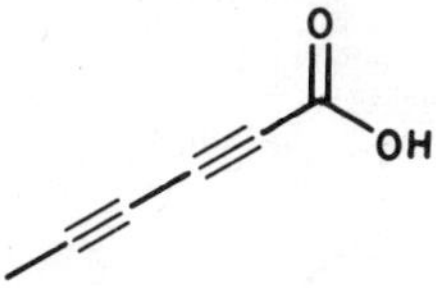

CCl_4 1688 39

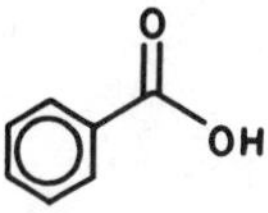

nuj. 1690 29
CCl_4 1730 48
$CHCl_3$ 1706 49

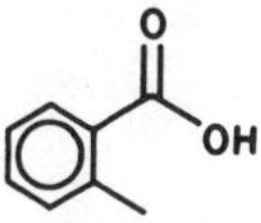

nuj. 1682 51
CCl_4 1740 50

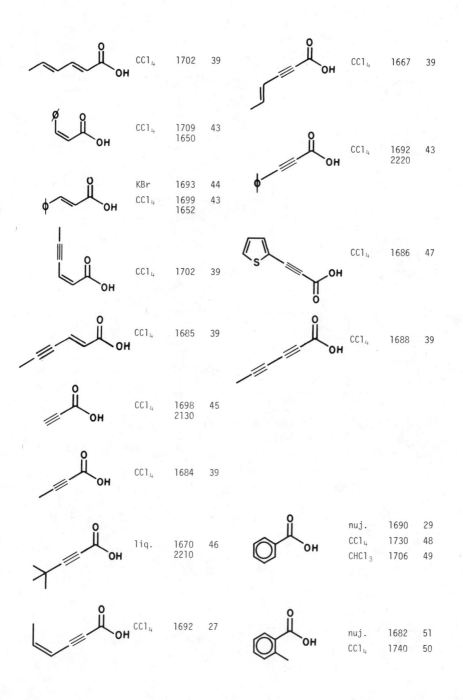

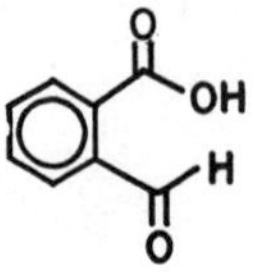

$CHCl_3$	1738	29

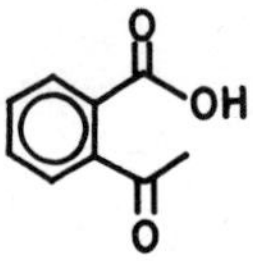

nuj.	1732	29

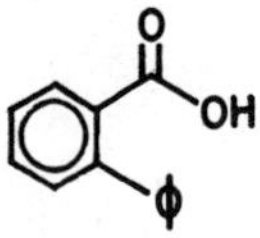

nuj.	1686	52

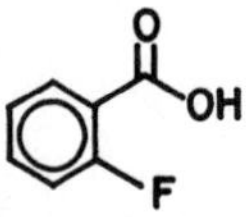

$CHCl_3$	1708	49

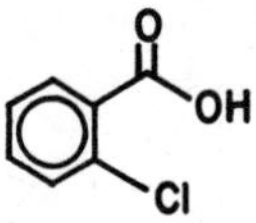

CCl_4	1755	50
$CHCl_3$	1713	49

CCl_4	1755	50
$CHCl_3$	1713	49

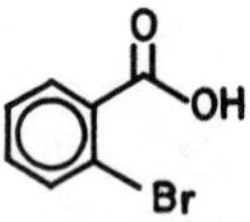

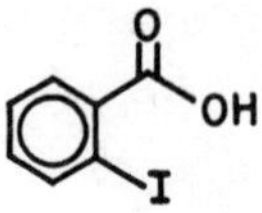

CCl_4	1752	50
$CHCl_3$	1712	49

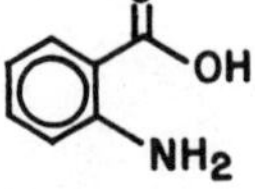

nuj.	1671	54
KBr	1668	53
CCl_4	1708	50
$CHCl_3$	1673	49

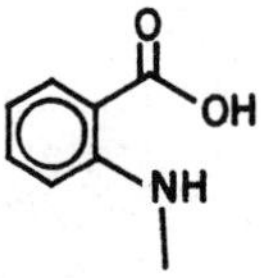

nuj.	1664	55

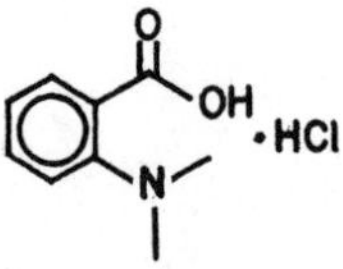

•HCl

	1770m 1672	56

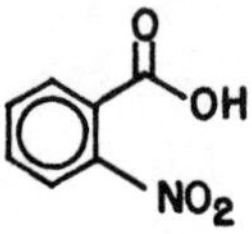

CCl_4	1759	50
$CHCl_3$	1719	49

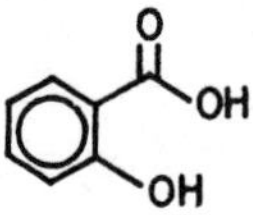

KBr	1667	58
CCl_4	1696	50
$CHCl_3$	1661	57

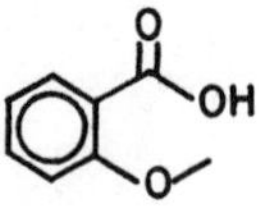

CCl_4	1750	50

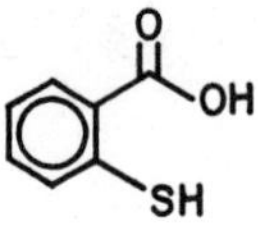

nuj.	1669	59

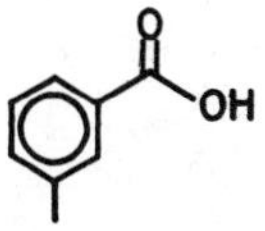

Medium	cm⁻¹	Ref.
nuj.	1686	60
CCl_4	1740	50
nuj.	1631	61
CCl_4	1705	62
nuj.	1689	63
nuj.	1667	64
KBr	1671	44
KBr	1704	44
KBr	1693	44
CCl_4	1693	62

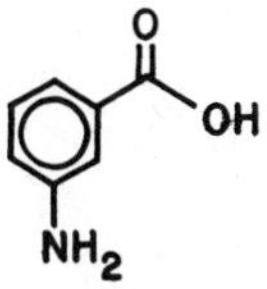

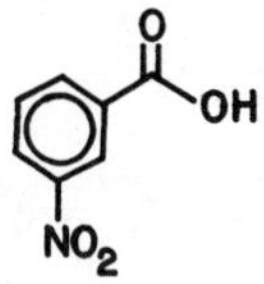

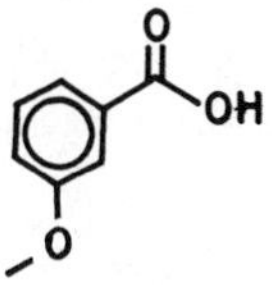

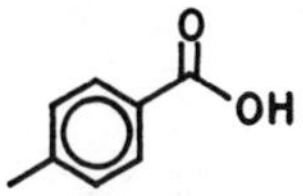

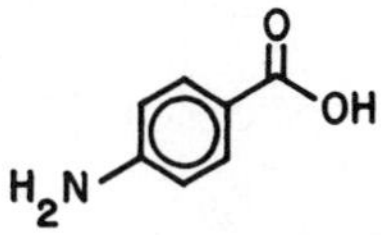

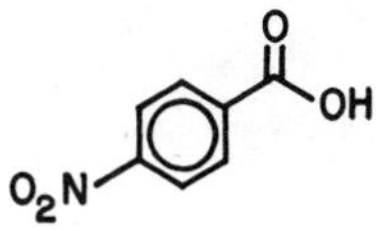

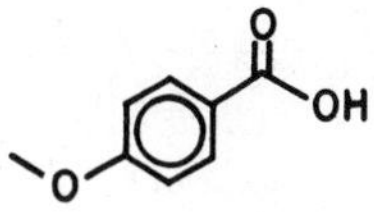

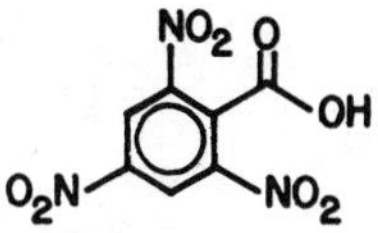

Medium	cm⁻¹	Ref.
KBr	1728	9
nuj.	1637	65
nuj.	1639 1610	66
KBr	1678	9
KBr	1690	9
KBr	1655	9

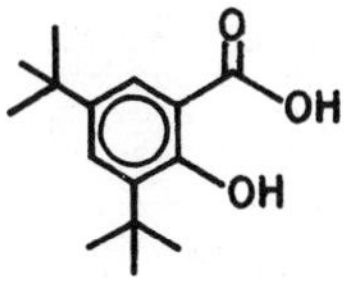

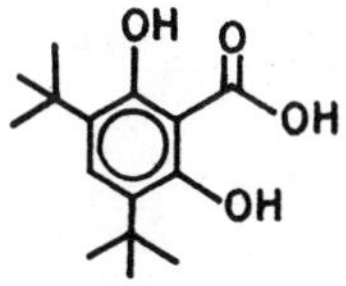

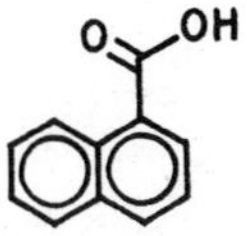

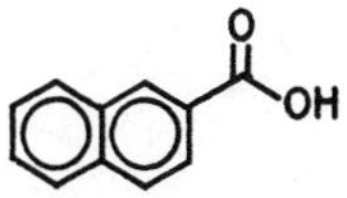

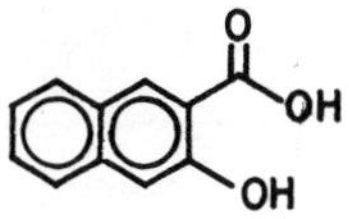

Solvent	ν	Ref.
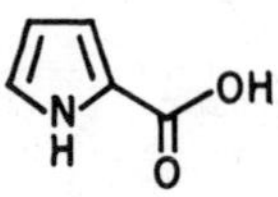 nuj.	1650 1629	67
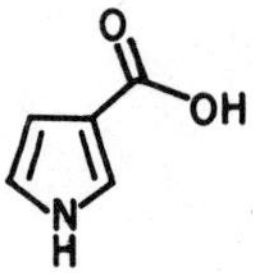 nuj.	1650 1630	67
nuj.	1709	69
KBr	1712 1678	68
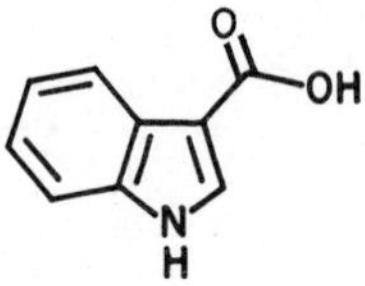 KBr	1642	68
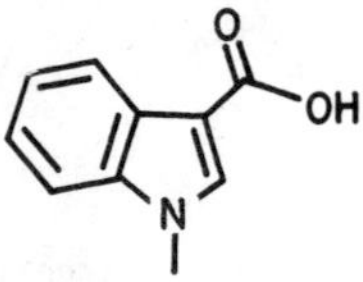 KBr	1639	68
nuj.	1672	70
KBr	1697	44
CCl_4	1684	47
$CHCl_3$	1679	71
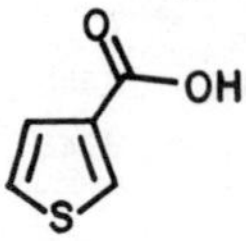 $CHCl_3$	1690	71
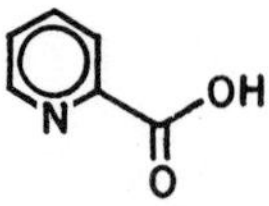 nuj.	1706b	72
nuj.	1704	73
KBr	1700	9
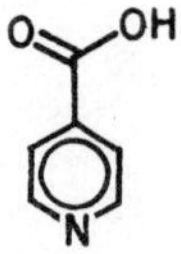 nuj.	1701b	74
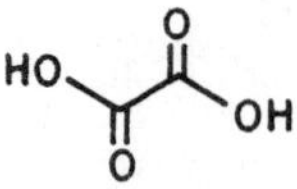 solid	1650	75
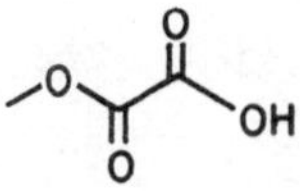 CCl_4	1808 1789 1758 1740	34
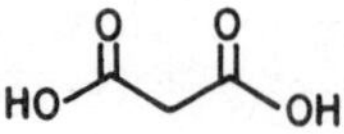 solid	1735w 1703	76

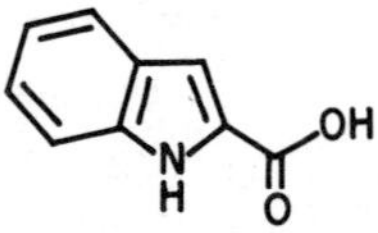

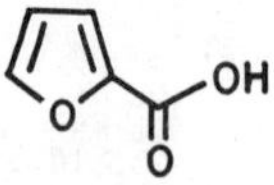

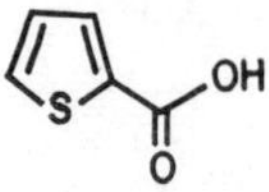

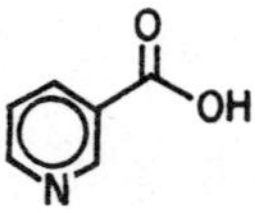

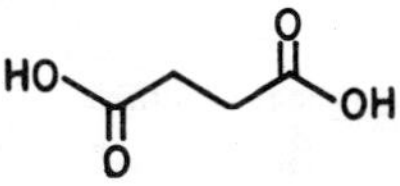

solid	1689	76
KBr	1724	19

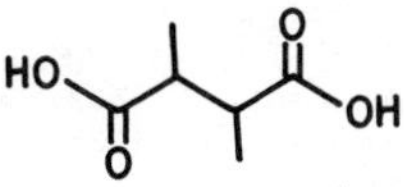

(±) KBr	1733 1704	77
meso KBr	1709	77

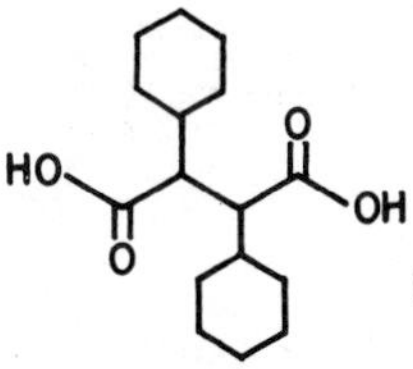

KBr	1700 1625*	78

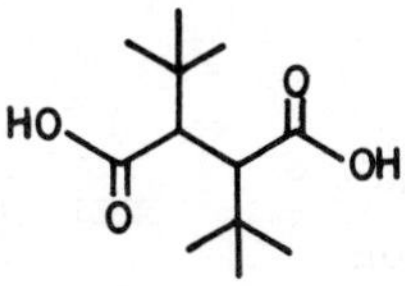

KBr	1695 1615w	78

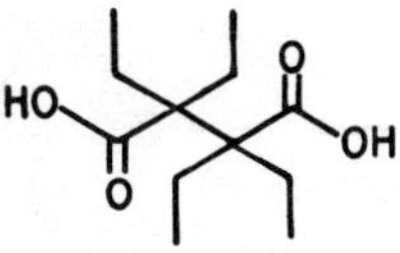

KBr	1683	78

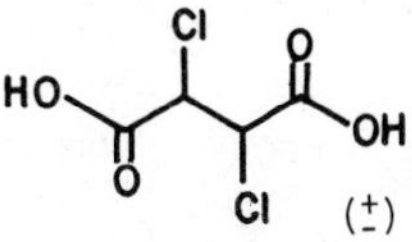

(±) nuj.	1754w 1724	79
meso nuj.	1712	79

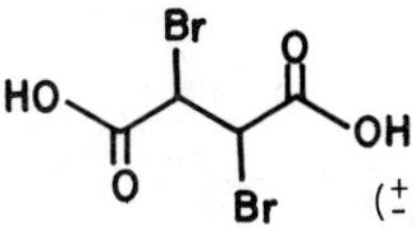

(±) nuj.	1736 1709	79
meso nuj.	1712	79

KBr	1709	44

KBr	1710	80

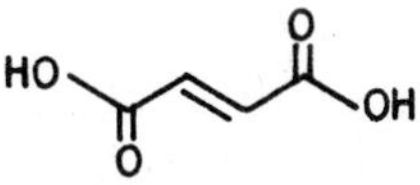

KBr	1670	80

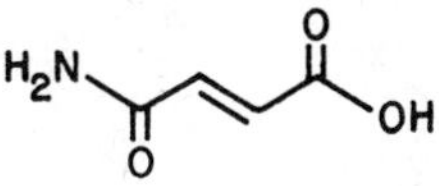

KBr	1705 1660	80

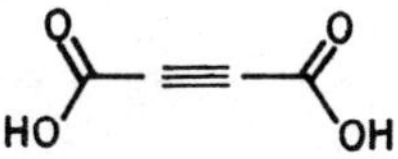

CS_2	1700	39

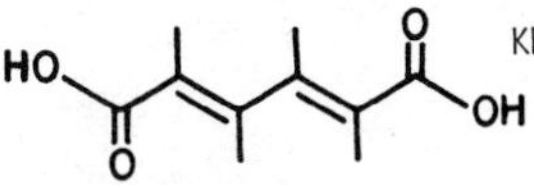

KBr	1678 1620	81

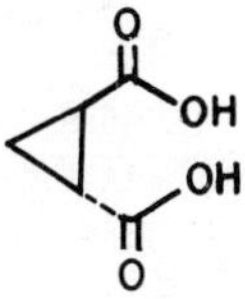

KBr	1695	82

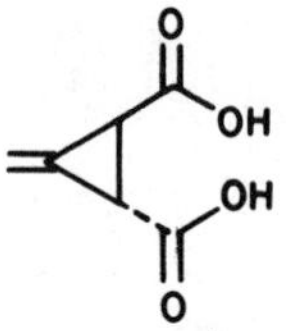

KBr	1701	82
nuj.	1698	83
nuj.	1701	84
nuj.	1748 1721 1678 1664*	85
nuj.	1686	86
nuj.	1681	87
nuj.	1689	88
solid	1685	76
KBr	1695	9

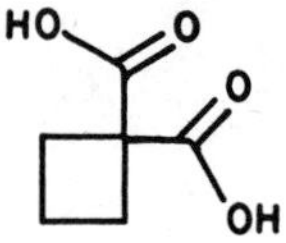

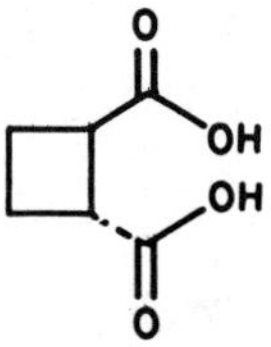

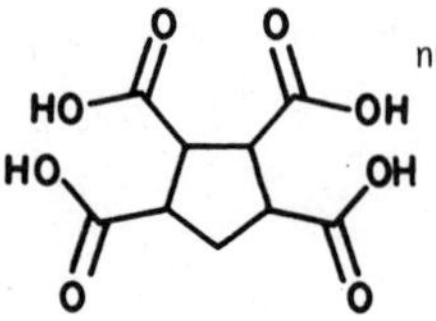

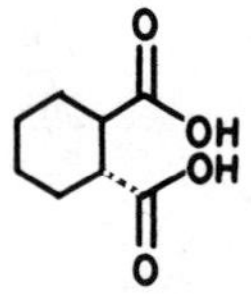

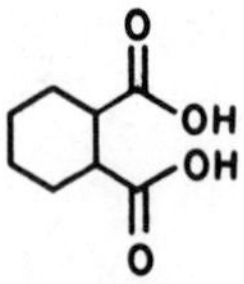

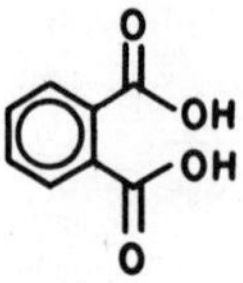

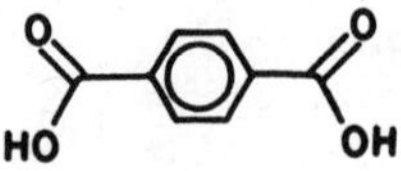

nuj.	1686	88
KBr	1690	9
nuj.	1652 1622	67
nuj.	1654 1628	67
nuj.	1644 1626	67
nuj.	1651	67
CCl_4	1788 1725	34
CCl_4	1785 1762w 1699w 1676	89

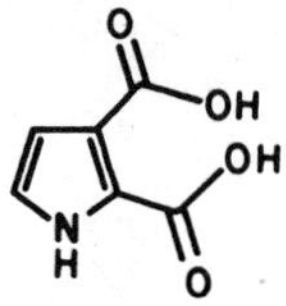

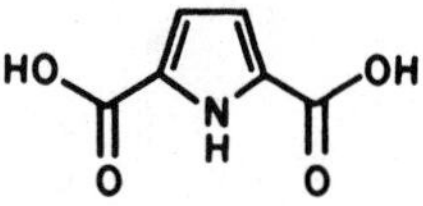

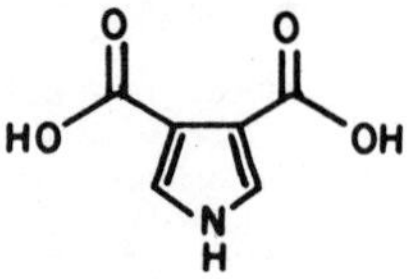

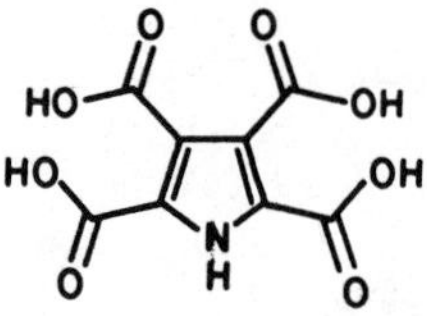

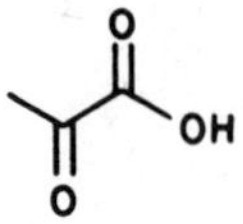

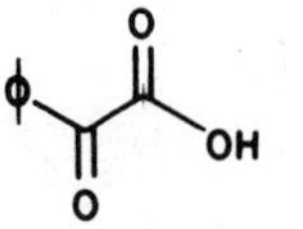

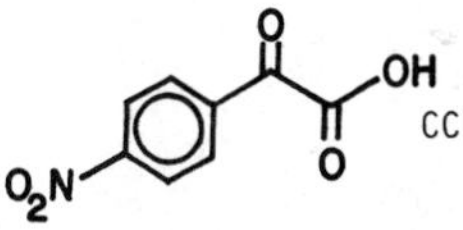

Solvent	ν (cm^{-1})	Ref.
CCl_4	1787 1759w 1707w 1689	89
CCl_4	1785 1758w 1689w 1659	89
KBr	1714 1621	68
nuj.	1675 1624w	90
$CHCl_3$	1689 1639w	90
$CHCl_3$	1690 1665 1630	91
KBr	1656	92
liq.	1739	93
liq.	1750	93
CCl_4	1748	94
liq.	1775	95
liq.	1775	95
CCl_4	1730	96
nuj.	1741 1713w	97
CCl_4	1750	97
nuj.	1739 1710	97
CCl_4	1735	97

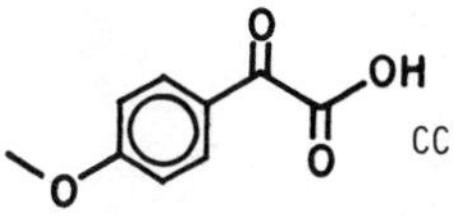

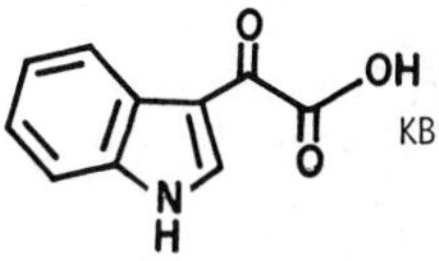

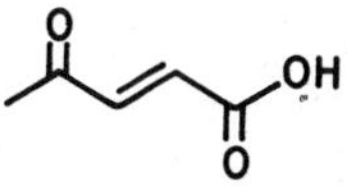

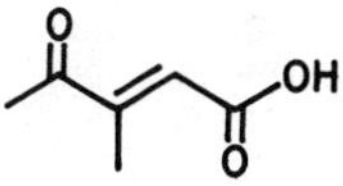

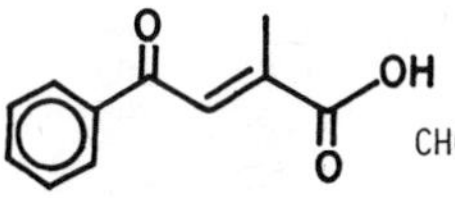

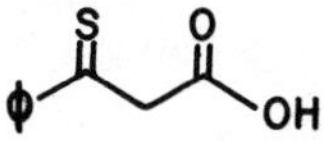

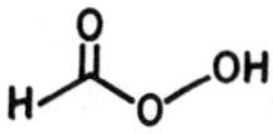

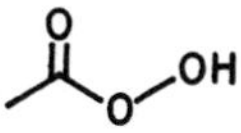

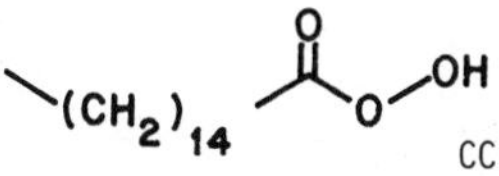

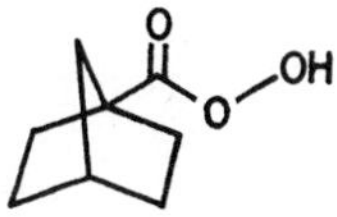

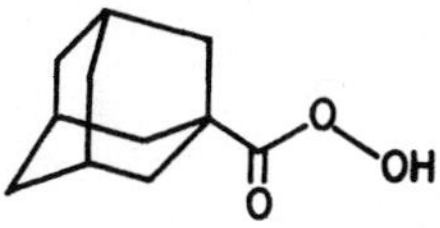

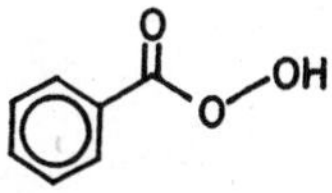

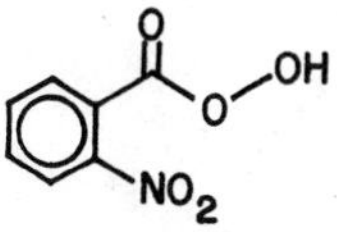

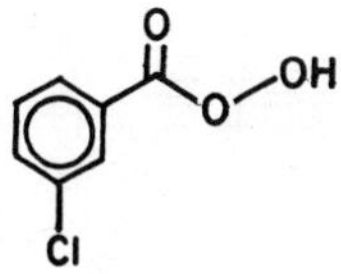

Compound	Solvent	ν (cm⁻¹)	Ref.
O_2N–C_6H_4–C(=O)–O–OH	nuj.	1738 1715	97
CH_3C(=O)SH	CCl_4	1712	98
Cl_3CC(=O)SH	CS_2	1726	99
$C_6H_5CH_2$C(=O)SH	CS_2	1708	98
C_6H_5C(=O)SH	CCl_4 CS_2	1690 1708	98 100
1-naphthyl–C(=O)SH	KBr	1685	100

Anion	Cation	Solvent	ν (cm⁻¹)	Ref.
HC(=O)O^-	Li^+	nuj.	1587 1410 1399	101
	Na^+	nuj.	1592 1381	101
	K^+	nuj.	1590 1385	101
CH_3C(=O)O^-	Na^+	KBr	1583 1421m	102
	Tl^+	nuj.	1412	103
	Pb^+	nuj.	1410	103
CH_3CH_2C(=O)O^-	Na^+	KBr	1573 1565 1553m 1429m	102
$(CH_3)_3C$C(=O)O^-	Na^+	KBr	1674m 1551 1413	102
$NCCH_2$C(=O)O^-	Na^+	KBr	1619 1602 1384	102
FCH_2C(=O)O^-	Na^+	KBr	1624 1448m	102
F_3CC(=O)O^-	Na^+	KBr	1689 1446w	102
$ClCH_2$C(=O)O^-	Na^+	KBr	1603 1418	102

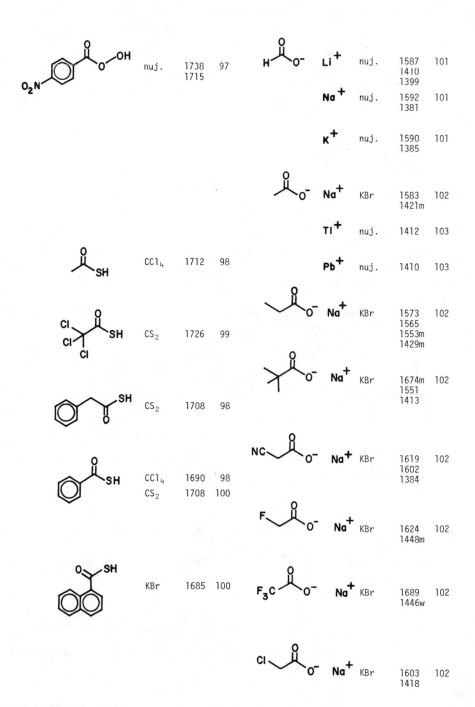

Structure	Cation	Medium	Frequencies	Ref.
$Cl_2CH{-}C(=O)O^-$	Na^+	KBr	1660w 1640 1399	102
$Cl_3C{-}C(=O)O^-$	Na^+	KBr	1677 1353	102
$BrCH_2{-}C(=O)O^-$	Na^+	KBr	1596 1415	102
$Br_3C{-}C(=O)O^-$	Na^+	KBr	1659 1381 1336	102
$ICH_2{-}C(=O)O^-$	Na^+	KBr	1583 1394m	102
$H{-}C{\equiv}C{-}C(=O)O^-$	Na^+	KBr	1619*w 1600 2140w 2095m	45
$C_6H_5{-}C(=O)O^-$	Li^+	nuj.	1561 1427	104
	Na^+	nuj.	1552 1413	104
	K^+	nuj.	1552 1395	104
$o\text{-}HO{-}C_6H_4{-}C(=O)O^-\ Na^+$		nuj.	1582 1407 1376	104
$p\text{-}H_2N{-}C_6H_4{-}C(=O)O^-$	Na^+	KBr	1538 1404	105
indole-2-$C(=O)O^-$ (N–H)	K^+	KBr	1529	68
indole-3-$C(=O)O^-$ (N–H)	K^+	KBr	1549	68
$^-O{-}C(=O){-}C(=O){-}O^-$	$2\ Na^+$	KBr	1616	44
	$2\ K^+$	solid	1627	75
$Na^+\ ^-O{-}C(=O){-}CH_2{-}C(=O){-}O^-\ Na^+$		KBr	1595	44
$Na^+\ ^-O{-}C(=O){-}CH_2CH_2{-}C(=O){-}O^-\ Na^+$		KBr	1575	44
$\phi{-}C(=O){-}S^-$	K^+	nuj.	1525	98

REFERENCES

1. S. Bratož, G. Hadži and N. Sheppard, *SCA* (1956) *8*, 249.
2. K. S. McCallum and W. D. Emmons, *JOC* (1956), 367.
3. E. J. Hartwell, R. E. Richards and H. W. Thompson, *JCS* (1948), 1436.
4. D. Mücke, G. Geppert and L. Kipke, *J Prakt Chem* (1959) *9*, 16.
5. P. J. Corish and D. Chapman, *JCS* (1957), 1746.
6. R. H. Gillette, *JACS* (1936), 1143.
7. O. W. Steward, J. E. Dziedzic and J. S. Johnson, *JOC* (1971), 3475.
8. L. J. Bellamy, R. F. Lake and R. J. Pace, *SCA* (1963), 443.
9. M. St. C. Flett, *JCS* (1951), 962.
10. C. J. Pouchert, *The Aldrich Library of Infrared Spectra*, Aldrich (1970), spectrum 693H.
11. C. J. Pouchert, *The Aldrich Library of Infrared Spectra*, Aldrich (1970), spectrum 240C.
12. J. P. Pete, *BSC France* (1967), 357.
13. E. F. Ullman and W. J. Fanshawe, *JACS* (1961), 2379.
14. J. Meinwald, J. W. Wheeler, A. A. Nimetz and J. S. Liu, *JOC* (1965), 1038.
15. E. Vogel, R. Erb, G. Lenz and A. A. Bothner-By, *A* (1965) *682*, 1.
16. C. J. Pouchert, *The Aldrich Library of Infrared Spectra*, Aldrich (1970), spectrum 246D.
17. C. J. Pouchert, *The Aldrich Library of Infrared Spectra*, Aldrich (1970), spectrum 246E.
18. C. J. Pouchert, *The Aldrich Library of Infrared Spectra*, Aldrich (1970), spectrum 241G.
19. R. J. Koegel, J. P. Greenstein, M. Winitz, S. M. Birnbaum and R. A. McCallum, *JACS* (1955), 5708.
20. C. J. Pouchert, *The Aldrich Library of Infrared Spectra*, Aldrich (1970), spectrum 247C.
21. W. Lwowski and T. W. Mattingly, Jr., *JACS* (1965), 1947.
22. C. J. Pouchert, *The Aldrich Library of Infrared Spectra*, Aldrich (1970), spectrum 248A.
23. C. J. Pouchert, *The Aldrich Library of Infrared Spectra*, Aldrich (1970), spectrum 248B.
24. P. E. Eaton and T. W. Cole, Jr., *JACS* (1964), 3157.
25. C. J. Pouchert, *The Aldrich Library of Infrared Spectra*, Aldrich (1970), spectrum 248F.

26. J. R. Barceló and C. Otero, *SCA* (1962), 1231.
27. J. Bellanato and J. R. Barceló, *SCA* (1960), 1333.
28. G. M. Barrow, *JACS* (1956), 5802.
29. J. F. Grove and H. A. Willis, *JCS* (1951), 877.
30. G. M. Barrow, *JACS* (1958), 86.
31. F. Micheel and B. Schleppinghoff, *B* (1955), 763.
32. N. Fuson, M.-L. Josien and R. L. Powell, *JACS* (1952), 1.
33. J. Bolard, *CR* (1963) *256*, 4388.
34. M. Öki and M. Hirota, *BSC Japan* (1961), 374.
35. C. J. Pouchert, *The Aldrich Library of Infrared Spectra*, Aldrich (1970), spectrum 695F.
36. C. J. Pouchert, *The Aldrich Library of Infrared Spectra*, Aldrich (1970), spectrum 696G.
37. A. Solladié-Cavallo and P. Vieles, *BSC France* (1967), 517.
38. M. I. Batuev, A. S. Onishchenko, A. D. Matveeva and N. I. Aronova, *J Gen Chem USSR* (1960), 657.
39. J. L. H. Allan, G. D. Meakins and M. C. Whiting, *JCS* (1955), 1874.
40. C. H. Eugster, L. Leichner and E. Jenny, *Helv* (1963), 543.
41. D. B. Kurland, *Harvard Ph.D. Thesis* (1967).
42. L. O. Ross, L. Goodman and B. R. Baker, *JOC* (1959), 1152.
43. J. Lecomte and J. Guy, *CR* (148) *227*, 54.
44. M. St.C. Flett, *SCA* (1962), 1537.
45. J. E. Katon and N. T. McDevitt, *SCA* (1965), 1717.
46. J. W. Wilson and V. S. Stubblefield, *JACS* (1968), 3423.
47. J. J. Peron, P. Saumagne and J. M. Lebas, *SCA* (1970), 1651.
48. L. J. Bellamy, *JCS* (1955), 4221.
49. D. G. O'Sullivan and P. W. Sadler, *JCS* (1957), 2839.
50. M.-L. Josien, D. Peltier and A. Pichevin, *CR* (1960) *250*, 1643.
51. C. F. Pouchert, *The Aldrich Library of Infrared Spectra*, Aldrich (1970), spectrum 712A.
52. C. F. Pouchert, *The Aldrich Library of Infrared Spectra*, Aldrich (1970), spectrum 734F.
53. A. E. Kellie, D. G. O'Sullivan and P. W. Sadler, *JOC* (1957), 29.
54. C. F. Pouchert, *The Aldrich Library of Infrared Spectra*, Aldrich (1970), spectrum 713B.
55. C. F. Pouchert, *The Aldrich Library of Infrared Spectra*, Aldrich (1970), spectrum 713C.
56. C. F. Pouchert, *The Aldrich Library of Infrared Spectra*, Aldrich (1970), spectrum 713D.
57. L. J. Bellamy and L. Beecher, *JCS* (1954), 4487.
58. H. Musso, *B* (1955), 1915.

59. C. F. Pouchert, *The Aldrich Library of Infrared Spectra*, Aldrich (1970), spectrum 712H.
60. C. F. Pouchert, *The Aldrich Library of Infrared Spectra*, Aldrich (1970), spectrum 713H.
61. C. F. Pouchert, *The Aldrich Library of Infrared Spectra*, Aldrich (1970), spectrum 714G.
62. M. St.C. Flett, *Trans Farad Soc* (1948), 767.
63. C. F. Pouchert, *The Aldrich Library of Infrared Spectra*, Aldrich (1970), spectrum 714E.
64. C. F. Pouchert, *The Aldrich Library of Infrared Spectra*, Aldrich (1970), spectrum 715B.
65. C. F. Pouchert, *The Aldrich Library of Infrared Spectra*, Aldrich (1970), spectrum 730D.
66. C. F. Pouchert, *The Aldrich Library of Infrared Spectra*, Aldrich (1970), spectrum 733H.
67. M. Scrocco and R. Nicolaus, *Atti Accad nazl Lincei, Rend., Classe sci fis., mat e nat* (1956) *20*, 795.
68. F. Millich and E. I. Becker, *JOC* (1958), 1096.
69. C. F. Pouchert, *The Aldrich Library of Infrared Spectra*, Aldrich (1970), spectrum 919D.
70. C. F. Pouchert, *The Aldrich Library of Infrared Spectra*, Aldrich (1970), spectrum 884B.
71. S. Gronowitz and A. Rosenberg, *Arkiv Kemi* (1955) *8*, 23.
72. C. F. Pouchert, *The Aldrich Library of Infrared Spectra*, Aldrich (1970), spectrum 980C.
73. C. F. Pouchert, *The Aldrich Library of Infrared Spectra*, Aldrich (1970), spectrum 980E.
74. C. F. Pouchert, *The Aldrich Library of Infrared Spectra*, Aldrich (1970), spectrum 980H.
75. H. Murata and K. Kawai, *JCP* (1956) *25*, 589.
76. C. Duval and C. Wadier, *Anal Chim Acta* (1960) *23*, 541.
77. A. Rosenberg and L. Schotte, *Acta Chem Scand* (1954), 867.
78. L. Eberson, *Acta Chem Scand* (1959), 224.
79. L. Schotte and A. Rosenberg, *Arkiv Kemi* (1956) *8*, 551.
80. M. K. Hargreaves and E. A. Stevinson, *SCA* (1965), 1681.
81. W. Adam, *B* (1964), 1811.
82. A. T. Blomquist and D. T. Longone, *JACS* (1959), 2012.
83. C. F. Pouchert, *The Aldrich Library of Infrared Spectra*, Aldrich (1970), spectrum 249E.
84. C. F. Pouchert, *The Aldrich Library of Infrared Spectra*, Aldrich (1970), spectrum 249F.

85. C. F. Pouchert, *The Aldrich Library of Infrared Spectra*, Aldrich (1970), spectrum 250F.
86. C. F. Pouchert, *The Aldrich Library of Infrared Spectra*, Aldrich (1970), spectrum 250B.
87. C. F. Pouchert, *The Aldrich Library of Infrared Spectra*, Aldrich (1970), spectrum 250A.
88. F. González-Sánchez, *SCA* (1958), *12*, 17.
89. G. Oehme, G. Fischer and A. Schellenberger, *B* (1967), 425.
90. R. Scheffold and P. Dubs, *Helv* (1967), 798.
91. G. Leclerc, C.-G. Wermuth and J. Schreiber, *BSC France* (1967), 1302.
92. Z. Reyes and R. M. Silverstein, *JACS* (1958), 6367.
93. P. A. Giguère and A. Weingartshofer Olmos, *Canad J Chem* (1952), 821.
94. D. Swern, L. P. Witnauer, C. R. Eddy and W. E. Parker, *JACS* (1955), 5537.
95. L. B. Humphrey, B. Hodgson and R. E. Pincock, *Canad J Chem* (1968), 3099.
96. E. Briner and E. Dallwigk, *Helv* (1956), 1446.
97. R. Kavčič, B. Plesničar and D. Hadži, *SCA* (1967), 2483.
98. R. A. Nyquist and W. J. Potts, *SCA* (1959) *15*, 514.
99. B. Sjöberg, *Acta Chem Scand* (1957), 945.
100. B. Sjöberg, *Acta Chem Scand* (1959), 1036.
101. J. D. Donaldson, J. F. Knifton and S. D. Ross, *SCA* (1964), 847.
102. E. Spinner, *JCS* (1964), 4217.
103. J. D. Donaldson, J. F. Knifton and S. D. Ross, *SCA* (1965), 275.
104. J. H. S. Green, W. Kynaston and A. S. Lindsey, *SCA* (1961), 486.
105. M. M. Stimson, *JCP* (1954), 1942.

4.8

acid halides

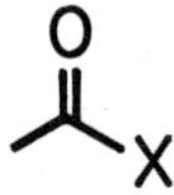

As might be expected, direct attachment of a halogen atom to a carbonyl group causes an even greater increase of the carbonyl stretching frequency than does α-halogen substitution. Thus carbonyl fluoride has the highest recorded carbonyl stretching frequency at 1928 cm^{-1} in the vapor phase.

In most respects substitution on the α-carbon atom has the same effect as with ketones (see Chapter 4.2, p. 175). Thus α-halogens cause an increase in frequency (often accompanied by multiple absorption due to conformation isomers), whereas alkenyl and aryl substituents cause a decrease in the carbonyl frequency. In addition, splitting into doublets, due to Fermi resonance, is frequently observed.

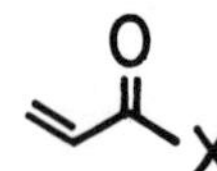

Only small changes in frequency occur between gaseous and condensed phases, and even smaller changes are observed between various solvents.

Structure (labels as printed)	Medium	Frequency	Ref.
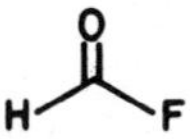 O; H, F	gas	1837	1
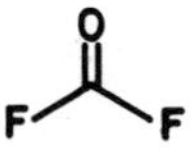 O; F, F	gas	1928	1
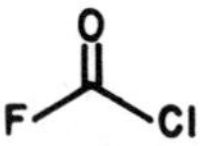 O; F, Cl	gas	1886	1
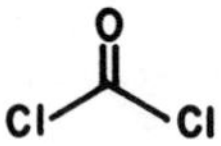 O; Cl, Cl	gas	1827	1
	$CHCl_3$	1801	2
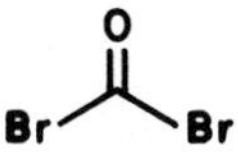 O; Br, Br	gas	1826	1
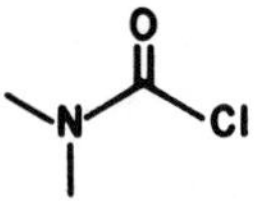 O; N, Cl	CCl_4	1744	3
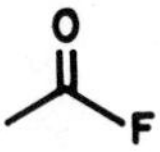 O; F	liq.	1840	4
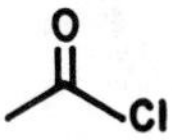 O; Cl	liq.	1806	5
	CCl_4	1799	6

Structure (labels as printed)	Medium	Frequency	Ref.
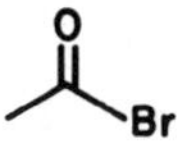 O; Br	liq.	1814	5
	CCl_4	1912w 1818 1751w	6
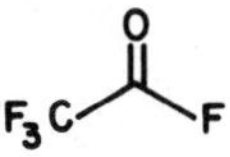 O; F_3C, F	gas	1901	1
	liq.	1905	7
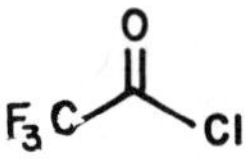 O; F_3C, Cl	gas	1810	1
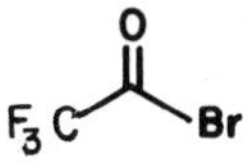 O; F_3C, Br	gas	1838	1
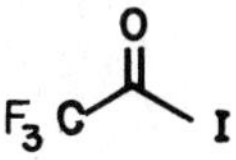 O; F_3C, I	gas	1812	1
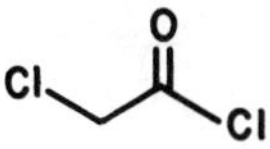 O; Cl, Cl	liq.	1809	8
	CCl_4	1821 1785	9
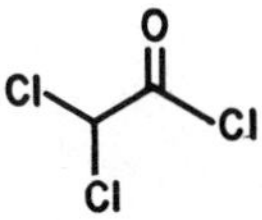 O; Cl, Cl, Cl	liq.	1803	4
	CCl_4	1810 1779	9
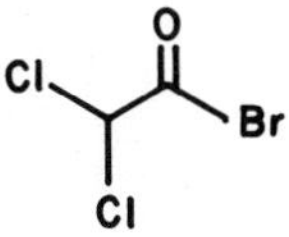 O; Cl, Br, Cl	liq.	1799	4

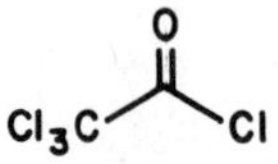 CCl_4	1805	10
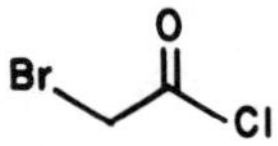 liq.	1802	8
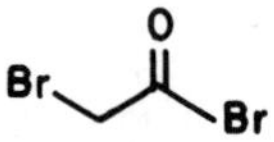 liq.	1807	8
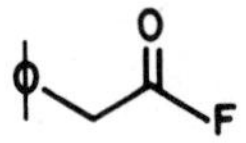 liq.	1843	11
liq.	1792	12
CCl_4	1799 1712w	6
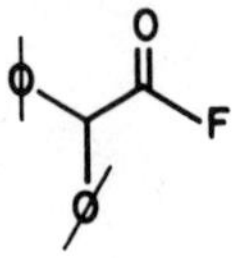 liq.	1847	11
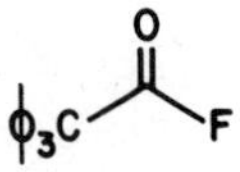 liq.	1831	11
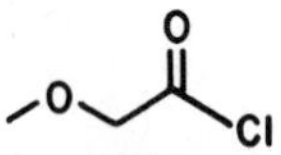 liq.	1848m* 1795	13

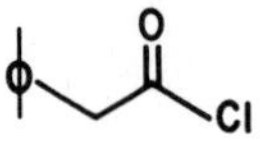

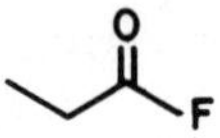 liq.	1845	11
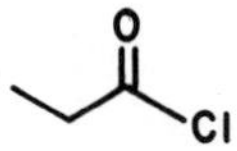 liq.	1832* 1789	14
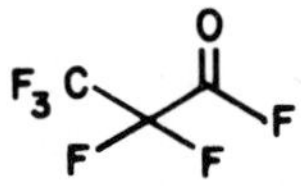 liq.	1890	7
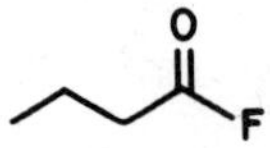 liq.	1824	11
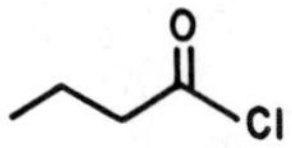 liq.	1792	15
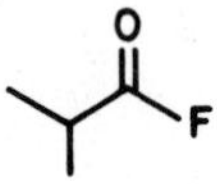 liq.	1840	11
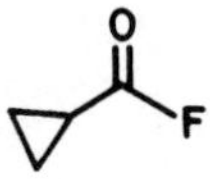 CCl_4	1840	16
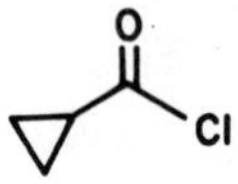 CCl_4	1785	16
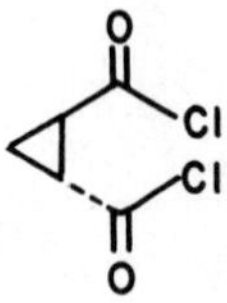 liq.	1783	17

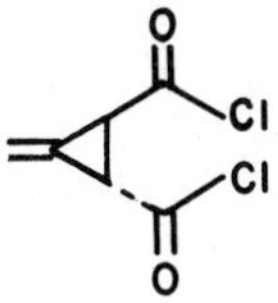

liq.	1786	17
liq.	1802	18
liq.	1823	11
liq.	1821 1779	19
liq.	1795	18
CS_2	1844	20
liq.	1776	21
CCl_4	1762	3
liq.	1786* 1757 1629	22
liq.	1754 1736*	23
CCl_4	1812	24
liq.	1773 1730	25
CCl_4	1773 1736	6
CS_2	1773 1736	26
CS_2	1779 1701w	26
CS_2	1761 1698w	26
liq.	1785	27
liq.	1764 1733	28
CS_2	1776 1745	26

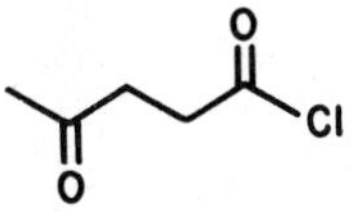

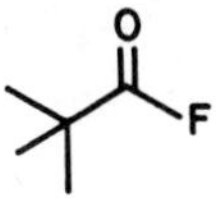

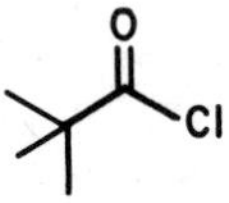

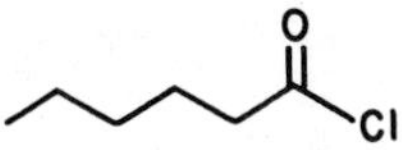

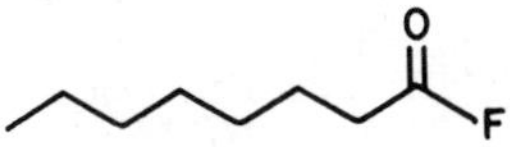

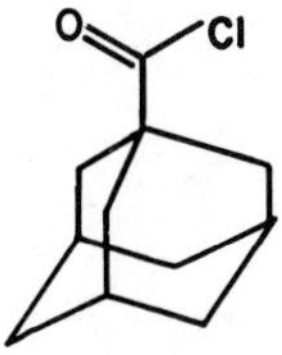

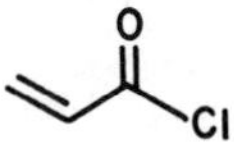

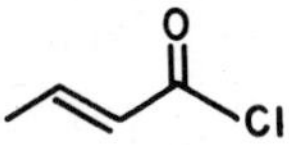

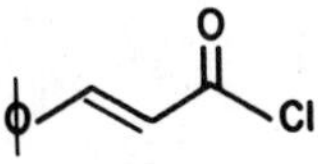

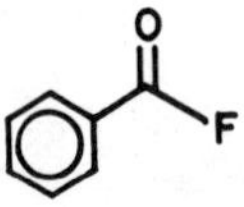

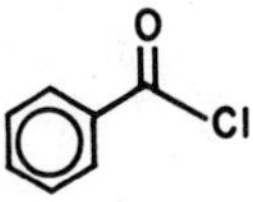

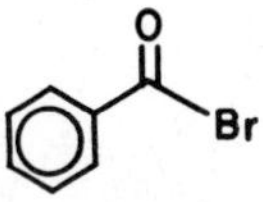

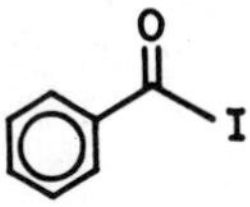

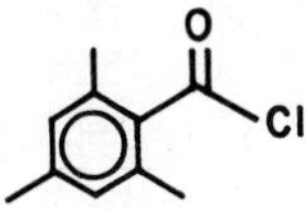

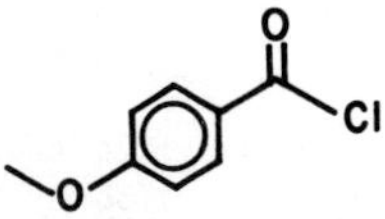

Solvent	ν (cm⁻¹)	Ref.
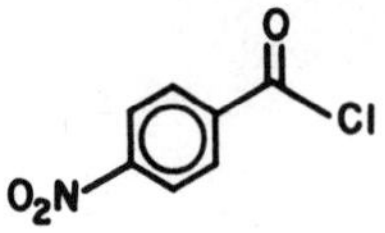 liq.	1779* 1756 1730*	29
CS_2	1792m 1767 1730w	26
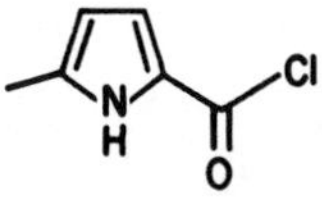 CH_2Cl_2	1752	30
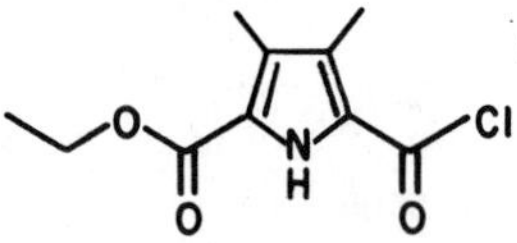 CH_2Cl_2	1755 1720	30
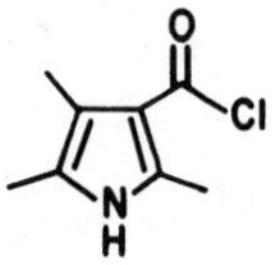 CH_2Cl_2	1695	30
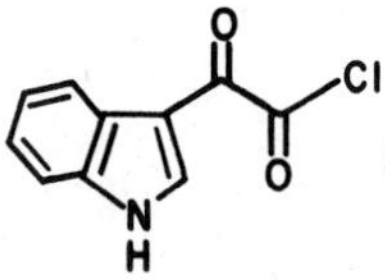 KBr	1792 1629	31
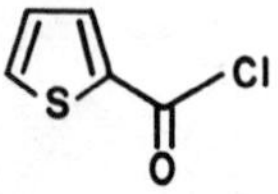 CCl_4	1753	32
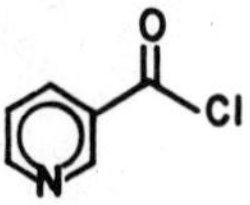 CH_2Cl_2	1775 1738	30

Solvent	ν (cm⁻¹)	Ref.
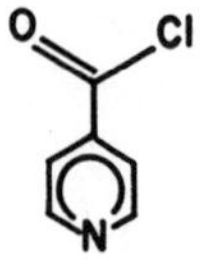 liq.	1780 1745	30
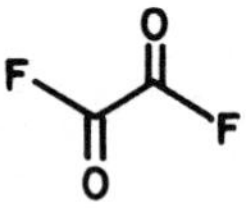 gas	1870	33
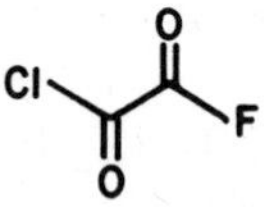 gas	1858 1790	33
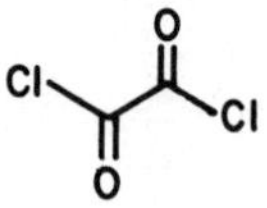 gas	1862m 1790	34
liq.	1832 1764b	35
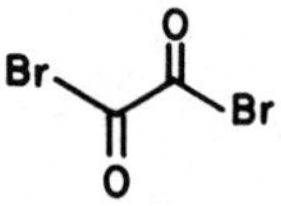 liq.	1799 1770*	36
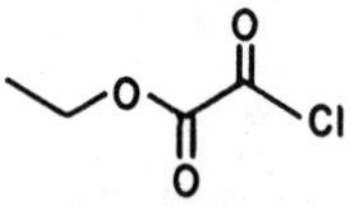 liq.	1786 1751	37
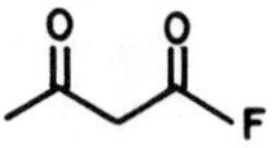 liq.	1832 1721	38
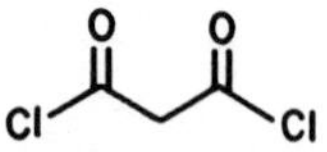 liq.	1812* 1783	39

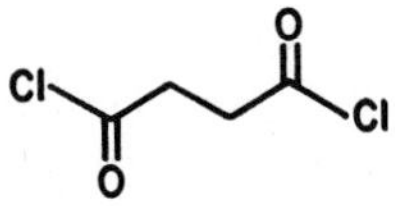	liq.	1786	18
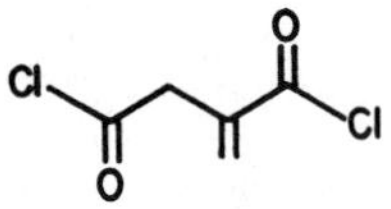	liq.	1795 1742	40
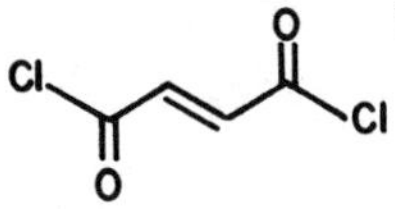	liq.	1751	41
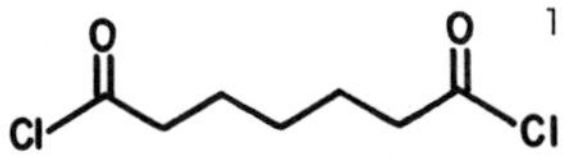	liq.	1792	18
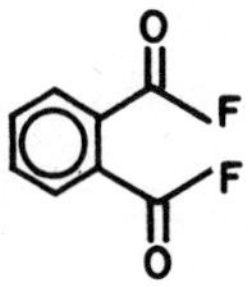	CCl_4	1830	24
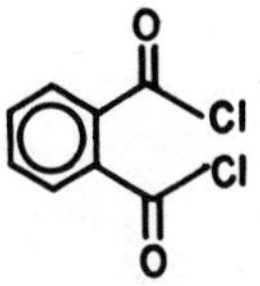	CCl_4	1785	24
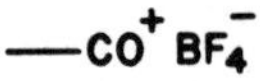 $—CO^+ BF_4^-$	nuj.	2299 1619w 1560w	42
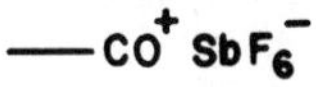 $—CO^+ SbF_6^-$	nuj.	2294 1621w 1554w	42
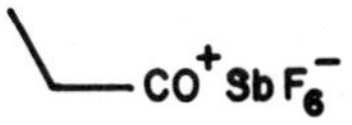 $CO^+ SbF_6^-$	nuj.	2290 1610w	42
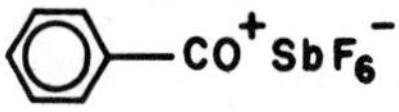 $CO^+ SbF_6^-$	nuj.	2212 1531w	42

REFERENCES

1. S. Forsén, *SCA* (1962), 595.
2. C. Black and D. Dolphin, unpublished results.
3. A. W. Baker and G. H. Harris, *JACS* (1960), 1923.
4. R. E. Kagarise, *JACS* (1955), 1377.
5. L. J. Bellamy and R. L. Williams, *JCS* (1957), 863.
6. R. S. Rasmussen and R. R. Brattain, *JACS* (1949), 1073.
7. R. N. Haszeldine, *JCS* (1954), 4026.
8. I. Nakagawa, I. Ichishima, K. Kuratani, T. Miyazawa, T. Shimanouchi and S. Mizushima, *JCP* (1952), 1720.
9. L. J. Bellamy and R. L. Williams, *JCS* (1958), 3465.
10. H. W. Thompson and D. A. Jameson, *SCA* (1958) *13* , 236.

11. G. A. Olah, W. S. Tolgyesi, S. J. Kuhn, M. E. Moffat, I. J. Bastien and E. B. Baker, *JACS* (1963), 1328.
12. C. F. Pouchert, *The Aldrich Library of Infrared Spectra*, Aldrich 1970, spectrum 778A.
13. C. F. Pouchert, *The Aldrich Library of Infrared Spectra*, Aldrich 1970, spectrum 326G.
14. C. F. Pouchert, *The Aldrich Library of Infrared Spectra*, Aldrich 1970, spectrum 322G.
15. C. F. Pouchert, *The Aldrich Library of Infrared Spectra*, Aldrich 1970, spectrum 323C.
16. M. Hanack and H. Eggensperger, *B* (1963), 1341.
17. A. T. Blomquist and D. T. Longone, *JACS* (1959), 2012.
18. J. Cason and E. J. Reist, *JOC* (1958), 1492.
19. C. F. Pouchert, *The Aldrich Library of Infrared Spectra*, Aldrich 1970, spectrum 323E.
20. E. S. Rothman, G. G. Moore and S. Serota, *JOC* (1969), 2486.
21. C. F. Pouchert, *The Aldrich Library of Infrared Spectra*, Aldrich 1970, spectrum 327H.
22. C. F. Pouchert, *The Aldrich Library of Infrared Spectra*, Aldrich 1970, spectrum 324F.
23. C. F. Pouchert, *The Aldrich Library of Infrared Spectra*, Aldrich 1970, spectrum 778G.
24. F. Seel and J. Langer, *B* (1958), 2553.
25. C. F. Pouchert, *The Aldrich Library of Infrared Spectra*, Aldrich 1970, spectrum 779A.
26. S. Yoshida, The 6th Symposium of Infrared and Raman Spectroscopy, 1959.
27. B. P. Susz and D. Cassimatis, *Helv* (1961), 395.
28. C. F. Pouchert, *The Aldrich Library of Infrared Spectra*, Aldrich 1970, spectrum 780F.
29. C. F. Pouchert, *The Aldrich Library of Infrared Spectra*, Aldrich 1970, spectrum 780G.
30. A. Wick, unpublished results.
31. F. Millich and E. I. Becker, *JOC* (1958), 1096.
32. J. J. Peron, P. Saumagne and J. M. Lebas, *SCA* (1970), 1651.
33. J. L. Hencher and G. W. King, *J Mol Spectr* (1965) *16*, 168.
34. J. L. Hencher and G. W. King, *J Mol Spectr* (1965) *16*, 158.
35. C. F. Pouchert, *The Aldrich Library of Infrared Spectra*, Aldrich 1970, spectrum 325G.
36. C. F. Pouchert, *The Aldrich Library of Infrared Spectra*, Aldrich 1970, spectrum 325H.

37. C. F. Pouchert, *The Aldrich Library of Infrared Spectra*, Aldrich 1970, spectrum 325C.

38. G. A. Olah and S. J. Kuhn, *JOC* (1961), 225.

39. C. F. Pouchert, *The Aldrich Library of Infrared Spectra*, Aldrich 1970, spectrum 326A.

40. C. F. Pouchert, *The Aldrich Library of Infrared Spectra*, Aldrich 1970, spectrum 327E.

41. C. F. Pouchert, *The Aldrich Library of Infrared Spectra*, Aldrich 1970, spectrum 326F.

42. G. A. Olah, S. J. Kuhn, W. S. Tolgyesi and E. B. Baker, *JACS* (1962), 2733.

4.9

anhydrides diacylperoxides

ANHYDRIDES

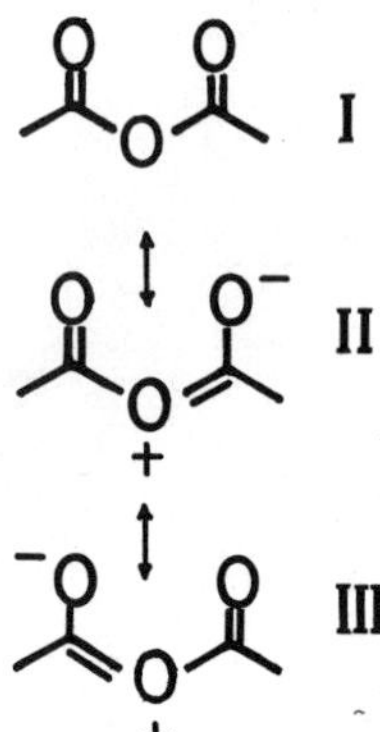

Contributions from the dipolar resonance forms II and III result in a planar structure for the anhydride group. This leads to coupled vibrations of the two carbonyl stretching vibrations, and two bands are seen in the carbonyl region. The higher frequency band is associated with the asymmetric mode and the lower with the symmetric mode.

With the exception of the mixed anhydrides of formic acid, these two bands are separated by 60 to 75 cm^{-1}. In acyclic anhydrides the higher frequency band is more intense, whereas in cyclic systems the situation is reversed and the lower frequency band is stronger.

Field effects of electronegative substituents, conjugations with alkenyl and aryl groups, and ring strain are similar to those seen with esters (Chapter 4.10, p. 332) and lactones (Chapter 4.11, p. 365). Even with unsymmetrically substituted anhydrides, both the symmetric and asymmetric modes are affected by the substitution pattern in the same manner.

C–O

Anhydrides, like esters, show strong C-O single-bond stretching frequencies. Acyclic systems absorb between 1180 and 1040 cm^{-1} and cyclic systems between 1320 and 1200 cm^{-1}, but neither range is characteristic.

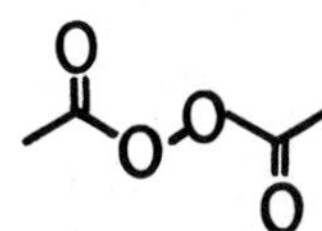

DIACYLPEROXIDES

Diacylperoxides exhibit two bands in the carbonyl region suggesting that these are coupled vibrations of the two carbonyl stretching modes. The separation of these two bands is usually between 20 and 30 cm^{-1}.

Aryl diacylperoxides, whether symmetrically or asymmetrically substituted, exhibit lower frequencies for both carbonyl bands in a manner analogous to that observed with anhydrides.

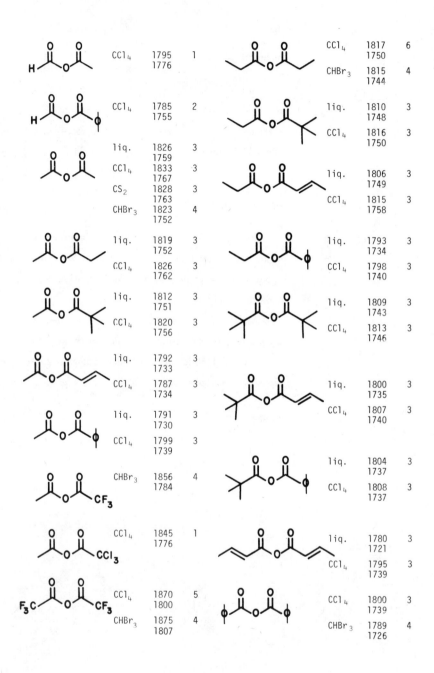

Solvent	cm⁻¹	Ref.
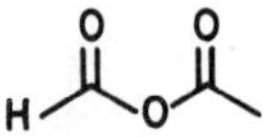 CCl_4	1795, 1776	1
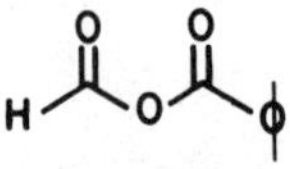 CCl_4	1785, 1755	2
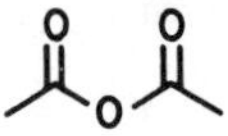 liq.	1826, 1759	3
CCl_4	1833, 1767	3
CS_2	1828, 1763	3
$CHBr_3$	1823, 1752	4
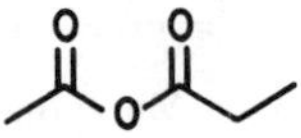 liq.	1819, 1752	3
CCl_4	1826, 1762	3
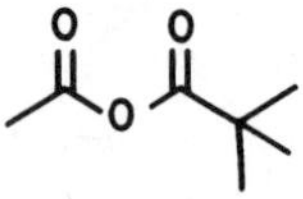 liq.	1812, 1751	3
CCl_4	1820, 1756	3
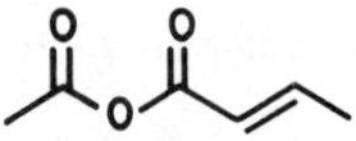 liq.	1792, 1733	3
CCl_4	1787, 1734	3
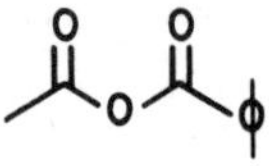 liq.	1791, 1730	3
CCl_4	1799, 1739	3
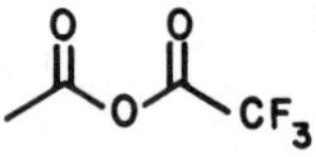 $CHBr_3$	1856, 1784	4
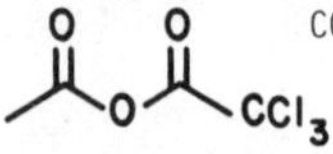 CCl_4	1845, 1776	1
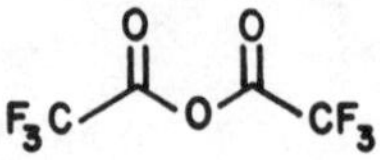 CCl_4	1870, 1800	5
$CHBr_3$	1875, 1807	4

Solvent	cm⁻¹	Ref.
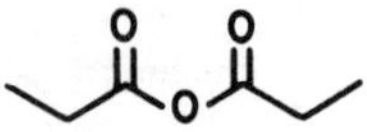 CCl_4	1817, 1750	6
$CHBr_3$	1815, 1744	4
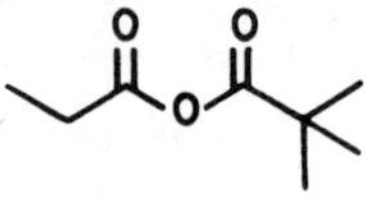 liq.	1810, 1748	3
CCl_4	1816, 1750	3
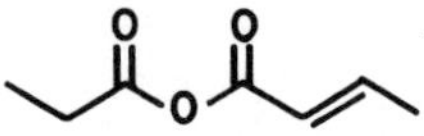 liq.	1806, 1749	3
CCl_4	1815, 1758	3
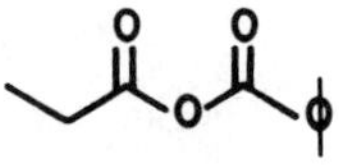 liq.	1793, 1734	3
CCl_4	1798, 1740	3
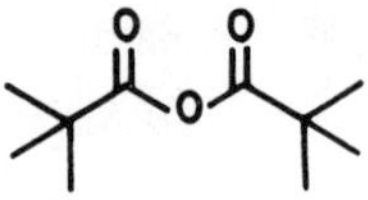 liq.	1809, 1743	3
CCl_4	1813, 1746	3
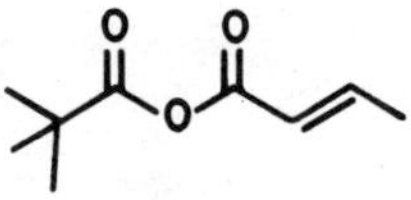 liq.	1800, 1735	3
CCl_4	1807, 1740	3
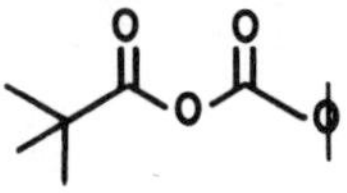 liq.	1804, 1737	3
CCl_4	1808, 1737	3
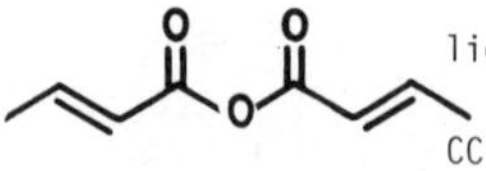 liq.	1780, 1721	3
CCl_4	1795, 1739	3
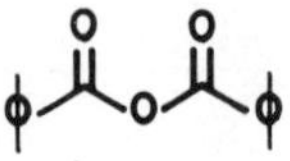 CCl_4	1800, 1739	3
$CHBr_3$	1789, 1726	4

Solvent	ν (cm⁻¹)	Ref.
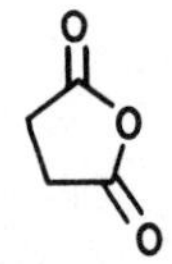 CCl_4	1872 1796	7
$CHBr_3$	1868 1786	4
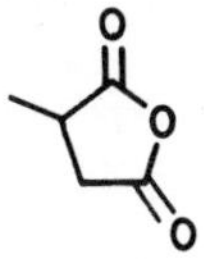 $CHBr_3$	1863 1784	4
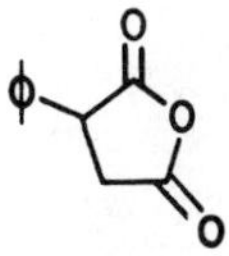 $CHBr_3$	1867 1791	4
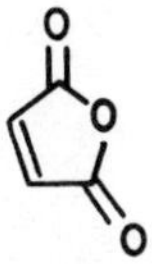 CCl_4	1856 1796* 1784	8
$CHCl_3$	1858 1796 1785	8
$CHBr_3$	1851 1780	4
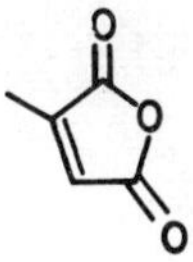 $CHBr_3$	1844 1775	4
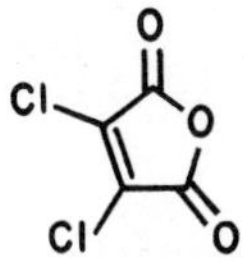 KBr	1876 1835 1786	9
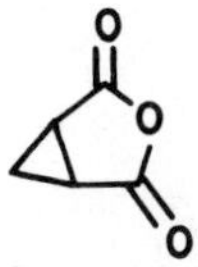 CCl_4	1862 1799	10
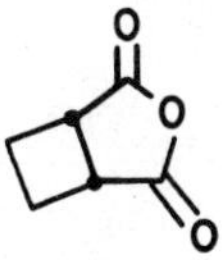 CCl_4	1840 1760	11
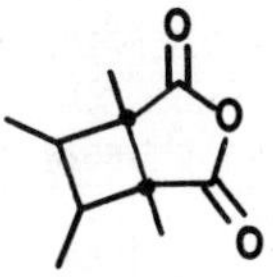 CCl_4	1850 1780	12
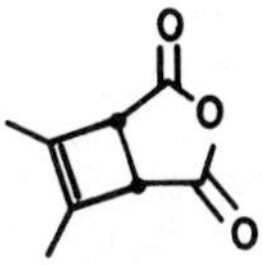 CCl_4	1860 1780	13
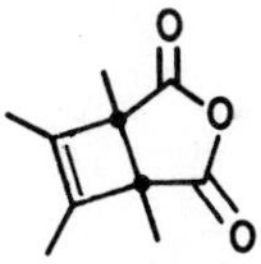 CCl_4	1840 1780	13
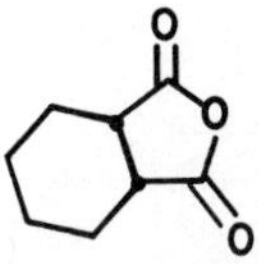 CCl_4	1858 1795	7
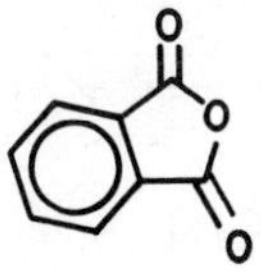 nuj.	1844 1791 1766	14
KBr	1855m 1770	15
CCl_4	1858 1793 1779	14
$CHCl_3$	1854 1789 1775	14
$CHBr_3$	1853 1789	4
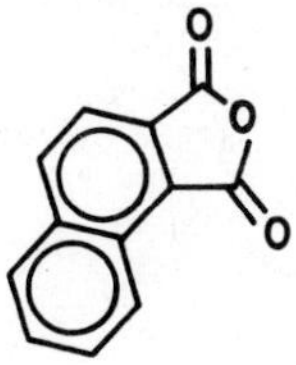 CCl_4	1848 1783	16

Solvent	Frequency	Ref.
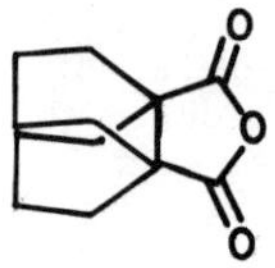 CCl_4	1859 1787	17
nuj.	1862w 1783	18
CCl_4	1860 1780	17
CCl_4	1802 1761	7
$CHBr_3$	1810 1765	4
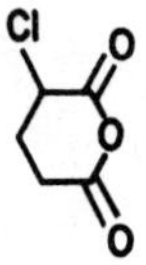 $CHBr_3$	1821 1773	4
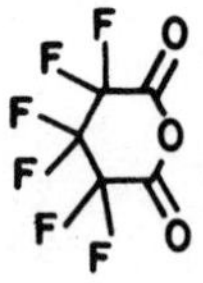 liq.	1852 1792	19
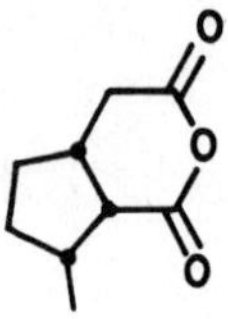 liq.	1755	20
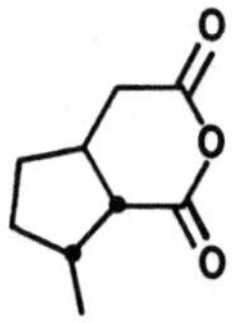 nuj.	1700	20
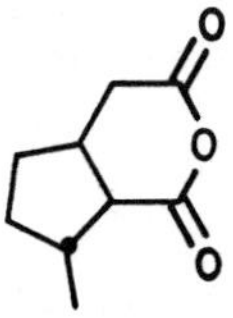 liq.	1805 1760	20
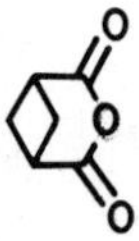 CCl_4	1721	21
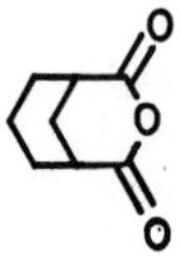 CCl_4	1811 1744	7
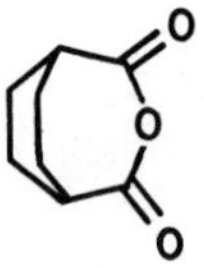 CCl_4	1811 1744	7

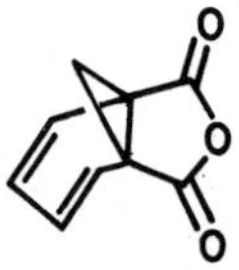

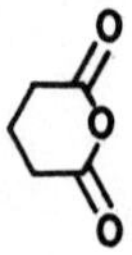

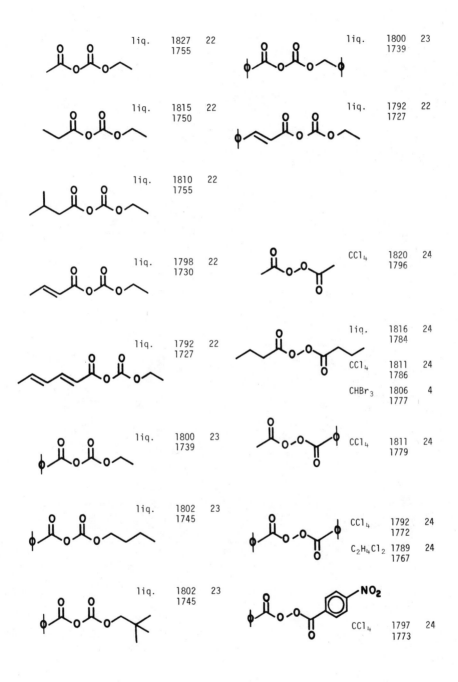

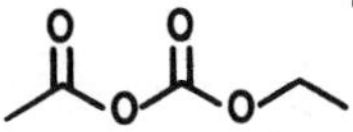

liq. 1827 22
1755

liq. 1815 22
1750

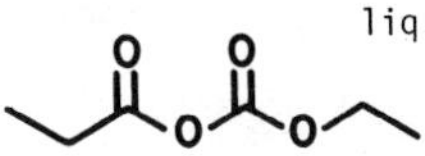

liq. 1810 22
1755

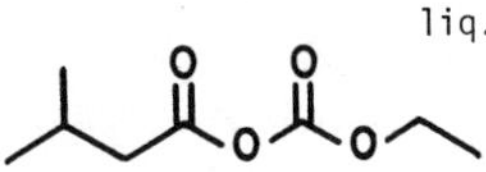

liq. 1798 22
1730

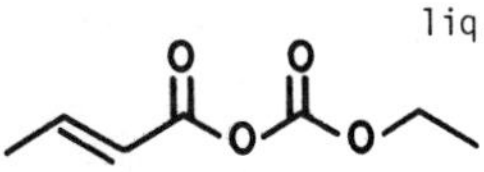

liq. 1792 22
1727

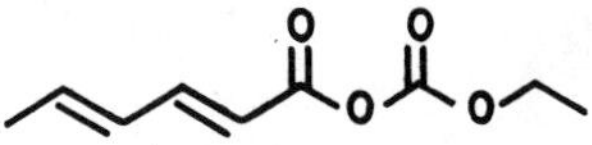

liq. 1800 23
1739

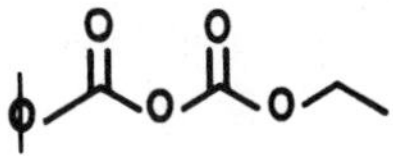

liq. 1802 23
1745

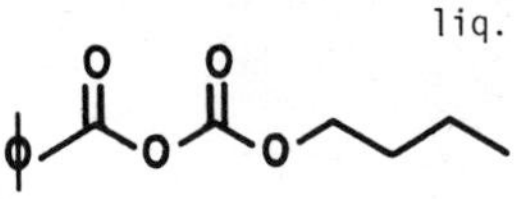

liq. 1802 23
1745

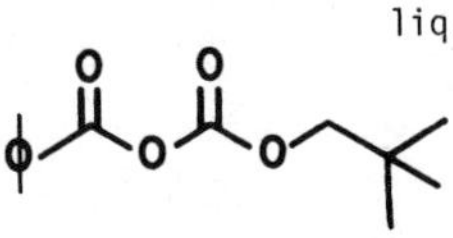

liq. 1800 23
1739

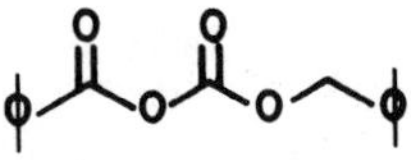

liq. 1792 22
1727

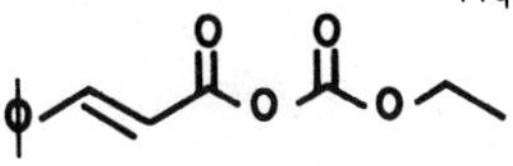

CCl_4 1820 24
1796

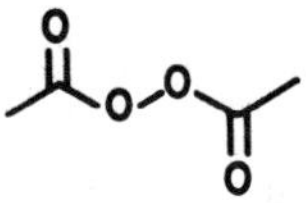

liq. 1816 24
1784

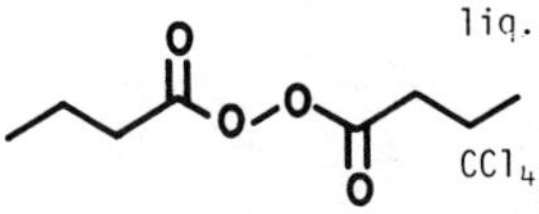

CCl_4 1811 24
1786

$CHBr_3$ 1806 4
1777

CCl_4 1811 24
1779

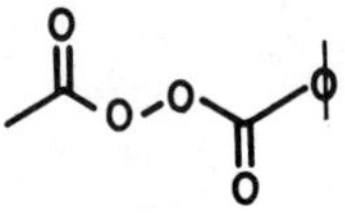

CCl_4 1792 24
1772

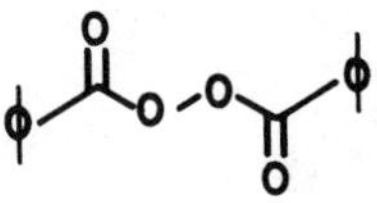

$C_2H_4Cl_2$ 1789 24
1767

NO_2

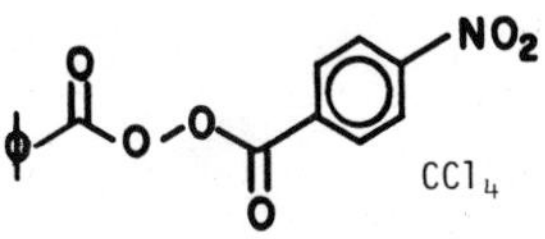

CCl_4 1797 24
1773

Solvent	cm⁻¹	Ref.
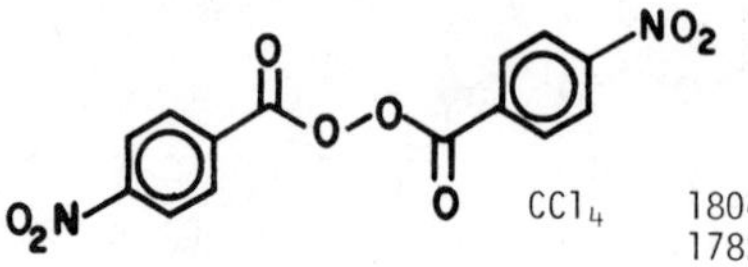 CCl_4	1804, 1782	24
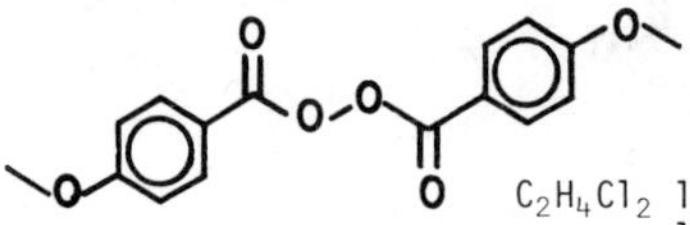 $C_2H_4Cl_2$	1780, 1758	24
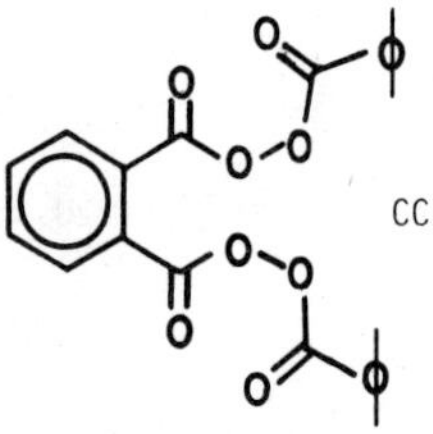 CCl_4	1801, 1776	24
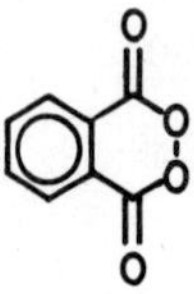 nuj.	1805, 1783	24
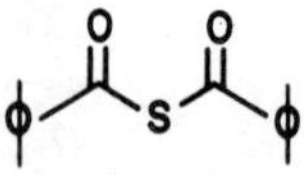 CCl_4	1739, 1709, 1680	25
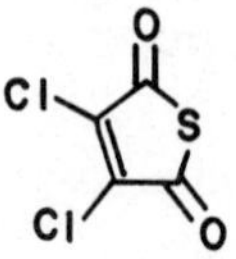 KBr	1876, 1789, 1702b	9
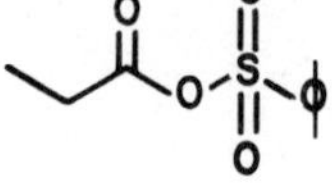 sng	1800	26
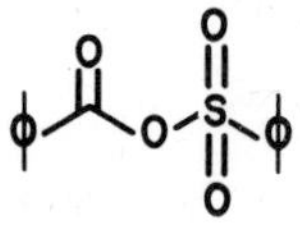 sng	1780	26
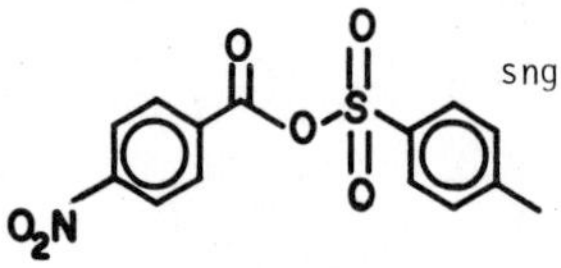 sng	1780	26
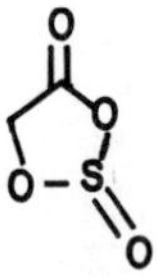 liq.	1825	27
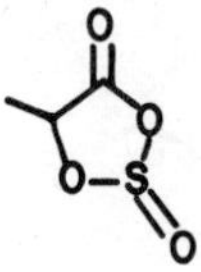 liq.	1815	27
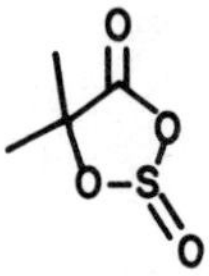 liq.	1825	27
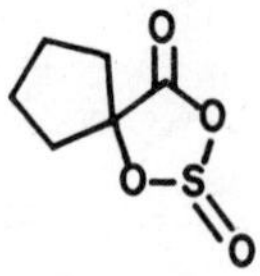 liq.	1818	27
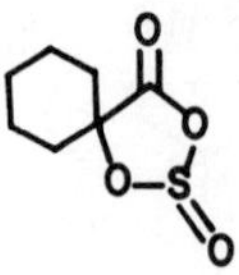 liq.	1810	27

REFERENCES

1. H. H. Wasserman and P. S. Wharton, *JACS* (1960), 1411.
2. G. F. Fanta, *JOC* (1964), 981.
3. D. Dolphin and R. Linn, unpublished results.
4. L. J. Bellamy, B. R. Connelly, A. R. Philpotts and R. L. Williams, *Z Elektrochem* (1960), 563.
5. T. G. Bonner, E. G. Gabb, P. McNamara and B. Smethurst, *T* (1965) *21*, 463.
6. J. H. Looker and D. M. Thatcher, *JOC* (1958), 403.
7. H. K. Hall, Jr. and R. Zbinden, *JACS* (1958), 6428.
8. P. Mirone and P. Chiorboli, *SCA* (1962), 1426.
9. O. Scherer and F. Kluge, *B* (1966), 1973.
10. W. G. Dauben and W. W. Epstein, *JOC* (1959), 1595.
11. E. Casadevall, C. Langeau and P. Moreau, *BSC France* (1968), 1514.
12. G. Maier and F. Seidler, *B* (1966), 1236.
13. R. Criegee, U. Zirngibl, H. Furrer, D. Seebach and G. Freund, *B* (1964), 2942.
14. R. G. Cooke, *Chem & Ind* (1955), 142.
15. A. R. Lepley and J. P. Thelman, *T* (1966) *22*, 101.
16. E. J. Modest and J. Smuszkovicz, *JACS* (1950), 577.
17. G. Snatzke and G. Zanati, *A* (1965) *684*, 62.
18. R. Darms, T. Threlfall, M. Pesaro and A. Eschenmoser, *Helv* (1963), 2893.
19. M. Hauptschein, C. S. Stokes and E. A. Nodiff, *JACS* (1952), 4005.
20. T. Sakan, F. Murai, Y. Hayashi, Y. Honda, T. Shono, M. Nakajima and M. Kato, *T* (1967), 4635.
21. H. Musso, K. Naumann and K. Grychtol, *B* (1967), 3614.
22. D. S. Tarbell and M. A. Leister, *JOC* (1958), 1149.
23. D. S. Tarbell and E. J. Longosz, *JOC* (1959), 774.
24. W. H. T. Davison, *JCS* (1951), 2456.
25. R. A. Nyquist and W. J. Potts, *SCA* (1959) *15*, 514.
26. C. G. Overberger and E. Sarlo, *JACS* (1963), 2446.
27. J. Pump, E. G. Rochow and U. Wannagat, *Monatsh Chem* (1963), 588.

4.10 esters

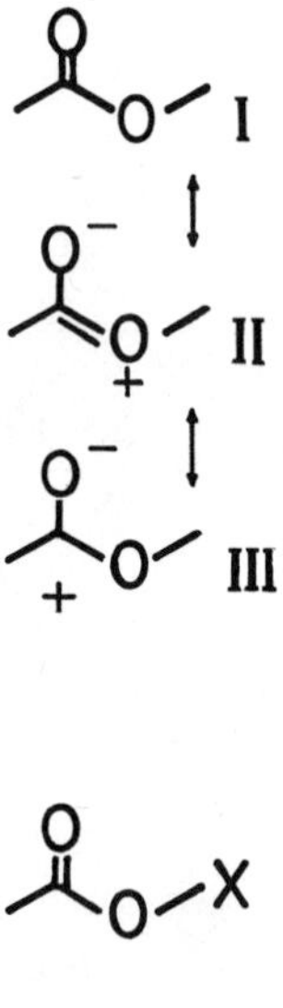

Delocalization of the oxygen lone pair electrons into the carbonyl group I ⟷ II will lower the frequency of the ester carbonyl compared to that of the corresponding ketone. This effect will be counterbalanced by the destabilization of the resonance form III by both the electronegativity of the ether oxygen and by the field effect of this oxygen. Since esters absorb a higher frequencies than ketones, it must be the latter effect that predominates over electron delocalization, which in the case of esters is considerably smaller than with amides.

As would be expected from the preceding arguments, electron withdrawing groups on the ether oxygen increase the frequency of the carbonyl stretching mode both by decreasing delocalization and by decreasing the apparent electronegativity of the oxygen. This effect is observed both with electron withdrawing groups attached to the carbon atom bonded to the ether oxygen, as well as with alkenyl and aryl groups which are bonded directly to the ether oxygen.

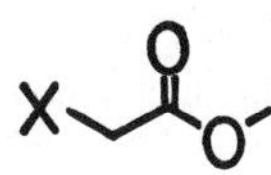

The effects of substitution on the α-carbon atom of esters cause changes in the carbonyl frequency that parallel those already seen with ketones (Chapter 4.2, p. 175).

Changes in state and solvent cause similar frequency changes as are seen with ketones. Thus the vapor phase frequencies are between 20 and 30 cm^{-1} higher than those of the condensed state, and the frequencies in $CHCl_3$ are usually from 10 to 20 cm^{-1} lower than those measured in CCl_4.

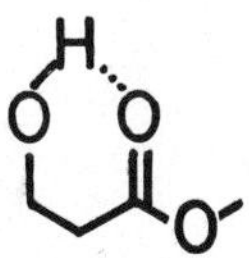

Intramolecular hydrogen bonding can cause a considerable lowering in the carbonyl stretching frequency. Thus a dilute solution in CCl_4 of methyl benzoate absorbs at 1727 cm^{-1}, whereas methyl anthranilate absorbs at 1684 cm^{-1}. Similarly, β-keto esters, when enolized, show not only the typical ester and ketone bands, but lower frequency bands that are associated with the olefinic double bond and the hydrogen-bonded ester carbonyl.

C—O

The C-O single-bond stretching vibration is always strong in esters and occurs between 1300 and 1000 cm^{-1}. The position of the band is of little diagnostic value.

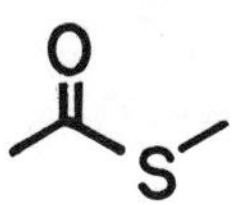

The larger covalent radius of sulfur compared to that of oxygen results in less delocalization of the nonbonding sulfur electrons into the carbonyl group. This is balanced by the lower electronegativity of sulfur, with the result that thiol esters absorb at frequencies lower than those of oxygen-containing esters. The effects of substitution on either side of the thiol ester are comparable to those observed with normal esters.

Anhydrides (Chapter 4.9, p. 324), carbonates (Chapter 4.14, p. 403), lactones (Chapter 4.11, p. 365), and carbamates (Chapter 4.13, p. 396) are covered separately.

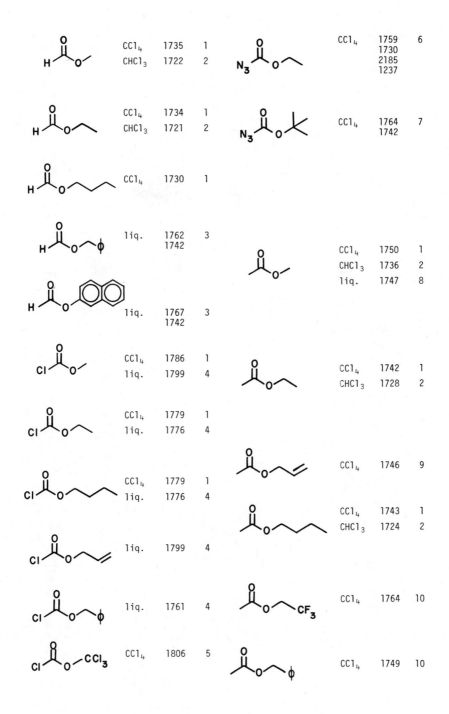

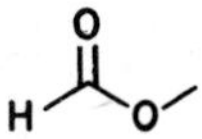

Solvent	Frequency	Ref.
CCl_4	1735	1
$CHCl_3$	1722	2
CCl_4	1734	1
$CHCl_3$	1721	2
CCl_4	1730	1
liq.	1762 1742	3
liq.	1767 1742	3
CCl_4	1786	1
liq.	1799	4
CCl_4	1779	1
liq.	1776	4
CCl_4	1779	1
liq.	1776	4
liq.	1799	4
liq.	1761	4
CCl_4	1806	5

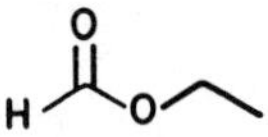

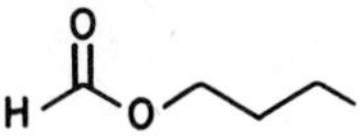

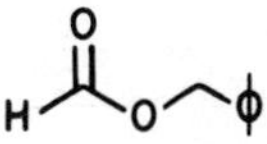

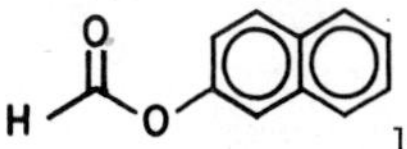

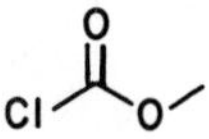

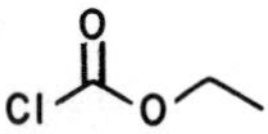

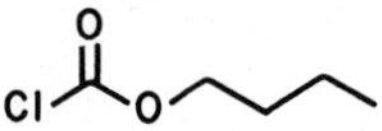

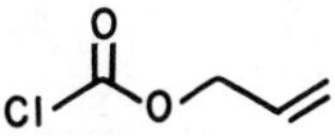

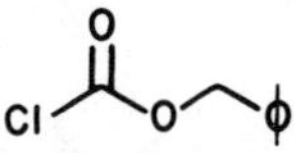

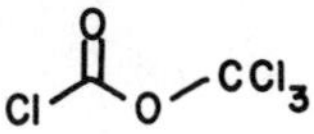

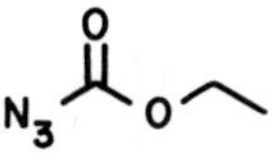

Solvent	Frequency	Ref.
CCl_4	1759 1730 2185 1237	6
CCl_4	1764 1742	7
CCl_4	1750	1
$CHCl_3$	1736	2
liq.	1747	8
CCl_4	1742	1
$CHCl_3$	1728	2
CCl_4	1746	9
CCl_4	1743	1
$CHCl_3$	1724	2
CCl_4	1764	10
CCl_4	1749	10

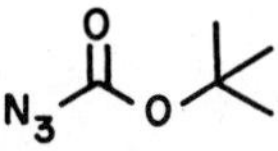

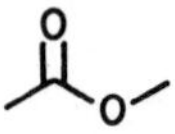

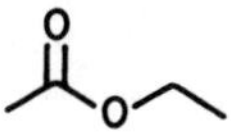

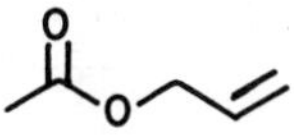

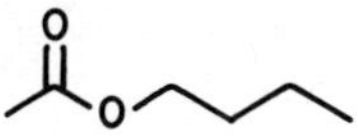

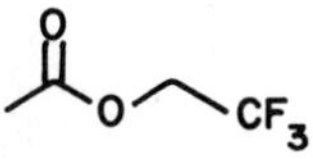

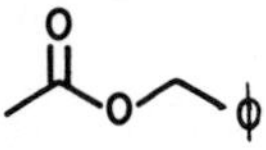

Solvent	ν (cm⁻¹)	Ref.
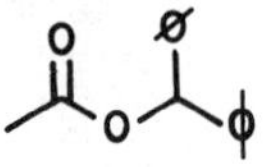 CCl_4	1741	10
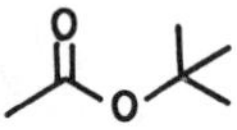 CCl_4	1736	10
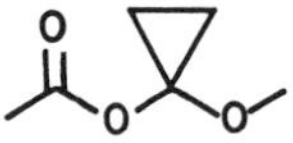 CCl_4	1740	11
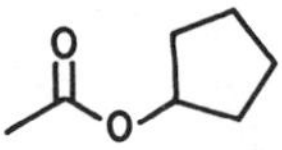 liq.	1748	12
CCl_4	1738	1
liq.	1740	8
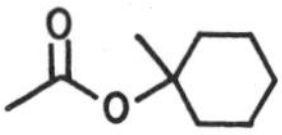 liq.	1750	12
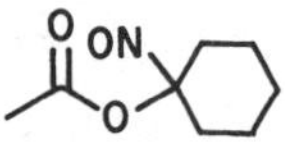 liq.	1745 1555	13
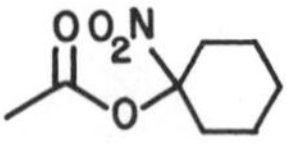 liq.	1760 1552 1359	13
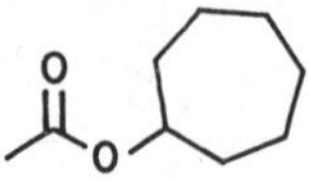 liq.	1740	12
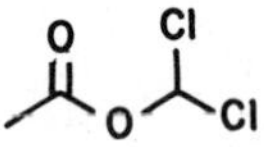 CCl_4	1780	14
CCl_4	1763	1
liq.	1762	8
CCl_4	1759 1678	15
liq.	1755	16
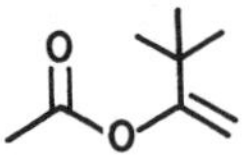 liq.	1755 1700w 1652	17
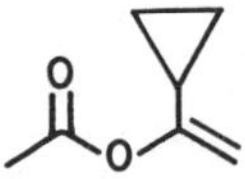 liq.	1760 1660	18
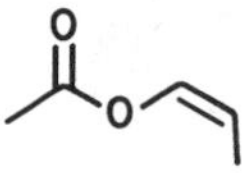 CCl_4	1758 1690	15
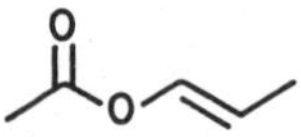 CCl_4	1752 1675	15
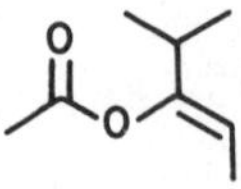 CCl_4	1755 1692	15
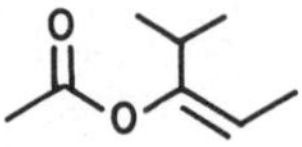 CCl_4	1755 1683	15

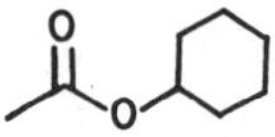

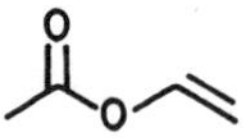

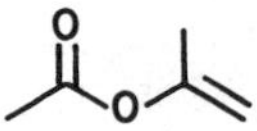

Solvent	Frequency	Ref.
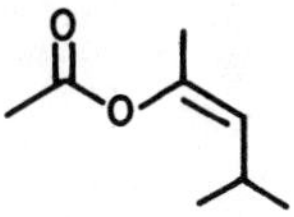 CCl_4	1752 1695	15
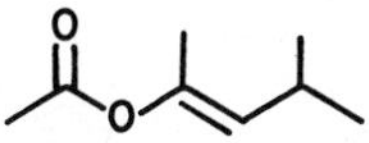 CCl_4	1758 1754 1690	15
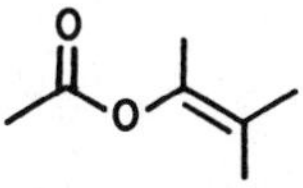 liq.	1765 1710	19
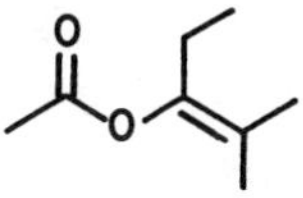 CCl_4	1748 1693	15
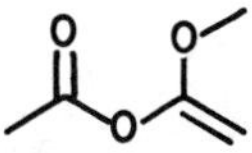 CCl_4	1681	20
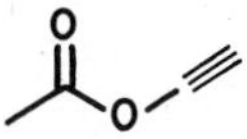 CCl_4	1764	10
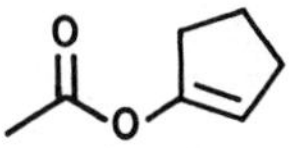 liq.	1761	21
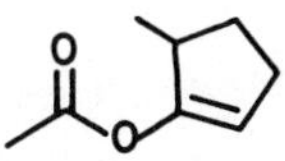 CCl_4	1755 1713	15
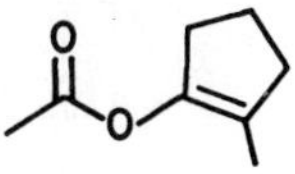 CCl_4	1750 1703	15
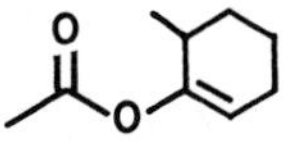 CCl_4	1750 1680	15
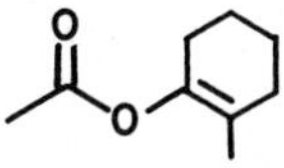 CCl_4	1750 1705	15
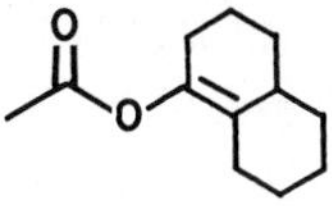 CCl_4	1750 1695	22
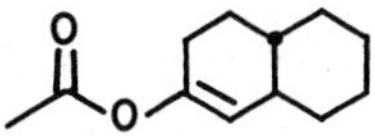 CCl_4	1750 1678	22
CCl_4	1769	1
$CHCl_3$	1764 1744	23
liq.	1766	8
$CHCl_3$	1770 1755	23
CCl_4	1775 1758m	23
$CHCl_3$	1774w 1757	23
CCl_4	1773 1763	23

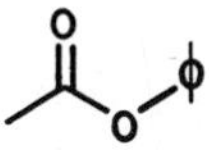

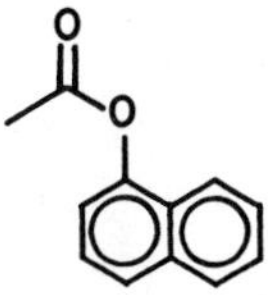

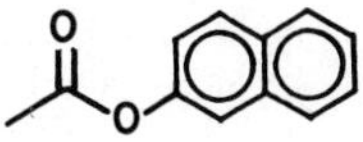

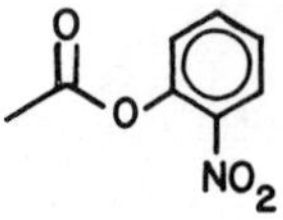

CCl_4	1786	24

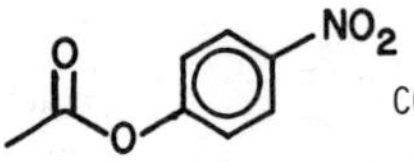

CCl_4	1779	25

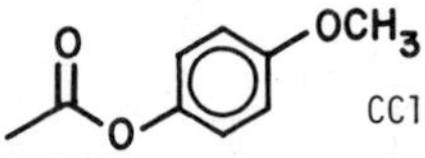

CCl_4	1770	25

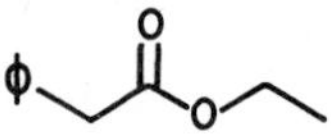

CCl_4	1741	26
$CHCl_3$	1730	27

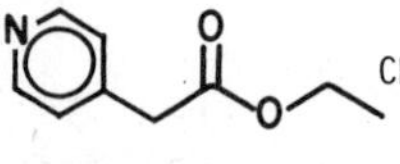

$CHCl_3$	1736	27

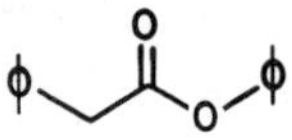

CCl_4	1755	28

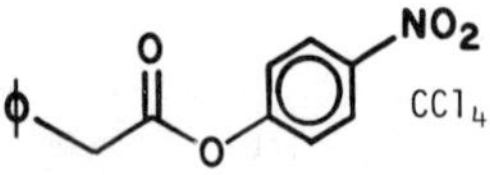

CCl_4	1749	10

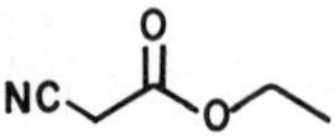

CCl_4	1754	29

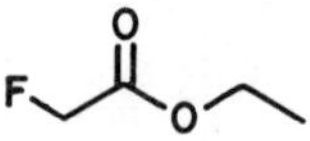

CCl_4	1775 1746	30
$CHCl_3$	1718	30
liq.	1778 1750	31

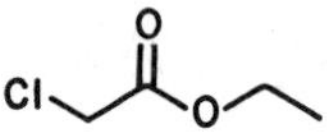

CCl_4	1768 1745	26

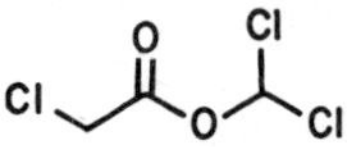

CCl_4	1775	14

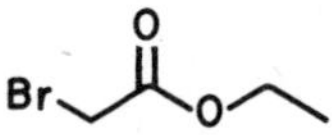

liq.	1740	31

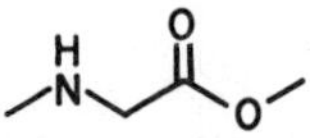

liq.	1742	24

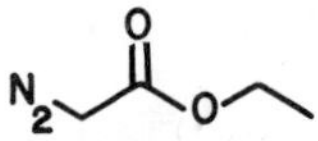

CH_2Cl_2	1695 2101	32

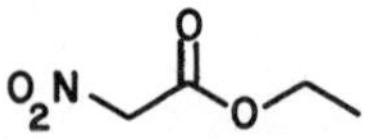

liq.	1565 1372	33

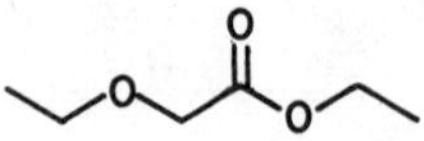

CCl_4	1758 1736	34

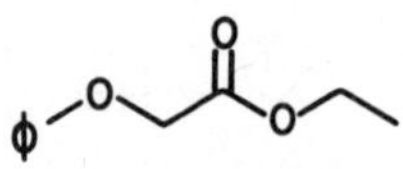

CCl_4	1767 1739	35

Compound	Solvent	ν (cm⁻¹)	Ref.
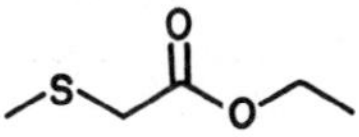 $CH_3SCH_2COOC_2H_5$	CCl_4	1747	36
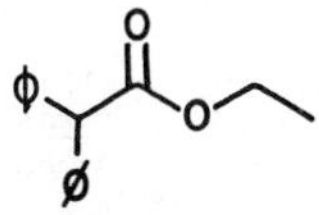 $\phi_2CHCOOC_2H_5$	$CHCl_3$	1727	24
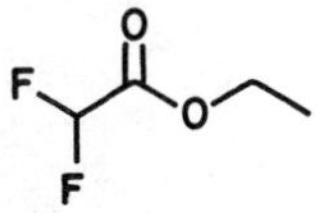 $F_2CHCOOC_2H_5$	CCl_4	1780 1765	30
	liq.	1770	31
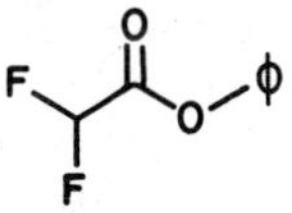 $F_2CHCOO\phi$	liq.	1790	37
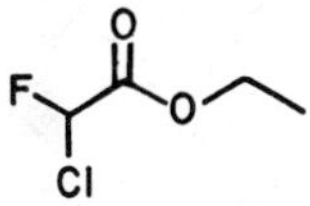 $FClCHCOOC_2H_5$	liq.	1754	31
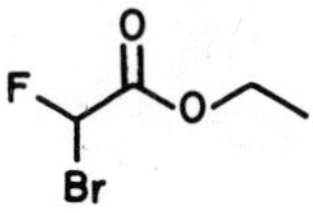 $FBrCHCOOC_2H_5$	liq.	1754	31
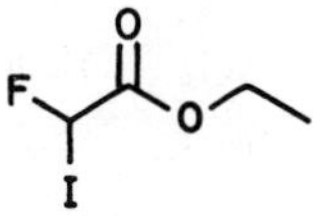 $FICHCOOC_2H_5$	liq.	1754	31
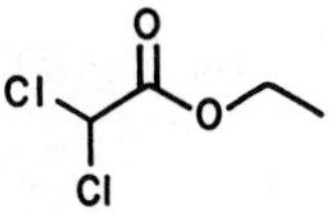 $Cl_2CHCOOC_2H_5$	liq.	1758	31
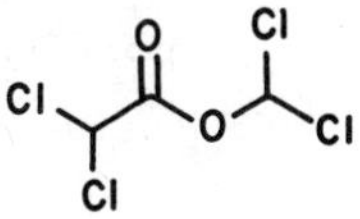 $Cl_2CHCOOCHCl_2$	CCl_4	1790 1770	14
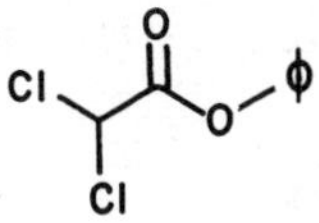 $Cl_2CHCOO\phi$	nuj.	1775	37
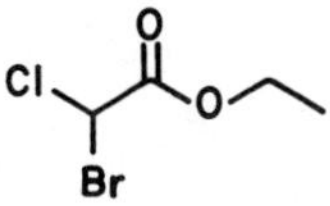 $ClBrCHCOOC_2H_5$	liq.	1758 1742	31
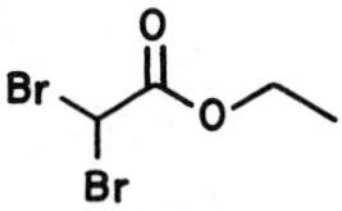 $Br_2CHCOOC_2H_5$	liq.	1753 1739	31
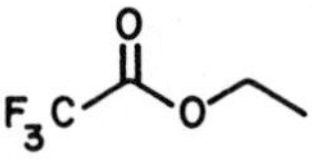 $F_3CCOOC_2H_5$	CCl_4	1789	30
	liq.	1789	31
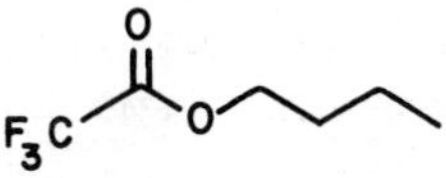 $F_3CCOO(CH_2)_3CH_3$	CCl_4	1787	10
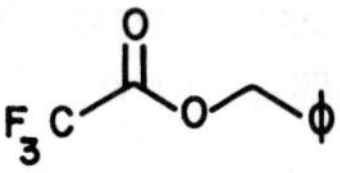 $F_3CCOOCH_2\phi$	CCl_4	1776	10
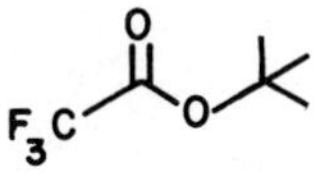 $F_3CCOOC(CH_3)_3$	CCl_4	1777	10

Solvent	Frequency	Ref.
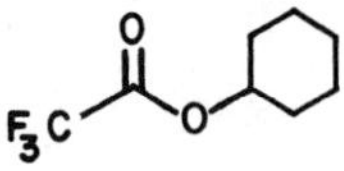 CCl_4	1777	10
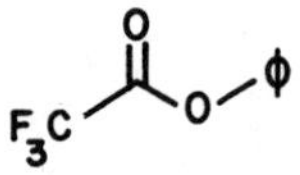 CCl_4	1795	10
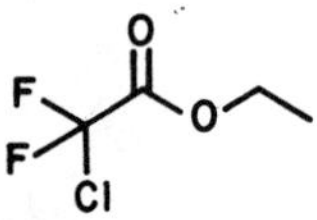 liq.	1770	31
CCl_4	1769	34
liq.	1768	8
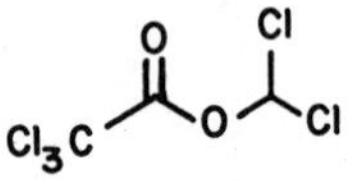 CCl_4	1780	14
$CHCl_3$	1728	2
CCl_4	1745	1
$CHCl_3$	1723	2
CCl_4	1736	1
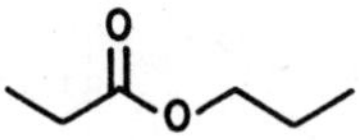 $CHCl_3$	1728	2

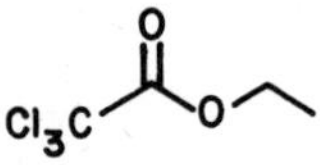

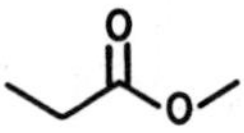

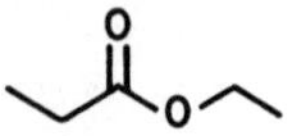

Solvent	Frequency	Ref.
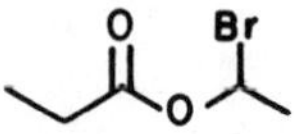 CCl_4	1747	38
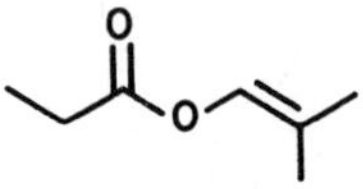 liq.	1748 1681	39
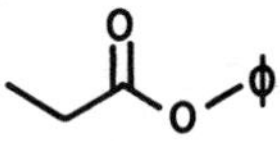 CCl_4	1756	28
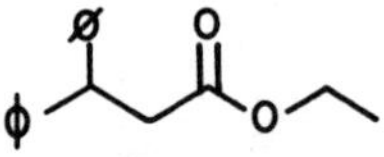 $CHCl_3$	1727	24
CCl_4	1747	30
$CHCl_3$	1741	30
CCl_4	1743	30
$CHCl_3$	1735	30
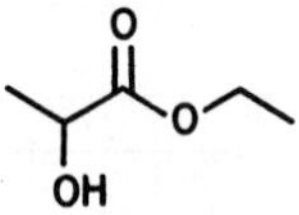 CCl_4	1733b	40
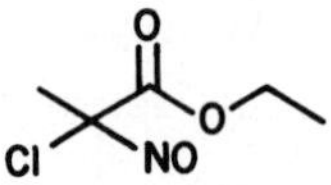 liq.	1751	41

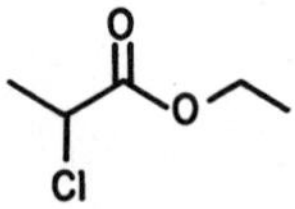

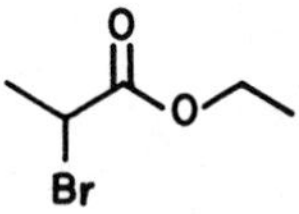

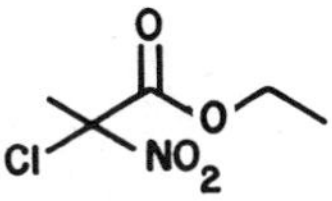

Solvent	Frequency	Ref.
liq.	1764	41
liq.	1739	24
CCl_4	1739	26
liq.	1739	42
liq.	1767	42
CCl_4	1746	28
CCl_4	1741	43
liq.	1786	44
liq.	1734	45
CCl_4	1736	30
$CHCl_3$	1725	30
liq.	1735	45
CCl_4	1738	30
$CHCl_3$	1731	30
nuj.	1721	46
CCl_4	1736	1
$CHCl_3$	1718	47
CCl_4	1728	30
$CHCl_3$	1718	30
CCl_4	1723	30
$CHCl_3$	1715	30
CCl_4	1770	14
CCl_4	1738	43
CCl_4	1745 1727	43

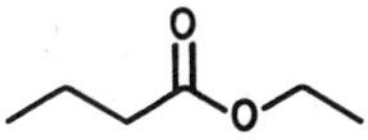

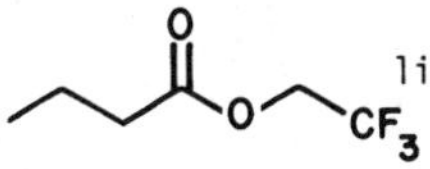

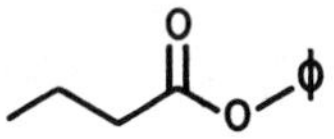

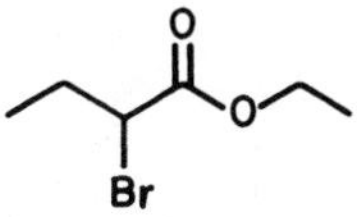

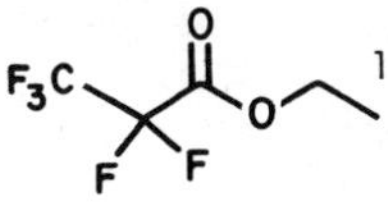

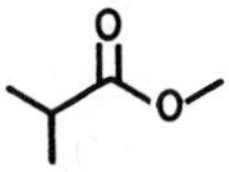

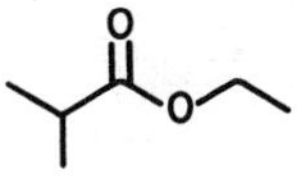

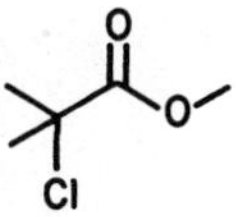

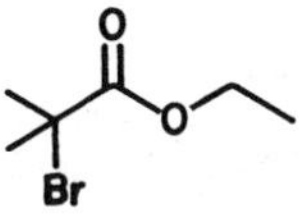

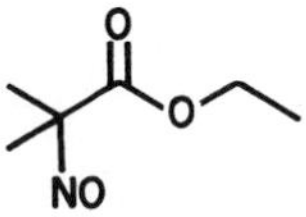

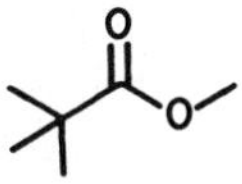

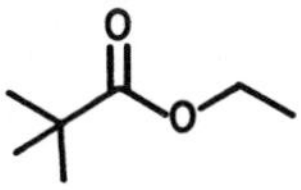

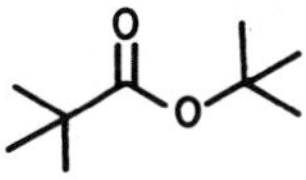

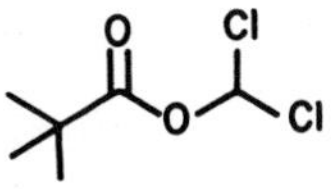

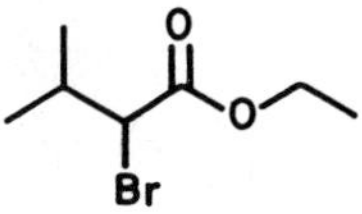

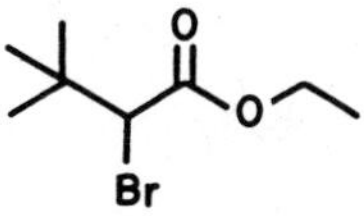

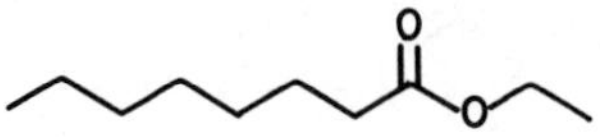

Solvent	Frequency	Ref.
liq.	1743	48
CCl_4	1738	49
CCl_4	1730	50
$CHCl_3$	1690	51
CCl_4	1725	50
CCl_4	1722	50
liq.	1786 1727	39
liq.	1750	52

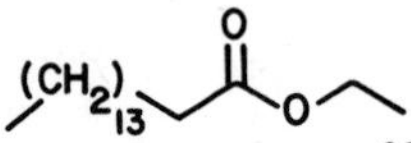

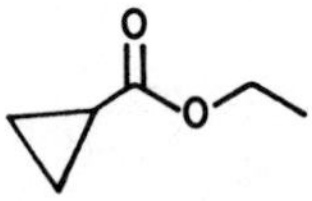

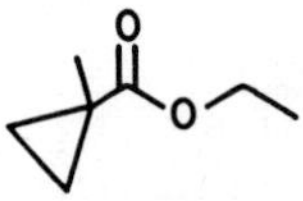

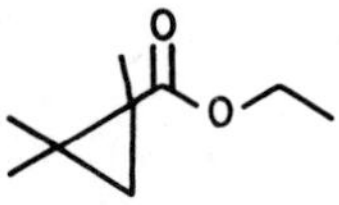

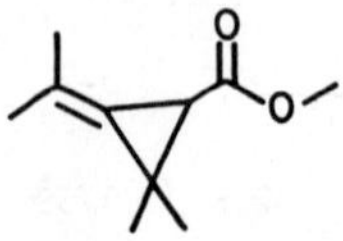

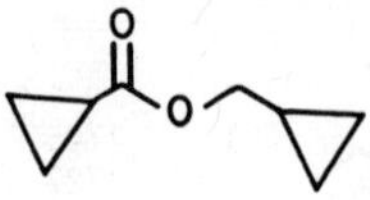

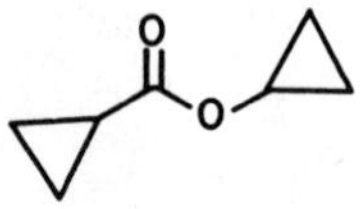

Solvent	Frequency	Ref.
CCl_4	1720	53
CCl_4	1691* 1683	54
CCl_4	1750 1726	54
CCl_4	1760 1728	54
CCl_4	1690*	54
$CHCl_3$	1700	51
liq.	1725	55
$CHCl_3$	1705	51

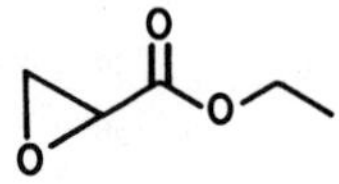

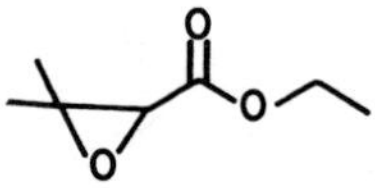

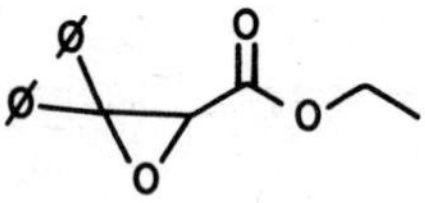

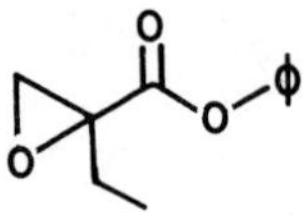

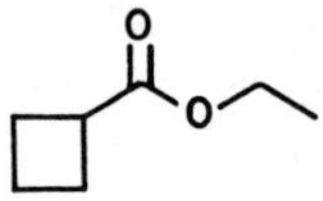

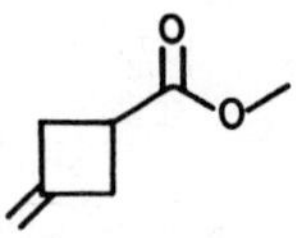

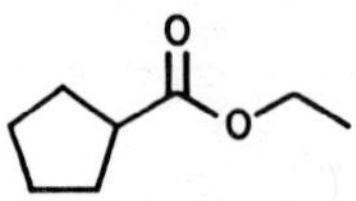

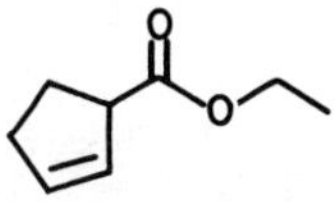

Solvent	Frequency (cm^{-1})	Ref.
liq.	1730 1616	56
CCl_4	1750 1730	54
$CHCl_3$	1705	51
CCl_4	1722 1650w	6
CCl_4	1733	28
CCl_4	1733	57
liq. CCl_4	1732 1735	49 49
CCl_4	1730	1

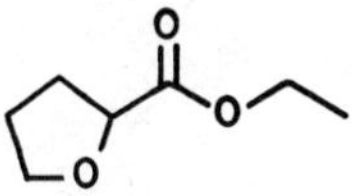

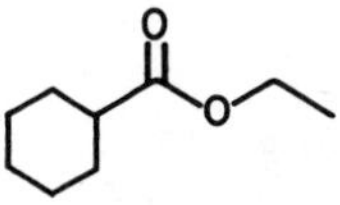

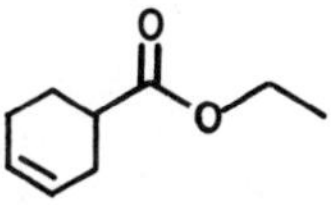

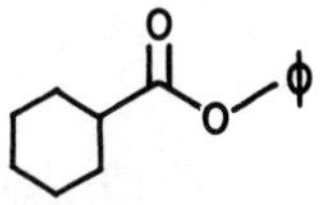

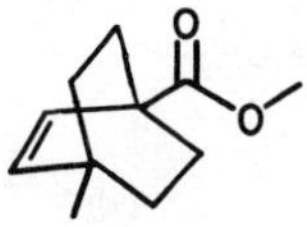

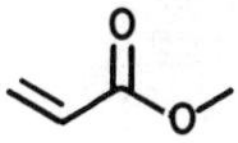

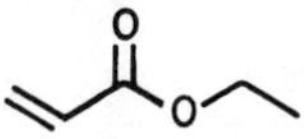

Solvent	Frequency (cm^{-1})	Ref.
CCl_4	1770	14
liq.	1735 1644	58
CS_2	1721	59
CS_2	1726	59
CCl_4 $CHCl_3$	1724 1712	29 29
liq.	1730 1658 1638	58
liq.	1721	42
liq.	1742	42
liq.	1739	42

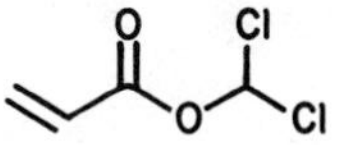

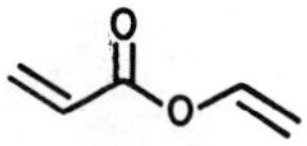

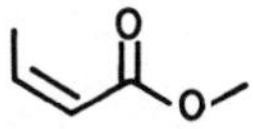

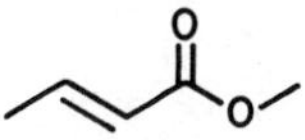

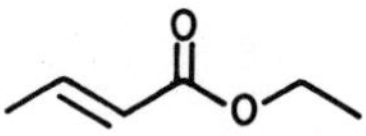

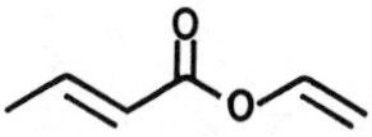

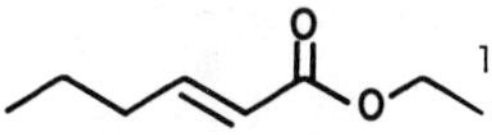

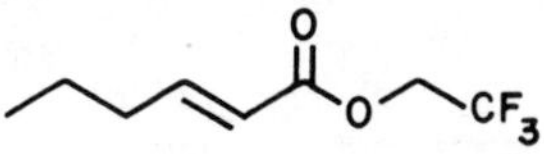

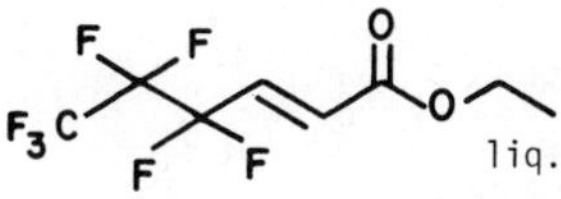

Solvent	Frequency	Ref.
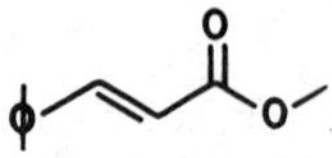 CCl_4	1724	1
CCl_4	1718	38
$CHCl_3$	1711	27
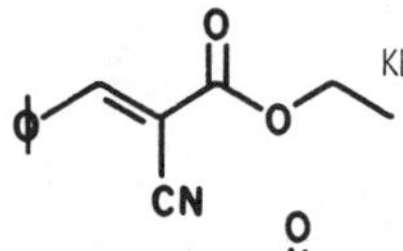 KBr	1732 1682 2247	60
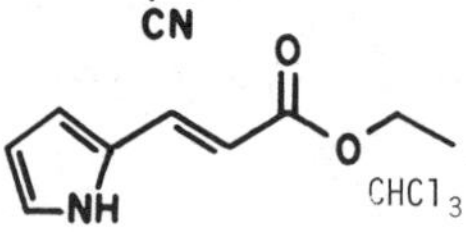 $CHCl_3$	1704* 1690	61
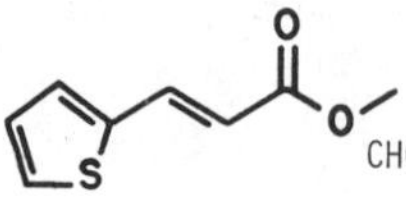 $CHCl_3$	1705 1631	62
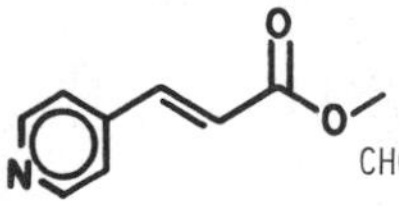 $CHCl_3$	1718	27
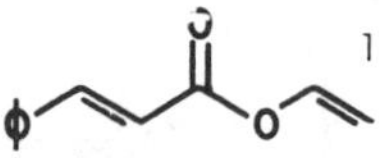 liq.	1723 1629 1597	58
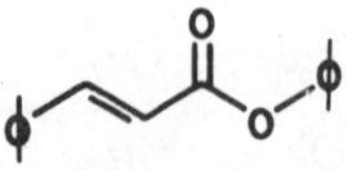 $CHCl_3$	1732	63
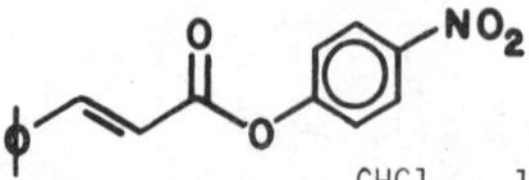 $CHCl_3$	1733	63

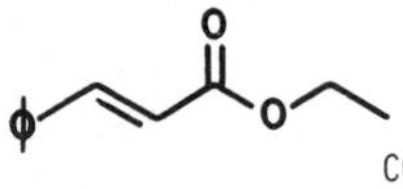

Solvent	Frequency	Ref.
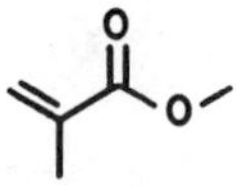 CCl_4	1727	1
CCl_4	1721	29
$CHCl_3$	1712	29
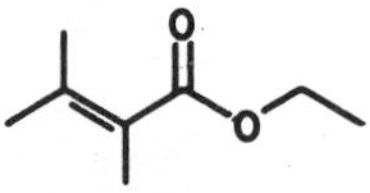 CCl_4	1715 1640	64
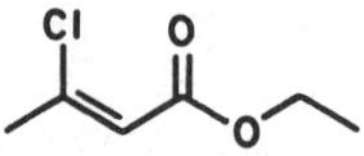 liq.	1712 1631	65
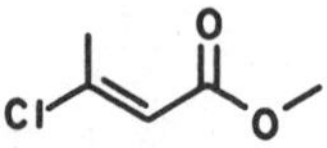 liq.	1721 1637	65
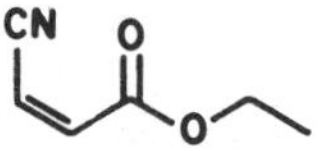 CCl_4	1738 1725	29
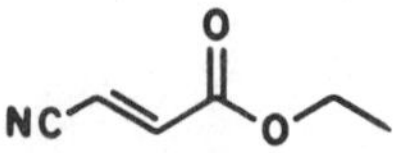 CCl_4	1728	29
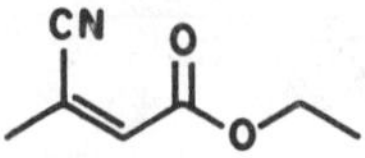 CCl_4	1738 1725	29
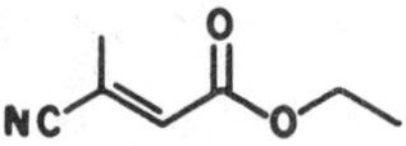 CCl_4	1728	29

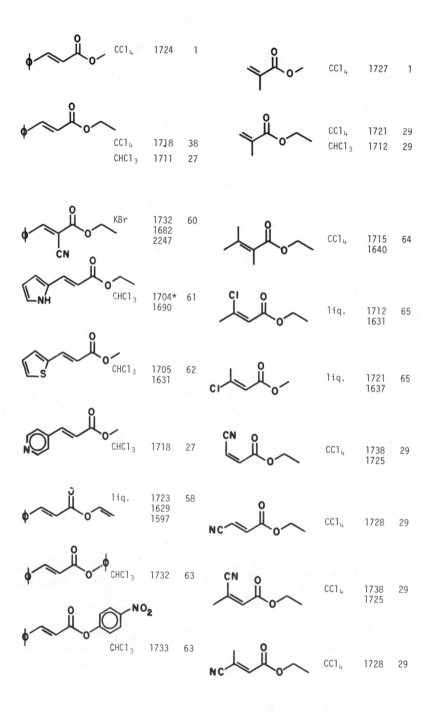

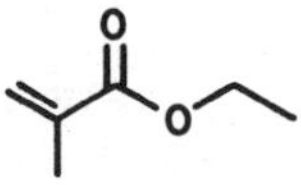

Solvent	Frequency	Ref.
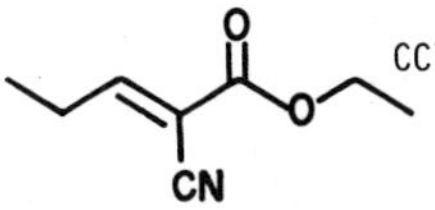 CCl_4	1737	29
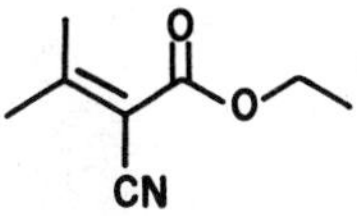 CCl_4	1729 1614 2222	66
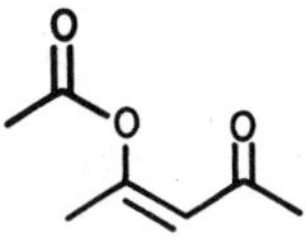 CCl_4	1762 1696 1670w	67
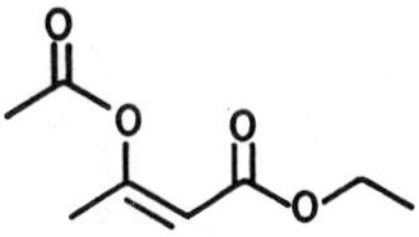 CCl_4	1760 1715 1660m	67
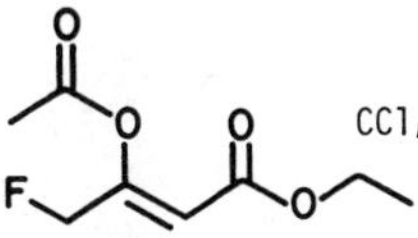 CCl_4	1780 1720 1678	67
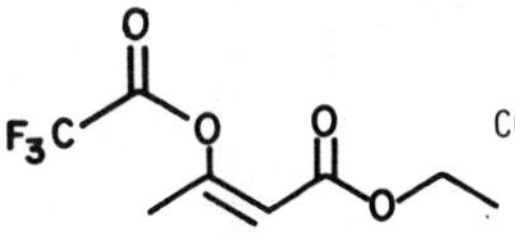 CCl_4	1785 1715m 1650w	67
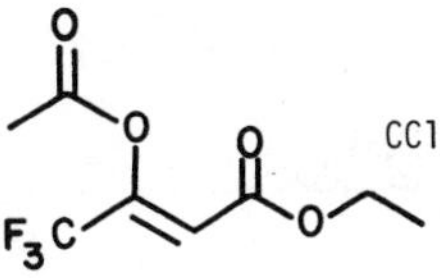 CCl_4	1795 1725 1675m	67
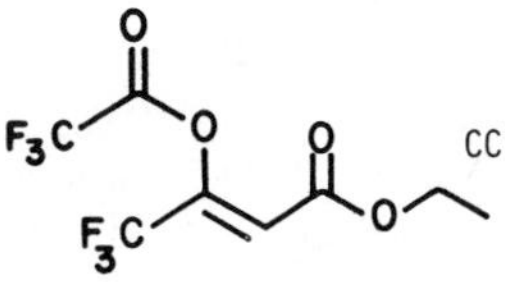 CCl_4	1790 1737m	67

Solvent	Frequency	Ref.
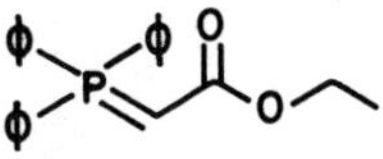 KBr	1620	68
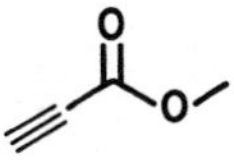 CS_2	1719	59
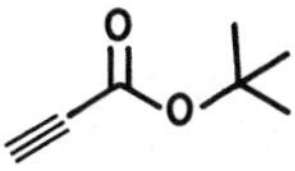 $CHCl_3$	1700 2105	69
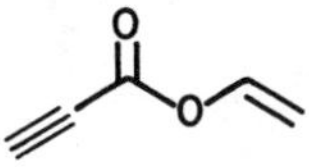 liq.	1721 1644 2120	58
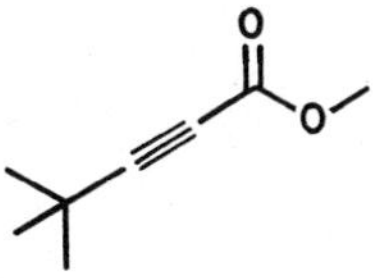 liq.	1705 2220	70
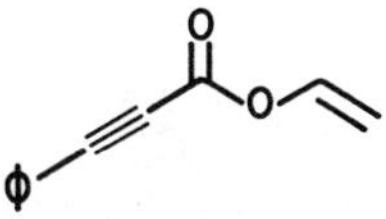 liq.	1714 1639 1592 2213	58
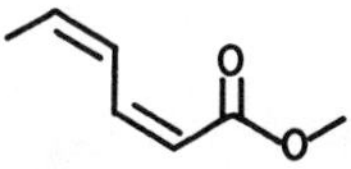 CS_2	1715	59
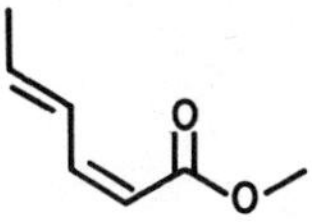 CS_2	1715	59
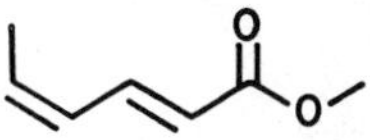 CS_2	1715	59

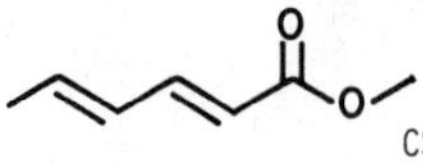

Solvent	cm⁻¹	Ref.
CS_2	1716	59
CCl_4	1705, 1620	71
CCl_4	1705, 1615	71
CS_2	1719	59
CS_2	1733, 1718	59
CS_2	1713	59
CS_2	1708	59
CS_2	1716	59
CS_2	1721	59
liq.	1710, 1620	56
liq.	1709, 1595	72
liq.	1715, 1650	73
CH_2Cl_2	1712, 1653w	74
liq.	1724	75
CCl_4	1727	1
$CHCl_3$	1724	27
CCl_4	1720	1

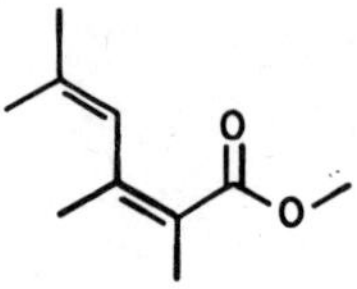

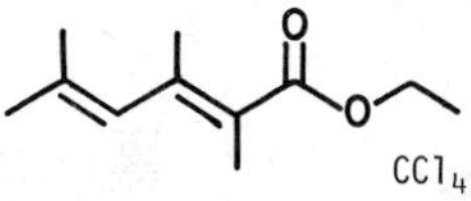

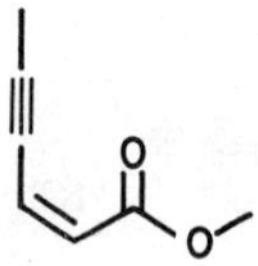

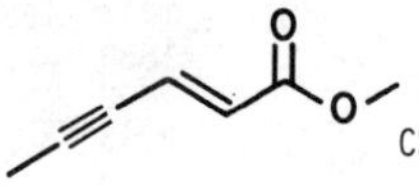

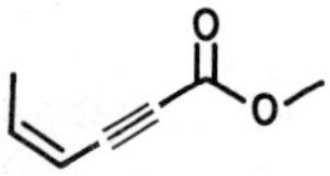

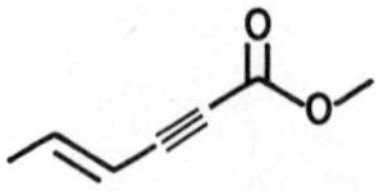

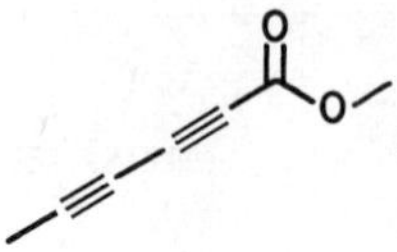

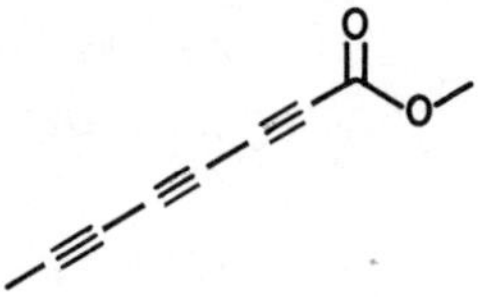

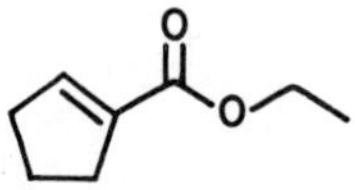

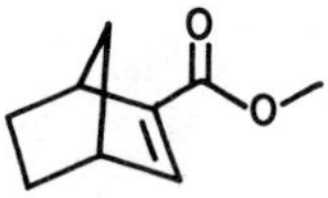

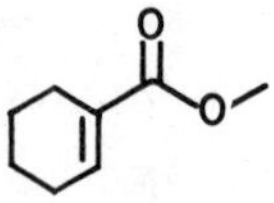

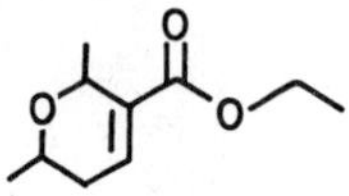

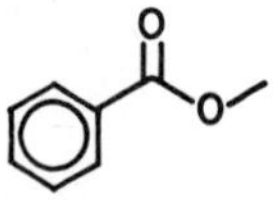

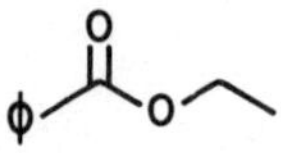

Solvent	ν (cm⁻¹)	Ref.
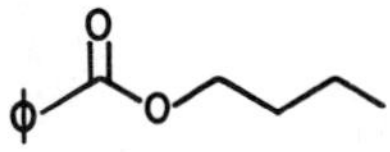 CCl_4	1726	1
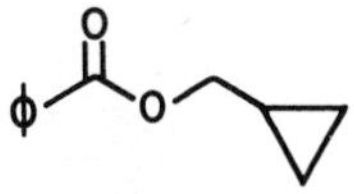 liq.	1718	76
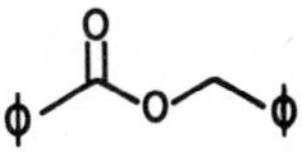 CCl_4	1726	10
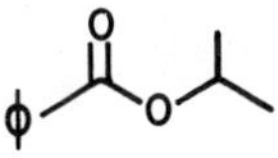 CCl_4	1708	28
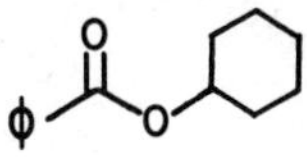 CCl_4	1708	28
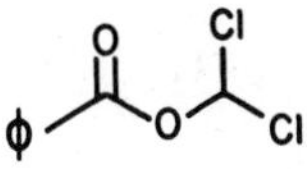 CCl_4	1750	14
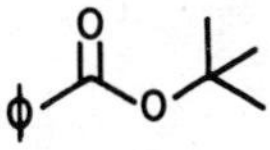 CCl_4	1717	77
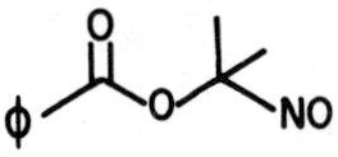 $CHCl_3$	1730 1577 1508	78
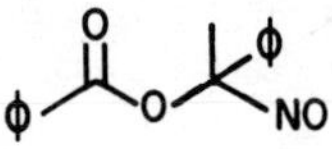 CH_2Cl_2	1733 1572	78

Solvent	ν (cm⁻¹)	Ref.
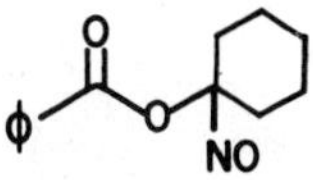 CH_2Cl_2	1724 1565 1499	78
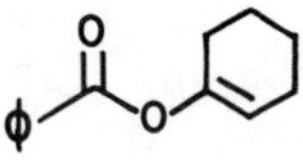 liq.	1739 1695 1608	79
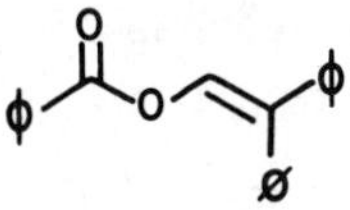 $CHCl_3$	1710	80
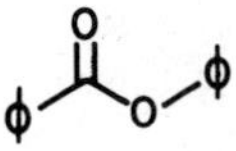 CCl_4	1743	1
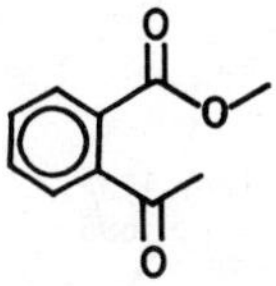 liq.	1725 1708	75
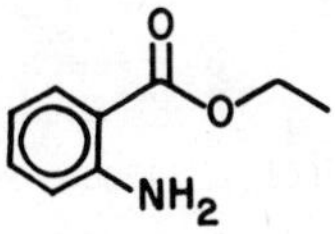 CCl_4	1697	81
liq.	1685	82
CCl_4	1684	24
liq.	1730	82
CCl_4	1727	24

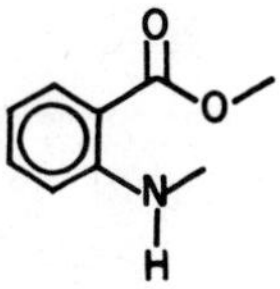

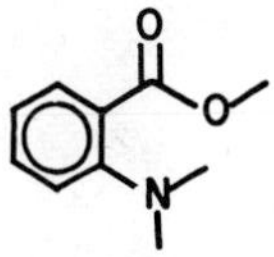

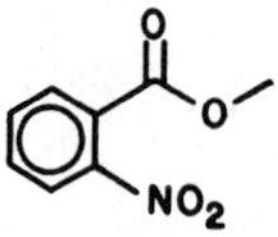

CCl_4 1747 83

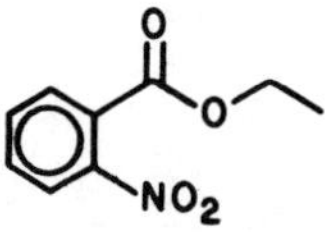

$CHCl_3$ 1724 84

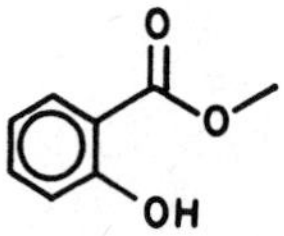

CCl_4 1684 77

$CHCl_3$ 1679 77

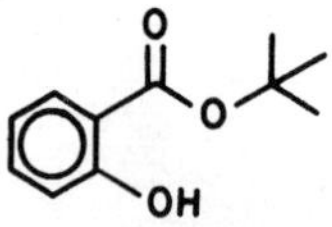

CCl_4 1674 77

$CHCl_3$ 1673 77

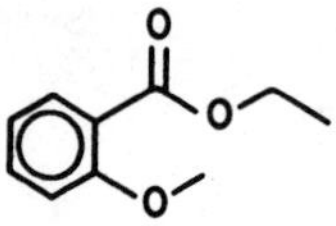

$CHCl_3$ 1704 84
1695

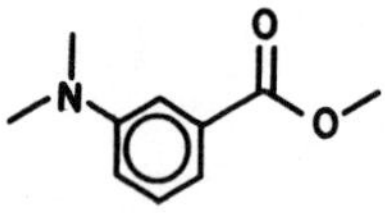

CCl_4 1730 85

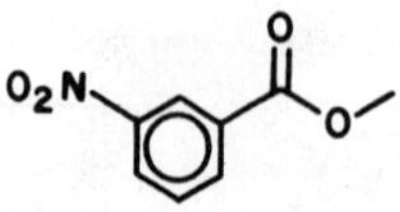

CCl_4 1741 85

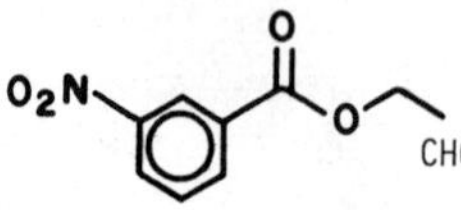

$CHCl_3$ 1724 84

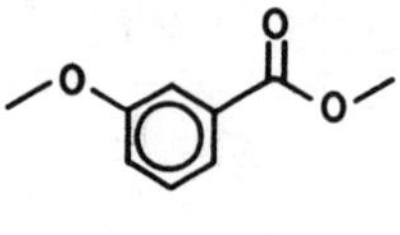

CCl_4 1731 85

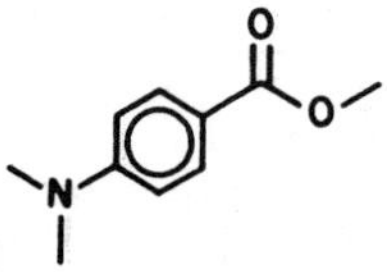

liq. 1720 86

CCl_4 1715 85

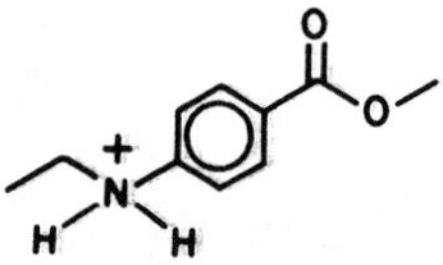

KBr 1730 87

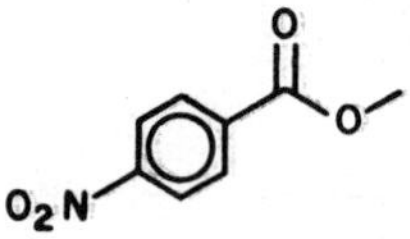

CCl_4 1737 85

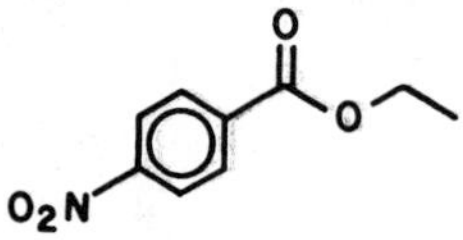

$CHCl_3$ 1718 84

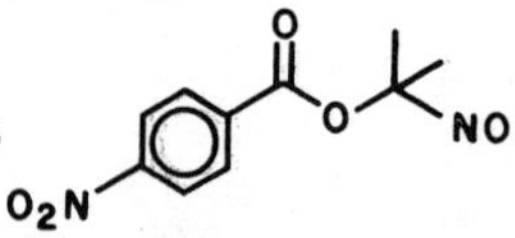

$CHCl_3$ 1730 78
1575

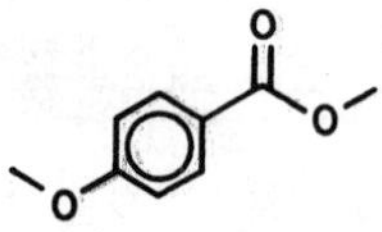

CCl_4 1721 85

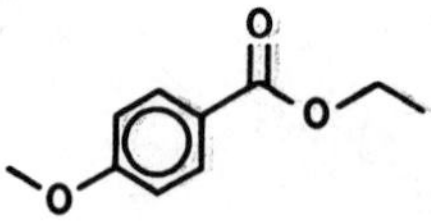

$CHCl_3$ 1698 84

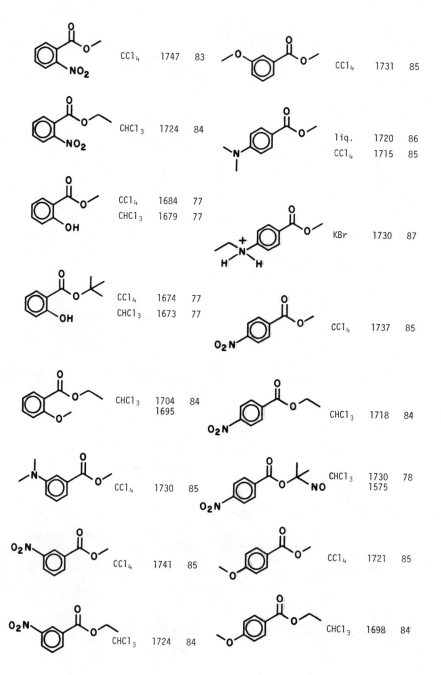

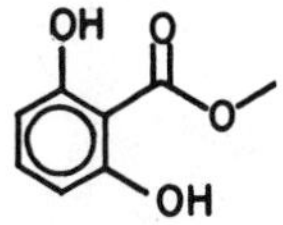

Solvent	Frequency	Ref.
CCl_4	1680	77
liq.	1720	86
CCl_4	1716 1700	88
$CHCl_3$	1721 1697	88
CCl_4	1726 1712 1702* 1669	89
CCl_4	1715 1685 1670	89
CCl_4	1715 1685 1670	89
$CHCl_3$	1669	90
KBr	1689	91
CCl_4	1732 1711	88
$CHCl_3$	1731 1702	88
CCl_4	1716 1691 1595 1525	92
CCl_4	1701 1683*	89
CCl_4	1711 1693	89
CCl_4	1702 1683*	89
KBr	1669	91
$CHCl_3$	1723 1704m	93

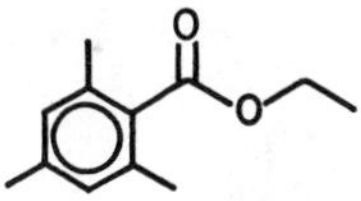

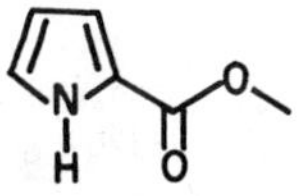

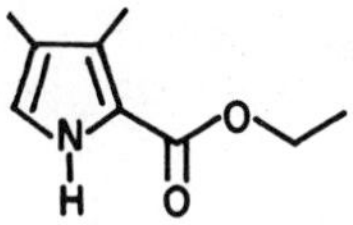

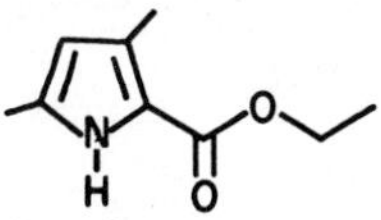

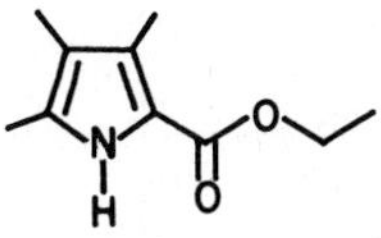

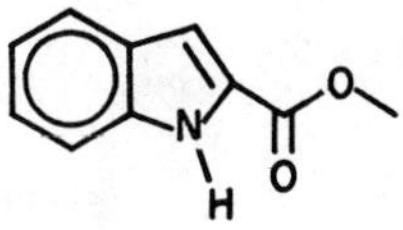

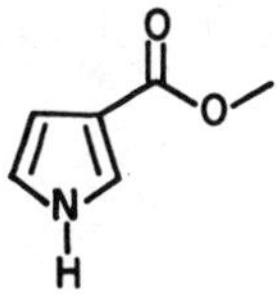

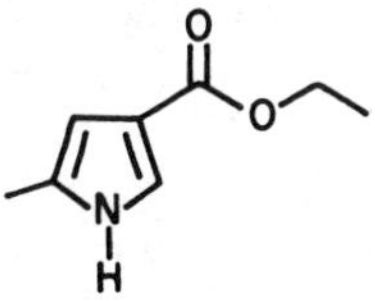

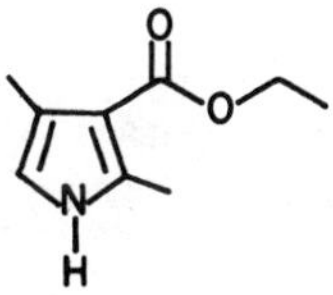

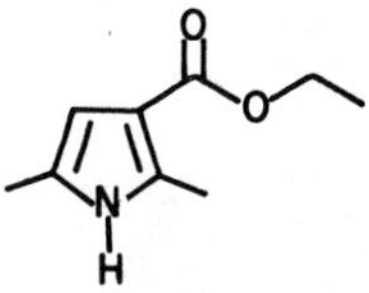

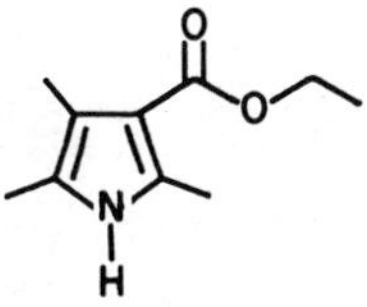

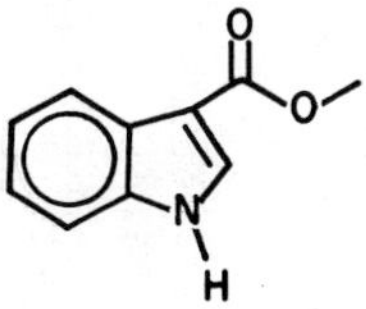

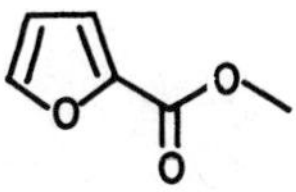

Solvent	Frequency	Ref.
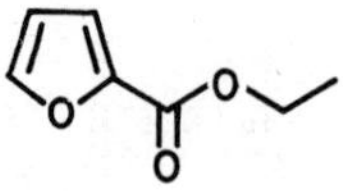 $CHCl_3$	1721 1715	93
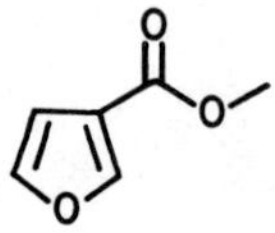 $CHCl_3$	1726 1703	94
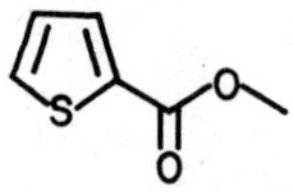 $CHCl_3$	1707	62
CCl_4	1699	95
$CHCl_3$	1707	62
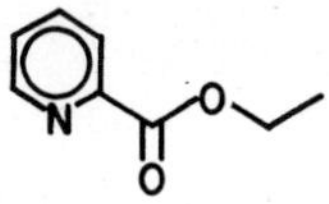 $CHCl_3$	1723	27
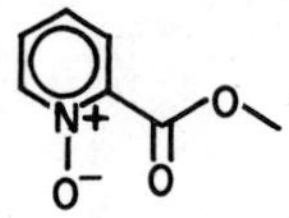 $CHCl_3$	1741	27
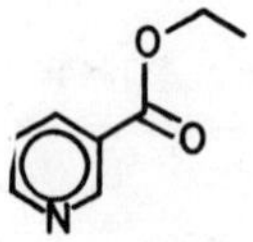 $CHCl_3$	1724	27
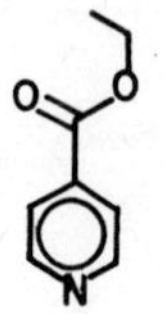 $CHCl_3$	1730	27

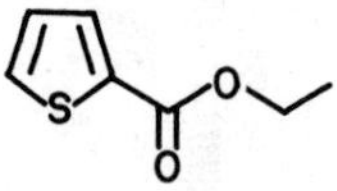

Solvent	Frequency	Ref.
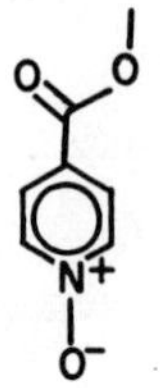 $CHCl_3$	1722	27
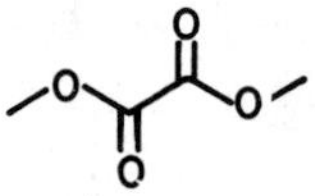 CCl_4	1780 1752	35
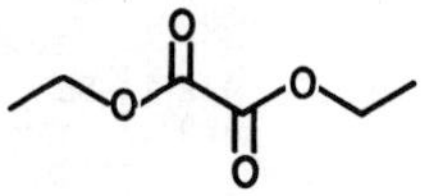 liq.	1774 1754	96
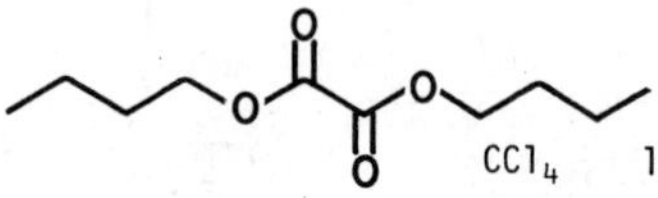 CCl_4	1746	49
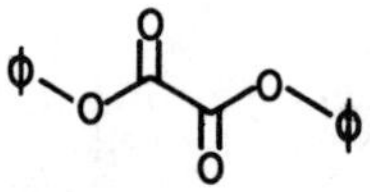 CCl_4	1788 1764	3
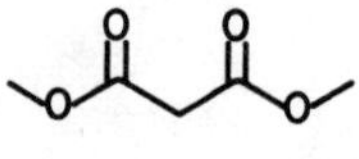 liq.	1768 1750	96
liq.	1760 1742	97
CCl_4	1757 1740	29
$CHCl_3$	1747 1731	97

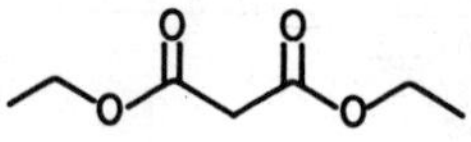

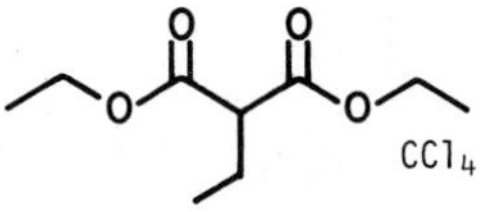

CCl_4 1760 97
1742

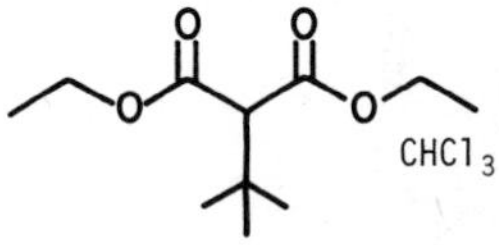

$CHCl_3$ 1761 97
1756

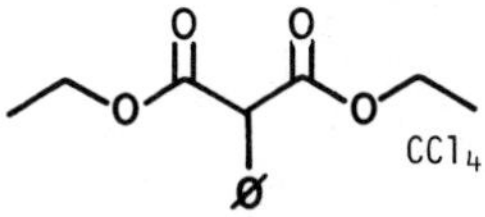

CCl_4 1764 97
1743
1727*

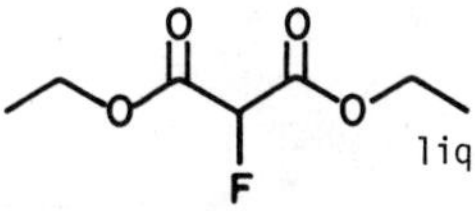

liq. 1780 98
1750

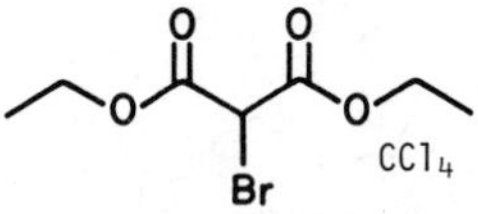

CCl_4 1776 97
1754

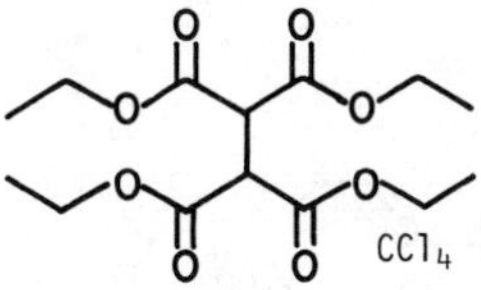

CCl_4 1751 97
1742

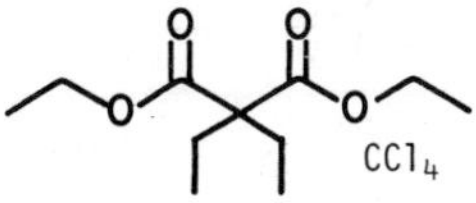

CCl_4 1757* 97
1742

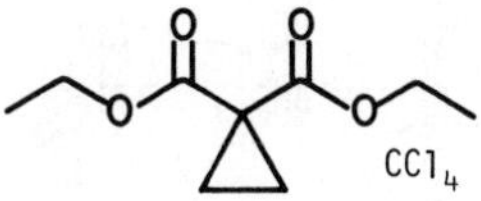

CCl_4 1731 50

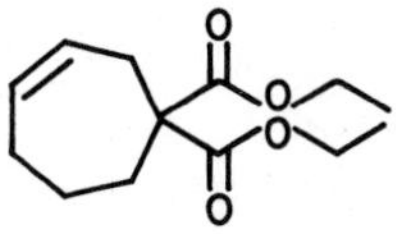

CCl_4 1729 99
CS_2 685 99

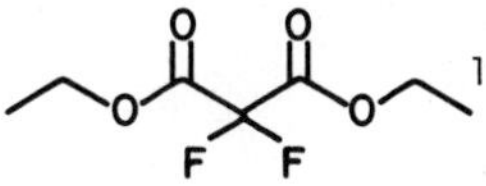

liq. 1750 98

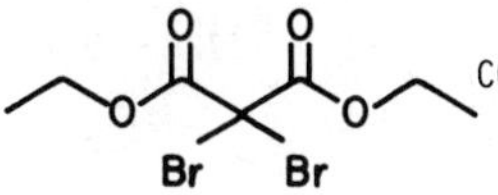

CCl_4 1767 97
1751

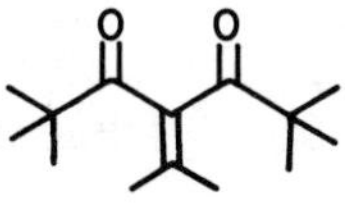

liq. 1720 69
1645w

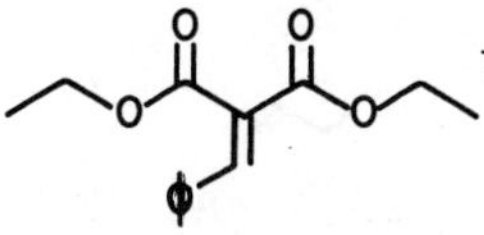

liq. 1747 60
1654

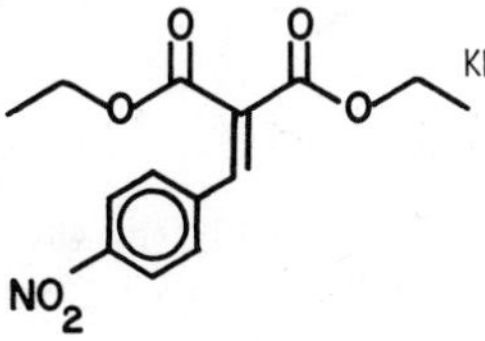

KBr 1740 60
1645
1620

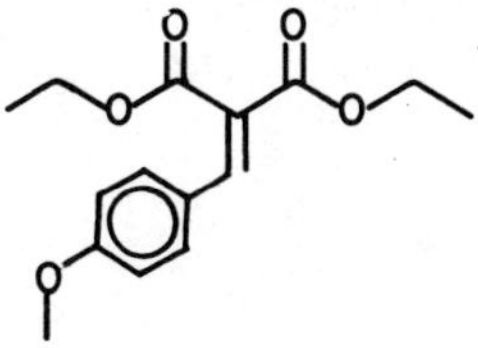

liq. 1740 60
1620

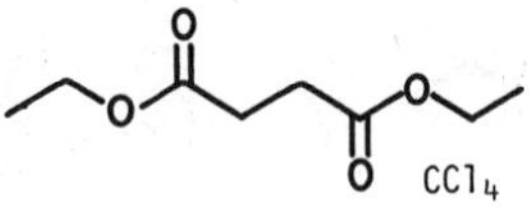

CCl4 1740 29

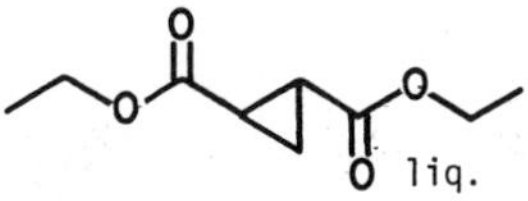

liq. 1730 100

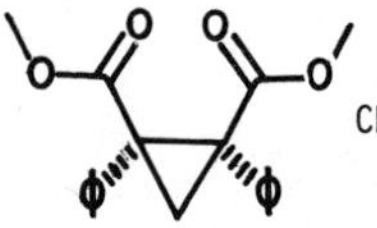

$CHCl_3$ 1725 101

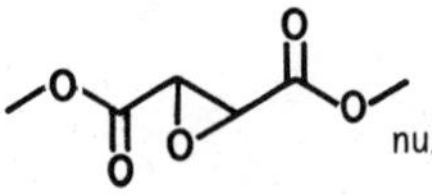

nuj. 1745 94

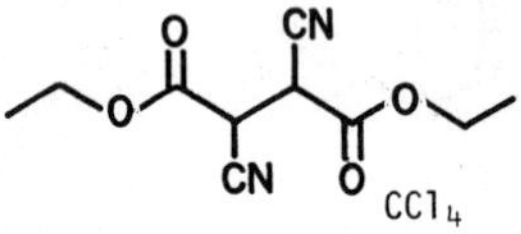

CCl_4 1755 29

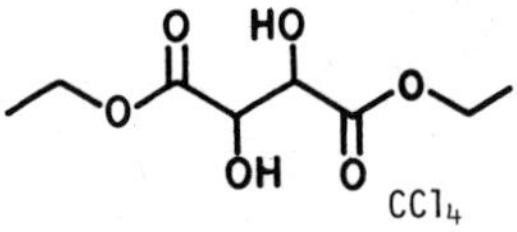

CCl_4 1744b 40

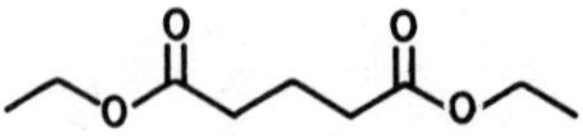

liq. 1749 96

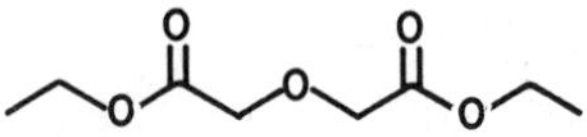

CCl_4 1750 102

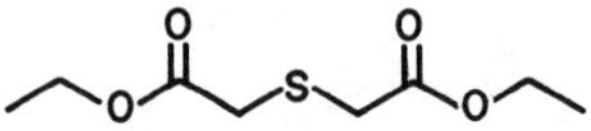

CCl_4 1730 102

CCl_4 1734 29
$CHCl_3$ 1724 29

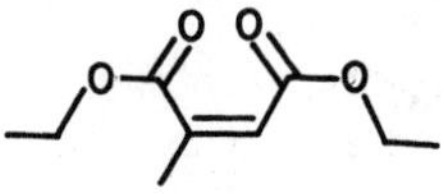

CCl_4 1732 29
$CHCl_3$ 1723 29

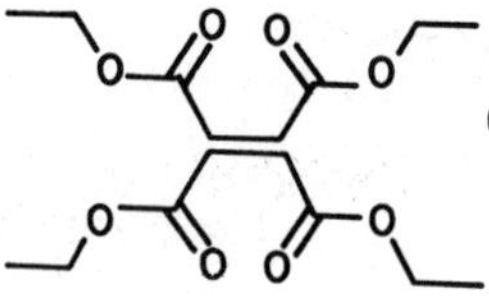

CCl_4 1750 29
1739

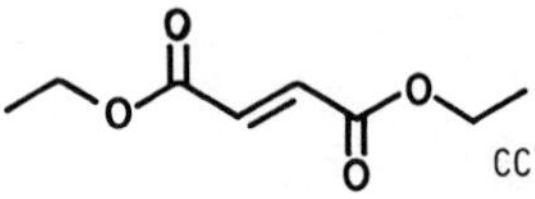

CCl_4 1727 29
$CHCl_3$ 1719 29

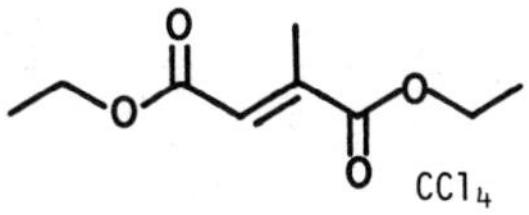

Solvent	cm⁻¹	Ref.
CCl_4	1725	29
$CHCl_3$	1716	29

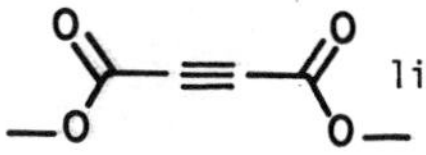

Solvent	cm⁻¹	Ref.
nuj.	1753	106
$CHCl_3$	1763	106

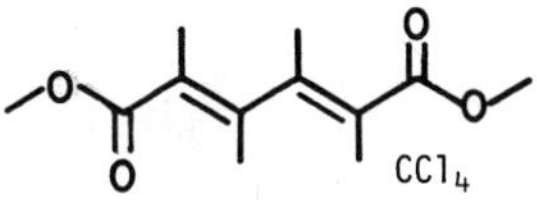

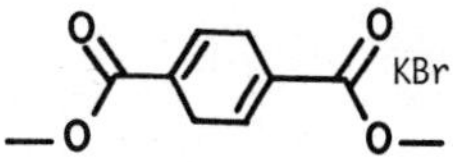

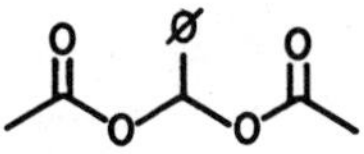

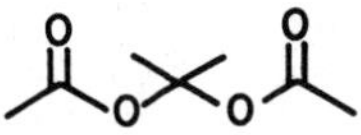

liq. 1722 59

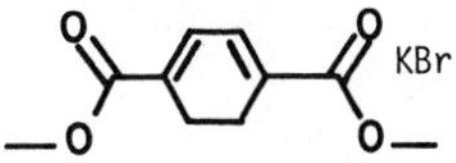

liq. 1761 24

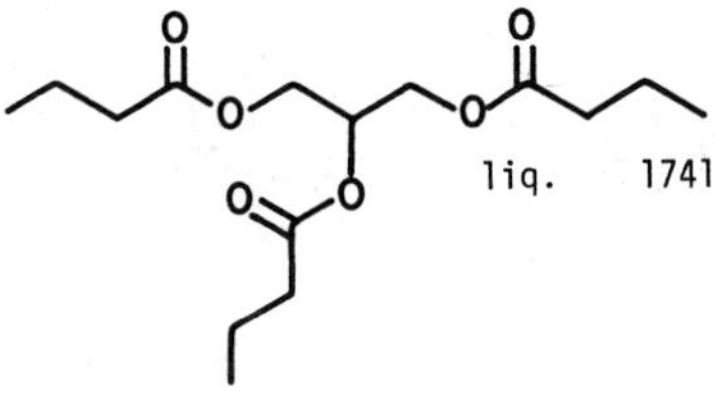

liq. 1741 107

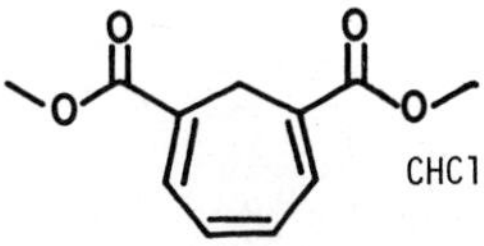

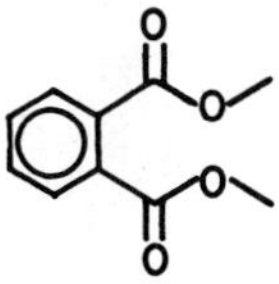

CCl_4 1710 1640 103

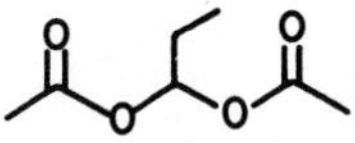

KBr 1710 1645 104

liq. 1733 108

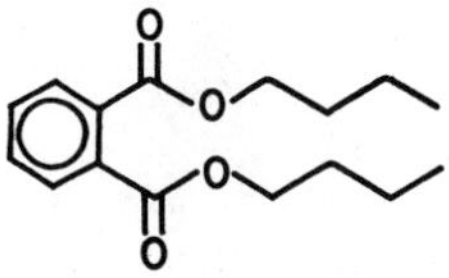

KBr 1705 1645 104

CCl_4 1746 49

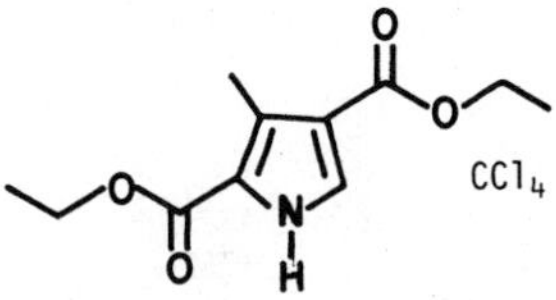

$CHCl_3$ 1710 1617 105

CCl_4 1724 1697 1676 89

Solvent	cm⁻¹	Ref.
liq.	1761	24
nuj.	1761	106

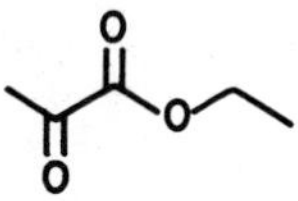

CCl_4 1761* 1735 43

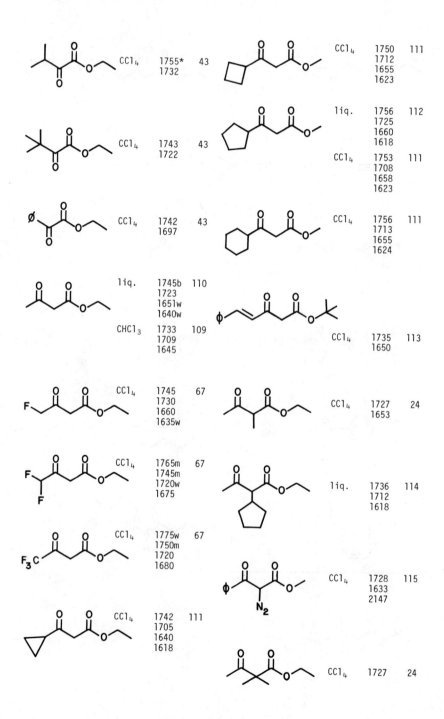

Solvent	Frequencies	Ref.
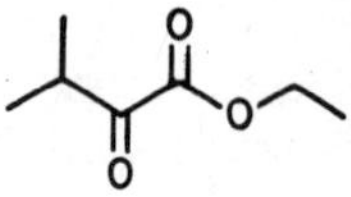 CCl_4	1755* 1732	43
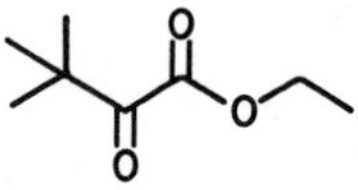 CCl_4	1743 1722	43
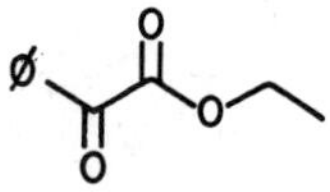 CCl_4	1742 1697	43
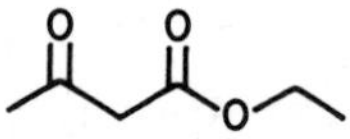 liq.	1745b 1723 1651w 1640w	110
$CHCl_3$	1733 1709 1645	109
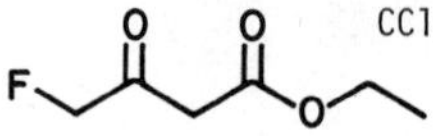 CCl_4	1745 1730 1660 1635w	67
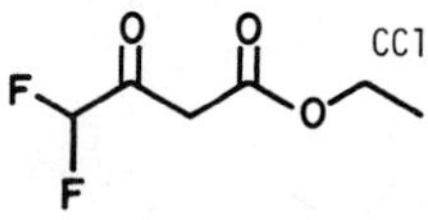 CCl_4	1765m 1745m 1720w 1675	67
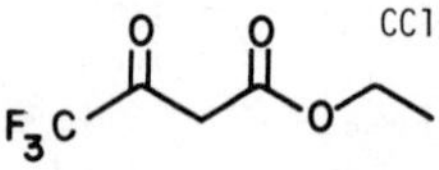 CCl_4	1775w 1750m 1720 1680	67
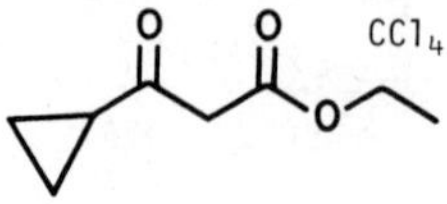 CCl_4	1742 1705 1640 1618	111
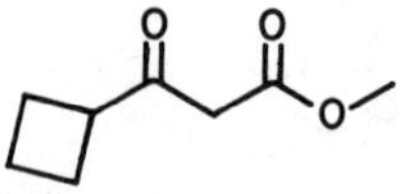 CCl_4	1750 1712 1655 1623	111
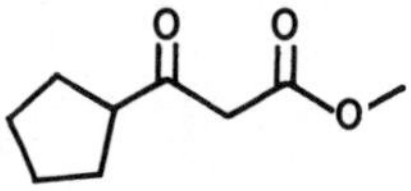 liq.	1756 1725 1660 1618	112
CCl_4	1753 1708 1658 1623	111
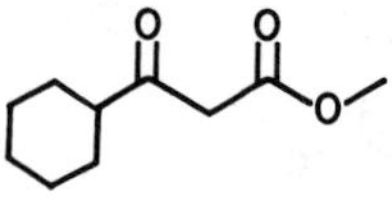 CCl_4	1756 1713 1655 1624	111
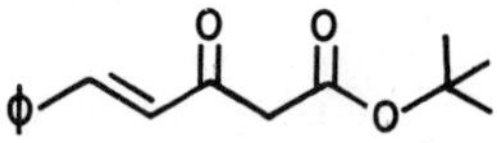 CCl_4	1735 1650	113
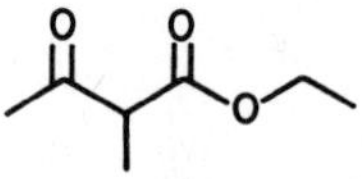 CCl_4	1727 1653	24
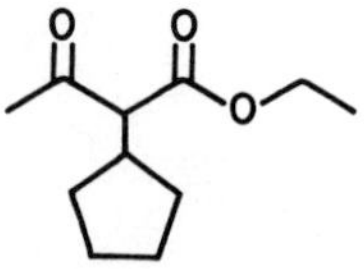 liq.	1736 1712 1618	114
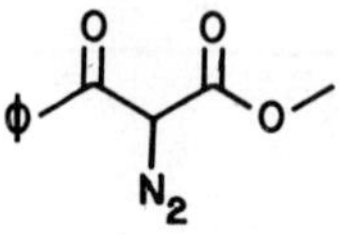 CCl_4	1728 1633 2147	115
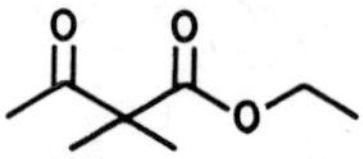 CCl_4	1727	24

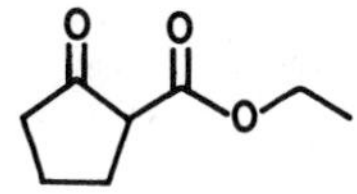

Solvent	Bands	Ref.
liq.	1757, 1730, 1661w, 1622wb	110
$CHCl_3$	1753, 1725, 1661w	110
liq.	1754, 1733	110
$CHCl_3$	1747, 1725	110
liq.	1744w, 1720w, 1656b, 1618m	110
$CHCl_3$	1735w, 1712m, 1657*, 1650, 1617m	110
liq.	1736*, 1721*, 1715	110
$CHCl_3$	1722*, 1710	110
liq.	1741, 1720, 1660, 1624	112
liq.	1739, 1709, 1639, 1587	116
CCl_4	1675	117

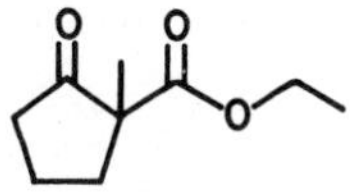

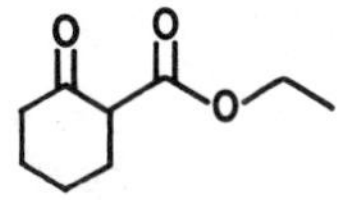

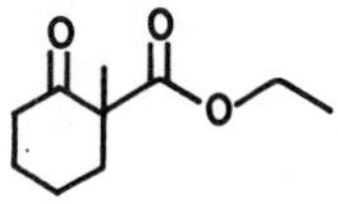

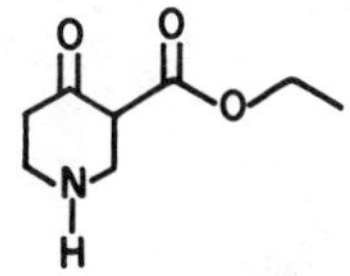

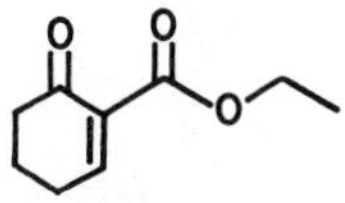

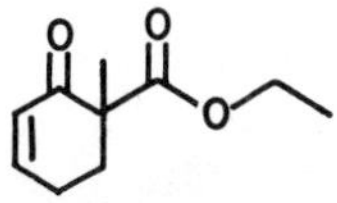

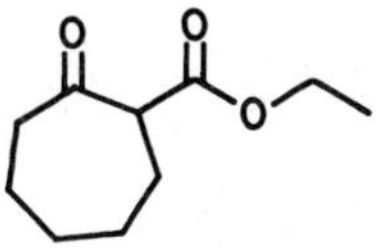

Solvent	Bands	Ref.
liq.	1740, 1708, 1638m, 1614wb	110
liq.	1738, 1710	110
liq.	1745m, 1709m, 1643, 1612mb	110
liq.	1738, 1710	110
CCl_4	1740, 1711	99
CCl_4	1732	24
liq.	1721	118

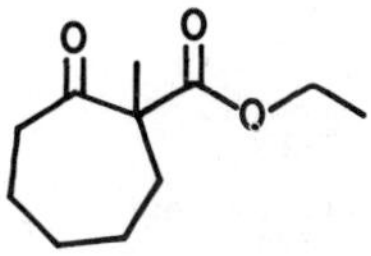

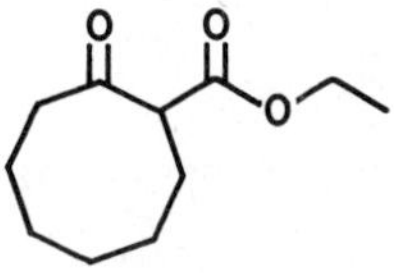

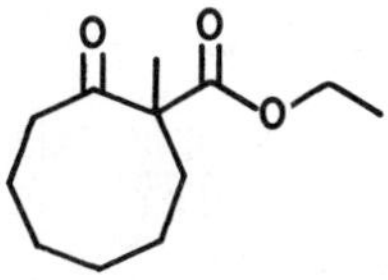

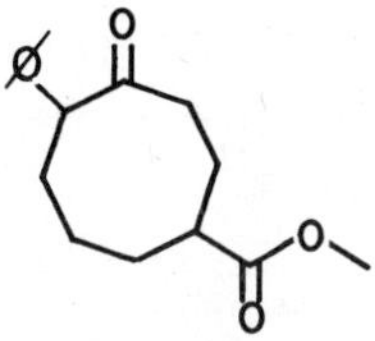

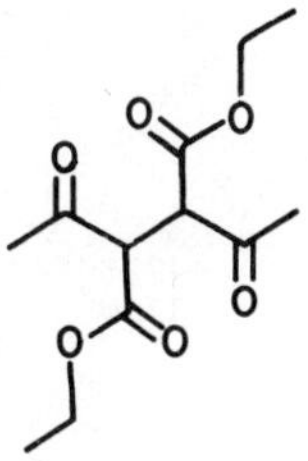

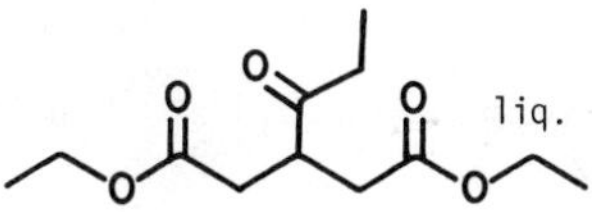

Solvent	ν (cm⁻¹)	Ref.
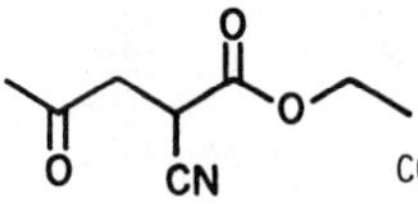 CCl_4	1755 1733 2260	92
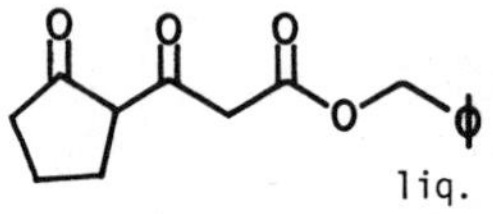 liq.	1745 1710 1660 1625	119
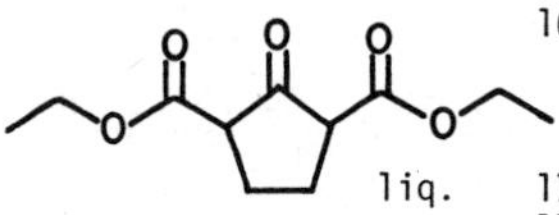 liq.	1763 1736b 1671m 1623	112
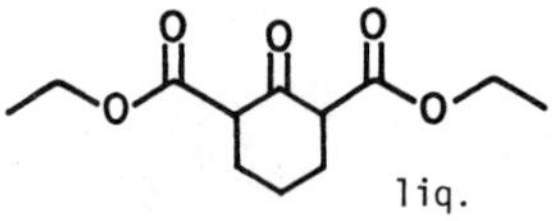 liq.	1738 1660 1621	112
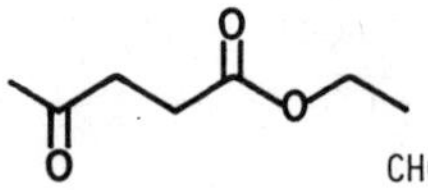 $CHCl_3$	1726 1712	47
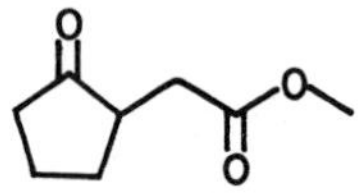 $CHCl_3$	1733	47
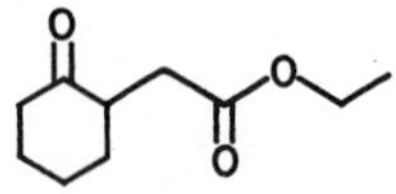	1724 1704	47
 liq.	1733 1695	120
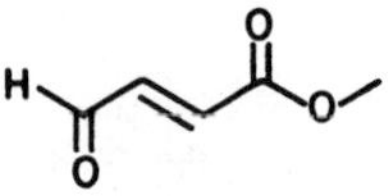 CS_2	1739 1733	120
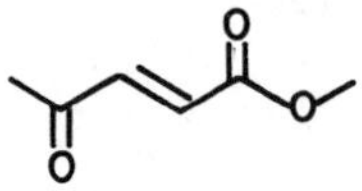 $CHCl_3$	1724 1695	121
CCl_4	1752	122
$CHCl_3$	1747 1733	47
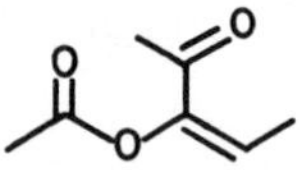 CCl_4	1762 1692 1670	67
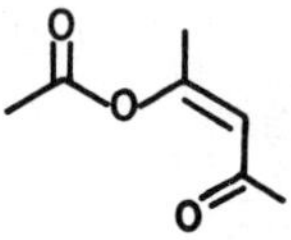 $CHCl_3$	1695 1664 1631	123
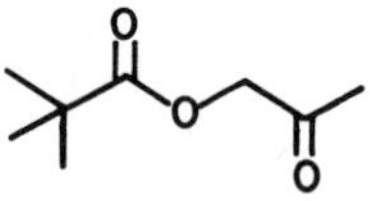 $CHCl_3$	1736 1723	47
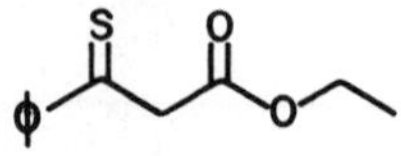 liq.	1739 1695* 1667	124
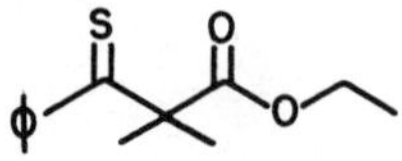 liq.	1727	125

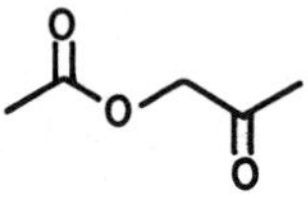

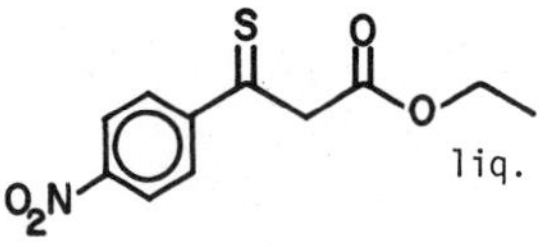

	Solvent	Frequencies	Ref.
	liq.	1739 1681	125
	liq.	1751 1701* 1681	125
	liq.	1757 1575	126
	liq.	1748 1630	127
	CCl_4	1705 1680 1600	128
	nuj.	1700 1670 1600 1580	127
	liq.	1681 1587	129
	liq.	1686 1592	129
	$CHCl_3$	1685 1595	128
	liq.	1658 1610	130
	nuj.	1650 1610	127
	CH_2Cl_2	1639 1608	131
	KBr	1639 1605	131
	CH_2Cl_2	1639 1603	131
	KBr	1647 1610 1592	132

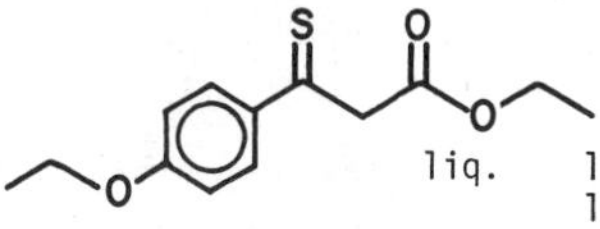

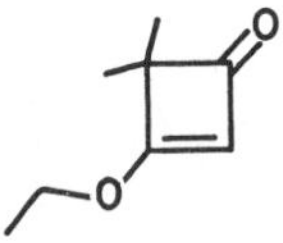

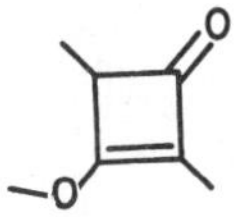

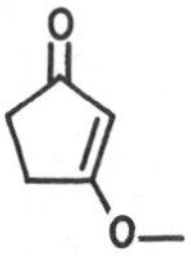

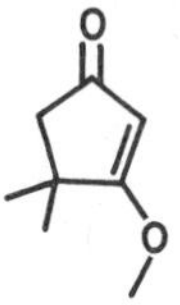

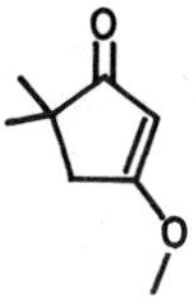

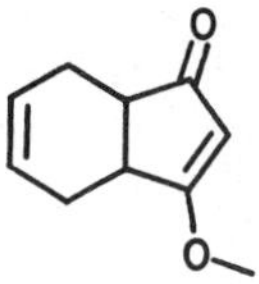

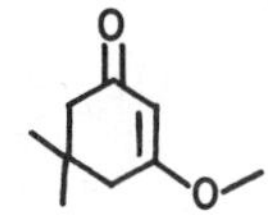

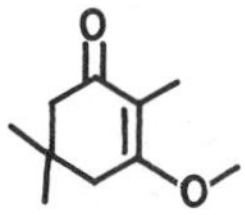

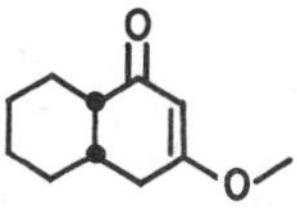

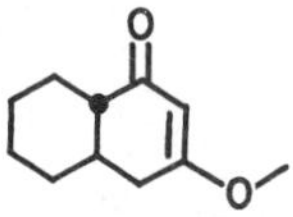

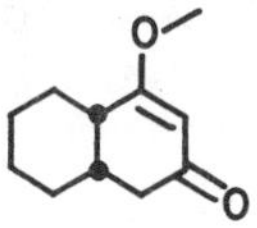

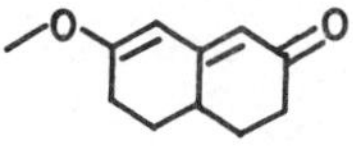

Solvent	Frequency	Ref.
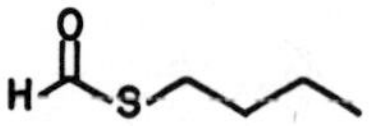 CCl_4	1675	133
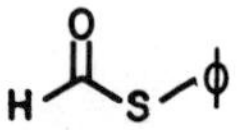 CCl_4	1693	133
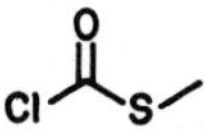 CCl_4	1766	134
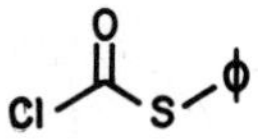 CCl_4	1772b	135
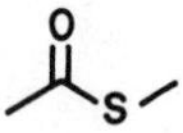 CCl_4	1698	134
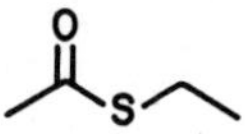 CCl_4	1669	134
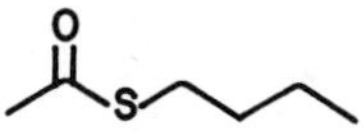 CCl_4	1695	133
CCl_4	1714	134
$CHCl_3$	1704	136
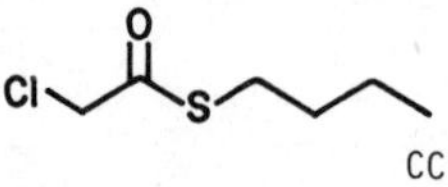 CCl_4	1699 1671	133

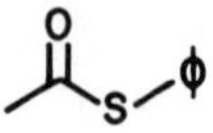

Solvent	Frequency	Ref.
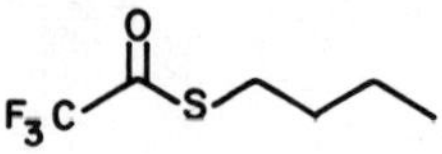 CCl_4	1710	133
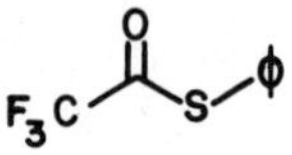 CCl_4	1722	133
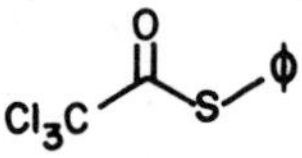 CCl_4	1711	133
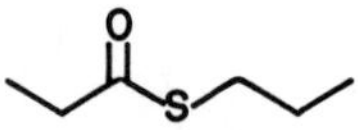 CCl_4	1691	133
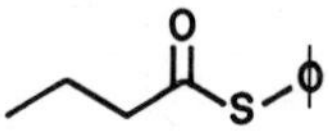 CCl_4	1710	133
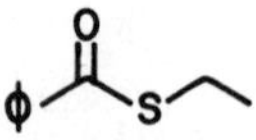 CCl_4	1669	134
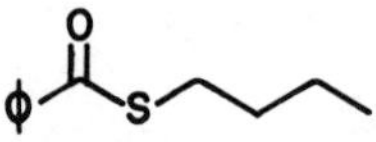 CCl_4	1665	133
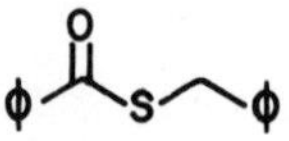 CCl_4	1665	133
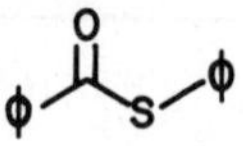 CCl_4	1685	133
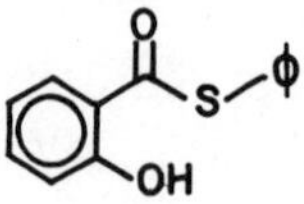 CCl_4	1640	133

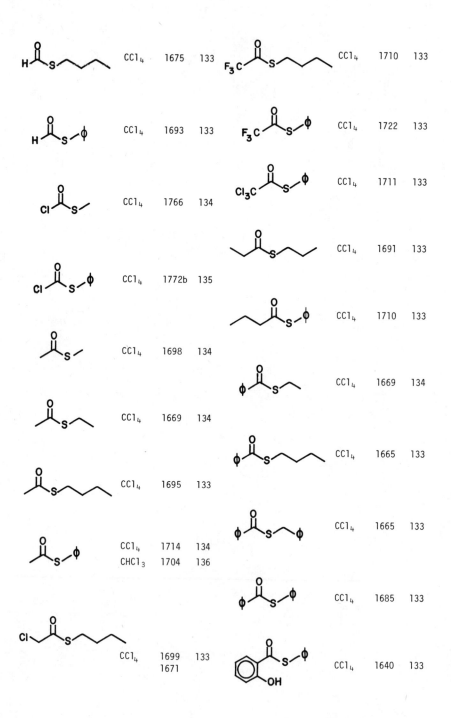

Solvent	ν	Ref.
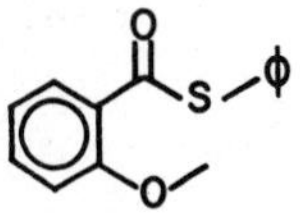 CCl_4	1700 1652	133
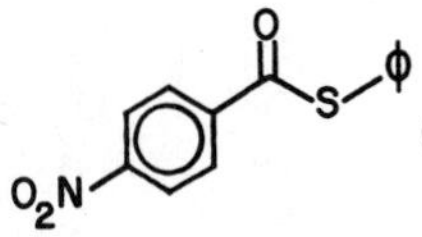 CCl_4	1683	133
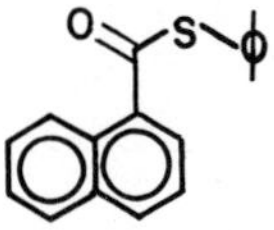 CCl_4	1690	133
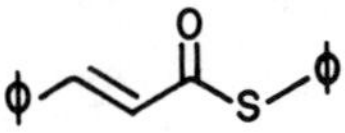 $CHCl_3$	1681	136
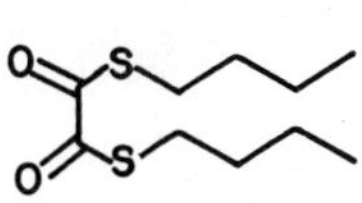 CCl_4	1680	133
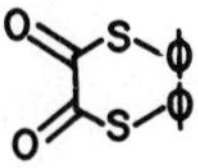 CCl_4	1698	133
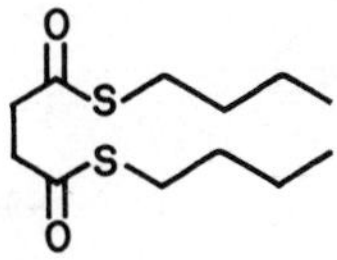 CCl_4	1690	133
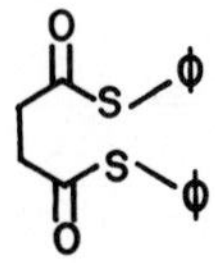 CCl_4	1705	133
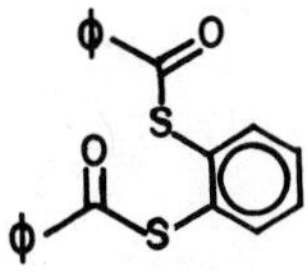 liq.	1670	137
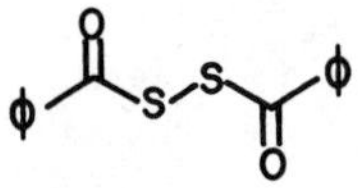 KBr	1705 1690 1600	138
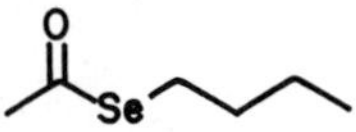 CCl_4	1712	139
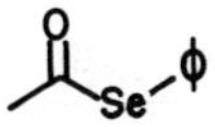 CCl_4	1723	139
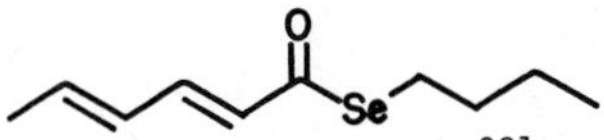 CCl_4	1681	139
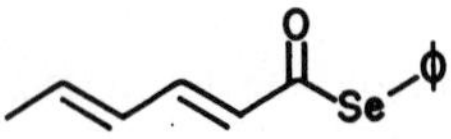 CCl_4	1695	139
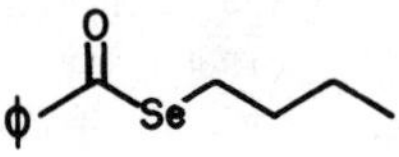 CCl_4	1675	139
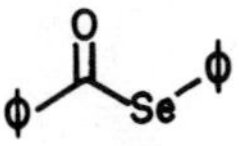 CCl_4	1686	139
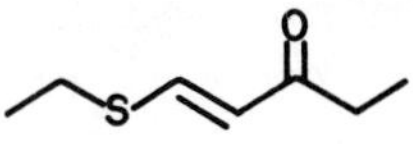 CCl_4	1682 1669	140

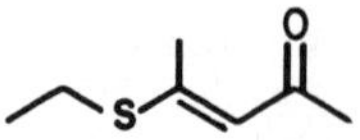

Solvent	ν	Ref.
CCl_4	1643 1540	141
KBr	1652	142
liq.	1655	143
liq.	1665	144

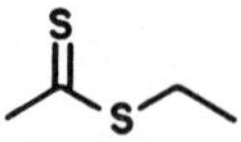

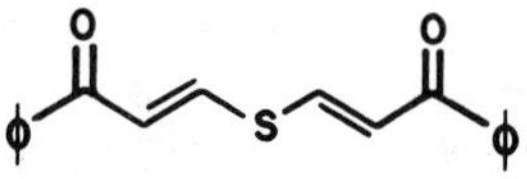

Solvent	ν	Ref.
liq.	1195	145
CCl_4	1197	146
CS_2	1195	146
CCl_4	1197	146
CS_2	1195	146
liq.	1187	145
liq.	1172 1153	145

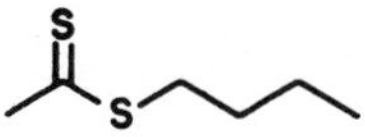

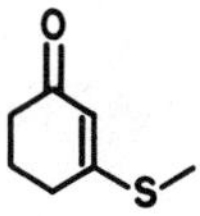

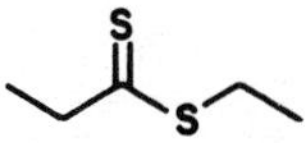

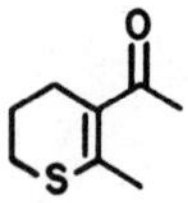

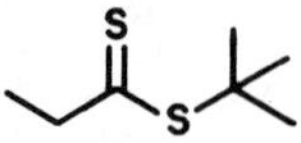

REFERENCES

1. H. W. Thompson and D. A. Jameson, *SCA* (1958) *13*, 236.
2. A. R. Katritzky, J. M. Lagowski and J. A. T. Beard, *SCA* (1960), 964.
3. W. M. Horspool and P. L. Pauson, *JCS* (1965), 5162.
4. H. A. Ory, *SCA* (1960), 1488.
5. J. L. Hales, J. I. Jones and W. Kynaston, *JCS* (1957), 618.
6. W. Lwowski and T. W. Mattingly, Jr., *JACS* (1965), 1947.
7. L. A. Carpino, C. A. Giza and B. A. Carpino, *JACS* (1959), 955.
8. E. J. Hartwell, R. E. Richards and H. W. Thompson, *JCS* (1948), 1436.
9. M.-L. Josien and R. Calas, *CR* (1955) *240*, 1641.
10. E. J. Bourne, M. Stacey, J. C. Tatlow and R. Worrall, *JCS* (1958), 3268.

11. N. J. Turro and W. B. Hammond, *T* (1968), 6017.
12. S. B. Kulkarni and Sukh Dev, *T* (1968), 561.
13. H. Kropf and R. Lambeck, *A* (1967) *700*, 1.
14. D. Seyferth and J. Yick-Pui Mui, *JACS* (1966), 4672.
15. H. O. House and V. Kramar, *JOC* (1963), 3362.
16. K. Nakanishi, *Zikken Kagaku Koza*, Maruzen, Tokyo (1959) *1*, 291.
17. F. Merényi and M. Nilsson, *Acta Chem Scand* (1967), 1755.
18. J. A. Landgrebe and L. W. Becker, *JACS* (1968), 395.
19. H. Nozaki, Z. Yamaguti, T. Okada, R. Noyori and M. Kawanisi, *T* (1967), 3993.
20. H. H. Wasserman and P. S. Wharton, *JACS* (1960), 661.
21. L. Goodman, A. Benitez, C. D. Anderson and B. R. Baker, *JACS* (1958), 6582.
22. H. O. House and H. W. Thompson, *JOC* (1961), 3729.
23. H. Lee and J. K. Wilmshurst, *JCS* (1965), 3590.
24. R. S. Rasmussen and R. R. Brattain, *JACS* (1949), 1073.
25. H. H. Freedman, *JACS* (1960), 2454.
26. L. Gutjahr, *SCA* (1960), 1209.
27. A. R. Katritzky, A. M. Monro, J. A. T. Beard, D. P. Dearnaley and N. J. Earl, *JCS* (1958), 2182.
28. M. F. Hawthorne, W. D. Emmons and K. S. McCallum, *JACS* (1958), 6393.
29. D. G. I. Felton and S. F. D. Orr, *JCS* (1955), 2170.
30. T. L. Brown, *SCA* (1962), 1615.
31. E. T. McBee and D. L. Christman, *JACS* (1955), 755.
32. P. Yates, B. L. Shapiro, N. Yoda and J. Fugger, *JACS* (1957), 5756.
33. N. Kornblum, H. E. Ungnade and R. A. Smiley, *JOC* (1956), 377.
34. T. L. Brown, *JACS* (1958), 3513.
35. M. Ōki and M. Hirota, *BCS Japan* (1961), 374.
36. A. Schönberg and K. Praefcke, *B* (1966), 2371.
37. A. Sami, A. S. Shawali and S. S. Biechler, *JACS* (1967), 3020.
38. L. J. Bellamy and R. J. Pace, *SCA* (1963), 1831.
39. J. Meinwald, J. W. Wheeler, A. A. Nimetz and J. S. Liu, *JOC* (1965), 1038.
40. N. Mori, Y. Tsuzuki and H. Tsubomura, *NKZ* (1956), 459.
41. L. W. Kissinger and H. E. Ungnade, *JOC* (1958), 1517.
42. R. Filler, *JACS* (1954), 1376.
43. G. Oehme and A. Schellenberger, *B* (1968), 1499.
44. G. Rappaport, M. Hauptschein, J. F. O'Brien and R. Filler, *JACS* (1953), 2695.
45. A. Kirrmann and F. Druesne, *BSC France* (1964), 1098.

46. M. Masui, H. Sayo and K. Kishi, *T* (1965), 2831.
47. Y. Mazur and F. Sondheimer, *Experientia* (1960), 181.
48. DMS 1245.
49. R. R. Hampton and J. E. Newell, *Anal Chem* (1949), 914.
50. G. W. Cannon, A. A. Santilli and P. Shenian, *JACS* (1959), 1660.
51. O. H. Wheeler, O. Chao, and J. R. Sánchez-Caldas, *JOC* (1961), 2505.
52. D. B. Denney and J. W. Hanifin, *JOC* (1964), 732.
53. F. P. B. van der Maeden, H. Steinberg and Th.J. de Boer, *TL* (1967), 4521.
54. H. O. House and J. W. Blaker, *JACS* (1958), 6389.
55. H. Musso, K. Naumann and K. Grychtol, *B* (1967), 3614.
56. H. D. Scharf and F. Korte, *B* (1964), 2425.
57. F. W. Baker and L. M. Stock, *JOC* (1967), 3344.
58. M. F. Shostakovskii, L. I. Komarova, A.Kh. Filippova and G. V. Ratovskii, *Izv Akad Nauk USSR, Ser Khim* (1967), 2526.
59. J. L. H. Allan, G. D. Meakins and M. C. Whiting, *JCS* (1955), 1874.
60. J. Zabicky, *JCS* (1961), 683.
61. R. A. Jones and J. A. Lindner, *Aust J C* (1965), 875.
62. A. R. Katritzky and A. J. Boulton, *JCS* (1959), 3500.
63. M. Yamaguchi, Y. Hayashi and S. Matsukawa, *Benseki Kagaku* (1961) *10*, 1106.
64. J. C. Sheehan and J. H. Beeson, *JACS* (1967), 362.
65. D. E. Jones, R. O. Morris, C. A. Vernon and R. F. M. White, *JCS* (1960), 2349.
66. C. H. Eugster, L. Leichner and E. Jenny, *Helv* (1963), 543.
67. R. Filler and S. M. Naqvi, *T* (1963) *19*, 879.
68. G. Aksnes, *Acta Chem Scand* (1961), 692.
69. W. Haefliger and T. Petrzilka, *Helv* (1966), 1937.
70. J. W. Wilson and V. S. Stubblefield, *JACS* (1968), 3423.
71. F. Korte and D. Scharf, *B* (1962), 443.
72. R. A. Finnegan and R. S. McNees, *JOC* (1964), 3234.
73. J. Klein, *T* (1964), 465.
74. F. E. Bader, *Helv* (1953), 215.
75. J. F. Grove and H. A. Willis, *JCS* (1951), 877.
76. R. A. Moss and F. C. Shulman, *T* (1968), 2881.
77. C. J. W. Brooks, C. Eglinton and J. F. Morman, *JCS* (1961), 661.
78. E. H. White and W. J. Considine, *JACS* (1958), 626.
79. M. Gorodetsky and Y. Mazur, *T* (1966), 3607.
80. H. O. House and D. J. Reif, *JACS* (1955), 6525.
81. A. N. Hambly and J. Bonnyman, *Aust J C* (1958), 529.

82. B. Witkop, *JACS* (1956), 2873.
83. C. J. W. Brooks, G. Eglinton and J. F. Morman, *JCS* (1961), 106.
84. M. Yamaguchi, *NKZ* (1959), 155.
85. J. L. Mateos, R. Cetina, E. Olivera and S. Meza, *JOC* (1961). 2494.
86. H. U. Brechbühler, *Thesis ETH* (1963).
87. M. St.C. Flett, *SCA* (1962), 1537.
88. M. K. A. Khan and K. J. Morgan, *T* (1965). 2197.
89. R. Grigg, *JCS* (1965), 5149.
90. U. Eisner and R. L. Erskine, *JCS* (1958), 971.
91. F. Millich and R. I. Becker, *JOC* (1958), 1096
92. F. Korte and K. Trautner, *B* (1962), 307.
93. A. R. Katritzky and J. M. Lagowski, *JCS* (1959), 657.
94. I. J. Cantlon, W. Cocker and T. B. H. McMurry, *T* (1961) *15*, 46.
95. J. J. Peron, P. Saumagne and J. M. Lebas, *SCA* (1970), 1651.
96. P. J. Corish and W. H. T. Davison, *JCS* (1958), 927.
97. R. A. Abramovitch, *Canad J Chem* (1959), 361.
98. E. D. Bergmann, S. Cohen, and I. Shahak, *JCS* (1959), 3286.
99. G. L. Buchanan, A. C. W. Curran, J. M. McCrae and G. W. McLay, *T* (1967), 4729.
100. A. T. Blomquist and D. T. Longone, *JACS* (1959), 2012.
101. W. Broser and D. Rahn, *B* (1967), 3472.
102. A. Solladié-Cavallo and P. Vièles, *BSC France* (1967), 517.
103. W. Adam, *B* (1964), 1811.
104. J. C. Kauer, R. E. Benson, and G. W. Parshall, *JOC* (1965), 1431.
105. R. Darms, T. Threlfall, M. Pesaro, and A. Eschenmoser, *Helv* (1963), 2893.
106. R. D. Guthrie, *JCS* (1961), 2525.
107. DMS 575.
108. DMS 1531.
109. L. J. Bellamy and L. Beecher, *JCS* (1954), 4487.
110. S. J. Rhoads, J. C. Gilbert, A. W. Decora, T. R. Garland, R. J. Spangler and M. J. Urbigkit, *T* (1963), 1625.
111. V. C. Petrus, *Thesis University of Montpelier* (1965).
112. N. J. Leonard, H. S. Gutowsky, W. J. Middleton and E. M. Peterson, *JACS* (1952), 4070.
113. J.-E. Dubois, F. Hennequin, and M. Durand, *BSC France* (1963), 791.
114. M. I. Kabachnik, S. T. Ioffe, E. M. Popov and K. V. Vatsuro, *T* (1961) *12*, 76.
115. J. H. Looker and D. N. Thatcher, *JOC* (1957), 1233.
116. J. E. Brenner, *JOC* (1961), 22.
117. S. Inayama, *Chem Pharm Bull* (1956), 198.

118. N. B. Mehta and W. E. McEwen, *JACS* (1953), 240.
119. A. W. Allan and R. P. A. Sneeden, *T* (1962) *18*, 821.
120. S. H. Schroeter, R. Appel, R. Brammer and G. O. Schenck, *A* (1966) *697*, 42.
121. N. L. Wender and H. L. Slates, *JOC* (1967), 849.
122. J. P. Freeman, *JACS* (1958), 5954.
123. R. S. Rasmussen, D. D. Tunnicliff and R. R. Brattain, *JACS* (1949), 1068.
124. Z. Reyeš and R. M. Silverstein, *JACS* (1958), 6367.
125. Z. Reyes and R. M. Silverstein, *JACS* (1958), 6373.
126. R. H. Hasek and J. C. Martin, *JOC* (1962), 3743.
127. A. Corbella, G. Jommi, G. Ricca and G. Russo, *Gazz Chim Ital* (1965), 948.
128. H. O. House and G. H. Rasmusson, *JOC* (1963), 27.
129. J. M. Landesberg and D. Kellner, *JOC* (1968), 3374.
130. T. A. Spencer, A. L. Hall and C. F. Von Reyn, *JOC* (1968), 3369.
131. H. Mühle and Ch. Tamm, *Helv* (1962), 1475.
132. J. A. Marshall and N. H. Andersen, *JOC* (1965), 1292.
133. R. A. Nyquist and W. J. Potts, *SCA* (1959) *15*, 514.
134. A. W. Baker and G. H. Harris, *JACS* (1960), 1923.
135. R. A. Nyquist and W. J. Potts, *SCA* (1961), 679.
136. M. Yamaguchi, *NKZ* (1957), 1236.
137. J. C. Sheehan and G. F. Holland, *JACS* (1956), 5631.
138. E. H. Hoffmeister and D. S. Tarbell, *T* (1965), 35 and 2857.
139. M. Renson and G. Draguet, *BSC Belge* (1962) *71*, 260.
140. P. D. Bartlett and M. Stiles, *JACS* (1955), 2806.
141. N. J. Leonard and J. A. Adamcik, *JACS* (1959), 595.
142. G. W. Fischer, *B* (1970), 3470.
143. L. Bateman and F. W. Shipley, *JCS* (1955), 1996.
144. G. W. Cannon, A. A. Santilli and P. Shenian, *JACS* (1959), 1660.
145. C. S. Marvel, P. de Radzitzky and J. J. Brader, *JACS* (1955), 5997.
146. L. J. Bellamy and P. E. Rogasch, *JCS* (1960), 2218.

4.11 lactones

Unstrained six-membered ring lactones exhibit the same carbonyl stretching frequencies as their corresponding acyclic counterparts. Substitution on either the carbonyl side or the ether-oxygen side of the chromophore has the same effect as with acyclic esters, and the comments made concerning esters (Chapter 4.10, p. 332) can be applied directly to lactones.

In addition to substituent effects, ring strain and its accompanying rehybridization of the carbonyl-carbon causes an increase in the carbonyl bond order (see Chapter 4.2, p. 175) and a consequent increase in the carbonyl stretching frequency.

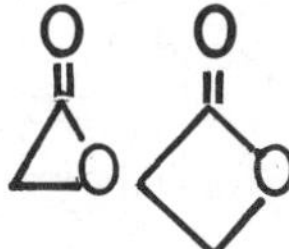

β-Propiolactone (1841 cm^{-1}) and γ-butyrolactone (1783 cm^{-1}) exhibit higher frequencies (compared to δ-valerolactone) than are observed with the corresponding cyclic ketones. This results because not only is there a change in the hybridization of the carbonyl carbon, but also of the ether oxygen. The greater s-character of the oxygen lone-pair orbitals (with increasing ring strain) causes less delocalization of the lone pairs into the carbonyl. This effect coupled with the field effect of the electronegative oxygen results in the observed increased frequencies.

Solvent	Frequency	Ref.
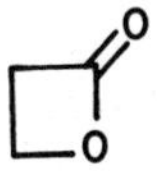 CCl_4	1841	1
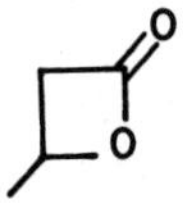 CCl_4	1818	2
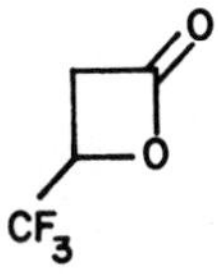 CCl_4	1890	3
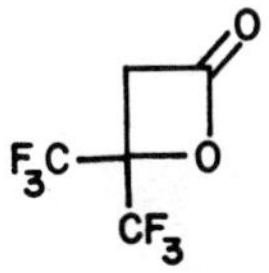 liq.	1949	4
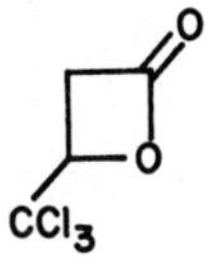 CCl_4	1855	3
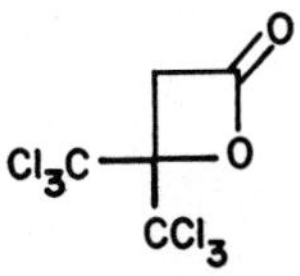 CCl_4	1905	3
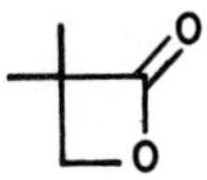 CCl_4	1830	5
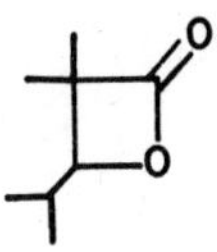 CCl_4	1825	5

Solvent	Frequency	Ref.
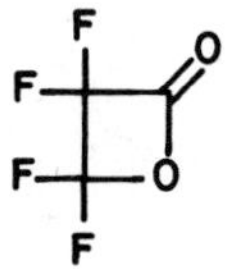 liq.	1873	6
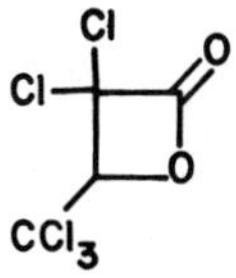 CCl_4	1890	3
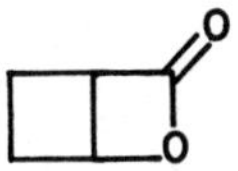 liq.	1818	7
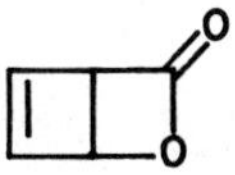 CCl_4	1848 1818 1543	7
CCl_4	1900 1867 1752 1708 1675	8
liq.	1897 1864 1744 1689	8
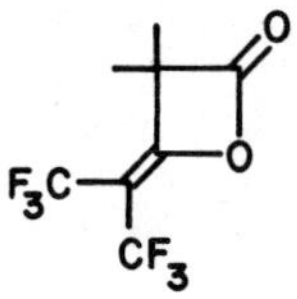 liq.	1905 1693	9
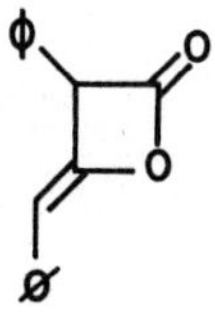 CCl_4	1910 1875 1850w 1710	10

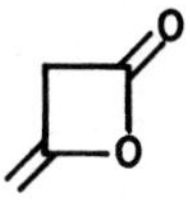

Compound	Solvent	ν (cm⁻¹)	Ref.
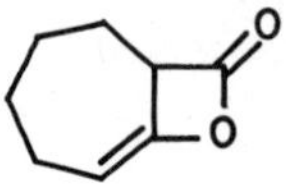 (structure)	CCl_4	1925w 1890 1855 1820 1725	11
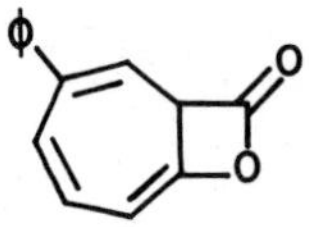 (structure, ϕ)	KBr	1818	12
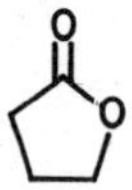 (structure)	liq.	1770	14
	CCl_4	1783	1
	$CHCl_3$	1793* 1774 1770	13
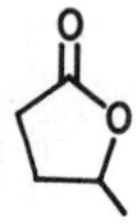 (structure)	CS_2	1790	15
	$CHCl_3$	1775	15
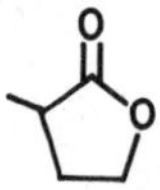 (structure)	liq.	1765	16
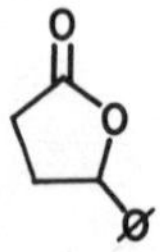 (structure, ϕ)	CCl_4	1795	17
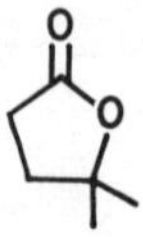 (structure)	CS_2	1785	15
	$CHCl_3$	1775	15

Compound	Solvent	ν (cm⁻¹)	Ref.
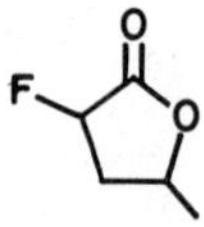 (structure, F)	CCl_4	1790	18
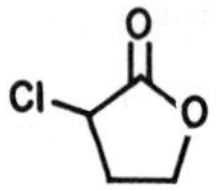 (structure, Cl)	liq.	1790	19
	CCl_4	1810	19
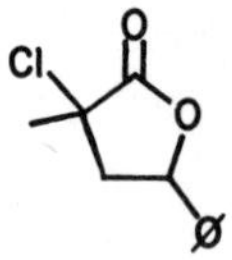 (structure, Cl, ϕ)	$CHCl_3$	1770	20
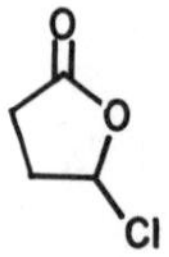 (structure, Cl)	liq.	1810	21
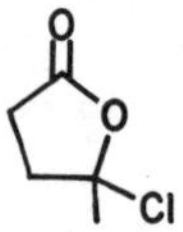 (structure, Cl)	liq.	1806	22
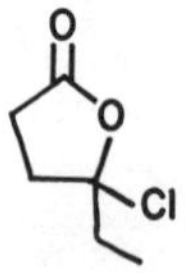 (structure, Cl)	liq.	1800	19
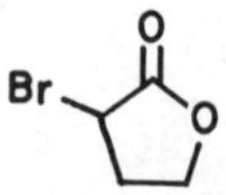 (structure, Br)	liq.	1785	19
	CCl_4	1800	19

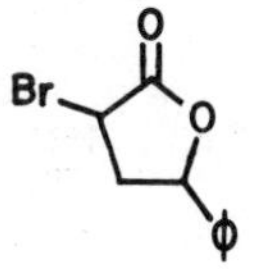

Solvent	ν (cm⁻¹)	Ref.
CCl_4	1793	17
liq.	1800	19
liq.	1773 1718 1656w	23
$CHCl_3$	1700	20
liq.	1783 1739	24
$CHCl_3$	1795 1748	23
liq.	1787	19
CCl_4	1799	25
nuj.	1762	26

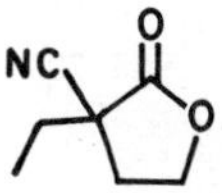

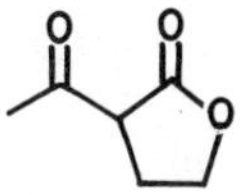

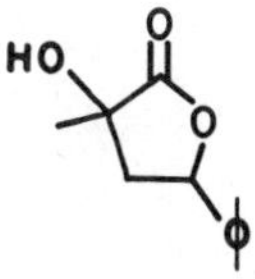

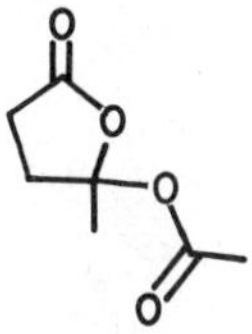

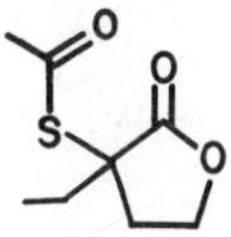

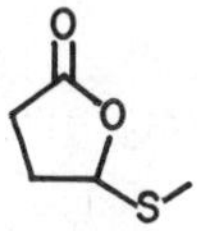

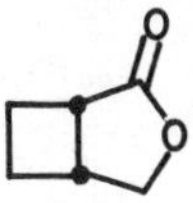

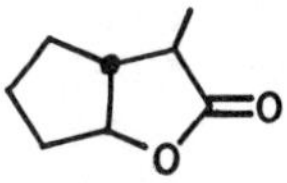

Solvent	ν (cm⁻¹)	Ref.
liq.	1756	27
CCl_4	1783 1775	13
$CHCl_3$	1757	
nuj.	1786	28
nuj.	1789 1774 1751	29
$CHCl_3$	1783	29
CCl_4	1787 1747	30
CCl_4	1768 1653w	31

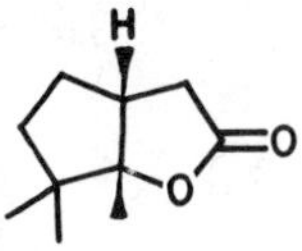

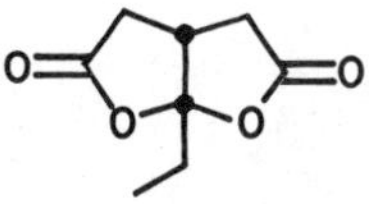

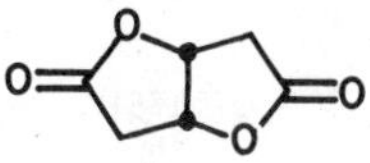

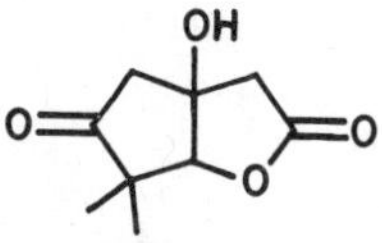

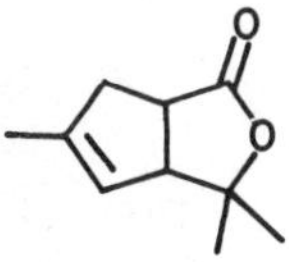

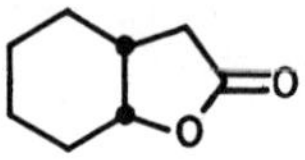

liq. 1775 32
nuj. 1778 26

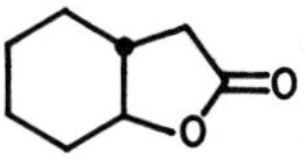

liq. 1785 32

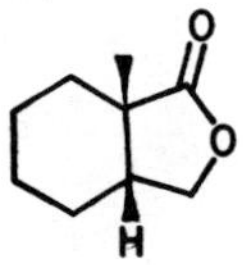

$CHCl_3$ 1770 26

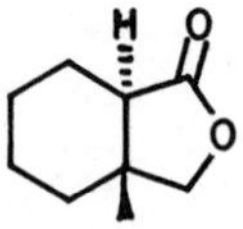

$CHCl_3$ 1775 26

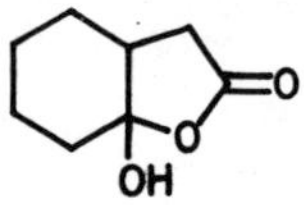

CCl_4 1751, 1661 33

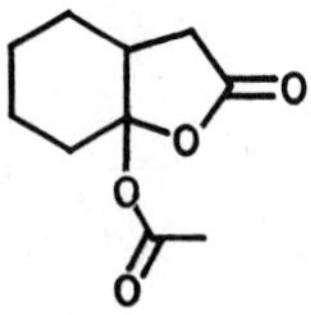

$CHCl_3$ 1761, 1658 33

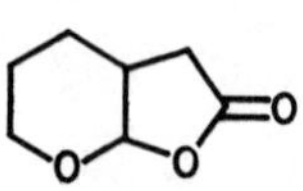

liq. 1795 34

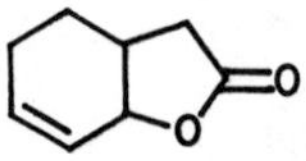

liq. 1765 32

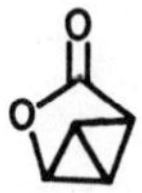

CCl_4 1818, 1795 35

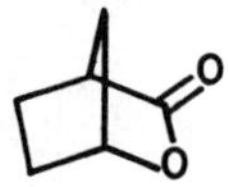

liq. 1770 36

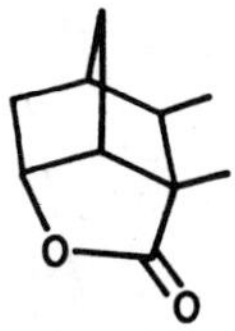

KBr 1769 37

$CHCl_3$ 1800, 1770 38

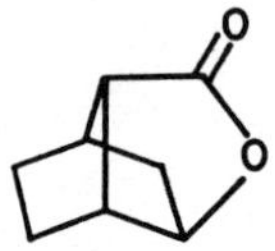

KBr 1775 37

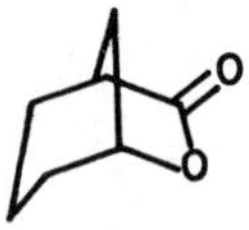

CCl_4 1764 39

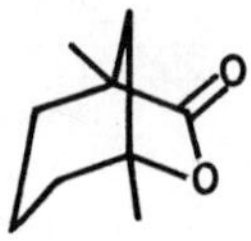

$CHCl_3$ 1764 40

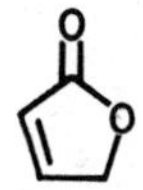

Solvent	cm⁻¹	Ref.
liq.	1775 1745	23
CCl_4	1785	15
$CHCl_3$	1778	15
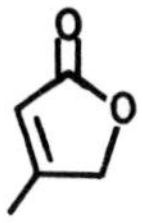 liq.	1750 1640	41
CCl_4	1782	15
$CHCl_3$	1783	15
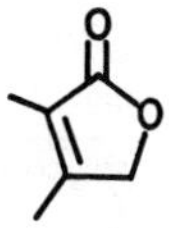 liq.	1745 1686	41
CCl_4	1776	15
$CHCl_3$	1774	15
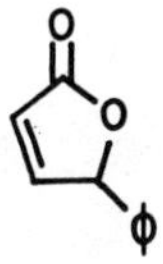 CCl_4	1818	42
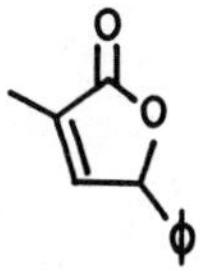 $CHCl_3$	1750 1650	20
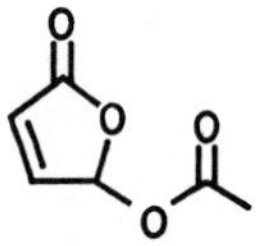 $CHCl_3$	1805 1779	42
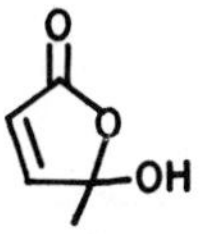 $CHCl_3$	1771 1650w	43
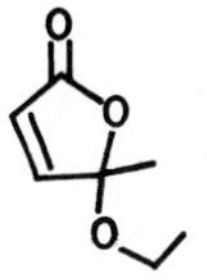 CCl_4	1802 1761	42
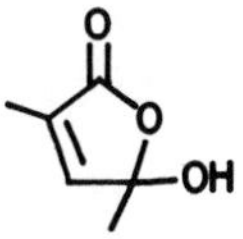 $CHCl_3$	1761	44
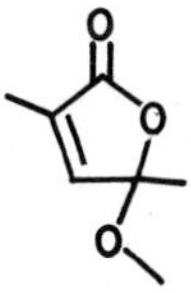 $CHCl_3$	1761	44
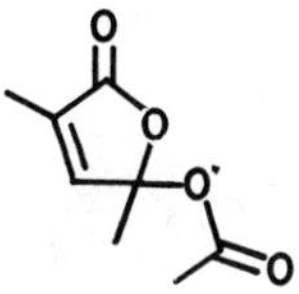 $CHCl_3$	1786 1748	44
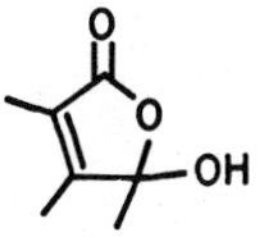 $CHCl_3$	1750 1690w	43
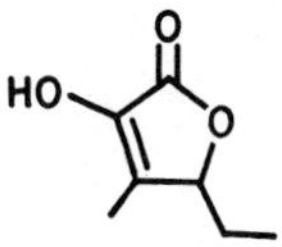 CCl_4	1770 1753	45
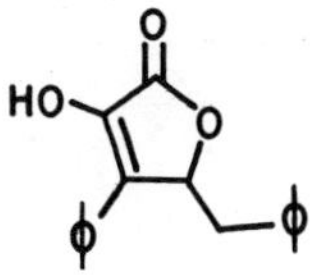 CCl_4	1770 1748*	45

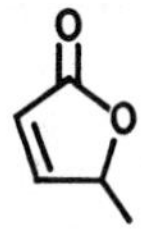

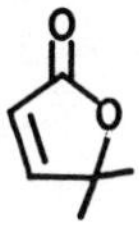

Solvent	Frequency	Ref.
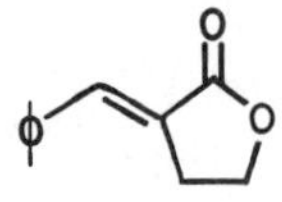 nuj.	1742	14
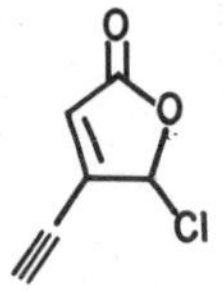 $CHCl_3$	1814	46
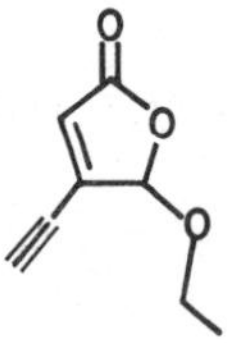 $CHCl_3$	1756	46
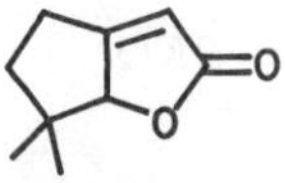 CCl_4	1783 1759	13
$CHCl_3$	1782 1752 1744	13
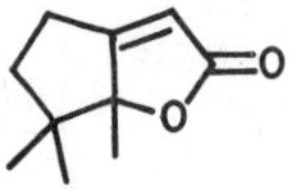 CCl_4	1783 1759	13
$CHCl_3$	1782 1752	13
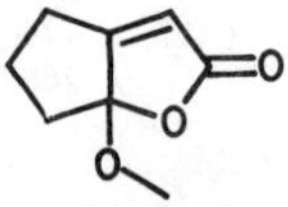 $CHCl_3$	1761 1664	33
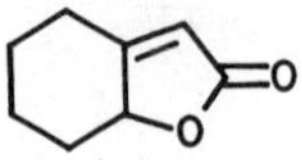 liq.	1748 1645	41
CCl_4	1779	15
$CHCl_3$	1755 1650	15
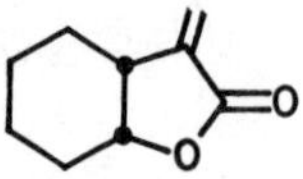 liq.	1767	47
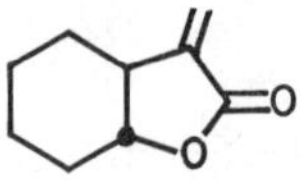 liq.	1779	47
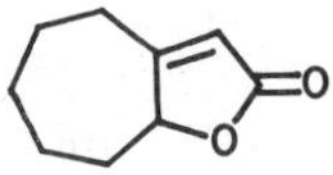 CCl_4	1760 1630	48
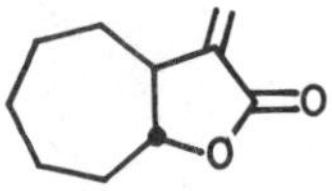 liq.	1764	47
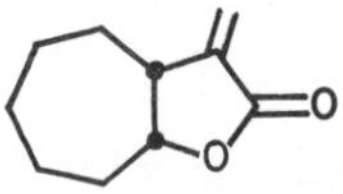 liq.	1761	47
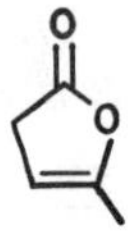 liq.	1799 1681	23
CCl_4	1806 1722	15
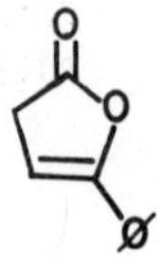 KBr	1800	20
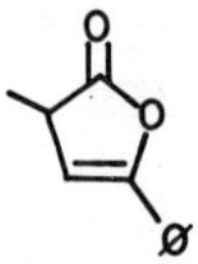 $CHCl_3$	1780 1640	20

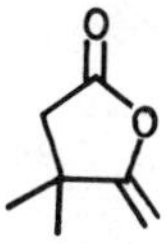

Solvent	ν	Ref.
$CHCl_3$	1800	49
liq.	1792	50
KBr	1754 1661 1600	33
CCl_4	1819 1708w	15
$CHCl_3$	1791 1709w	15
$CHCl_3$	1827 1665	51
CCl_4	1800 1680	48
liq.	1776	52
liq.	1776	52
CCl_4	1769 1749	30

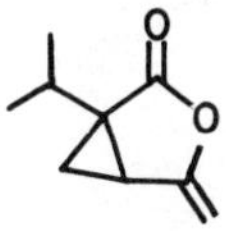

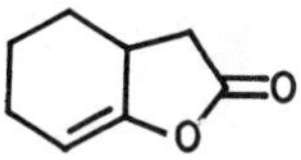

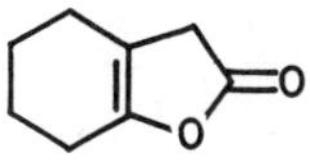

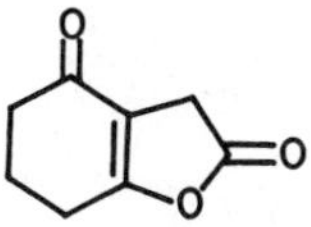

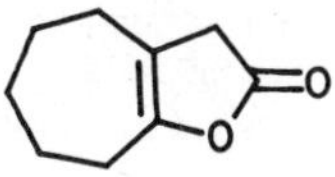

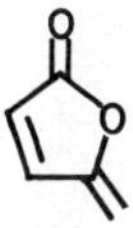

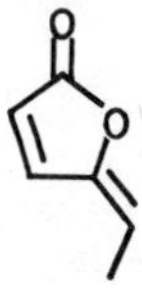

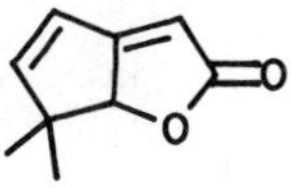

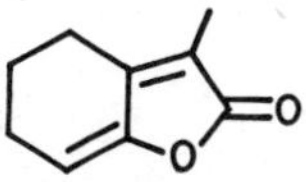

Solvent	ν	Ref.
$CHCl_3$	1780 1755 1670	53
nuj.	1768 1745	53
KBr	1780	54
KBr	1770	54
CCl_4	1761	55
$CHCl_3$	1761	15
KBr	1750	56
CCl_4	1761	55
nuj.	1768	57

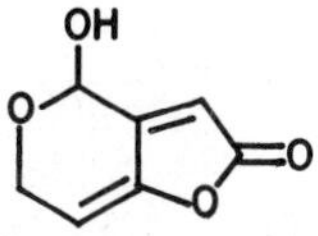

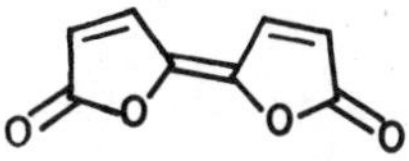

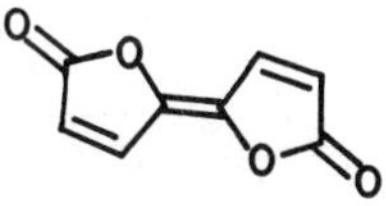

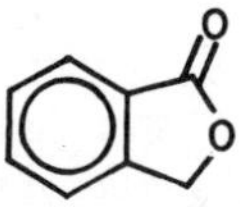

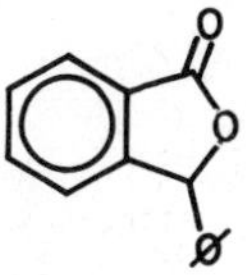

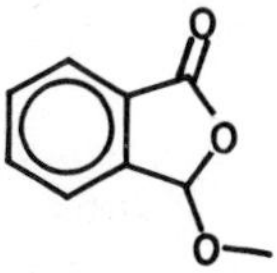

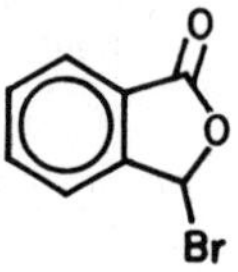

Solvent	Frequency	Ref.
CCl_4	1786	55
$CHCl_3$	1780	57
CCl_4	1751	55
CCl_4	1802	55
CCl_4	1773	55
$CHCl_3$	1805 1618	60
$CHCl_3$	1738	58
$CHCl_3$	1821 1626	60
CCl_4 $CHCl_3$	1749 1738	58 58
CCl_4 $CHCl_3$	1782 1754	58 58
$CHCl_3$	1818 1621	60
nuj.	1780	59
nuj.	1833 1740	61

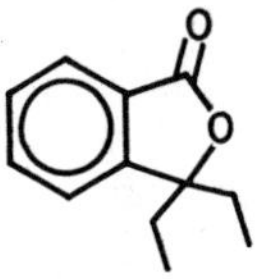

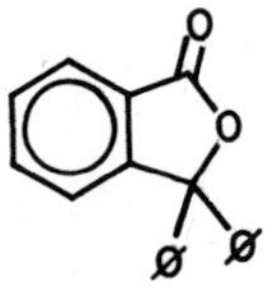

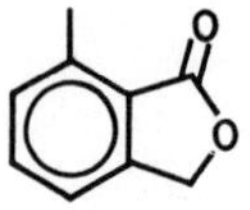

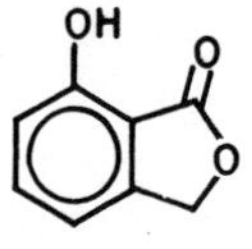

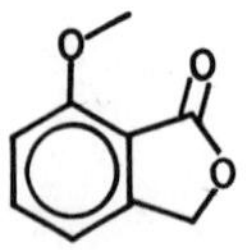

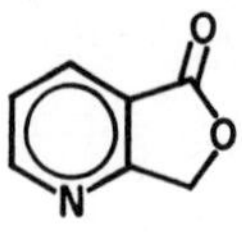

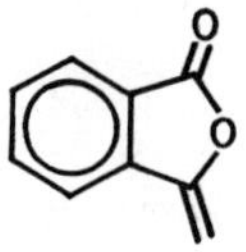

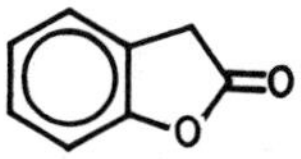

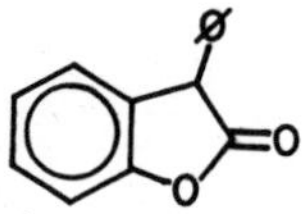

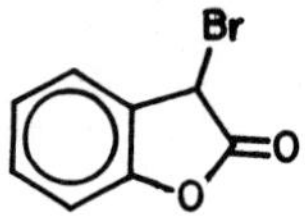

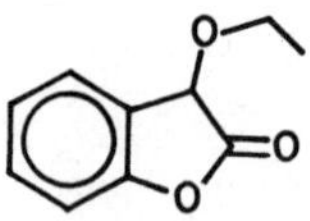

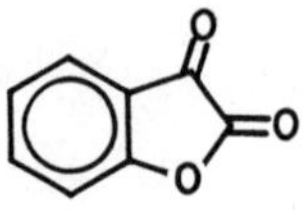

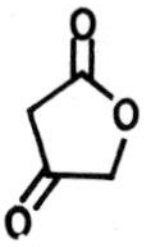

Solvent	Frequency	Ref.
nuj.	1690	62
	1636m	
$CHCl_3$	1820w	62
	1778*	
	1738	
	1632	

Solvent	Frequency	Ref.
nuj.	1720	62
	1634m	
	1575	
$CHCl_3$	1810w	62
	1758m	
	1739	
	1632	

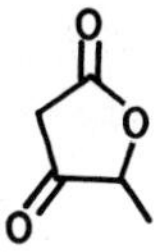

Solvent	Frequency	Ref.
nuj.	1720m	62
	1650	
	1580*	
$CHCl_3$	1744	62
	1675	

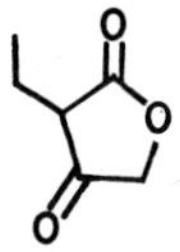

Solvent	Frequency	Ref.
nuj.	1758	62
	1675	
	1600	
$CHCl_3$	1772	62
	1705	
	1675m	
	1610	

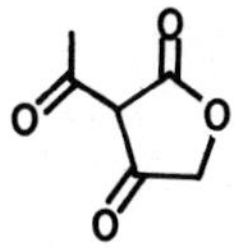

Solvent	Frequency	Ref.
liq.	1730	16
CCl_4	1748	1
CS_2	1750	15
$CHCl_3$	1732	15

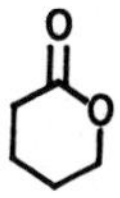

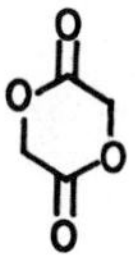

Solvent	Frequency	Ref.
liq.	1745	63

Solvent	Frequency	Ref.
liq.	1730	16

Solvent	Frequency	Ref.
liq.	1733	14

Solvent	Frequency	Ref.
liq.	1725	16

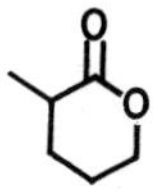

Solvent	Frequency	Ref.
CCl_4	1732	64

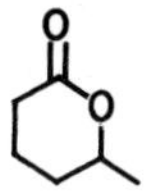

Solvent	Frequency	Ref.
CCl_4	1725	64

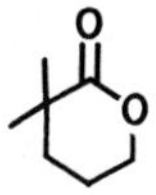

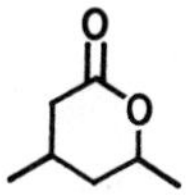

Solvent	Frequency	Ref.
CCl_4	1737	65

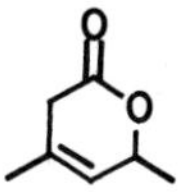

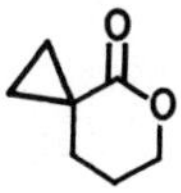

Solvent	Frequency	Ref.
CCl_4	1740	65

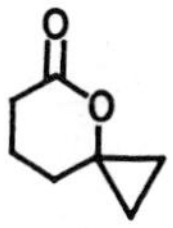

Solvent	Frequency	Ref.
CCl_4	1730	66

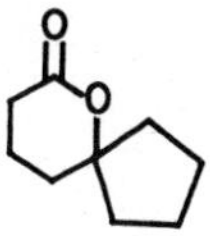

Solvent	Frequency	Ref.
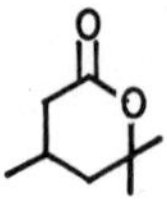 CCl_4	1730	64
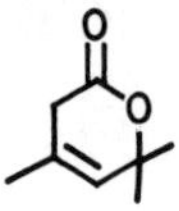 CCl_4	1725	64
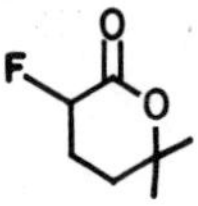 CCl_4	1765	5
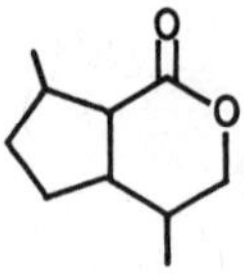 liq.	1727	67
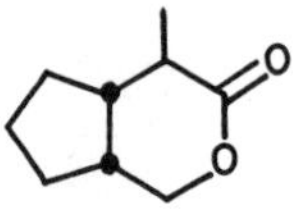 CCl_4	1733	68
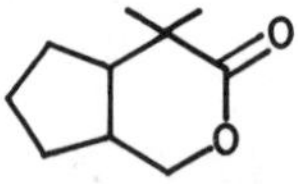 liq.	1736	68
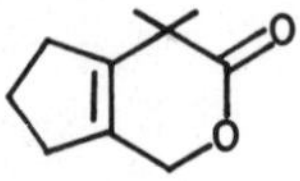 CCl_4	1738	68
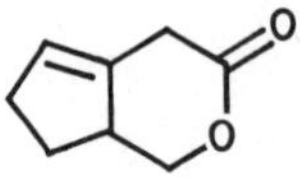 CCl_4	1743 1653	69
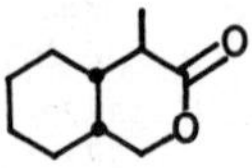 CCl_4	1736	68
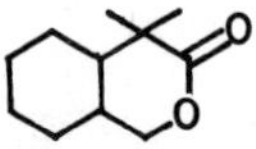 CCl_4	1736	68
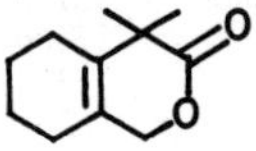 CCl_4	1740 1655	68
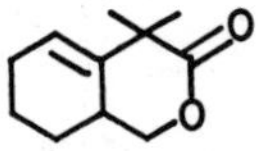 CCl_4	1746 1662	69
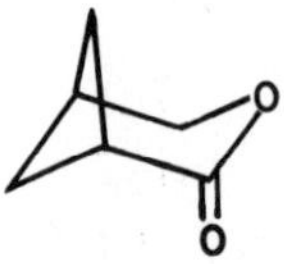 KBr	1738	36
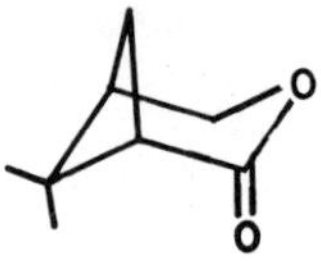 liq.	1742	70
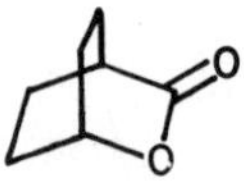 CCl_4	1739	39
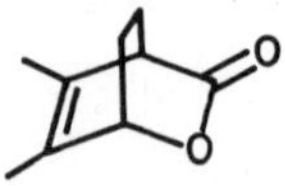 $CHCl_3$	1743	71
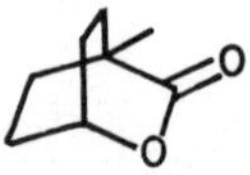 liq.	1715	72

 liq.	1730	73
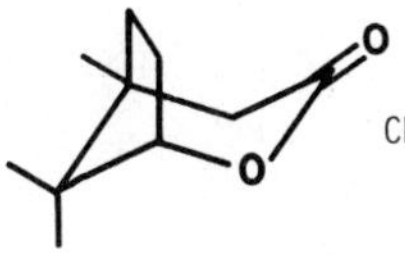 $CHCl_3$	1743	74
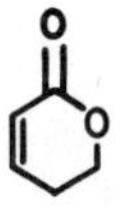 CCl_4	1743	15
$CHCl_3$	1729	15
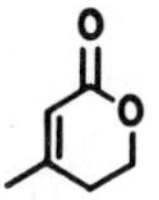 CCl_4	1710	64
$CHCl_3$	1722	75
	1670	
	1640	
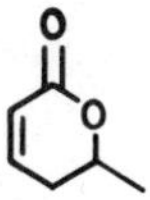 CCl_4	1710	64
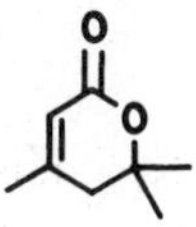 CCl_4	1715	76
	1665	
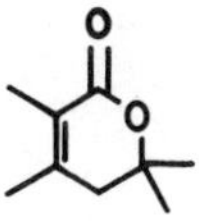 CCl_4	1725	76
	1655	
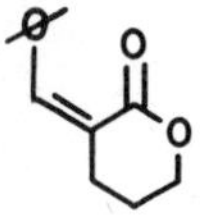 nuj.	1712	14
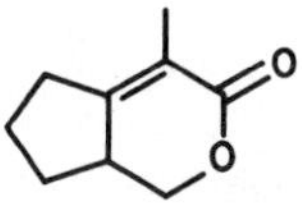 CCl_4	1709	68
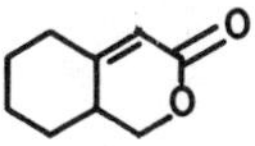 CCl_4	1708	64
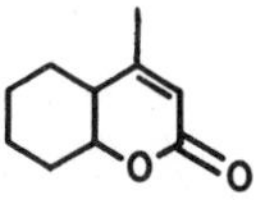 CCl_4	1716	68
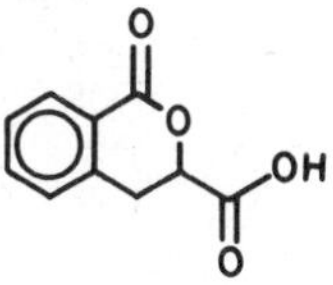 KBr	1754	77
	1709	
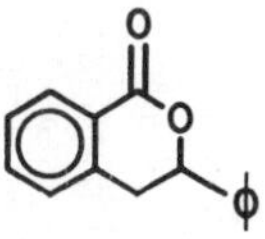 KBr	1730	78
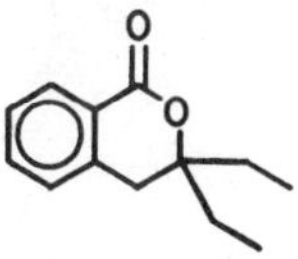 $CHCl_3$	1710	79

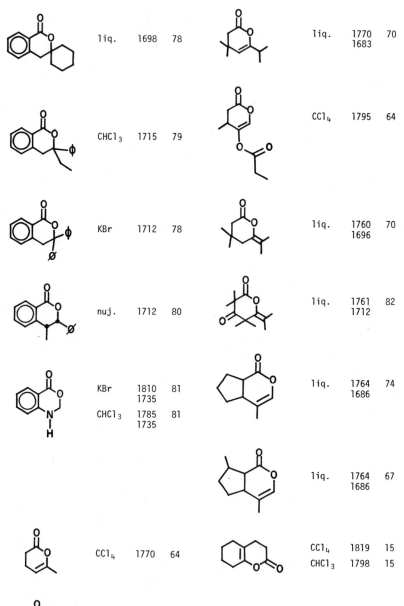

Medium	Frequency	Ref.
liq.	1698	78
$CHCl_3$	1715	79
KBr	1712	78
nuj.	1712	80
KBr	1810 1735	81
$CHCl_3$	1785 1735	81
CCl_4	1770	64
CCl_4	1773	64

Medium	Frequency	Ref.
liq.	1770 1683	70
CCl_4	1795	64
liq.	1760 1696	70
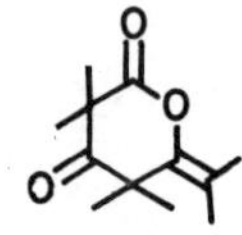 liq.	1761 1712	82
liq.	1764 1686	74
liq.	1764 1686	67
CCl_4	1819	15
$CHCl_3$	1798	15

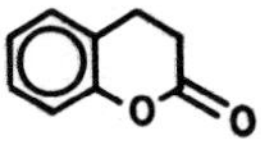

Solvent	cm⁻¹	Ref.
liq.	1780	83
CCl_4	1775	64
liq.	1770	83
KBr	1760	81
$CHCl_3$	1775	81
KBr	1715	81
$CHCl_3$	1715	81
CCl_4	1752	15
$CHCl_3$	1738 1722 1708	84
CCl_4	1753	15
$CHCl_3$	1750	15
CCl_4	1738	64
KBr	1712 1645 1536	85
KBr	1685	86
KBr	1725	86
nuj.	1707 1660 1620 1585 1545	87
CCl_4	1736	55
CCl_4	1745	55
KBr	1761 1695 1618	77
KBr	1705	88
CCl_4	1757	15
$CHCl_3$	1754	15

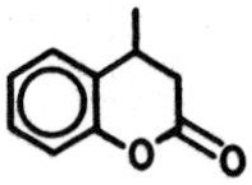

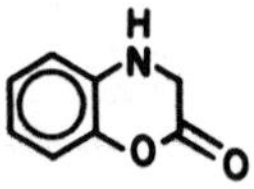

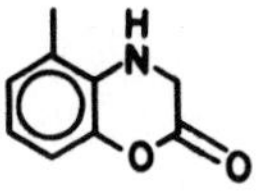

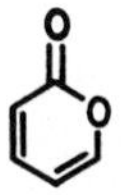

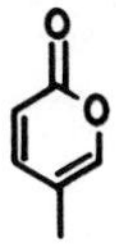

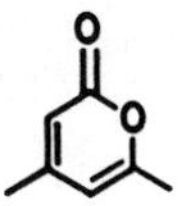

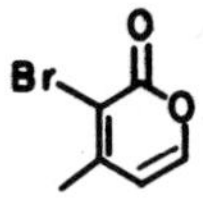

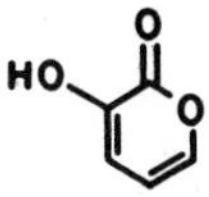

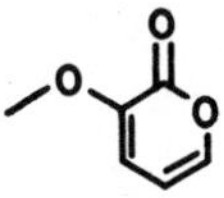

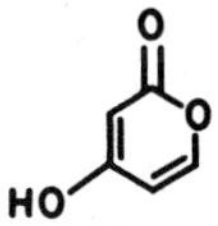

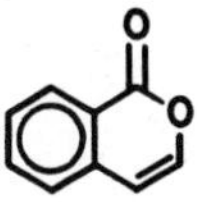

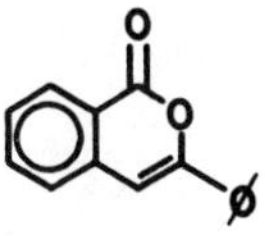

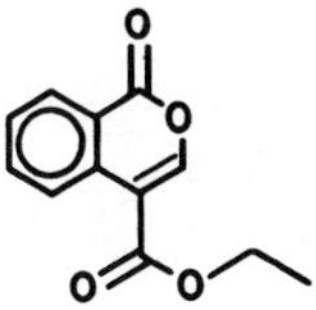

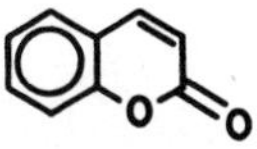

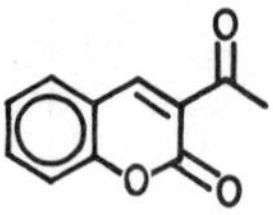 KBr	1730 1688m	88
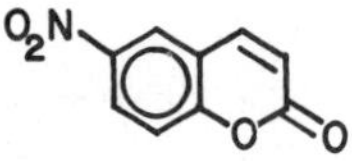 KBr	1738	88
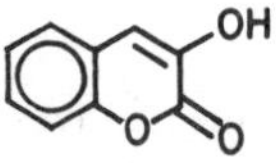 KBr	1701	88
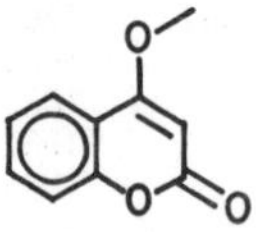 CCl_4	1730	89
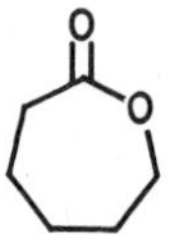 CCl_4	1727	39
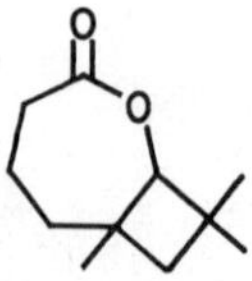 liq.	1754	90
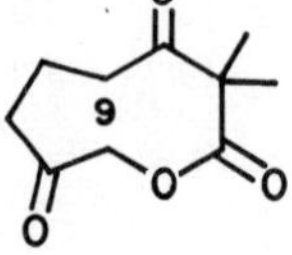 KBr	1770 1733	91
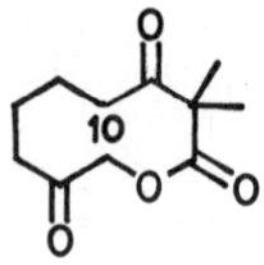 KBr	1754 1730	91
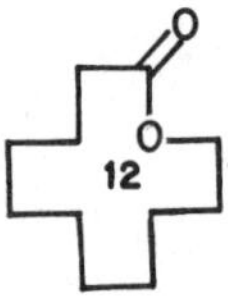 liq.	1727	92
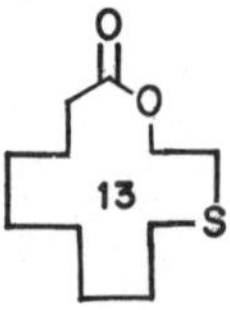 CS_2	1750	93
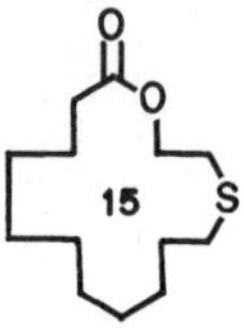 CS_2	1752	93
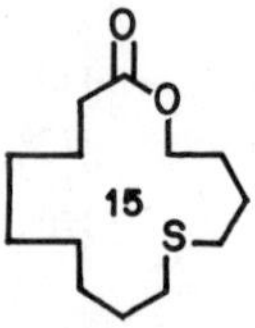 CS_2	1740	93
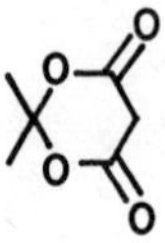 $CHCl_3$	1778 1749	94

Solvent	ν (cm^{-1})	Ref.
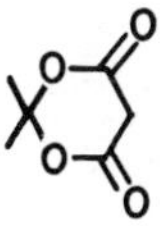 $CHCl_3$	1787 1760	95
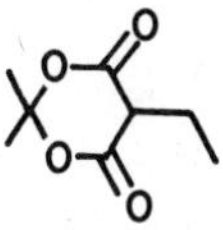 $CHCl_3$	1783 1747	94
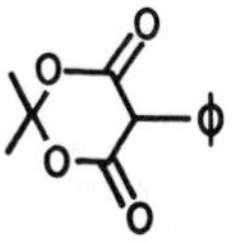 KBr	1792 1754	95
CCl_4	1787 1749	97
$CHCl_3$	1770 1735	94
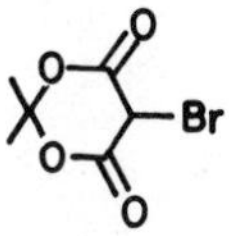 $CHCl_3$	1780 1767	95
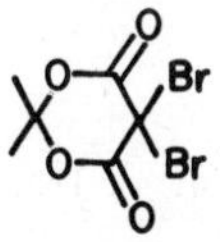 $CHCl_3$	1790 1768	95
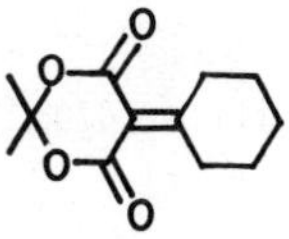 $CHCl_3$	1755 1730	95

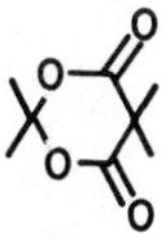

Solvent	ν (cm^{-1})	Ref.
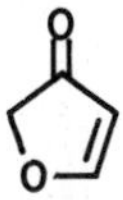 CCl_4	1733 1706 1572	96
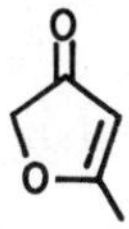 CCl_4	1742 1707 1645 1608	97
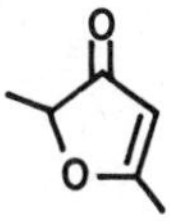 CCl_4	1712 1610	97
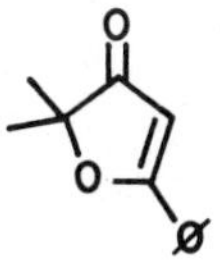 CCl_4	1701 1610	97
CCl_4	1678	15
$CHCl_3$	1674 1661 1637* 1634 1613	98
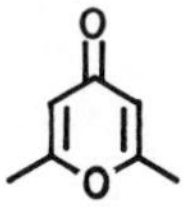 $CHCl_3$	1670 1620* 1615	98
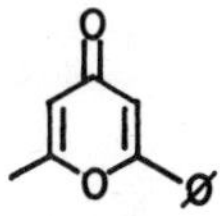 KBr	1667 1616	99
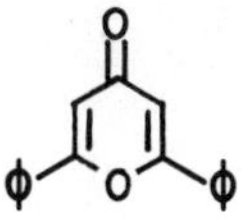 KBr	1653 1621w 1592w 1572w	99

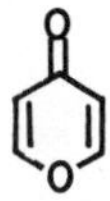

Solvent	ν (cm⁻¹)	Ref.
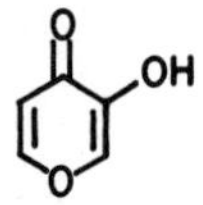 KBr	1647 1615 1563	100
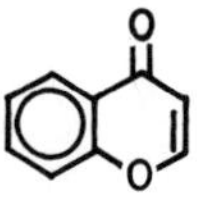 KBr	1617	84
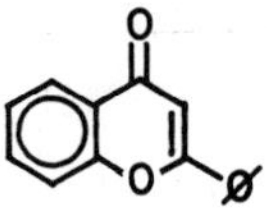 CCl_4	1619	101
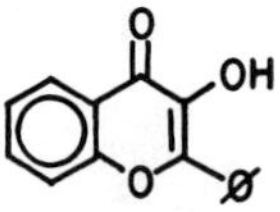 CCl_4	1619	101
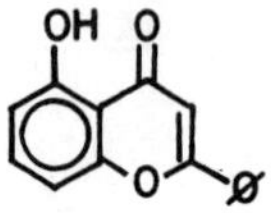 CCl_4	1648	101
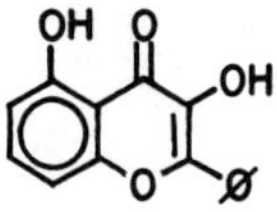 CCl_4	1638	101
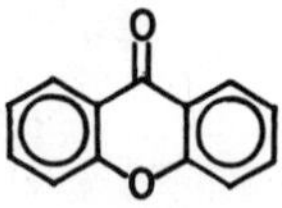 nuj.	1661 1616* 1608 1567w	102
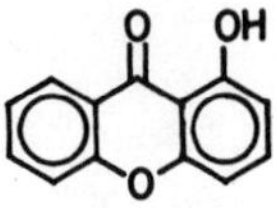 nuj.	1645 1610 1570	102
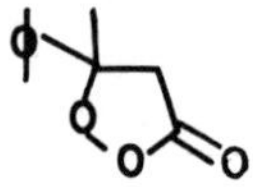 CCl_4	1810	103
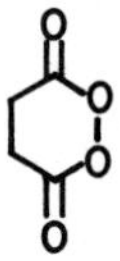 nuj.	1745	103
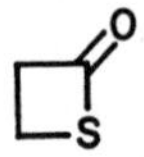 $CHCl_3$	1776	105
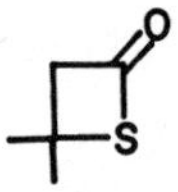 CCl_4	1772	105
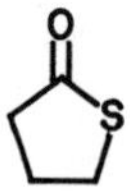 liq.	1705	106
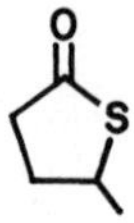 liq.	1705	107
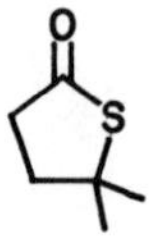 liq.	1705	107

Solvent	Frequency	Ref.
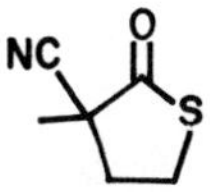 liq.	1695 2245	108
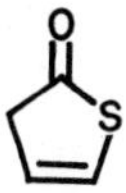 CCl_4	1669	109
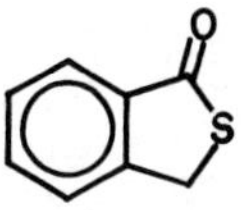 CCl_4	1686	55
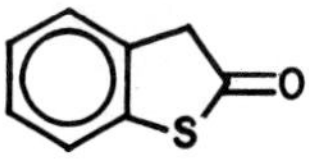 KBr CCl_4	1745 1723	111 110
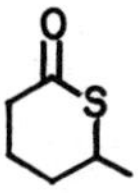 liq.	1665	107
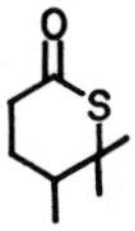 liq.	1665	107
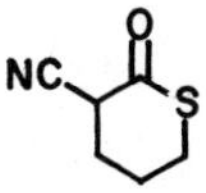 liq.	1665 2250	108
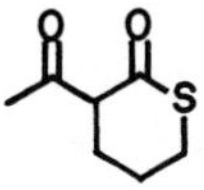 liq.	1715 1665	112
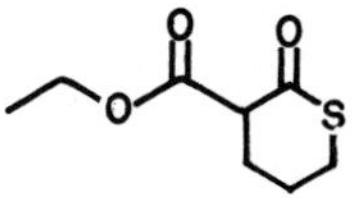 liq.	1735 1675	112
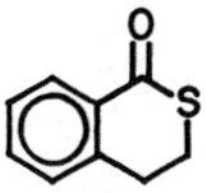 CCl_4	1715w 1655	110
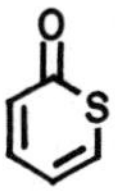 $CHCl_3$	1736	84
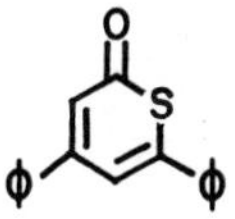 KBr	1640	113
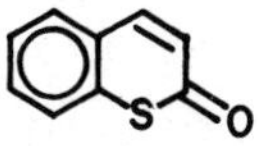 CCl_4	1683	55
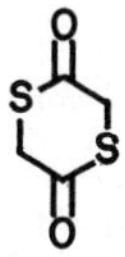 liq.	1678	114
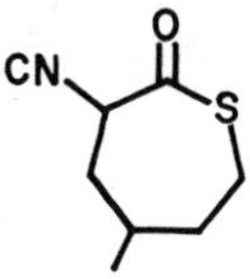 liq.	1665 2250	108

Structure	Medium	Frequency	Ref.
4H-thiopyran-4-one	nuj.	1609b	84
4H-thiochromen-4-one	KBr	1624	84
methyl 4-oxo-4H-thiopyran-3-carboxylate	nuj.	1607 1682	115

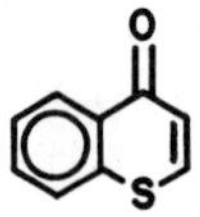

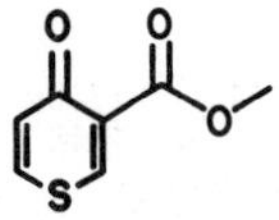

REFERENCES

1. S. Searles, M. Tamres, and G. M. Barrow, *JACS* (1953), 71.
2. L. J. Bellamy, *Infrared Spectra of Complex Molecules* (1958), 2nd edition, Methuen, London, p. 188.
3. D. Borrmann and R. Wegler, *B* (1966), 1245.
4. I. L. Knunyants and Yu. A. Cheburkow, *Izvest Akad Nauk USSR* (1960), 678.
5. F. Merger, *B* (1968), 2413.
6. M. Hauptschein, C. S. Stokes and A. V. Grosse, *JACS* (1952), 1974.
7. E. J. Corey and J. Streith, *JACS* (1964), 950.
8. F. A. Miller and G. L. Carlson, *JACS* (1957), 3995.
9. D. C. England and C. G. Krespan, *JOC* (1970), 3322.
10. J. E. Baldwin and J. D. Roberts, *JACS* (1963), 2444.
11. J. E. Baldwin, *JOC* (1964), 1882.
12. S. Sarel and E. Breuer, *JACS* (1959), 6522.
13. R. P. M. Bond, T. Cairns, J. D. Connolly, G. Eglinton and K. H. Overton, *JCS* (1965), 3958.
14. A. R. Pinder, *JCS* (1952), 2236.
15. R. N. Jones, C. L. Angell, T. Ito and R. J. D. Smith, *Canad J Chem* (1959), 2007.
16. J. Falbe, N. Huppes and F. Korte, *B* (1964), 863.
17. N. H. Cromwell, P. L. Creger and K. E. Cook, *JACS* (1956), 4412.
18. R. Filler and H. A. Leipold, *JOC* (1962), 4440.
19. W. Brügel, G. Stengel, F. Reicheneder and H. Suter, *Angew* (1956), 441.
20. G. Leclerc, C.-G. Wermuth and J. Schreiber, *BSC France* (1967), 1302.
21. H. G. Kuivila, *JOC* (1960), 284.
22. J. Cason and E. J. Reist, *JOC* (1958), 1492.

23. R. S. Rasmussen and R. R. Brattain, *JACS* (1949), 1073.
24. E. E. van Tamelen, F. M. Strong and U. C. Quark, *JACS* (1959), 750.
25. L. Birkofer and I. Hartwig, *B* (1954), 1189.
26. J. J. Bloomfield and S. L. Lee, *JOC* (1967), 3919.
27. T. Tanaka, *BCS Japan* (1959), 1320.
28. N. B. Mehta and W. E. McEwen, *JACS* (1953), 240.
29. Md. E. Ali and L. N. Owen, *JCS* (1958), 1074.
30. J. D. Connolly and K. H. Overton, *Proc Chem Soc* (1959), 188.
31. U. A. Huber and A. S. Dreiding, *Helv* (1970), 495.
32. J. Klein, *JACS* (1959), 3611.
33. A. Mondon, H. U. Menz, and J. Zander, *B* (1963), 826.
34. F. Korte and K. H. Büchel, *B* (1960), 1025.
35. E. J. Corey and W. H. Pirkle, *TL* (1967), 5255.
36. H. Musso, K. Naumann and K. Grychtol, *B* (1967), 3614.
37. S. Beckmann and H. Geiger, *B* (1959), 2411.
38. G. Snatzke and D. Marquarding, *B* (1967), 1710.
39. H. K. Hall, Jr. and R. Zbinden, *JACS* (1958), 6428.
40. J. Meinwald and H. C. Hwang, *JACS* (1957), 2910.
41. W. W. Epstein and A. C. Sonntag, *JOC* (1967), 3390.
42. A. N. Sagredos and J. D. von Mikusch, *A* (1966), *697*, 111.
43. R. Scheffold and P. Dubs, *Helv* (1967), 798.
44. N. L. Wendler and H. L. Slates, *JOC* (1967), 849.
45. T. Sakan, M. Kato and T. Miwa, *BCS Japan* (1964), 1171.
46. M. S. Newman and G. R. Kahle, *JOC* (1958), 666.
47. J. A. Marshall and N. Cohen, *JOC* (1965), 3475.
48. W. Herz and L. A. Glick, *JOC* (1963), 2970.
49. R. Scheffold, *Thesis ETH* Nr. 3356 (1963).
50. H. E. Smith and R. H. Eastman, *JACS* (1957), 5500.
51. B. M. Goldschmidt, B. L. Van Duuren and C. Mercado, *JCS* (1966), *C*, 2100.
52. W. Brügel, K. Dury, G. Stengel and H. Suter, *Angew* (1956), 440.
53. G. Singh, *Harvard Ph.D. Thesis* (1949).
54. J. C. Sauer, R. D. Cramer, V. A. Engelhardt, T. A. Ford, H. E. Holmquist and B. W. Howk, *JACS* (1959), 3677.
55. V. Prey, B. Kerres and H. Berbalk, *Monatsh Chem* (1960), 774.
56. W. H. Puterbaugh and C. R. Hauser, *JOC* (1964), 853.
57. J. F. Grove and H. A. Willis, *JCS* (1951), 877.
58. L. A. Duncanson, J. F. Grove and J. Zealley, *JCS* (1953), 1331.
59. Y. Sato, T. Iwashige and T. Miyadera, *Chem Pharm Bull* (1960), 427.

60. W. H. Washburn, *Appl Spectr* (1964) *18*, 61.
61. J. F. Grove, *JCS* (1951), 883.
62. L. A. Duncanson, *JCS* (1953), 1207.
63. T. Miyazawa, *J Mol Spectr* (1960) *4*, 155.
64. F. Korte, K. H. Büchel and KL. Göhring, *Angew* (1959), 523.
65. J. K. Crandall and R. J. Seidewand, *JOC* (1970), 697.
66. H. Christol, F. Plénat and C. Reliaud, *BSC France* (1968), 1566.
67. J. Meinwald, *JACS* (1954), 4571.
68. F. Korte, J. Falbe and A. Zschocke, *T* (1959) *6*, 201.
69. J. Falbe, H. Weitkamp and F. Korte, *T* (1963) *19*, 1479.
70. L. R. Subramanian and G. S. Krisha Rao, *T* (1967), 4167.
71. A. E. Wick, *Thesis ETH* Nr. 3617 (1964).
72. T. Tanaka, *BCS Japan* (1962), 1890.
73. R. K. Hill and A. G. Edwards, *T* (1965) *21*, 1501.
74. R. R. Sauers, *JACS* (1959), 925.
75. E. Fetz, B. Böhner, and Ch. Tamm, *Helv* (1965), 1669.
76. F. Korte and D. Scharf, *B* (1962), 443.
77. J. Aknin and D. Molho, *BSC France* (1967), 1813.
78. R. L. Vaulx, W. H. Puterbaugh and C. R. Hauser, *JOC* (1964), 3514.
79. T. A. Foglia, L. M. Gregory and G. Maerker, *JOC* (1970), 3779.
80. G. Berti and F. Mancini, *Gazz Chim Ital* (1958), 714.
81. D. G. O'Sullivan and P. W. Sadler, *JCS* (1957), 2916.
82. R. D. Clark, *JOC* (1967), 399.
83. K. Sato, T. Amakaxu and S. Abe, *JOC* (1964), 2971.
84. A. Schönberg and R. von Ardenne, *B* (1968), 346.
85. R. H. Wiley and J. G. Esterle, *JOC* (1957), 1257.
86. Y. Yamada, *Kagaku-no-Ryoiki (special issue)* (1959) *37*, 97.
87. A. W. Allan and R. P. A. Sneeden, *T* (1962) *18*, 821.
88. P. Bassignana and C. Cogrossi, *T* (164) *20*, 2859.
89. I. Chmielewska and J. Cieslak, *T* (1958) *4*, 135.
90. E. J. Corey, J. D. Bass, R. LeMahieu and R. B. Mitra, *JACS* (1964), 5570.
91. J. Falbe and F. Korte, *B* (1963), 919.
92. R. Huisgen and H. Ott, *T* (1959) *6*, 253.
93. V. N. Belov, N. P. Solov'eva, T. A. Rudol'fi and I. A. Voronina, *JOC USSR* (1965), 551.
94. R. A. Abramovitch, *Canad J Chem* (1959), 361.
95. H. R. Snyder and C. W. Kruse, *JACS* (1958), 1942.
96. A. Hofmann, W. v. Philipsborn and C. H. Eugster, *Helv* (1965), 1322.
97. R. E. Rosenkranz, K. Allner, R. Good, W. v. Philipsborn and C. H. Eugster, *Helv* (1963), 1259.

98. A. R. Katritzky and R. A. Jones, *SCA* (1961), 64.
99. R. J. Light and C. R. Hauser, *JOC* (1960), 539.
100. T. Shimanouchi, Y. Mashika, K. Nakanishi and S. Hayao, ed., *I. R. Spectra* (1959) *7*, Nankodo Tokyo, 97.
101. B. L. Shaw and T. H. Simpson, *JCS* (1955), 655.
102. F. Scheinmann, *T* (1962) *18*, 853.
103. S. Mizushima, T. Shimanouchi, I. Ichishima, T. Miyazawa, I. Nakagawa and T. Araki, *JACS* (1956), 2038.
104. F. D. Green, W. Adam and G. A. Knudsen, Jr., *JOC* (1966), 2087.
105. P. Y. Johnson and G. A. Berchtold, *JOC* (1970), 584.
106. F. Korte and K. H. Büchel, *B* (1960), 1021.
107. F. Korte and H. Christoph, *B* (1961), 1966.
108. F. Korte and H. Wamhoff, *B* (1964), 1970.
109. C. D. Hurd and K. L. Kreuz, *JACS* (1950), 5543.
110. W. C. Lumma, Jr., G. A. Dutra and C. A. Voeker, *JOC* (1970), 3442.
111. K. Gewald and G. Neumann, *B* (1968), 1933.
112. F. Korte and F.-F. Wiese, *B* (1964), 1963.
113. H. Behringer and A. Grimm, *A* (1965) *682*, 188.
114. A. Schöberl and G. Wiehler, *A* (1955) *595*, 101.
115. D. S. Tarbell and P. Hoffman, *JACS* (1954), 2451.

4.12
ureas

—NH

Although detailed assignments of the N-H modes of ureas have not been made, both the stretching frequencies in the 3500 to 3000 cm^{-1} region and the deformation and coupled deformation and C-N stretching modes in the 1600 to 1300 cm^{-1} region parallel those found for amides (see Chapter 4.3, p. 241).

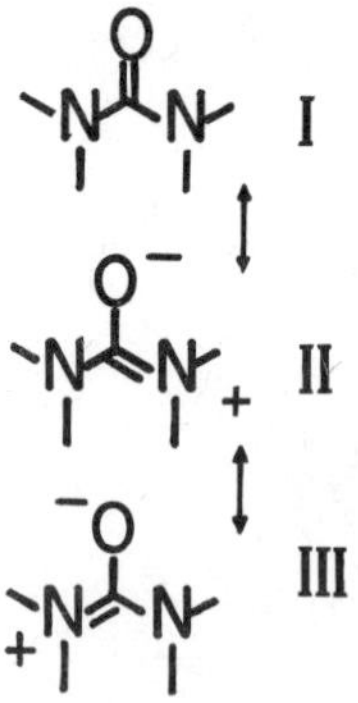

I ⟷ II ⟷ III

As with amides the dipolar resonance forms II and III make a major contribution to the structure, and the carbonyl stretching vibration is found at frequencies lower than those of ketones but slightly higher than those of amides. Substitution on nitrogen of electronegative groups and conjugated unsaturated groups cause an increase in the carbonyl stretching frequency. These changes are not as large with ureas as those seen with amides.

Association in the solid state or solute solvent interactions in polar solvents causes a lowering of the carbonyl stretching frequency, and in concentrated solution bands arising from both associated and nonassociated molecules may be seen.

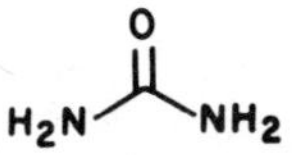

Medium	Frequencies	Ref.
nuj.	1679 1627	1
KBr	1688 1624 1608w 1466	1
nuj.	1700 1642 1625 1550 1475m	1
nuj.	1645 1567 1418	2
KBr	1672	3
nuj.	1653 1605 1563	2
KBr	1645	4
nuj.	1653 1600 1550 1420	2
nuj.	1661 1618 1563	5

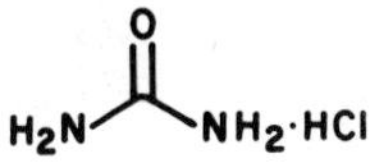

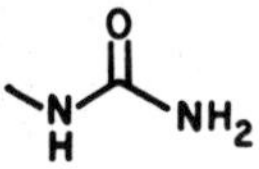

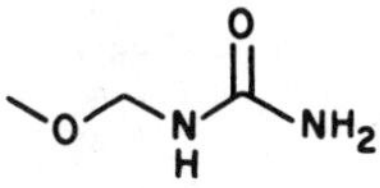

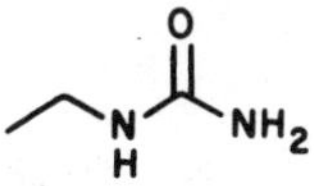

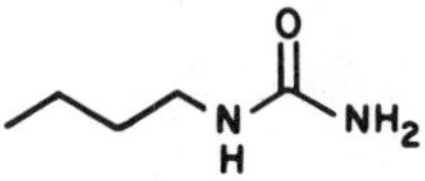

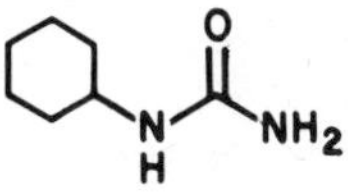

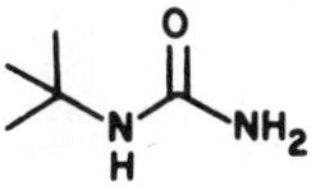

Medium	Frequencies	Ref.
nuj.	1656 1610 1511	5
KBr	1613 1575	6
nuj.	1622 1580* 1530	5
CCl_4	1717w 1695	
$CHCl_3$	1663 1548	8
KBr	1640	3
nuj.	1618 1590	9
nuj.	1620 1582	10

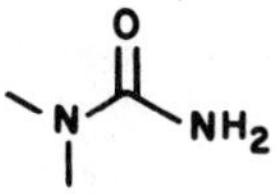

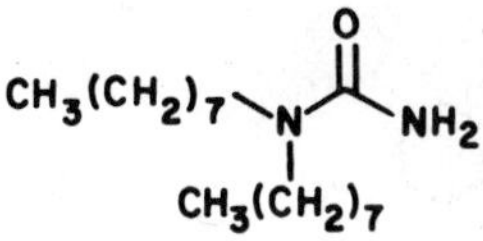

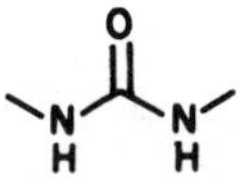

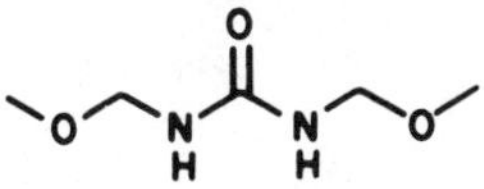

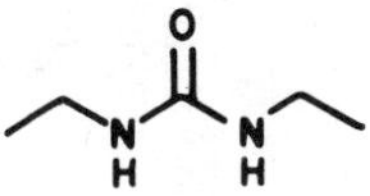

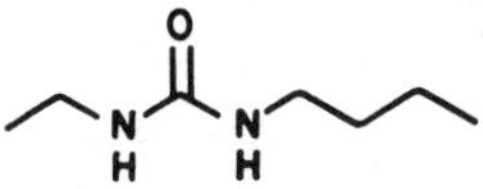

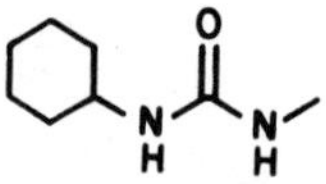

nuj. 1661 1645 1626 1587 1534 1418 2

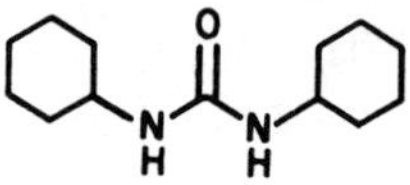

nuj. 1629 1575 1537 9

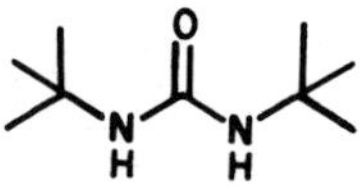

CCl_4 1680 1650 11

$CHCl_3$ 1684 1549 8

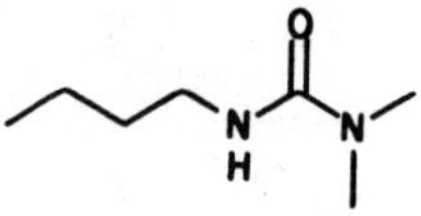

CCl_4 1655 4

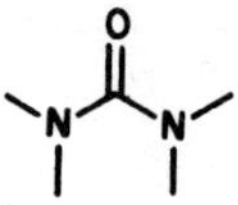

liq. 1875w 1645 1560w 1497 12

CCl_4 1654 13

$CHCl_3$ 1675 8

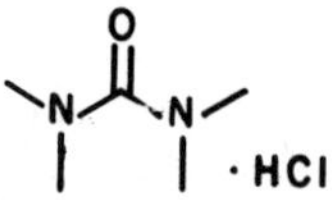

·HCl liq. 1755 1635 1538 1405m 12

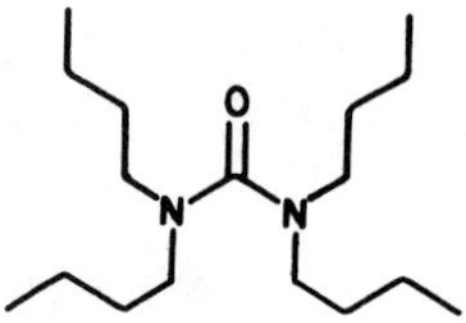

liq. 1637 6

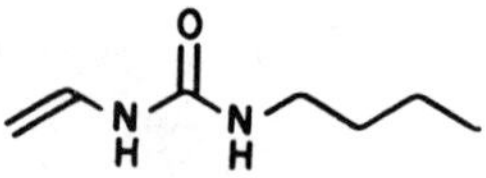

nuj. 1666 1638 1583 10

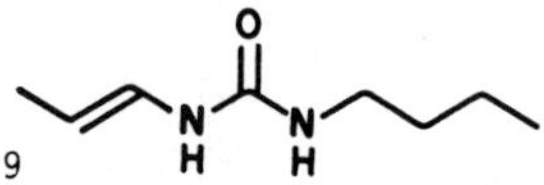

nuj. 1697w 1643 1582 10

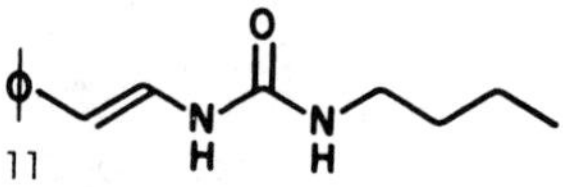

nuj. 1681 1640 1575 10

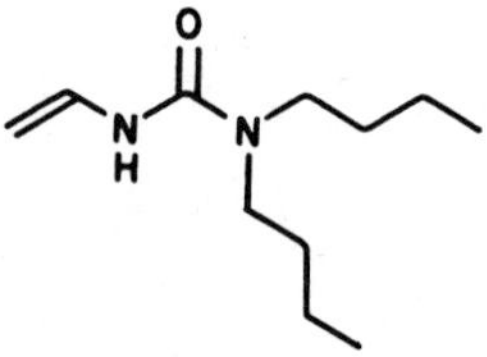

liq. 1663* 1625 1530* 10

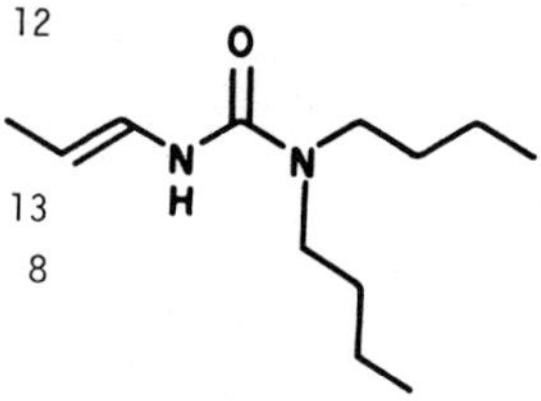

liq. 1680w 1628 1530 10

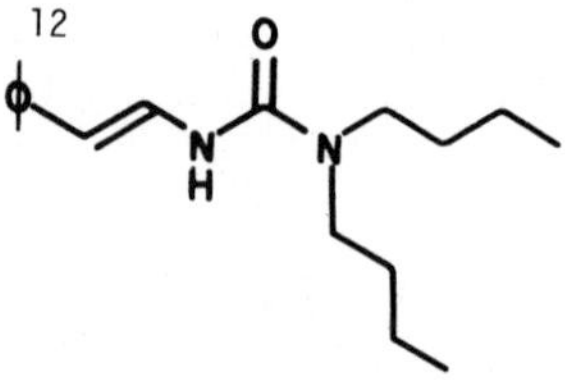

nuj. 1670 1632 10

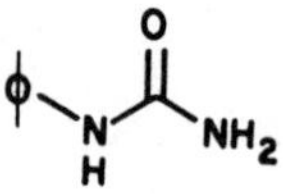

NH_2 nuj. 1661 1621 1600 1553 1499 1359 2

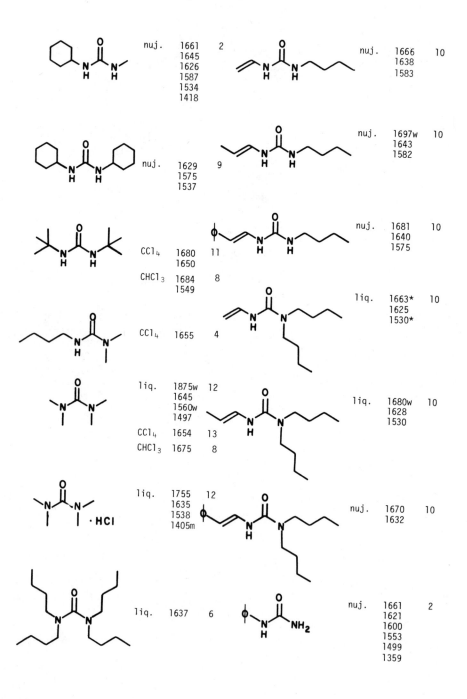

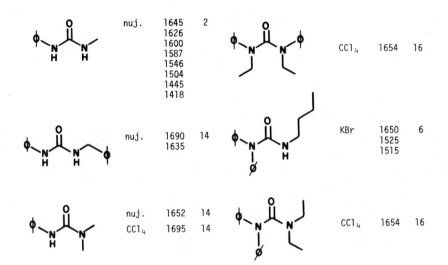

Solvent	Frequency	Ref.
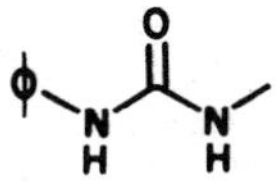 nuj.	1645 1626 1600 1587 1546 1504 1445 1418	2
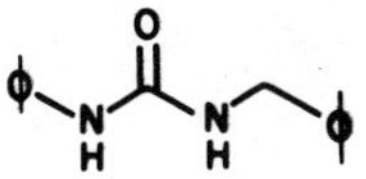 nuj.	1690 1635	14
nuj.	1652	14
CCl_4	1695	14
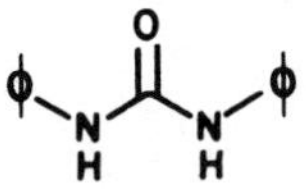 nuj.	1650 1600 1500	15
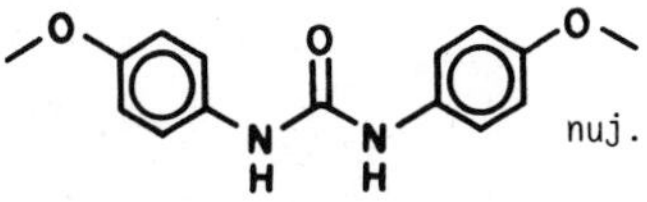 nuj.	1630 1570	15
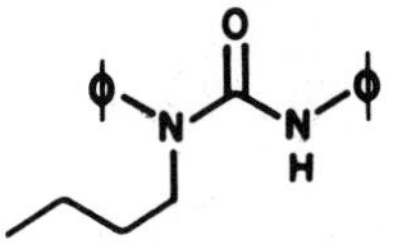 KBr	1650 1530	6

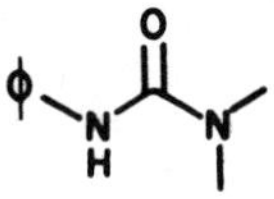

Solvent	Frequency	Ref.
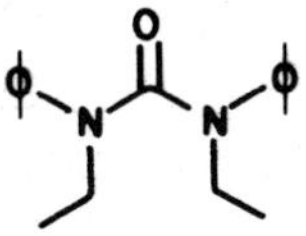 CCl_4	1654	16
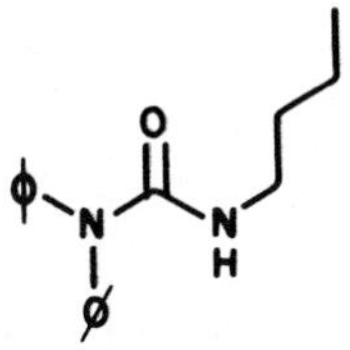 KBr	1650 1525 1515	6
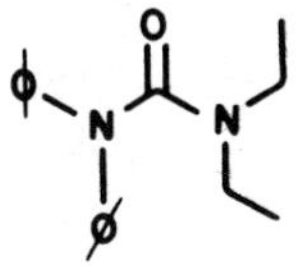 CCl_4	1654	16
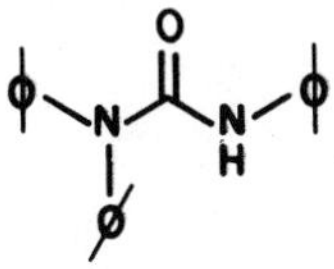 KBr	1676 1515	6
nuj.	1730 1650	14
CCl_4	1740	14
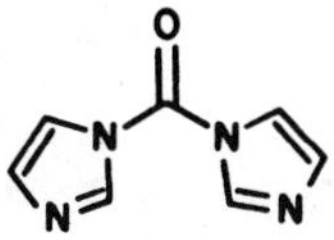 $CHCl_3$	1745	17

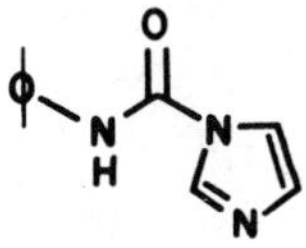

Solvent	cm⁻¹	Ref.
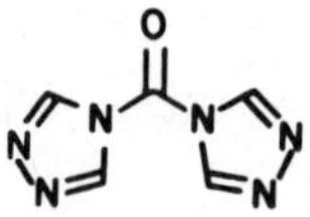 $CHCl_3$	1781	17
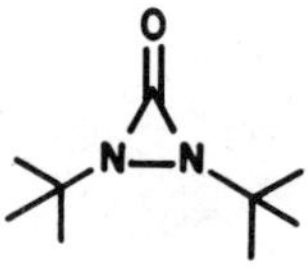 CCl_4	1926* 1880 1860* 1800*	11
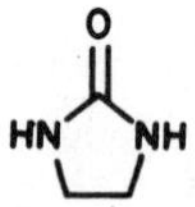 CCl_4	1735w 1718	7
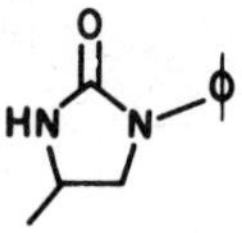 nuj.	1700	18
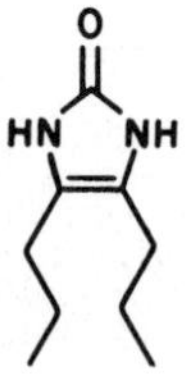 KBr	1684	19
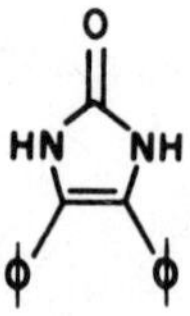 KBr	1677 1631	19
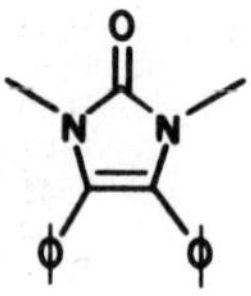 KBr	1684 1640*	19
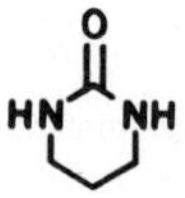 CCl_4	1718	7
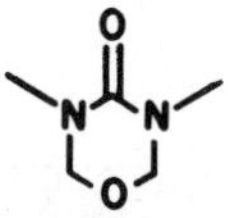 liq.	1635	20
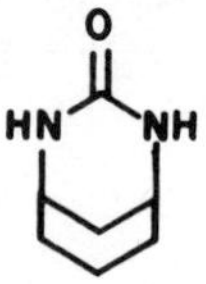 CCl_4	1712 1655	7
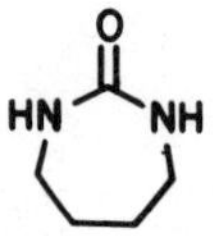 CCl_4	1689	7
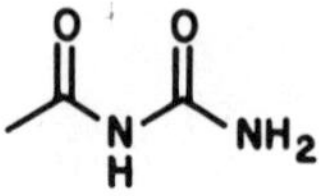 nuj.	1667 1634	21

Medium	Frequencies	Ref.
CCl_4	1720 1690	22
KBr	1724* 1703 1674	23
nuj.	1776 1697	18
nuj.	1730 1694	18
nuj.	1712 1676	21
nuj.	1748 1706	21
nuj.	1748 1706	18
nuj.	1761 1692	18
nuj.	1767 1718 1703	21
nuj.	1767 1695	21
nuj.	1718 1664	21
KBr	1768 1737 1716 1673 1653	23
KBr	1754* 1746 1723* 1711 1694	23
nuj.	1730 1698	21

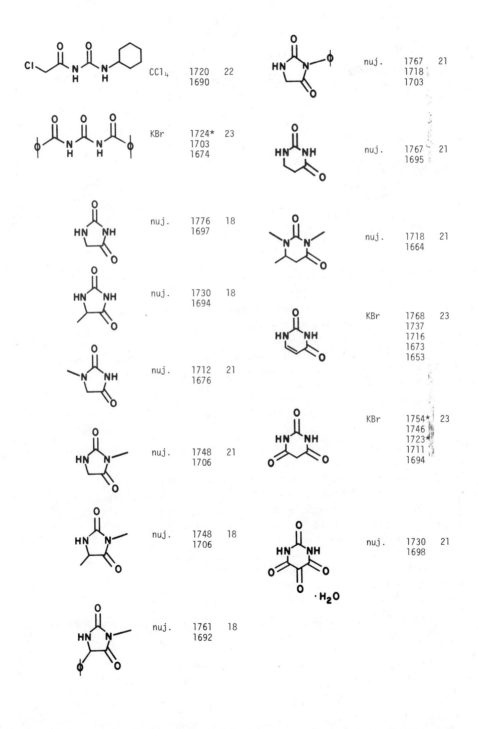

Structure	Solvent	cm⁻¹	Ref.
$H_2N-C(=O)-NH-NO_2$	nuj.	1724 1613 1314	24
$H_2N-C(=O)-N(NO)-CH_2CH_3$	CCl_4	1742 1506	25
$H_2N-C(=O)-N(NO)-CH_3$	CCl_4	1754 1497	25

REFERENCES

1. E. Spinner, *SCA* (1959) *15*, 95.
2. J. L. Boivin and P. A. Boivin, *Canad J Chem* (1954), 561.
3. H. J. Becher and F. Griffel, *B* (1958), 2025.
4. C. Collard-Charon and M. Renson, *BSC Belge* (1963), 149.
5. C. I. Jose, *SCA* (1969), 111.
6. S. E. Forman, C. A. Erickson and H. Adelman, *JOC* (1963), 2653.
7. H. K. Hall, Jr. and R. Zbinden, *JACS* (1958), 6428.
8. C. N. Rao, G. C. Chaturvedi and R. K. Gosavi, *J Mol Spectr* (1968) *28*, 526.
9. H. G. Khorana, *Canad J Chem* (1954), 261.
10. M. Sato, *JOC* (1961), 770.
11. F. D. Greene, J. C. Stowell and W. R. Bergmark, *JOC* (1969), 2254.
12. M. J. Janssen, *SCA* (1961), 475.
13. A. W. Baker and G. H. Harris, *JACS* (1960), 1923.
14. J. Denkosch, K. Schlögl and H. Woidich, *Monatsh Chem* (1957) *88*, 35.
15. P. A. Boivin, W. Bridgeo and J. L. Boivin, *Canad J Chem* (1954), 242.
16. H. W. Thompson and D. A. Jameson, *SCA* (1958) *13*, 236.
17. H. A. Staab, *A* (1957) *609*, 83.
18. L. Crombie and K. C. Hooper, *JCS* (1955), 3010.
19. R. Gompper and H. Herlinger, *B* (1956), 2825.
20. H. J. Becher and F. Griffel, *B* (1958), 2032.
21. H. M. Randall, R. G. Fowler, N. Fuson and J. R. Dangl, *Infrared Determinations of Organic Structure*, 1949, D. Van Nostrand Company, Inc., New York.

22. C. I. Jose and P. R. Pabrai, *SCA* (1967), 734.
23. R. Gompper, *B* (1960), 198.
24. E. Lieber, D. R. Levering and L. J. Patterson, *Anal Chem* (1951), 1594.
25. J. R. Dyer, R. D. Randall, Jr. and H. M. Deutsch, *JOC* (1964), 3423.

4.13

carbamates urethanes

–NH

The N-H stretching frequencies and the amide II band occur in the same region as for amides (Chapter 4.3, p. 241) and suffer the same changes in frequency with change in state, solvent, and concentration as do amides.

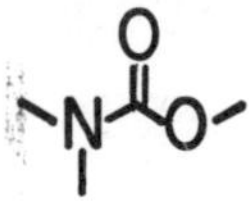

Delocalization of the nitrogen lone pair into the carbonyl group will lower the carbonyl stretching frequency, whereas the electronegativity of the ether oxygen will raise the frequency. In general, these .two opposing effects balance each other, and the carbonyl stretching frequency for urethanes and carbamates fall between those of the corresponding amides and esters.

Electron withdrawing groups, including alkenyl, aryl, and halogens, cause an increase in the carbonyl stretching frequency, as does increased ring strain in cyclic systems.

As with amides, these systems show a lowering of the carbonyl frequency on hydrogen bonding, and such changes in frequency are functions of state, solvent, and concentration.

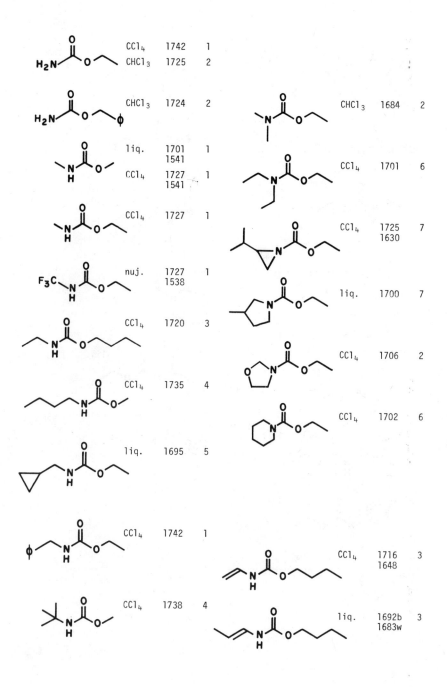

Solvent	Frequency	Ref.
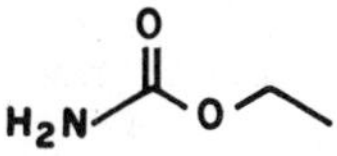 CCl_4	1742	1
$CHCl_3$	1725	2
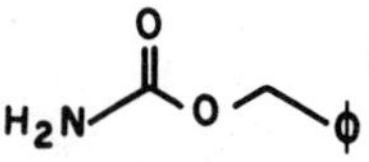 $CHCl_3$	1724	2
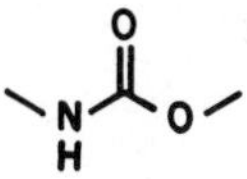 liq.	1701 1541	1
CCl_4	1727 1541	1
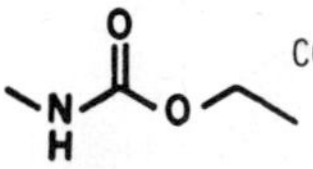 CCl_4	1727	1
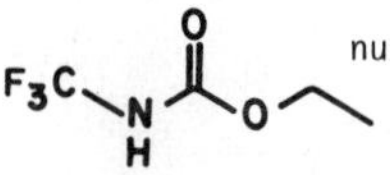 nuj.	1727 1538	1
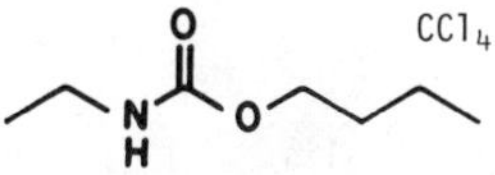 CCl_4	1720	3
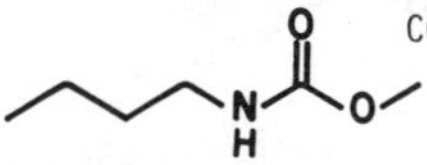 CCl_4	1735	4
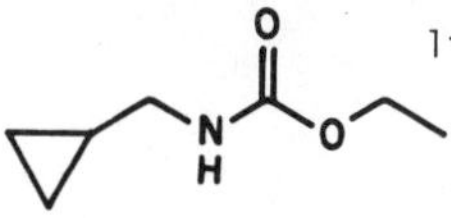 liq.	1695	5
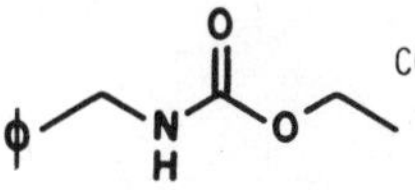 CCl_4	1742	1
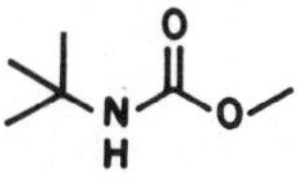 CCl_4	1738	4

Solvent	Frequency	Ref.
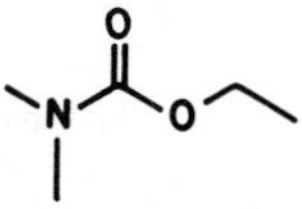 $CHCl_3$	1684	2
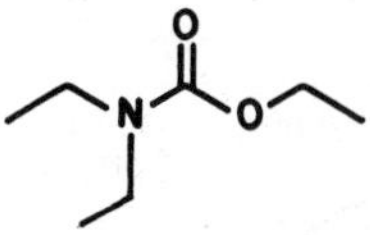 CCl_4	1701	6
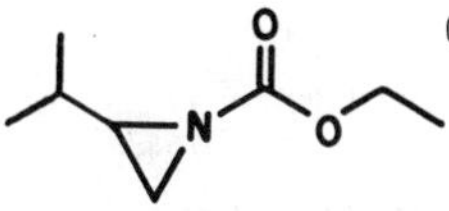 CCl_4	1725 1630	7
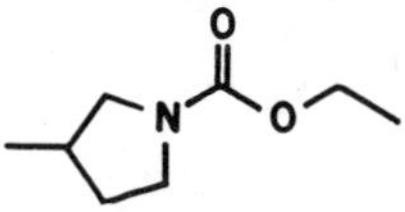 liq.	1700	7
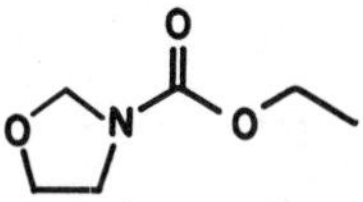 CCl_4	1706	2
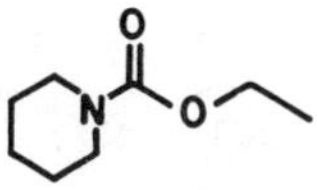 CCl_4	1702	6
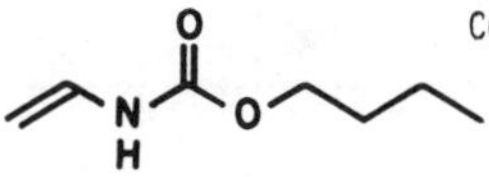 CCl_4	1716 1648	3
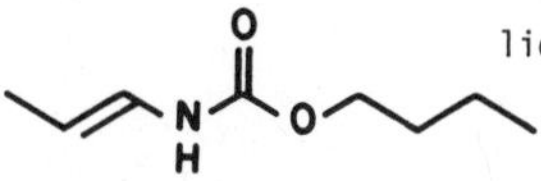 liq.	1692b 1683w	3

Solvent	Frequency	Ref.
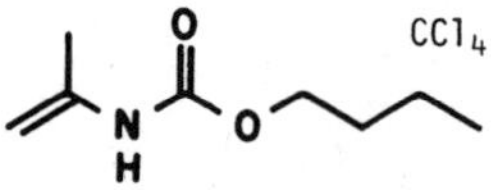 CCl_4	1710 1681w 1654	3
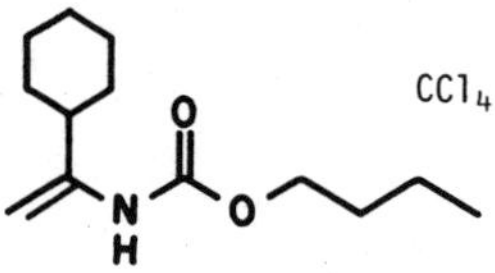 CCl_4	1730 1673w 1648w	3
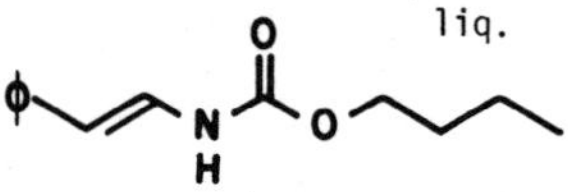 liq.	1710 1658	3
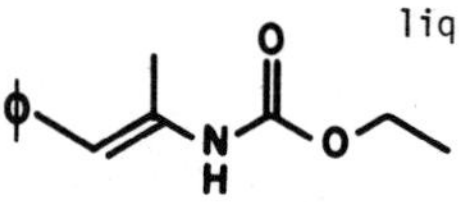 liq.	1723b 1647	3
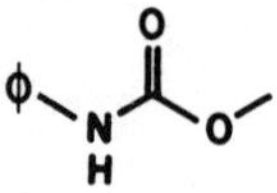 CCl_4	1748	4
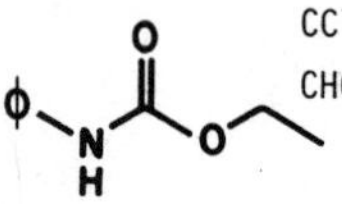 CCl_4 $CHCl_3$	1742 1730	1 8
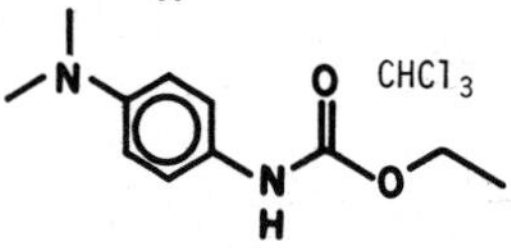 $CHCl_3$	1719	8
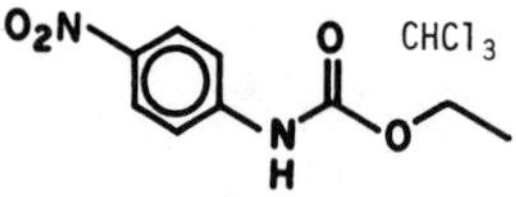 $CHCl_3$	1736	8
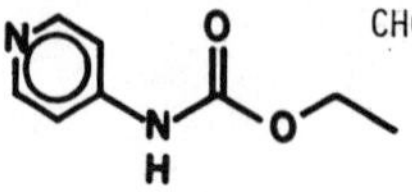 $CHCl_3$	1733	8
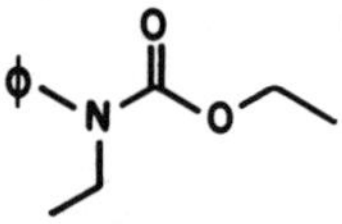 CCl_4	1706	6
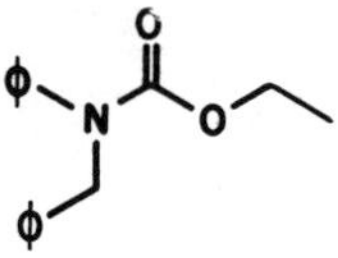 CCl_4	1706	6
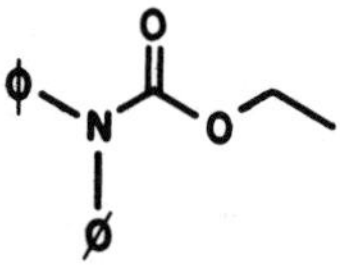 CCl_4	1719	6
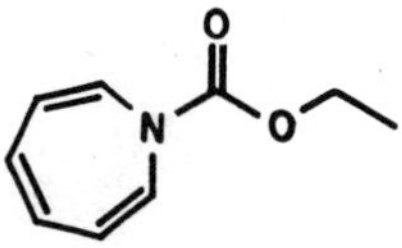 CCl_4	1717 1651	7
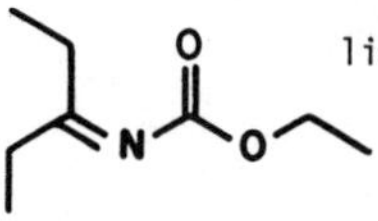 liq.	1703	3
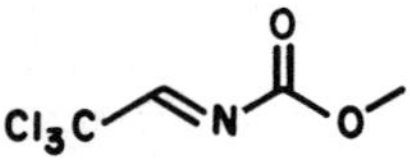 $CHCl_3$	1745 1669	9
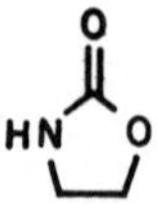 CCl_4 $CHCl_3$	1783 1760	8 2
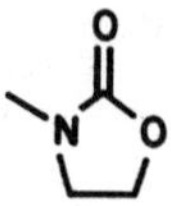 $CHCl_3$	1746	2

Solvent	ν (cm^{-1})	Ref.
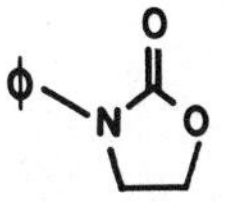 $CHCl_3$	1753	2
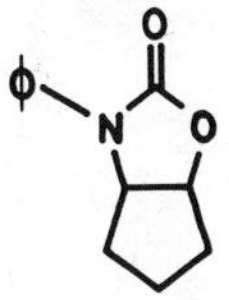 KBr	1733	10
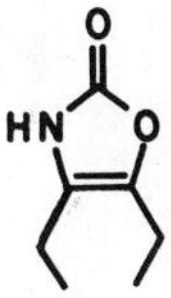 KBr	1753 1728 1704	11
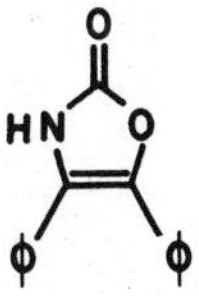 KBr	1750 1688*	11
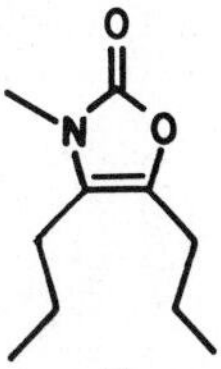 KBr	1754 1695	11
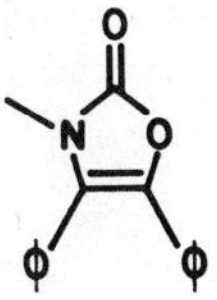 KBr	1749 1709	11
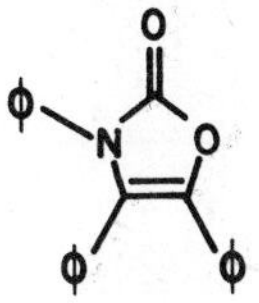 KBr	1750 1715* 1671	11
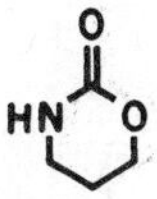 CCl_4	1743	12

Solvent	ν (cm^{-1})	Ref.
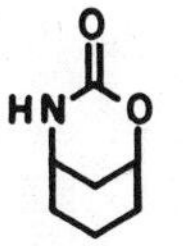 CCl_4	1725	12
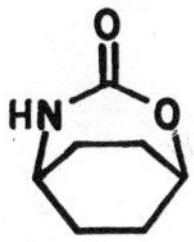 CCl_4	1717	12
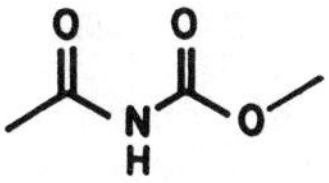 liq.	1706	13
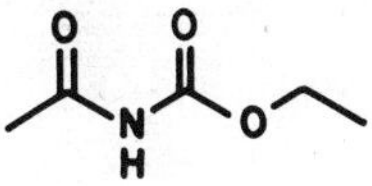 $CHBr_3$	1754 1704 1488	14
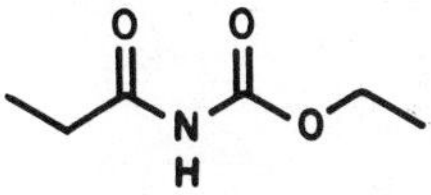 KBr	1761 1698	15
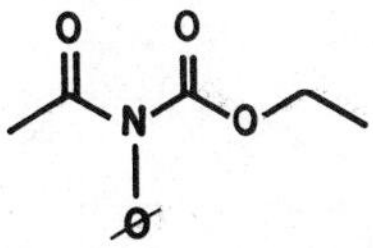 nuj.	1748	13
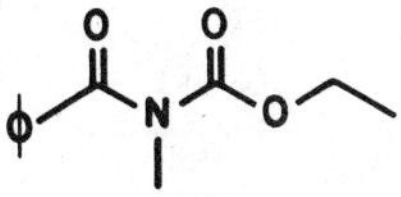 liq.	1736 1672	13
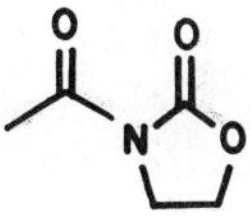 KBr	1776 1698	15
CCl_4	1795 1711	12
$CHCl_3$	1785 1679	2

Solvent	cm⁻¹	Ref.
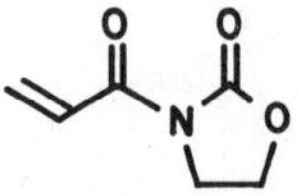 KBr	1773 1681	15
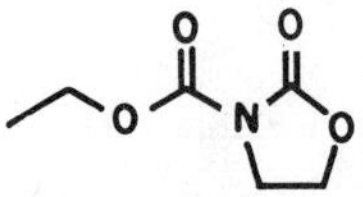 $CHCl_3$	1722	2
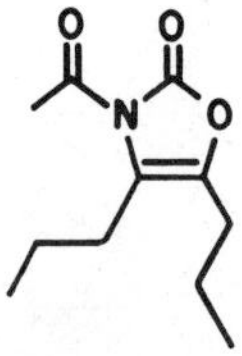 KBr	1782 1725 1703	11
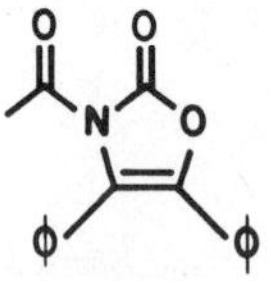 KBr	1784 1728 1680	11
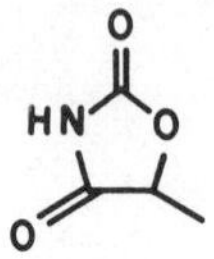 $CHBr_3$	1826 1748 1732	14
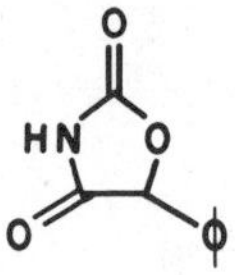 $CHBr_3$	1860 1825 1750	14
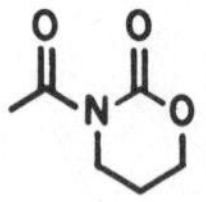 CCl_4	1748 1707	12
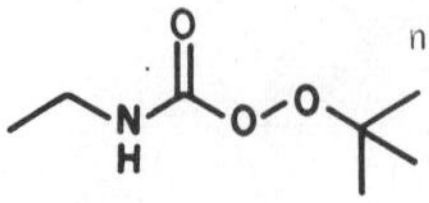 nuj.	1730	16
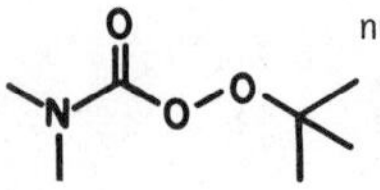 nuj.	1745	16
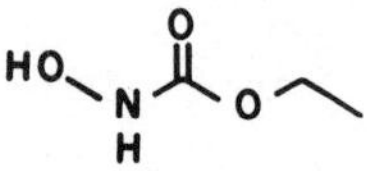 liq.	1718	13
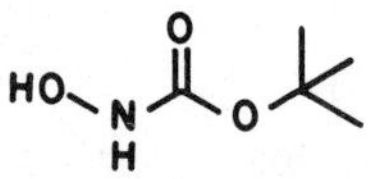 nuj.	1727	17
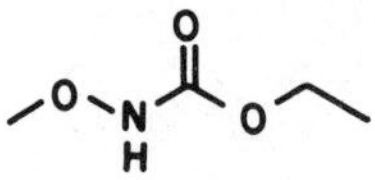 liq.	1727	13
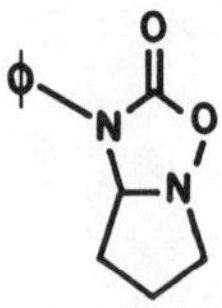 KBr	1764b	18
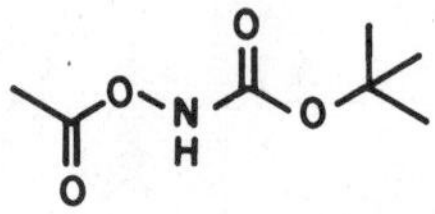 liq.	1754	17
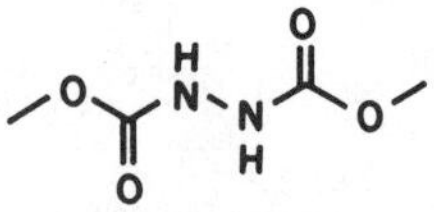 KBr	1770 1740b 1560	19
$CHCl_3$	1739	20
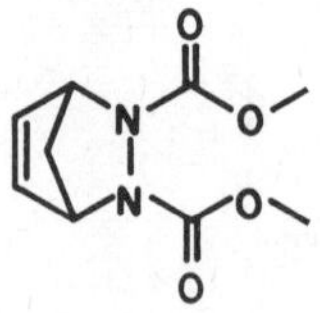 CCl_4	1754 1713	20
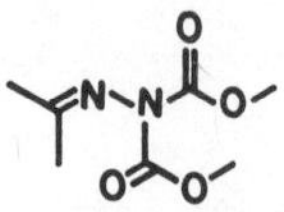 KBr	1782 1643	19

Solvent	ν	Ref.
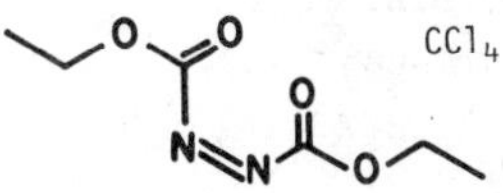 CCl_4	1750	21
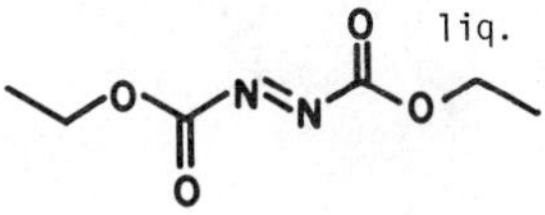 liq.	1783	21
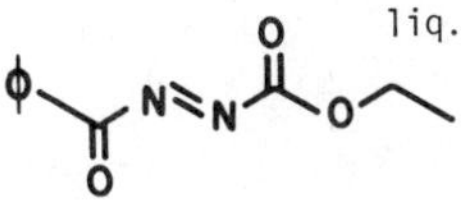 liq.	1783 1733	21
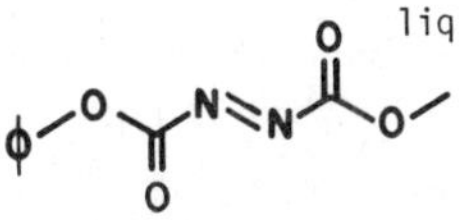 liq.	1789	21
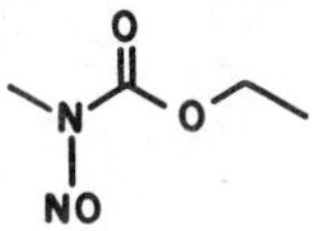 CCl_4	1751	1
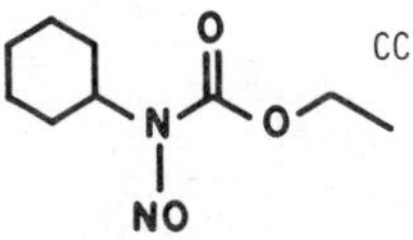 CCl_4	1754	22
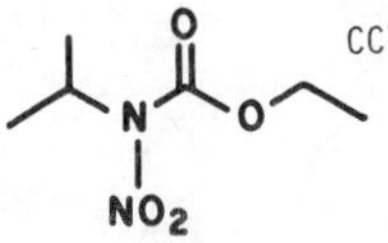 CCl_4	1773 1739	23
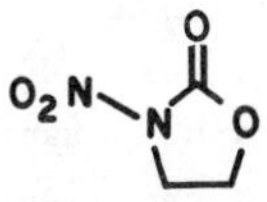 CCl_4	1835	24
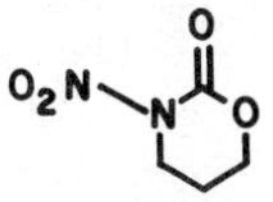 CCl_4	1795	24
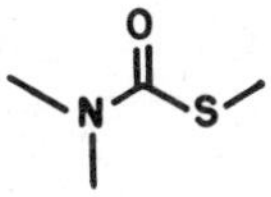 CCl_4	1662	25
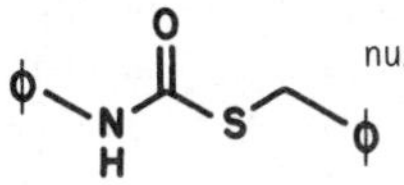 nuj.	1660 1515	26
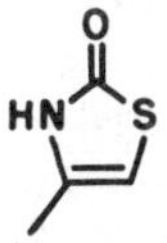 $CHCl_3$	1639	27
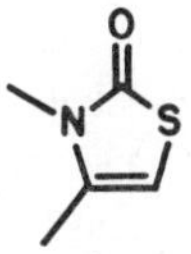 $CHCl_3$	1639 1592	27
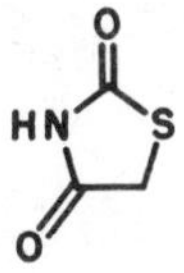 nuj.	1703 1658	13

REFERENCES

1. D. A. Barr and R. N. Haszeldine, *JCS* (1956), 3428.
2. S. Pinchas and D. B. Ishai, *JACS* (1957), 4099.
3. M. Sato, *JOC* (1961), 770.
4. R. A. Nyquist, *SCA* (1963), 509.
5. R. A. Moss and F. C. Shulman, *T* (1968), 2881.
6. L. J. Bellamy and R. J. Pace, *SCA* (1963), 1831.
7. A. Mishra, S. N. Rice and W. Lwowski, *JOC* (1968), 481.

8. A. R. Katritzky and R. A. Jones, *JCS* (1960), 676.
9. H. Ulrich, B. Tucker and A. A. Sayigh, *JOC* (1968), 2887.
10. H. De Pooter and N. Schamp, *BSC Belge* (1968), 377.
11. R. Gompper and H. Herlinger, *B* (1956), 2825.
12. H. K. Hall and R. Zbinden, *JACS* (1958), 6428.
13. H. M. Randall, R. G. Fowler, N. Fuson and J. R. Dangl, *Infrared Determinations of Organic Structures* 1949, D. Van Nostrand Company, Inc., New York.
14. M. Pianka and D. J. Polton, *JCS* (1960), 983.
15. P. Bassignana, C. Cogrossi, S. Franco and G. Polla Mattiot, *SCA* (1965), 677.
16. C. J. Pedersen, *JOC* (1958), 255.
17. L. A. Carpino, C. A. Giza and B. A. Carpino, *JACS* (1959), 955.
18. J. Thesing and W. Sirrenberg, *B* (1959), 1748.
19. J.-C. Bloch, *T* (1969), 619.
20. R. M. Moriarty, Sr., M. R. Murphy, S. J. Druck and L. May, *TL* (1967), 1603.
21. H. Bock and J. Kroner, *B* (1966), 2039.
22. E. H. White, *JACS* (1955), 6008.
23. E. H. White and D. W. Grisley, Jr., *JACS* (1961), 1191.
24. E. H. White, M. C. Chen and L. A. Dolak, *JOC* (1966), 3038.
25. A. W. Baker and G. H. Harris, *JACS* (1960), 1923.
26. P. Baudet, M. Calin and E. Cherbuliez, *Helv* (1965), 2023.
27. M. G. Ettlinger, *JACS* (1950), 4699.

4.14 carbonates

Delocalization of the nonbonding ether-oxygen electrons into the carbonyl groups, which will decrease the frequency of the carbonyl stretching vibration, and the electronegativity of the two oxygen atoms, which will increase the carbonyl stretching frequency, compete with each other in a manner similar to that seen with esters. As in the case of esters the latter effect predominates, and carbonates absorb at higher frequencies than ketones or esters.

Electronegative groups, as well as alkenyl and aryl groups, bonded to the ether oxygens cause an increase in the carbonyl stretching frequency by decreasing the extent of lone-pair delocalization.

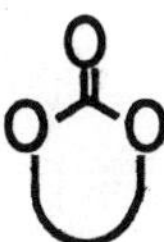

Ring strain and consequent increase in the bond order of the carbonyl group causes an increase in the frequency of the cyclic carbonate compared to the acyclic analogs.

Only small decreases in frequency (10 cm^{-1}) are seen between $CHCl_3$ and CCl_4 solutions.

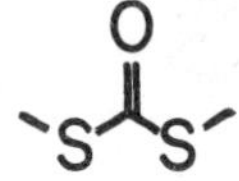

THIOCARBONATES

For the same reasons as those discussed for thiol esters (Chapter 4.10, p. 332), replacement of one of the ether-oxygen atoms by sulfur causes a decrease in the carbonyl stretching frequency, and replacement of both oxygen atoms by sulfur causes a further decrease in frequency.

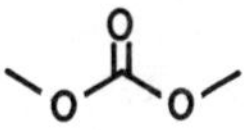

Solvent	cm⁻¹	Ref.
CCl_4	1758	1
$CHCl_3$	1751	1

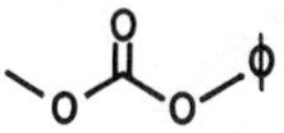

Solvent	cm⁻¹	Ref.
CCl_4	1739	5

ϕ

Solvent	cm⁻¹	Ref.
nuj.	1757	3

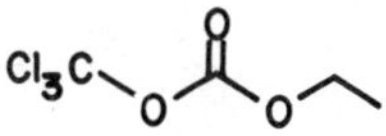

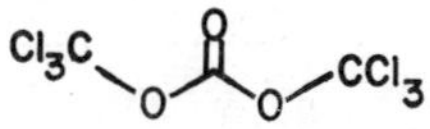

ϕ ϕ

Solvent	cm⁻¹	Ref.
CCl_4	1786	1
CS_2	1775	4

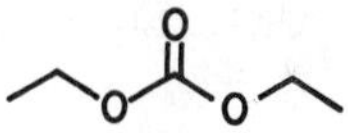

Cl_3C

Solvent	cm⁻¹	Ref.
liq.	1780	4

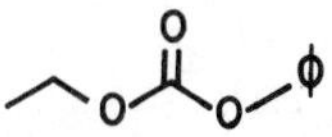

Solvent	cm⁻¹	Ref.
CCl_4	1779	5

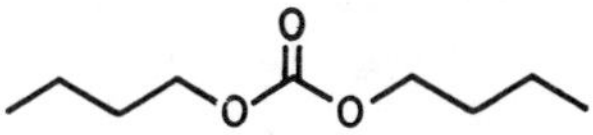

Cl_3C CCl_3

Solvent	cm⁻¹	Ref.
CCl_4	1832	3

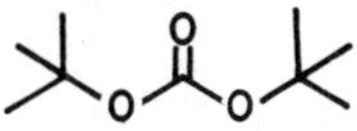

Solvent	cm⁻¹	Ref.
CCl_4	1783	5

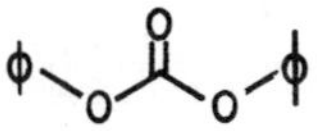

Solvent	cm⁻¹	Ref.
CCl_4	1748	1

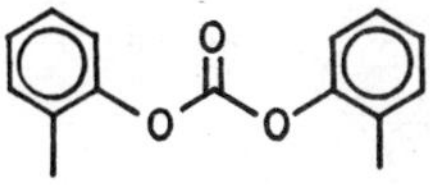

Solvent	cm⁻¹	Ref.
CCl_4	1822m 1748	6
$CHCl_3$	1808 1779	7

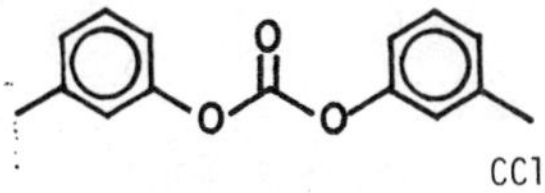

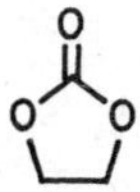

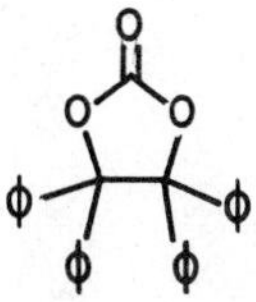

ϕ

Solvent	cm⁻¹	Ref.
liq.	1757	4

Solvent	cm⁻¹	Ref.
CCl_4	1746	1
$CHCl_3$	1738	1

ϕ ϕ ϕ ϕ

Solvent	cm⁻¹	Ref.
KBr	1786	8

Solvent	Frequency	Ref.
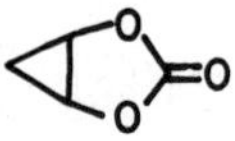 CCl_4	1859 1844 1822	9
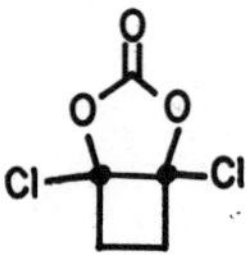 CCl_4	1885	10
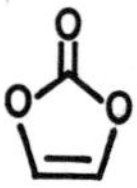 CCl_4	1833 1770w	4
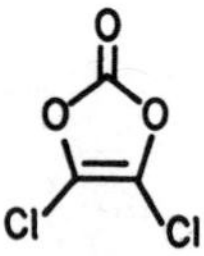 liq.	1890	10
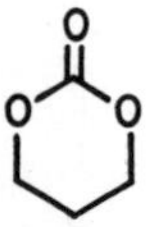 CCl_4	1777 1753w	6
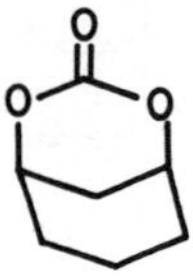 CCl_4	1762	6
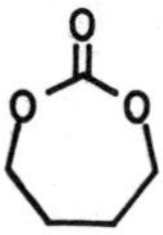 liq.	1730 1718	11
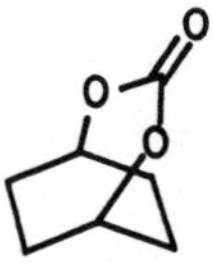 CCl_4	1753 1718w	6

Solvent	Frequency	Ref.
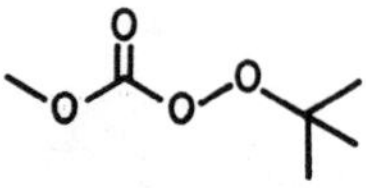 CCl_4	1795 1764	5
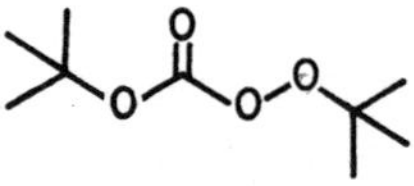 liq.	1789	12
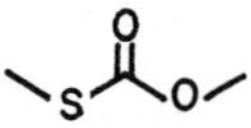 CCl_4	1719	13
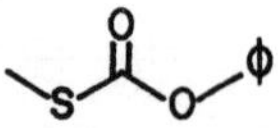 CCl_4	1734	13
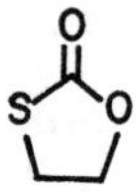 CCl_4 $CHCl_3$	1757 1739	14 7
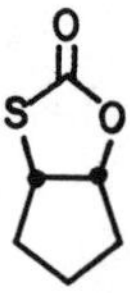 liq.	1736	15
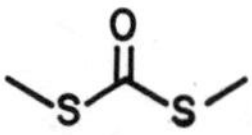 CCl_4	1653	13

Compound	Solvent	Frequency	Ref.
φ–S–C(=O)–S–φ	nuj.	1779	3
1,3-dithiolan-2-one (ring S–C(=O)–S)	CCl_4	1718 1677 1640	14
	$CHCl_3$	1672 1637	7
2,2-dimethyl-1,3-dithietan-4-one (ring S–C(=O)–S)	CS_2	1780 1710	16
1,3-dithiolane-2-thione (ring S–C(=S)–S)	CCl_4	1083	17
	CS_2	1079	17

REFERENCES

1. H. W. Thompson and D. A. Jameson, *SCA* (1958) *13*, 236.
2. E. H. Hoffmeister and D. S. Tarbell, *T* (1965) *35*, 2857.
3. B. M. Gatehouse, S. E. Livingstone and R. S. Nyholm, *JCS* (1958), 3137.
4. J. L. Hales, J. Idris Jones and W. Kynaston, *JCS* (1957), 618.
5. H. Minato, *BSC Japan* (1963), 1020.
6. H. K. Hall, Jr. and R. Zbinden, *JACS* (1958), 6428.
7. F. N. Jones and S. Andreades, *JOC* (1969), 3011.
8. W. A. Mosher, F. W. Steffgen and P. T. Lansbury, *JOC* (1961), 670.
9. F. W. Breitbeil, D. T. Dennerlein, A. E. Fiebig and R. E. Kuznicki, *JOC* (1968), 3389.
10. H.-D. Scharf, W. Droste and R. Liebig, *Angew* (1968), 194.
11. S. Sarel, L. A. Pohoryles and R. Ben-Shoshan, *JOC* (1959), 1873.
12. P. D. Bartlett and H. Sakurai, *JACS* (1962), 3269.
13. A. W. Baker and G. H. Harris, *JACS* (1960), 1923.
14. R. A. Nyquist and W. J. Potts, *SCA* (1961), 679.
15. L. Goodman, A. Benitez, C. D. Anderson and B. R. Baker, *JACS* (1958), 6582.
16. J. I. Jones, W. Kynaston and J. L. Hales, *JCS* (1957), 614.
17. L. J. Bellamy and P. E. Rogasch, *JCS* (1960), 2218.

4.15

ketenes ketimines

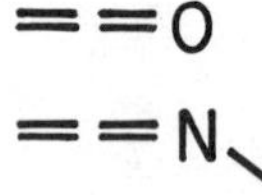

Like allenes, both ketenes and ketenimines show coupled vibrations. In both cases the asymmetric vibration occurs as a strong, diagnostically useful band in the range of 2200 to 2000 cm^{-1}. The symmetric vibration occurs around 1100 cm^{-1} and is of little diagnostic value.

Structure	Phase	cm⁻¹	Ref.
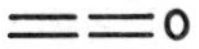 ==O	gas	2151 1120	1
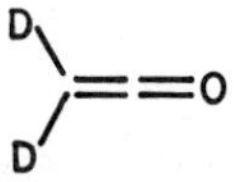 D, D ==O	gas	2119 870	1
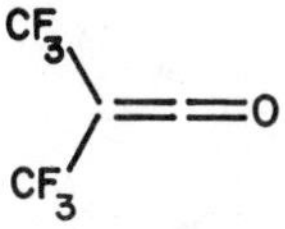 CF_3, CF_3 ==O	gas	2195	2
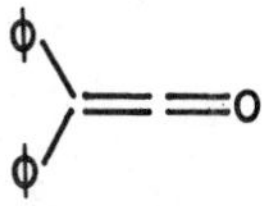 ϕ, ϕ ==O	CCl_4	2105	2
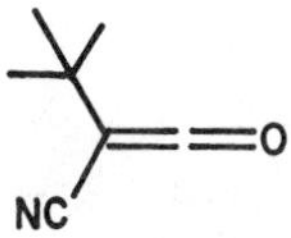 NC ==O	liq.	2220w 2130	3
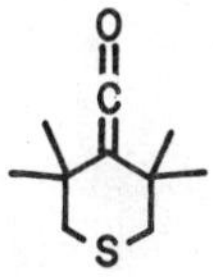 O=C, S	liq.	2200	4
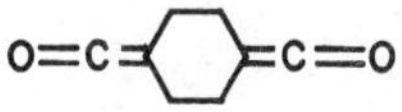 O=C= =C=O	nuj.	2105	5
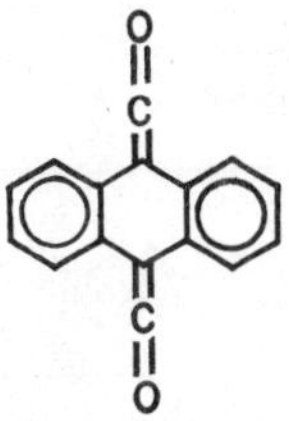 O=C, C=O	nuj.	2083	6
 O=C=C=C=O	gas	2400 2378 2151 1579 1568 1573	7
	solid	2250 1585	7
 S=C=C=C=S	solid	2135 2080 2054m 1030m 1023 1015m	8
	CS_2	2065 1019	8
 ==N	liq.	2140	9
	solid	2152	9
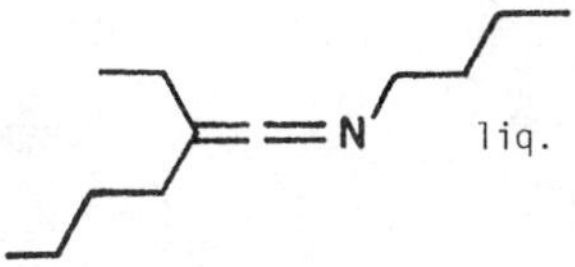 ==N	liq.	2045	10
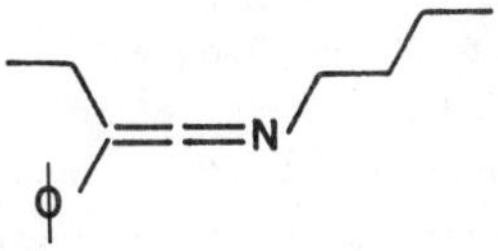 ϕ ==N	liq.	2000	11
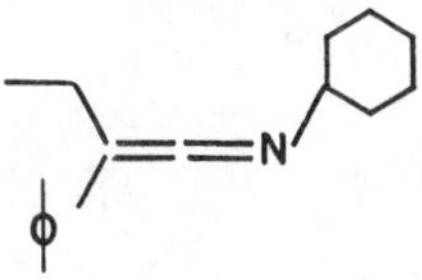 ϕ ==N	liq.	2000	11

Phase	Frequency	Ref.
liq.	2020	10
liq.	2009	10
liq.	2012	10
CS_2	2000	10
KBr	2273	12
CH_2Cl_2	2160	12
KBr	2137	12
KBr	2075	12

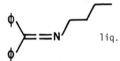

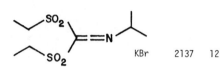

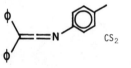

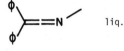

REFERENCES

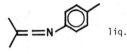

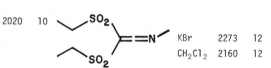

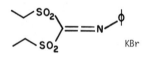

1. W. F. Arendale and W. H. Fletcher, *JCP* (1957) *26*, 793.
2. Sh. Nadzhimudtinov, N. A. Slovokhotova and V. A. Kargin, *Zh Fiz Khim* (1966), 893.
3. H. W. Moore and W. Weyler, Jr., *JACS* (1970), 4132.
4. A. E. DeGroot, J. A. Boerma, J. DeVolk and H. Wynberg, *JOC* (1968), 4025.
5. W. R. Hatchard and A. K. Schneider, *JACS* (1957), 6261.
6. A. T. Blomquist and Y. C. Meinwald, *JACS* (1957), 2021.
7. F. A. Miller and W. G. Fateley, *SCA* (1964), 253.
8. J. Bates and W. H. Smith, *SCA* (1971), 409.
9. G. Rapi and G. Sbrana, *JACS* (1971), 5213.
10. C. L. Stevens and J. C. French, *JACS* (1954), 4398.
11. W. S. Wadsworth, Jr. and W. D. Emmons, *JOC* (1964), 2816.
12. K. Hartke, *B* (1966), 3163.

4.16

isocyanates isothiocyanates

ISOCYANATES

$-N{=}C{=}O$ I

$\updownarrow$

$-\overset{+}{N}{\equiv}C{-}\overset{-}{O}$ II

Organic isocyanates may be represented by the resonance forms I and II. The predominant contribution comes from I with the result that isocyanates show a high-frequency asymmetric coupled mode and a symmetrically coupled vibration at lower frequencies. In practice, the asymmetric mode appears in the narrow range of 2280 to 2250 cm^{-1}. The symmetric mode occurs as a weak absorption between 1460 and 1340 cm^{-1} and is of little diagnostic value. Carboxylic acid isocyanates absorb in the same general region, but frequencies as low as 2225 cm^{-1} may be observed. On rare occasions two bands are seen in the region of the asymmetric mode.

Changes in state or solvent have little or no effect on the frequency of the asymmetric mode.

ISOTHIOCYANATES

−N=C=S

In the case of isothiocyanates a greater contribution occurs from the resonance form II, which would tend to increase the frequency of the asymmetric mode. This is, however, balanced by difference in reduced mass such that, in general, isothiocyanates absorb at frequencies slightly lower than those of isocyanates. Nevertheless, electron withdrawing groups, including alkenyl and aryl, decrease the contribution from II and lower the asymmetric stretching frequency. The symmetric mode occurs as an absorption of variable intensity in the range of 1250 to 1080 cm^{-1} and is of no diagnostic value.

Aliphatic isothiocyanates, unlike their aromatic counterparts, exhibit two, and occasionally more, bands in the region of the asymmetric vibration. The bands are sensitive to changes in both state and solvent. Thus in changing to a more polar solvent or a condensed phase, the band close to 2100 cm^{-1} may increase in frequency (up to 50 cm^{-1}), whereas the other band in this region will decrease in frequency. Aromatic and alkenyl isothiocyanates generally show only one band in the asymmetric region, and once again this band increases in frequency as the polarity of the solvent increases, or in going to a condensed phase.

Compound	Phase	cm⁻¹	Ref.
HNCO	gas	3531 2274 1327 797	1
KNCO	solid	3133 2246 1326	1
 NCO	liq.	2280	2
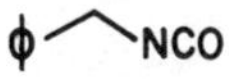 ϕ NCO	CCl_4	2265	3
 O_2N NCO	$CHCl_3$	2268	4
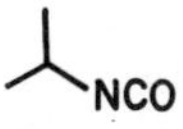 NCO	liq.	2270	2
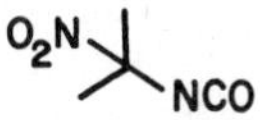 O_2N NCO	$CHCl_3$	2252	4
 NCO	CCl_4	2273	3
 OCN NCO	CCl_4	2270	5

Compound	Phase	cm⁻¹	Ref.
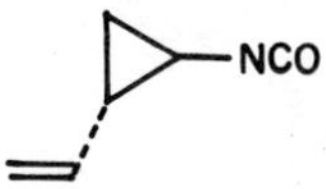 NCO	liq.	2265 1637	6
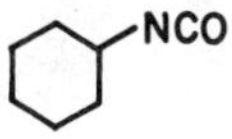 NCO	CCl_4	2257	3
 NCO	CCl_4	2256 1629	7
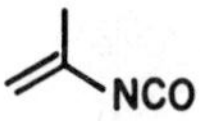 NCO	CCl_4	2262 1651	7
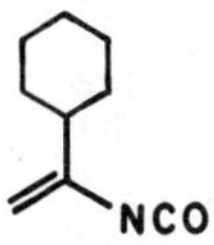 NCO	CCl_4	2275 2254 1639	7
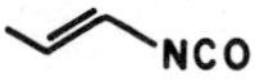 NCO	CCl_4	2265 1660	7
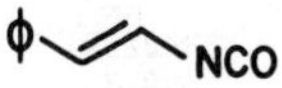 ϕ NCO	CCl_4	2258 1646	7

Structure	Solvent	Frequency	Ref.
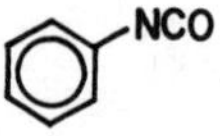 NCO	CCl_4	2267	8
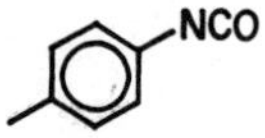 NCO	CCl_4	2263	3
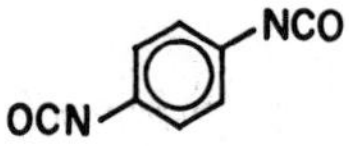 OCN, NCO	CCl_4	2257	3
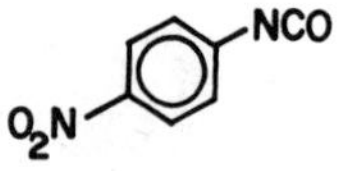 O_2N, NCO	CCl_4	2261	3
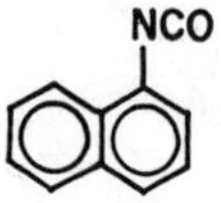 NCO	CCl_4	2275	5
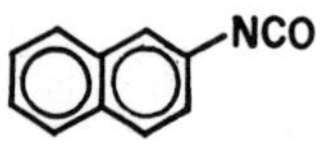 NCO	CCl_4	2267	5
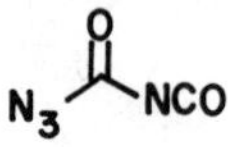 O, N_3, NCO	CCl_4	2255 2170 1754 1718	9

Structure	Solvent	Frequency	Ref.
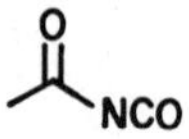 O, NCO	CCl_4	2246	3
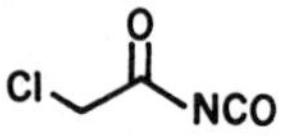 Cl, O, NCO	liq.	2250	10
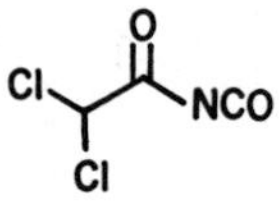 Cl, Cl, O, NCO	liq.	2250	10
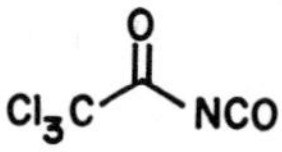 Cl_3C, O, NCO	liq.	2250	10
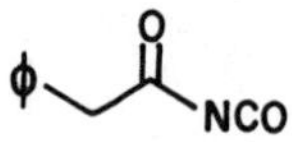 ϕ, O, NCO	liq.	2250	10
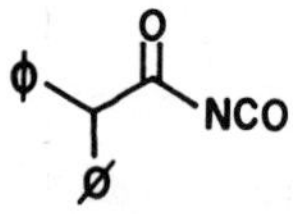 ϕ, ϕ, O, NCO	liq.	2225	10
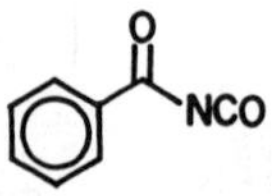 O, NCO	liq.	2225	10
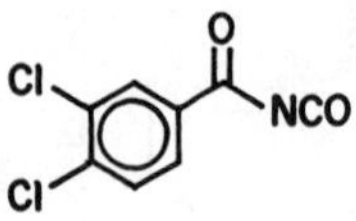 Cl, Cl, O, NCO	liq.	2275	10

Compound	Medium	Frequency	Ref.
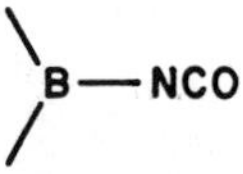 B—NCO	liq.	2280 2230	11
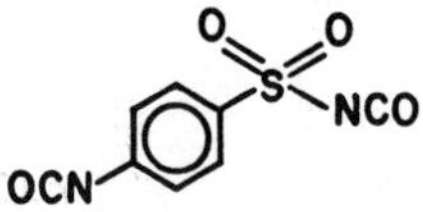 OCN, O, O, S, NCO	$CHCl_3$	2273 2232	12
HNCS	CCl_4	1980 3469	15
KNCS	nuj.	2020	2
 NCS	liq.	2206 2126	13
	CCl_4	2221 2106 2077	3
 NCS	liq.	2183 2114	13
	CCl_4	2103	8
 NCS	liq.	2180 2100	14
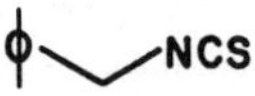 ϕ NCS	liq.	2180m 2100	4
 NCS	CCl_4	2173 2097 2068	3

Compound	Medium	Frequency	Ref.
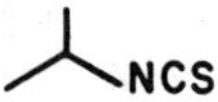 NCS	liq.	2149 2093	13
	CCl_4	2083	8
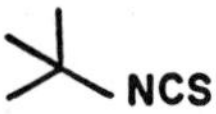 NCS	liq.	2090 1976w	13
	CCl_4	2090	15
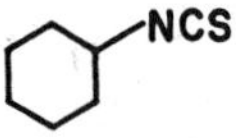 NCS	CCl_4	2180 2100 2067 2053	3
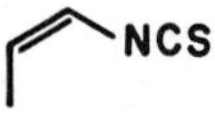 NCS	liq.	2105	16
 NCS	liq.	2062	16
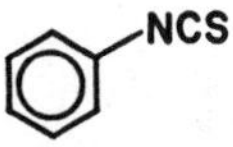 NCS	liq.	2090	14
	CCl_4	2065	14
	$CHCl_3$	2112	14
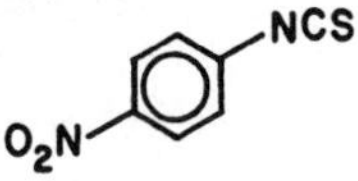 O_2N, NCS	$CHCl_3$	2045	5
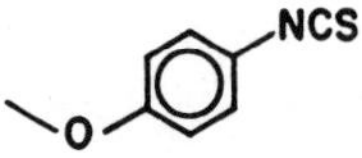 O, NCS	liq.	2100	4
	CCl_4	2087	5
	$CHCl_3$	2130	5

Compound	Solvent	cm⁻¹	Ref.
1-naphthyl NCS	CCl_4	2190 2150 2080 2015	3
2-pyridyl NCS	$CHCl_3$	2017	6
benzoyl NCS	CCl_4	1990	5
	$CHCl_3$	2000	5
4-methoxybenzoyl NCS	$CHCl_3$	1998	5
phenyl NCSe	liq.	2127 2048w	7
4-methoxyphenyl NCSe	KBr	2135 2068w	7

REFERENCES

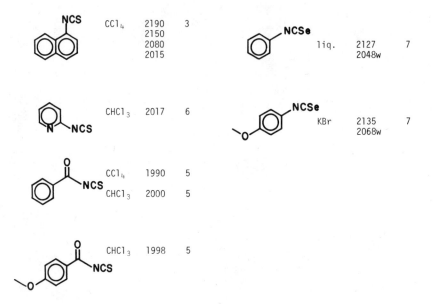

1. G. Herzberg and C. Revil, *Disc Farad Soc* (1950) *9*, 92.
2. R. P. Hirschmann, R. N. Kniseley and V. A. Fassel, *SCA* (1965), 2125.
3. H. Hoyer, *B* (1956), 2677.
4. H. E. Ungnade and L. W. Kissinger, *JOC* (1957), 1662.
5. W. H. T. Davison, *JCS* (1953), 3712.
6. E. Vogel, R. Erb, G. Lenz and A. A. Bothner-By, *A* (1965) *682*, 1.
7. M. Sato, *JOC* (1961), 770.
8. G. L. Caldow and H. W. Thompson, *SCA* (1958) *13*, 212.
9. H. Roesky and O. Glemser, *B* (1964), 1710.
10. A. J. Speziale and L. R. Smith, *JOC* (1962), 3742.
11. J. Goubeau and H. Gräbner, *B* (1960), 1379.
12. H. Ulrich, B. Tucker and A. A. Sayigh, *JOC* (1966), 2568.
13. R. N. Kniseley, R. P. Hirschmann and V. A. Fassel, *SCA* (1967), 109.
14. N. S. Ham and J. B. Willis, *SCA* (1960), 279.
15. E. Lieber, C. N. R. Rao and J. Ramachandran, *SCA* (1958) *13*, 296.
16. M. G. Ettlinger and J. E. Hodgkins, *JACS* (1955), 1831.

4.17

thiocarbonyl compounds

I

II

The π-orbital overlap between first- and second-row elements is less than with elements of the same periodic row. Consequently, there is a greater contribution from the dipolar form II in the structure of the C=S bond compared to that of the C=O bond. This affects both the chemistry of such compounds (thio aldehydes and ketones frequently undergo dipolar additions and usually exist as trimers) as well as their infrared group frequencies.

In those cases where the C=S-containing molecule is monomeric there appears to be little correlation between the frequency of the C=S chromophore and the electronegativities of the neighboring groups. Nonetheless all these systems exhibit a strong band in the region of 1250 to 1100 cm^{-1}.

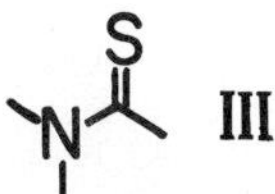

Systems containing the N-C=S grouping also have a large contribution from the dipolar form IV, which gives rise to bands in the region of 1600 to 1450 cm^{-1}. The bands cannot be assigned to a particular vibration of one

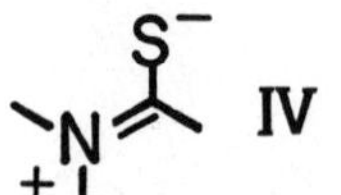

group and are best designated as symmetric and asymmetric vibrations of the NCS system.[a]

[a]M. Davis and W. J. Jones, *JCS* (1958), 955.

Compound	Medium	Frequency	Ref.
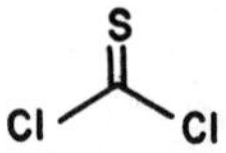 $Cl_2C{=}S$	liq.	1124	1
O = C = S	gas	1048m 859m 2064	2
S = C = S	liq.	1510	3
Se = C = S	liq.	1408	3
Te = C = S	CS_2	1347	3
 pyridine-4(1H)-thione (S, N, H)	KBr	1122 1111	6
	$CHCl_3$	1126	6
 N-methylpyridine-4-thione (S, N)	KBr	1111	6
	CS_2	1127	6
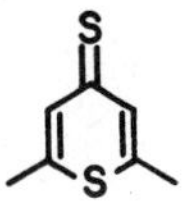 2,6-dimethylthiopyran-4-thione (S, S)	$CHCl_3$	1089	7
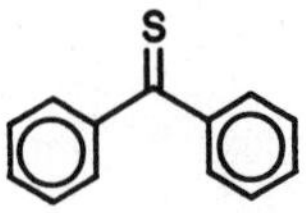 thiobenzophenone (S)	nuj.	1207	4
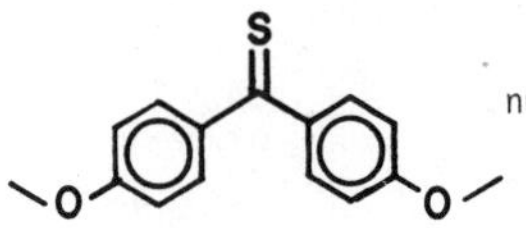 4,4'-dimethoxythiobenzophenone (S, O, O)	nuj.	1224	4
 pyran-4-thione (S, O)	$CHCl_3$	1156	5
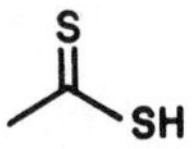 thioacetic acid (S, SH)	liq.	1216	8
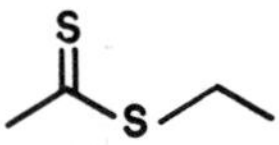 ethyl dithioacetate (S, S)	liq.	1195	9
	CCl_4	1197	10
	CS_2	1195	10
butyl dithioacetate (S, S)	CCl_4	1197	10
	CS_2	1195	10

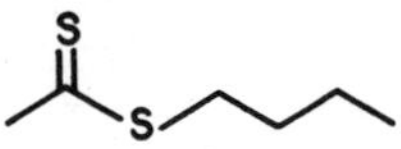

Compound	Phase	cm⁻¹	Ref.
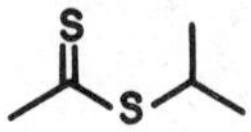 S-isopropyl dithioacetate	liq.	1195	9
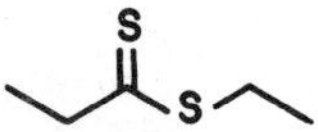 S-ethyl dithiopropionate	liq.	1187	9
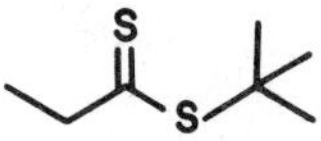 S-tert-butyl dithiopropionate	liq.	1172	9
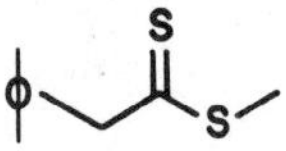 methyl phenyldithioacetate	KBr	1215	11
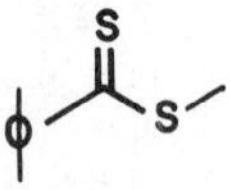 methyl dithiobenzoate	KBr	1226	11
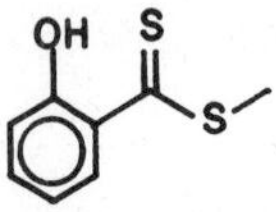 methyl 2-hydroxydithiobenzoate	KBr	1219	11
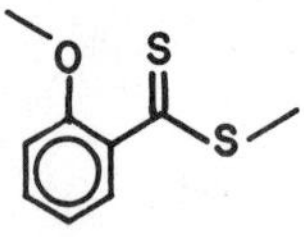 methyl 2-methoxydithiobenzoate	KBr	1218	11
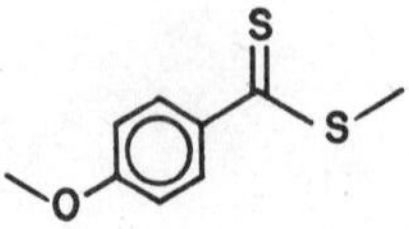 methyl 4-methoxydithiobenzoate	KBr	1215	11
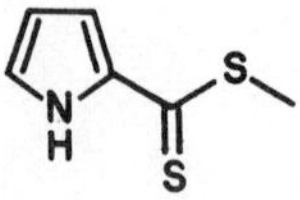 methyl pyrrole-2-dithiocarboxylate	KBr	1206	11
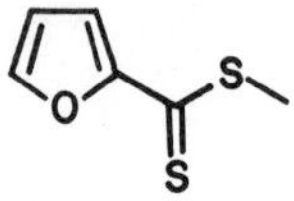 methyl furan-2-dithiocarboxylate	KBr	1203	11
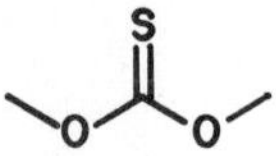 dimethyl thionocarbonate	liq.	1127	5
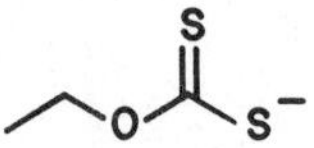 sodium ethyl xanthate (Na^+)	nuj.	1179	12
potassium ethyl xanthate (K^+)	nuj.	1143	12
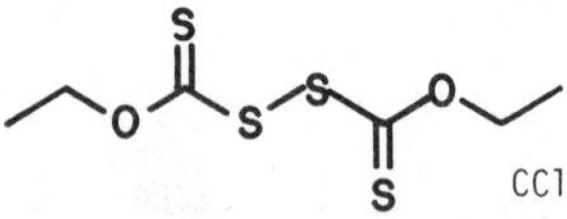 diethyl dixanthogen	CCl_4	1262 1242	12
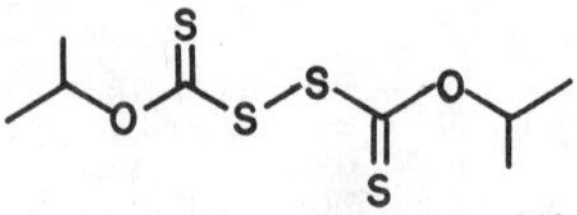 diisopropyl dixanthogen	CCl_4	1258	12

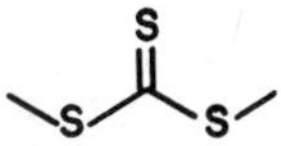

Medium	Frequencies	Ref.
liq.	1076	5
KBr	1171 1533 1290	13
KBr	1058	5
CCl_4	1081 1033	14
liq.	1125 1443 1325	15
liq.	1136 1537 1479 1443 1297	16
liq.	1020 1400b 1302	17
CCl_4	966 1367 1316	17
liq.	1100 1547 1360 1242	16
$CHCl_3$	1099 1534 1390	16
liq.	1181 1542	9
CCl_4	1122	5
solid	1260 1504	9

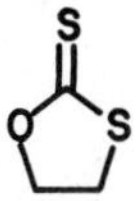

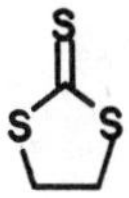

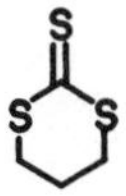

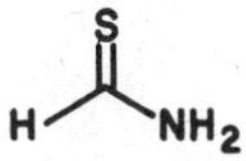

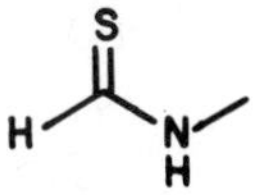

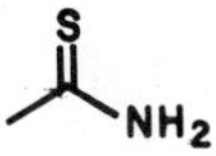

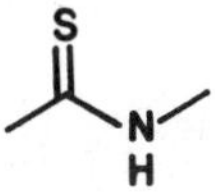

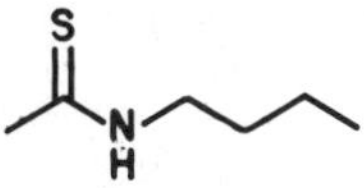

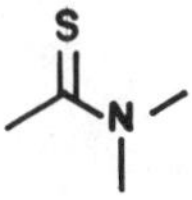

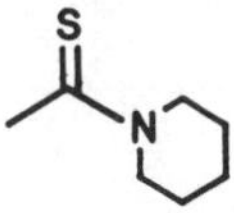

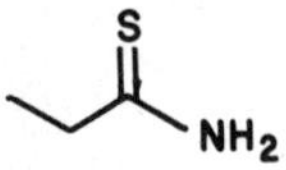

Solvent	Frequency	Ref.
KBr	1097	5
liq.	1122	5
KBr	1109 1533 1290	13
CCl_4	1115 1504 1292	13
$CHCl_3$	1112 1510 1290	13
KBr	1115 1572 1315	13
$CHCl_3$	1112 1545 1317	13
KBr	1113 1550 1333	13
$CHCl_3$	1117 1545 1317	13

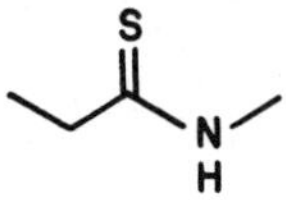

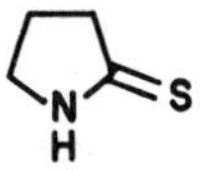

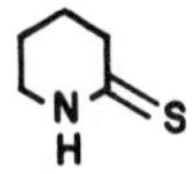

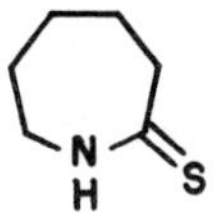

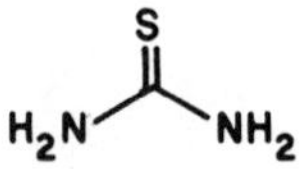

Solvent	Frequency	Ref.
KBr	1080m 1585 1460	18
KBr	1155m 1115m 1555 1255	18
KBr	1183	5
CCl_4	1155 1110m 1525 1230m	18
KBr	1201 1506 1273	13
KBr	1204 1550 1312	13
KBr	1209 1565 1322	13

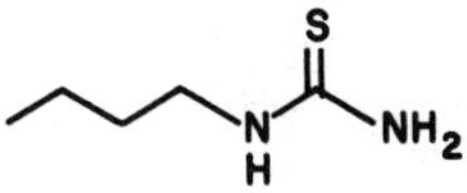

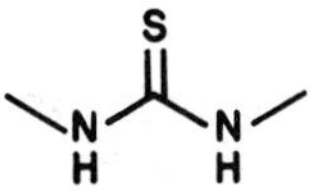

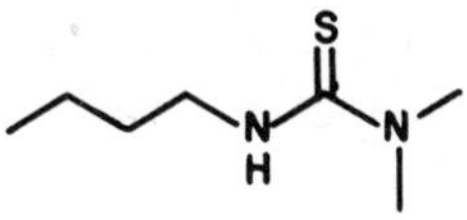

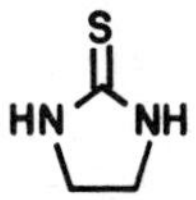

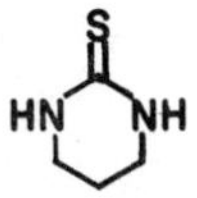

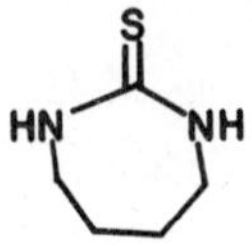

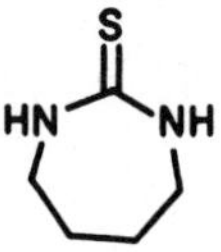

Solvent	Frequency	Ref.	Solvent	Frequency	Ref.
KBr	1209 1565 1322	13	KBr	1167 1590	5
KBr	1047 1515 1290	13	$CHCl_3$	1104 1581	20
nuj.	1155 1045	19	$CHCl_3$	1088	20
$CHCl_3$	1150 1050	19	$CHCl_3$	1169 1570	20
CS_2 $CHCl_3$	1144 1143	6 6	$CHCl_3$	1176	20
CCl_4 CS_2	1143 1117 1142 1114	6 6	solid	1207	21
KBr	1186 1567	20	$CHCl_3$	1165 1127 1070	22

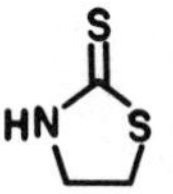

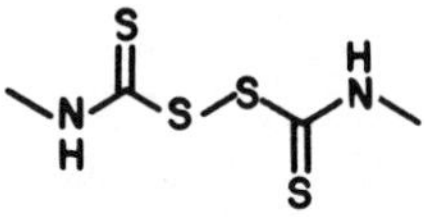

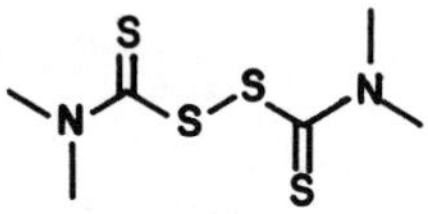

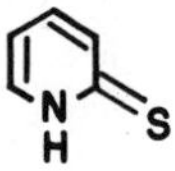

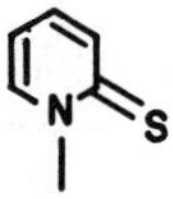

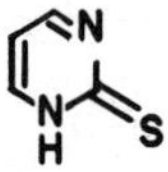

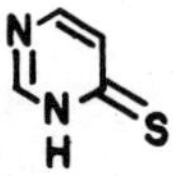

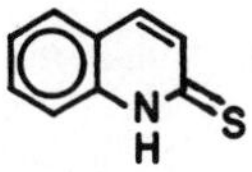

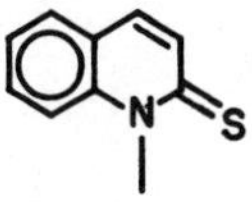

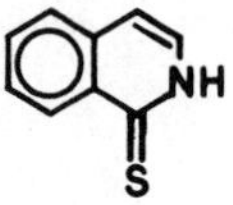

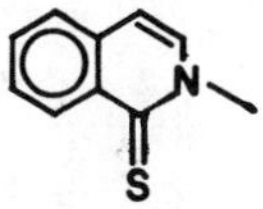

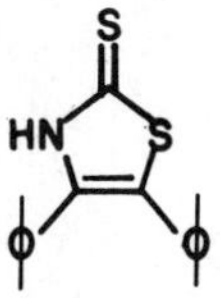

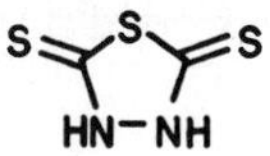

Compound	Medium	ν	Ref.	Compound	Medium	ν	Ref.
Se=C=Se	liq.	1267	3	n-$C_4H_9NHC(=Se)NH_2$	KBr	1135m 1110m 1550 1285	18
$H_2NC(=Se)NH_2$	KBr	1082m 1600 1400	18	n-$C_4H_9NHC(=Se)N(CH_3)_2$	CCl_4	1118m 1111m 1535 1238	18

REFERENCES

1. C. F. Pouchert, *The Aldrich Library of Infrared Spectra*, Aldrich 1970, spectrum 325 F.
2. H. J. Callomon, D. C. McKean and H. W. Thompson, *Proc Roy Soc* (1951), *208A*, 341.
3. T. Wentink, Jr., *JCP* (1958) *29*, 188.
4. N. Lozach and G. Guillouzo, *BSC France* (1957), 1221.
5. R. Mecke, R. Mecke and A. Lüttringhaus, *B* (1957), 975.
6. E. Spinner, *JOC* (1958), 2037.
7. A. R. Katritzky and R. A. Jones, *SCA* (1961), 64.
8. R. Mecke and H. Spiesecke, *B* (1956), 1110.
9. C. S. Marvel, P. De Radzitzky and J. J. Brader, *JACS* (1955), 5997.
10. L. J. Bellamy and P. E. Rogash, *JCS* (1960), 2218.
11. B. Bak, L. Hansen-Nygaard and C. Pedersen, *Acta Chem Scand* (1958), 451.
12. M. L. Shankaranarayana and C. C. Patel, *Canad J Chem* (1961), 1633.
13. R. Mecke, Jr. and R. Mecke, Sr., *B* (1956), 343.
14. S. M. Iqbal and L. N. Owen, *JCS* (1960), 1030.
15. I. Suzuki, *BCS Japan* (1962), 1286.
16. I. Suzuki, *BCS Japan* (1962), 1456.
17. I. Suzuki, *BCS Japan* (1962), 1449.
18. C. Collard-Charon and M. Renson, *BSC Belge* (1963), 149.
19. C. D. Thorn, *Canad J. Chem* (1960), 2349.
20. E. Spinner, *JOC* (1960), 1237.
21. R. Gompper and H. Herlinger, *B* (1956), 2825.
22. C. D. Thorn, *Canad J Chem* (1960),1438.

5.1

azo compounds

The N-N double-bond stretching vibration is weak in all azo compounds so far studied, and differentiation between this mode and those associated with aromatic systems is difficult.

Too few examples of azo-group frequencies have been reported to enable characteristic trends to be discerned. Cis-azobenzene absorbs at a frequency higher than that of the trans isomer, and substitution with either electron withdrawing or donating groups causes little change in the azo-group stretching frequency of azobenzenes.

Structure	Phase	Frequency	Ref.
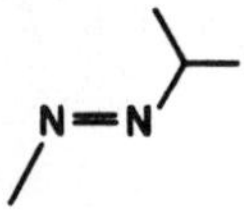 N=N (methyl/isopropyl)	liq.	1563	1
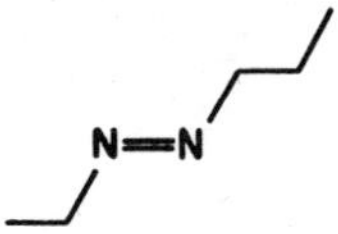 N=N (ethyl/propyl)	liq.	1563	1
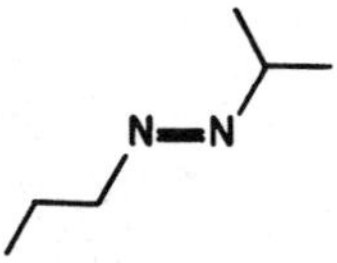 N=N (propyl/isopropyl)	liq.	1563	1
 N=N (ring)	CCl_4	1545	2
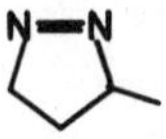 N=N (ring, methyl)	CCl_4	1540	2
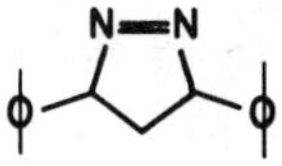 N=N (ring, ϕ, ϕ)	KBr	1548	3
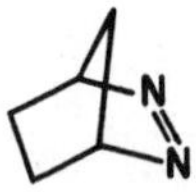 N=N (bicyclic)	KBr	1502 1449	4
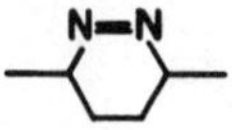 N=N (ring, dimethyl)	$CHCl_3$	1608w 1458m	4
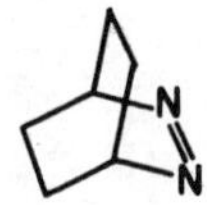 N=N (bicyclic)	KBr	1527 1508 1453	4
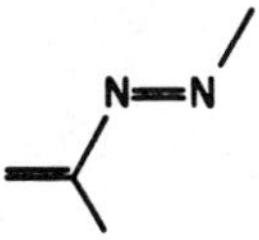 N=N	sng	1500 1640	5
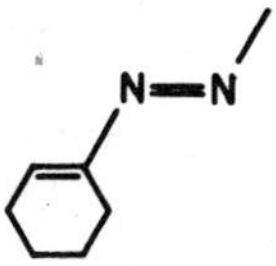 N=N	sng	1500 1640	5
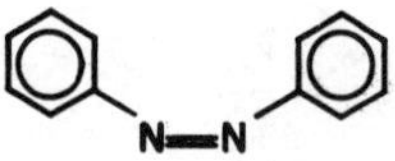 N=N	KBr	1511	6
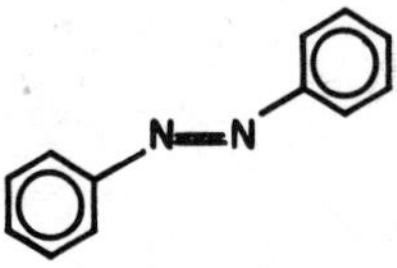 N=N	KBr	1455 1395w	7
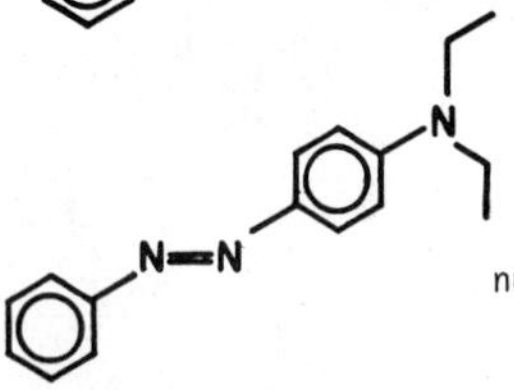 N=N, N	nuj.	1403 1336	8
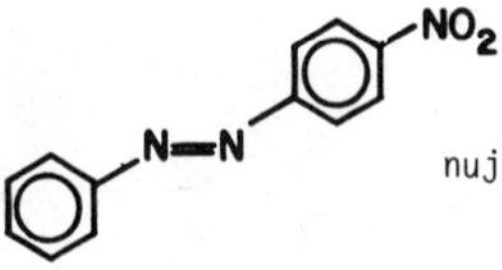 N=N, NO_2	nuj.	1442	8

Structure	Medium	Frequency	Ref.
4-methoxyazobenzene (N=N)	nuj.	1455 1448	8
4,4'-dinitroazobenzene (N=N, NO_2, O_2N)	KBr	1467 1408	7
4,4'-diaminoazobenzene (N=N, NH_2, H_2N)	KBr	1462 1420	7
4,4'-dimethoxyazobenzene (N=N)	KBr	1499 1250	9

REFERENCES

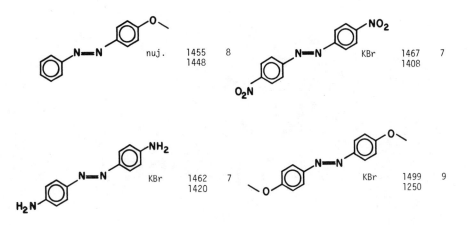

1. L. Spialter, D. H. O'Brien, G. L. Untereiner and W. A. Rush, *JOC* (1965), 3278.
2. R. J. Crawford, A. Mishra and R. J. Dummel, *JACS* (1966), 3959.
3. G. C. Overberger, J.-P. Anselme and J. R. Hall, *JACS* (1963), 2752.
4. S. G. Cohen and R. Zand, *JACS* (1962), 586.
5. B. T. Gillis and J. D. Hagarty, *JACS* (1965), 4576.
6. R. Kübler, W. Lüttke and S. Weckherlin, *Z Electrochem* (1960), 650.
7. K. Ueno, *JACS* (1957), 3205.
8. R. J. W. Le Fèvre and R. L. Werner, *Aust J Chem* (1957), 26.
9. W. Maier and G. Englert, *Z Phys Chem* (Frankfurt) (1959) *19*, 168.

5.2 diazo compounds

$$\rangle \overset{+}{C}-\overset{-}{\bar{N}}\equiv N \quad \text{I}$$
$$\updownarrow$$
$$\rangle C=\overset{+}{N}=\overset{-}{\bar{N}} \quad \text{II}$$

Organic diazo compounds have contributions from the resonance forms I and II and exhibit a strong band in the region from 2200 to 2000 cm^{-1}. Electron withdrawing groups, especially phenyl substituents, stabilize the resonance form I and cause a decrease in the vibrational frequency of diazo chromophores. Within phenyldiazo compounds electron withdrawing groups raise, and electron donating groups lower, the frequency.

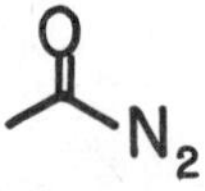

Diazoketones show an increase in frequency, compared to the simple aliphatic diazocompounds, which reflects a greater contribution from the resonance form I, as a result of delocalization into the carbonyl group. This in turn leads to a lowering of the carbonyl stretching frequency compared to that of simple ketones (Chapter 4.2, p. 175).

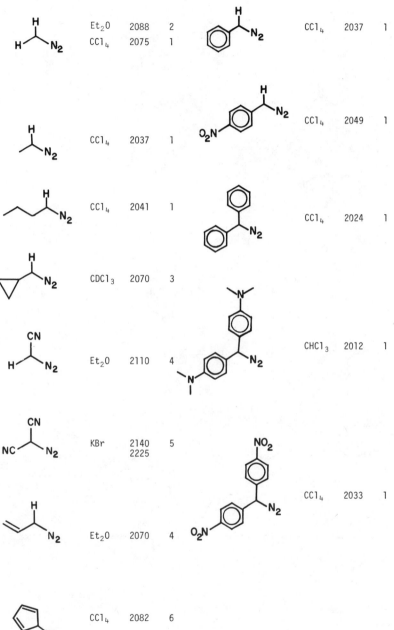

Structure	Solvent	ν	Ref.
CH_2N_2 (H, H, N_2)	Et_2O	2088	2
	CCl_4	2075	1
CH_3CHN_2 (H, N_2)	CCl_4	2037	1
$CH_3CH_2CH_2CHN_2$ (H, N_2)	CCl_4	2041	1
cyclopropyl-CHN_2 (H, N_2)	$CDCl_3$	2070	3
H, CN, N_2	Et_2O	2110	4
NC, CN, N_2	KBr	2140	5
		2225	
allyl-CHN_2 (H, N_2)	Et_2O	2070	4
cyclopentadienyl-N_2	CCl_4	2082	6
phenyl-CHN_2 (H, N_2)	CCl_4	2037	1
O_2N-phenyl-CHN_2 (H, N_2)	CCl_4	2049	1
diphenyl-CN_2 (N_2)	CCl_4	2024	1
bis(4-dimethylaminophenyl)-CN_2 (N, N_2)	$CHCl_3$	2012	1
bis(4-nitrophenyl)-CN_2 (NO_2, O_2N, N_2)	CCl_4	2033	1

Phase	Frequency	Ref.
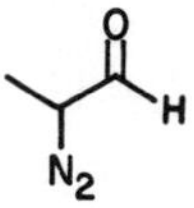 liq.	2092	7
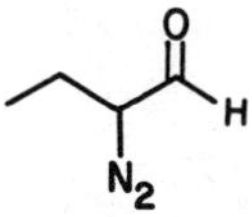 liq.	2083	7
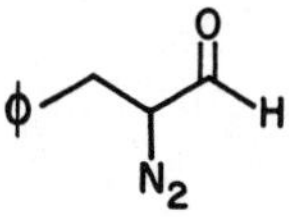 liq.	2092	7
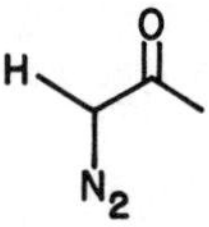 CCl_4	2108 1662	8
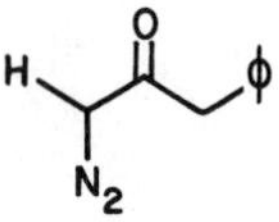 $CHCl_3$	2101 1634	9
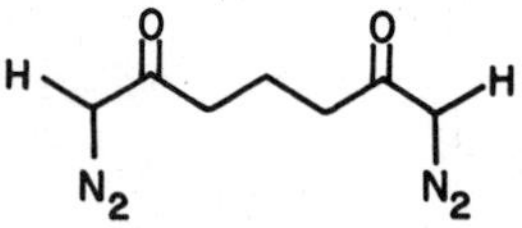 KBr	2232w 2119	10
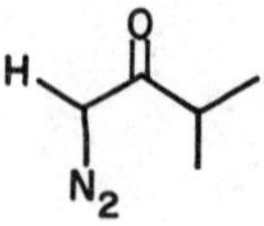 liq.	2198w 2105	7
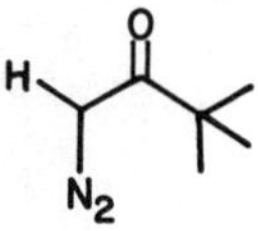 liq.	2227w 2110	7
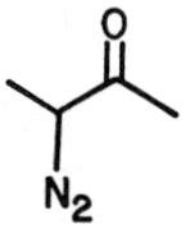 liq.	2128w 2075	7
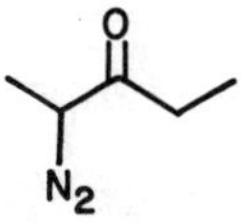 liq.	2128	7
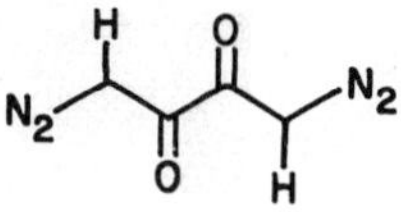 KBr	2217 2128 1621	10
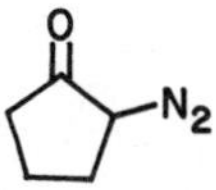 liq.	2086	11
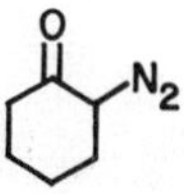 liq.	2083	11
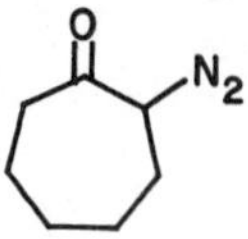 liq.	2079	11

liq.	2081	11

CH_2Cl_2	2058	1

KBr	2075	11

nuj.	2014 1618	12

KBr	2257w 2193m 2151 1672m 1639	10

KBr	2128	13

KBr	2108	8
CCl_4	2108 1633 1622*	8
$CHCl_3$	2111	8

KBr	2165w 2110	1
CCl_4	2112	2
CH_2Cl_2	2101	1
C_2Cl_4	1642	2

KBr	2105 2179	1
CCl_4	2105	2
CH_2Cl_2	2101	1
C_2Cl_4	1631	2

liq.	2169w 2075	1
$CHCl_3$	2062	1

CCl_4	2073 1630 1621*	8
CH_2Cl_2	2062	1

CH_2Cl_2	2088	1

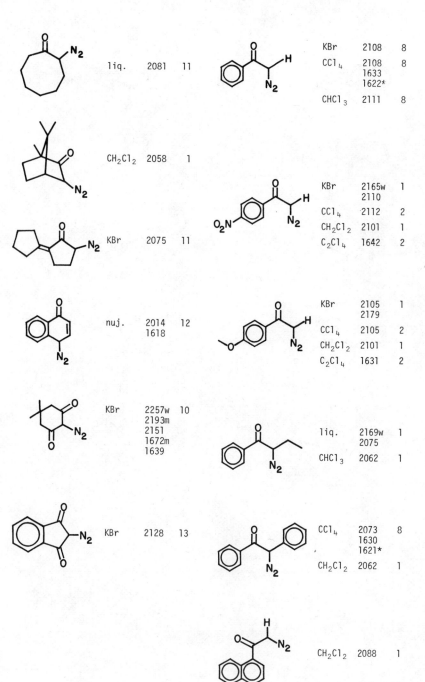

Solvent	Frequency	Ref.
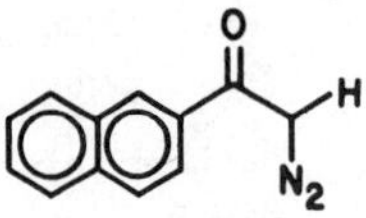 CH_2Cl_2	2092	1
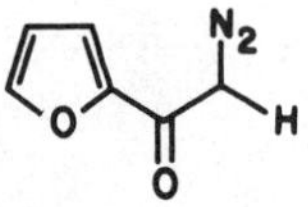 liq.	2169m 2114	7
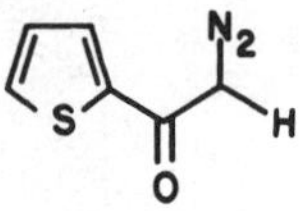 KBr	2110	7
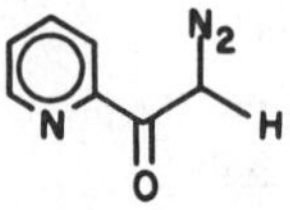 KBr	2112	7
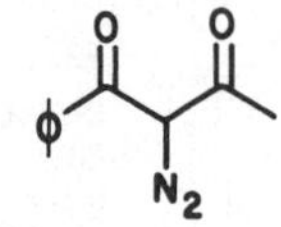 KBr	2123 1650 1629	4
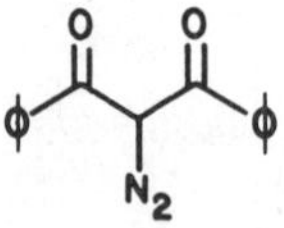 KBr	2128 1645 1631	4
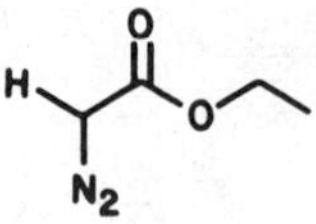 CH_2Cl_2	2101	1
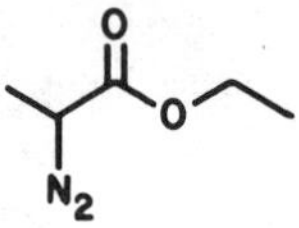 liq.	2083	7
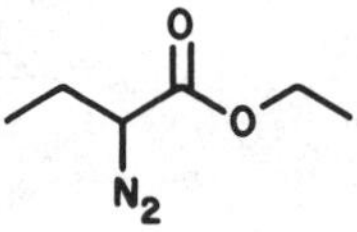 liq.	2088	7
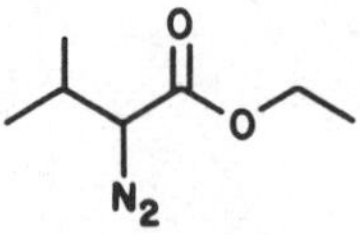 liq.	2092	7
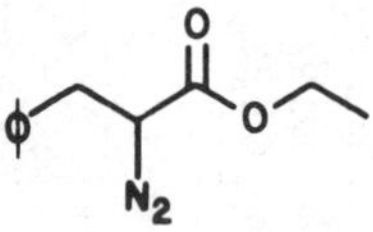 liq.	2088	7
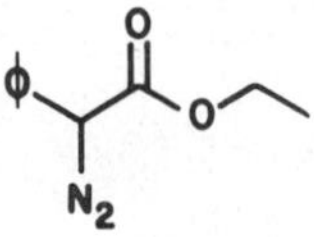 liq.	2092	7
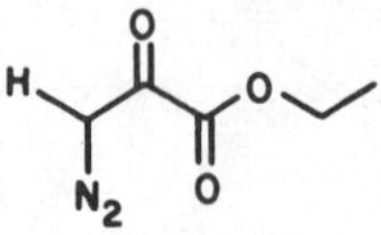 KBr	2188w 2165 2110	10

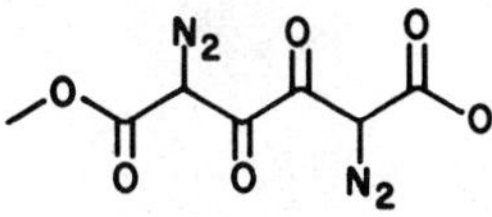	KBr	2203 2160	10
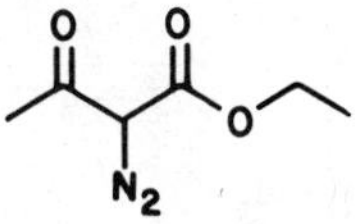	liq.	2146 1724 1661	4
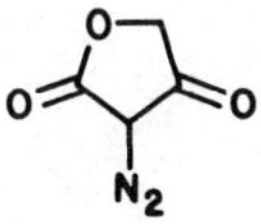	KBr	2222 2169	10
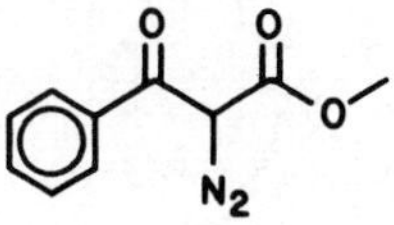	nuj. KBr CCl_4	2162 2165 2147	15 10 14
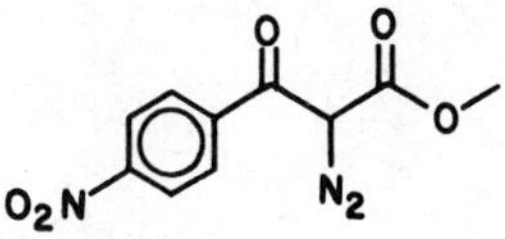	nuj.	2160	15

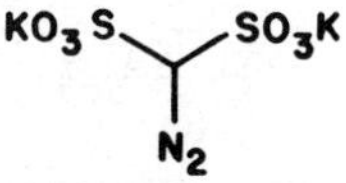	KBr	2105	4
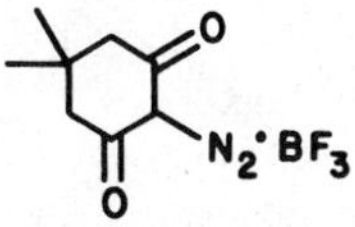	nuj.	2268 2232	16
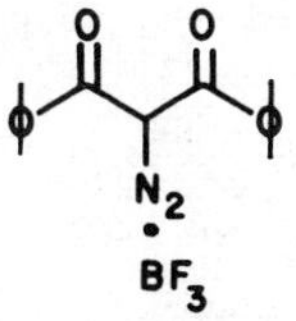	nuj.	2212	16
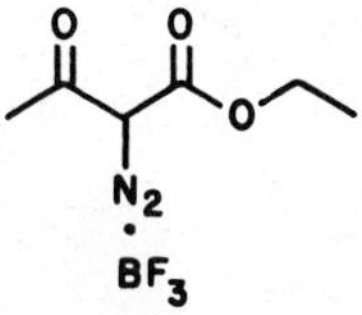	nuj.	2242	16
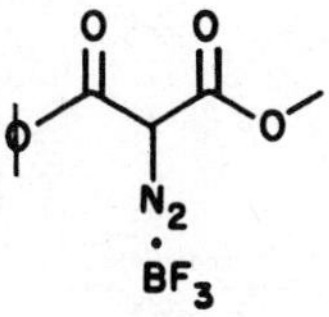	nuj.	2237	16

REFERENCES

1. P. Yates, B. L. Shapiro, N. Yoda and J. Fugger, *JACS* (1957), 5756.
2. A. Foffani, C. Pecile and S. Ghersetti, *T* (1960) *11*, 285.
3. R. A. Moss and F. C. Shulman, *T* (1968), 2881.
4. E. Fahr, *A* (1960) *638*, 1.
5. E. Ciganek, *JOC* (1965), 4198.
6. W. von E. Doering and C. H. DePuy, *JACS* (1953), 5955.
7. M. Regitz and F. Menz, *B* (1968), 2622.
8. C. Pecile, A. Foffani and S. Ghersetti, *T* (1964), 823.
9. H. Dahn and H. Gold, *Helv* (1963), 983.
10. E. Fahr. *A* (1958) *617*, 11.
11. M. Regitz and J. Rüter, *B* (1968), 1263.
12. R. J. W. Le Fèvre, J. B. Sousa and R. L. Werner, *JCS* (1954), 4686.
13. M. Regitz, H. Schwall, G. Heck, B. Eistert and G. Bock, *A* (1965) *690*, 125.
14. J. H. Looker and D. N. Thatcher, *JOC* (1957), 1233.
15. J. H. Looker and C. H. Hayes, *JOC* (1963), 1342.
16. E. Fahr and W. D. Hörmann, *A* (1965) *682*, 48.

5.3 diazonium compounds

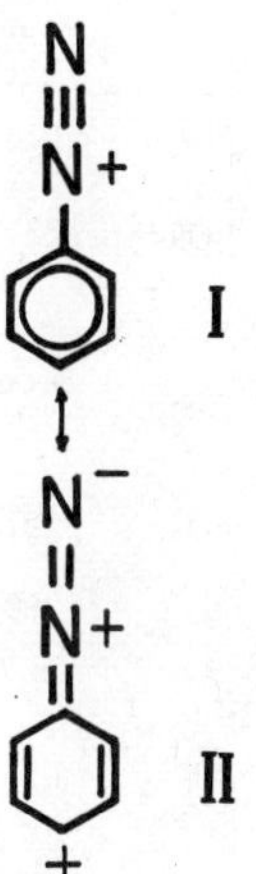

Aryl diazonium salts can be represented by resonance structures I and II. Clearly the principal contribution comes from the first form where each atom maintains an octet, and this results in diazonium groups absorbing in the triple-bond region between 2200 and 2100 cm^{-1}. Electron donating groups either ortho or para to the diazonium group increase the contribution from the resonance form II and consequently decrease the diazonium stretching frequency, whereas electron withdrawing groups decrease the contribution from II and increase the frequency. Small changes in frequency can result from association with the counter ion.

$C_6H_5N_2^+$

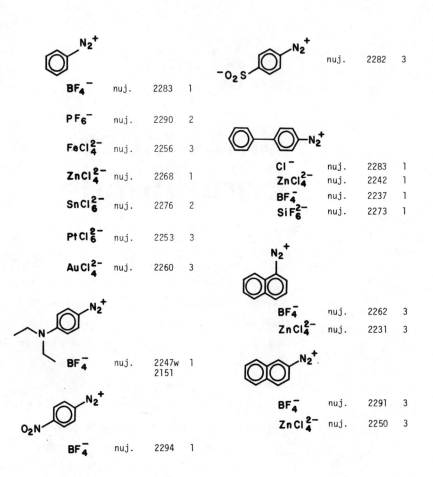

Anion	Medium	Frequency	Ref.
BF_4^-	nuj.	2283	1
PF_6^-	nuj.	2290	2
$FeCl_4^{2-}$	nuj.	2256	3
$ZnCl_4^{2-}$	nuj.	2268	1
$SnCl_6^{2-}$	nuj.	2276	2
$PtCl_6^{2-}$	nuj.	2253	3
$AuCl_4^{2-}$	nuj.	2260	3

$p\text{-}(C_2H_5)_2N\text{-}C_6H_4N_2^+$

Anion	Medium	Frequency	Ref.
BF_4^-	nuj.	2247w 2151	1

$p\text{-}O_2N\text{-}C_6H_4N_2^+$

Anion	Medium	Frequency	Ref.
BF_4^-	nuj.	2294	1

$p\text{-}^-O_2S\text{-}C_6H_4N_2^+$

Medium	Frequency	Ref.
nuj.	2282	3

$p\text{-}C_6H_5\text{-}C_6H_4N_2^+$

Anion	Medium	Frequency	Ref.
Cl^-	nuj.	2283	1
$ZnCl_4^{2-}$	nuj.	2242	1
BF_4^-	nuj.	2237	1
SiF_6^{2-}	nuj.	2273	1

1-Naphthyl-N_2^+

Anion	Medium	Frequency	Ref.
BF_4^-	nuj.	2262	3
$ZnCl_4^{2-}$	nuj.	2231	3

2-Naphthyl-N_2^+

Anion	Medium	Frequency	Ref.
BF_4^-	nuj.	2291	3
$ZnCl_4^{2-}$	nuj.	2250	3

REFERENCES

1. K. B. Whetsel, G. F. Hawkins and F. E. Johnson, *JACS* (1956), 3360.
2. R. H. Nuttall, E. R. Roberts, and D. W. A. Sharp, *SCA* (1961), 947.
3. M. Aroney, R. L. W. Le Fèvre and R. L. Werner, *JCS* (1955), 276.

5.4
azides

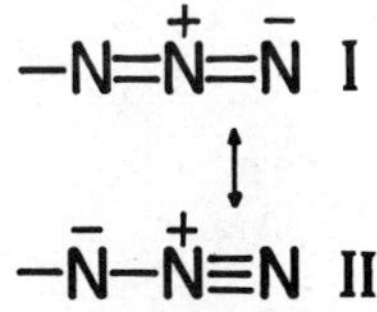

Organic azides can be represented by the resonance forms I and II. The predominant contribution comes from I, and as with allenes and ketenes, coupled vibrations occur leading to a strong and diagnostically useful band in the 2100 cm^{-1} region arising from the asymmetric mode. The symmetric mode gives rise to a weaker band in the region from 1350 to 1170 cm^{-1}, which is of little diagnostic value. The variations of the asymmetric mode with substitution are collected in the tables that follow.

Electron withdrawing groups and the carbonyl group of carboxylic acid azides increase the contribution of the resonance form II and cause an increase in the frequency of the coupled asymmetric vibration.

The symmetric ionic azide ion does not show the coupled symmetric vibration, and association with counter ions can cause changes in the frequency of the asymmetric mode.

Compound	Phase	cm⁻¹	Ref.
HN_3	gas	2140	1
	solid	2162	1
LiN_3	nuj.	2092	4
NaN_3	nuj.	2128	2
	KBr	2140	3
KN_3	nuj.	2041	4
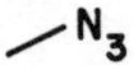 N_3	CCl_4	2090	5
 ϕ N_3	liq.	2103	6
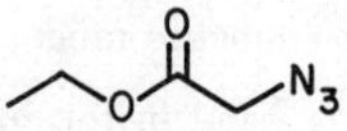 O, O, N_3	CCl_4	2110	5
	liq.	2108 1744	6
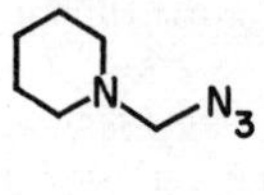 N, N_3	liq.	2110	5
 O, N_3	liq.	2100	5
 S, N_3	liq.	2080	5
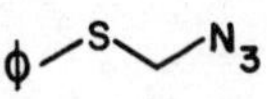 ϕ, S, N_3	liq.	2090	5
 N_3, S, N_3	liq.	2120	5
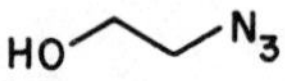 HO, N_3	liq.	2105	6
 N_3	liq.	2107	6
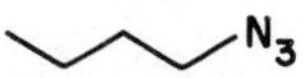 N_3	sng	2083	7
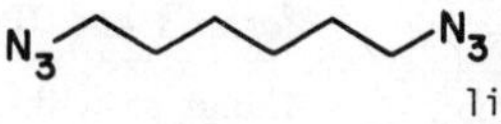 N_3, N_3	liq.	2100	5
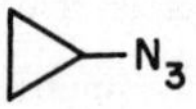 N_3	sng	2083	7
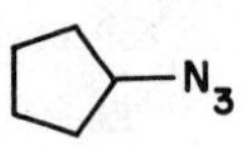 N_3	sng	2083	7
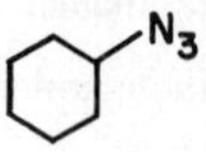 N_3	liq.	2110	5
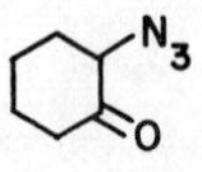 N_3, O	liq.	2104	6
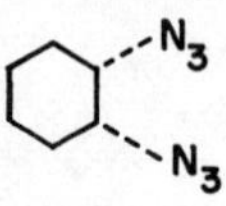 N_3, N_3	liq.	2075	8

Compound	Solvent	Frequency	Ref.
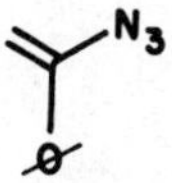 N_3, O	sng	2202w 2142 2115	9
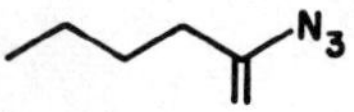 N_3	sng	2220* 2120 2100	9
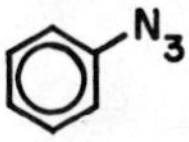 N_3	CCl_4	2100	5
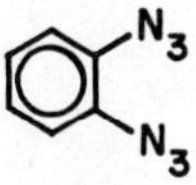 N_3, N_3	liq.	2120	10
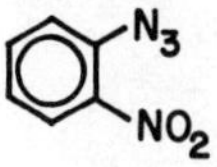 N_3, NO_2	nuj.	2130	6
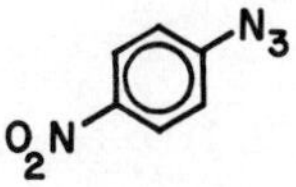 N_3, O_2N	nuj.	2132	6
NCN_3	CCl_4	2240 2199 2143 2090	11

Compound	Solvent	Frequency	Ref.
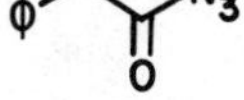 φ, N_3, O	sng	2143	9
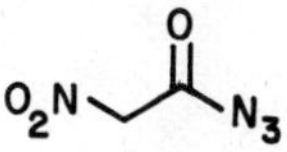 O, O_2N, N_3	$CHCl_3$	2155	12
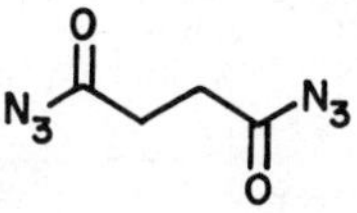 O, N_3, N_3, O	sng	2148	9
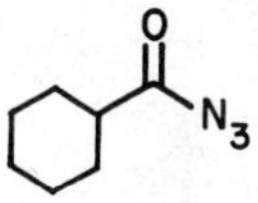 O, N_3	sng	2146	9
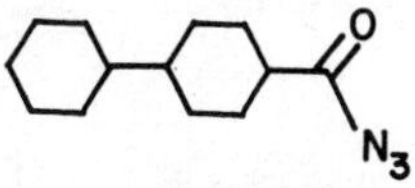 O, N_3	CH_2Cl_2	2165	13
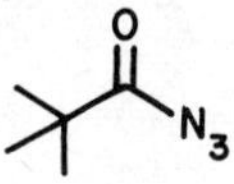 O, N_3	CCl_4	2137 1712	14
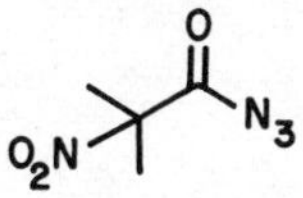 O, O_2N, N_3	$CHCl_3$	2146	12
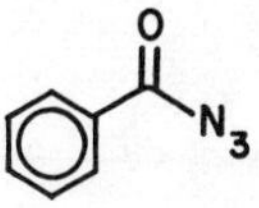 O, N_3	nuj.	2179m 2141 1709	15
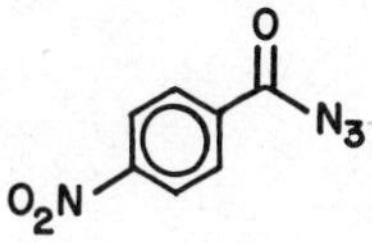 O, O_2N, N_3	nuj.	2193 2146 1701	15

Solvent	Frequency	Ref.
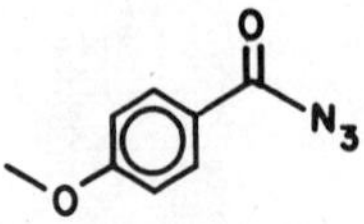 sng	2193w 2137	9
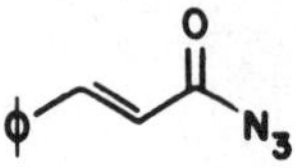 sng	2150 2110w	9
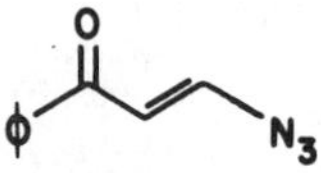 KBr	2140 1662	16
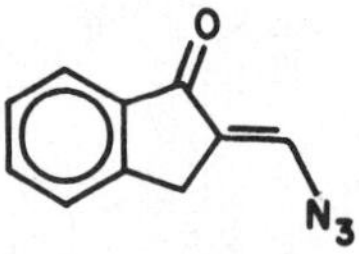 KBr	2131 1703	16
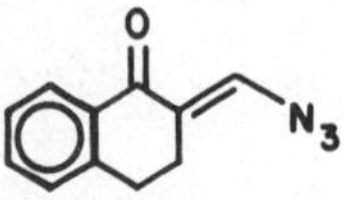 KBr	2115 1671	16
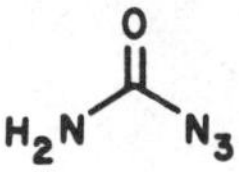	2160	9
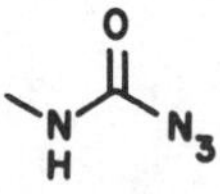 CCl_4	2150	5
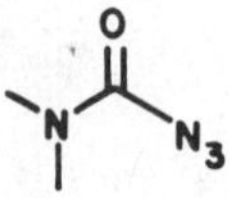 CCl_4	2160	5
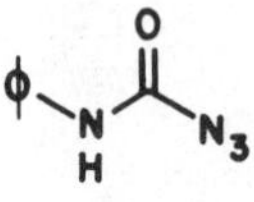 CCl_4	2150	5
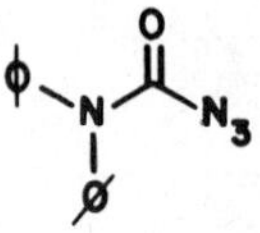 sng	2200w 2162	9
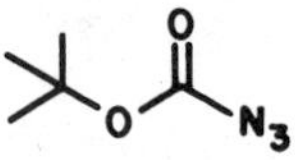 CCl_4	2193 2141 1764 1742	17
liq.	2130	6
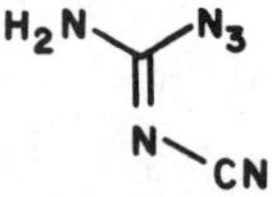 KBr	2170	5
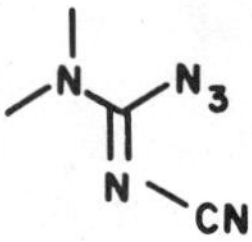 $CHCl_3$	2180	5
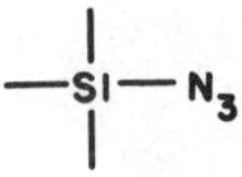 liq.	2139	18
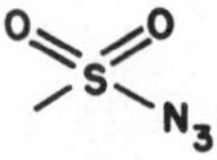 $CHCl_3$	2336w 2137	19
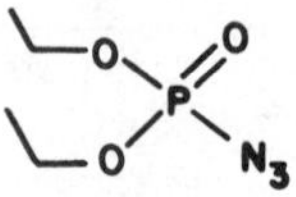 CCl_4	2166	20

REFERENCES

1. D. A. Dows and G. C. Pimentel, *JCP* (1955) *23*, 1258.
2. E. Lieber, D. R. Levering and L. J. Patterson, *Anal Chem* (1951), 1594.
3. R. T. M. Frazer, *Anal Chem* (1959), 1602.
4. P. Gray and T. C. Waddington, *Trans Farad Soc* (1957), 901.
5. W. R. Carpenter, *Appl Spectr* (1963), 70.
6. Yu. N. Sheinker, L. B. Senyavina and V. N. Zheltova, *Doklady Chem* (1968) *160*, 1339.
7. E. Lieber, C. N. R. Rao, T. S. Chao and C. W. W. Hoffman, *Anal Chem* (1957), 916.
8. G. Swift and D. Swern, *JOC* (1967), 511.
9. E. Lieber, C. N. R. Rao, A. E. Thomas, E. Oftedeahl, R. Minnis and C. V. N. Nambury, *SCA* (1963), 1135.
10. J. H. Hall, *JACS* (1965), 1147.
11. F. D. Marsh and M. E. Hermes, *JACS* (1964), 4506.
12. H. E. Ungnade and L. W. Kissinger, *JOC* (1957), 1662.
13. A. E. Wick, unpublished data.
14. G. T. Tissue, S. Linke and W. Lwowski, *JACS* (1967), 6303.
15. E. Lieber and E. Oftedahl, *JOC* (1959), 1014.
16. G. W. Fischer, *B* (1970), 3470.
17. L. A. Carpino, C. A. Giza and B. A. Carpino, *JACS* (1959), 955.
18. N. Wiberg and B. Neruda, *B* (1966), 740.
19. J. H. Boyer, C. H. Mack, N. Goebol and L. R. Morgan, Jr., *JOC* (1958), 1051.
20. G. Rappaport, M. Hauptschein, J. F. O'Brien and R. Filler, *JACS* (1953), 2695.

5.5

azoxy compounds

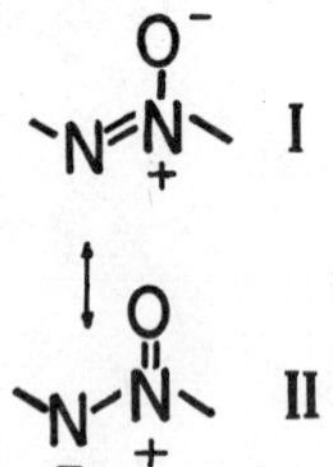

Two bands are observed for the azoxy group. The higher frequency band is associated with the N-N double-bond stretching vibration, whereas the lower frequency band arises from the N-O single-bond stretch.

Too few examples have been reported to enable any definitive generalizations to be made. Nevertheless, although the major contribution to the azoxy group comes from a resonancy structure of type I, some contribution also arises from type II. Thus electron stabilizing groups on the non-oxygenated nitrogen can be expected to decrease the frequency of the N-N double-bond stretching vibration, while at the same time increasing the frequency of the N-O single-bond vibration. Similarly, electron withdrawing groups on the oxygenated nitrogen will increase the frequency of the N-N vibration and lower the frequency of the N-O vibration.

	liq.	1527 1342	1		nuj.	1513 1479 1276	4
F_3C, CF_3	gas	1570 1282 1259	2		solid	1490 1335	5
	liq.	1506 1316	1	F, F	KBr	1480 1325	5
	liq.	1500 1295	1	Cl, Cl	solid	1465 1320	5
	liq.	1493 1336	3	I, I	KBr	1460 1342	6
	liq.	1495 1285	1	I, I	KBr	1471 1325	6
	nuj.	1479 1312 1302	4		KBr	1473 1331	5

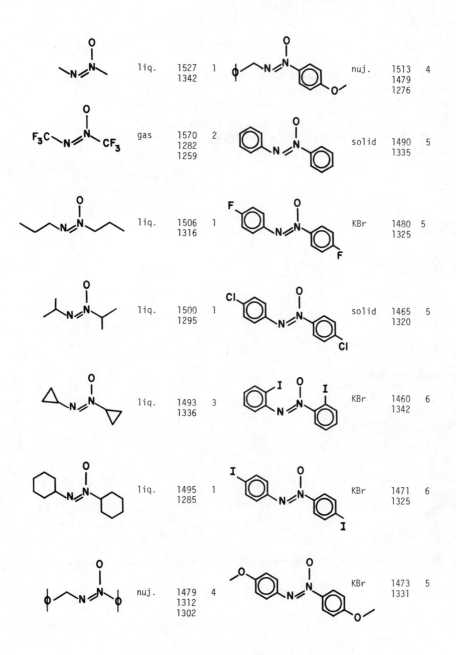

REFERENCES

1. B. W. Langley, B. Lythgoe and L. S. Rayner, *JCS* (1952), 4191.
2. J. Jander and R. N. Haszeldine, *JCS* (1954), 919.
3. P. A. Iversen, *B* (1971), 2195.
4. B. Witkop and H. M. Kissman, *JACS* (1953), 1975.
5. W. Maier and G. Englert, *Z Electrochem* (1958) *62*, 1020.
6. M. V. King, *JOC* (1961), 3323.

6.1

nitroso compounds

$$\overset{+}{N}=\overset{+}{N}\ (O^-,\ O^-)\ \text{(cis)}$$

$$\overset{+}{N}=\overset{+}{N}\ (O^-,\ O^-)\ \text{(trans)}$$

The majority of nitroso compounds associate in solution to give a mixture of the cis and trans dimers. When monomeric, aliphatic nitroso compounds exhibit a strong band in the region of 1620 to 1550 cm^{-1}, which is associated with the N-O double-bond stretching vibration. Aromatic nitroso compounds show a similar strong band at somewhat lower frequencies.

Upon dimerization the trans isomers exhibit bands between 1300 and 1100 cm^{-1}, whereas the cis isomers absorb in the higher range between 1430 and 1330 cm^{-1}.

Nitrosyl halides absorb at considerably higher frequencies than organic nitroso compounds, whereas the nitrosonium cation absorbs in the triple-bond region, reflecting the dominant contribution form $N\equiv\overset{+}{O}:$.

Nitrosyl derivatives of transition metal complexes are listed in Chapter 8.2, p. 493. Other nitrogen oxides are to be found in Chapter 8.1, p. 485.

Compound	Solvent	Frequency	Ref.
CCl_3NO	liq.	1616	1
$(CH_3)_3C$–NO	CCl_4	1548	2
O_2N–$C(CH_3)_2$–NO	$CHCl_3$	1563	3
NC–$C(CH_3)_2$–NO	C_2Cl_4	1570	4
Cl–$C(CH_3)_2$–NO	liq.	1587 1570	1
1-chloro-1-nitrosocyclopentane	liq.	1578 1568	1
1-chloro-1-nitrosocyclohexane	liq.	1577 1564	1
1,4-dichloro-1,4-dinitrosocyclohexane	KBr	1570	4
nitrosobenzene	CCl_4	1513	2
p-nitroso-N,N-dimethylaniline	CCl_4 $CHBr_3$	1537 1443	5 6
p-nitronitrosobenzene	CCl_4	1505	5

DIMERS

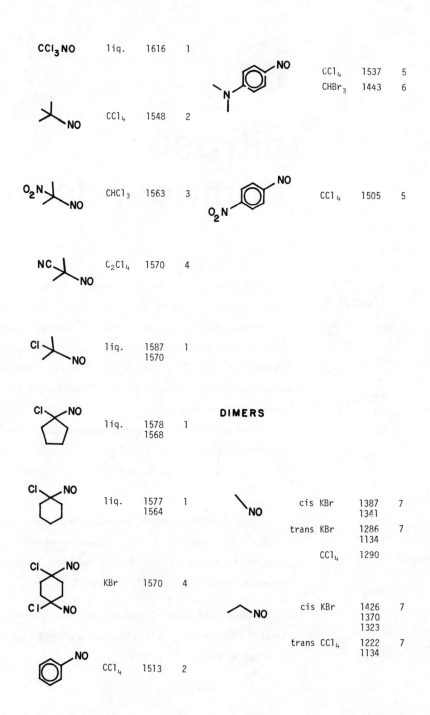

Compound	Form	Solvent	Frequency	Ref.
CH_3–NO	cis	KBr	1387 1341	7
	trans	KBr	1286 1134	7
		CCl_4	1290	
C_2H_5–NO	cis	KBr	1426 1370 1323	7
	trans	CCl_4	1222 1134	7

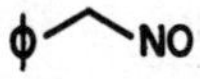

	State	Frequency	Ref.
ϕ NO	trans KBr	1176	8
NO (n-butyl)	cis KBr	1426 1385 1336	7
	trans KBr	1212 1144	7
	CCl_4	1211 1148 1116	7
NO (isopropyl)	cis KBr	1426 1408 1382 1330	7
	trans CCl_4	1212 1140	7
NO (cyclohexyl)	trans KBr	1199	8
O_2N NO	trans KBr	1297	8
NO (tert-butyl)	trans KBr	1262 1233 1181	7
NC NO	trans KBr	1295	8
Cl N=N O O Cl	KBr	1399 1350	8
NO (phenyl)	KBr	1409 1397	8
NO (p-tolyl)	KBr	1252	8
O_2N NO	KBr	1259	8
O NO	KBr	1248	8
NO (mesityl)	KBr	1261	8
NO (naphthyl)	KBr	1409 1389	8

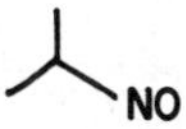

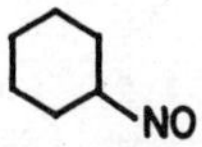

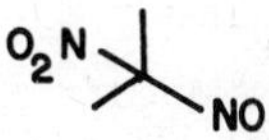

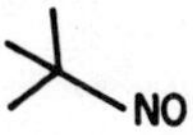

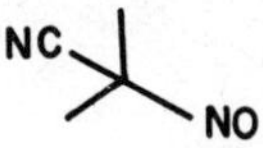

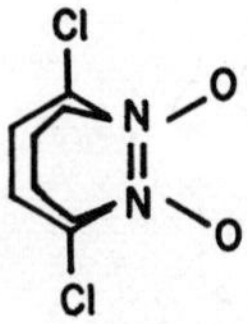

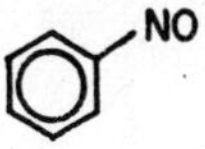

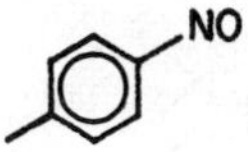

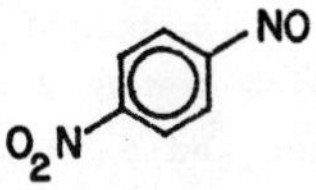

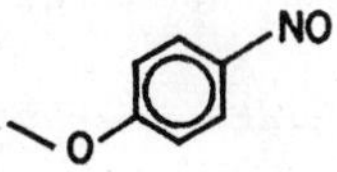

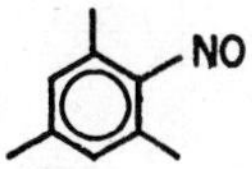

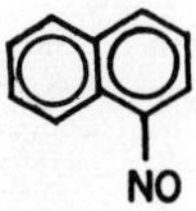

Compound	Phase	Frequency	Ref.
2,2'-Biphenyl cyclic N,N'-dioxide (N(O)=N(O) bridge)	KBr	1399 1342	8
NOCl	gas	1799	9
NOBr	gas	1801	9
N_2O_2	cis solid	1862 1768	10
	trans solid	1740	10
$NO^+\ AlCl_4^-$	solid	2236	11
$NO^+\ ClO_4^-$	solid	2313	11

REFERENCES

1. P. Tarte, *BSC Belge* (1954), 525.
2. R. N. Haszeldine and B. J. H. Mattinson, *JCS* (1955), 4172.
3. L. W. Kissinger and H. E. Ungnade, *JOC* (1958), 1517.
4. W. Lüttke, *Z Elektrochem* (1957), 302.
5. K. Nakamoto and R. E. Rundle, *JACS* (1956), 1113.
6. M. Piskorz and T. Urbański, *Bull Acad Polon* (1963) *11*, 607.
7. B. G. Gowenlock, H. Spedding, J. Trotman and D. H. Whiffen, *JCS* (1957), 3927.
8. W. Lüttke, *Z Elektrochem* (1957), 976.
9. W. G. Burns and H. J. Bernstein, *JCP* (1950) *18*, 1669.
10. W. G. Fateley, H. A. Bent and B. Crawford, Jr., *JCP* (1959) *31*, 204.
11. H. Gerding and H. Houtgraaf, *Rec Trav Chim* (1953), 21.

6.2

nitrosamines

Compound	Medium	Frequency	Ref.
N—NO	liq.	1475	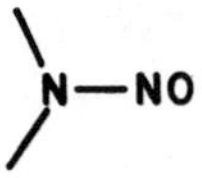 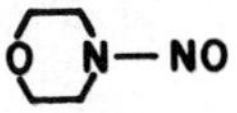 1
	CCl_4	1470	1
	$CHCl_3$	1480	1
	$CHBr_3$	1443	2
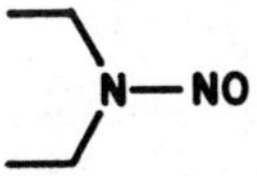 N—NO	CCl_4	1470	1 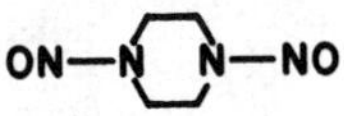
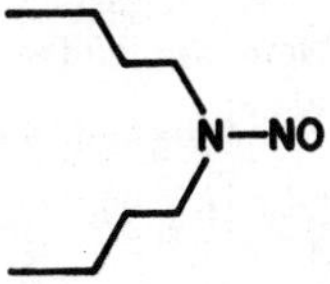 N—NO	CCl_4	1465	1 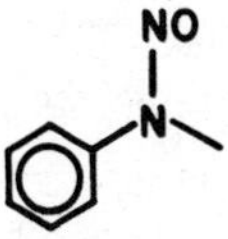
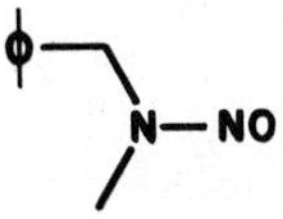 ϕ N—NO	CCl_4	1455	1 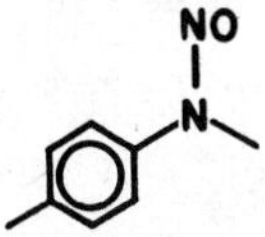
N—NO	$CHCl_3$	1429	3
O N—NO	CCl_4	1465	1
ON—N N—NO	KBr	1429	4
NO N	liq.	1470	1
NO N	liq.	1460	1
ϕ N NO ϕ	$CHBr_3$	1443 1321m	2

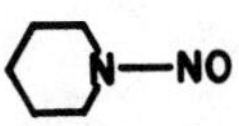

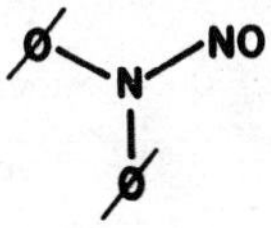

REFERENCES

1. C. E. Looney, W. D. Phillips and E. L. Reilly, *JACS* (1957), 6136.
2. M. Piskorz and T. Urbański, *Bull Acad Polon* (1963) *11*, 607.
3. R. N. Haszeldine and J. Jander, *JCS* (1954), 691.
4. M. V. George and G. F. Wright, *JACS* (1958), 1200.

6.3

nitro compounds

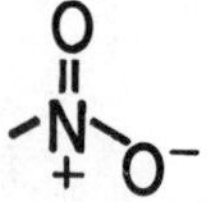

The polar nitro group shows two strong bands, the higher frequency band arising from the asymmetric mode and the lower frequency band from the symmetric mode.

Conjugation to alkenyl and aryl groups causes a lowering of the frequency of both the symmetric and asymmetric modes. Further substitution on a conjugated phenyl group causes predictable change in both modes. Thus ortho and para electron withdrawing groups raise the frequencies, whereas electron donating groups lower them.

This parallel behavior of the two modes is not observed when electronegative groups are attached to the same carbon as the nitro-group. In these cases the frequency of the asymmetric vibration is raised, whereas that of the symmetric mode is lowered.

The dipolar nature of the nitro group enables it to form strong hydrogen bonds, but this results in only small changes of the nitro-group frequency with change in state or solvent, and only in the cases where intramolecular hydrogen bonding can occur is an appreciable lowering of the asymmetric mode observed.

α-Nitro carbanions show bands in the regions associated with the nitro group frequencies. These absorptions, however, cannot be assigned to the nitro group but rather to the delocalized chromophore that is represented by I and II.

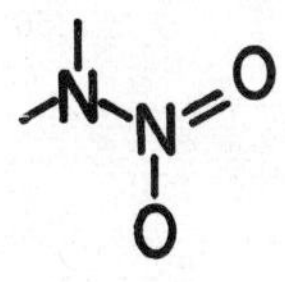

Delocalization of the nitrogen lone pair into the nitro group of nitroamines causes a lowering of the frequency of both the symmetric and asymmetric modes; similar effects are seen with nitroamides and nitroureas.

The uncoordinated nitrite anion absorbs at 1328 and 1261 cm^{-1} in nujol and 1375 and 1270 cm^{-1} in KBr.

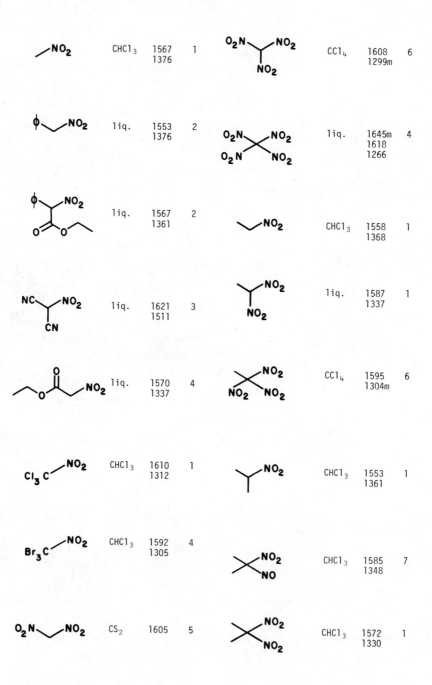

Structure labels	Solvent	Frequencies	Ref.
NO_2	$CHCl_3$	1567 1376	1
φ NO_2	liq.	1553 1376	2
φ NO_2 O O	liq.	1567 1361	2
NC NO_2 CN	liq.	1621 1511	3
O O NO_2	liq.	1570 1337	4
Cl_3C NO_2	$CHCl_3$	1610 1312	1
Br_3C NO_2	$CHCl_3$	1592 1305	4
O_2N NO_2	CS_2	1605	5
O_2N NO_2 NO_2	CCl_4	1608 1299m	6
O_2N NO_2 O_2N NO_2	liq.	1645m 1618 1266	4
NO_2	$CHCl_3$	1558 1368	1
NO_2 NO_2	liq.	1587 1337	1
NO_2 NO_2 NO_2	CCl_4	1595 1304m	6
NO_2	$CHCl_3$	1553 1361	1
NO_2 NO	$CHCl_3$	1585 1348	7
NO_2 NO_2	$CHCl_3$	1572 1330	1

Structure	Medium	Frequencies	Ref.
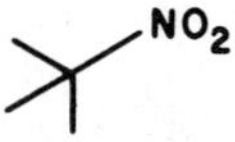 NO_2	liq.	1543 1346	1
NO_2, O, O	liq.	1560 1368	2
NO_2	$CHCl_3$	1553 1383	1
 NO_2	liq.	1538 1357	1
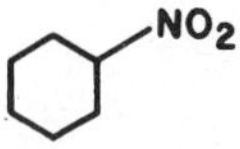 NO_2	liq.	1550 1379	2
NO_2	liq.	1536 1343	2
NO_2, NO_2	$CHCl_3$	1572 1323	1

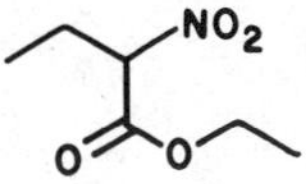

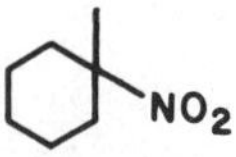

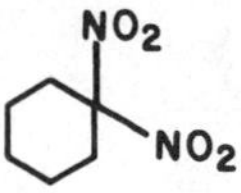

Structure	Medium	Frequencies	Ref.
 NO_2	liq.	1527 1353	1
 ϕ, NO_2	liq.	1517 1340	1
 NO_2	liq.	1520 1348	1
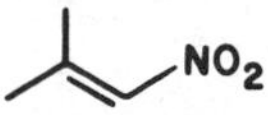 NO_2	$CHCl_3$	1520 1355	1
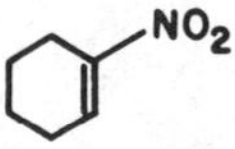 NO_2	liq.	1513 1337	1
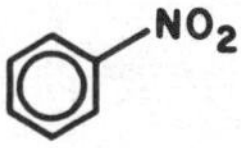 NO_2	liq.	1534 1354	9
	$CHCl_3$	1534 1348	8
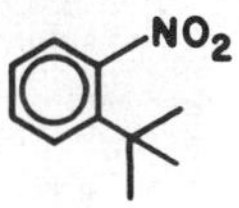 NO_2	$CHCl_3$	1527 1357	8
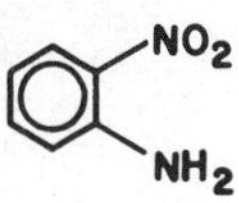 NO_2, NH_2	$CHCl_3$	1516 1342	8

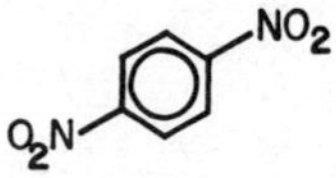

Solvent	cm^{-1}	Ref.
$CHBr_3$	1553	10
$CHCl_3$	1362	10
$CHBr_3$	1514	10
$CHCl_3$	1340	10
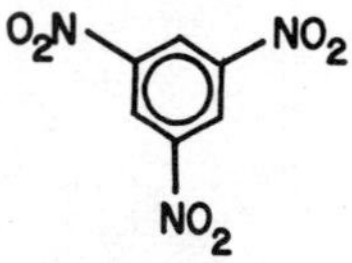 $CHCl_3$	1557 1344	8
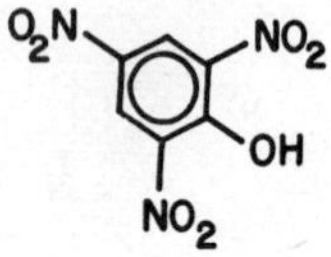 $CHCl_3$	1554 1347	8
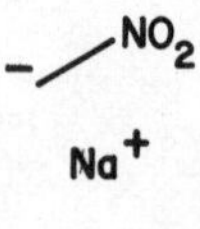 nuj.	1587 1272 1258	11
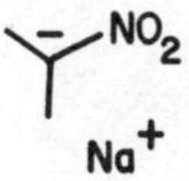 nuj.	1605 1295	11

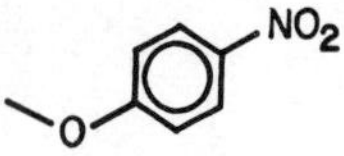

Solvent	cm^{-1}	Ref.
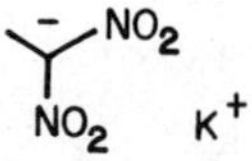 KBr	1495 1256	12
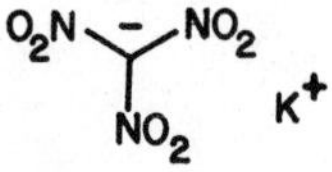 KBr	1516b 1283b	12
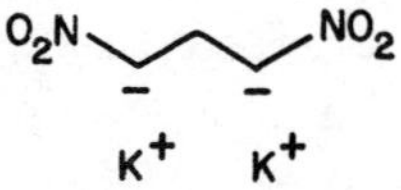 nuj.	1610 1241 1130	11
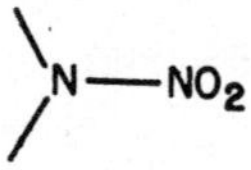 KBr	1529w 1297	13
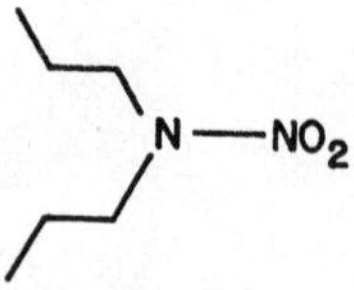 KBr	1504w 1304	13
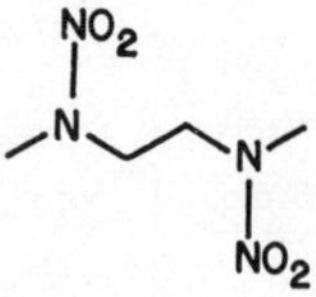 KBr	1504 1282	13

Compound	Phase	Frequencies	Ref.
N—NO_2	liq.	1500 1250	14
O_2N—N N—NO_2	KBr	1548 1287w	13
O N—NO_2	KBr	1520 1285	13

REFERENCES

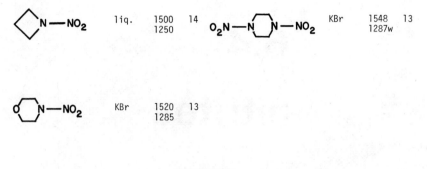

1. J. F. Brown, Jr., *JACS* (1955), 6341.
2. N. Kornblum, H. E. Ungnade and R. A. Smiley, *JOC* (1956), 377.
3. H. Hellmann, G. Hallmann and F. Lingens, *B* (1953), 1346.
4. R. N. Haszeldine, *JCS* (1953), 2525.
5. K. Singh, *SCA* (1967), 1089.
6. H. E. Ungnade and L. W. Kissinger, *T* (1963) *Suppl 19*, 121.
7. L. W. Kissinger and H. E. Ungnade, *JOC* (1958), 1517.
8. C. P. Conduit, *JCS* (1959), 3273.
9. R. R. Randle and D. H. Whiffen, *JCS* (1952), 4153.
10. R. D. Kross and V. A. Fassel, *JACS* (1956), 4225.
11. H. Feuer, C. Savides and C. N. R. Rao, *SCA* (1963), 431.
12. M. J. Kamlet, R. E. Oesterling and H. G. Adolph, *JCS* (1965), 5838.
13. M. V. George and G. F. Wright, *JACS* (1958), 1200.
14. C. L. Bumgardner, K. S. McCallum and J. P. Freeman, *JACS* (1961), 4417.

6.4 nitrites

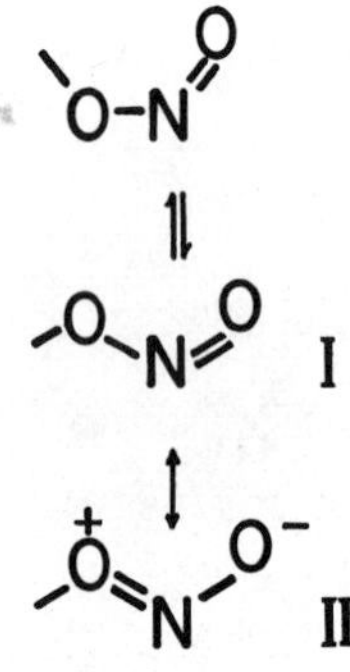

Nitrites exist in cis and trans configurations. The activation energy for isomerization is about 8 Kcal $mole^{-1}$, suggesting that the resonance form II contributes significantly to the overall structure.

The dipolar nature of the chromophore gives rise to intense bands for both the cis and trans isomers, with the trans isomer absorbing at higher frequencies. Only small decreases are observed in the stretching frequencies between the gaseous and condensed phases.

The nitrite anion absorbs at 1328 and 1261 cm^{-1}.

ONO (methyl nitrite)

cis gas	1631	1
	1621	
trans gas	1692	1
	1672	

ONO (ethyl nitrite)

cis gas	1629	1
	1621	
cis CCl_4	1613	1
trans gas	1684	1
	1669	
trans CCl_4	1653	1

ONO (n-butyl nitrite)

cis gas	1621	1
cis liq.	1608	1
cis CCl_4	1608	1
trans gas	1678	1
	1672	
trans liq.	1647	1
trans CCl_4	1650	1

ONO (allyl nitrite)

gas	1681	2
	1623	

ONO (isoamyl nitrite)

cis gas	1623	1
cis liq.	1610	1
cis CCl_4	1608	1
trans gas	1675	1
trans liq.	1653	1
trans CCl_4	1653	1

ϕ ONO (benzyl nitrite)

gas	1678	2
	1621	

ONO (isopropyl nitrite)

gas	1667	2
	1615	

ONO (tert-butyl nitrite)

gas	1665	2
	1610	

ONO (cyclopentyl nitrite)

gas	1664	2
	1613	

ONO (cyclohexyl nitrite)

gas	1664	2
	1615	

REFERENCES

1. R. N. Haszeldine and B. J. H. Mattinson, *JCS* (1955), 4172.

2. P. Tarte, *JCP* (1952) *20,* 1570.

6.5

nitrates

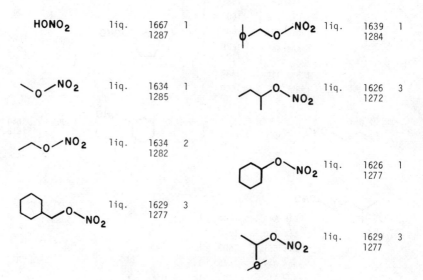

Compound	State	Bands	Ref.
$HONO_2$	liq.	1667 1287	1
CH_3ONO_2	liq.	1634 1285	1
$C_2H_5ONO_2$	liq.	1634 1282	2
cyclohexyl-CH_2ONO_2	liq.	1629 1277	3
$\phi CH_2CH_2ONO_2$	liq.	1639 1284	1
sec-butyl-ONO_2	liq.	1626 1272	3
cyclohexyl-ONO_2	liq.	1626 1277	1
$CH_3CH(\phi)ONO_2$	liq.	1629 1277	3

REFERENCES

1. J. F. Brown, Jr., *JACS* (1955), 6341.
2. J. Jander and R. N. Haszeldine, *JCS* (1954), 912.
3. N. Kornblum, H. E. Ungnade and R. A. Smiley, *JOC* (1956), 377.

7.1 sulfoxy compounds

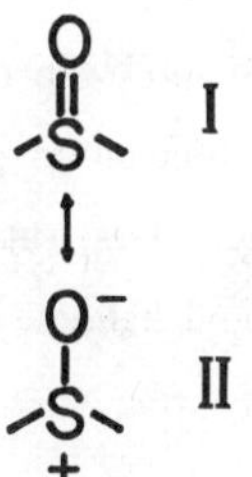

Sulfoxides can be represented by the two resonance forms I and II. The length of the S-O bond and the infrared stretching frequency indicates that the pyramidal (at sulfur) sulfoxide is predominantly represented by II. This suggests that electronegative groups which will destabilize II with respect to I will increase the frequency of the S=O stretching vibration and that alkenyl and aryl substituents bonded to sulfur should have little effect on the stretching frequency. Moreover, the change in bond angle when the sulfoxide group is part of a cyclic system should have little effect on the stretching frequency. This is observed in practice, there being a steady increase in the frequency of the medium-intensity band of the S=O stretching vibration of sulfoxides, sulfinic esters, sulfinyl halides, sulfites, and thionyl halides, and only very small changes being observed as a result of unsaturation or ring size.

Small changes are observed in the stretching frequency with change in state and solvent polarity. In general, condensed samples show the lowest frequency, which are comparable to those in $CHCl_3$, whereas CCl_4 solutions absorb at higher frequencies. This is not, however, the case with cyclic sulfoxides or sulfites where the condensed phase has a higher frequency than a CCl_4 solution.

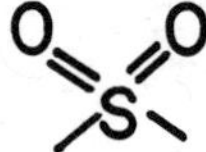

Compounds containing the SO_2 group show two strong bands: the higher frequency arising from the asymmetric vibration and the lower frequency from the symmetric vibration. The frequencies of both the symmetric and asymmetric modes depend principally on the electronegativities of the groups bonded to sulfur. Increases in the frequencies of these bands are observed in going from sulfones, sulfonamides, sulfonic acid esters, sulfonyl halides, and dialkylsulfates, through sulfonyl halides. Once again, conjugated unsaturated groups and ring size have little effect on the frequencies. Condensed phases, polar solvents, and hydrogen-bonded systems cause decreases in the frequencies of the absorptions.

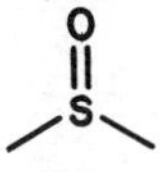

Phase	ν	Ref.
liq.	1057	1
CCl_4	1072	2
$CHCl_3$	1055	1
nuj.	1231	3
KBr	1232	4
liq.	1018	1
$CHCl_3$	1020	1
liq.	1031	5
KBr	1018	1
$CHCl_3$	1012	1
KBr	1016	1
$CHCl_3$	1012	1
CCl_4	1061 1047	6
CCl_4	1047	6

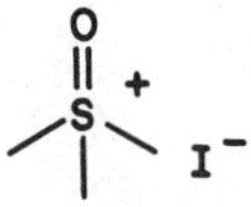

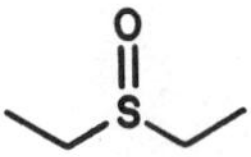

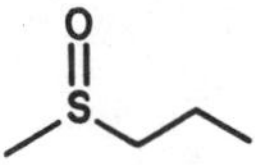

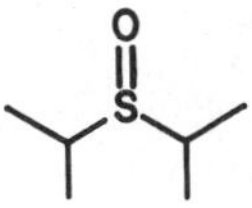

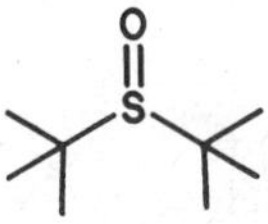

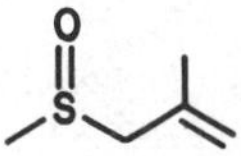

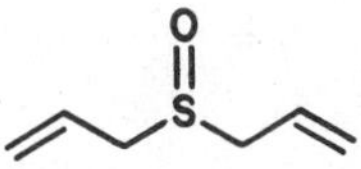

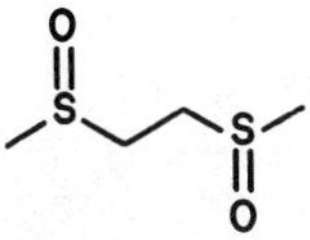

Phase	ν	Ref.
KCl	1017	7
$CHCl_3$	1045	7
liq.	1040	6
CCl_4	1055	6
liq.	1070	5
liq.	1044	6
CCl_4	1055	6
$CHCl_3$	1045	8
$CHCl_3$	1054	8
$CHCl_3$	1042	8
KCl	1035	7
$CHCl_3$	1044	7

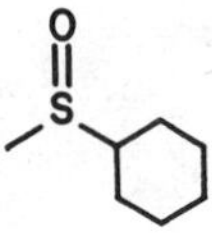

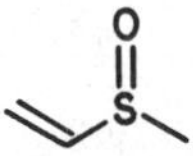

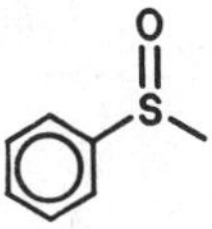

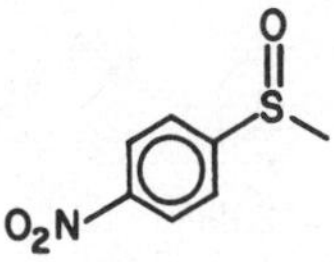

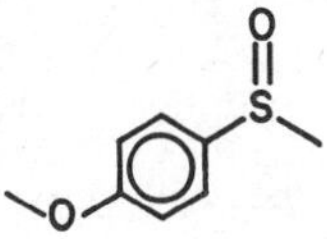

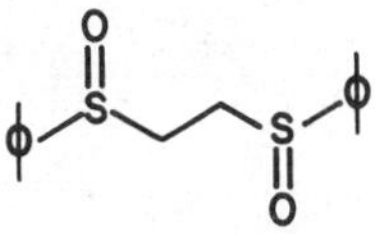

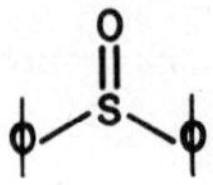

liq.	1044	6
KBr	1037	1
CCl_4	1051	9
$CHCl_3$	1041	9
KBr	1037	1
CCl_4	1034	2
$CHCl_3$	1020	1
KBr	1067	10
$CHCl_3$	1060	10
KBr	1035	1
$CHCl_3$	1029	1
KBr	1019	1
$CHCl_3$	1044	1
nuj.	1020 1010	11
KCl	1023	7
$CHCl_3$	1036	7
nuj.	1047 1032 1012	11
$CHCl_3$	1055	7

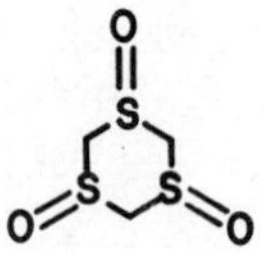

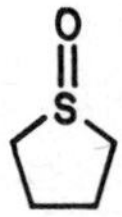

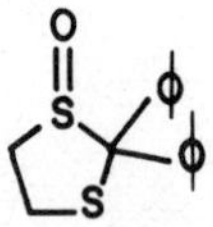

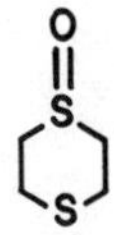

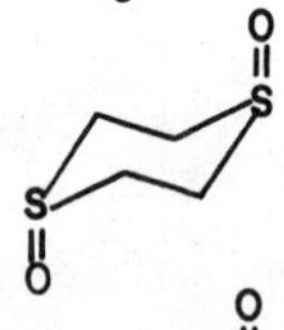

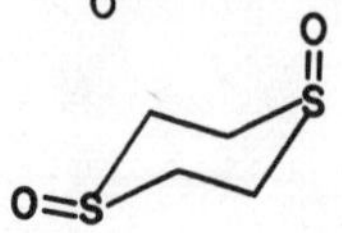

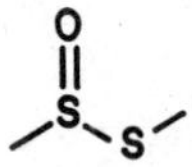

KCl	1053	7
CCl_4	1097	12
$CHCl_3$	1075	12
CCl_4	1101	12
$CHCl_3$	1078	12
gas	1269	13
liq.	1241 1218 1125	14
liq.	1244 1220 1195 1167 1798	14
liq.	1250 1234 1203 1149 1817	14

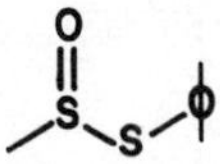

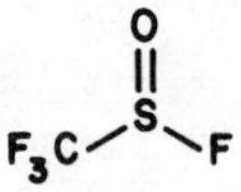

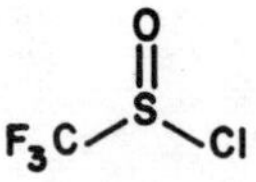

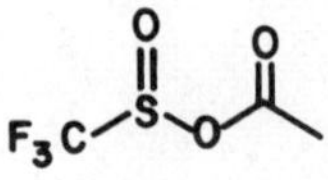

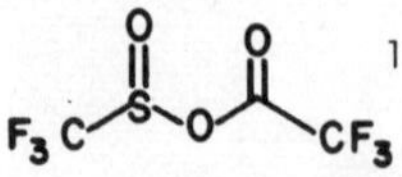

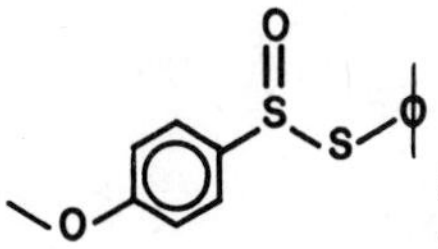

Phase	Frequency	Ref.
liq.	1150	15
nuj.	1090	15
liq.	1126	2
CCl_4	1104	12
$CHCl_3$	1088	12
CCl_4	1108	12
$CHCl_3$	1091	12
CCl_4	1110	12
$CHCl_3$	1104	12
CCl_4	1106	12
$CHCl_3$	1090	12
CCl_4	1114	12
$CHCl_3$	1098	12

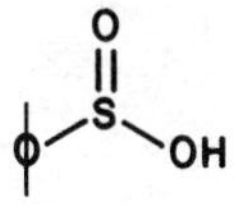

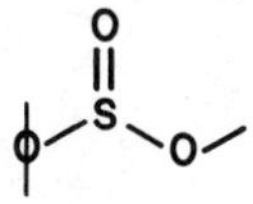

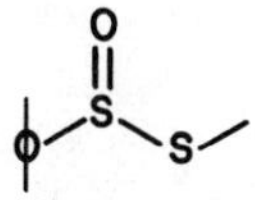

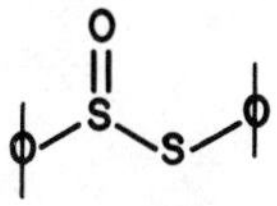

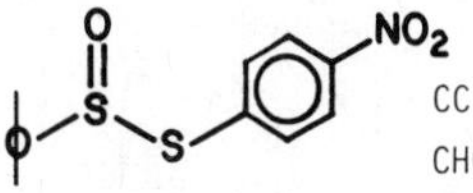

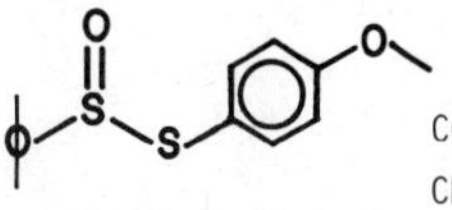

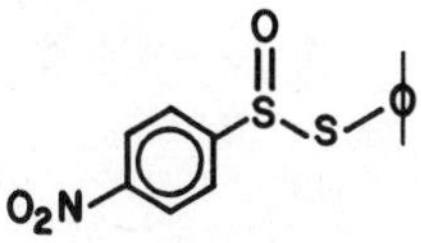

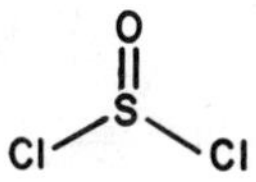

Phase	Frequency	Ref.
CCl_4	1099	12
$CHCl_3$	1090	12
liq.	1251	16
CCl_4	1350w 1238 1200*	17
CS_2	1224	18
nuj.	1493 1265 1166	19
liq.	1200	20
CCl_4	1220 1205	21

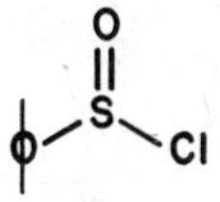

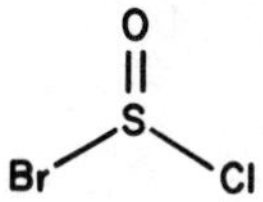

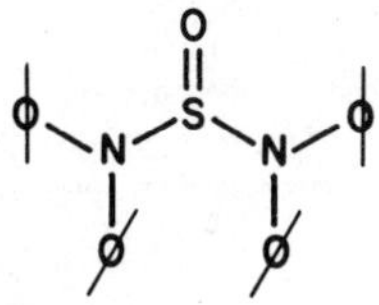

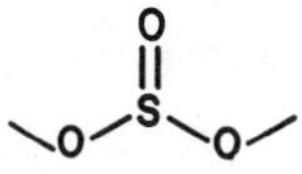

Compound	Phase	Frequency	Ref.
$CH_3O-S(=O)-OC_2H_5$	liq.	1205	21
$C_2H_5O-S(=O)-OC_2H_5$	CCl_4	1210 1192	2
$C_3H_7O-S(=O)-OC_3H_7$	CCl_4	1211 1193	2
$(CH_3)_2CHO-S(=O)-OCH(CH_3)_2$	CCl_4	1203	22
$\phi O-S(=O)-O\phi$	CCl_4	1245	22
five-membered cyclic sulfite ($-OCH_2CH_2O-$ ring)	liq. CCl_4	1214 1220	23 22
five-membered cyclic sulfite with methyl substituent	liq.	1215	23
six-membered cyclic sulfite ($-OCH_2CH_2CH_2O-$ ring)	liq. CCl_4	1234m 1190 1224	23 22
$Cl_2C=S=O$	liq.	1156 1052	24
F−N=S=O	liq.	1230	25
Cl−N=S=O	liq.	1221	25
Br−N=S=O	liq.	1214	25
I−N=S=O	liq.	1247	25
$C_6H_5-N=S=O$	liq.	1303 1288	26

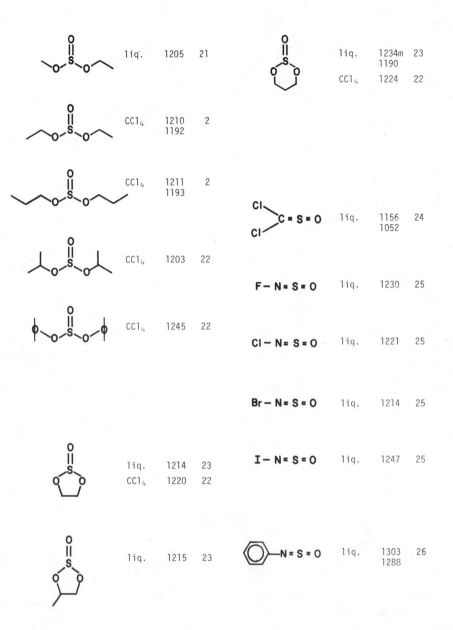

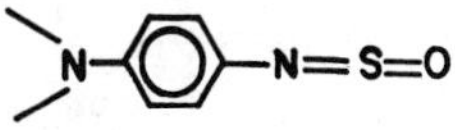

Compound	Medium	ν	Ref.
$(CH_3)_2N$–C₆H₄–N=S=O	CCl_4	1309	26
		1292	
		1272	
O_2N–C₆H₄–N=S=O	CCl_4	1303	26
		1293	
	nuj.	1311	27
		1137	
	CCl_4	1324	27
		1152	
	liq.	1307	5
		1136	
	CCl_4	1323	28
		1138	
	nuj.	1312	27
		1136	
	CCl_4	1322	27
		1136	
	nuj.	1305	27
		1127	
	CCl_4	1308	27
		1130	

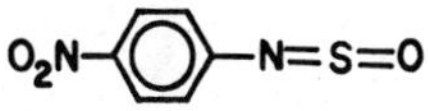

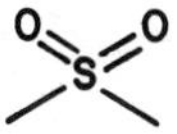

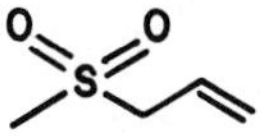

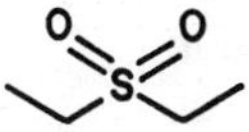

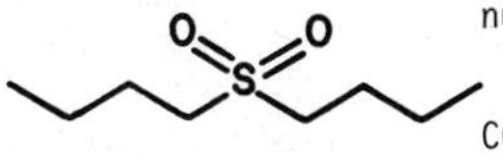

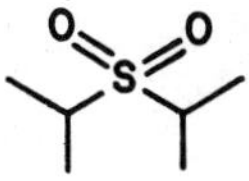

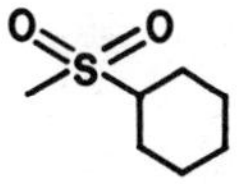

Compound	Medium	ν	Ref.
	liq.	1309	6
		1138	
	CCl_4	1321	6
		1144	
	CCl_4	1305	28
		1134	
	liq.	1303	27
		1135	
	CCl_4	1327	27
		1144	
	liq.	1313	27
		1134	
	CCl_4	1325	27
		1136	
	CCl_4	1332	28
		1139	
	liq.	1318	6
		1135	
	KBr	1084	8
	CCl_4	1334	6
		1160	
	liq.	1363	29
		1188	

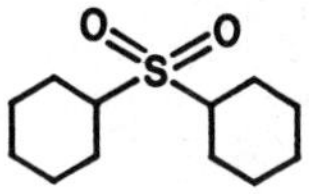

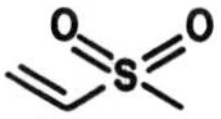

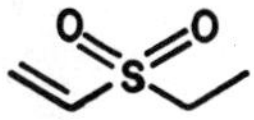

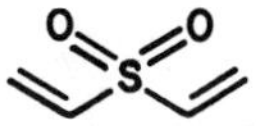

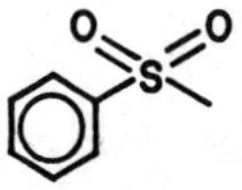

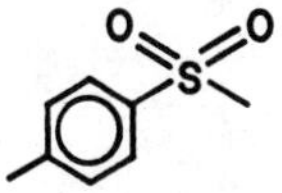

Solvent	Frequency	Ref.
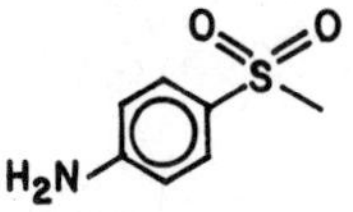 nuj.	1282 1140	30
nuj.	1192	29
KBr	1088	8
$CHCl_3$	1377	29
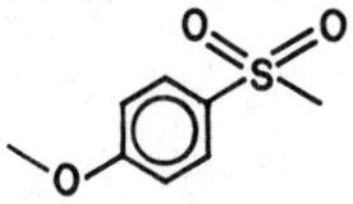 KBr	1082	8
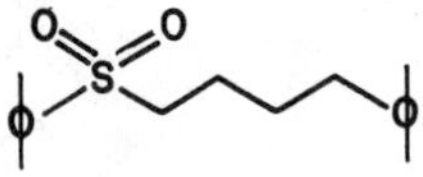 CCl_4	1324 1151	31
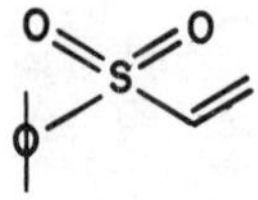 $CHCl_3$	1325 1153	32
nuj.	1319 1155	27
CCl_4	1327 1309m 1295w 1162	9
CS_2	1326 1161	33
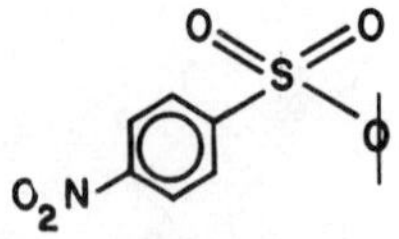 CS_2	1341 1163	33
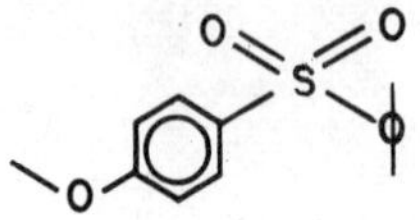 nuj.	1150	34

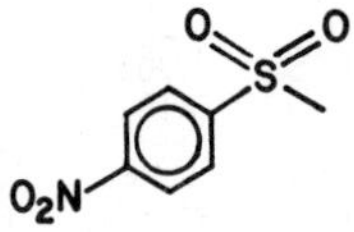

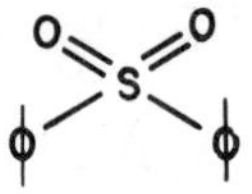

Solvent	Frequency	Ref.
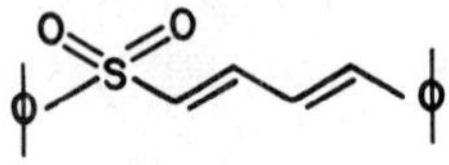 CCl_4	1325 1149	31
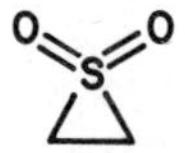 liq.	1310 1168	35
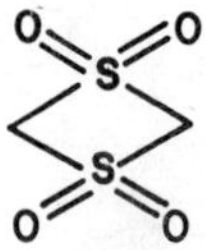 KBr	1340 1200 1090	36
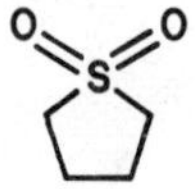 liq.	1298 1143	27
nuj.	1343 1125	27
CCl_4	1328 1142	27
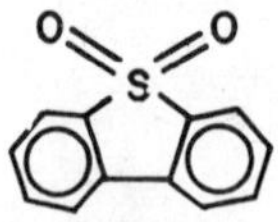 CS_2	1332 1138	33

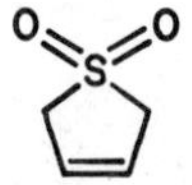

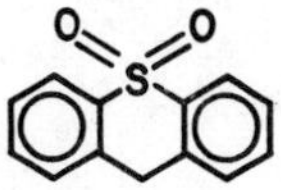

Phase	Frequencies	Ref.
CCl_4	1321 1167	33
CS_2	1314 1175	33
CCl_4	1324 1157	31
CCl_4	1321 1164	31
CCl_4	1327 1153	31
liq.	1412b 1213b	37
CCl_4	1426 1402 1214	37

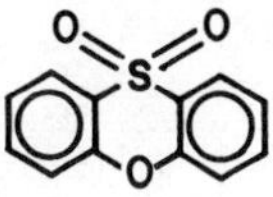

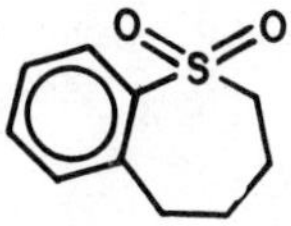

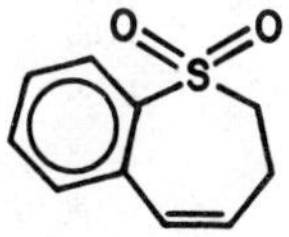

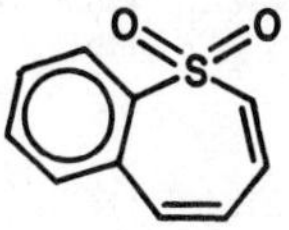

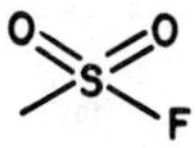

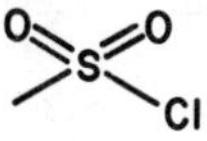

Phase	Frequencies	Ref.
liq.	1367 1175	38
CCl_4	1382 1179	39
gas	1439 1239	20
solid	1430 1220	40
liq.	1389 1205	41
nuj.	1404 1183	20
CCl_4	1406 1188	20
liq.	1372 1180	42
CCl_4	1390 1179 1171	43
liq.	1377 1163	6
nuj.	1371 1185w	6
liq.	1408 1209	41

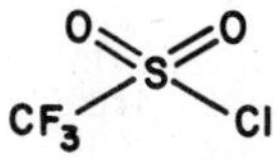

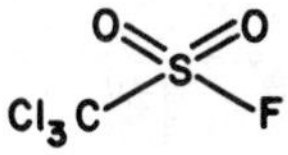

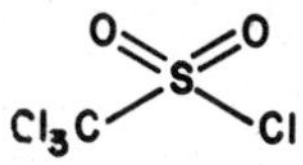

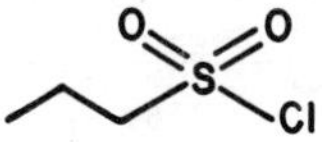

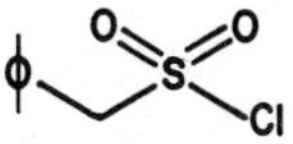

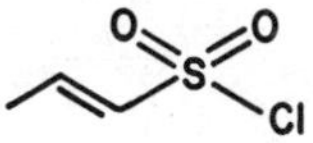

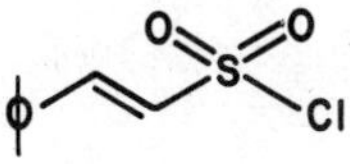

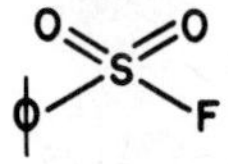

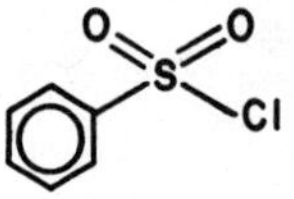

	liq.	1380b 1180b	37
	CCl_4	1396 1195 1185	43
	CCl_4	1410 1400 1204 1185	43
	CCl_4	1391 1195 1175	43
	nuj.	1336 1169	44
	CCl_4	1362 1162	44
	nuj.	1351 1179	41
	nuj.	1315 1155 1131	44
	CCl_4	1333 1159	44
	$CHCl_3$	1328 1152	45
	$CHCl_3$	1340 1151	45
	liq.	1383 1183	41

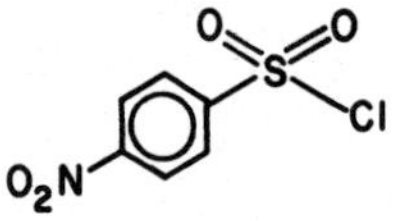

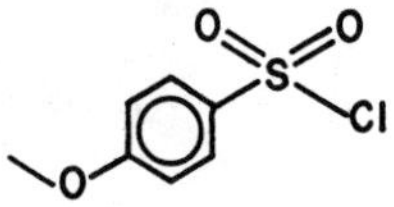

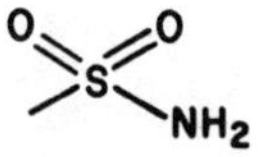

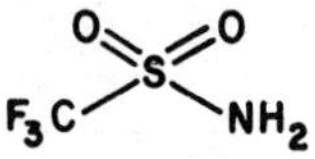

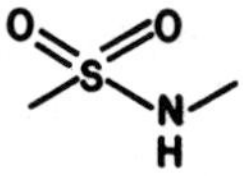

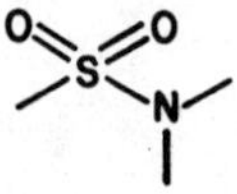

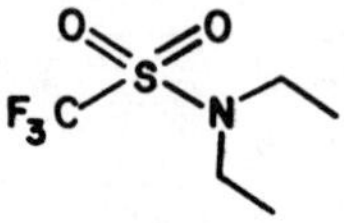

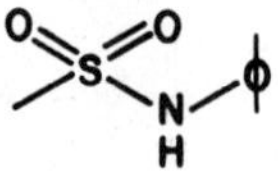

	CCl_4	1348 1160	44
	$CHCl_3$	1349 1329 1158 1154	45
	$CHCl_3$	1347 1326 1156 1146	45
	liq.	1200 2143	46
	nuj.	1339 1315 1156	30
	CCl_4	1159	44
	nuj.	1330 1155	44
	CCl_4	1363 1161	44
	nuj.	1341 1324 1317 1148	30
	nuj.	1322 1161	44
	CCl_4	1336 1166	44
	nuj.	1333 1160	44
	CCl_4	1348 1333 1166	44

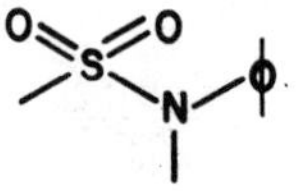

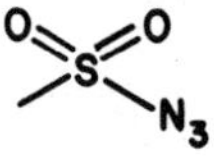

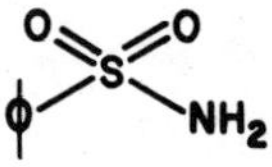

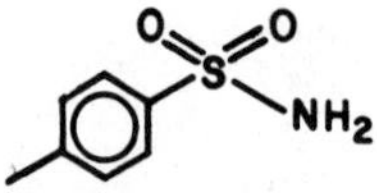

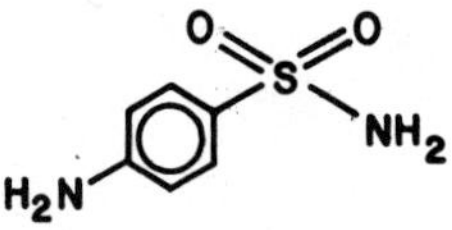

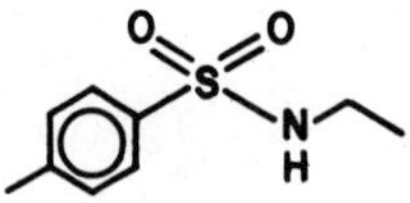

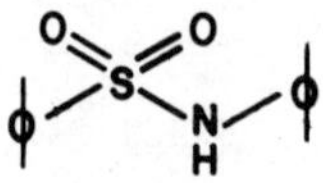

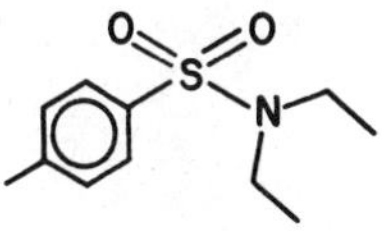

	Phase	cm^{-1}	Ref.
	nuj.	1337 1161	44
	CCl_4	1358 1343 1161	44
	liq.	1190 1050	20
Na^+	nuj.	1195 1060	41
Ag^+	nuj.	1186 1058 1028	41
Na^+	nuj.	1279 1042	41
Ag^+	nuj.	1266 1026	41
	liq.	1471 1460 1131	47
	CCl_4	1368 1182	39
	liq.	1412 1205 1144	41
	CCl_4	1371 1351 1176	48
	liq.	1414 1205 1144	41
	CCl_4	1380 1172	39
	CCl_4	1370 1351 1176	48
	CCl_4	1370 1351 1176	48
	solid	1345 1150	49
Na^+	nuj.	1020	20

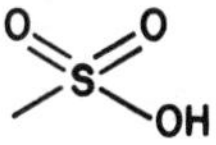

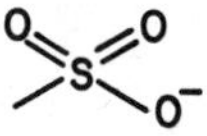

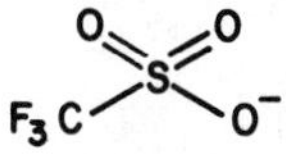

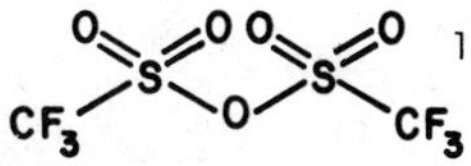

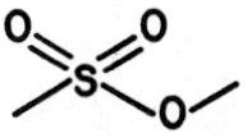

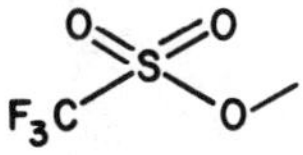

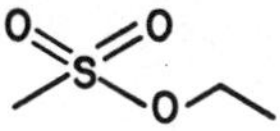

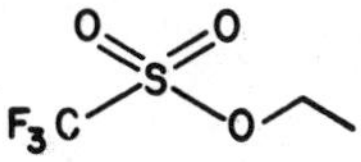

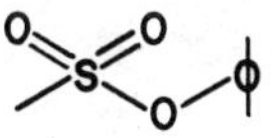

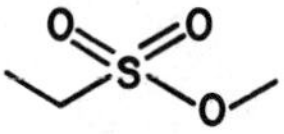

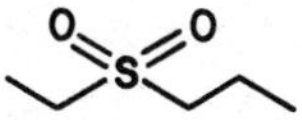

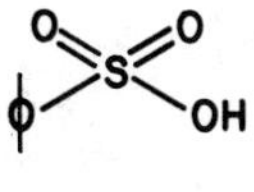

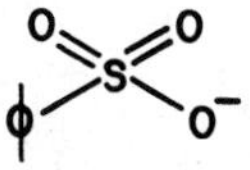

O O S ϕ O	liq.	1365b 1189	37
O O S OH	solid	1342 1160	49
O O S $OH \cdot H_2O$	nuj.	1240 1065 1030	50
O O S O	CCl_4	1375 1315w 1198 1185 1100	17
O O S O ϕ	nuj.	1375 1192 1170	50
O O S O O O S	nuj.	1375 1174	49
O O S S^- Na^+	KBr	1200 1099	51
O O S S	liq.	1310 1130	52
O O S S^- Na^+	KBr	1192 1094	51

O O S S	liq.	1310 1125	52
O O ϕ S S ϕ	nuj.	1331 1144	53
NH_2 O O S S H_2N	nuj.	1342 1154	53
O O Cl S Cl	CCl_4	1419 1196 1160 1128m	17
O O O S Cl	CCl_4	1415 1205	39
O O ϕ O S Cl	CCl_4	1422 1211	39
O O O S O	CCl_4	1415 1187 1161m	17

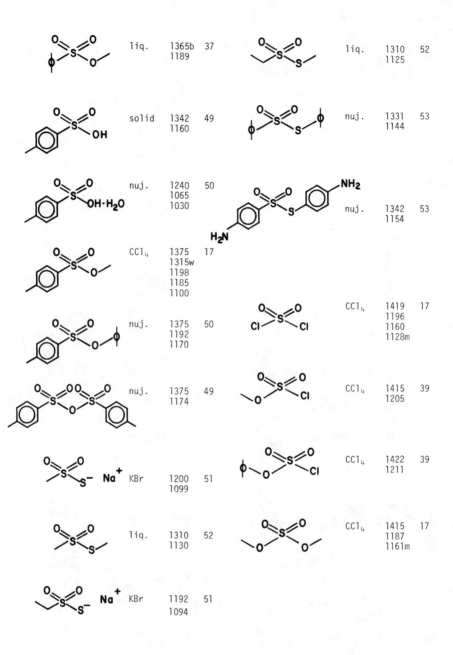

Compound	Phase	Frequency	Ref.
SO_2	gas	1361 1151	54
	solid	1316 1330 1308m 1147* 1142	54
SO_3	gas	1391 529	55
	liq.	1330 1065 653	55
$S_2O_3^{=}\cdot 5H_2O$ 2Na^+	nuj.	1000	56
HSO_4^- Na^+	nuj.	1175m 1075 1045 865	56
$SO_4^{=}$ 2$(NH_4)^+$	nuj.	1105b	56
2Na^+	nuj.	1110b	56
2K^+	nuj.	1110b	56
$SO_3^{=}$ 2Na^+	nuj.	960b	56
$SO_3^{=}$ 2H_2O 2K^+	nuj.	943b	56
$SO_5^{=}$ 2$(NH_4)^+$	solid	1190 1052 976m	57
2K^+	solid	1254 1059 900m	57

REFERENCES

1. T. Cairns, G. Eglinton and D. T. Gibson, *SCA* (1964), 31.
2. T. Gramstad, *SCA* (1963), 829.
3. J. A. Creighton, J. H. S. Green and D. J. Harrison, *SCA* (1969), 1314.
4. A. Blaschette and H. Burger, *Inorg & Nucl Chem Letters* (1969), 639.
5. C. C. Price and R. G. Gillis, *JACS* (1953), 4750.
6. D. Barnard, J. M. Fabian and H. P. Koch, *JCS* (1949), 2442.
7. T. Cairns, G. Eglinton and D. T. Gibson, *SCA* (1964), 159.
8. G. Kresze, E. Ropte and B. Schrader, *SCA* (1965), 1633.
9. S. Pinchas, D. Samuel and M. Weiss-Broday, *JCS* (1962), 3968.

10. W. Otting and F. A. Neugebauer, *B* (1962), 540.
11. P. B. D. de la Mare, D. J. Millen, J. G. Tillett and D. Watson, *JCS* (1963), 1619.
12. S. Ghersetti and G. Modena, *SCA* (1963), 1809.
13. E. W. Lawless and L. D. Harman, *Inorg Chem* (1968), 391.
14. D. T. Dauer and J. M. Shreeve, *Inorg & Nucl Chem Letters* (1970), 501.
15. S. Detoni and D. Hadži, *JCS* (1955), 3163.
16. R. J. Gillespie and E. A. Robinson, *Canad J Chem* (1961), 2171.
17. K. C. Schreiber, *Anal Chem* (1949), 1169.
18. R. Steudel and D. Lautenbach, *Z Naturf* (1969) *24b*, 350.
19. R. D. Peacock and I. N. Rozhkov, *JCS* (1968) *A*, 107.
20. R. N. Haszeldine and J. M. Kidd, *JCS* (1955), 2901.
21. A. Simon, H. Kriegsmann and H. Dutz, *B* (1956), 2390.
22. H. H. Szmant and W. Emerson, *JACS* (1956), 454.
23. P. B. D. de la Mare, W. Klyne, D. J. Millen, J. G. Pritchard and D. Watson, *JCS* (1956), 1813.
24. J. Šilhánek and M. Zbirovský, *Chem Comm* (1969), 878.
25. H. H. Eysel, *J Mol Structure* (1970) *5*, 275.
26. G. Kresze and A. Maschke, *B* (1961), 450.
27. W. R. Feairheller, Jr. and J. E. Katon, *SCA* (1964), 1099.
28. E. S. Waight, *JCS* (1952), 2440.
29. D. E. Freeman and A. N. Hambly, *Aust J Chem* (1957), 239.
30. T. Momose, Y. Ueda, T. Shoji and H. Yano, *Chem Pharm Bull* (1958), 669.
31. V. J. Traynelis and R. F. Love, *JOC* (1961), 2728.
32. C. C. Price and H. Morita, *JACS* (1953), 4747.
33. N. Marziano and G. Montaudo, *Gazz Chim Ital* (1961), 587.
34. E. D. Amstutz, I. M. Hunsberger and J. J. Chessick, *JACS* (1951), 1220.
35. G. Hesse, E. Reichold and S. Majmudar, *B* (1957), 2106.
36. G. Opitz and H. R. Mohl, *Angew* (1969), 36.
37. R. A. McIvor, G. A. Grant and C. E. Hubley, *Canad J Chem* (1956), 1611.
38. L. J. Bellamy and L. Beecher, *JCS* (1952), 475.
39. F. K. Butcher, J. Charalambous, M. J. Frazer and W. Gerrard, *SCA* (1967), 2399.
40. H. W. Roesky, *Inorg & Nucl Chem Letters* (1970), 795.
41. T. Gramstad and R. N. Haszeldine, *JCS* (1956), 173.
42. G. Geiseler and K. O. Bindernagel, *Z Elektrochem* (1960), 421.
43. G. Malewski and H.-J. Weigmann, *SCA* (1962), 725.

44. J. N. Baxter, J. Cymerman-Craig and J. B. Willis, *JCS* (1955), 669.
45. A. R. Katritzky and R. A. Jones, *JCS* (1960), 4497.
46. Yu. N. Sheinker, L. B. Senyavina and V. N. Zheltova, *Doklady Chem* (1968) *160*, 1339.
47. T. Gramstad and R. N. Haszeldine, *JCS* (1957), 4069.
48. A. Simon, H. Kriegsmann and H. Dutz, *B* (1956), 2378.
49. S. Detoni and D. Hadži, *SCA* (1958) *11*, 601.
50. R. S. Tipson, *JACS* (1952), 1354.
51. A. Simon and D. Kunath, *Z Anorg Chem* (1961) *308*, 321.
52. I. B. Douglass and B. S. Farah, *JOC* (1959), 973.
53. J. Cymerman and J. B. Willis, *JCS* (1951), 1332.
54. R. N. Wiener and E. R. Nixon, *JCP* (1956) *25*, 175.
55. R. W. Lovejoy, J. H. Colwell, D. F. Eggers, Jr. and G. D. Halsey, Jr., *JCP* (1962) *36*, 612.
56. F. A. Miller and C. H. Wilkins, *Anal Chem* (1952), 1253.
57. P. Pascal, C. Duval, J. Lecomte and A. Pacault, *CR* (1951) *233*, 118.

7.2

phosphoryl compounds

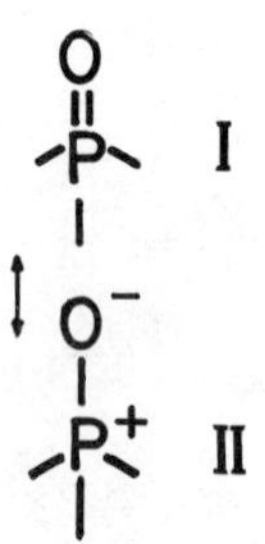

The P-O double-bond of pentavalent phosphorus compounds may be represented by the resonance forms I and II. dπ-pπ Overlap between a first- and second-row element is less than for two elements in the same row, and a large contribution to the phosphoryl group comes from II. This, coupled with the mass of phosphorus compared to carbon, is exemplified in the infrared spectrum where a strong absorption is seen in the 1300 to 1150 cm^{-1} region. Those few compounds which absorb at the high end of the region always have electronegative groups bonded to the phosphorus.

In general, the frequency of the P=O stretching vibration is related directly to the sum of the electronegativities of the phosphorus substituents. Thus trimethylphosphine oxide and tri-(*p*-methoxyphenyl) phosphine oxide absorb at the lower end of the scale. Substitution of a P-C bond by successive P-O bonds causes increases in the frequencies, which are further increased by phosphorus-halogen substitutions.

The phosphoryl group can form strong hydrogen bonds, and such hydrogen bonding causes a considerable lowering of the P=O stretching frequency. This frequency lowering is usually accompanied by a broadening and intensifying of the band. Not only do changes in state and solvent cause these frequency shifts, but intermolecular hydrogen bonding and solute-solvent interactions have the same effect. Thus phosphoryl compounds containing -NH or -OH groups will associate and show frequencies lower than one might anticipate on the basis of group electronegativities.

Unlike cyclic carbonyl-containing compounds, cyclic phosphoryl compounds do not show changes in frequency with change in ring size.

P=S Systems show a decrease in their stretching frequencies consistent with their increased reduced mass.

Inorganic phosphoryl-containing ions show very broad absorptions between 1100 and 1000 cm^{-1}.

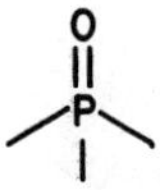

nuj.	1176	1
KBr	1174	2
KBr	1220	3
nuj.	1190	1
KBr	1176b	4
KBr	1175	4
KBr	1176	4
liq.	1236	1
KBr	1220	3
KBr	1240	5
KBr	1200	5
liq.	1185	3

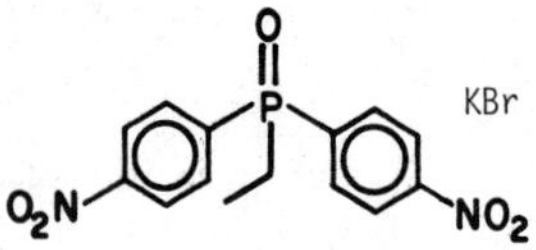

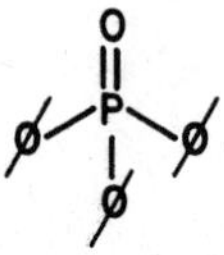

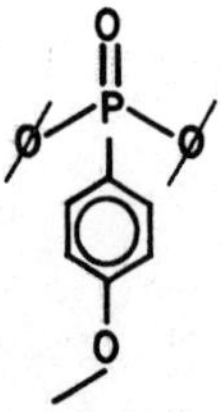

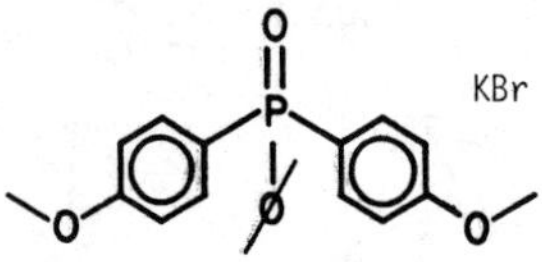

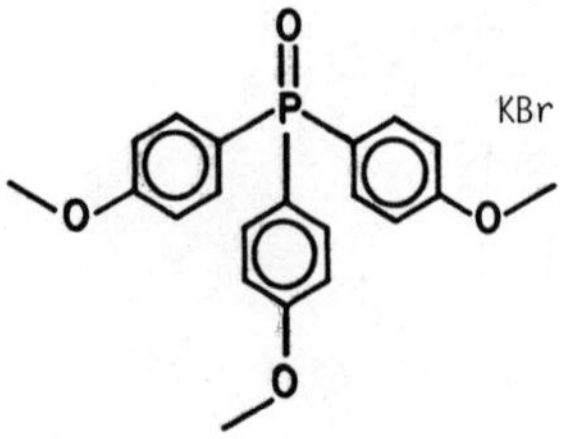

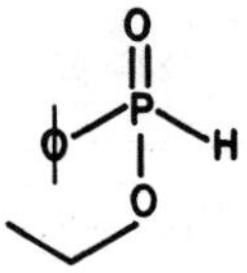

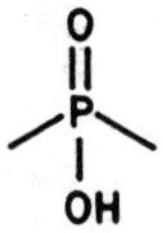

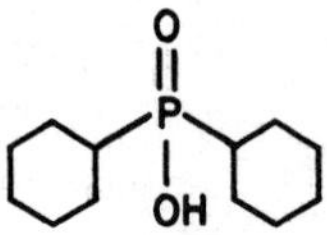

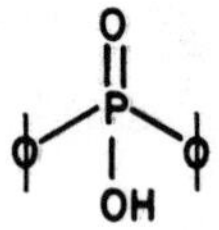

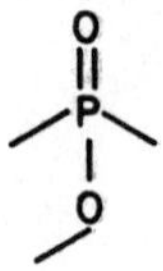

KBr	1294	3
liq.	1236	6
liq.	1212 1190	7
liq.	1212	7
liq.	1266	8
liq.	1265	6
liq.	1262	8
liq.	1250	9
liq.	1205	10
liq.	1234	10
CCl_4	1272	6
nuj.	1200b 1140 1060	11
liq.	1200	5

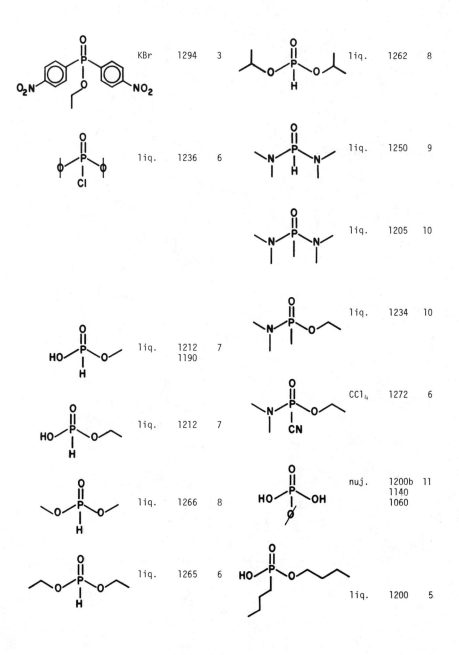

Medium	Frequency	Ref.
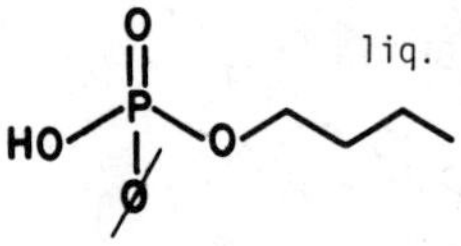 liq.	1215b	5
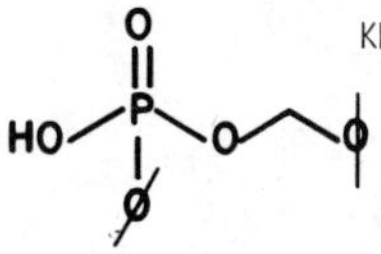 KBr	1220	5
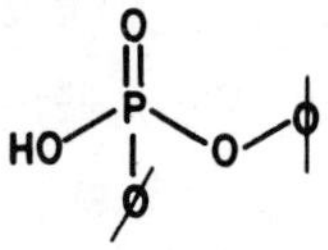 KBr	1215	5
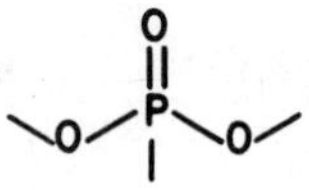 liq.	1241	8
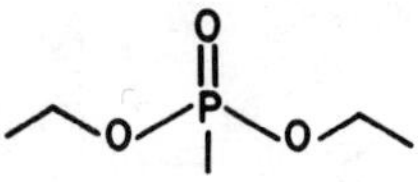 liq.	1250	8
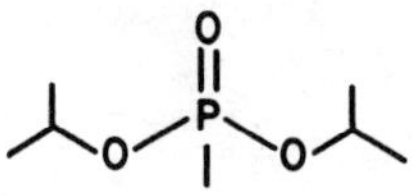 liq.	1248	8
liq.	1230	8
CS_2	1250	12
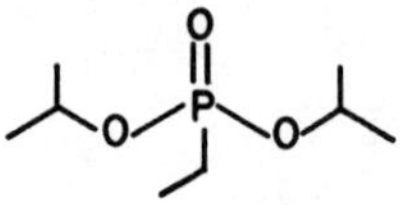 liq.	1224	8
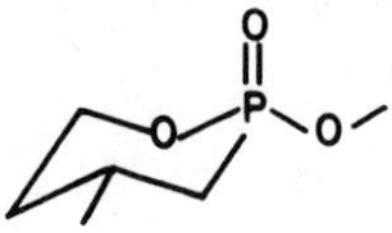 liq.	1245	13
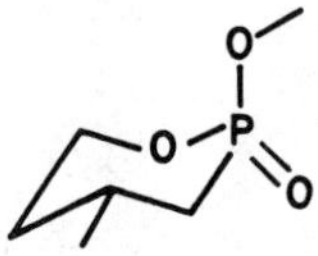 liq.	1255 1240	13
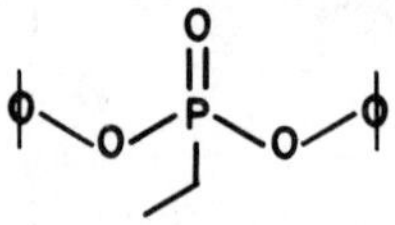 CS_2	1314 1297	12
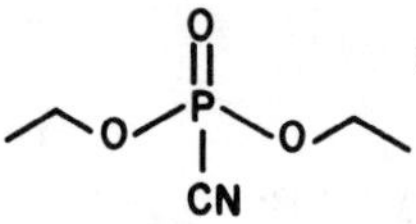 CCl_4	1290	6
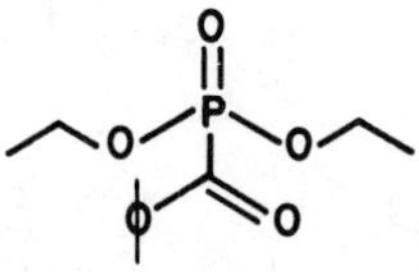 liq.	1243	3
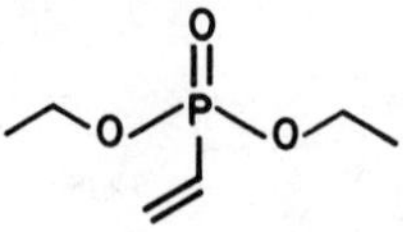 liq.	1250	14

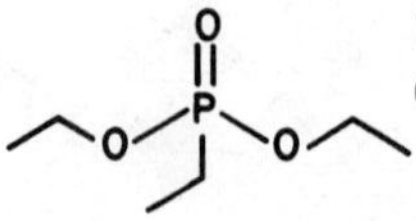

Phase	Frequency	Ref.
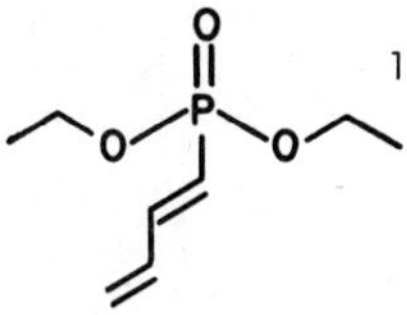 liq.	1250	14
liq.	1257	6
CS_2	1290 1279	12
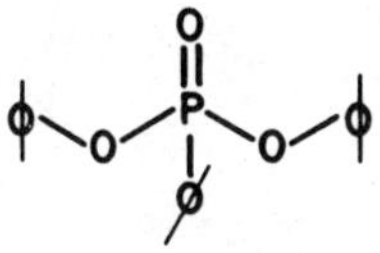 CS_2	1311 1299	12
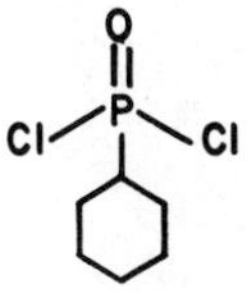 CS_2	1242	12
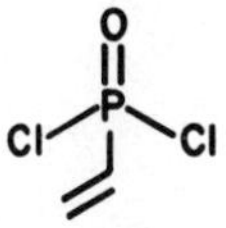 liq.	1270	14
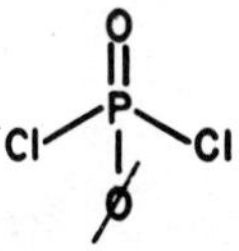 liq.	1275	6
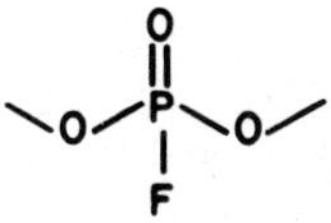 liq.	1305	6
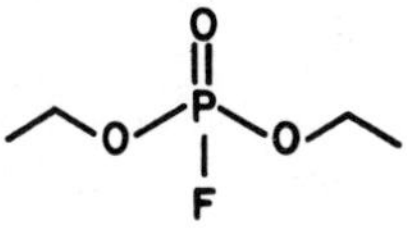 liq.	1309	6
CCl_4	1292	6
CS_2	1311	12
$CHCl_3$	1261	15
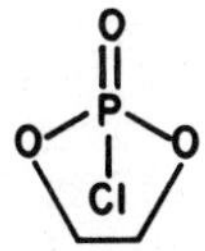 $CHCl_3$	1287	15
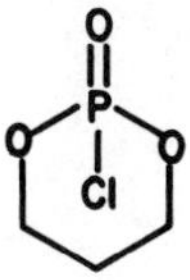 $CHCl_3$	1290	15
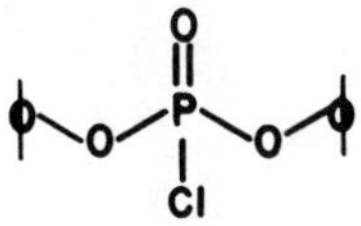 CS_2	1305 1297	12
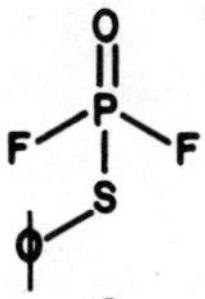 KBr	1330	16
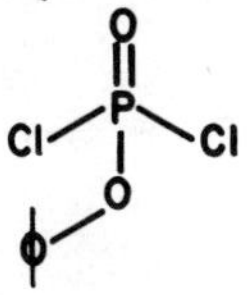 CS_2	1305	12

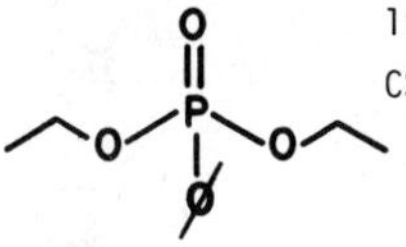

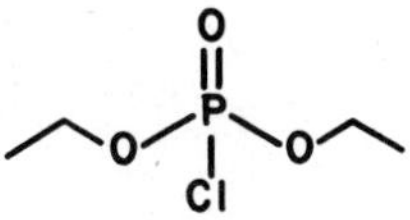

Compound	Phase	Frequency	Ref.
OPF_3	gas	1415	17
$OPCl_2F$	gas	1384	17
	liq.	1331	18
$OPCl_3$	gas	1321	17
	liq.	1290	19
$OPBr_3$	liq.	1261	19
$OP(NH_2)_3$	KBr	1200	20
$OP(NMe_2)_3$	CCl_4	1211	21
	$CHCl_3$	1202	21
$OP(NEt_2)_3$	CCl_4	1247	21
	$CHCl_3$	1244	21
$OP(NCS)_3$	gas	1307	22
	liq.	1280	18
$(Me_2N)_2P(O)OEt$	liq.	1220	10
$(Me_2N)_2P(O)Cl$	liq.	1241	10
$(EtO)_2P(O)NH_2$	CCl_4	1294m	6
$(EtO)_2P(O)NMe_2$	liq.	1250	10
$(EtO)_2P(O)NH\phi$	CS_2	1290	12
$Cl(EtO)P(O)NMe_2$	CCl_4	1269	6
$Cl_2P(O)NMe_2$	liq.	1268	10
	CCl_4	1269	6

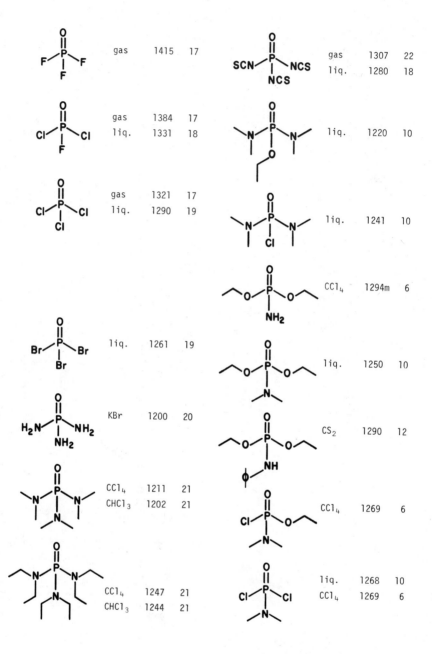

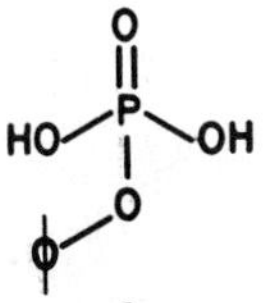

Medium	Frequency	Ref.
nuj.	1215 1150	11
liq.	1220	23
nuj.	1250	24
liq.	1290 1275	25
CS_2	1275	6
liq.	1270 1260	25
CS_2	1280 1264	12
CCl_4	1257	26
$CHCl_3$	1261	15
liq.	1274	12
$CHCl_3$	1287	15
$CHCl_3$	1290	15
liq.	1282	12
KBr	1297	3
liq.	1314 1298	25
CCl_4	1311 1297	21
$CHCl_3$	1309 1295	21
KBr	1298	3
CCl_4	1200	21
$CHCl_3$	1205	21
liq.	1234	10

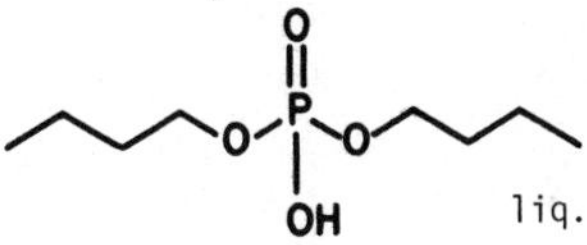

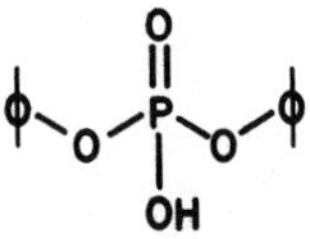

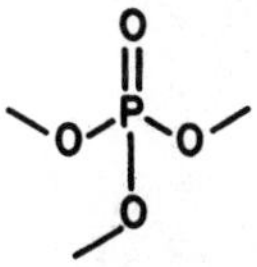

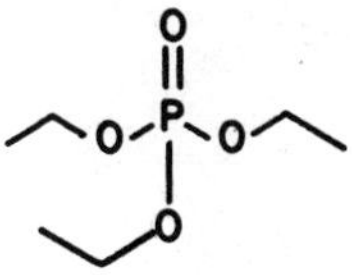

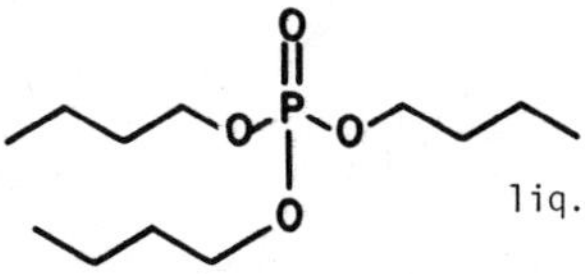

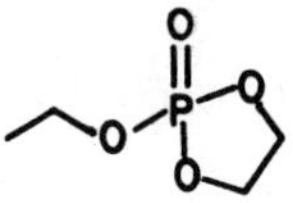

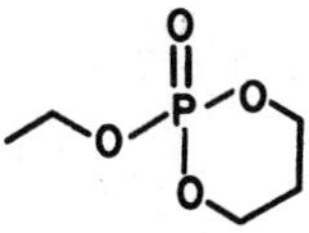

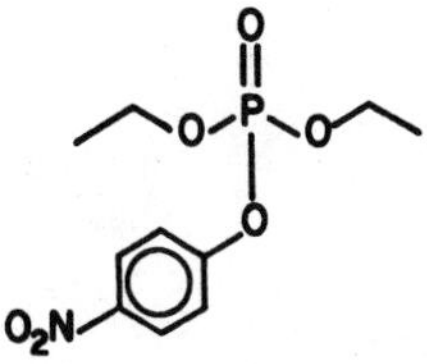

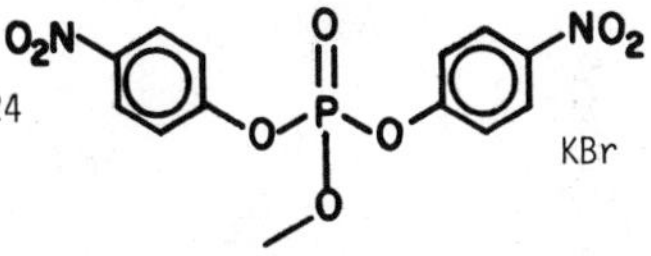

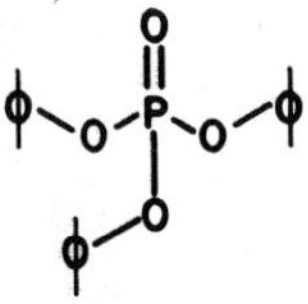

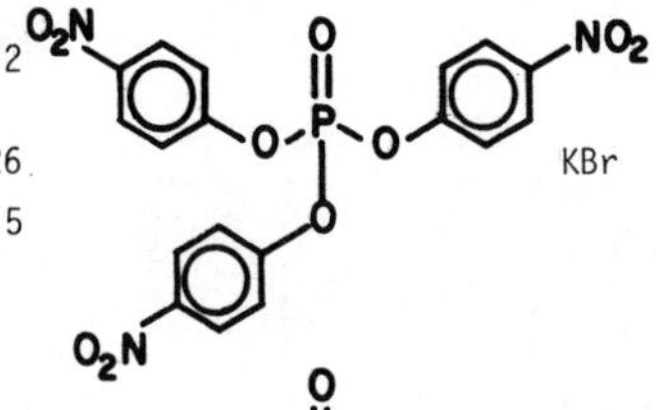

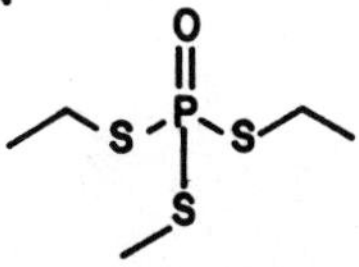

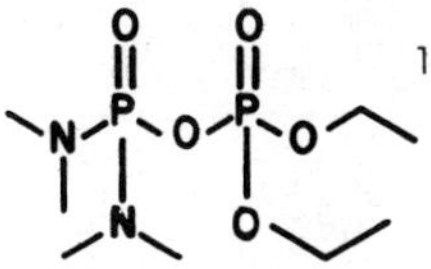

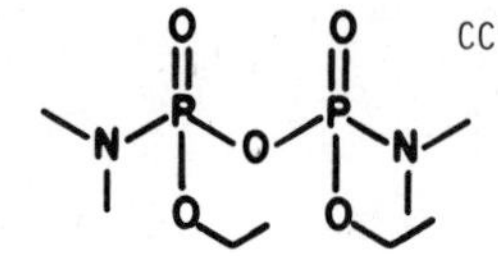

CCl_4	1266	6
liq.	1268	10
liq.	1280	26
CCl_4	1284	12
liq.	794b	27
KBr	860	20
liq.	790	27
liq.	780b	27
liq.	818b	27
CS_2	763	12
CS_2	766	12
liq.	687	27
liq.	782	27
liq.	834	27

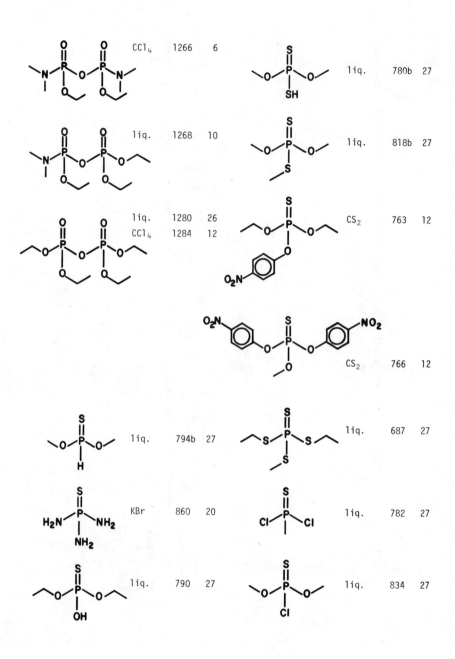

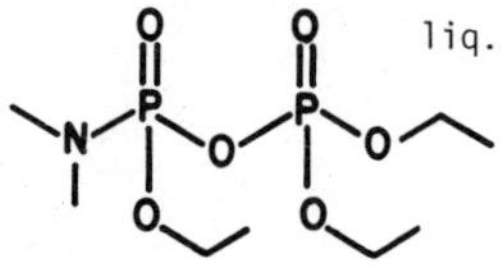

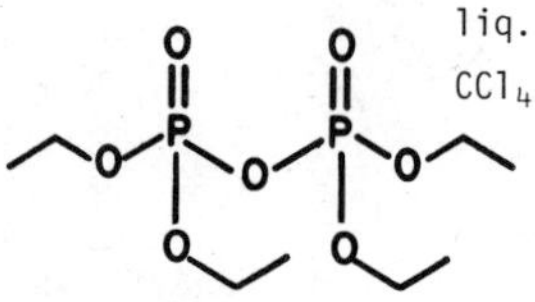

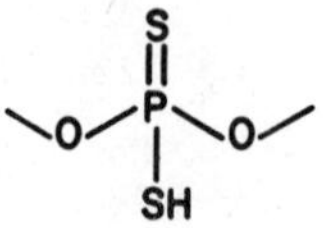

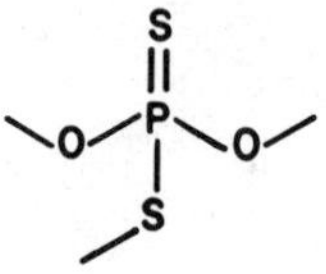

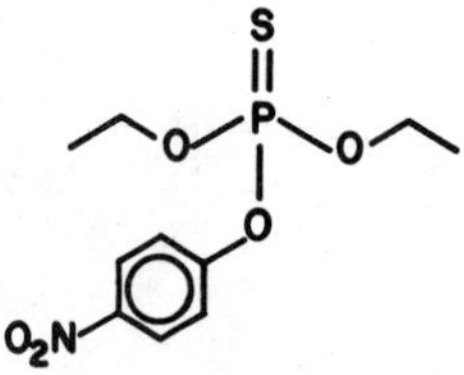

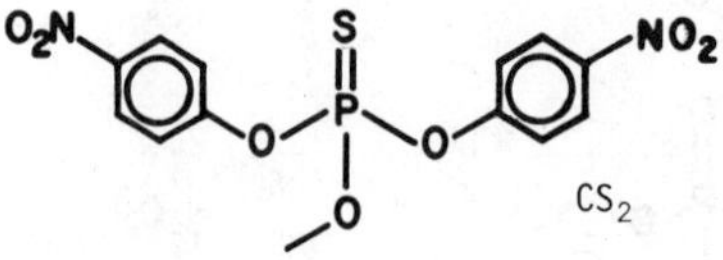

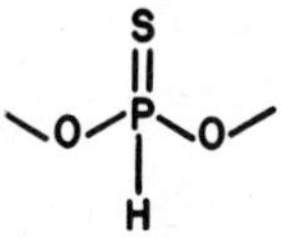

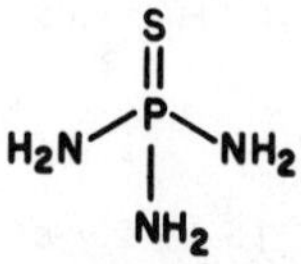

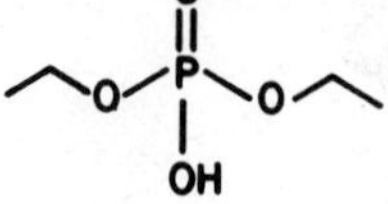

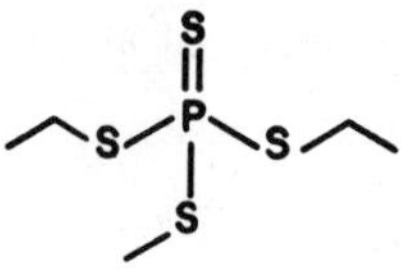

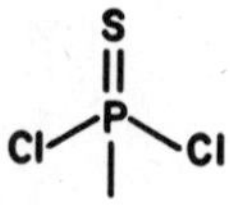

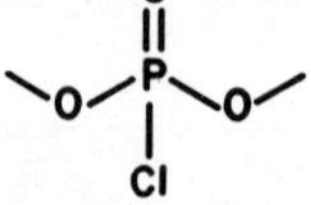

Compound	Phase	cm⁻¹	Ref.
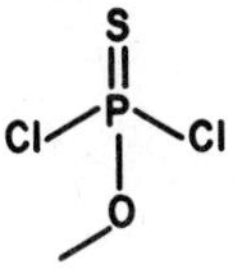 (S=P(Cl)(Cl)–O–CH₃ structure)	liq.	830	27
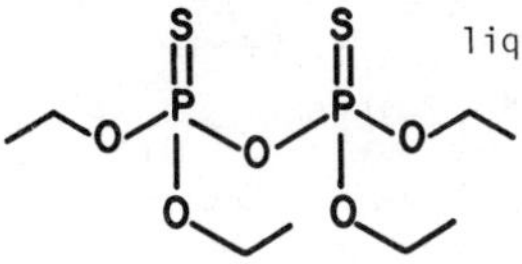 (tetraethyl thiopyrophosphate structure)	liq.	824	27
$[(CH_3O)_4P]^+ SbCl_6^-$	KBr	no P=O stretch	28
	CH_2Cl_2	1270m	28

Compound	Phase	cm⁻¹	Ref.
$(NH_4)_2HPO_4$	nuj.$_F$	1070b	29
$Na_2HPO_4 \cdot 12H_2O$	nuj.	1070b	29
K_2HPO_4	nuj.$_F$	1110b	29
$Na_3PO_4 \cdot 12H_2O$	nuj.$_F$	1070b	29
K_3PO_4	nuj.	1000b	29

REFERENCES

1. L. W. Daasch and D. C. Smith, *Anal Chem* (1951), 853.
2. F. A. Cotton, R. D. Barnes and E. Bannister, *JCS* (1960), 2199.
3. J. A. A. Ketelaar and H. R. Gersmann, *Rec Trav Chim* (1959), 190.
4. A. E. Senear, W. Valient and J. Wirth, *JOC* (1960), 2001.
5. D. F. Peppard, J. R. Ferraro and G. W. Mason, *JINC* (1961) *16*, 246.
6. B. Holmstedt and L. Larsson, *Acta Chem Scand* (1951), 1179.
7. L. C. Thomas and K. P. Clark, *Nature* (1963) *198*, 855.
8. C. I. Meyrick and H. W. Thompson, *JCS* (1950), 225.
9. R. Burgada and J. Roussel, *BSC France* (1970), 192.
10. R. G. Harvey and J. E. Mayhood, *Canad J Chem* (1955), 1552.
11. J. R. Ferraro, D. F. Peppard and G. W. Mason, *JINC* (1965), 2055.
12. L. J. Bellamy and L. Beecher, *JCS* (1952), 475.

13. K. Bergesen, *Acta Chem Scand* (1970), 2019.
14. N. A. Slovokhotova, K. N. Anisimov, G. M. Kunitskaya and N. E. Kolobova, *Izvest Akad Nauk Khim* (1961), 71.
15. R. A. Y. Jones and A. R. Katritzky, *JCS* (1960), 4376.
16. H. W. Roesky, *B* (1968), 636.
17. A. Müller, E. Niecke and O. Glemser, *Z Anorg Chem* (1967) *350*, 246.
18. H. W. Roesky and A. Müller, *Z Anorg Chem* (1967) *353*, 265.
19. M.-L. Delwaulle and F. François, *CR* (1945) *220*, 817.
20. E. Steger, *Z Elektrochem* (1957), 1004.
21. M. W. Hanson and J. B. Bouck, *JACS* (1957), 5631.
22. K. Oba, F. Watari and K. Aida, *SCA* (23A), 1515.
23. L. J. Bellamy and L. Beecher, *JCS* (1953), 728.
24. L. J. Bellamy and L. Beecher, *JCS* (1952), 1701.
25. F. S. Mortimer, *SCA* (1957) *9*, 270.
26. E. D. Bergmann, U. Z. Littauer and S. Pinchas, *JCS* (1952), 847.
27. R. A. McIvor, G. A. Grant and C. E. Hubley, *Canad J Chem* (1956), 1611.
28. J. S. Cohen, *JACS* (1967), 2543.
29. F. A. Miller and C. H. Wilkins, *Anal Chim Acta* (1952), 1253.

8.1

main group compounds

The diverse structures of the inorganic ions compiled here do not allow for any generalizations. However, the symmetry of these ions plays an important role in determining the bands that are observed in the infrared. These topics are covered in considerable detail by Nakamoto,[a] Cotton,[b] and Siebert.[c]

Many of the fundamental bands of inorganic ions occur in the far infrared. This region is one that is not normally available to the organic chemist, and such modes have not been included in this survey.

In general, inorganic ions show broad absorptions whose position and intensity depend considerably on the physical state and degree of solvation of the sample, and the counter ion.

[a] K. N. Nakamoto, *Infrared Spectra of Inorganic and Coordination Compounds,* Wiley, New York, 2nd edition, 1970.

[b] F. A. Cotton, *Chemical Applications of Group Theory,* Interscience, New York, 1963.

[c] H. Siebert, *Anwendunger Der Schwingungsspektroskopie In Der Anorganischen Chemie,* Springer-Verlag, Berlin, 1966.

Compound	Medium	Frequency	Ref.
BN^-	nuj.	1390 810w	1
BO_2^-	nuj.	1310 1175m 925b	1
BO_3^-	nuj.	1175 1075m 934	1
H_3BO_4	nuj.	1450b 1197 882w 824b	2
$B_4O_7^{2-}$	nuj.	1000 943 828	1
BF_4^-	KBr	1128 1107 1088 1077 1063 1038	3
Al_2O_3	KBr	741b	4
HCN	gas	2097	5
	CCl_4	2096	5
	$CHCl_3$	2095	5
FCN	gas	2290 1077	6
ClCN	gas	2219 784 714w	7
	liq.	2206 730	8

Compound	Medium	Frequency	Ref.
BrCN	liq.	2191 586	8
	solid	2194 573m	7
CN^-	nuj.	2085	9
$(CN)_2^{2-}$			
$2Na^+$	nuj.	2120	10
$2Hg^+$	nuj.	2075	10
Hg^{2+}	nuj.	2100	10
CO_2	gas	2349 1343	11
	H_2O	2342	12
HCO_3^-	nuj.	1295 1035 1000 838	1
CO_3^{2-}	KBr	1460m 1414 1156m 1138m	13
HCNO	gas	2190 1251	14
NCO^-	KBr	2170	13
HNCS	CCl_4	1980 2601	15
NCS^-	nuj.	2020	16

Compound	State	Frequencies	Ref.
SiO_3^{2-}	nuj.	1165m 1125m 980 832	1
SILICA GEL	KBr	3440b 1080b 955m 795m	17
SILICONE GREASE	liq.	2980w 1265 1090b 1020b 800b	17
N_2	liq.	2350	18
Li_3N	KBr	1471* 1429 862	4
N_3^-	nuj.	2128	19
NH_3	gas	3448m 3337 1626 950	20
NH_4^+	KBr	3125b 1401	21
NH_2OH	nuj.	3331 3268 3207 3088m 1657m 1502m 1187 926	22

Compound	State	Frequencies	Ref.
$NH_2OH \cdot HCl$	solid	3080 3025m 1876 1563 1482 1200 1176 1000	22
NH_2NH_2	liq.	3338 3200 1280 1038 885	23
N_2O	gas	2224 1286	24
Cis-N_2O_2	solid	1862 1768	25
Trans-N_2O_2	solid	1742	25
N_2O_3	gas	1863 1589 1297 983	26
NO_2	gas	1618 1318 750	27
N_2O_4	gas	1748 1261 750	25
	liq.	1737 1253 743	25
NOF	gas	1844 766	28
NOCl	gas	1800	29

Cis−HONO	gas	1696	29
		1260	
		794	
Trans−HONO	gas	1640	29
		1292	
		856	
$N_2O_2^{2-}$	nuj.	1160w	30
		1060s	
		865w	
NO_2^+	H_2SO_4	2360m	31
NO_2F	gas	1792	32
		1310	
NO_2Cl	gas	1685	33
		1318	
		1293	
		1160	
NO_2^-	nuj.	1328	34
		1261	
	KBr	1375	13
		1270	
NO_3^-	nuj.	1049	35
		974	
		828	
PF_6^-	KBr	913m	17
		830	
		557	
POF_3	gas	1415	36
$POCl_3$	gas	1321	36
	liq.	1290	37
HPO_4^{2-}	nuj.	1110*w	1
		990b	
		934	
		837	
$HPO_4^{2-}\cdot 12H_2O$	nuj.	1070	1
		985	
		958*	
		865	
$H_2PO_4^-$	nuj.	1090b	1
		900b	
$H_2PO_4^-\cdot H_2O$	nuj.	1125*m	1
		1050	
		983	
		925	
		873m	
PO_4^{3-}	nuj.	1000b	1
$PO_4^{3-}\cdot 12H_2O$	nuj.	1000b	1
SbF_6^-	KBr	695m	17
		661	
O_2	liq.	1557	38
O_3	gas	1110	39
		1043	
		705	

Compound	Medium	Frequency	Ref.
H_2O	gas	3756	40
		3657	
		1595	
	liq.	3445	40
		3219	
		1627	
H_2O_2	liq.	3360	41
		2780m	
		1350	
		878m	
H_2S	gas	2627	42
		2615	
		1183	
SO_2	gas	1361	43
		1151	
	solid	1316	43
		1330	
		1308m	
		1147*	
		1142	
SO_3	gas	1391	44
	liq.	1330	44
$SOCl_2$	liq.	1251	45
SO_2Cl_2	CCl_4	1419	46
		1196	
		1160	
		1128m	
$CF_3SO_3^-$	KBr	1270vb	62
		1237	
		1167	
		1043	
		760m	
		647	
SO_3^{2-}	nuj.	1215w	1
		1135w	
		960b	

Compound	Medium	Frequency	Ref.
HSO_4^-	nuj.	1175m	1
		1075	
		1045	
		865	
SO_4^{2-}	nuj.	1110	1
SO_5^{2-}	solid	1254	47
		1059	
		900m	
$S_2O_3^{2-}$	nuj.	1165	1
		1125	
		1000	
$S_2O_8^{2-}$	nuj.	1300	1
		1270	
		1060	
KHF_2	nuj.	1455b	48
		1230	
	KBr	1527	48
		1257	
Cl_2O	gas	969	49
ClO_2	gas	1108	49
		945	
ClO_2^-	H_2O	840	50
		790	
ClO_3^-	nuj.	991	51
		974	
		938	
$HClO_4 \cdot H_2O$	solid	1577	52
		1175	

ClO_4^-	nuj.	1110b 935	53
Br_2O	solid	925 821	54
BrO_3^-	nuj.	820 800	55
IO_3^-	nuj.	800m 775 767	1
IO_4^-	nuj.	848	1
$VOCl_3$	liq.	1035	56
$VOCl_4^{2-}$	nuj.	912	57
CrO_2Cl_2	gas	1000 990	58

Cr_2O_3	KBr	1111w	4
CrO_3	nuj.	969	59
CrO_4^{2-}	nuj.	915 890 855b 820b	1
Cr_2O_7	nuj.	940b 905m 885m 795m 760	1
MnO_4^-	nuj.	900b 845w	1
OsO_4	CCl_4	1921 1910 954	60
RuO_4	liq.	1792 913	61
	CCl_4	1820 1794 914	60

REFERENCES

1. F. A. Miller and C. H. Wilkins, *Anal Chem* (1952), 1253.
2. D. E. Bethell and N. Sheppard, *Trans Farad Soc* (1955) *51*, 9.
3. N. N. Greenwood, *JCS* (1959), 3811.
4. E. G. Brame, Jr., J. L. Margrave and V. W. Meloche, *JINC* (1957) *5*, 48.
5. G. L. Caldow and H. W. Thompson, *Proc Roy Soc* (1960) *A254*, 1.
6. R. E. Dodd and R. Little, *SCA* (1960), 1083.
7. W. O. Freitag and E. R. Nixon, *JCP* (1956) *24*, 109.
8. J. Wagner, *Z Physik Chem* (1943) *193A*, 55.
9. M. F. Amr El-Sayad and R. K. Sheline, *JINC* (1958) *6*, 187.
10. S. K. Deb and A. D. Yoffe, *Trans Farad Soc* (1959) *55*, 106.
11. J. H. Taylor, W. S. Benedict and J. Strong, *JCP* (1952) *20*, 1884.
12. L. H. Jones and E. McLaren, *JCP* (1958) *28*, 995.
13. A. Maki and J. C. Decius, *JCP* (1959) *31*, 772.
14. W. Beck and K. Feldl, *Angew* (1966), 746.
15. T. M. Barakat, N. Legge and A. D. E. Pullin, *Trans Farad Soc* (1963), 1764.
16. P. C. H. Mitchell and R. J. P. Williams, *JCS* (1960), 1912.
17. J. E. Bulkowski and D. Dolphin, unpublished results.
18. A. L. Smith, W. E. Keller and H. L. Johnston, *Phys Rev* (1950) *79*, 728.
19. E. Lieber, D. R. Levering and L. J. Pullenson, *Anal Chem* (1951), 1594.
20. D. Dolphin and A. Wick, unpublished results.
21. W. A. Pliskin and R. P. Eischens, *JPC* (1955) *59*, 1156.
22. P. A. Giguère and I. D. Liu, *Canad J Chem* (1952), 948.
23. P. A. Giguère and I. D. Liu, *JCP* (1952) *20*, 136.
24. G. M. Begun and W. H. Fletcher, *JCP* (1958) *28*, 414.
25. W. G. Fateley, H. A. Bent and B. L. Crawford, *JCP* (1959) *31*, 204.
26. I. C. Hisatsune and J. P. Devlin, *SCA* (1960) *16*, 401; (1961) *17*, 218.
27. E. T. Arakawa and A. H. Nielsen, *J Mol Spectr* (1958) *2*, 413.
28. P. J. H. Woltz, E. A. Jones and A. H. Neilsen, *JCP* (1952) *20*, 378.
29. L. Landau, *J Mol Spectr* (1960) *4*, 276.
30. M. N. Hughes, *JINC* (1967), 1376.
31. R. A. Marcus and J. M. Fresco, *JCP* (1957) *27*, 564.
32. P. J. Bruna, J. E. Sicre and H. J. Schumacher, *Chem Comm* (1970), 1542.
33. R. Ryason and M. K. Wilson, *JCP* (1954) *22*, 2000.
34. R. E. Weston and T. F. Brodasky, *JCP* (1957) *27*, 683.
35. M. Anbar, M. Halmann and S. Pinchas, *JCS* (1960) *1242*, 1246.
36. A. Müller, E. Niecke and O. Glemser, *Z Anorg Chem* (1967) *350*, 246.
37. M. L. Delwaulle and F. Francois, *CR* (1945) *220*, 817.

38. A. L. Smith and H. L. Johnston, *JCP* (1952) *20*, 1972.
39. M. K. Wilson and R. M. Badger, *JCP* (1948) *16*, 741.
40. W. S. Benedict, N. Gailar and E. K. Plyler, *JCP* (1956) *24*, 1139.
41. O. Bain and P. A. Giguère, *Canad J Chem* (1955), 527.
42. H. C. Allen and E. K. Plyler, *JCP* (1956) *25*, 1132.
43. R. N. Wiener and E. R. Nixon, *JCP* (1956) *25*, 175.
44. R. W. Lovejoy, J. H. Colwell, D. F. Eggers and G. D. Halsey, *JCP* (1962) *36*, 612.
45. R. J. Gillespie and E. A. Robinson, *Canad J Chem* (1961), 2171.
46. K. C. Schreiber, *Anal Chem* (1949), 1169.
47. P. Pascal, C. Duval, J. Lecompte and A. Pacault, *CR* (1951) *233*, 118.
48. J. A. A. Ketelaar, C. Haas and J. van der Elsken, *JCP* (1956) *24*, 624.
49. K. Hedberg, *JCP* (1951) *19*, 509.
50. J. P. Mathieu, *CR* (1952) *234*, 2272.
51. J. L. Hollenberg and D. A. Dows, *SCA* (1960) *16*, 1155.
52. R. C. Taylor and G. L. Vidale, *JACS* (1956), 5999.
53. H. Colm, *J Chem Soc* (1952), 4282.
54. C. Campbell, J. P. M. Jones and J. J. Turner, *Chem Comm* (1968), 888.
55. G. Djega-Mariadassou, R. Kircher, R. Diament, J. Breiss, F. Gans and G. Pannetier, *BSC France* (1968), 2726.
56. F. A. Miller and L. R. Cousins, *JCP* (1957) *26*, 329.
57. A. Feltz, *Z Anorg Chem* (1957) *355*, 120.
58. W. E. Hobbs, *JCP* (1958) *28*, 1220.
59. R. Mattes, *Z Naturf* (1969) *24b*, 772.
60. R. E. Dodd, *Trans Farad Soc* (1959), 1480.
61. M. H. Ortner, *JCP* (1961) *34*, 556.
62. H. Bürger, K. Burczyk and A. Blaschette, *Monatsh Chem* (1970), 102.

8.2

transition metal complexes

M—H

M—CO

M—NO

M—N≡N

M—(C=C)

M—(C≡C)

The considerable increase in the use of organometallic intermediates in organic syntheses has prompted us to include a chapter for such compounds. We have limited ourselves to those ligands that show characteristic group frequencies in the infrared. Those are principally H^-, alkenes, alkynes, CO, NO, and N_2.

It is not possible to make any brief generalizations, since in most cases the bonding of the ligand to the metal is synergic and involves interactions between filled π-orbitals on the ligand with empty metal d-orbitals and empty ligand π^*-orbitals with filled metal d-orbitals. The overall result is that the frequency of the ligand or metal-ligand bond depends not only on the metal to which the ligand is bonded, but the oxidation state of the metal and the other ligands bonded to the metal, as well as the symmetry of the complex. Detailed discussions of these topics can be found in *Infrared Spectra of Inorganic and Coordination Compounds,* K. Nakamoto, Wiley, New York, 2nd edition, 1970, and *Metal-Ligand and Related*

Vibrations, D. M. Adams, E. Arnold Ltd., London, 1967.

Compounds in this chapter are ordered by metal and then by ligand. The ligands appear in the following order: H^-, CO, alkene, alkyne, NO, and N_2.

Main group hydrides are covered in Chapter 1.5, p. 17 , and inorganic ions in Chapter 8.1, p. 485.

Compound	Solvent	ν (cm⁻¹)	Ref.
$Ti(CO)_2(\pi-C_5H_5)_2$	C_6H_6	1964 1883	1
$[V(CO)_3(\pi-C_5H_5)]^{2-}$	nuj.	1748 1645	2
$[V(CO)_2(\pi-C_5H_5)_2]^{+}$	nuj.	2050 2010	3
$V(CO)_4(\pi-C_5H_5)$	C_6H_{14}	2031 1931 1901	4
$[V(CO)_4(\pi-C_6H_6)]^{+}$	THF	2068 2018 1986	5
$V(CO)_5(NO)$	C_6H_{12}	2108m 2064m 1992 1700	6
$V(CO)_6$	C_6H_{12}	1976	6
$[V(CO)_6]^{-}$	THF	1895w 1859	7
$V_2(CO)_5(\pi-C_5H_5)_2$	C_6H_{14}	2006 1953 1905 1869 1828	4
$Nb(CO)_4(\pi-C_5H_5)$	CS_2	2000 1901	8
$TaH_3(\pi-C_5H_5)_2$	nuj.	1735	9
$Ta(CO)_4(\pi-C_5H_5)$	CS_2	2020 1900	8
$[Ta(CO)_6]^{-}$	diglyme	1850	10
$Cr\,H(CO)_3(\pi-C_5H_5)$	sng	1828	11
$[Cr(CO)(NO)_2(\pi-C_5H_5)]^{+}$	nuj.	1873 1779	12

cis-$Cr(CO)_2[P(CH_3)_3]_4$

$C_{16}H_{34}$	1846 1785	13

cis-$Cr(CO)_2[P(OCH_3)_3]_4$

$C_{16}H_{34}$	1901 1847	13

$Cr(CO)_3(\pi\text{-}C_6H_6)$

C_6H_{12}	1987 1917	15
CCl_4	1980 1909	14

fac-$Cr(CO)_3[P(CH_3)_3]$

$C_{16}H_{34}$	1935 1842	13

mer-$Cr(CO)_3[P(CH_3)_3]_3$

$C_{16}H_{34}$	1943 1844* 1839	13

fac-$Cr(CO)_3[P(OCH_3)_3]_3$

$C_{16}H_{34}$	1966 1888* 1879	13

mer-$Cr(CO)_3[P(OCH_3)_3]_3$

$C_{16}H_{34}$	1981 1891* 1878	13

$[Cr(CO)_4(\pi\text{-}C_5H_5)]^+$

sng	2114 2037	16

$Cr(CO)_4(py)_2$	CH_3CN	2020 1899 1878 1837	17

cis-$Cr(CO)_4[P(CH_3)_3]_2$

$C_{16}H_{34}$	2006 1911 1892 1881	13

trans-$Cr(CO)_4[P(CH_3)_3]_2$

$C_{16}H_{34}$	1881	13

$Cr(CO)_4(P\phi_3)_2$	CCl_4	2012w 1949w 1897	18

cis $Cr(CO)_4[P(OCH_3)_3]_2$

$C_{16}H_{34}$	2026 1947 1939 1913	13

trans-$Cr(CO)_4[P(OCH_3)_3]_2$

$C_{16}H_{34}$	1914	13

$Cr(CO)_5(py)$	$CHCl_3$	2073 1986 1938 1905	17
$Cr(CO)_5(PH_3)$	C_6H_{14}	2074	19
$Cr(CO)_5(P(CH_3)_3)$	$C_{16}H_{34}$	2063 1949 1938	13
$Cr(CO)_5(P(OCH_3)_3)$	$C_{16}H_{34}$	2073 1985 1963 1948	13
$Cr(CO)_5P\phi_3$	CCl_4	2066m 1988w 1942	18
	C_6H_{14}	2065 1947 1944	20
$Cr(CO)_5Cl^-$	$CHCl_3$	2058 1921 1856	21

Compound	Medium	ν	Ref.
$Cr(CO)_5I$	CH_2Cl_2	2100 2025 1988	22
$[Cr(CO)_5Cl]^-$	KBr	2056w 1912 1875m	23
$[Cr(CO)_5I]^-$	CH_2Cl_2	2054 1920 1867m	22
	$CHCl_3$	2045 1915 1856	21
$Cr(CO)_6$	solid	1988 1961m	24
	C_6H_{12}	1900	15
	CCl_4	1977 1912	15
$Cr(NO)_4$	C_5H_{12}	1716	25
$Cr(NO)_2Cl(\pi-C_5H_5)$	$CHCl_3$	1823 1715	26
$[Cr_2(CO)_{10}]^{2-}$	THF	1903m 1880 1851m 1801w 1776w 1756w	27
$Cr_2(CO)_{10}I$	C_6H_{12}	2021 1999 1986	28
$[CrCo(CO)_9]^-$	nuj.	2055w 2000w 1936 1910*w 1864m 1854m	29

Compound	Medium	ν	Ref.
$MoH(CO)_3(\pi-C_5H_5)$	CS_2	2027 2020* 1940 1904m 1790	30
$MoH_2(\pi-C_5H_5)_2$	CS_2	1847	9
$[Mo(CO)(\pi-C_5H_5)(\pi-C_6H_6)]^+$	nuj.	2013	31
$Mo(CO)_2(\pi-C_5H_5)(P\,nBu_3)Cl$	C_5H_{12}	1977 1884	32
$Mo(CO)_3(\pi-C_6H_6)$	CCl_4	1982 1910	14
$Mo(CO)_3(py)_3$	nuj.	1888 1850* 1818* 1746b	33
$Mo(CO)_3(P\,nBu_3)_3$	C_5H_{12}	1939 1845	32
$Mo(CO)_3(P\phi_3)_3$	nuj.	1949 1908w 1891w 1835	33
$[Mo(CO)_4(\pi-C_5H_5)]^+$	sng	2128 2041 1980	34

Compound	Medium	Frequency	Ref.
$Mo(CO)_4(P\,nBu_3)_2$	C_5H_{12}	2015 1918 1903 1892	32
cis $Mo(CO)_4(P\phi_3)_2$	CCl_4	2020 1926 1908 1893	35
$Mo(CO)_4(P\phi_3)_2$	CCl_4	1902	18
$Mo(CO)_4Cl_2$	nuj.	2100w 2050m 1980 1956w	36
$[Mo(CO)_4Cl_2]^-$	KBr	2098 2052 2010 1939	36
$[Mo(CO)_4I_3]^-$	nuj.(F)	2065 2003 1994 1961 1941	37
$Mo(CO)_3(\pi\text{-}C_6H_6)$	nuj.	1972 1855	38
	CCl_4	1988 1912	38
$Mo(CO)_5(py)$	$CHCl_3$	2079 1987 1944 1890	17
$Mo(CO)_5(PH_3)$	C_6H_{14}	2081	19
$Mo(CO)_5(PEt_3)$	$C_{16}H_{34}$	2069 1980 1947 1941	39
$Mo(CO)_5(P\phi_3)$	CCl_4	2074m 1988w 1946	18
$Mo(CO)_5(DMF)$	DMF	2068w 1924 1847m	40
$[Mo(CO)_5Cl]^-$	KBr	2064w 1913 1871m	23
$Mo(CO)_6$	solid	1990 1962	24
	CCl_4	1980	18
	C_6H_{14}	1989	20
trans-$[Mo(N_2)_2(diphos)_2]$		2200w 1970	41
$WH_2(\pi\text{-}C_5H_5)_2$	nuj.	1896	9
$[W(CO)(\pi\text{-}C_5H_5)(\pi\text{-}C_6H_6)]^+$	nuj.	2009	31
$[W(CO)_4(\pi\text{-}C_5H_5)]^+$	sng	2128 2028 1965	34

Compound	Medium	Frequency	Ref.
$[W(CO)_4I]^-$	nuj.(F)	2067 1990b 1949 1930	37
$W(CO)_4Cl_2$	KBr	2100 2014 1988 1939	36
$[W(CO)_4Cl_2]^-$	KBr	2099 2042 1990 1920	36
$W(CO)_5(P\phi_3)$	CCl_4	2075m 1980* 1938	18
$W(CO)_5(py)$	$CHCl_3$	2076 1980 1933 1895	17
$W(CO)_5(DMF)$	DMF	2067w 1917 1847m	40
$[W(CO)_5Cl]^-$	KBr	2061w 1904 1869	23
$W(CO)_6$	solid	1982 1949m	24
	CCl_4	1980	18
$MnH(CO)_3(P\phi_3)_2$	nuj.	1912 1898 1827w	42
$MnH(CO)_5$	C_6H_{14}	2118w 2113* 2046w 2017 1982w 1967w 1780w	43
$[Mn(CO)(NO)(\pi-C_5H_5)]_2$	KBr	1949 1920 1779 1735 1700	44
$Mn(CO)(NO)_3$	C_6H_{12}	2088 1823 1734	45
$Mn(CO)_3(diphos)$	nuj.	1990 1900	46
$[Mn(CO)_2(diphos)_2]^-$	nuj.	1880	46
$[Mn(CO)_3(\pi-C_6H_6)]^+$	KBr	2083 2024	47

Compound	Solvent	Frequency	Ref.
cis-$Mn(CO)_4(CH_3)(P\phi_3)$	C_6H_{14}	2055w	48
		1983m	
		1969	
		1939m	
cis $Mn(CO)_4(COCH_3)(P\phi_3)$	C_6H_{14}	2067w	48
		1994m	
		1964	
		1957	
$Mn(CO)_4(NO)$	C_2Cl_4	2095m	49
		2019	
		1972	
		1759	
$[Mn(CO)_4(P\phi_3)_2]^+$	THF	2003	50
$Mn(CO)_4I$	CCl_4	2125w	51
		2044	
		2016w	
		2003	
$[Mn(CO)_4Cl_2]^-$	$C_2H_4Cl_2$	2098w	52
		2026	
		1986w	
		1936	
$Mn(CO)_5(CH_3)$	C_2Cl_4	2108m	53
		2102w	
		2010	
		1989	
$Mn(CO)_5(CF_3)$	C_6H_{12}	2144m	54
		2050	
		2025	
$Mn(CO)_5(COCH_3)$	C_6H_{12}	2116m	54
		2053*m	
		2009	
		1994	
$Mn(CO)_5\phi$	CCl_4	2123	54
		2028	
		2007	

Compound	Solvent	Frequency	Ref.
$[Mn(CO)_5(\pi-C_2H_4)]^+$	sng	2165	5
		2083	
		2062	
		1522	
$[Mn(CO)_5(P\phi_3)]^+$	THF	2138m	50
		2086w	
		2046	
$Mn(CO)_5Cl$	CCl_4	2138w	51
		2054	
		2022w	
		1999w	
	$CHCl_3$	2138	21
		2053	
		2001	
$Mn(CO)_5Br$	$CHCl_3$	2135w	55
		2052	
		2001m	
$Mn(CO)_5I$	$CHCl_3$	2125	21
		2044	
		2007	
$[Mn(CO)_6]^+$	THF	2090	50
$[Mn(CO)_4Cl]_2$	CCl_4	2104w	56
		2045	
		2012	
		1977	
$[Mn(CO)_6]_2^{2+}$	CH_3COCH_3	2091	50
$[Mn_2(CO)_8Cl_2]^{2-}$	$CHCl_3$	2022	57
		1931	
$Mn_2(CO)_{10}$	C_6H_{12}	2044m	58
		2013	
		1983m	

Compound	Medium	Frequency	Ref.
$[MnFe(CO)_9]^-$	CH_2Cl_2	2066w 2024w 1968 1937*m 1870m	29
$[TcH_2(\pi-C_5H_5)_2]^+$	KBr	1984w	59
$[Tc(CO)_4Cl]_2$	CCl_4	2119w 2048 2011m 1972m	56
$Tc(CO)_5Cl$	CCl_4	2153w 2057 2028w 1991w	51
$Tc(CO)_5I$	CCl_4	2146w 2055 2024 2000m	51
$ReH(\pi-C_5H_5)_2$	CS_2	2030w	60
$ReH(CO)_5$	nuj.	1832	30
$[ReH_2(\pi-C_5H_5)_2]^+$	nuj.	2360w 2330w 2250w 2070	60
$ReH_3(P\phi_3)_3$	sng	2000	61
$ReH_3(P\phi_3)_4$	sng	2050	61
$ReH_5(P\phi_3)_3$	nuj.	2000 1961w 1934m 1912 1890*	62
$ReH_7(P\phi_3)_2$	nuj.	1961w 1894m	62
$[Re(CO)_4Cl]_2$	CCl_4	2114w 2032 2000m 1959m	56
$[Re(CO)_5]^-$	THF	1910 1864	63
$Re(CO)_5(COCH_3)$	CCl_4	2131m 2068m 2061m 2045m 2018 2001m 1976m	63
$Re(CO)_5Cl$	KBr	2156w 2062 2035* 1970	64
	CCl_4	2156w 2045 2016w 1982	51
$Re(CO)_5Br$	KBr	2156w 2064 2032* 1965	64

Compound	Medium	Frequency	Ref.
$Re(CO)_5I$	CCl_4	2145w 2042 2013 1987m	51
$Re(N_2)(diphos)_2Cl$	$CHCl_3$	1980	65
$FeH(CO)_2(\pi-C_5H_5)$	CS_2	2014 1960 1930* 1900w 1835	30
$FeH_2(CO)_4$	gas	2071w 2061m 2056 2050m	66
$FeH_2(N_2)(PEt\phi_2)_3$	nuj.	2058 1955w 1859m	67
$Fe(CN)(NO)_2(P\phi_3)$	KBr	2082 1825 1734	68
$[Fe(CN)_2(NO)_2]^-$	KBr	2079 1951 1808 1730	68
$Fe(CO)(NO)_2(P\phi_3)$	C_2Cl_4	2009 1764 1722	69

Compound	Medium	Frequency	Ref.
$Fe(CO)(PF_3)_4$	C_6H_{12}	2038m 2009m	70
$Fe(CO)_2(CH_3)(\pi-C_5H_5)$	CS_2	2016 2005 1925	30
$Fe(CO)_2(C\equiv CH)(\pi-C_5H_5)$	nuj.	2050 1998 1965	71
$[Fe(CO)_2(\pi-C_2H_4)(\pi-C_5H_5)]^+$	sng	2083 2049 1527	34
$[Fe(CO)_2(\pi-C_5H_5)(P\phi_3)]^+$	nuj.	2066 2030	72
$Fe(CO)_2(\pi-C_5H_5)Cl$	C_6H_{12}	2054 2013	73
$Fe(CO)(NO)_2(P\phi_3)$	Toluene	2010 1766 1725	74
$Fe(CO)_2(NO)_2$	C_6H_{12}	2087 2034 1810 1756	69
$Fe(CO)_3(\pi-C_4H_4)$	sng	2055 1985	75

Compound	Solvent	ν (cm⁻¹)	Ref.
$Fe(CO)_3(\pi\text{-}C_3H_5)Cl$	KBr	2091 2013 1985*w	76
(π-allyl structure) $Fe(CO)_3$	sng	2064 1998	77
$Fe(CO)_3(\pi\text{-}C_4H_7)Cl$	KBr	2090 2026 2000	76
$[Fe(CO)_3(\pi\text{-}C_5H_5)]^+$	nuj.	2120 2070	72
$Fe(CO)_3(COT)$	CCl_4	2050 1996 1972	78
$[Fe(CO)_3(NO)]^-$	CH_3COCH_3	1984m 1881 1658 1651 1647*	79
trans $Fe(CO)_3(P\phi_3)_2$	CCl_4	1886	78
$[Fe(CO)_4]^{2-}$	H_2O	1788	80
	DMF	1730	81
$[FeH(CO)_4]^-$	DMF	2008w 1914m 1880	81
$Fe(CO)_4(\pi\text{-}C_2H_4)$	liq.	2088 2013* 2007 1986 1511m	82
$Fe(CO)_4(\pi\text{-}C_4H_8)$	KBr	2090 2026 2000 1522w	83
$Fe_2(CO)_4(\pi\text{-}C_5H_5)_2$	C_7H_{16}	2005 1961	84
$Fe(CO)_4(NH_3)$	nuj.	2050m 2023m 2003 1960* 1941 1900*	85
$Fe(CO)_4(PF_3)$	C_6H_{12}	2101w 2094w 2021 2004 1996 1970w 1960w	70
$Fe(CO)_4(PEt_3)$	$C_{16}H_{34}$	2049 1975 1936	86
$Fe(CO)_4(P\phi_3)$	CCl_4	2059 1984 1946	87
	C_6H_{14}	2052 1979 1946	88
$Fe(CO)_4Cl_2$	$CHCl_3$	2167w 2126 2082	89

Compound	Medium	Frequency	Ref.
cis $Fe(CO)_4Br_2$	C_6H_{12}	2149m 2107 2097 2074	90
cis $Fe(CO)_4I_2$	C_6H_{12}	2129m 2084 2060m	90
$Fe(CO)_5$	liq.	2222m 2020b	91
	C_5H_{12}	2022 2000	92
$Fe(NO)(\pi\text{-}C_5H_5)I_2$	KBr	1835	93
$Fe(NO)_2(P\phi_3)_2$	C_2Cl_4	1724 1678	99
$Fe(NO)_3Cl$	CCl_4	1826 1763	94
$[Fe(CO)_2(\pi\text{-}C_5H_5)]_2$	C_6H_{12}	2006m 1961 1793	95
$[Fe(CO)_3(NH)]_2$	CCl_4	2074 2032 1987	85
$[Fe(CO)_3S]_2$	CCl_4	2081 2042 2005	85
$[Fe_2H(CO)_8]^-$	DMF	1998w 1980m 1930 1906m 1873w	81
$[Fe_2(CO)_8]^{2-}$	DMF	1916m 1866 1842w	81

Compound	Medium	Frequency	Ref.
$Fe_2(CO)_9$	solid	2080m 2034 1829	91
$[FeCo(CO)_8]^-$	CH_2Cl_2	2056w 1989m 1953 1895w 1776m	29
$[Fe(NO)_2(NH)]_2$	CCl_4	1767 1735	85
$[Fe(NO)_2I]_2$	CCl_4	1819 1771	85
$RuH(CO)(P\phi_3)_3Cl$	nuj.	2020	97
$RuH(CO)_2(\pi\text{-}C_5H_5)$	CS_2	1853	98
$RuH_2(CO)(P\phi_3)_3$	nuj.	1960 1900	99
$RuH_2(P\phi_3)_4$	nuj.	2080	99
	KBr	2080	100
$RuH_2(CO)_2(P\phi_3)_2$	nuj.(F)	2011 1974 1878m 1823m	101
$RuH_2(CO)_4$	gas	2082w 2074m 1980w	66

$Ru\ H_2(N_2)(P\phi_3)_3$
KBr 2147 100
2027

$Ru(CO)(NO)(P\phi_3)_2Cl$
nuj. 1925 102
1592

$Ru(CO)_2(CH_3)(\pi-C_5H_5)$
CS_2 2028 30
1960
1933*

$Ru(CO)_2(\pi-C_5H_5)Cl$
C_6H_{12} 2056 73
2008

$[Ru(CO)_2(P\phi_3)I_2]_2$
CH_2Cl_2 2061 103
2005

trans-$Ru(CO)_2(P\phi_3)_2I_2$
CH_2Cl_2 1987 104

$Ru(CO)_3(\pi-C_4H_4)$
sng 2061 75
1995

$Ru(CO)_3(COT)$ C_6H_{12} 2070 105
2009
1996

$Ru(CO)_3(P\phi_3)_2$ nuj. 1900 99

$Ru(CO)_4P\phi_3$ nuj. 2060 103
1980m
1940
1906w

cis-$Ru(CO)_4Br_2$
$CHCl_3$ 2180 66
2127
2111m
2088

$Ru(CO)_4I_2$ CCl_4 2161m 106
2119w
2106
2097
2068

$Ru(CO)_5$ C_5H_{12} 2035 92
1999

$[Ru(NH_3)_5(N_2)]^{2+}$
nuj. 2114 107

$[Ru(NH_3)_5(NO)]^{3+}$ nuj. 1903 108

$[Ru(NO)_2(P\phi_3)_2Cl]^+$
KBr 1845 109
1687

cis-$[Ru(CO)_3Cl_2]_2$
$CHCl_3$ 2138 66
2063
2021w

$OsH(CO)(P\phi_3)_3Cl$
nuj. 2000 96
1933
1912
1899

$OsH_2(CO)_2(P\phi_3)_2$
nuj.(F) 2014 101
1996
1928m
1873m

$OsH_2(CO)_3(P\phi_3)$ C_7H_{16} 2079 110
2027*
2018
1959w
1922w

$Os\ H_2(CO)_4$ C_7H_{16} 2141w 110
2067m
2055
2048
2016w
1942w

Compound	Medium	Frequency	Ref.
$OsH_2(N_2)(PEt\phi_2)_3$	nuj.	2085 1925	111
$OsH_4(P\phi_3)_3$	nuj.	2080w 2020m 1890	99
$Os(CO)_5$	C_5H_{12}	2034 1991	92
$[Os(N_2)(NH_3)_5]^{2+}$	nuj.	2028	112
$Os_2(CO)_9$	C_7H_{16}	2080 2038 2024 2013 2000w 1778m	113
$CoH(CO)(P\phi_3)_3$	sng	1910 1960	115
$CoH(CO)_2(P\phi_3)_2$	sng	1920b 1990b	115
$CoH(CO)_4$	nuj.	1934	116
	C_7H_{16}	2119 2070 2053 2030 1996	116
$CoH(N_2)(P\phi_3)_3$	nuj.	2085 2105	117
$CoH_3(P\phi_3)_3$	C_6H_6	1947m 1760	118
$Co(CN)(NO)_2P\phi_3$	KBr	2098 1852 1790	68
$[Co(CN)_2(NO)_2]^-$	KBr	2096 2074 1837 1748	68
$Co(CO)(\pi-C_5H_5)P\phi_3$	C_6H_{14}	1937	119
$Co(CO)(NO)(P\phi_3)_2$	CCl_4	1959 1714	120
$Co(CO)_2(\pi-C_5H_5)$	liq.	2020 1986	121
$Co(CO)_2(P\phi_3)_2Cl$	$CHCl_3$	1984 1919	122
$Co(CO)_3(NO)$	gas	2108 2047 2010 1822	123
$[Co(CO)_3(P\phi_3)_2]^+$	$CHCl_3$	2073 2013 2006	124
$Co(CO)_3(P\phi_3)I$	THF	2066 1966	122

Compound	Medium	Frequency	Ref.
$Co(CO)_4(CH_3)$	C_6H_{14}	2105 2036 2019	88
$Co(CO)_4(COCF_3)$	liq.	2127m 2041	125
$Co(N_2)(P\phi_3)_3$	THF	2093	126
$Co(NO)(P\phi_3)_3$	nuj.	1633	127
$Co(NO)_2(P\phi_3)Cl$	nuj.	1820 1765 1735	127
$Co(NO)_3$	C_6H_{12}	1860 1795	128
$[Co_2(CO)_4(diphos)_3]^{2+}$	nuj.	2000 1950	46
$Co_2(CO)_8$	C_5H_{12}	2071* 2069 2044* 2042 2031m 2022	129
$[Co(NO)_2Cl]_2$	CCl_4	1859 1790	94
$[RhH(NH_3)_5]^{2+}$			
$2Br^-$	nuj.	2015	130
SO_4^{2-}	nuj.	2079	130
$RhH(P\phi_3)_4$	nuj.	2140	99
	KBr	2140	100
$RhH(CO)(P\phi_3)_3$	nuj.	2040 1923m	131
$RhH(CO)_2(P\phi_3)_2$	C_6H_{14}	2038 1980 1939	131
$[RhH(CN)_5]^{3+}$		2138 2123 2110 1970	132
$RhH(P\phi_3)_3$	sng	2020	133
$Rh(\pi-C_2H_4)_2(\pi-C_5H_5)$	KBr	1490	134
$Rh(\pi-CH_2=C=CH_2)(P\phi_3)_2Cl$	nuj.	1730	135
$Rh(CO)_2(\pi-C_5H_5)$	CH_3COCH_3	2041 1947	136
$Rh(NO)(P\phi_3)_3$	nuj.	1610	99
	KBr	1610	137
$[Rh(NO)_2Cl]_4$	nuj.	1703 1605	114

Compound	Medium	Frequency	Ref.
$Rh(N_2)(P\phi_3)_2Cl$	nuj.	2152	138
$IrH(PEt_3)Cl_2$	nuj.	2112	139
$IrH(P\phi_3)_3Cl_2$	nuj.	2197	99
$[IrH(diphos)_2Cl]^+$	nuj.(F)	2216w	140
$IrH(CO)(P\phi_3)_3$	CH_2Cl_2	2070 1925	141
$IrH(CO)(P\phi_3)_2Cl_2$	nuj.	2245 2030	142
$IrH(NO)(P\phi_3)_2Cl$	KBr	1550	143
$[IrH(NO)P\phi_3]^+$	nuj.	1715	144
$Ir(CH_3)(NO)(P\phi_3)_2I$	KBr	1525	143
$IrH_2(CO)(P\phi_3)_2Cl$	nuj.	2190 2100 1970	142
$IrH_2(P\phi_3)_3Cl$	nuj.	2210 2130	145

Compound	Medium	Frequency	Ref.
$[IrH_2(diphos)]^+$	nuj.(F)	2091 2080	140
trans-$IrH_3(P\phi_3)_3$	nuj.	2100m 1771 1750*	139
$Ir(CO)(\sigma-CH{=}C{=}CH_2)(P\phi_3)_2Cl_2$	KBr	2050 1920w	146
$Ir(CO)(\sigma-HC{=}{<})(P\phi_3)_2Cl_2$	KBr	2050 1925w	146
$Ir(CO)(\pi-C_5H_5)P\phi_3$	C_6H_{14}	1947	119
$Ir(CO)(P\phi_3)_2Cl$	nuj.	1950	99
$Ir(CO)(NO)(P\phi_3)_2$	nuj.	1950 1920 1660 1645	144
$[Ir(CO)(NO)(P\phi_3)_2Cl]^+$	$CHCl_3$	2060 1690	147
$[Ir(CO)_2I_2]^-$	nuj.	2020 1941	148
$[Ir(CO)_2I_4]^{2-}$	nuj.	2109 2074	148
$[Ir(CO)_3(P\phi_2)_2]^+$	$CHCl_3$	2080w 2018 2010	124

Compound	Medium	ν	Ref.
$Ir(CO)_3I_3$	nuj.	2178 2114	148
$Ir(N_2)(P\phi_3)_2Cl$	$CHCl_3$	2105	149
$Ir(NO)(P\phi_3)_3$	nuj.	1600	144
	KBr	1600	137
$[Ir(NO)(P\phi_3)_2Cl_3]^+$	nuj.	1945	147
$[Ir(NO)Cl_5]^-$	nuj.	2006	108
$Ni(\sigma\text{-}C\equiv CH)_3(\pi\text{-}C_5H_5)P\phi_3$	nuj.	2220 2160 2090 2010	71
$Ni(CN)(NO)(P\phi_3)_2$	THF	2109 1760	68
$Ni(CO)_3P\phi_3$	C_6H_{12}	2070 2000	151
$[Ni(CO)_3I]^-$	DMF	2049 1958	152
$Ni(CO)_4$	gas	2057	153
	liq.	2039	153
	solid	2038 2022	154
$Ni(NO)(\pi\text{-}C_5H_5)$	CCl_4	1833	155
$[Ni(\pi\text{-}C_3H_5)]_2$	nuj.	1449	156
$NiH(N_2)(PEt_3)_2$	nuj.	2075 1911	150
$Ni(\sigma\text{-}C\equiv CH)(\pi\text{-}C_5H_5)P\phi_3$	nuj.	1965	71
$Ni(\sigma\text{-}C\equiv C\text{-}C\equiv CH)(\pi\text{-}C_5H_5)P\phi_3$	nuj.	2125 1980	71
$PdH(PEt_3)_2Cl$	nuj.	2035	157
$Pd(CO)(P\phi_3)_3$	KBr	1955	158
$Pd(NO)(\pi\text{-}C_5H_5)$	KBr	1789	159
$Pd(NO)_2Cl_2$	nuj.	1833 1818	114

Compound	Medium	Frequency	Ref.
$[Pd(\pi\text{-}H_2C{=}CH_2)Cl_2]_2$	nuj.	1525	160
$[Pd(\pi\text{-}C_3H_5)Cl]_2$	KBr	1458m	161
$[Pd(\pi\text{-}C_3H_5)Br]_2$	CS_2	1384m 1017	162
$[Pd(\pi\text{-}C_4H_7)Cl]_2$	KBr	1461	161
trans-$[PtH(CO)(PEt_3)_2]^+$	$CHCl_3$	2167 2064	163
trans-$PtH(PMe_3)_2Cl$	CCl_4	2182	164
trans-$[PtH(PEt_3)_2(py)]^+$	$CHCl_3$	2216	163
trans-$PtH(P\phi_3)_2Cl$	CCl_4	2186	164
$PtH_2(PEt_3)_2$	nuj.	1670	165
$PtH_2(P\phi_3)_2$	nuj.	1670	166
$PtH_2(PEt_3)_2Cl_2$	nuj.	2254 2265*	164
$[Pt(\pi\text{-}C_2H_4)Cl_3]^-$	KBr	1527w	167
$Pt(NBD)Cl_2$	nuj.	1436	168
$Pt(\sigma\text{-}CH{=}C{=}CH_2)(P\phi_3)_2Cl$	KBr	1910w	146
$[Pt(\sigma\text{-}CH{\equiv}C)_4]^{2-}$	KBr	1931	169
$[Pt(CH_3\text{-}C{\equiv}C)_4]^{2-}$	KBr	2116	169
$[Pt(CO)(PEt_3)Cl]^+$	$CHCl_3$	2109	163
$[Pt(CO)Cl_3]^-$	nuj.	2126 2101w	108
$Pt(CO)_2Cl_2$	CCl_4	2200 2162	170
$Pt(NO)(\pi\text{-}C_5H_5)$	KBr	1739	159

$[Pt(\pi-CH_2{=}CH_2)Cl_2]_2$	nuj.	1239w 1234w	172
	KBr	1511w	171
$[Pt(CO)Cl_2]_2$	nuj.	2152	173
$[Pt(CO)I_2]_2$	nuj.	2112	173
$Pt_2(CO)_2Cl_4$	nuj.	2114 2079	174
$[Ag(\pi-C_3H_6)]^+$	nuj.	1595	175
	$CHCl_3$	1580	175
$[Ag(\pi-cis-C_4H_8)]^+$	nuj.	1604	175
	$CHCl_3$	1595	175
$[Ag(\pi-trans-C_4H_8)]^+$	nuj.	1615	175
$[Ag_2(\pi-NBD)]^+$	nuj.	1470	168
$Ag_2C{=}C{=}O$	sng	2060	176

REFERENCES

1. J. G. Murray, *JACS* (1961), 1287.
2. R. D. Fisher, *B* (1960), 165.
3. F. Calderazzo and S. Bacciarelli, *Inorg Chem* (1963), 721.
4. E. O. Fischer and R. J. J. Schneider, *Angew* (1967), 537.
5. E. O. Fischer and K. Őfele, *Angew* (1961), 581.
6. W. Hieber, J. Peterhans and E. Winter, *B* (1961), 2572.
7. F. Calderazzo, *Inorg Chem* (1964), 810.
8. R. P. M. Werner, A. H. Filbey and S. A. Manastyrskyj, *Inorg Chem* (1964), 298.
9. M. L. H. Green, J. A. McCleverty, L. Pratt and G. Wilkinson, *JCS* (1961), 4854.
10. R. P. M. Werner and H. E. Dodall, *Chem & Ind* (1961), 144.
11. H. P. Fritz, *Advances in Organometallic Chemistry*, Vol. 1, p. 285, eds. F. G. A. Stone and R. West, Academic Press, New York, 1964.
12. E. O. Fischer and P. Kuzel, *Z Anorg Chem* (1962) *317*, 226.
13. R. Mathieu, M. Lenzi and R. Poilblanc, *Inorg Chem* (1970), 2030.
14. D. A. Brown and F. J. Hughes, *JCS* (1968) *A*, 1519.

15. H. P. Fritz and J. Manchot, *SCA* (1962), 171.
16. E. O. Fischer and K. Ulm, *Z Naturf* (1961) *16b*, 757.
17. C. S. Kraihanzel and F. A. Cotton, *Inorg Chem* (1963), 533.
18. T. A. Magee, C. N. Matthews, T. S. Wang and J. H. Wotiz, *JACS* (1961), 3200.
19. E. O. Fischer and L. Knauss, *B* (1969), 223.
20. T. L. Brown and D. J. Darensbourg, *Inorg Chem* (1967), 971.
21. E. W. Abel and I. S. Butler, *Trans Farad Soc* (1967), 45.
22. E. Lindner and H. Behrens, *SCA* (1967) *23A*, 3025.
23. E. W. Abel, I. S. Butler and I. G. Reid, *JCS* (1963), 2068.
24. R. L. Amster, R. B. Hannan and M. C. Tobin, *SCA* (1963), 1489.
25. B. I. Swanson and S. K. Satija, *Chem Comm* (1973), 40.
26. T. S. Piper and G. Wilkinson, *JINC* (1956) *2*, 38.
27. W. C. Kaska, *JACS* (1968), 6340.
28. H. Behrens and R. Schwab, *Z Naturf* (1964) *19b*, 768.
29. J. K. Ruff, *Inorg Chem* (1968), 1818.
30. A. Davison, J. A. McCleverty and G. Wilkinson, *JCS* (1963), 1133.
31. E. O. Fischer and F. J. Kohl, *Z Naturf* (1963) *18b*, 504.
32. R. J. Mawby and C. White, *Chem Comm* (1968), 312.
33. E. W. Abel, M. A. Bennett and G. Wilkinson, *JCS* (1959), 2323.
34. E. O. Fischer and K. Fichtel, *B* (1961), 1200.
35. R. Poilblanc and M. Bigorgne, *CR* (1960) *250*, 1064.
36. J. A. Bowden and R. Colton, *Aust J Chem* (1968), 2657.
37. R. B. King, *Inorg Chem* (1964), 1039.
38. H. P. Fritz and J. Manchot, *Z Naturf* (1962) *17b*, 711.
39. J. Dalton, I. Paul, J. G. Smith and F. G. A. Stone, *JCS* (1968) *A*, 1195.
40. I. W. Stolz, G. R. Dobson and R. K. Sheline, *Inorg Chem* (1963), 323.
41. M. Hidai, K. Tominari, Y. Uchida and A. Misono, *Chem Comm* (1969), 1392.
42. R. Ugo and F. Bonati, *J Organometallic Chem* (1967) *8*, 189.
43. W. J. Miles, Jr. and R. G. Clark, *Inorg Chem* (1968), 1801.
44. T. A. James and J. A. McCleverty, *JCS* (1970) *A*, 850.
45. C. G. Barraclough and J. Lewis, *JCS* (1960), 4842.
46. A. Sacco, *Gazz Chim Ital* (1963), 698.
47. G. Winkhaus, L. Pratt and G. Wilkinson, *JCS* (1961), 3807.
48. C. S. Kraihanzel and P. K. Maples, *Inorg Chem* (1968), 1806.
49. P. M. Treichel, E. Pitcher, R. B. King and F. G. A. Stone, *JACS* (1961), 2593.
50. T. Kruck and M. Noack, *B* (1963), 3028.
51. J. C. Hileman, D. K. Huggins and H. D. Kaesz, *Inorg Chem* (1962), 933.
52. R. J. Angelici, *Inorg Chem* (1964), 1099.
53. E. Pitcher and F. G. A. Stone, *SCA* (1962), 585.

54. W. Beck, W. Hieber and H. Tengler, *B* (1961), 862.
55. B. F. G. Johnson, J. Lewis, J. R. Miller, B. H. Robinson, D. W. Robinson and A. Wojcicki, *JCS* (1968) *A*, 522.
56. M. A. El-Sayed and H. D. Kaesz, *Inorg Chem* (1963), 158.
57. E. W. Abel and I. S. Butler, *JCS* (1964), 434.
58. N. Flitcroft, D. K. Huggins and H. D. Kaesz, *Inorg Chem* (1964), 1123.
59. E. O. Fischer and M. W. Schmidt, *Angew* (1967), 99.
60. M. L. H. Green, L. Pratt and G. Wilkinson, *JCS* (1958), 3916.
61. M. Freni and V. Valenti, *Gazz Chim Ital* (1961), 1357.
62. J. Chatt and R. S. Coffey, *JCS* (1969) *A*, 1963.
63. W. Hieber, G. Braun and W. Beck, *B* (1960), 901.
64. W. A. McAllister and A. L. Marston, *SCA* (1971), 523.
65. J. Chatt, J. R. Dilworth and G. J. Leigh, *Chem Comm* (1969), 687.
66. J. D. Cotton, M. I. Bruce and F. G. A. Stone, *JCS* (1968) *A*, 2162.
67. A. Sacco and M. Aresta, *Chem Comm* (1968), 1223.
68. W. Hieber and H. Führling, *Z Anorg Chem* (1970) *373*, 48.
69. D. W. McBride, S. L. Stafford and F. G. A. Stone, *Inorg Chem* (1962), 386.
70. R. J. Clark, *Inorg Chem* (1964), 1395.
71. P. J. Kim, H. Masai, K. Sonogashira and N. Hagihara, *Inorg & Nucl Chem Letters* (1970), 181.
72. A. Davison, M. L. H. Green and G. Wilkinson, *JCS* (1961), 3172.
73. T. Blackmore, J. D. Cotton, M. I. Bruce and F. G. A. Stone, *JCS* (1968) *A*, 2931.
74. D. E. Morris and F. Basolo, *JACS* (1968), 2531.
75. R. G. Amiet, P. C. Reeves and R. Pettit, *Chem Comm* (1967), 1208.
76. H. D. Murdoch and E. Weiss, *Helv* (1962), 1927.
77. J. S. Ward and R. Pettit, *Chem Comm* (1970), 1419.
78. F. Faraone, F. Zingales, P. Uguagliati and U. Belluco, *Inorg Chem* (1968), 2362.
79. W. Beck, *B* (1961), 1214.
80. W. F. Edgell, J. Huff, J. Thomas, H. Lehman, C. Angell and G. Asato, *JACS* (1960), 1254.
81. W. F. Edgell, M. T. Yang, B. J. Bulkin, R. Bayer and N. Koizumi, *JACS* (1965), 3080.
82. H. D. Murdoch and E. Weiss, *Helv* (1963), 1588.
83. H. D. Murdoch and E. Weiss, *Helv* (1962), 1156.
84. K. Noack, *JINC* (1963) *25*, 1383.
85. W. Hieber and H. Beutner, *Z Anorg Chem* (1962) *317*, 63.
86. J. Dalton, I. Paul, J. G. Smith and F. G. A. Stone, *JCS* (1968) *A*, 1199.
87. F. A. Cotton and R. V. Parish, *JCS* (1960), 1440.
88. G. Bor, *Inorg Chem Acta* (1967), 81.

89. R. C. Taylor and W. D. Horrocks, Jr., *Inorg Chem* (1964), 584.
90. B. F. G. Johnson, J. Lewis, P. W. Robinson and J. R. Miller, *JCS* (1968) *A*, 1043.
91. R. K. Sheline and K. S. Pitzer, *JACS* (1950), 1107.
92. F. Calderazzo and F. L'Eplattenier, *Inorg Chem* (1967), 1220.
93. H. Brunner and H. Wachsmann, *J Organometallic Chem* (1968) *15*, 409.
94. W. Hieber and A. Jahn, *Z Naturf* (1958) *13b*, 195.
95. R. J. Haines, A. L. DuPreez and G. T. W. Wittmann, *Chem Comm* (1968), 611.
96. J. Chatt and B. L. Shaw, *Chem & Ind* (1961), 290.
97. L. Vaska and J. W. DiLuzio, *JACS* (1961), 1262.
98. A. Davison, J. A. McCleverty and G. Wilkinson, *JCS* (1963), 1133.
99. J. J. Levison and S. D. Robinson, *JCS* (1970) *A*, 2947.
100. T. Ito, S. Kitazume, A. Yamamoto and S. Ikeda, *JACS* (1970), 3011.
101. F. L'Eplattenier and F. Calderazzo, *Inorg Chem* (1968), 1290.
102. K. R. Laing and W. R. Roper, *JCS* (1970) *A*, 2149.
103. F. Piacenti, M. Bianchi, E. Benedetti and G. Braca, *Inorg Chem* (1968), 1815.
104. P. John, *B* (1970), 2178.
105. M. I. Bruce, M. Cooke and M. Green, *J Organometallic Chem* (1968) *13*, 227.
106. E. R. Corey, M. V. Evans and L. F. Dahl, *JINC* (1962) *24*, 926.
107. A. D. Allen, F. Bottomley, R. O. Harris, V. P. Reinsalu and G. V. Senoff, *JACS* (1967), 5595.
108. M. J. Cleare and W. P. Griffith, *JCS* (1969) *A*, 372.
109. C. G. Pierpont, D. G. Van Derveer, W. Durland and R. Eisenberg, *JACS* (1970), 4760.
110. F. L'Eplattenier and F. Calderazzo, *Inorg Chem* (1967), 2092.
111. B. Bell, J. Chatt and G. J. Leigh, *Chem Comm* (1970), 576.
112. A. D. Allen and J. R. Stevens, *Chem Comm* (1967), 1147.
113. J. R. Moss and W. A. G. Graham, *Chem Comm* (1970), 835.
114. W. P. Griffith, J. Lewis and G. Wilkinson, *JCS* (1959), 1775.
115. A. Misono, Y. Uchida, M. Hidai and T. Kuse, *Chem Comm* (1968), 981.
116. R. A. Friedel, I. Wender, S. L. Shufler and H. W. Sternberg, *JACS* (1955), 3951.
117. J. H. Enemark, B. R. Davis, J. A. McGinnety and J. A. Ibers, *Chem Comm* (1968), 96.
118. M. Rossi and A. Sacco, *Chem Comm* (1969), 471.
119. A. J. Hart-Davis and W. A. G. Graham, *Inorg Chem* (1970), 2659.
120. G. Cardaci, S. M. Murgia and G. Reichenbach, *Inorg Chim Acta* (1970), 118.

121. D. J. Cook, J. L. Dawes and R. D. W. Kemmitt, *JCS* (1967) *A*, 1547.
122. W. Hieber and H. Duchatsch, *B* (1965), 2530.
123. R. S. McDowell, W. D. Horrocks, Jr. and J. T. Yates, *JCP* (1961) *34*, 530.
124. M. J. Church and M. J. Mays, *Chem Comm* (1968), 435.
125. W. Hieber, W. Beck and E. Lindner, *Z Naturf* (1961) *16b*, 229.
126. G. Speier and L. Marko, *Inorg Chim Acta* (1969), 126.
127. T. Bianco, M. Rossi and L. Ova, *Inorg Chim Acta* (1969), 443.
128. I. H. Sabherwal and A. B. Burg, *Chem Comm* (1970), 1001.
129. K. Noack, *SCA* (1963), 1925.
130. K. Thomas, J. A. Osborn, A. R. Powell and G. Wilkinson, *JCS* (1968) *A*, 1801.
131. D. Evans, G. Yagupsky and G. Wilkinson, *JCS* (1968) *A*, 2660.
132. K. Korgmann and W. Binder, *J Organometallic Chem* (1968) *11*, P27.
133. W. Keim, *J Organometallic Chem* (1967) *8*, P25.
134. R. B. King, *Inorg Chem* (1963), 528.
135. J. A. Osborn, *Chem Comm* (1968), 1231.
136. J. L. Dawes and R. D. W. Kemmitt, *JCS* (1968) *A*, 1072.
137. J. P. Collman, N. W. Hoffman and D. E. Morris, *JACS* (1969), 5659.
138. L. Y. Ukhin, A. Y. Shvetsov and M. L. Khidekel, *Izvest Akad Nauk Khim* (1967), 957.
139. J. Chatt, R. S. Coffey and B. L. Shaw, *JCS* (1965), 7391.
140. L. Vaska and D. L. Catone, *JACS* (1966), 5324.
141. J. G. Harrod, D. F. F. Gilson and R. Charles, *Canad J Chem* (1969), 1431.
142. L. Vaska and J. W. DiLuzio, *JACS* (1962), 679.
143. C. A. Reed and W. R. Roper, *Chem Comm* (1969), 155.
144. C. A. Reed and W. R. Roper, *JCS* (1970) *A*, 3054.
145. R. G. Hayter, *JACS* (1961), 1259.
146. J. P. Collman, J. N. Cawse and J. W. Kang, *Inorg Chem* (1969), 2574.
147. C. A. Reed and W. R. Roper, *Chem Comm* (1969), 1459.
148. L. Maletesta, L. Naldini and F. Cariati, *JCS* (1964), 961.
149. J. P. Collman, M. Kubota, F. D. Vastine, J. Y. Sun and J. W. Kang, *JACS* (1968), 5430.
150. S. C. Srivastava and M. Bigorgne, *J Organometallic Chem* (1969) *18*, 30.
151. L. S. Meriwether and M. L. Fiene, *JACS* (1959), 4200.
152. L. Cassar and M. Foà, *Inorg & Nucl Chem Let* (1970), 291.
153. L. H. Jones, *JCP* (1958) *28*, 1215.
154. L. H. Jones, *SCA* (1963), 1899.
155. T. S. Piper, F. A. Cotton and G. Wilkinson, *JINC* (1955) *1*, 165.
156. E. O. Fischer and O. Bürger, *B* (1961), 2409.
157. J. Chatt, L. A. Duncanson and B. L. Shaw, *Chem & Ind* (1958), 859.

158. A. Misono, Y. Uchida, M. Hidai and K. Kudo, *J Organometallic Chem* (1969) *20*, P7.
159. E. O. Fischer and H. Schuster-Woldan, *Z Naturf* (1964) *19b*, 766.
160. H. P. Fritz and C. G. Kreiter, *B* (1963), 1672.
161. H. P. Fritz, *B* (1961), 1217.
162. D. M. Adams and A. Squire, *JCS* (1970) *A*, 1808.
163. M. J. Church and M. J. Mays, *JCS* (1968) *A*, 3074.
164. J. Chatt and B. L. Shaw, *JCS* (1962), 5075.
165. J. A. Chopoorian, J. Lewis and R. S. Nyholm, *Nature* (1961) *190*, 528.
166. L. Malatesta and R. Ugo, *JCS* (1963), 2080.
167. H. B. Jonassen and W. B. Kirsch, *JACS* (1957), 1279.
168. E. W. Abel, M. A. Bennett and G. Wilkinson, *JCS* (1959), 3178.
169. R. Nast and W.-D. Heinz, *B* (1962), 1478.
170. R. J. Irving and E. A. Magnusson, *JCS* (1958), 2283.
171. H. B. Jonassen and J. E. Field, *JACS* (1957), 1275.
172. J. Hiraishi, *SCA* (1969), 749.
173. R. J. Irving and E. A. Magnusson, *JCS* (1956), 1860.
174. L. Malatesta and L. Naldini, *Gazz Chim Ital* (1960), 1505.
175. H. W. Quinn, J. S. McIntyre and D. J. Peterson, *Canad J Chem* (1965), 2896.
176. E. T. Blues, D. Bryce-Smith, H. Hirsch and M. J. Simons, *Chem Comm* (1970), 699.

9

spectra of common solvents

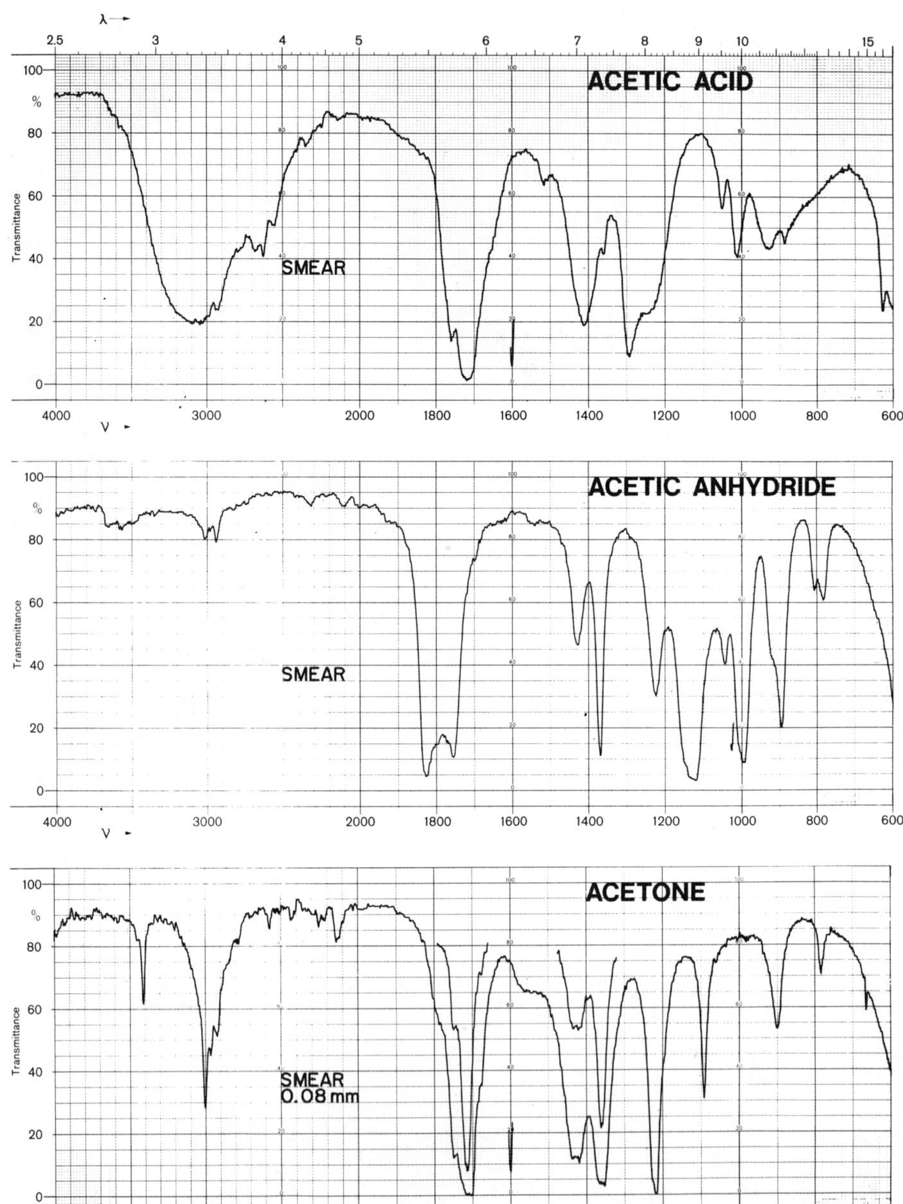
λ→
2.5
3
4
5
6
7
8
9
10
15
100
%
80
60
40
20
0
Transmittance
ACETIC ACID
SMEAR
4000
3000
2000
1800
1600
1400
1200
1000
800
600
ν
ACETIC ANHYDRIDE
SMEAR
ACETONE
SMEAR
0.08 mm

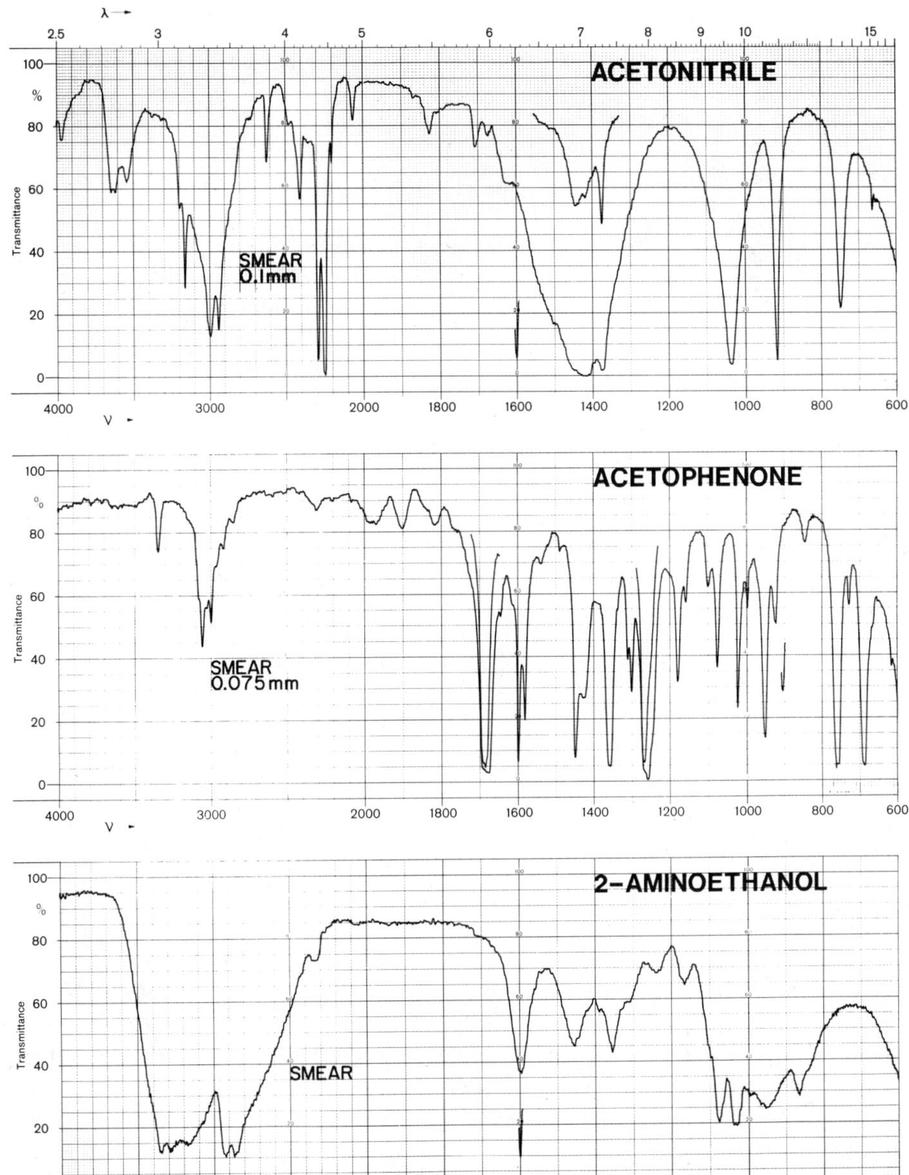
λ →
2.5
3
4
5
6
7
8
9
10
15
100
%
80
60
40
20
0
Transmittance
ACETONITRILE
SMEAR
0.1mm
4000
3000
2000
1800
1600
1400
1200
1000
800
600
ν →
ACETOPHENONE
SMEAR
0.075mm
2-AMINOETHANOL
SMEAR

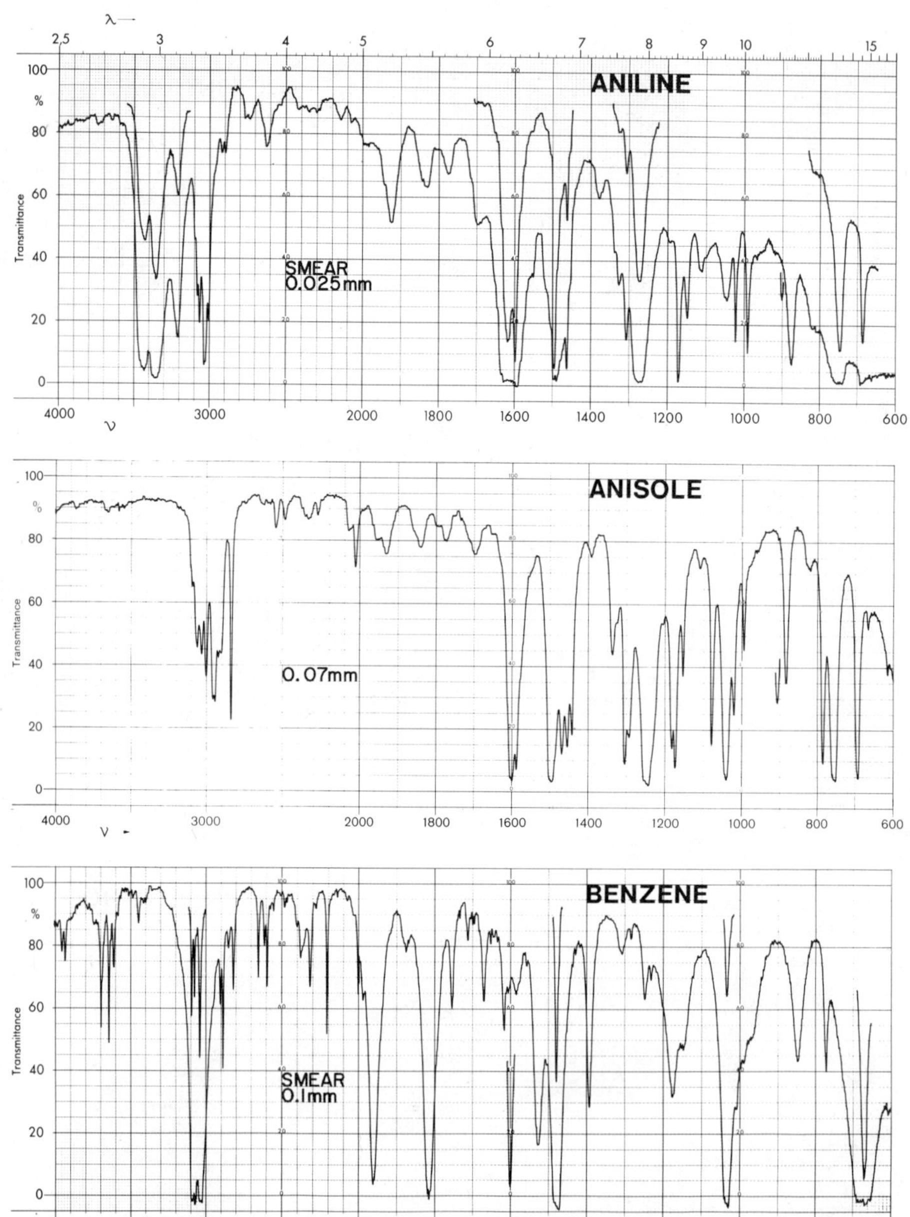
λ→
2.5
3
4
5
6
7
8
9
10
15
100
%
80
60
40
20
0
Transmittance
ANILINE
SMEAR
0.025mm
4000
3000
2000
1800
1600
1400
1200
1000
800
600
ν
ANISOLE
0.07mm
BENZENE
SMEAR
0.1mm

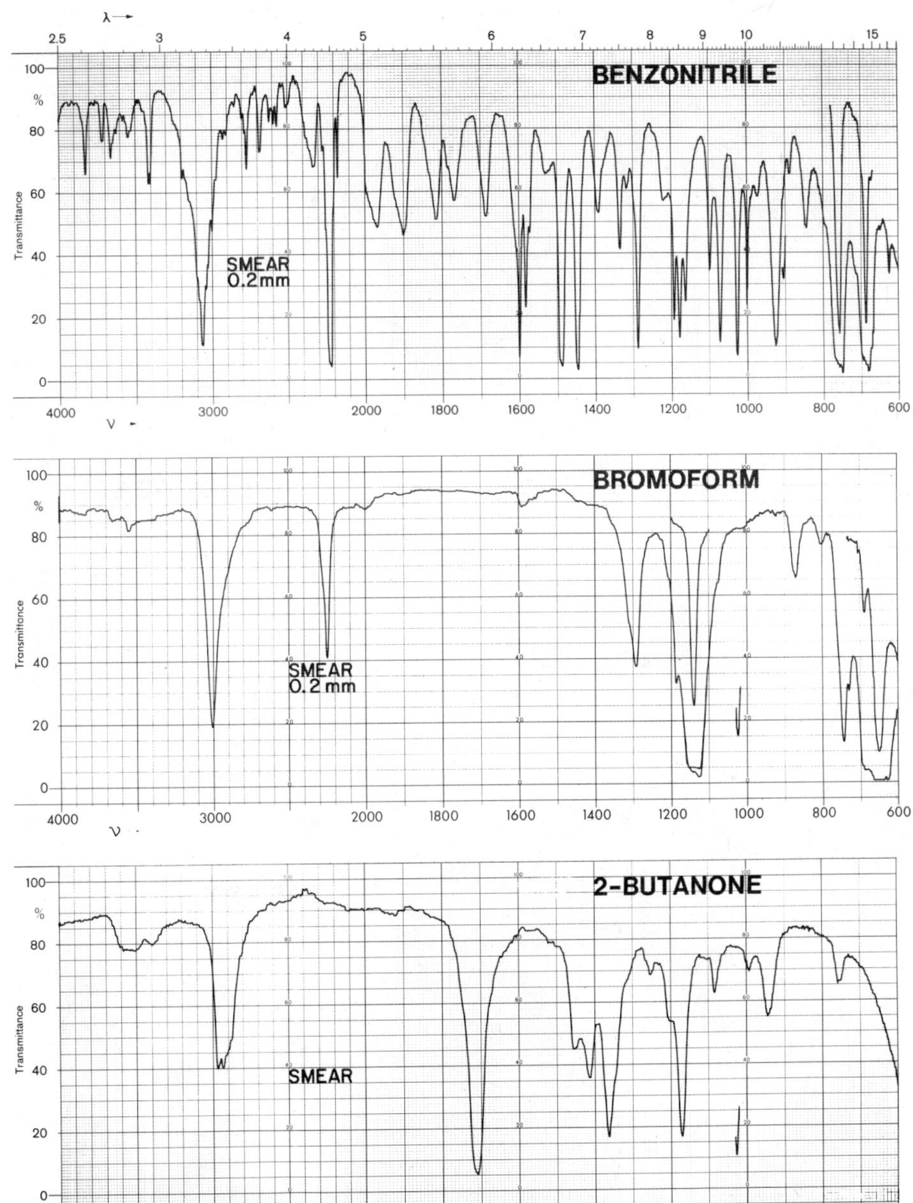
λ→
2.5
3
4
5
6
7
8
9
10
15
BENZONITRILE
100
%
80
60
40
20
0
Transmittance
SMEAR
0.2mm
4000
3000
2000
1800
1600
1400
1200
1000
800
600
ν→
BROMOFORM
SMEAR
0.2 mm
2-BUTANONE
SMEAR

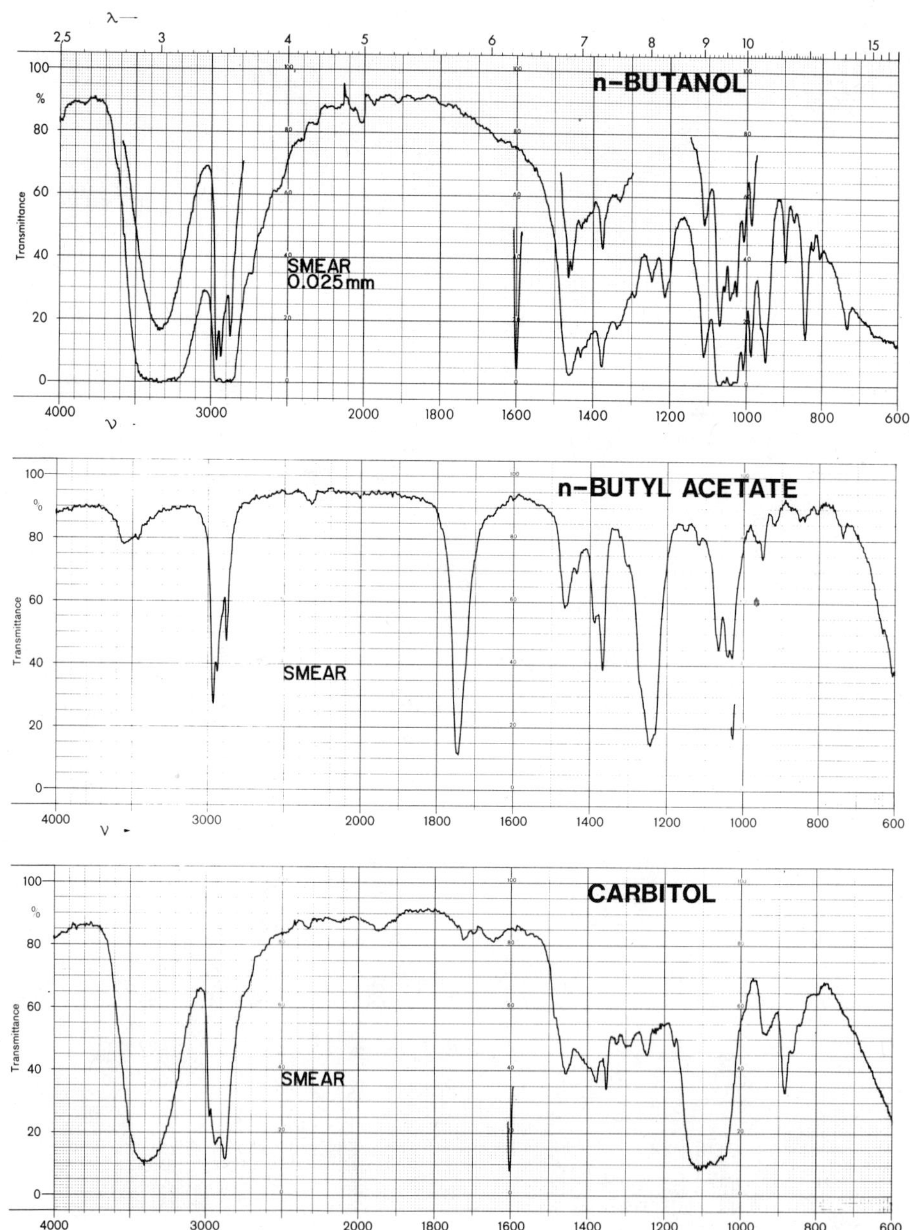

λ→
2.5
3
4
5
6
7
8
9
10
15
100
%
80
60
40
20
0
Transmittance
n-BUTANOL
SMEAR
0.025 mm
4000
3000
2000
1800
1600
1400
1200
1000
800
600
ν
n-BUTYL ACETATE
SMEAR
CARBITOL
SMEAR

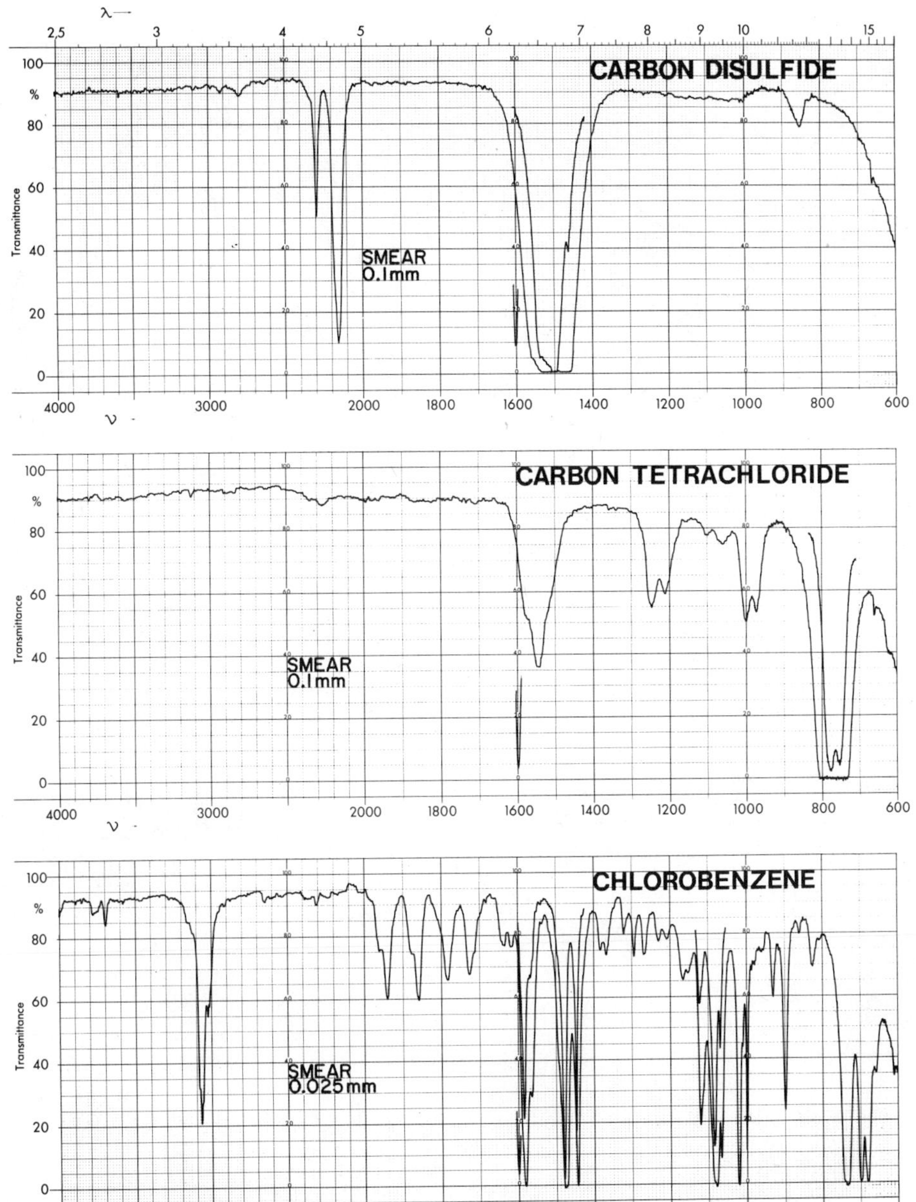
λ→
2.5
3
4
5
6
7
8
9
10
15
100
%
80
60
40
20
0
Transmittance
CARBON DISULFIDE
SMEAR
0.1mm
4000
3000
2000
1800
1600
1400
1200
1000
800
600
ν
CARBON TETRACHLORIDE
SMEAR
0.1mm
CHLOROBENZENE
SMEAR
0.025mm

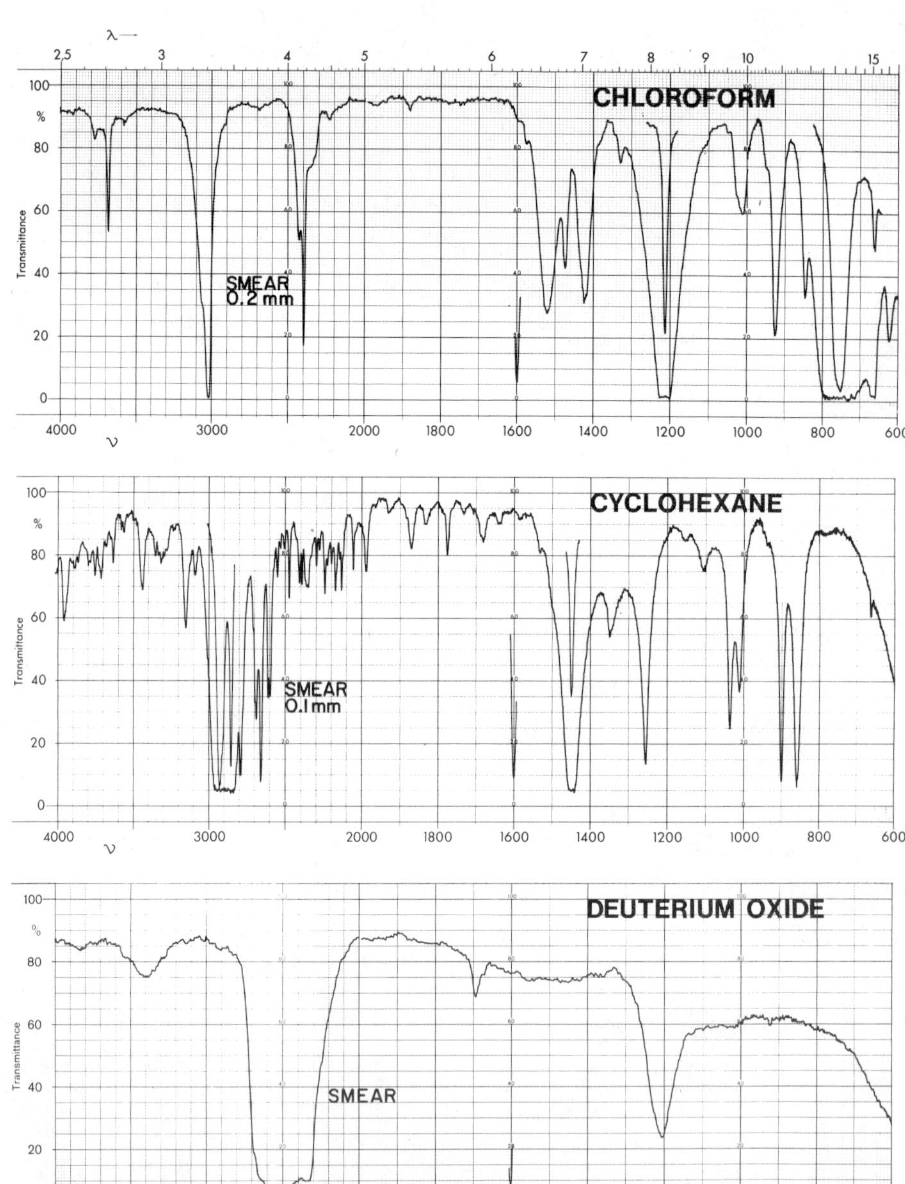
λ→
2.5
3
4
5
6
7
8
9
10
15
100
%
80
60
40
20
0
Transmittance
CHLOROFORM
SMEAR
0.2 mm
4000
3000
2000
1800
1600
1400
1200
1000
800
600
ν
CYCLOHEXANE
SMEAR
0.1mm
DEUTERIUM OXIDE
SMEAR

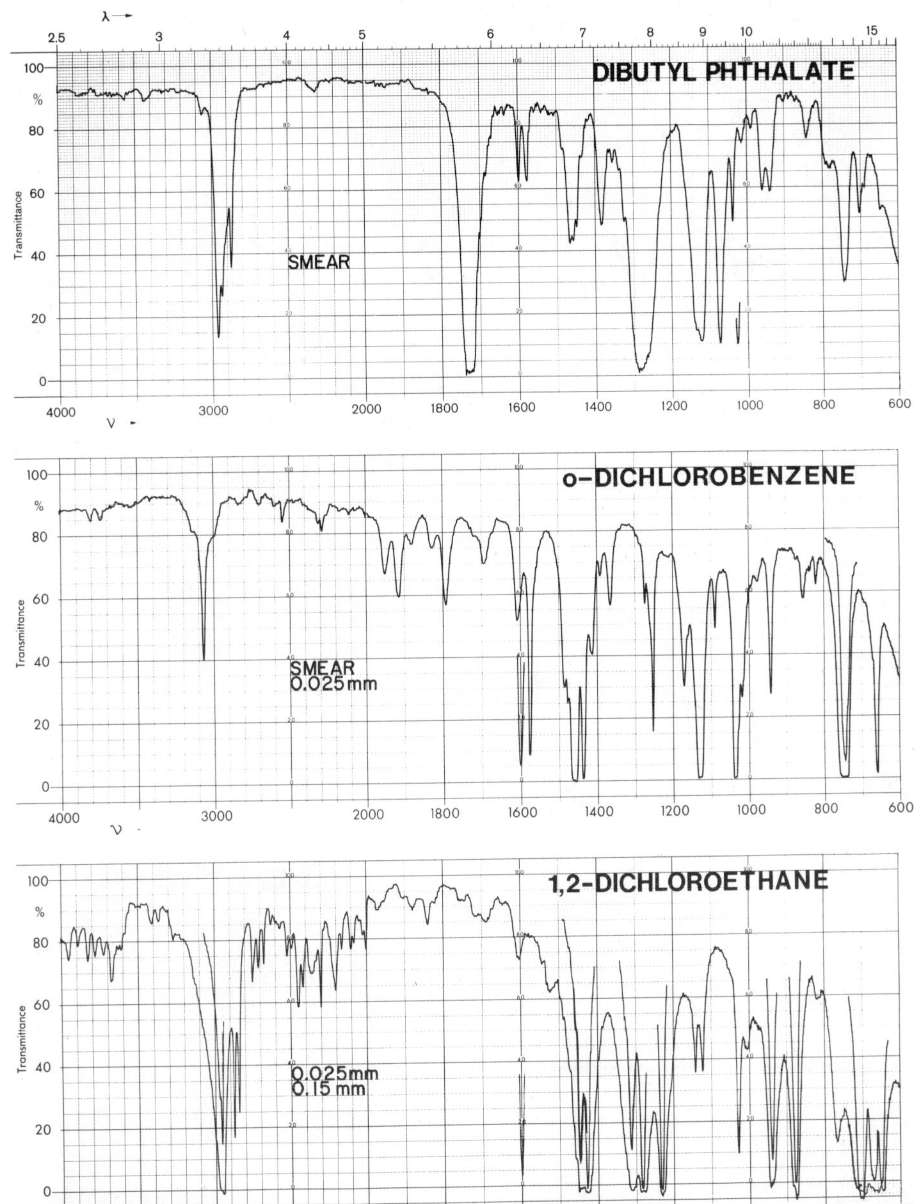
DIBUTYL PHTHALATE
SMEAR
o-DICHLOROBENZENE
SMEAR
0.025mm
1,2-DICHLOROETHANE
0.025mm
0.15mm
Transmittance
%
λ
ν

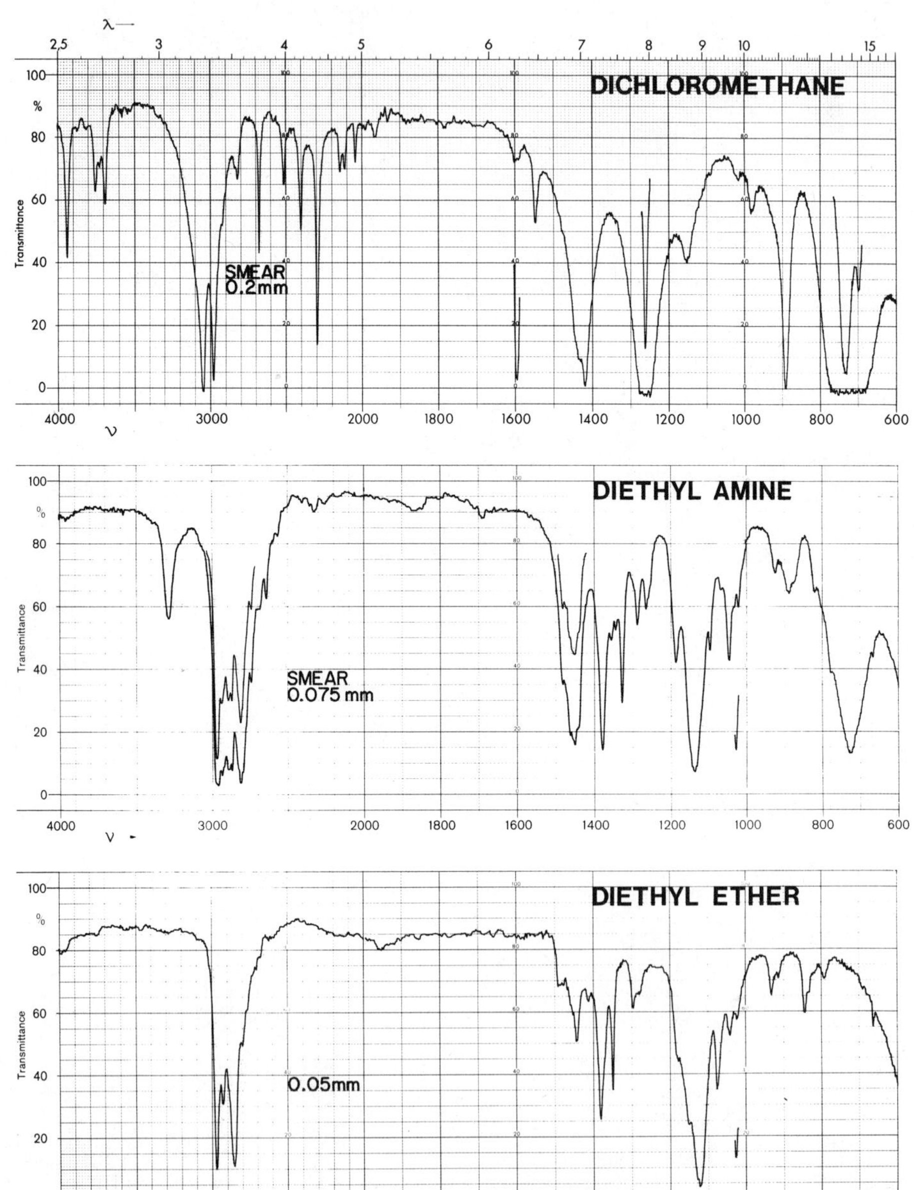
λ—
2.5
3
4
5
6
7
8
9
10
15
100
%
80
60
40
20
0
Transmittance
DICHLOROMETHANE
SMEAR
0.2mm
4000
3000
2000
1800
1600
1400
1200
1000
800
600
ν
DIETHYL AMINE
SMEAR
0.075 mm
DIETHYL ETHER
0.05mm

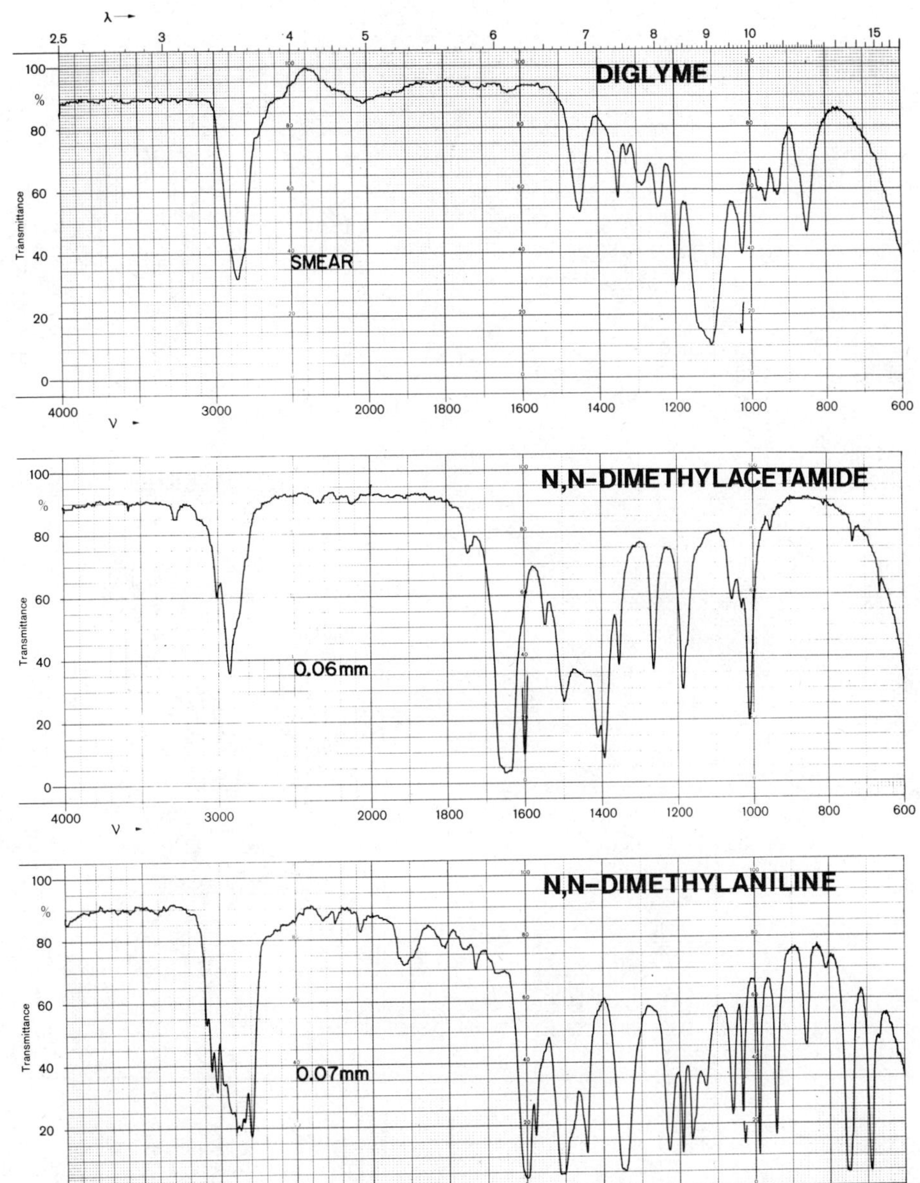
λ
2.5
3
4
5
6
7
8
9
10
15
DIGLYME
SMEAR
100
%
80
60
40
20
0
Transmittance
4000
3000
2000
1800
1600
1400
1200
1000
800
600
ν
N,N-DIMETHYLACETAMIDE
0.06mm
N,N-DIMETHYLANILINE
0.07mm

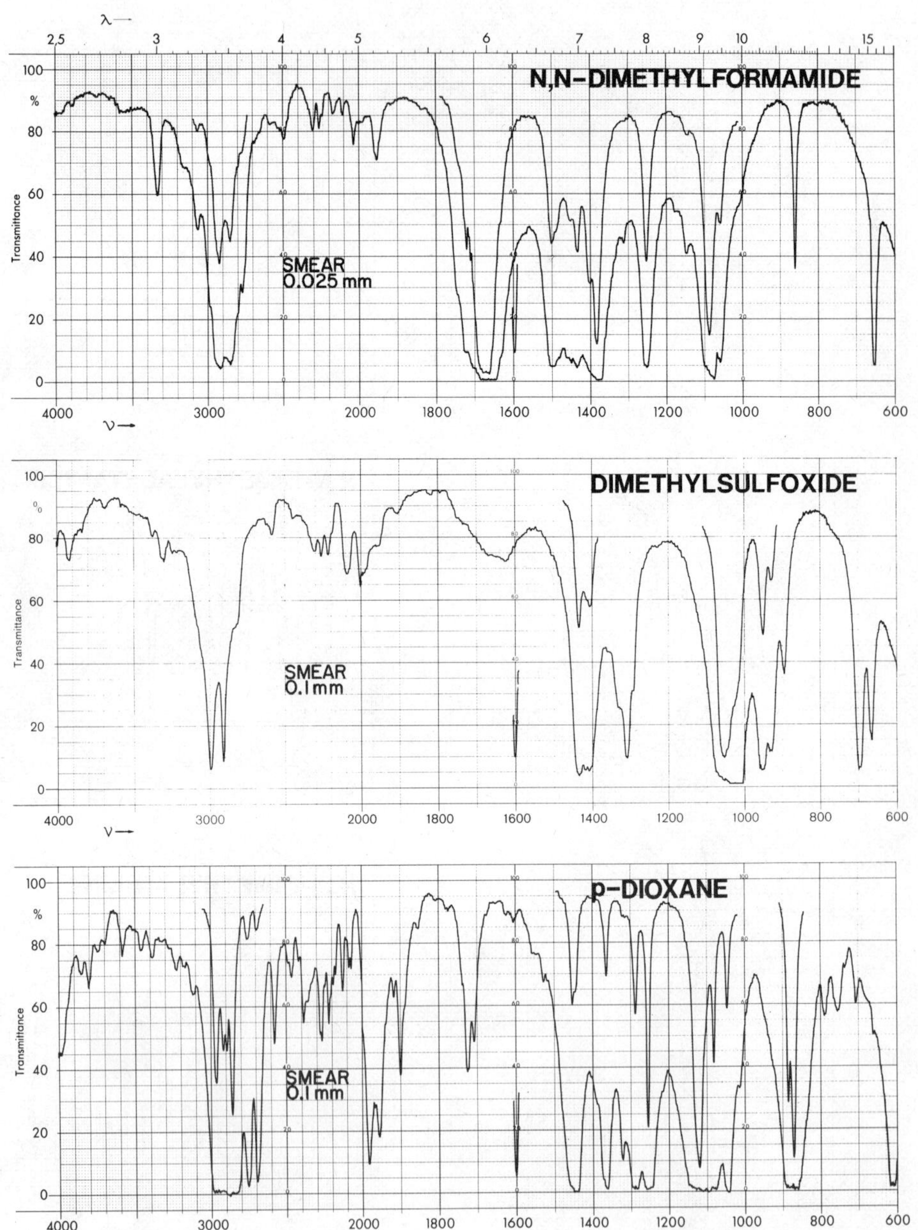
N,N-DIMETHYLFORMAMIDE
SMEAR
0.025 mm
DIMETHYLSULFOXIDE
SMEAR
0.1 mm
p-DIOXANE
SMEAR
0.1 mm
Transmittance
%
λ→
ν→

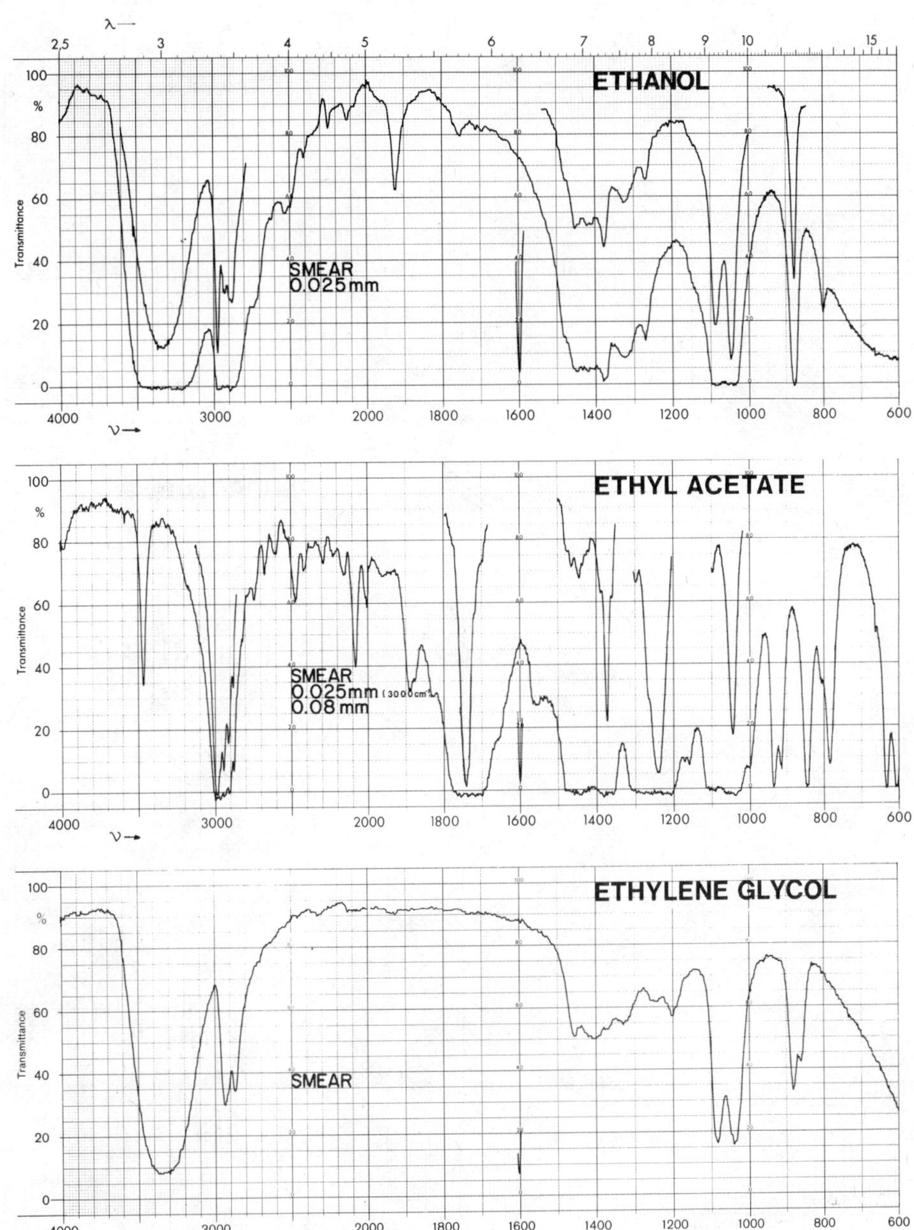
ETHANOL
SMEAR
0.025mm
ETHYL ACETATE
SMEAR
0.025mm (3000cm⁻¹)
0.08 mm
ETHYLENE GLYCOL
SMEAR
Transmittance
%
λ→
ν→
2.5 3 4 5 6 7 8 9 10 15
4000 3000 2000 1800 1600 1400 1200 1000 800 600
0 20 40 60 80 100

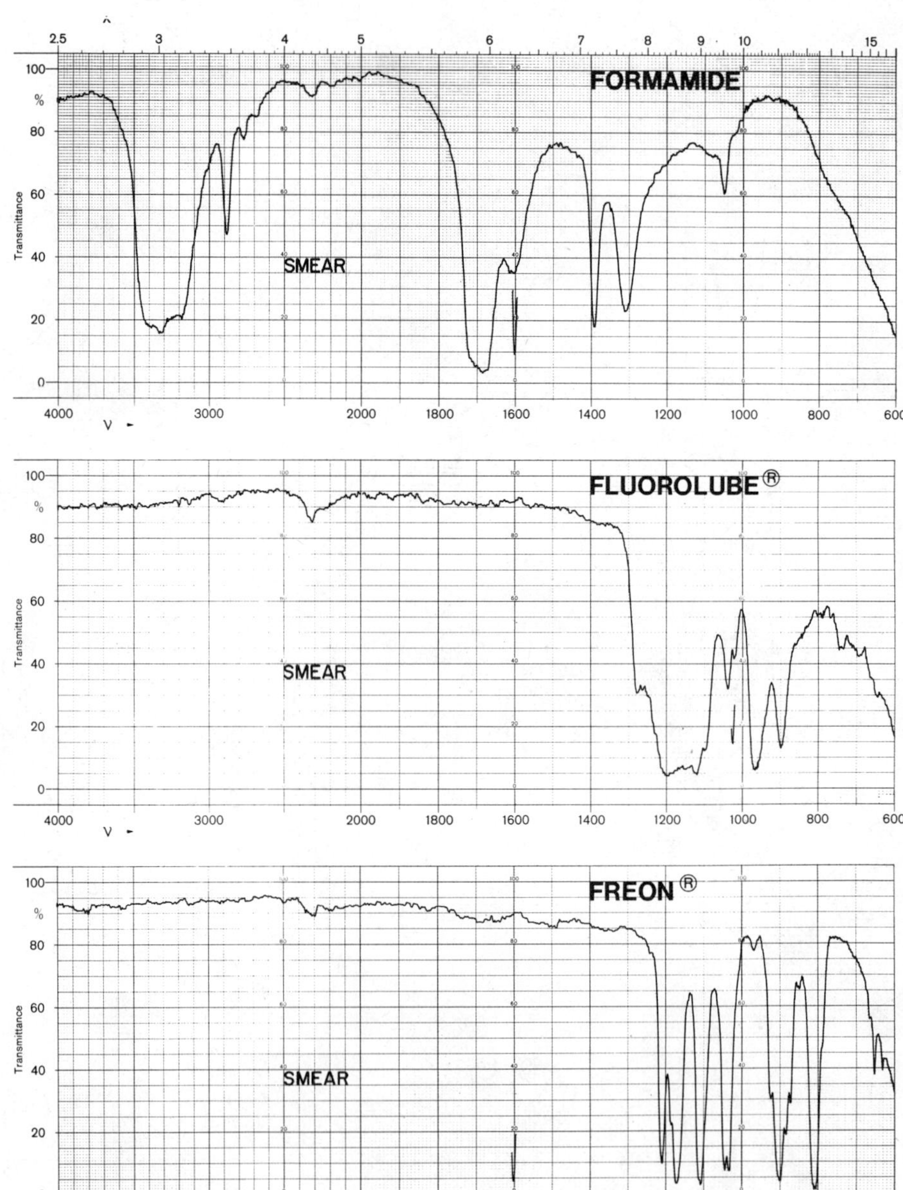
FORMAMIDE
SMEAR
FLUOROLUBE®
SMEAR
FREON®
SMEAR
Transmittance
%
ν

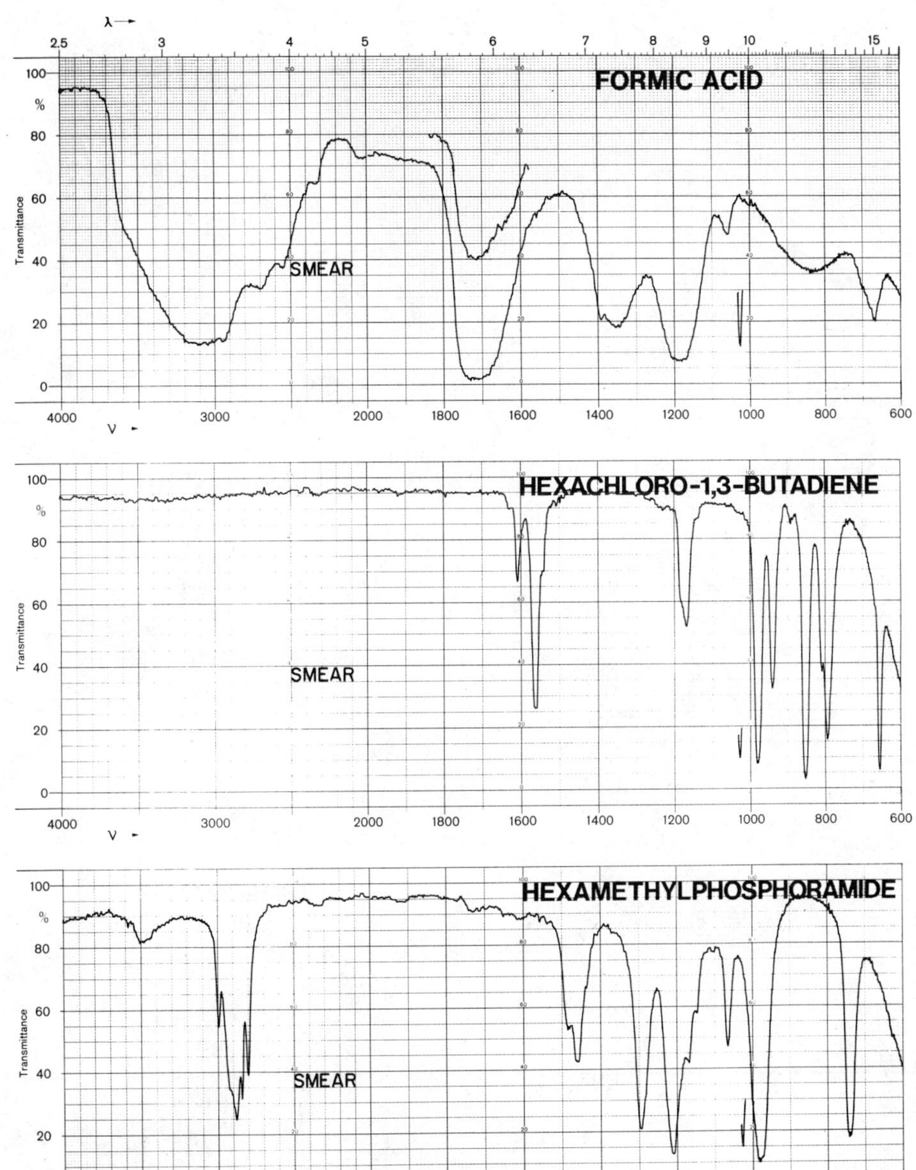
FORMIC ACID
SMEAR
HEXACHLORO-1,3-BUTADIENE
SMEAR
HEXAMETHYLPHOSPHORAMIDE
SMEAR
Transmittance
%
λ
ν

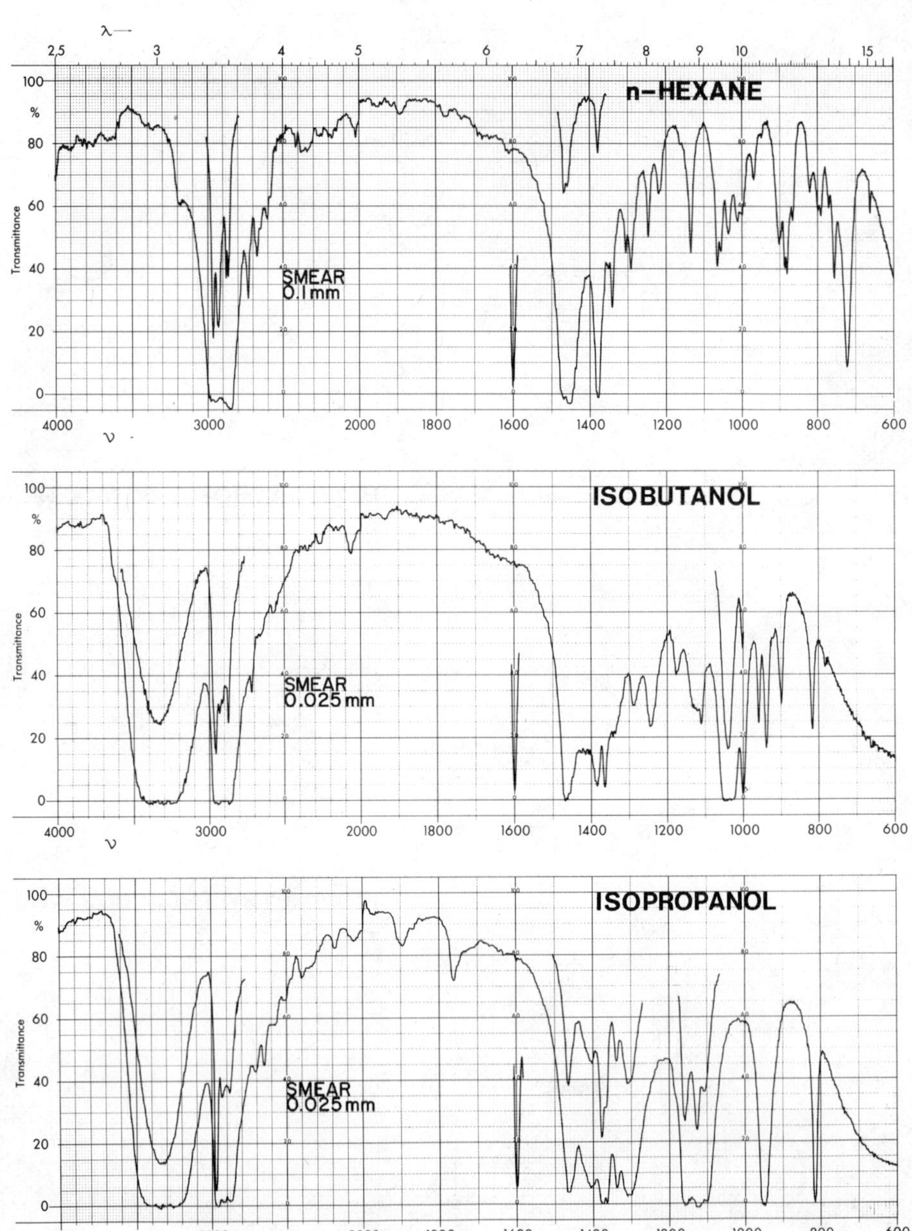
n-HEXANE
SMEAR
0.1mm
ISOBUTANOL
SMEAR
0.025mm
ISOPROPANOL
SMEAR
0.025mm
Transmittance
%
λ
ν

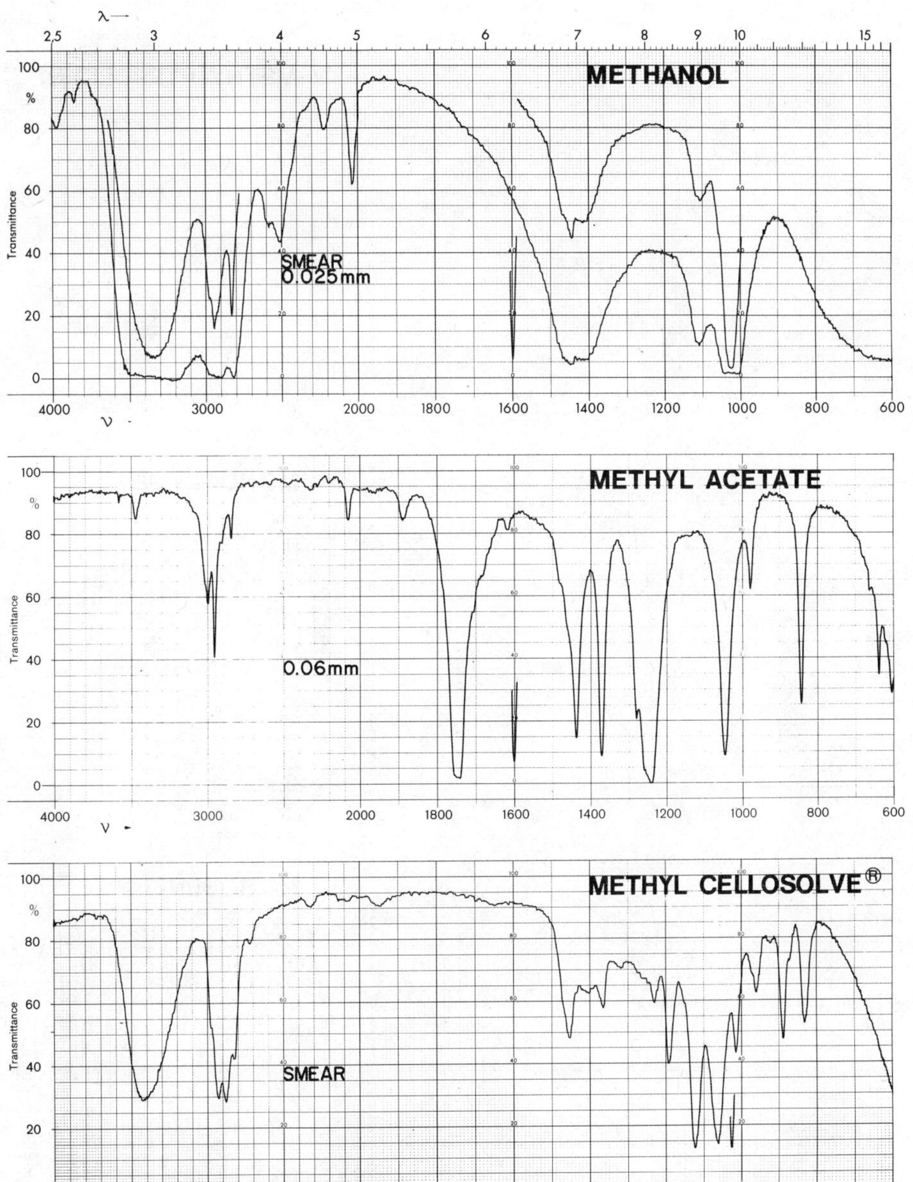
λ→
2.5
3
4
5
6
7
8
9
10
15
METHANOL
100
%
80
60
40
20
0
Transmittance
SMEAR
0.025mm
4000
3000
2000
1800
1600
1400
1200
1000
800
600
ν
METHYL ACETATE
0.06mm
METHYL CELLOSOLVE®
SMEAR

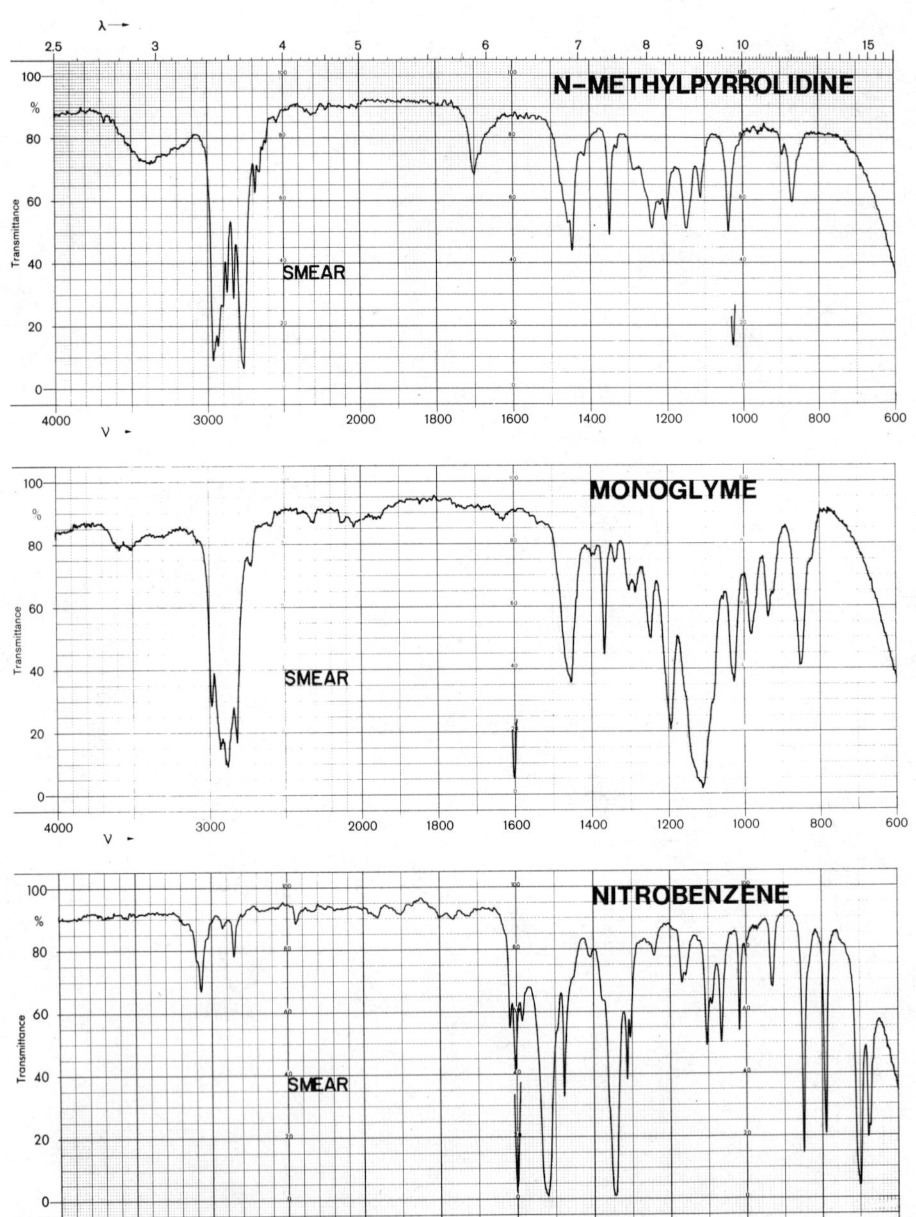
N-METHYLPYRROLIDINE
SMEAR
MONOGLYME
SMEAR
NITROBENZENE
SMEAR
Transmittance
%
λ
ν

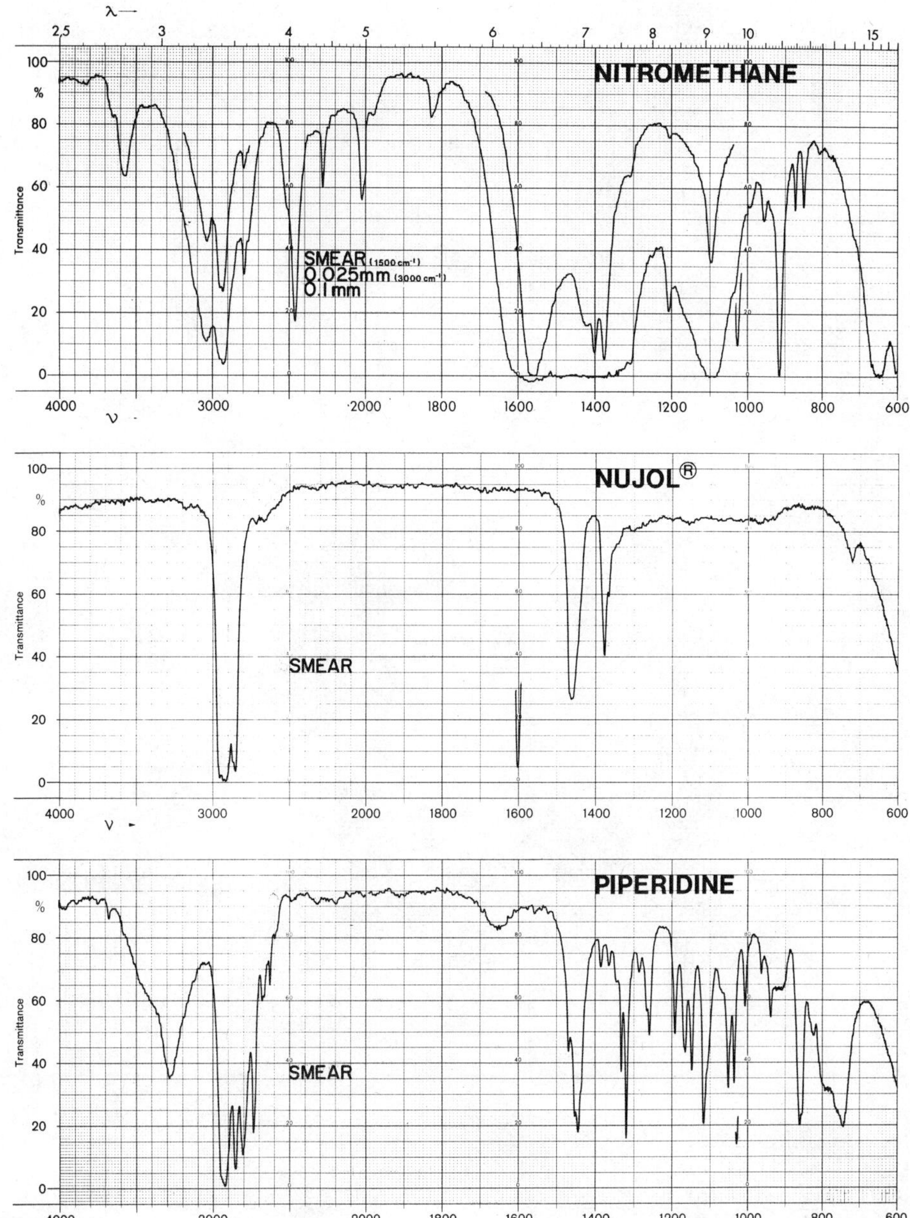
NITROMETHANE
SMEAR (1500 cm⁻¹)
0.025mm (3000 cm⁻¹)
0.1mm
Transmittance
NUJOL®
SMEAR
PIPERIDINE
SMEAR

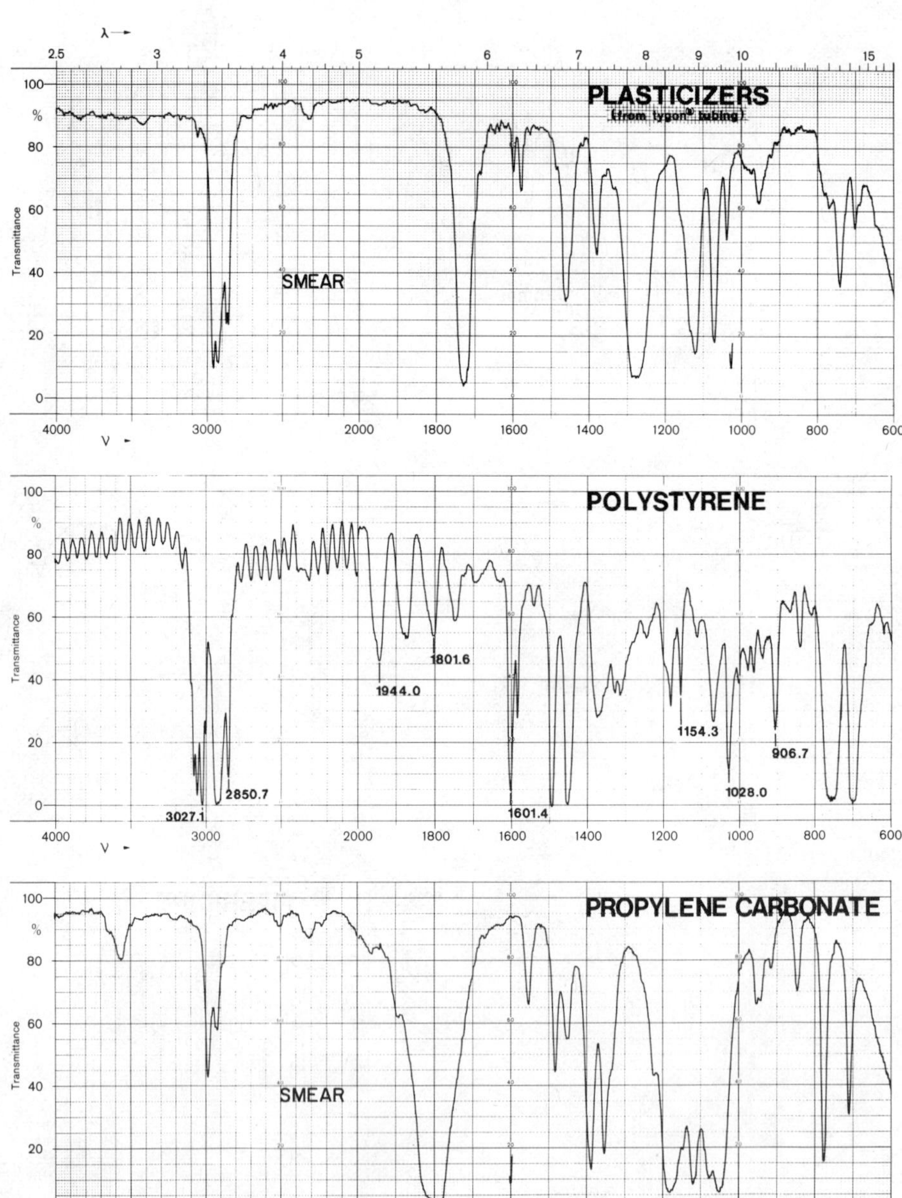
PLASTICIZERS
(from tygon® tubing)
SMEAR
POLYSTYRENE
3027.1
2850.7
1944.0
1801.6
1601.4
1154.3
1028.0
906.7
PROPYLENE CARBONATE
SMEAR
Transmittance
λ
ν

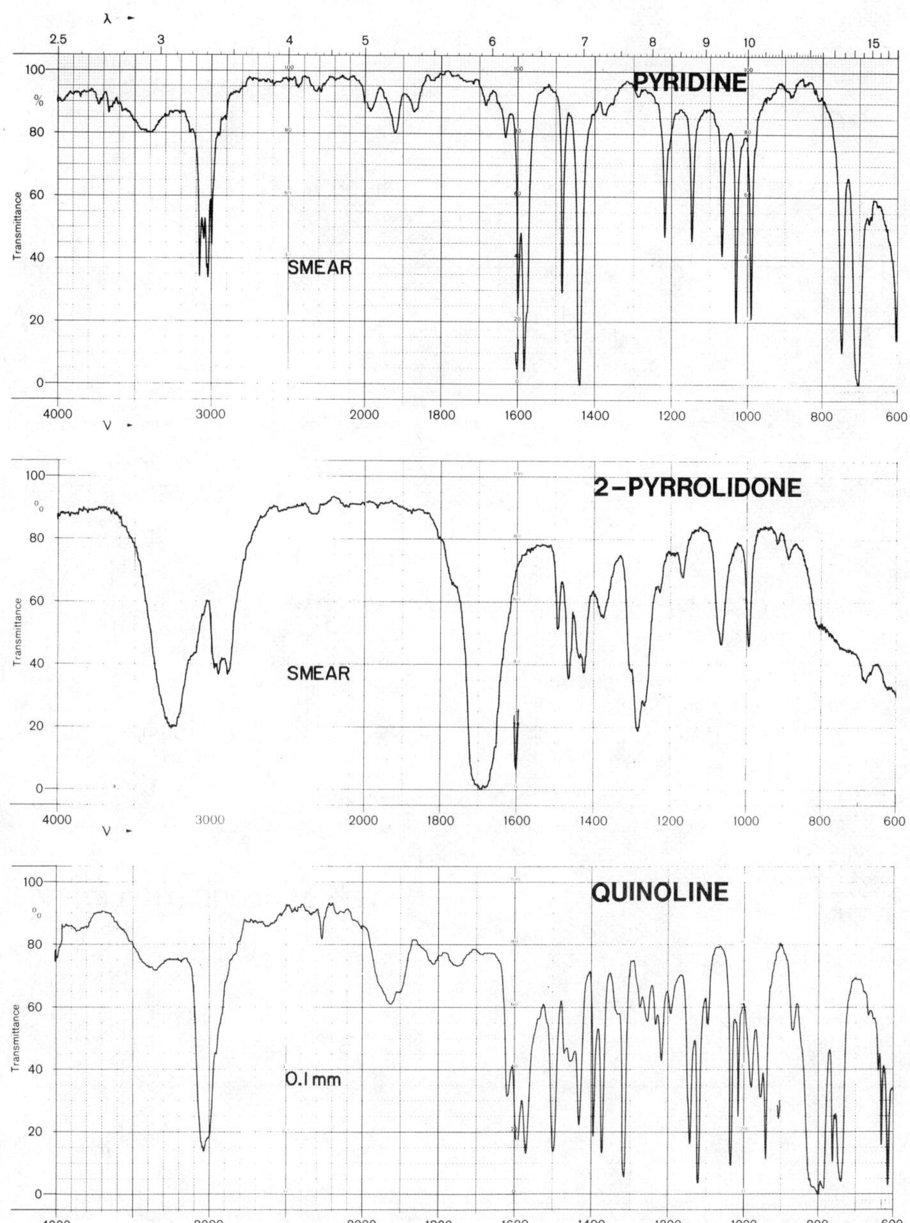
λ
2.5
3
4
5
6
7
8
9
10
15
100
%
80
60
40
20
0
Transmittance
PYRIDINE
SMEAR
4000
3000
2000
1800
1600
1400
1200
1000
800
600
ν
2-PYRROLIDONE
SMEAR
QUINOLINE
0.1mm

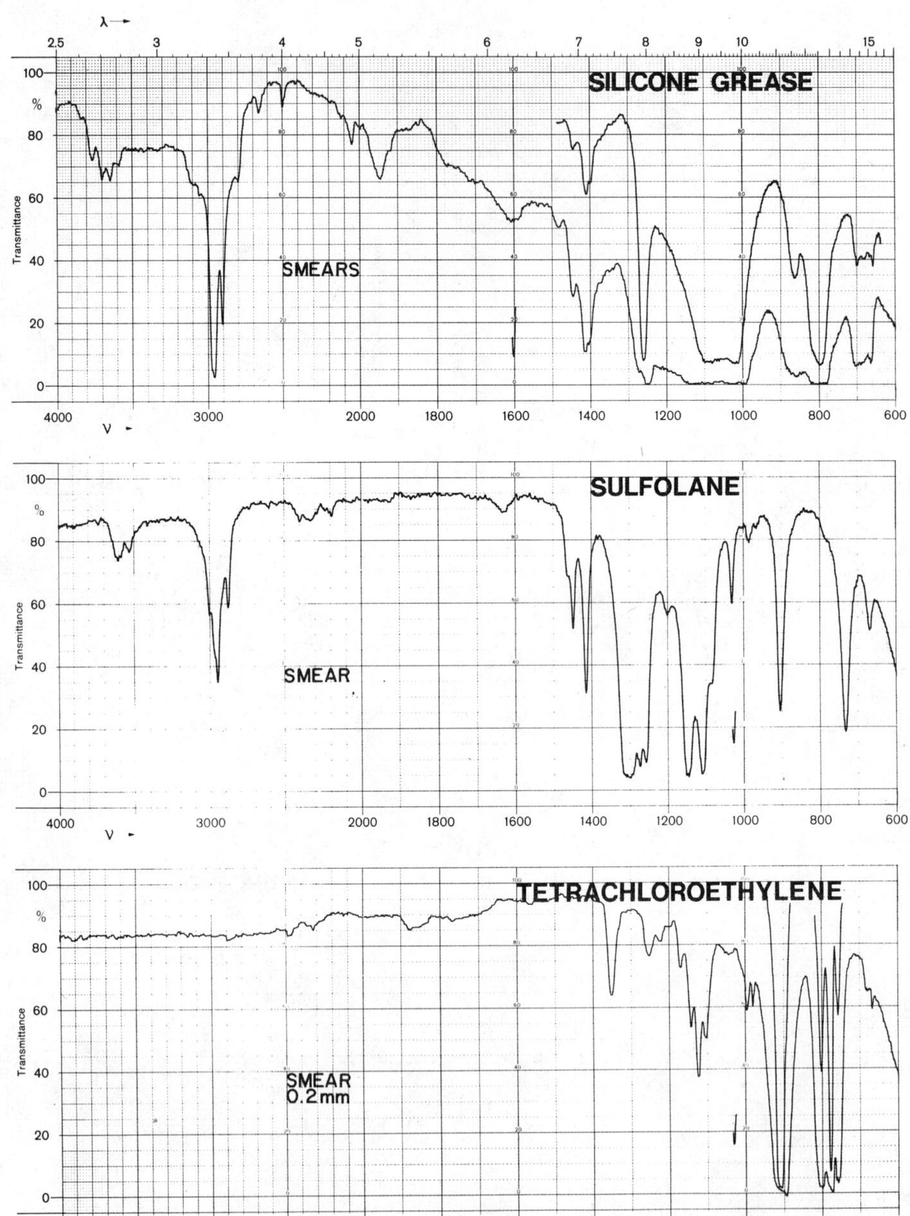
λ
2.5
3
4
5
6
7
8
9
10
15
100
%
80
60
40
20
0
Transmittance
SILICONE GREASE
SMEARS
4000
3000
2000
1800
1600
1400
1200
1000
800
600
ν
SULFOLANE
SMEAR
TETRACHLOROETHYLENE
SMEAR
0.2mm

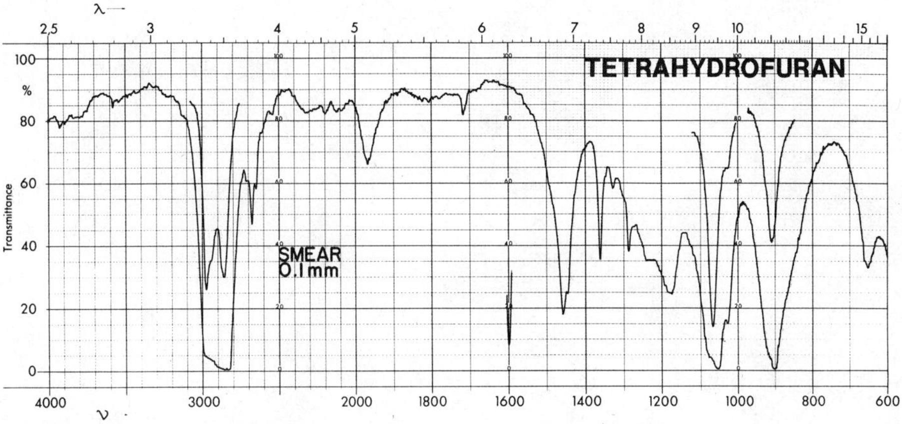

TETRAHYDROFURAN

SMEAR
0.1mm

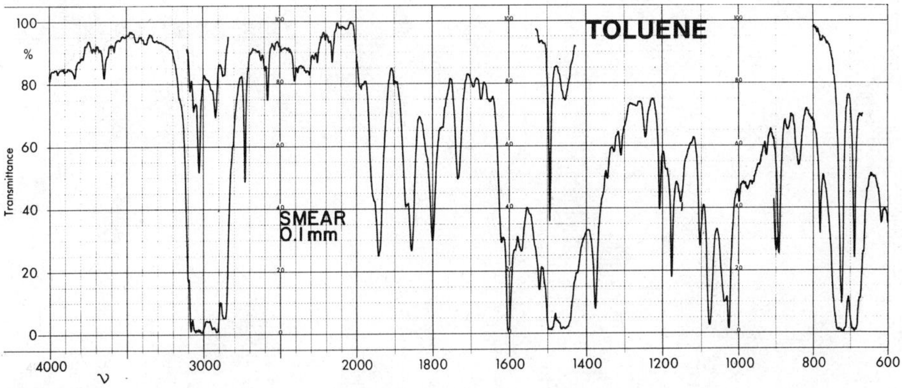

TOLUENE

SMEAR
0.1mm

TRI-n-BUTYL PHOSPHATE

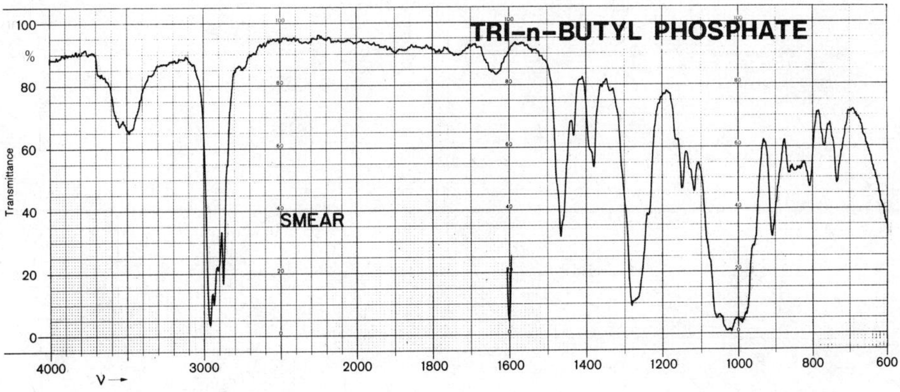

SMEAR

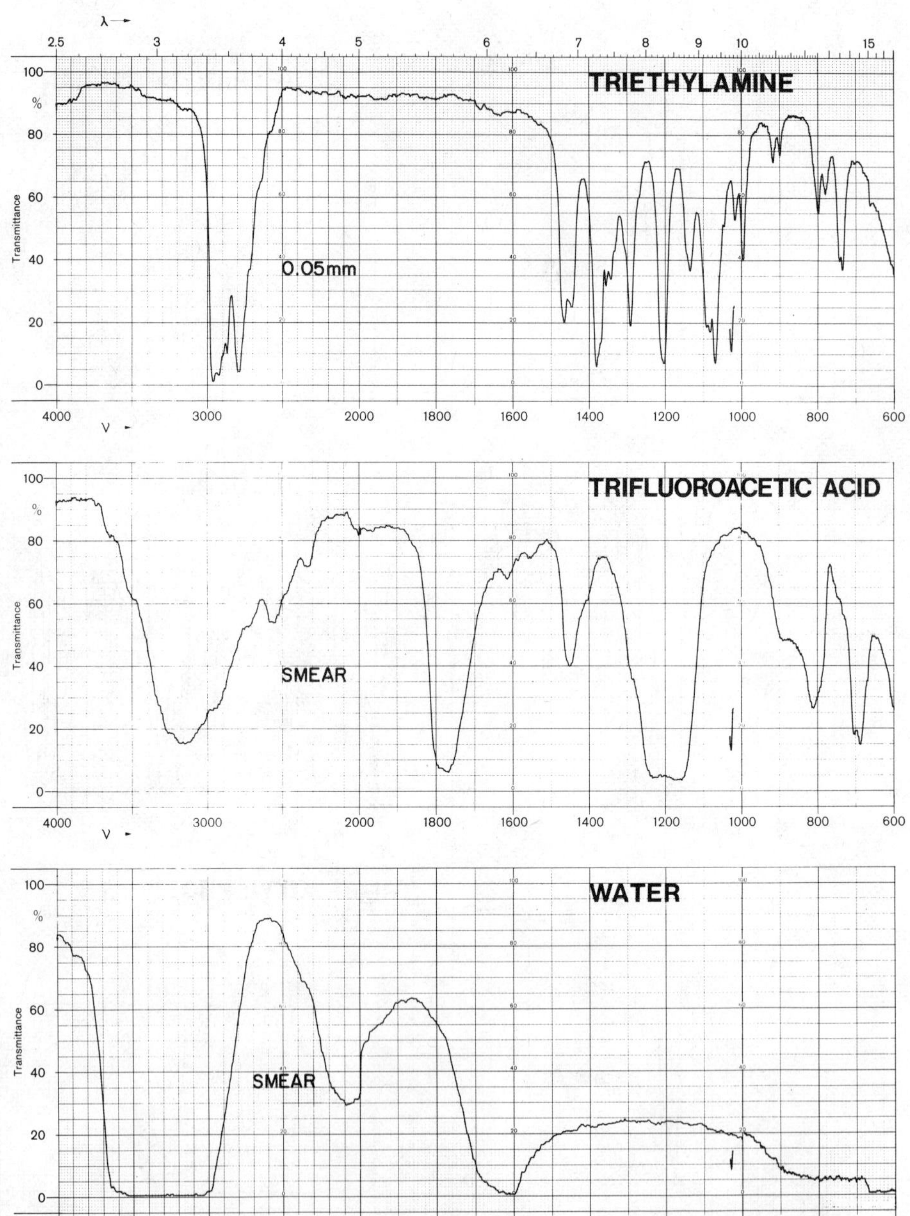
TRIETHYLAMINE
0.05mm
TRIFLUOROACETIC ACID
SMEAR
WATER
SMEAR
Transmittance
%
λ
ν

index

A

D

I

J

K

L

U

V

W